Lecture Notes in Computer Science 15156

Advanced Research in Computing and Software Science

Subline of Lecture Notes in Computer Science

More information about this series at https://link.springer.com/bookseries/558

Guido Schäfer · Carmine Ventre
Editors

Algorithmic Game Theory

17th International Symposium, SAGT 2024
Amsterdam, The Netherlands, September 3–6, 2024
Proceedings

Springer

Editors
Guido Schäfer
Centrum Wiskunde & Informatica (CWI)
Amsterdam, The Netherlands

Carmine Ventre
King's College London
London, UK

ISSN 0302-9743 ISSN 1611-3349 (electronic)
Lecture Notes in Computer Science
ISBN 978-3-031-71032-2 ISBN 978-3-031-71033-9 (eBook)
https://doi.org/10.1007/978-3-031-71033-9

This Springer imprint is published by the registered company Springer Nature Switzerland AG
The registered company address is: Gewerbestrasse 11, 6330 Cham, Switzerland

Preface

This volume contains the papers and extended abstracts presented at the 17th International Symposium on Algorithmic Game Theory (SAGT 2024), held during September 3–6, 2024, at Centrum Wiskunde & Informatica (CWI), Amsterdam, The Netherlands. The purpose of SAGT is to bring together researchers from computer science, economics, mathematics, operations research, psychology, physics, and biology to present and discuss original research at the intersection of algorithms and game theory.

This year, we received a record number of 84 submissions, which were rigorously peer-reviewed by the program committee (PC). Each paper was reviewed in an open peer-review process by at least three PC members, and evaluated on the basis of originality, soundness, significance, and exposition. The PC eventually decided to accept 32 papers to be presented at the conference, for an acceptance ratio of 38%.

The works accepted for publication in this volume cover most of the major aspects of algorithmic game theory, including auction theory, mechanism design, markets and matching, computational aspects of games, resource allocation problems, and computational social choice. To accommodate the publishing traditions of different fields, authors of accepted papers could ask that only a one-page abstract of the paper appeared in the proceedings. Among the 32 accepted papers, the authors of three papers chose this option.

Due to the financial support by Springer, we were able to provide two Best Paper Awards, one for the best student paper and one for a the best regular paper. The PC decided to give the Best Student Paper Award to the paper "Playing Repeated Games with Sublinear Randomness" by Farid Arthaud. The Best Paper Award was given to the paper "Swim Till You Sink: Computing the Limit of a Game" by Rashida Hakim, Jason Milionis, Christos Papadimitriou, and Georgios Piliouras.

The program also included three invited talks by distinguished researchers in algorithmic game theory, namely Paul Duetting (Google Research, Switzerland), Katrina Ligett (Hebrew University of Jerusalem, Israel), and Vasilis Gkatzelis (Drexel University, USA). In addition, SAGT 2024 featured two tutorials, which were given by Jan Maly (WU Wien, Austria) and Rebecca Reiffenhäuser (University of Amsterdam, Netherlands).

We would like to thank all the authors for their interest in submitting their work to SAGT 2024, as well as the 35 PC members and the external reviewers for their great work in evaluating the submissions. We also want to thank the local organizers Ulle Endriss, Sophie Klumper, Artem Tsikiridis, and Susanne van Dam, as well as several people from ITF and COM at CWI, for their help with the organization of the conference.

We also want to thank our sponsors G-Research, IOG, CWI, Google, ILLC, Springer, Networks, NWO, and VVSOR for their generous financial support. Finally, we would like to thank Springer for their work on the proceedings, as well as the EasyChair conference management system for facilitating the peer-review process.

July 2024

Guido Schäfer
Carmine Ventre

Organization

Program Committee

Georgios Amanatidis	University of Essex, UK
Vittorio Bilò	University of Salento, Italy
Ioannis Caragiannis	Aarhus University, Denmark
Riccardo Colini Baldeschi	Meta, UK
Andrés Cristi	EPFL, Switzerland
Bart de Keijzer	King's College London, UK
Argyrios Deligkas	Royal Holloway University of London, UK
Tomer Ezra	Harvard University, USA
Diodato Ferraioli	Università di Salerno, Italy
Federico Fusco	Sapienza University of Rome, Italy
Martin Gairing	University of Liverpool, UK
Thekla Hamm	Utrecht University, The Netherlands
Kristoffer Arnsfelt Hansen	Aarhus University, Denmark
Tobias Harks	University of Passau, Germany
Martin Hoefer	RWTH Aachen University, Germany
Alexandros Hollender	University of Oxford, UK
Simon Jantschgi	University of Oxford, UK
Panagiotis Kanellopoulos	University of Essex, UK
Pieter Kleer	Tilburg University, The Netherkands
Max Klimm	Technische Universität Berlin, Germany
Piotr Krysta	Augusta University, USA and University of Liverpool, UK
Pascal Lenzner	Hasso-Plattner-Institute Potsdam, Germany
Bo Li	Hong Kong Polytechnic University, China
Minming Li	City University of Hong Kong, China
David Manlove	University of Glasgow, UK
Evangelos Markakis	Athens University of Economics and Business, Greece
Gianpiero Monaco	University of Chieti-Pescara, Italy
Paolo Penna	IOG, Switzerland
Daniel Schoepflin	Rutgers University, USA
Guido Schäfer (Chair)	CWI and University of Amsterdam, The Netherlands
Alkmini Sgouritsa	Athens University of Economics and Business and Archimedes/Athena RC, Greece
Tami Tamir	Reichman University, Israel
Yixin Tao	Shanghai University of Finance and Economics, China
Zoi Terzopoulou	University Lyon - Saint-Étienne, France

Carmine Ventre (Chair)	King's College London, UK
Adrian Vetta	McGill University, Canada
Anaëlle Wilczynski	CentraleSupélec, Université Paris-Saclay, France

Organizing Committee

Ulle Endriss	University of Amsterdam, The Netherlands
Sophie Klumper	CWI and University of Amsterdam, The Netherlands
Guido Schäfer	CWI and University of Amsterdam, The Netherlands
Artem Tsikiridis	CWI, The Netherlands
Susanne van Dam	CWI, The Netherlands

Steering Committee

Elias Koutsoupias	University of Oxford, UK
Marios Mavronicolas	University of Cyprus, Cyprus
Dov Monderer	Technion, Israel
Burkhard Monien	University of Paderborn, Germany
Christos Papadimitriou	Columbia University, USA
Giuseppe Persiano	University of Salerno, Italy
Paul Spirakis (Chair)	University of Liverpool, UK

Additional Reviewers

Aziz, Haris
Bilò, Davide
Bérczi-Kovács, Erika Renáta
Cembrano, Javier
Chionas, Giorgos
Christoforidis, Vasilis
Cleveland, Colin
de la Haye, Merlin
Dogan, Battal
Dupré La Tour, Max
Dósa, György
Fehrs, Karl
Gawendowicz, Hans
Ghosh, Abheek
Goldsmith, Tiger-Lily
Greger, Matthias
Gualà, Luciano
Huth, Lars
Knaack, Martin
Kraiczy, Sonja
Krogmann, Simon
Lam, Alexander
Li, Weian
Lock, Edwin
Luo, Junjie
Machaira, Paraskevi
Michini, Carla
Narayan, Vishnu
Ndiaye, Ndiamé
Neubert, Stefan
Papasotiropoulos, Georgios
Patsilinakos, Panagiotis
Romen, René
Schecker, Conrad
Schierreich, Šimon
Schmalhofer, Marco
Shyam, Sudarshan
Springham, Drew
Sun, Ankang
Troyan, Peter
Tsikiridis, Artem
Turrini, Paolo

Varloot, Estelle
Verma, Paritosh
Vinci, Cosimo
Vlachos, Ioannis
Wang, Chenhao
Warode, Philipp
Wu, Xiaowei
Wulf, Lasse
Zhou, Houyu

Invited Talks

Ambiguous Contracts

Paul Dütting

Google Research, Zurich, Switzerland
duetting@google.com

Contract theory captures situations where two parties—a principal and an agent—can benefit from mutual cooperation. The prototypical situation is one in which the principal seeks to delegate the execution of a job to an agent. The agent can take different costly actions, and his choice of action entails a stochastic outcome (with attached reward) for the principal. The principal cannot directly observe the agent's choice of action but can influence the agent's decision through a contract that specifies outcome-contingent payments. Given a contract, the agent aims to maximize his expected payment minus cost. The goal of the principal is to maximize her expected utility given by expected reward minus payment, under the action chosen by the agent.

One feature of real-life contracts that is not captured (or explained) by this classic model is that practical contracts are often ambiguous. For example, the promotion guidelines of a university may require a candidate to demonstrate "research productivity and excellence." Similarly, a professional services contract may require that a provider exert "due diligence." This ambiguity may be due to an inability of the two parties to provide more precise specification. In contrast, we explore the deliberate infusion of ambiguity as a tool to enhance the principal's contracting power over the agent.

Towards this goal we propose an extension of the classic (hidden action) principal-agent model. In this model, an ambiguous contract consists of a set of classic contracts, which the agent evaluates by considering the minimum utility a given action yields against any contract in the support of the ambiguous contract. At the same time, we require that the principal's utility—under the action chosen by the agent—is the same for all contracts in the support of the ambiguous contract. We show that this expands the set of actions that the principal can implement, and that the principal's gain from using an ambiguous contract can be arbitrarily large. We further characterize the structure of optimal ambiguous contracts, showing that ambiguity drives optimal contracts towards simplicity. We also provide a characterization of ambiguity-proof classes of contracts, where the principal cannot gain by infusing ambiguity. Finally, we show that when the agent can engage in mixed actions, the advantages of ambiguous contracts disappear.

Based on joint work with Michal Feldman, Daniel Peretz, and Larry Samuelson (EC 2023 and arXiv).

Learning-Augmented Mechanism Design

Vasilis Gkatzelis

Drexel University, Philadelphia, PA, 19104, USA
gkatz@drexel.edu
https://www.cs.drexel.edu/

Abstract. For more than half a century, the dominant approach for the mathematical analysis of algorithms in computer science has been worst-case analysis. While worst-case analysis provides a useful signal regarding the robustness of an algorithm, it can be overly pessimistic, and it often leads to uninformative bounds or impossibility results that may not reflect real-world obstacles. Meanwhile, advances in machine learning have led to very practical algorithms, most of which do not provide any non-trivial worst-case performance guarantees. Motivated by the tension between worst-case analysis and machine learning, a surge of recent work focuses on the design algorithms that are guided by machine-learned predictions, aiming to perform better in practice, while maintaining their robustness. Specifically, the goal of this literature on **"learning-augmented algorithms"** is to design algorithms that simultaneously provide two types of guarantees: **"robustness"** (which corresponds to the classic worst-case guarantees, even if the predictions that the algorithm is provided with are arbitrarily bad) and **"consistency"** (i.e., stronger performance guarantees when the predictions are accurate). This "learning-augmented framework" has been used successfully in a variety of settings, e.g., toward a refined analysis of competitive ratios in online algorithms and running times in traditional algorithms.

A recent line of work on **"learning-augmented mechanism design"** has deployed this learning augmented framework in **settings involving strategic agents**. In such settings, the designer often faces additional obstacles which further limit their ability to reach desired outcomes. For example, some of the input that the designer needs may be private information held by the participating agents, and the agents could strategically misreport this information, aiming to maximize their own utility. In other settings, the agents may even have direct control over some aspects of the outcome. The long literature on mechanism design has proposed a variety of solutions for these types of problems, aiming to align the incentives of the agents with those of the designer, but the worst-case guarantees of these solutions are often underwhelming from a practical perspective. This talk will introduce the "learning-augmented mechanism design" model and provide an overview of some of the results in this line of work.

Actually, data is a rival good

Katrina Ligett

The Hebrew University of Jerusalem
katrina@cs.huji.ac.il

Abstract. There is a tendency in many fields, including computer science, economics, and industry, to model data as a non-rival good, meaning that one entity using a particular piece of data doesn't impinge on its use by others. Food is a classic rival good (if I eat the apple, you cannot); digital music is a classic non-rival good (my listening to the song has no effect on your listening experience). Data might, at first blush, seem more like digital music than like an apple. In this talk, I will give arguments from three fields—economics, privacy, and statistics—for why modeling data as non-rival is problematic, and will argue that we need a new paradigm.

The core of the economic argument is that generative AI has transformed the market for data, making competition (and rivalrousness) for and around data newly central. The privacy and statistical validity arguments rely on mathematical frameworks that help us understand how repeated uses of a dataset accumulate and interact. All of these arguments suggest new models and metaphors for data, and directions for further work.

Acknowledgments. This talk will touch on work that was supported in part by a gift to the McCourt School of Public Policy and Georgetown University, Simons Foundation Collaboration 733792, Israel Science Foundation (ISF) grants 1044/16 and 2861/20, a grant from the Israeli Ministry of Education, and the Dieter Schwarz Foundation's TUM-HUJI Joint AI Research Hub.

Disclosure of Interests. The speaker works part-time for Google. This work was not done at Google nor subject to pre-publication review. All views presented are the speaker's own.

Contents

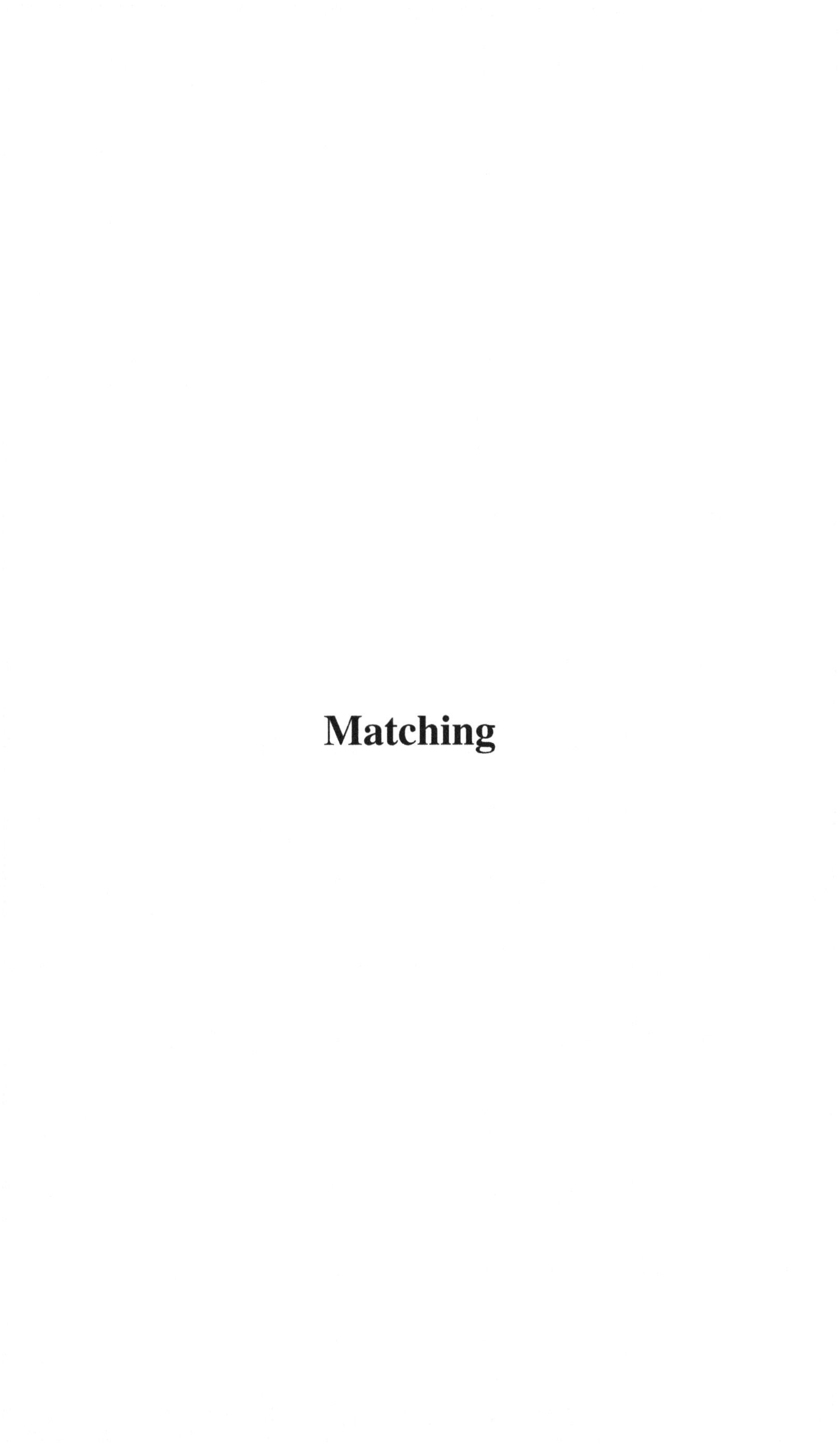

Matching

Structural and Algorithmic Results for Stable Cycles and Partitions in the Roommates Problem

Frederik Glitzner[(✉)] and David Manlove

School of Computing Science, University of Glasgow, Glasgow G12 8QQ, UK
f.glitzner.1@research.gla.ac.uk, david.manlove@glasgow.ac.uk

Abstract. In the Stable Roommates problem, we seek a *stable matching* of the agents into pairs, in which no two agents have an incentive to deviate from their assignment. It is well known that a stable matching is unlikely to exist, but a *stable partition* always does and provides a succinct certificate for the *unsolvability* of an instance. Furthermore, apart from being a useful structural tool to study the problem, every stable partition corresponds to a *stable half-matching*, which has applications, for example, in sports scheduling and time-sharing applications. We establish new structural results for stable partitions and show how to enumerate all stable partitions and the cycles included in such structures efficiently. We also adapt known fairness and optimality criteria from stable matchings to stable partitions and give complexity and approximability results for the problems of computing such "fair" and "optimal" stable partitions.

Keywords: Stable Roommates Problem · Stable Partition · Enumeration · Optimal Stable Partitions

1 Introduction

1.1 Background

The Stable Roommates problem (sr) is a classical combinatorial problem with applications to computational social choice. Consider a group of friends that want to play one hour of tennis, where everyone has preferences over who to play with. Can we match them into pairs such that no two friends prefer to play with each other rather than their assigned partners? If the problem instance I admits such a *stable matching* M which does not admit a *blocking pair* of agents which would rather be matched to each other than their partners in M, then we call I *solvable*. Otherwise, we call I *unsolvable*. Even an instance with as few as 4 agents may be unsolvable as Gale and Shapley [9] showed.

There are many practical applications of the sr model. As the name suggests, it can model campus housing allocation where two students either share a room or a flat. Furthermore, sr can model pairwise kidney exchange markets, with

G. Schäfer and C. Ventre (Eds.): SAGT 2024, LNCS 15156, pp. 3–20, 2024.
https://doi.org/10.1007/978-3-031-71033-9_1

centralised matching schemes existing, for example, in the US[1] and UK[2]. SR can also model peer-to-peer networks and pair formation in chess tournaments.

In SR, a problem instance I and a matching M can be represented in the form of preference lists and a set of pairs or as a complete graph and a set of edges; these representations are equivalent. The SR problem is a non-bipartite extension of the classical STABLE MARRIAGE problem and is well-studied in its own right [14,19]. Irving [15] presented an algorithm to find a stable matching or decide that none exists in linear time. Furthermore, SR now has a wide range of algorithms and structural results relating to special kinds of stable matchings, including algorithms for finding stable matchings that satisfy further optimality constraints on the structure of the matching [4,6,23]. For example, the *profile* is often used in fairness measures of a matching, which is a vector $(p_1 \dots p_n)$ where p_i captures the number of agents that are matched to their ith choice, and n is the number of agents. However, naturally, these optimality properties cannot be realised if the instance is unsolvable. Mertens [20] presented empirical evidence suggesting that, as n grows large, stable matchings are unlikely to be a good solution concept in practice as they are unlikely to exist. In the past, many alternative solution concepts have been proposed, such as *stable partitions* [24], *maximum stable matchings* [25], *almost-stable matchings* [1], *popular matchings* [11], and more, some of which leave agents unassigned or are NP-hard to compute.

One of the fundamental SR-related questions posed by Gusfield and Irving [14] asked about the existence of a succinct certificate for the unsolvability of an instance. The question was answered positively by Tan [24], who generalised the notion of a stable matching to a new structure called a *stable partition* that always exists. Over 20 years after their initial publication, Manlove [19] described the work on stable partitions in the 1990s as a key landmark in the progress made on the SR problem after 1989. Although a stable partition gives a succinct certificate for the unsolvability of an SR instance, this notion was adopted as a solution concept in itself. Remaining in the tennis analogy, a stable partition can be interpreted as an assignment of the friends into half-hour sessions, or a weekly-alternating schedule. The stability definition leads to a solution in which every friend plays either for exactly one hour with some other friend or two half-hour sessions with two different friends, at most one agent does not play at all, and no two friends who would like to play more time with each other. Note that any stable partition can be transformed into a *reduced stable partition* (i.e., stable partitions with no cycles of even length longer than 2) in linear time, which corresponds to a stable matching in the case of solvable instances, while some agents are guaranteed to be in half-time partnerships in the case of unsolvable instances.

Some have highlighted the advantages of considering truly fractional stable matchings, whereas stable matchings are inherently integral, and stable partitions can be interpreted as stable half-integral matchings [3]. While stable frac-

[1] http://www.paireddonation.org.

[2] http://www.organdonation.nhs.uk.

tional matchings might be advantageous in applications such as time-sharing, it is unlikely to be a suitable solution concept for other applications such as sports tournament scheduling.

1.2 Our Contributions

We prove new structural results for stable partitions, allowing us to construct algorithms to enumerate stable partitions and to compute or approximate various types of optimal stable partitions.

Specifically, we prove a bijective correspondence between the set of reduced stable partitions and the set of stable matchings of a solvable sub-instance. With this, we show how the set of all reduced stable partitions can be enumerated efficiently. We build on this by considering all stable partitions, not necessarily reduced. We show that any non-reduced stable partition can be constructed from two reduced stable partitions. However, a deeper structural investigation yields that any predecessor-successor pair can be part of at most one cycle of length longer than 2. Therefore, the union of all stable partitions of any instance with n agents contains at most $O(n^2)$ cycles, and we show how these can be enumerated in $O(n^4)$ time. With this, we improve the enumeration algorithm for all stable partitions, denoted $P(I)$, and present an algorithm that runs in $O(|P(I)|n^3+n^4)$ time. Finally, we adapt six natural fairness measures and optimality criteria from stable matchings to stable partitions. We find that a stable partition that minimises the maximum predecessor rank (called *minimum-regret stable partition*) always exists and is the only tractable problem variant of these. Furthermore, although two of the NP-hard optimisation problem variants are polynomial-time approximable within a factor of 2, one of them is not approximable within any constant factor.

All of these results are consistent with those previously known about stable matchings, strengthening the tight correspondence between stable matchings and stable partitions. Overall, our results shed more light on the complexity-theoretic and structural properties of stable partitions.

1.3 Related Work

Irving [16] and Gusfield [12] published an extensive collection of early structural results, leading to a book discussing the structure and algorithms of the SM and SR problems [14]. They proved, for example, that the stable matchings of a given instance form a semi-lattice structure and that, of a given SR instance I with n agents, all stable pairs (union of all pairs part of some stable matching) and all stable matchings, denoted $S(I)$, can be found in $O(n^3 \log n)$ and $O(|S(I)|n^2+n^3 \log n)$ time, respectively. Later, Feder [7] improved this to $O(n^3)$ and $O(|S(I)|n+n^2)$, respectively.

With regards to the complexity of profile-optimal stable matching problems, *minimum-regret* is one of few efficiently solvable ones, with a linear-time algorithm presented by Gusfield and Irving [14]. On the other hand, Feder [6] proved that finding an egalitarian (minimum sum of ranks of agents assigned) stable matching for SR is NP-hard. Later, Feder [7], and Gusfield and Pitt [13] gave 2-approximation algorithms for the problem. Cooper [4] showed that for an instance with regret r, the problems of finding *rank-maximal* (lexicographically maximal profile), *generous* (lexicographically minimal inverse profile), *first-choice maximal* (maximum number of first choices achieved), and *regret-minimal* (minimum number of agents assigned to r) stable matchings, if they exist, are all NP-hard. Simola and Manlove [23] extended these results to short preference lists and presented some approximability results.

In his original paper, Tan [24] provided a linear-time algorithm to compute a stable partition and showed that every SR instance admits at least one stable partition. Alternative proofs for the existence of stable half-matchings in more general settings using Scarf's Lemma exist [2]. Tan and Hsueh [26] considered the online version of the problem of finding a stable partition, in which a new agent arrives and the preference lists are updated, and constructed an exact algorithm that runs in linear time (for each newly arriving agent).

Pittel [21,22] derived a range of probabilistic results about the algorithm by Tan and stable partitions in general. He showed, for example, that at most $O(\sqrt{(n \log n)})$ members are likely to be involved in the cycles of odd length three or more, that the expected number of stable partitions is $O(n^{\frac{1}{2}})$, which reduces to $O(n^{\frac{1}{4}})$ for reduced stable partitions. Stable partitions are also used to show various other results with regards to the SR problem. For example, in connection with their study of so-called "almost-stable" matchings, Abraham et al. [1] make use of stable partitions as a structural tool.

Over time, many variations of SR have been studied, such as SRT (where ties are allowed in the preferences) [17], SRI (where incomplete preferences are permitted) [23], or both, denoted SRTI [17]. Fleiner [8] also showed how, under some assumptions, the strict linearly-ordered preference lists can be replaced with choice functions, while keeping the problem polynomial-time solvable. Irving and Scott [18] introduced a many-to-many extension of SR called STABLE FIXTURES and presented a linear-time algorithm to find a stable matching or report that none exists. Dean and Munshi [5] investigated the even more general STABLE ALLOCATION problem and used transformations between non-bipartite and bipartite instances to derive efficient algorithms.

1.4 Structure of the Paper

In Sect. 2, we present a mixture of existing and new formal definitions and known results relevant to this paper. In Sect. 3, we characterise and exploit the structure of reduced stable partitions, longer cycles, and all stable partitions. Finally, in Sect. 4, we adapt common profile-based optimality criteria from stable matchings to stable partitions and investigate the complexity and approximability of these problems. We finish with a discussion of the results and some related open problems in Sect. 5. All omitted proofs and some other details and examples can be found in the full version of this paper [10].

2 Formal Definitions and Preliminary Results

We begin this section by defining SR instances formally as follows.

Definition 1 (SR Instance). *Let $I = (A, \succ)$ be an* SR instance *where $A = \{a_1, a_2, \ldots, a_n\}$, also denoted $A(I)$, is a set of $n \in 2\mathbb{N}$ agents and every agent $a_i \in A$ has a strict preference ranking or* preference relation *$\succ_i$ over all other agents $a_j \in A \setminus \{a_i\}$. For stable partitions, we will assume that $\succ$ is extended such that every agent ranks themselves last to allow self-assigned agents.*

Note that in this work, we assume an even number of agents. However, it has been shown [14] that most concepts and results transfer over to the case where the number of agents is odd and we accept an unmatched (but non-blocking) agent. Furthermore, there will be cases where we transform an SR instance with *complete preferences* (i.e. a ranking $\succ$ in which all agents rank all other agents but themselves) into an instance with *incomplete preferences* (technically denoted by SRI), but all existing techniques that we use still apply in the SRI case. A solution to an SR instance is now defined.

Definition 2 (Stable Matchings). *Let $I = (A, \succ)$ be an* SR *instance. M is a* matching *of I if it is an assignment of some agents in A into pairs such that no agent is contained in more than one pair of M. A* blocking pair *of a matching M is a pair of two agents $a_i, a_j \in A$ such that either a_j is unassigned in M or $a_i \succ_j M(a_j)$, and either a_i is unassigned or $a_j \succ_i M(a_i)$, where $M(a_i)$ is the partner of a_i in M. If M does not admit any blocking pair, then it is called* stable. *A stable matching is* complete *if it contains all $|A|$ agents.*

Any stable matching of an instance with complete preferences and an even number of agents will be complete, as any two unmatched agents would rather be matched to each other than to be unmatched.

Definition 3 (Solvability and Sub-Instances). *Let $I = (A, \succ)$ be an* SR *instance. I is* solvable *if it admits at least one stable matching, otherwise it is* unsolvable. *$I' = (A', \succ')$ is a* sub-instance *or a* restriction *of I if $A' \subseteq A$ and $a_i \succ'_k a_j$ for all agents $a_i, a_j, a_k \in A'$ where $a_i \succ_k a_j$.*

As previously mentioned, stable partitions Π are permutations on the subset of agents. Therefore, they can be written in cyclic notation where every agent is part of exactly one cycle. A formal definition follows.

Definition 4 (Stable Partition). *Let $I = (A, \succ)$ be an* SR *instance. Then a partition Π is* stable *if it is a permutation of A and*

1. *$\forall a_i \in A$ we have $\Pi(a_i) \succeq_i \Pi^{-1}(a_i)$, and*
2. *$\nexists . a_i, a_j \in A$, $a_i \neq a_j$, such that $a_j \succ_i \Pi^{-1}(a_i)$ and $a_i \succ_j \Pi^{-1}(a_j)$,*

where $\Pi(a_i) \succeq_i \Pi^{-1}(a_i)$ means that either a_i's successor in Π is equal to its predecessor, or the successor has a better rank than the predecessor in the preference list of a_i.

Each stable partition Π corresponds to a stable half-matching in the sense that each successor-predecessor pair in Π is assigned a half-integral match. In our study of stable partitions, it is important to differentiate between different kinds of cycles.

Definition 5 (Cycles). *Let I be an* SR *instance and let $C = (a_{i_1}\ a_{i_2}\ \dots\ a_{i_k})$ be an ordered collection of one or more agents in I. Then C is a* cycle *and we can* apply *cycle C to some agent $a_{i_j} \in C$ to get its successor in C, denoted by $C(a_{i_j})$. Similarly, we can apply the inverse $C^{-1} = (a_{i_1}\ a_{i_k}\ \dots\ a_{i_2})$ to a_{i_j} to get its predecessor in C, denoted by $C^{-1}(a_{i_j})$. The same holds for sets of cycles.*

We will implicitly assume that two disjoint cycles or sets of cycles C_1, C_2 can be added under concatenation $C_1 +\!\!+\ C_2$, simply be denoted by C_1C_2 when obvious. C_1 can also be removed from a collection of cycles C, denoted by $C \backslash C_1$.

Definition 6 (Stable Cycles). *Let I be an* SR *instance and let C be a cycle. If there exists some stable partition Π of I such that $C \in \Pi$, then C is a* stable cycle *of I. A stable cycle of odd length is called an* odd cycle, *of length 2 a* transposition, *and of even length longer than 2 an* even cycle *of I. Clearly, for every stable partition Π of I, $\Pi = \mathcal{T}\mathcal{E}\mathcal{O}_I$, where $\mathcal{T}$ denotes the transpositions, $\mathcal{E}$ the even cycles, and $\mathcal{O}_I$ the odd cycles of Π. Let $A(C)$ denote the set of agents contained in a single cycle or a collection of cycles C. Unless specified otherwise, let $n_1 = |A(\mathcal{T}\mathcal{E})|$ and $n_2 = |A(\mathcal{O}_I)|$.*

Definition 7 (Reduced Stable Partition). *A cycle is called* reduced *if its length is either 2 or odd and* non-reduced *if not. Similarly, a stable partition Π is* reduced *if it consists only of reduced cycles, and* non-reduced *if not.*

We will also need a definition for sub-sequences of cycles.

Definition 8 (Partial Cycles). *Let $C = (a_{i_1}\ a_{i_2} \dots a_{i_k} \dots)$ be a cycle where only some agents $a_{i_1} \dots a_{i_k}$ and their position in the cycle are known. Then we call C a* partial cycle *and if there exist agents $a_{i_{k+1}} \dots a_{i_r}$ such that cycle $(a_{i_1}\ a_{i_2} \dots a_{i_k}\ a_{i_{k+1}} \dots a_{i_r})$ is stable, then C is a* partial stable cycle *and the sequence of agents $a_{i_{k+1}} \dots a_{i_r}$ is a* completion *of C.*

Now we can introduce the notion of blocking agents and pairs for stable partitions. Note that the definition is very close to that of blocking pairs of stable matchings, and it should be clear from the context which definition applies.

Definition 9 (Blocking Pairs and Agents). *Let $I = (A, \succ)$ be an* SR *instance and let Π be a partition of A. Then an agent $a_i \in A$* blocks Π *(is a* blocking agent*) if $\Pi^{-1}(a_i) \succ_i \Pi(a_i)$. Two agents a_i, a_j* block *(form a* blocking pair*) of Π if $a_j \succ_i \Pi^{-1}(a_i)$ and $a_i \succ_j \Pi^{-1}(a_j)$. In the second case, we can also say that (wlog) a_i* blocks Π with a_j.

Note that a stable matching $M = \{\{a_{i_1}, a_{i_2}\}, \dots, \{a_{i_{2k-1}}, a_{i_{2k}}\}\}$ consisting of k pairs can be used interchangeably with its induced collection of transpositions $(a_{i_1}\ a_{i_2}) \dots (a_{i_{2k-1}}\ a_{i_{2k}})$, which is a stable partition of I. The following has been shown and will also be assumed throughout this work.

Theorem 1 ([24,25]). *The following properties hold for any* SR *instance I.*

- *I admits at least one stable partition and any non-reduced stable partition can be transformed into a reduced one by breaking down its longer even length cycles into collections of transpositions.*
- *Any two stable partitions of I contain the same odd cycles.*
- *I admits a complete stable matching if and only if no stable partition of I contains an odd cycle.*

3 Structure of Stable Cycles, Transpositions, and Partitions

First, we will formalise and characterise the structure of reduced stable partitions in Sect. 3.1. Then, naturally, we will extend the study to all stable partitions. This requires the study of stable cycles. After showing various results about the structure of even cycles in Sect. 3.2, we focus on algorithmic questions related to the enumeration of all reduced and non-reduced stable cycles in Sect. 3.3, finishing this section with an enumeration algorithm for all stable partitions in Sect. 3.4 which ties together most of our results.

3.1 Structure of Reduced Stable Partitions

Let $I = (A, \succ)$ be an instance of SR and let Π_i be any reduced stable partition of I. We know that its odd cycles $\mathcal{O}_I$ are invariant under all stable partitions of I, while the transpositions might not be. Let $I_E = (A', \succ')$, where $A' = A \backslash A(\mathcal{O}_I)$ and $\succ'$ is the restriction of $\succ$ to the agents in A', be the instance I restricted to agents in transpositions.

Lemma 1. *Let Π_i be any reduced stable partition of an* SR *instance I. Then Π_i is the union $M_i \mathcal{O}_I$ of its odd cycles $\mathcal{O}_I$ and a perfect stable matching M_i of the sub-instance I_E of I constructed above as a cyclic permutation of transpositions.*

On the other hand, we are not guaranteed a bijection between the set $P(I)$ of all reduced stable partitions of I and the set $S(I_E)$ of all stable matchings of I_E. To deal with this case, we can reduce I_E further. Specifically, all pairs of agents $\{a_i, a_j\} \subseteq A'$ need to be deleted (made mutually unacceptable) from the preference relation $\succ'$ whenever there exists an agent $a_r \in A(\mathcal{O}_I)$ such that $a_i \succ_r \mathcal{O}_I^{-1}(a_r)$ and $a_r \succ_i a_j$. This is equivalent to truncating the preference list of a_i at the first agent $a_s \in A'$ such that $a_s \succ_i a_r$, for the best-ranked such $a_r \in A(\mathcal{O}_I)$ of a_i, and removing a_i from the preference list of every agent in A' that is less preferred by a_i. The deletion operations do not reduce the set of agents, just the preference relation, so we let the resulting sub-instance of I_E be denoted by $I_T = (A', \succ'')$. We assume that preference comparison and partner lookup can be done in constant time, such that the procedure takes at most linear time overall.

Although these deletions might destroy some stable matchings admitted by I_E, we must ensure that I_T still admits at least one.

Lemma 2. *Let $\Pi_i = M_i \mathcal{O}_I$ be any reduced stable partition of an* SR *instance I and let I_T be its sub-instance as constructed above. Then M_i is a perfect stable matching of I_T.*

We can now show the correspondence between the stable matchings of I_T and the reduced stable partitions of I.

Theorem 2. *M_i is a perfect stable matching of I_T if and only if $M_i \mathcal{O}_I$ is a reduced stable partition of I, where I_T is constructed as above and $\mathcal{O}_I$ are the invariant odd cycles of I.*

Corollary 1. *The number of reduced stable partitions that an* SR *instance I with $n \in 2\mathbb{N}$ agents admits is equal to the number of stable matchings that its sub-instance I_T with $n - |A(\mathcal{O}_I)|$ agents admits.*

With these results, we now have a procedure to enumerate all reduced stable permutations by enumerating $S(I_T)$ using the algorithm by Feder [6].

Theorem 3. *We can enumerate all reduced stable partitions of an* SR *instance $I = (A, \succ)$ in $O(|S(I_T)|n + n^2)$ time, where I_T is the sub-instance of I as constructed above, $S(I_T)$ is the set of all stable matchings of I_T, and $n = |A|$.*

3.2 Structure of Longer Even Cycles

Reduced stable partitions are canonical in the sense that they cannot be broken down into smaller cycles. However, while Tan [25] shows that any stable cycle of even length greater than two can be decomposed into a stable collection of transpositions, it may be of theoretical interest to look at how longer cycles can be constructed from a collection of transpositions. Naively, we could try any combination of transpositions to construct longer even cycles. However, this would quickly lead to a combinatorial explosion due to the potential number of transpositions involved. Instead, we look at the different ways to decompose an even cycle into transpositions.

Lemma 3. *Any even cycle $C = (a_{i_1}\ a_{i_2}\ \dots\ a_{i_{2k}})$ of some stable partition Π where $k \geq 2$ can be broken into two distinct collections of transpositions $C_1 = (a_{i_1}\ a_{i_2})(a_{i_3}\ a_{i_4})\dots(a_{i_{2k-1}}\ a_{i_{2k}})$ and $C_2 = (a_{i_1}\ a_{i_{2k}})(a_{i_2}\ a_{i_3})\dots(a_{i_{2k-2}}\ a_{i_{2k-1}})$ such that both partitions $\Pi_1 = (\Pi \backslash C) +\!\!+\ C_1$ and $\Pi_2 = (\Pi \backslash C) +\!\!+\ C_2$ are stable.*

However, there might be more ways to rearrange the agents of an even cycle into a stable collection of disjoint transpositions.

Theorem 4. *Any non-reduced stable partition Π can be constructed from two reduced stable partitions Π_a, Π_b.*

This gives a naive way to construct all stable partitions from the set of reduced stable partitions.

Corollary 2. *We can construct all stable partitions, $P(I)$, of an* SR *instance I from its set of reduced stable partitions, $RP(I)$, in $O(|RP(I)|^2 n^2)$ time without repetitions.*

As $|RP(I)|$ may be exponential in n and some combinations of reduced stable partitions might not to lead to a non-reduced stable partition, we focus the study on stable cycles themselves and how they could be useful in enumerating all stable partitions directly. First, to refine the search for longer even cycles, we can show restrictions on the combinatorics of agents. We want to show that it suffices to know a predecessor-successor pair of agents in a partial even cycle $C = (a_{i_1}\ a_{i_2}\ \dots)$ to find a completion if one exists, and that we can find such a completion in polynomial time. For this, we show that the correspondence between C and $C\backslash\{a_{i_2}\}$, where the latter is a potential odd cycle of some related instance. Specifically, the procedure takes as input the instance and a predecessor-successor pair a_{i_1}, a_{i_2} of the partial (not necessarily stable) cycle in question and creates an instance I_S related to I without agent a_{i_2} in which the preference lists of all agents preferred by a_{i_2} over a_{i_1} in I are modified in I_S such that a_{i_1} takes the spot of a_{i_2} and its previous entry is deleted, and these agents, if not preferred by a_{i_1} over a_{i_2}, are promoted in the preference list of a_{i_1} in I_S to before the rank position of a_{i_2} in a_{i_1}'s preference list in I while maintaining the relative preference order given by a_{i_2} in I.

Lemma 4. *Let I be an* SR *instance, let $(a_{i_1}\ a_{i_2} \dots)$ be a partial cycle of I, and let I_S be the instance constructed from I as above. If $(a_{i_1}\ a_{i_2} \dots)$ has a completion to an even stable cycle, then a_{i_1} belongs to an odd stable cycle in I_S.*

Due to odd cycles being invariant, we can use the statement above to reason about the structural restrictions of longer even cycles implied by a predecessor-successor pair.

Lemma 5. *If $C = (a_{i_1}\ a_{i_2}\ a_{i_3}\ \dots\ a_{i_{2k}})$ and $C' = (a_{i_1}\ a_{i_2}\ a'_{i_3}\ \dots\ a'_{i_{2l}})$ are both even cycles of I, then $k = l$ and $a_{i_j} = a'_{i_j}$ for $3 \leq j \leq 2k$.*

Now note that this also implies that no instance can admit, for example, the following stable partitions: $\Pi = (a_1\ a_2\ a_3\ a_4\ a_5\ a_6)$, $\Pi' = (a_1\ a_2\ a_3\ a_4)(a_5\ a_6)$.

Lemma 6. *If some stable partition Π of* SR *instance I contains an even cycle $C = (a_{i_1}\ \dots\ a_{i_k})$, then C cannot be contained in a longer even cycle $C' = (a_{i_1}\ \dots\ a_{i_k}\ a_{j_1}\ \dots\ a_{j_l})$ of some other stable partition Π' of I.*

The results also greatly restrict the number of even cycles that an instance can admit.

Lemma 7. *Any* SR *instance with n agents admits at most $O(n_1^2 + n_2) = O(n^2)$ cycles of length not equal to two.*

Finally, Lemma 5 also establishes the following powerful result.

Theorem 5. *Let* $(a_{i_1}\ a_{i_2} \dots)$ *be a partial cycle in some* SR *instance* I*. In linear time, we can determine whether this partial cycle has a completion* $C = (a_{i_1}\ a_{i_2}\ a_{i_3} \dots a_{i_{2k}})$ *to an even cycle of some stable partition of* I*, or report that no such completion exists. If a completion* C *to a stable even cycle does exist, then* C *must be unique.*

Proof. We can construct I_S from I using the procedure described above and find some stable partition Π' of I_S, all in linear time. Now, if a_{i_1} does not belong to an odd stable cycle C' of Π', then by Lemma 4, $(a_{i_1}\ a_{i_2} \dots)$ has no completion to a stable even cycle of I.

Otherwise, suppose that $C' = (a_{i_1}\ a_{i_3} \dots a_{i_{2k}})$ is an odd stable cycle of Π'. Then the cycle $C = (a_{i_1}\ a_{i_2}\ a_{i_3}\ \dots\ a_{i_{2k}})$ is an even stable cycle of I unless $\{a_{i_1}, a_{i_3}\}$ block in $(\Pi' \backslash C') +\!\!+\ C$.

Thus, we can either output C, which is unique by Lemma 5, or it must be the case that "no" is the correct answer, as by the same Lemma, if $(a_{i_1}\ a_{i_2} \dots)$ has a completion to an even stable cycle in I, then $C = (a_{i_1}\ a_{i_2}\ a_{i_3} \dots a_{i_{2k}})$ is the only candidate. □

We have shown above that if C is a partial even stable cycle of the considered instance I, then we must have a corresponding invariant odd cycle C' of some related instance of I, and that therefore, C has a unique completion. Furthermore, Theorem 5 establishes that we only need to consider a potential successor-predecessor pair to see whether that pair is indeed part of some longer unique stable even cycle, and we can do so in linear time. Furthermore, given an instance I and a (non-partial) candidate cycle C, we can also efficiently verify whether it is in some stable partition as it is, by Theorem 5, sufficient to consider any successor-precessor pair in it. If so, we can also explicitly construct a stable partition Π containing C using the method described in the proof, where $\Pi = (\Pi' \backslash C') +\!\!+\ C$.

Corollary 3. *Given an* SR *instance* I *and a candidate (not necessarily stable) cycle* C *of even length, we can verify whether* C *is stable and, if so, construct a stable partition* Π *containing* C*, all in linear time.*

3.3 All Stable Cycles

With stable matchings, we can consider the problem of finding all stable pairs, denoted $SP(I)$, which is defined as the union of all pairs contained in some stable matching of an SR instance I. Similarly, one can ask about the fixed pairs, which are all pairs contained in all stable matchings of I. Analogously, we introduce the corresponding questions for stable partitions: given an SR instance I, what are all the stable cycles, denoted $SC(I)$, admitted by some stable partition of the instance? Similarly, what are the fixed stable cycles, denoted $FC(I)$, admitted by all its stable partitions?

The enumeration of reduced stable partitions might naturally run in time exponential in the number of agents, similar to the enumeration of all stable matchings of a solvable instance. In contrast, when only looking at stable pairs rather than matchings, Feder [7] presented an $O(n^3)$ algorithm to find them in a SR instance with n agents. Therefore, it is natural to ask whether we can construct a similar efficient algorithm for finding all stable cycles of a stable partition. Given an SR instance I, we are looking to find stable cycles $SC(I) = \{C \in \Pi \mid \Pi \in P(I)\}$ where $P(I)$ are all stable partitions admitted by I. Now note that the odd cycles $\mathcal{O}_I$ are invariant and can be computed efficiently, so we require to find the union of all stable cycles of even length. Observe that $\mathcal{O}_I \subseteq FC(I) \subseteq RSC(I) \subseteq SC(I)$, where $RSC(I)$ are the reduced stable cycles (i.e. stable transpositions and odd cycles) of I. This is because for any stable cycle C of even length longer than 2, we can break it into two collections of stable transpositions by Lemma 3, such that C cannot be fixed. We can find all reduced fixed cycles directly in $O(n^2)$ time by constructing the odd cycles $\mathcal{O}_I$, which are fixed, and the sub-instance I_T. Then, we can derive the fixed pairs (i.e. fixed stable transpositions) as described by Gusfield and Irving [14].

Theorem 6. *Given an* SR *instance I with n agents, we can find all its fixed cycles in linear time.*

Theorem 2 states that any reduced stable partition Π_i is the union of some stable matching M_i of the sub-instance I_T of I, and $\mathcal{O}_I$. Therefore, in order to find all stable transpositions, we can apply the known all stable pairs algorithm to I_T.

Lemma 8. *Given an* SR *instance I with n agents, we can find all reduced stable cycles, $RSC(I)$, i.e. odd cycles and stable transpositions, in $O(n^2 + n_1^3)$ time.*

This lets us extend Lemma 7 to include transpositions.

Theorem 7. *Any* SR *instance with n agents admits at most $O(n_1^2 + n_2) = O(n^2)$ stable cycles.*

Now to compute all stable cycles, not necessarily reduced or fixed, we can use our previous work on reconstructing stable cycles of even length longer than 2 to show the following.

Theorem 8. *Given an* SR *instance I with n agents, we can find all its stable cycles, $SC(I)$, in $O(n_1^2(n^2 + n_1)) = O(n^4)$ time.*

Proof. Due to Lemma 8, we can find all reduced stable cycles, i.e. all stable transpositions and odd cycles, in $O(n^2 + n_1^3)$ time.

Now with regards to the stable cycles of even length longer than 2, we can consider all pairs of agents that could be part of a longer even stable cycle. There are $\binom{n}{2} = O(n^2)$ such subsets in general, but we can limit ourselves to the stable transpositions we found in the previous step due to Theorem 4, in which we showed that any even cycle can be broken into at least two collections of stable transpositions, therefore no agent pair that is not a stable transposition will lead to an even stable cycle.

For each pair $\{a_i, a_j\}$, we need to consider each of the orderings $(a_i\ a_j\ \ldots)$ and $(a_j\ a_i\ \ldots)$. Clearly, there are $O(n_1^2)$ such pairs. Then, for each ordering, we can try to find its unique completion in $O(n^2)$ time due to Theorem 5 and finally verify that it is a stable cycle, also in $O(n^2)$ time due to Corollary 3. Altogether, we consider $O(n_1^2)$ candidates and need $O(n^2)$ time for each, therefore requiring $O(n_1^2 n^2)$ time. As any consecutively ordered pair of agents will only be part of at most one stable cycle of length longer than 2 by Theorem 5, we can eliminate all such pairs from the candidate set once detected. This will ensure that the even cycles are enumerated exactly once.

Altogether, the steps take $O(n_1^2(n^2 + n_1))$, or $O(n^4)$ time. □

3.4 Enumerating Stable Partitions

Now that we know how to find all stable cycles of our SR instance I efficiently, we show that this leads to an alternative approach to enumerating all stable partitions, $P(I)$, compared to the naive method from Corollary 2. In this procedure, we will recursively build up all stable partitions, starting with the invariant odd cycles and adding one additional cycle with each recursive call. Simultaneously, at each recursive step, we fix an arbitrary agent not yet in any cycle of the partial partition and branch on all its cycles (as given by $SC(I)$) that, together with the current partial partition, lead to at least one stable partition. This way, the recursive tree consists of paths leading from the origin (a partial stable partition containing only the odd cycles) to all leaf nodes (all stable partitions).

The procedure $\texttt{EnumP}_I$ described in Algorithm 1 implements the logic described in the previous paragraph to enumerate all stable partitions. The parameter "cycles" contains a partial stable partition and grows with each recursive call until it is a full stable partition. The parameter "agents" contains all agents in I not yet in any cycle of "cycles" and is naturally decreasing in size as "cycles" grows. Finally, the parameter SC contains all cycles of some stable partition of I containing some agent in "agents", not necessarily compatible with "cycles" (in the sense that not every cycle in SC belongs to a stable partition with "cycles"). Note that by this behaviour, if "agents" is empty, then "cycles" is a stable partition and we print it. If not, then the algorithm proceeds to return and remove an arbitrary agent a_i from "agents". Then, the algorithm checks whether there is a unique cycle c containing a_i such that "cycles" $+\!\!+$ c is part of some stable partition of I using the omitted procedure FC which uses Theorem 6. If there is a unique such c, then we reduce the set of stable cycles SC to SC_{red} which does not contain any cycles with agents in c and recurse with the increased partial stable partition "cycles" $+\!\!+$ c, the reduced set of stable cycles to choose from SC_{red}, and the reduced set of remaining agents "agents"$\setminus A(c)$. If there is no unique choice c, then the procedure loops through all candidate cycles $c' \in SC$ that contain a_i, at least two of which must correspond to a partial stable partition "cycles" $+\!\!+$ c' by construction. Whether "cycles" $+\!\!+$ c' does correspond to a partial stable partition or only a collection of cycles that is not a subset of any stable partition can be determined with the procedure `verify` which is omitted but uses the result from Corollary 3 to check whether the collection of

Algorithm 1 $\texttt{EnumP}_I(SC$, cycles, agents), recursively enumerates all stable partitions $P(I)$

Input: SC: all cycles of some stable partition of I containing some agent in "agents" and no agent in A(cycles); cycles: a collection of cycles which is a subset of some stable partition of I and includes all odd cycles of I; agents: all agents in $A(I)\backslash A$(cycles)

```
if agents = ∅ then
    print(cycles)
    return
end if
a_i ← agents.pop()
fixed ← FC(I, cycles, a_i) ▷ finds the fixed cycle compatible with "cycles" that a_i in or returns null
if fixed ≠ null then
    SC_red ← {(a_j1 ... a_jk) ∈ SC | ∀ 1 ≤ s ≤ k, a_js not in fixed}
    EnumP_I(SC_red, cycles ++ fixed, agents\A(fixed))
else
    for cycle in SC containing a_i do
        if verify(cycles ++ cycle) then ▷ checks whether the cycle combination is part of some stable partition
            SC_red ← {(a_j1 ... a_jk) ∈ SC | ∀ 1 ≤ s ≤ k, a_js not in cycle}
            EnumP_I(SC_red, cycles ++ cycle, agents\A(cycle))
        end if
    end for
end if
```

cycles completes to a stable partition. If so, we, again, create the appropriate SC_{red} and recurse as above.

Overall, this gives a method to enumerate $P(I)$ by first finding all $O(n_1^2+n_2)$ stable cycles, $SC(I)$, in $O(n_1^2(n^2+n_1))$ time as previously shown and then calling the procedure $\texttt{EnumP}_I$ described in Algorithm 1 with stable cycles $SC(I)\backslash \mathcal{O}_I$ as parameter SC, the invariant odd cycles $\mathcal{O}_I$ as parameter "cycles", and agents $A(I)\backslash A(\mathcal{O}_I)$ as parameter "agents".

Theorem 9. *Let I be an instance of* SR *with n agents. Then Algorithm 1 enumerates all stable partitions $P(I)$ of I without repetition in $O(|P(I)|n^3 + n^4)$ time, or more specifically, $O((|P(I)| + n_1)n_1n^2 + n_1^3)$ time.*

4 Complexity of Profile-Optimal Stable Partitions

We have shown how to efficiently enumerate all reduced stable partitions admitted by an SR instance I. However, as the size of the set $S(I_T)$ of all stable matchings of the sub-instance I_T of I can, in general, be exponential in the number of agents n, this might not be a good solution to finding a reduced stable partition with some special property. Therefore, we might be interested in computing some "fair" or "optimal" stable partitions directly. As previously noted, there are different measures of fairness and optimality for stable matchings considered in the literature. The main measure is the *profile* $p(M)$ of a matching M, a vector $(p_1 \dots p_{n-1})$, where p_i counts the number of agents assigned to their ith choice in M. We define a similar measure for stable partitions, accounting for all half-assignments in the associated half-matching.

Definition 10. *Let I be an* SR *instance with n agents and Π a stable partition of I. An agent a_i has* rank *i for an agent a_j if it appears at position $i-1$ in a_j's preference list. We define the* successor profile, *denoted $p^s(\Pi) = (p_1^s \dots p_n^s)$, to capture the number of agents p_i^s whose successor in Π has rank i. Analogously, let $p^p(\Pi)$ denote the* predecessor profile *of Π. Finally, let $p(\Pi) = (p_1 \dots p_n)$ denote the combined* profile *of Π, where $p_i = p_i^s + p_i^p$. Similarly, for a stable cycle C, let $p^s(C), p^p(C)$, and $p(C)$ denote the profiles containing the number and positions of predecessors, successors, or both achieved in C. The* regret *of Π, denoted $r(\Pi)$, is the rank of the worst-ranked agent assigned in Π, which is the index of the last positive entry in $p(\Pi)$. Similarly, for C some cycle of agents, let $r(C)$ be the rank of the worst-ranked agent assigned in C. The* cost *of Π, denoted $c(\Pi)$, is the sum of ranks achieved in Π, averaged over the predecessor and successor assignments. Specifically, we define $c(\Pi) = \frac{1}{2}\sum_{1 \leq i < n} p_i(\Pi) * i$.*

As the odd cycles of SR instances are invariant, there is no point in seeking odd cycles with special properties for a given instance. However, we can consider the set of stable partitions of a given instance and seek, for example, a reduced stable partition with minimum regret. In this case, it is not enough to simply leverage the result from Theorem 2, as we are deleting entries in the preference lists during the construction of I_T, which makes the resulting profile entries in the larger instance inconsistent. For this, we use a *padding method* by replacing deleted agents in the preference lists by dummy agents which are guaranteed to be matched to each other by construction. Therefore, a solvable instance similar to I_T is created in which the ranks are maintained and known algorithms can be applied.

First, using the padding method, we can show that the problem of finding a minimum-regret stable partition is tractable.

Lemma 9. *Let I be an* SR *instance, let $\Pi = M_0\mathcal{O}_I$ be a reduced stable partition of I, where M_0 is a collection of transpositions and $\mathcal{O}_I$ are the odd cycles in Π, and I_P be the transformed instance using the padding method. If M_0 is a stable matching of I_P with minimum regret, then Π is a reduced stable partition with minimum regret.*

Note that the converse does not necessarily hold, because a reduced stable partition $\Pi = M_0\mathcal{O}_I$ as above could have $r(M_0) < r(\Pi) = r(\mathcal{O}_I)$.

We can furthermore show that there is no gap in minimum regret between the set of reduced stable partitions and the set of stable partitions.

Lemma 10. *Let Π be a minimum-regret reduced stable partition of some* SR *instance I and let $r(\Pi)$ be its regret. Then for any stable partition Π' of I, we have $r(\Pi) \leq r(\Pi')$.*

Theorem 10. *A stable partition with minimum regret always exists and can be computed in linear time.*

Consistent with stable matchings, minimum regret remains an exception concerning the tractability of profile-based optimal stable partitions.

With our definitions of profile, regret, and cost for stable partitions, we can define analogous optimality measures for stable partitions (and thus stable half-matchings). For some, we need an order on profiles as defined below.

Definition 11. *Let* $p = (p_1 \dots p_n), p' = (p'_1 \dots p'_n)$ *be two profile vectors. Then* $p = p'$ *if* $p_i = p'_i$ *for all* $1 \leq i \leq n$*. If* $p \neq p'$*, let* k *be the first position in which they differ. We define* $p \succ p'$ *if* $p_k > p'_k$*. Furthermore, we define* $p \succeq p'$ *if either* $p \succ p'$ *or* $p = p'$*. Finally, we will call* $p^{rev} = (p_n \dots p_1)$ *the* reverse profile *of* p*.*

We now also consider the following decision problems.

Definition 12. *Given an* SR *instance* I *with* n *agents and regret* r*, an integer* k*, and a vector* $\sigma \in \{0, 1, \dots, n\}^n$*, let*

- FC-Dec-SR-SP *denote the problem of deciding whether* I *admits a stable partition* Π *where* $p_1(\Pi) \geq k$*,*
- Rank-Dec-SR-SP *denote the problem of deciding whether* I *admits a stable partition* Π *where* $p(\Pi) \succeq \sigma$*,*
- RM-Dec-SR-SP *denote the problem of deciding whether* I *admits a minimum regret stable partition with at most* k r*th choices,*
- Gen-Dec-SR-SP *denote the problem of deciding whether* I *admits a stable partition* Π *where* $\sigma \succeq p^{rev}(\Pi)$*, and*
- Egal-Dec-SR-SP *denote the problem of deciding whether* I *admits a stable partition with cost at most* k*.*

We can show for all of these problems that no non-reduced stable partition is "better" with regards to our optimality criteria than the best reduced stable partition, which allows us to only consider reduced stable partitions. Knowing the NP-completeness of the associated problems above for stable matchings, we can show that all of these problems are NP-complete for stable partitions, even if the instance is solvable. One might ask whether the same result holds for unsolvable instances, which have some different structural properties. Using a reduction from solvable instances, we can show that this is indeed the case.

Theorem 11. *The problems* FC-Dec-SR-SP, Rank-Dec-SR-SP, RM-Dec-SR-SP, Gen-Dec-SR-SP, *and* Egal-Dec-SR-SP *are all NP-complete, even if the instance is solvable and even if it is unsolvable.*

Furthermore, the following approximation bounds apply reduced from the approximation bounds for stable matchings due to Simola and Manlove [23].

Theorem 12. *The problem of finding a stable partition* Π *of some* SR *instance* I *with a maximal number of first choices does not admit a polynomial-time approximation algorithm with any constant-factor performance guarantee, even if* I *is solvable and even if* I *is unsolvable, unless* $P = NP$*.*

However, for two other problems, the approximation method for OPTIMAL SR [14] due to Teo and Sethuraman [27] applies.

Theorem 13. *The problems of finding a stable partition Π of some* SR *instance I with minimal cost or minimal number of $r(I)$ choices admits polynomial-time approximation algorithms with a multiplicative performance guarantee of 2.*

These results are consistent with the approximability results of stable matchings, which fits the theme of compatibility and close correspondence throughout this research. Note that the NP-hardness arguments assume a solvable (sub-)instance that is computationally difficult in the considered context. However, the smaller the sub-instance, the easier to compute an optimal solution, as there is no choice involved in the odd cycles. In the extreme case, all agents are part of odd cycles, which means that there exists a unique stable partition for the instance which is automatically optimal regardless of the objective.

5 Discussion and Open Problems

In this paper, we have shown various new insights into the structure of stable partitions. Specifically, that there is a close correspondence between reduced stable partitions of unsolvable instances and stable matchings of certain solvable sub-instances, how even cycles can be constructed from transpositions and partial cycles, with the highlight being a result proving that a predecessor-successor pair is sufficient to find the unique stable cycle of even length longer than 2 (if one exists). Using these insights, we also provided the first algorithms to efficiently enumerate all reduced and non-reduced stable cycles and partitions of a given problem instance, which could be useful for a variety of cases. Furthermore, we showed how the stable matching notions of profile- and cost-optimality carry over to stable partitions and which of the problems are efficiently computable. This bridges the gap between stable matchings, which do not always exist, and fractional matchings, which always exist but may not be useful in certain practical applications. For a comparative overview of the complexity and approximability results previously known and newly established, see Table 1.

Overall, we have shown a close correspondence between stable matchings and stable partitions in every way, but many open questions remain. One direction for future work is to improve on the complexity results presented here, for example by investigating whether a technique similar to the 2-SAT reduction by Feder [7] or the transformations between non-bipartite to bipartite instances presented by Dean and Munshi [5] can speed up the enumeration of $SC(I)$ and $P(I)$. Currently, it is not clear how to approximate rank-maximal- and generous-profile problems, neither for matchings nor for partitions. On the structural side, it would be interesting to study whether stable half-matchings form a semi-lattice similar to stable matchings. It would also be interesting to confirm that the results presented here hold in the cases of incomplete preference lists and an odd number of agents. There might also be a closer relationship than currently discovered between the structure of stable partitions and approximate or exact

Table 1. Complexity results. Results from existing literature or not applicable are highlighted in grey, whilst new results are highlighted in white. n is the number of agents in the instance, n_1 of which are not in invariant odd cycles.

	Stable Matching	**Stable Partition**
Always Exists	No	Yes
Find Any (if exists)	$O(n^2)$ [15]	$O(n^2)$ [24]
Find $SP(I)/SC(I)$	$O(n^3)$ [7]	$O(n_1^2(n^2+n_1))$ (Theorem 8)
Find $FP(I)/FC(I)$	$O(n^2)$ [14]	$O(n^2)$ (Theorem 6)
Find $S(I)/RP(I)$	$O(\|S(I)\|n+n^2)$ [7]	$O(\|S(I_T)\|n+n^2)$ (Theorem 3)
Find $P(I)$		$O(\|P(I)\|n^3+n^4)$ (Theorem 9)
Min Regret	$O(n^2)$ [14]	$O(n^2)$ (Theorem 10)
Rank-Max	NP-h [4]	NP-h (Theorem 11)
Generous	NP-h [4]	NP-h (Theorem 11)
Egalitarian	NP-h [6], but 2-approx. [7]	NP-h (Theorem 11), but 2-approx. (Theorem 13)
Regret-Min	NP-h [4], but 2-approx. [23]	NP-h (Theorem 11), but 2-approx. (Theorem 13)
First-Choice Max	NP-h [4], and Inapprox. within $c \in \mathbb{R}$ [23]	NP-h (Theorem 11), and Inapprox. within $c \in \mathbb{R}$ (Theorem 12)

solutions to the almost-stable matching problem. Finally, it could be investigated how stable partitions can be generalised to the many-to-many setting, such as the STABLE FIXTURES problem [18].

Acknowledgments. Frederik Glitzner is supported by a Minerva Scholarship from the School of Computing Science, University of Glasgow. David Manlove is supported by the EPSRC, grant number EP/X013618/1. We would like to thank the anonymous MATCH-UP and SAGT reviewers for their helpful suggestions. The authors have no competing interests to declare that are relevant to the content of this article.

References

1. Abraham, D.J., Biró, P., Manlove, D.F.: "Almost stable" matchings in the Roommates problem. In: Erlebach, T., Persinao, G. (eds.) WAOA 2005. LNCS, vol. 3879, pp. 1–14. Springer, Heidelberg (2006). https://doi.org/10.1007/11671411_1
2. Aharoni, R., Fleiner, T.: On a lemma of Scarf. J. Comb. Theory Ser. B **87**(1), 72–80 (2003)
3. Chen, J., Roy, S., Sorge, M.: Fractional matchings under preferences: stability and optimality. In: Proceedings of the IJCAI 2021, vol. 1, pp. 89–95 (2021)
4. Cooper, F.: Fair and large stable matchings in the stable marriage and student-project allocation problems. Ph.D. thesis, University of Glasgow (2020)
5. Dean, B.C., Munshi, S.: Faster algorithms for stable allocation problems. Algorithmica **58**(1), 59–81 (2010)
6. Feder, T.: A new fixed point approach for stable networks and stable marriages. J. Comput. Syst. Sci. **45**(2), 233–284 (1992)

7. Feder, T.: Network flow and 2-satisfiability. Algorithmica **11**, 291–319 (1994)
8. Fleiner, T.: The stable roommates problem with choice functions. Algorithmica **58**, 82–101 (2010)
9. Gale, D., Shapley, L.S.: College admissions and the stability of marriage. Am. Math. Mon. **69**, 9 (1962)
10. Glitzner, F., Manlove, D.: Structural and algorithmic results for stable cycles and partitions in the roommates problem (2024). https://doi.org/10.48550/arXiv.2406.00437. arXiv:2406.00437
11. Gupta, S., et al.: Popular matching in roommates setting is NP-hard. ACM Trans. Comput. Theory **13**, 1–20 (2021)
12. Gusfield, D.: The structure of the stable roommate problem: efficient representation and enumeration of all stable assignments. SIAM J. Comput. **17**, 742–769 (1988)
13. Gusfield, D., Pitt, L.: A bounded approximation for the minimum cost 2-sat problem. Algorithmica **8**, 103–117 (1992)
14. Gusfield, D., Irving, R.W.: The Stable Marriage Problem: Structure and Algorithms. MIT Press, Cambridge (1989)
15. Irving, R.W.: An efficient algorithm for the "stable roommates" problem. J. Algorithms **6**, 577–595 (1985)
16. Irving, R.W.: On the stable roommates problem. Research report CSC/86/R5, Department of Computing Science, University of Glasgow (1986)
17. Irving, R.W., Manlove, D.: The stable roommates problem with ties. J. Algorithms **43**, 85–105 (2002)
18. Irving, R.W., Scott, S.: The stable fixtures problem-a many-to-many extension of stable roommates. Discrete Appl. Math. **155**, 2118–2129 (2007)
19. Manlove, D.: Algorithmics of Matching Under Preferences. Series on Theoretical Computer Science, vol. 2. World Scientific (2013)
20. Mertens, S.: Random stable matchings. J. Stat. Mech: Theory Exp. **2005**, P10008 (2005)
21. Pittel, B.: On a random instance of a 'stable roommates' problem: likely behavior of the proposal algorithm. Comb. Probab. Comput. **2**, 53–92 (1993)
22. Pittel, B.: On random stable partitions. Int. J. Game Theory **48**, 433–480 (2019)
23. Simola, S., Manlove, D.: Profile-based optimal stable matchings in the roommates problem (2021). https://doi.org/10.48550/arXiv.2110.02555. arXiv:2110.02555
24. Tan, J.J.: A necessary and sufficient condition for the existence of a complete stable matching. J. Algorithms **12**, 154–178 (1991)
25. Tan, J.J.: Stable matchings and stable partitions. Int. J. Comput. Math. **39**, 11–20 (1991)
26. Tan, J.J., Hsueh, Y.C.: A generalization of the stable matching problem. Discrete Appl. Math. **59**, 87–102 (1995)
27. Teo, C.P., Sethuraman, J.: LP based approach to optimal stable matchings. In: Proceedings of SODA 1997, pp. 710–719 (1997)

Online Matching with High Probability

Milena Mihail and Thorben Tröbst(✉)

Department of Computer Science, University of California, Irvine, USA
{mihail,t.troebst}@uci.edu

Abstract. We study the classical, randomized RANKING algorithm, which is known to be $(1 - \frac{1}{e})$-competitive in expectation for the Online Bipartite Matching Problem. We give a tail inequality bound (Theorem 1), namely that RANKING is $(1 - \frac{1}{e} - \alpha)$-competitive with probability at least $1 - e^{-2\alpha^2 n}$ where n is the size of the maximum matching in the instance. Building on this, we show similar concentration results for several generalizations of the Online Bipartite Matching Problem, including the Fully Online Matching Problem and the Online Vertex-Weighted Bipartite Matching Problem.

Keywords: Online Algorithms · Concentration of Measure · Online Matching · Randomized Algorithms

1 Introduction

In the Online Bipartite Matching Problem, we have an undirected, bipartite graph $G = (S, B, E)$ with a set S of *goods* and a set B of *buyers*. The buyers arrive online in adversarial order and every time a buyer i arrives, all of its neighbors $N(i)$ are revealed. We must then decide immediately and irrevocably which as of yet unmatched neighbor of i should get matched to i. The goal is to maximize the number of edges in the final matching M relative to the size of the maximum matching in G (in the worst case over all choices of G and arrival orders of the buyers), i.e. the so-called competitive ratio.

A matching M is considered *maximal* if there is no edge in E which can be added to M while preserving the matching property. Due to the well-known fact that every maximal matching contains at least half of the edges of any *maximum* matching, it is easy to see that any algorithm which matches arriving buyers whenever possible must be $\frac{1}{2}$-competitive. Moreover, because of the adversarial arrival of buyers and choice of the underlying graph, this is best possible for deterministic algorithms. In their seminal work, Karp et al. [13] defined the randomized RANKING algorithm (see Algorithm 1) and showed that it is $(1 -$

M. Mihail—The author was supported in part by a UCI-faculty startup grant.
T. Tröbst—The author was supported in part by NSF grant CCF-1815901. Part of this work was done while the author attended the "Trimester Program on Discrete Optimization" at the Hausdorff Institute for Mathematics in Bonn, Germany.

G. Schäfer and C. Ventre (Eds.): SAGT 2024, LNCS 15156, pp. 21–34, 2024.
https://doi.org/10.1007/978-3-031-71033-9_2

$\frac{1}{e}$)-competitive in expectation. They also showed that this is the best possible competitive ratio for any randomized algorithm.

Algorithm 1: RANKING

```
Sample a uniformly random permutation π on S.
for each buyer i who arrives do
    Match i to the first unmatched buyer in N(i) wrt. to π.
```

Over the years, online matching problems have received a large amount of interest due to the vast number of applications created by the internet and mobile computing. Online advertising alone poses the AdWords Problem [9,17,19] that lies at the heart of a multi-billion dollar market. Another interesting application, the Fully Online Matching Problem [10,11], came about due to the rise of ride-sharing and/or ride-hailing apps such as Uber and Lyft where riders and drivers come online and need to be matched almost instantaneously while minimizing some function of latency and distance traveled to the rider. For a more complete overview of online matching and its place in matching-based market design, we refer to [6].

For many online matching problems, there are extensions of RANKING which achieve competitive ratios of $1 - \frac{1}{e}$ or at the very least strictly greater than $\frac{1}{2}$. Often, these are best-known for their respective problems. However, to the best of our knowledge, all results on RANKING-like online matching algorithms in the literature only establish the competitive ratio *in expectation* without guaranteeing any form of concentration beyond the trivial bounds that follow from Markov's inequality. Under the restriction that the graph is d-regular, Cohen and Wajc [3] proposed the MARKING algorithm for Online Bipartite Matching and showed that it has a competitive ratio of $1 - O(\sqrt{\log d}/\sqrt{d})$ in expectation and $1 - O(\log n/\sqrt{d})$ with high probability. They remark that this is the first high probability guarantee $> 1/2$ for Online Bipartite Matching, though only in this restricted setting. Accordingly, our result is the first such bound without additional assumptions on the problem instances.

The analysis of concentration bounds for randomized algorithms goes back to the 1970s with classic results such as the second moment bound for QUICKSORT [18]. See [5] for an extensive overview of the field. However, it has remained the case that in the analysis of algorithms, results are usually quantified in terms of expected solution quality only.

In some sense this is due to the well-known fact that, as a consequence of standard Chernoff bounds, any randomized algorithm which is good in expectation can be boosted to be good with probability $1 - \frac{1}{n}$ by simply repeating it $O(\log n)$ many times. But it is precisely in the case of online algorithms where this argument fails due to the fact that online algorithms, by definition, can not be repeated. Despite this, the literature on high probability bounds for online algorithms is relatively sparse (for some exceptions, see e.g. [14,15]). Given the impact that RANKING has had over the last 30 years, it is quite remarkable that such a fundamental aspect of it had been left unanswered.

1.1 Our Results and Techniques

Our results concern RANKING type algorithms in three different settings: the classic Online Bipartite Matching Problem (see Sect. 2), the Fully Online Matching Problem inspired by ride-sharing (see Sect. 3) and the Online Vertex-Weighted Bipartite Matching Problem inspired by the internet advertising markets (see Sect. 4).

In Sect. 2 we will show the following result, complementing the classic $(1-\frac{1}{e})$-competitiveness result of RANKING for the Online Bipartite Matching Problem [13].

Theorem 1. *Let $G = (S, B, E)$ be an instance of the Online Bipartite Matching Problem which admits a matching of size n. Then for any $\alpha > 0$ and any arrival order,*

$$\mathbb{P}\left[|M| < \left(1 - \frac{1}{e} - \alpha\right)n\right] < e^{-2\alpha^2 n}$$

where M is the random variable denoting the matching generated by RANKING.

The key technical ingredient for this result is a bounded differences property of the random variable $|M|$ (see Lemma 2). We prove this via structural properties of matchings (see Lemma 3) similar to ones which have been used in previous analyses of RANKING [2,8]. Together with McDiarmid's inequality shown below (a consequence of Azuma's inequality) this gives rise to a particularly natural proof of Theorem 1.

Lemma 1 (McDiarmid's Inequality[16]). *Let $c_1, \ldots, c_n \in \mathbb{R}_+$ and consider some function $f : [0,1]^n \to \mathbb{R}$ satisfying*

$$|f(x_1, \ldots, x_{i-1}, x_i', x_{i+1}, \ldots, x_n) - f(x_1, \ldots, x_n)| \leq c_i$$

for all $x \in [0,1]^n$, $i \in [n]$ and $x_i' \in [0,1]$. Moreover let Δ^n be the uniform distribution on $[0,1]^n$. Then for all $t > 0$, we have

$$\mathbb{P}_{x \sim \Delta^n}[f(x) < \mathbb{E}_{y \sim \Delta^n}[f(y)] - t] < e^{-\frac{2t^2}{\sum_{i=1}^n c_i^2}}.$$

We want to contrast this technique briefly with two related results. The analysis of MARKING by Cohen and Wajc [3] uses the d-regularity of the graph in an essential way. They are able to show directly that the probabilities that the offline vertices are unmatched are negatively correlated and apply a Chernoff bound. In fact, they even show that the probability that any given offline vertex is matched goes to 1 as $d \to \infty$ which is certainly not the case for RANKING.

The technique by Komm et al. [14] can be used to show concentration bounds for several problems which are loosely related to online matching such as the Online k-Server Problem. Their key idea is to use a repeating strategy where any existing randomized algorithm is used and simply restarted periodically when certain conditions are met. This improves the in expectation guarantee of the original algorithm to a high probability guarantee similar to typical re-running technique for non-online algorithms. However, this only works if one can indeed cheaply restart the algorithm without harming the analysis which is the case in the Online k-Server Problem but not in the Online Bipartite Matching Problem.

In Sect. 3 we will define the Fully Online Matching Problem and the natural extension of RANKING for this setting. We remark that we allow for non-bipartite graphs here and we will give a similar concentration bound as in Theorem 1.

Theorem 2. *Let G be an instance of the Fully Online Matching Problem which admits a matching of size n. Then for any $\alpha > 0$,*

$$\mathbb{P}\left[|M| < (\rho - \alpha)\, n\right] < e^{-\alpha^2 n}$$

where M is the random variable denoting the matching generated by RANKING and ρ is the competitive ratio of RANKING.

We remark that by [10], we know $\rho > 0.521$ and for the special case where G is bipartite, we have $\rho = W(1) \approx 0.567$.

Lastly, in Sect. 4 we will consider the Online Vertex-Weighted Bipartite Matching Problem. In this setting, a generalization of RANKING was shown to be $(1 - \frac{1}{e})$-competitive by Aggarwal et al. [1]. We will modify this algorithm to show the following.

Theorem 3. *For any $\alpha > 0$, there exists a variant of RANKING such that for any instance $G = (S, B, E)$ with weights $w : S \to \mathbb{R}_+$ of the Online Vertex-Weighted Bipartite Matching, any arrival order of B and any matching M^*,*

$$\mathbb{P}\left[w(M) < \left(1 - \frac{1}{e} - \alpha\right) w(M^*)\right] < e^{-\frac{1}{50}\alpha^4 \frac{w(M^*)^2}{||w||_2^2}}$$

where M denotes the matching generated by RANKING and

$$w(M) := \sum_{\{i,j\} \in M} w_j.$$

Lastly, we argue that this bound also applies to the Online Single-Valued Bipartite Matching Problem which is a variant of the vertex-weighted problem in which goods can be matched multiple times.

2 Online Bipartite Matching

In order to analyze RANKING, it is common to replace the sampling of the permutation π in Algorithm 1 by sampling an independent, uniform $x_j \in [0, 1]$ for every $j \in S$ called the *rank* of j. Then, sorting S by the values of x_j yields a uniformly random permutation. Formally, this is Algorithm 2.

Algorithm 2: RANKING

for $j \in S$ **do**
  Sample a uniformly random $x_j \in [0, 1]$.
for *each buyer i who arrives* **do**
  Match i to an unmatched $j \in N(i)$ with minimum x_j.

In the following, consider a fixed graph $G = (S, B, E)$ with a fixed arrival order. Assume that $|S| = n$ and that G has a matching of size n. We define a function $f : [0,1]^S \to \mathbb{R}$ by letting $f(y)$ be the size of the matching M generated by Algorithm 2 if $x_j = y_j$ for all $j \in S$. Our goal will then be to show the following Lemma which is a different perspective on a structural property that appears under various forms in the online matching literature (e.g. Lemma 2 in [2]).

Lemma 2 (Bounded Differences). *Let $x \in [0,1]^S$, $j^\star \in S$ and $\theta \in [0,1]$ be arbitrary. Define x'_j to be θ if $j = j^\star$ and x_j otherwise. Then $|f(x) - f(x')| \le 1$.*

Note that Lemma 2 implies Theorem 1 via McDiarmid's inequality (Lemma 1). Specifically, by applying McDiarmid to the function f with $c \equiv 1$ we get

$$\mathbb{P}\left[|M| < \left(1 - \frac{1}{e} - \alpha\right) n\right] \le \mathbb{P}_{x \sim \Delta^S}[f(x) < \mathbb{E}_{y \sim \Delta^S}[f(y)] - \alpha n] \le e^{-\alpha^2 n}$$

where we used that $(1 - \frac{1}{e})n \le \mathbb{E}_{y\sim\Delta^S}[f(y)]$ since RANKING is $(1-\frac{1}{e})$-competitive. It remains to prove Lemma 2.

Lemma 3. *Let $j \in S$, then we can define the graph G_{-j} which contains all vertices of G except for j. For some fixed values of $x \in [0,1]^S$, we let M be the matching produced by RANKING in G and let M_{-j} be the matching produced by RANKING in G_{-j}. Then $|M_{-j}| \le |M| \le |M_{-j}| + 1$.*

Proof. For any buyers $i, i' \in B$, let $N^{(i)}(i')$ be the set of neighbors of i' in G which are unmatched by the time that i arrives in the run of RANKING with the fixed values of x. Likewise, let $N^{(i)}_{-j}(i')$ be the set of unmatched neighbors of i' in the run of RANKING on G_{-j} when i arrives. We claim that for all $i \in B$ there exists some $j' \in S$ such that for all $i' \in B$ we have $N^{(i)}(i') = N^{(i)}_{-j}(i')$ or $N^{(i)}(i') = N^{(i)}_{-j}(i') \cup \{j'\}$.

Let us show this claim via induction on $i \in B$ in order of arrival. Note that when the first buyer arrives, this holds for $j' = j$ because we have removed only j' from the graph and nobody has been matched yet. Now assume that the statement holds when i arrives, we need to see that it still holds after i has been matched. Clearly, if i gets matched to the same vertex in G and in G_{-j}, then the inductive step follows trivially.

So now assume that i gets matched to different vertices in G and in G_{-j}. By the inductive hypothesis this can only happen if i gets matched to j' in G and it gets matched to some other j'' (potentially $j'' = \perp$, i.e. it is not matched at all) in G_{-j}. But then $N^{(i+1)}(i') = N^{(i+1)}_{-j}(i')$ or $N^{(i+1)}(i') = N^{(i+1)}_{-j}(i') \cup \{j''\}$ for all $i' \in B$. Thus the claim holds by induction.

Finally, let us see that the claim implies the lemma. First note that since i always has more unmatched neighbors in G than in G_{-j}, we have $|M| \ge |M_{-j}|$.

But on the other hand, if at some time in the algorithm i is matched to j' in G and not matched at all in G_{-j}, then we have that $N^{(i+1)}(i') = N^{(i+1)}_{-j}(i')$ for all $i' \in B$. Thus the two runs will be identical from that point onward and $|M| = |M_{-j}| + 1$. □

Finally, we can show that this implies the bounded differences property of f that we claimed in Lemma 2.

Proof of Lemma 2. By Lemma 3 we know that removing a good from the graph can decrease the size of the matching computed by RANKING by at most one assuming that the values of the x_j are fixed. But of course if we are removing $j^\star \in S$, the matching $M_{-j^\star}$ computed by RANKING in $G_{-j^\star}$ does not depend on the value of $x_{j^\star}$ or $x'_{j^\star}$. So we have

$$|M_{-j^\star}| \le f(x) \le |M_{-j^\star}| + 1$$

and

$$|M_{-j^\star}| \le f(x') \le |M_{-j^\star}| + 1$$

which implies $|f(x) - f(x')| \le 1$ as claimed. □

As we have already seen, this is enough to prove Theorem 1 in the case where $|S| = n$. To prove the general case we can use a simple reduction. In particular, assuming that there is a matching M of size n but $|S| > n$, let S_M be the goods covered by M and let $G_M = (S_M, B, E)$. We have seen in Lemma 3 that for any fixed $x \in [0,1]^S$, RANKING will produce a matching in G that is not smaller than the matching it produces in G_M when run with x restricted to S_M. Therefore, Theorem 1 on G_M implies Theorem 1 on G which establishes the general case.

3 Fully Online Matching

In the Fully Online Matching Problem we have a not necessarily bipartite graph G the vertices of which arrive and depart online in adversarial order. When a vertex arrives, it reveals all of its edges to vertices that have already arrived. By the time it departs we are guaranteed to have been revealed its entire neighborhood.

This problem was introduced by Huang et al. [10] and is motivated by ride-sharing. Each vertex represents a rider who, upon arrival, is willing to wait only for a certain amount of time. Two riders can only be matched if the time that they spend on the platform overlaps, even in the offline solution. This additional condition allows Huang et al. to show that the generalization of RANKING shown in Algorithm 3 is 0.521-competitive in general and 0.567-competitive on bipartite graphs.

Note that for bipartite graphs, another algorithm called BALANCED RANKING is known to be 0.569-competitive [12], showing that RANKING is not optimal for the Fully Online Matching Problem. However, it is not clear whether one can carry out a similar analysis for BALANCED RANKING as well.

Algorithm 3: FULLY ONLINE RANKING

for *vertex* i *who arrives* **do**
  Sample a uniformly random $x_i \in [0,1]$.
for *vertex* i *who departs* **do**
  Match i to an unmatched $j \in N(i)$ with minimum x_j.

In order to show a concentration bound, we can apply similar techniques as in Sect. 2. Let $G = (V, E)$ be a graph which admits a perfect matching of size n. Then let $f : [0,1]^V \to \mathbb{R}$ represent once again the size of the matching generated by Algorithm 3 when given the x_i values. The corresponding bounded differences condition then becomes:

Lemma 4 (Bounded Differences). *Let $x \in [0,1]^V$, $i^\star \in V$ and $\theta \in [0,1]$ be arbitrary. Define x'_i to be θ if $i = i^\star$ and x_i otherwise. Then $|f(x) - f(x')| \leq 1$.*

This implies Theorem 2 as before though note that this time we will lose a factor of 2 since we now have $2n$ variables. We remark that this follows directly from Lemma 2.3 in [10] but for completeness we will give a short proof sketch.

Lemma 5. *Using the notation from Lemma 3, we have $|M_{-j}| \leq |M| \leq |M_{-j}| + 1$ for any $j \in V$ and fixed values of $x \in [0,1]^V$.*

Proof. As in the proof of Lemma 3, let $N^{(i)}(i')$ (or $N^{(i)}_{-j}(i')$) be the set of neighbors of i' in G (or G_{-j}) which is unmatched by the time that i departs in the run of FULLY ONLINE RANKING with the fixed values of x. We claim that for all $i \in V$, there exists some $j' \in V$ such that for all $i' \in V$, we have $N^{(i)}(i') = N^{(i)}_{-j}(i')$ or $N^{(i)}(i') = N^{(i)}_{-j}(i') \cup \{j'\}$.

This claim follows via an almost identical induction as in Lemma 3. Then, since i always has more unmatched neighbors in G than in G_{-j}, we have $|M| \geq |M_{-j}|$. And if at some time in the algorithm, i is matched to j' in G and not matched at all in G_{-j}, then we have that $N^{(i+1)}(i') = N^{(i+1)}_{-j}(i')$ for all $i' \in V$. Thus the two runs will the identical from that point onward and $|M| = |M_{-j}|+1$. □

Since Lemma 5 implies Lemma 4, this yields Theorem 2 for graphs which contain a perfect matching. But as in Sect. 2, we may drop this condition by reducing a graph G with a matching M to the subgraph induced by the vertices covered by M. Adding the vertices back in only increases the performance of FULLY ONLINE RANKING by Lemma 5.

4 Online Vertex-Weighted Bipartite Matching

In this section we will consider a weighted extension of the Online Bipartite Matching Problem which has been inspired by online advertising markets. In the Online Vertex-Weighted Bipartite Matching Problem, we have a bipartite graph $G = (S, B, E)$ with vertex weights $w : S \to \mathbb{R}_+$ on the offline vertices. Here S represents the advertisers and B represents website impressions or search queries which should get matched to ads from the advertisers. The vertices B arrive online in adversarial order and should get matched to a neighbor j such that the total weight of the matched vertices in S is maximized. This problem can be seen as a special case of the AdWords Problem which instead imposes edge-weights and budgets on the offline vertices.

Perhaps somewhat surprisingly it took 20 years for RANKING to be extended for the unweighted to the vertex-weighted setting by Aggarwal et al. [1]. This is because in the presence of weights, it is no longer enough to pick a uniformly random permutation over the offline vertices. Instead, one has to skew the permutation so that heavier vertices are more likely to appear first. This is done elegantly in Algorithm 4 by ordering the vertices not by their x_j but rather by the careful chosen quantity $w_j(1 - e^{x_j - 1})$.

Algorithm 4: VERTEX-WEIGHTED RANKING

---

1 **for** $j \in S$ **do**

2   Sample a uniformly random $x_j \in [0, 1]$.

3 **for** *each buyer i who arrives* **do**

4   Match i to an unmatched $j \in N(i)$ with maximum $w_j \left(1 - e^{x_j - 1}\right)$.

However, Algorithm 4 does not lend itself to a straight-forward analysis via the method of bounded differences. This is because a vertex with small weight, which should have little impact on the total weight of the matching, can sometimes be chosen over a vertex with much larger weight. See the example shown in Fig. 1.

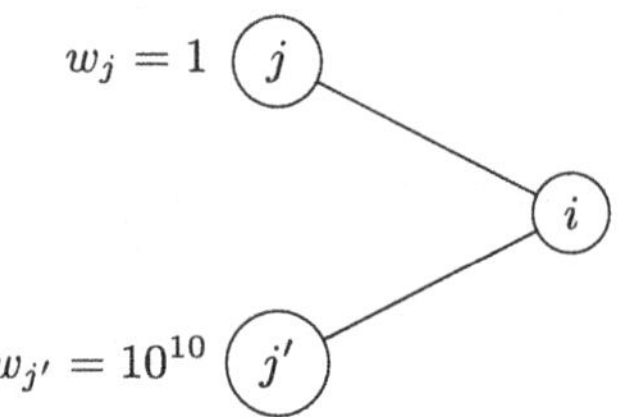

Fig. 1. Shown is a simple instance in which the value of x_j can have a large impact on the final matching despite the fact that w_j is small. If $x_{j'} \gg 1 - 10^{-10}$, i will choose j in line 4 for sufficiently small values of x_j.

In particular, the problem lies with the fact that $w_j(1 - e^{x_j - 1})$ can get arbitrarily close to 0 if x_j gets close to 1. We will overcome this problem by changing the function slightly. For any $\epsilon > 0$ we consider ϵ-RANKING as shown in Algorithm 5.

Algorithm 5: ϵ-RANKING

1 **for** $j \in S$ **do**
2   Sample a uniformly random $x_j \in [0, 1]$.
3 **for** *each buyer* i *who arrives* **do**
4   Match i to an unmatched $j \in N(i)$ with maximum $w_j\left(1 - e^{x_j-1-\epsilon}\right)$.

In the following fix some instance $G = (S, B, E)$ with vertex-weights w and some $\epsilon > 0$. Then we let $f : [0, 1]^S \to \mathbb{R}$ represent the total weight of the matching generated by Algorithm 5 with fixed samples x_j. We will show that ϵ-RANKING is still $(1 - \frac{1}{e} - \epsilon)$-competitive while also allowing us to give a concentration bound.

To give a concise proof of the $(1 - \frac{1}{e} - \epsilon)$-competitiveness we will use the economic analysis by Eden et al. [7] which is itself based on the primal-dual viewpoint due to Devanur et al. [4]. This analysis associates random variables r_j with all $j \in S$ and u_i with $i \in B$. The idea is that the value $w_j e^{x_j-1-\epsilon}$ represents the *price* of j and whenever a match between i and j is made, this is a *sale*. We will then set r_j (the *revenue*) to be $w_j e^{x_j-1-\epsilon}$ and u_i (the *utility*) to be $w_j e^{x_j-1-\epsilon}$. If a vertex is never matched, its revenue/utility will be zero.

Lemma 6. *Using the notation from Lemma 3, we have that for all $j \in S$ and fixed samples x,*

$$w(M_{-j}) - \frac{2}{\epsilon} w_j \leq w(M) \leq w(M_{-j}) + w_j.$$

Additionally, for any $i \in B$, its utility u_i in the run on G will be no less than in the run on G_{-j}.

Proof. For any buyers $i, i' \in B$, let $N^{(i)}(i')$ be the set of neighbors of i' in G which are unmatched by the time that i arrives in the run of Algorithm 5 with the fixed values of x. Likewise, let $N^{(i)}_{-j}(i')$ be the set of unmatched neighbors of i' in the run of ϵ-RANKING on G_{-j} when i arrives. We claim that for all $i \in B$ there exists some $j' \in S$ such that

$$w_{j'}(1 - e^{x_{j'}-1-\epsilon}) \leq w_j(1 - e^{x_j-1-\epsilon})$$

and for all $i' \in B$, we have $N^{(i)}(i') = N^{(i)}_{-j}(i')$ or $N^{(i)}(i') = N^{(i)}_{-j}(i') \cup \{j'\}$.

This claim is almost the same as in the proof of Lemma 3 and may likewise be shown via induction. Note that the extra condition on $w_{j'}$ holds at the beginning where $j' = j$ and whenever i matches to j', it frees up a vertex j'' with

$$w_{j''}(1 - e^{x_{j''}-1-\epsilon}) \leq w_{j'}(1 - e^{x_{j'}-1-\epsilon})$$

due to the fact that j' was picked over j'' in line 4. If i was not even matched in G_{-j}, we can simply set $j' = j$ for the induction.

Now note that since $N^{(i)}_{-j}(i) \subseteq N^{(i)}(i)$ for all $i \in B$, we always maximize over a larger set in line 4. Thus the utility of i will be no smaller in the run on G compared to the run on G_{-j}.

On the other hand, let $T \subseteq S$ be the set of goods matched in the run on G and let $T_{-j} \subseteq S \backslash \{j\}$ be the set of goods matched in the run on G_{-j}. Then we observe that $T \backslash T_{-j} \subseteq \{j\}$ because for all $j' \neq j$, if j' gets matched to i in M, then either $j' \in N^{(i)}_{-j}(i)$ implying that i will match to j' in M_{-j}, or j' was already matched to some other vertex. In both cases, if $j' \in T$ then $j' \in T_{-j}$. This implies that $w(M) \leq w(M_{-j}) + w_j$.

We also have that $|T_{-j} \backslash T| \leq 1$. Simply imagine a buyer i^* that arrives after all other buyers and has edges to all goods. Then by the claim, there exists some $j' \in S$ such that

$$(S \backslash \{j\}) \backslash T_{-j} = N^{(i^*)}_{-j}(i^*) \subseteq N^{(i^*)}(i^*) \cup \{j'\} = S \backslash (T \cup \{j'\})$$

and so $T_{-j} \subseteq T \cup \{j'\}$. This implies that $w(M) \geq w(M_{-j}) - w_{j'}$.

Finally, we also know by the claim that $w_{j'}(1 - e^{x_{j'}-1-\epsilon}) \leq w_j(1 - e^{x_j-1-\epsilon})$ which implies

$$w_{j'} \leq \frac{1}{1 - e^{-\epsilon}} w_j \leq \frac{1}{\left(1 - \frac{1}{e}\right)\epsilon} w_j \leq \frac{2}{\epsilon} w_j.$$

Thus we have shown $w(M_{-j}) - \frac{2}{\epsilon} w_j \leq w(M) \leq w(M_{-j}) + w_j$ as required. □

Lemma 7 (Bounded Differences). *Let $x \in [0,1]^S$, $j^\star \in S$ and $\theta \in [0,1]$ be arbitrary. Define x'_j to be θ if $j = j^\star$ and x_j otherwise. Then $|f(x) - f(x')| \leq \left(1 + \frac{2}{\epsilon}\right) w_{j^\star}$.*

Proof. As in the proof of Lemma 2, we can simply remove $j^\star$ and apply Lemma 6. Then

$$w(M_{-j^\star}) - \frac{2}{\epsilon} w_{j^\star} \leq f(x) \leq w(M_{-j^\star}) + w_{j^\star},$$
$$w(M_{-j^\star}) - \frac{2}{\epsilon} w_{j^\star} \leq f(x') \leq w(M_{-j^\star}) + w_{j^\star}$$

which implies the result. □

Lemma 8. *For any $\{i, j\} \in E$, we have $\mathbb{E}[r_j + u_i] \geq (1 - \frac{1}{e} - \epsilon) w_j$.*

Proof. Fix all samples x except for x_j. Then we can define u^* to be the utility of i when ϵ-Ranking is ran on G_{-j}. By Lemma 6, we know that $u_i \geq u^*$, regardless of the value of x_j.

On the other hand, if x_j is small enough that $w_j(1 - e^{x_j-1-\epsilon}) > u^*$, then j will definitely get matched because if j is not yet matched by the time that i arrives, then clearly j will be chosen in line 4 of the algorithm and so it gets matched to i. Now if u^* is very small, this may be the case for all values of x_j and in that case

$$\mathbb{E}[r_j \mid x_{-j}] \geq \int_0^1 w_j e^{t-1-\epsilon} \, \mathrm{d}t = \left(1 - \frac{1}{e}\right) e^{-\epsilon} w_j \geq \left(1 - \frac{1}{e} - \epsilon\right) w_j.$$

Otherwise there will be some value $z \in [0,1]$ such that $w_j(1 - e^{z-1-\epsilon}) = u^*$ and then we can compute

$$\mathbb{E}[r_j \mid x_{-j}] \geq \int_0^z w_j e^{t-1-\epsilon}\,\mathrm{d}t = \left(1 - \frac{1}{e}\right) w_j - u^*.$$

But clearly, in both cases we have

$$\mathbb{E}[r_j + u_i \mid x_{-j}] \geq \mathbb{E}[r_j \mid x_{-j}] + u^* \geq \left(1 - \frac{1}{e} - \epsilon\right) w_j$$

and so in particular $\mathbb{E}[r_j + u_i] \geq (1 - \frac{1}{c} - \epsilon)w_j$ as claimed. □

Lemma 9. *ϵ-RANKING is $(1 - \frac{1}{e} - \epsilon)$-competitive.*

Proof. Let M^* be a maximum weight matching and let M be the matching output by ϵ-RANKING. Notice that every time we match an edge in the algorithm, we increase $\sum_{j \in S} r_j + \sum_{i \in B} u_i$ by exactly the weight of the edge. Thus by Lemma 8,

$$\begin{aligned}\mathbb{E}[w(M)] = \mathbb{E}\left[\sum_{j \in S} r_j + \sum_{i \in B} u_i\right] &\geq \sum_{\{i,j\} \in M^*} \mathbb{E}[r_j + u_i] \\ &\geq \sum_{\{i,j\} \in M^*} \left(1 - \frac{1}{e} - \epsilon\right) w_j = \left(1 - \frac{1}{e} - \epsilon\right) w(M^*)\end{aligned}$$

and therefore ϵ-RANKING is $(1 - \frac{1}{e} - \epsilon)$-competitive. □

Finally, we have the tools necessary to show Theorem 3 by combining Lemma 7 with Lemma 9.

Proof of Theorem 3. Given some $\alpha > 0$, we consider the algorithm $\frac{\alpha}{2}$-RANKING which we know to be $(1 - \frac{1}{e} - \frac{\alpha}{2})$-competitive by Lemma 9. We apply Lemma 1 (McDiarmid's inequality) with Lemma 7 (bounded differences). This gives us

$$\begin{aligned}\mathbb{P}\left[w(M) < \left(1 - \frac{1}{e} - \alpha\right) w(M^*)\right] &< e^{-2\frac{\alpha^2}{2}\frac{w(M^*)^2}{(1+4/\alpha)^2 ||w||_2^2}} \\ &\leq e^{-\frac{\alpha^4}{50}\frac{w(M^*)^2}{||w||_2^2}}\end{aligned}$$

where we use that $\alpha < 1$ since otherwise the bound holds trivially. □

The results of this section may also be extended to a generalization of the Online Vertex-Weighted Bipartite Matching Problem which is called the Online Single-Valued Bipartite Matching Problem. The setup is almost identical in that we still have a bipartite graph $G = (S, B, E)$ with vertex weights $w : S \to \mathbb{R}_+$

on the offline vertices. However, now each offline vertex j also has a capacity $c_j \in \mathbb{N}$ that represents how often it is allowed to be matched.

Clearly, Theorem 3 can be extended to this setting by simply creating c_j many copies of each offline vertex j. This can be done implicitly and in a capacity-oblivious way by simply sampling a new x_j every time j is matched during the RANKING (or ϵ-RANKING) algorithm.

Recently, Vazirani [19] showed that this "resampling" is in fact not necessary, i.e. that the same value of x_j can be used for every copy of j while still achieving $(1 - \frac{1}{e})$-competitiveness of RANKING; see Algorithm 6.

Algorithm 6: SINGLE-VALUED RANKING

```
for j ∈ S do
    Sample a uniformly random x_j ∈ [0, 1].
for each buyer i who arrives do
    Match i to a j ∈ N(i) which has been matched less than c_j times,
    with maximum w_j (1 − e^(x_j − 1)).
```

The main benefit of Algorithm 6 is that it uses fewer random bits than running RANKING on the reduced instance with c_j many copies of each offline vertex j. However, it will accordingly be less tightly concentrated which leads to a version of Theorem 3 in which the bound depends not on $||w||_2^2$ but rather on $\sum_j (c_j w_j)^2$.

5 Discussion

We have shown that RANKING and its many variants achieve their competitive ratios with high probability rather than just in expectation. This leaves several interesting open problems. The first is to show a concentration bound for the original weighted version of RANKING rather than ϵ-RANKING. As mentioned, the bounded differences approach fails due to the large influence that vertices with small weight can have on the matching. However, this should happen rarely and so a more fine-grained analysis may be able to overcome this challenge.

A second interesting prospect is to consider the AdWords problem. Vazirani [19] showed that a variant of RANKING can be used for AdWords with small bids under the assumption of the so-called *no-surpassing property* which tends to hold in practice though the bound is once again given in terms of expectation. An advantage of this approach over the classic MSVV algorithm [17] is that RANKING does not need to know about the budgets. It may be possible to show a concentration bound for this algorithm as well. However, this is made more challenging by the fact that the setting is edge-weighted.

Acknowledgements. We would like to thank Vijay Vazirani for helpful comments and feedback.

References

1. Aggarwal, G., Goel, G., Karande, C., Mehta, A.: Online vertex-weighted bipartite matching and single-bid budgeted allocations. In: Proceedings of the 22nd Annual ACM-SIAM Symposium on Discrete Algorithms, pp. 1253–1264. SIAM (2011)
2. Birnbaum, B., Mathieu, C.: On-line bipartite matching made simple. SIGACT News **39**(1), 80–87 (2008). https://doi.org/10.1145/1360443.1360462
3. Cohen, I.R., Wajc, D.: Randomized online matching in regular graphs. In: Proceedings of the Twenty-Ninth Annual ACM-SIAM Symposium on Discrete Algorithms, pp. 960–979. SIAM (2018)
4. Devanur, N.R., Jain, K., Kleinberg, R.D.: Randomized primal-dual analysis of ranking for online bipartite matching. In: Proceedings of the 24th Annual ACM-SIAM Symposium on Discrete Algorithms, pp. 101–107. SIAM (2013)
5. Dubhashi, D.P., Panconesi, A.: Concentration of Measure for the Analysis of Randomized Algorithms. Cambridge University Press, Cambridge (2009). https://doi.org/10.1017/CBO9780511581274
6. Echenique, F., Immorlica, N., Vazirani, V.: Online and Matching-Based Market Design. Cambridge University Press, Cambridge (2023)
7. Eden, A., Feldman, M., Fiat, A., Segal, K.: An economics-based analysis of RANKING for online bipartite matching. In: Le, H.V., King, V. (eds.) 4th Symposium on Simplicity in Algorithms, SOSA 2021, pp. 107–110. SIAM (2021). https://doi.org/10.1137/1.9781611976496.12
8. Goel, G., Mehta, A.: Online budgeted matching in random input models with applications to AdWords. In: Proceedings of the 19th Annual ACM-SIAM Symposium on Discrete Algorithms, pp. 982–991 (2008)
9. Huang, Z., Zhang, Q., Zhang, Y.: Adwords in a panorama. In: 2020 IEEE 61st Annual Symposium on Foundations of Computer Science (FOCS), Los Alamitos, CA, USA, pp. 1416–1426. IEEE Computer Society (2020). https://doi.org/10.1109/FOCS46700.2020.00133
10. Huang, Z., Kang, N., Tang, Z.G., Wu, X., Zhang, Y., Zhu, X.: How to match when all vertices arrive online. In: Proceedings of the 50th Annual ACM SIGACT Symposium on Theory of Computing, pp. 17–29 (2018)
11. Huang, Z., Peng, B., Tang, Z.G., Tao, R., Wu, X., Zhang, Y.: Tight competitive ratios of classic matching algorithms in the fully online model. In: Proceedings of the 30th Annual ACM-SIAM Symposium on Discrete Algorithms, pp. 2875–2886. SIAM (2019)
12. Huang, Z., Tang, Z.G., Wu, X., Zhang, Y.: Fully online matching ii: beating ranking and water-filling. In: Proceedings of the 51st Annual IEEE Symposium on Foundations of Computer Science. IEEE (2020)
13. Karp, R.M., Vazirani, U.V., Vazirani, V.V.: An optimal algorithm for on-line bipartite matching. In: Proceedings of the 22nd Annual ACM Symposium on Theory of Computing, pp. 352–358 (1990)
14. Komm, D., Královic, R., Královic, R., Mömke, T.: Randomized online algorithms with high probability guarantees. In: Mayr, E.W., Portier, N. (eds.) 31st International Symposium on Theoretical Aspects of Computer Science (STACS 2014). Leibniz International Proceedings in Informatics (LIPIcs), vol. 25, pp. 470–481. Schloss Dagstuhl–Leibniz-Zentrum fuer Informatik, Dagstuhl (2014). https://doi.org/10.4230/LIPIcs.STACS.2014.470
15. Leonardi, S., Marchetti-Spaccamela, A., Presciutti, A., Rosén, A.: On-line randomized call control revisited. SIAM J. Comput. **31**(1), 86–112 (2001). https://doi.org/10.1137/S0097539798346706

16. McDiarmid, C.: On the Method of Bounded Differences. London Mathematical Society Lecture Note Series, pp. 148–188. Cambridge University Press, Cambridge (1989). https://doi.org/10.1017/CBO9781107359949.008
17. Mehta, A., Saberi, A., Vazirani, U., Vazirani, V.: Adwords and generalized online matching. J. ACM (JACM) **54**(5), 22-es (2007)
18. Sedgewick, R.: Quicksort. Ph.D. thesis, Stanford University (1975)
19. Vazirani, V.V.: Towards a practical, budget-oblivious algorithm for the adwords problem under small bids. In: Bouyer, P., Srinivasan, S. (eds.) 43rd IARCS Annual Conference on Foundations of Software Technology and Theoretical Computer Science (FSTTCS 2023). Leibniz International Proceedings in Informatics (LIPIcs), vol. 284, pp. 21:1–21:14. Schloss Dagstuhl – Leibniz-Zentrum für Informatik, Dagstuhl (2023). https://doi.org/10.4230/LIPIcs.FSTTCS.2023.21

The Team Order Problem: Maximizing the Probability of Matching Being Large Enough

Haris Aziz[1], Jiarui Gan[2], Grzegorz Lisowski[3(✉)], and Ali Pourmiri[1]

[1] UNSW Sydney, Sydney, Australia
haris.aziz@unsw.edu.au
[2] University of Oxford, Oxford, UK
jiarui.gan@cs.ox.ac.uk
[3] AGH University of Science and Technology, Kraków, Poland
glisowski@agh.edu.pl

Abstract. We consider a matching problem, which is meaningful in team competitions, as well as in information theory, recommender systems, and assignment problems. In the competitions which we study, each competitor in a team order plays a match with the corresponding opposing player. The team that wins more matches wins. We consider a problem where the input is the graph of probabilities that a team 1 player can win against the team 2 player, and the output is the optimal ordering of team 1 players given the fixed ordering of team 2. Our central result is a polynomial-time approximation scheme (PTAS) to compute a matching whose winning probability is at most ε less than the winning probability of the optimal matching. We also provide tractability results for several special cases of the problem, as well as an analytical bound on how far the winning probability of a maximum weight matching of the underlying graph is from the best achievable winning probability.

1 Introduction

Bipartite matching underpins several impactful problems in allocation and market design problems including kidney allocation, adword auctions, on demand taxi allocation, refugee assignment, or school choice (see, e.g., [11]). We consider a fundamental matching problem with an underlying weighted bipartite graph where each edge weight has weight between 0 and 1. Instead of focusing on the classical objective of maximizing the total weight of the matching, we focus on a different objective with a probabilistic interpretation: We want to compute a matching that maximizes the probability of reaching a target size. This problem can model several scenarios, including that of the so called *team order problem*.

One of the most relevant applications of our setting is the rivalry between teams of contestants. Consider team competitions in which both teams put forward an ordering of their players. The contestants then play matches against the corresponding contestants from the opposing team. The team that wins more

G. Schäfer and C. Ventre (Eds.): SAGT 2024, LNCS 15156, pp. 35–52, 2024.
https://doi.org/10.1007/978-3-031-71033-9_3

matches wins the overall competition. Such competitions are not only held in various inter-club tennis competitions, the same format is also used in international table tennis and badminton competitions, such as the Corbillon Cup, Swaythling Cup, Thomas Cup, and the Olympic Games. We focus on the problem in which one team's order is fixed (as is the case in many situations where the home team commits to an ordering) and the other team wants to compute the optimal ordering. As the ordering of one team is fixed, the problem of computing the other team's ordering is essentially a competitor matching problem.

The problem of finding a way to maximize the number of achieved goals by setting an appropriate line-up is not limited to sport competitions. Indeed, it admits several other motivations in competitive contexts such as politics (fielding political candidates in different constituencies against candidates of a rival party). Our problem also provides a perspective into finding *durable matchings.* Suppose that we are given the probability of success of various partnerships. For example a partnership could represent a job placement or allocation of refugee family to a council (see, e.g., [2,7]). A typical objective could be maximizing the expected number of partnerships. However, another meaningful objective that is centred around a particular target could be to maximize the probability of having a target number of successful partnerships, which maps to the objective that we study. Another potential application of our research relates to *information networks* (see, e.g., [21]). Suppose that we are given such a network, represented by a flow network. There, each edge has a reliability probability of a message reaching the other side, and we want to find a flow maximizing probability of delivering a target number of messages. Finally, our research is motivated by its applications in *recommendation systems* (see, e.g., [25]). Suppose that a ranked list of recommendations needs to be displayed with each item having a probability of being clicked depending on its position in the ranking list. One may want to maximize the probability of having a target number of items being clicked, which can be captured by our problem. We explore the following questions.

> *How hard is the team order problem? Under what conditions is it easy to solve? What are reasonable approximation approaches for the problem?*

We note that the problem that we study in this paper is closely related to the maximum-weight matching problem. There, we are given a bipartite graph, where each edge is assigned a weight, and the objective is to find a matching with the maximum sum of weights. In fact, our results reflect that finding the solution to that problem provides a good approximation of the optimal solution. However, the problem we study is substantially more complex. Indeed, for an instance of the team order problem to be positive, we require that the weights in a selected matching are large enough for some subset of edges, instead of maximizing their global sum. Furthermore, given the strategic games interpretation of our setting, our results concern the computation of the optimal response to the opponent choice, which is an important step towards the study of equilibria in this setting.

Contributions. We first show that the winning probability of a given matching (line-up) can be computed in polynomial time (Proposition 1). Subsequently, we show that in certain settings computing an optimal line-up is tractable. In particular, when the input winning probability of each partnership takes its value from a size-three set $\{\alpha, \beta, 0\}$ we show that the optimal matching can be computed in polynomial time (Theorem 1). While we conjecture that the team order problem is hard in the general case, we show that it is tractable for practical purposes. Our central result is a *polynomial-time approximation scheme (PTAS)*[1] to compute a matching whose winning probability is at most ε less than the winning probability of the optimal matching (Theorem 3). Although the winning probability is not a linear objective, we show that the general problem of computing an optimal matching can be solved via integer linear programming. Also, we provide an analytical bound on how far the winning probability of a maximum weight matching is from the best achievable winning probability.

2 Related Work

Our results are relevant to a number of research direction in multi-agent systems.

Matching Theory. Matching problems have been widely studied in combinatorial optimization. The standard objectives typically focus on maximizing the weight of the matching (see, e.g., [9,24]). In our context, maximizing the weight of the underlying weighted bipartite graph gives us a matching maximizing the expected number of matches won. Our objective is different as we want to maximize the probability of winning a target number of matches. The paper most relevant to our work is by Tang et al. [28], which concerns the same setting but considered different problems. It takes an economic design approach and presents necessary and sufficient conditions, ensuring that truthful reporting and maximal effort in matches are equilibrium strategies. We note that the probabilistic approach in matching has been previously studied. E.g., Aziz et al. [6] studied the stable matching problem with uncertain preferences.

Manipulation of Competitions. Within the wider topic of manipulations in competitions, there have been several papers on identifying conditions or manipulations under which a certain team or player can win. A notable example is manipulating the draw of a balanced knockout tournament to maximize the probability of a certain player winning, i.e., the *tournament fixing problem* [5,30,31]. Similarly, there has also been algorithmic research on round-robin formats to understand which teams have a chance to win the overall tournament [4,17].

Colonel Blotto Game. Furthermore, the team line-up setting bears resemblance to Colonel Blotto Games which are two-player zero-sum games in which two armies fight in n battle fields with each battle being won by the army that had more troops in the battle (see, e.g., [26,27]). The armies are interested in maximizing a weighted sum of utilities from the battlefields where they gain victories.

[1] A PTAS is a scheme which, for every instance of a problem and $\varepsilon > 0$, provides an approximate solution based on ε.

Although the team-line-up setting is similar in that each battle corresponds to a match, in Colonel Blotto games, the armies have more flexibility in shuffling their troops around. Secondly, in Colonel Blotto games the outcome of a battle depends on the *number* of troops of each army whereas in the team line up setting, the outcome of a match depends on the identities of the respective players. Independent of our work, Gaonkar et al. [15] considered a version of Blotto games in which every resource is unique and non-interchangeable which makes it close to our setting. They motivate the problem as *derby games* in which teams assign each resource to a particular round and wins a payoff corresponding to that round if they win the round. We note, however, that our work differs significantly from their results. In particular, they examine Nash equilibria, which are not the focus of our study. Furthermore, they do not take the information on winning probabilities into account and do not focus on algorithmic issues.

Sequential Games. Games between teams of players in which the ordering of contestants matters gained a substantial interest in recent literature. Fu et al. [14] studied the scenario in which teams compete in a number of games between pairs of players. Within this setting they investigated how the sequencing of those matches impacts the result. We note that, in contrast to our study, the games they considered are also based on private rewards for the individual players. Furthermore, Konishi et al. [18] studied the problem of whether the equilibrium winning probability in such games depends on whether matches are held simultaneously, or sequentially. Also, Fu and Lu [13] explored the topic of how teams can strategically assign contestants to time-slots of a sequential competition. Let us further note that in contrast to our work the discussed papers on sequential games do not focus on computational complexity.

Nominee Selection. Our setting is also related to the literature on strategic selection of group members participating in a competition. In social choice theory, this problem relates to the process of selecting representative for the elections (see, e.g., [3,12]). Regarding sport competitions, our problem relates to choosing a coalition member to participate in a tournament (see, e.g., [22,23]).

3 The Team Order Problem

We consider the following problem setting.

- Two teams T_1 and T_2 are to play a team competition.
- Each team T_i has n contestants $t_i^1, \dots, t_i^n$.
- We have information about the winning probability $p(t_i^a, t_j^b)$ of any contestant t_i^a against any other contestant t_j^b. The instance is said to be *degenerate* if all the winning probabilities are 0 or 1.

In the competition each team is required to report a line-up, i.e., an ordering $i_1, \dots, i_n$ of its contestants, which is a permutation of $1, \dots, n$. Then each contestant $t_i^{i_k}$ plays a match with the corresponding contestant $t_j^{j_k}$. The team that wins at least $\lfloor \frac{n}{2} \rfloor + 1$ matches wins the competition. All of our results hold equally well if the target $\lfloor \frac{n}{2} \rfloor + 1$ is replaced by some generic target L that is higher or lower than $\lfloor \frac{n}{2} \rfloor + 1$.

We will consider computational problems related to strategic aspects of deciding on a line-up of players of a team. Our primary consideration is the following problem of computing the best response to a given line-up of the opposing team.

Team Order	
Input:	A target probability $q \in [0,1]$ and a finite set Team Order instance, and a (deterministic) line-up of team T_2.
Question:	Does there exists a line-up for team T_1 under which the probability of T_1 winning against T_2 is at least q?

Without loss of generality, we can assume that the line-up of T_2 is fixed to $t_2^1, \ldots, t_2^n$ when dealing with the Team Order problem. From a graph theoretic perspective, it can be captured by a weighted and complete bipartite graph $G = (T_1 \cup T_2, E, p)$. The weight of an edge (t_i^a, t_j^b) is winning probability $p(t_i^a, t_j^b)$ of any contestant t_i^a against any other contestant t_j^b. We will call G the *corresponding graph*. The line-ups of the two teams correspond to a perfect matching in G, which pairs up every player in T_1 with a unique player in T_2. Assuming that matches are independent, we are interested in computing a perfect matching M whose edge weights maximize the winning probability:

$$\sum_{\substack{S \subseteq \{1,\ldots,n\} \\ |S| \geq \lfloor \frac{n}{2} \rfloor + 1}} \prod_{i \in S}^{n} p(t_1^i, t_2^{M(i)}) \prod_{i \notin S}^{n} \left(1 - p(t_1^i, t_2^{M(i)})\right),$$

where $M(i)$ denotes the index of the player in T_2 who is matched with t_1^i, and each S is an outcome of the competition represented as the set of players in T_1 who win against their opponents. For simplicity, we will also write the probabilities as $p_{i,j} = p(t_1^i, t_2^j)$.

In fact, even when the line-ups of both teams are given, it is not immediately clear that the winning probability of M can be computed efficiently, since there are exponentially (in n) many possible outcomes of the competition. One way that leads to a polynomial-time algorithm to compute this probability is via dynamic programming, which results in the proposition below.

Proposition 1. *Given the line-ups of T_1 and T_2, the winning probability of each team can be computed in time $O(n^2)$.*

We present an example below to illustrate the problem.

Example 1. Take an instance with the input winning probabilities as in Table 1. Also, Team T_1 has 3! different line-ups $O_1, \ldots, O_6$ as illustrated in Fig. 1.

Table 1. Each entry (i, j) is the probability $p(t_1^i, t_2^j)$.

	t_2^1	t_2^2	t_2^3
t_1^1	0.9^*	$\underline{1}$	1
t_1^2	0.5	0.9^*	$\underline{1}$
t_1^3	$\underline{0}$	0.5	0.9^*

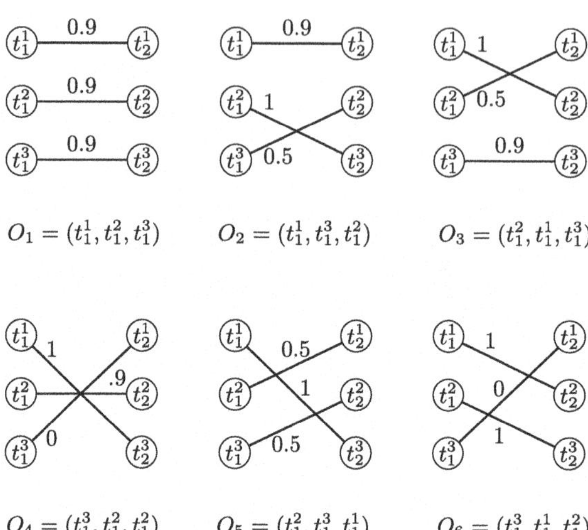

Fig. 1. Graph theoretic view of Example 1. There are 3! different line-ups for T_1 and each line-up is a perfect matching and has its own winning probabilities illustrated on the edges.

Suppose that T_2 uses the line-up (t_2^1, t_2^2, t_2^3). If T_1 responds with (t_1^3, t_1^1, t_1^2) (underlined entries), the probability that they beat T_2 is 1, as they will win two matches with certainty. On the other hand, if T_1 responds with (t_1^1, t_1^2, t_1^3) (starred entries), their winning probability becomes

$$\underbrace{0.9 \times 0.9 \times 0.9}_{\text{prob. of winning all the matches}} + \underbrace{0.9 \times 0.9 \times (1 - 0.9) \times 3}_{\text{prob. of winning exactly two matches}} = 0.972.$$

Indeed, in the above example, the line-up (t_1^1, t_1^2, t_1^3) corresponds to the perfect matching with the maximum total weight in this instance. This demonstrates that weight maximizing matchings may not be optimal solutions to TEAM ORDER. The next example shows that such matchings fail to even provide any approximation guarantee to TEAM ORDER.

Example 2. Suppose that $n = 7$ and the input winning probabilities are given in Table 2. The maximum weight matching gives the guarantee of winning three matching with certainty but losing all the others, and hence probability 0 of winning the competition. On the other hand, the matching that gives probability

Table 2. Each entry (i, j) is the probability $p(t_1^i, t_2^j)$.

	t_2^1	t_2^2	t_2^3	t_2^4	t_2^5	t_2^6	t_2^7
t_1^1	0	0	0	0.5	1	1	1
t_1^2	0	0	0	0	0.5	1	1
t_1^3	0	0	0	0	0	0.5	1
t_1^4	0	0	0	0	0	0	0.5
t_1^5	0	0	0	0	0	0	0
t_1^6	0	0	0	0	0	0	0
t_1^7	0	0	0	0	0	0	0

0.5 of winning four matches wins the competition with a non-zero probability. The example also shows that the maximum weight matching cannot approximate the highest winning probability within any multiplicative factor.

In the above example, the better solution has more balanced winning probabilities over the matches. In view of this, one may conjecture that a *leximin-maximizing* matching is optimal for the TEAM ORDER problem.[2] However, the next example disproves this conjecture: a leximin-maximizing matching may not be optimal, even when it is also maximum weight matchings.

Table 3. Each entry (i, j) is the probability $p(t_1^i, t_2^j)$.

	t_2^1	t_2^2	t_2^3
t_1^1	0.9	0.5	1
t_1^2	0.5	0.1	1
t_1^3	0	0	1

Example 3. Suppose that $n = 3$ and one match is guaranteed to be won as shown in Table 3. The edge weights of the maximum weight matchings are (1) $0.5, 0.5, 1$, or (2) $0.1, 0.9, 1$, and the first one is a leximin-maximizing matching. However, the winning probabilities of these two matchings are $1 - 0.25 = 0.75$ and $1 - 0.09 = 0.89$, respectively.

4 Tractable Variants

In this section we show that TEAM ORDER is tractable if there are only two values of probabilities which are greater than 0 in an instance. Moreover, we

[2] A vector x is leximin-greater than a vector y if x and y are in non-decreasing order and x is lexicographically greater than y.

ALGORITHM 1: Iterative Algorithm

Input: a Team Order instance $G = (T_1 \cup T_2, E, p)$ where $p_{i,j} \in \{\alpha, \beta, 0\}$, $\alpha > \beta > 0$.
Output: an optimal solution to Team Order.

```
Remove all zero-weight edges of G;
opt ← 0;
for s = ⌊n/2⌋ + 1, ... n do
    M_s ← maximum weight matching of size s; // polynomial-time solvable
    if M_s ≠ ∅ then
        p_s ← winning probability of line-up M_s; // see Proposition 1
        if p_s > opt then
            opt ← p_s;
            M* ← M_s;
        end
    end
end
return M*.
```

demonstrate that checking if a team can win with a non-zero probability can be done in polynomial time. Finally, we show that finding the line-up maximizing winning all the matches is tractable. Our reasoning in this section is closely related to the Maximum Weight Matching problem. We note that it can be solved in $O(n^3)$ time via the Hungarian algorithm [19].

Maximum Weight Matching	
Input:	A bipartite graph G, weight $w(e) \in \mathbb{R}_+$ for each edge e on G.
Question:	Compute a perfect matching M of G that maximizes $w(M) := \sum_{e \in M} w(e)$.

4.1 When Input Probabilities Have Three Values (Including 0)

Let us consider the case in which the input probabilities are from a set $\{\alpha, \beta, 0\}$ and, without loss of generality, assume that $\alpha > \beta > 0$. We note that the problem appears closely connected to a Colored Bipartite Matching problem with two types of colors: given a bipartite graph with red and blue edges, does there exists a matching with (exactly) a certain number of red edges? Although the complexity of this red-blue matching problem is open [32], we show that the optimal line-up problem can be solved in polynomial-time via Algorithm 1. We also remark that with this probability set $\{\alpha, \beta, 0\}$ the problem still remains different from Maximum Weight Matching, as we demonstrated via Example 1.

Theorem 1. *Suppose that $G = (T_1 \cup T_2, E, p)$ is a* Team Order *instance with $p_{i,j} \in \{\alpha, \beta, 0\}$ for all $i, j \in \{1, \ldots, n\}$. Then an optimal line-up can be computed in polynomial time.*

Proof. Suppose that M^* denotes an optimal line-up. Let X denote a random variable counting the number of games won by T_1 corresponding to M^*. Then X follows a Poisson Binomial (PB) distribution:

$$X \sim PB(\underbrace{\alpha, \ldots, \alpha}_{x}, \underbrace{\beta, \ldots, \beta}_{y}, \underbrace{0, \ldots, 0}_{z}) = PB(\underbrace{\alpha, \ldots, \alpha}_{x}, \underbrace{\beta, \ldots, \beta}_{y}),$$

where x, y and z are non-negative integers. Let us remove all 0-weight edges from G and call the resulting graph G'. Then, M^* is a matching of size $x + y$ in G'. Also, any maximum weight matching of size $x + y$, say M, has at least x α-weight edges. Notice that if M has at least $x + 1$ α-weight edges, then Poisson binomial random variable Y corresponding to M stochastically dominates X contradicting the fact that M^* is an optimal line-up. The argument also suggests that searching through all matchings of various sizes will hit the optimal line-up. Note that finding a maximum weight matching of a given size is polynomially solvable. For example, the Hungarian algorithm computes a maximum weight matching of a bipartite graph for each target size [20].

Similar approaches based on Maximum Weight Matching also lead to efficient algorithms for two variants of Team Order. First, if the goal is to decide whether T_1 can beat T_2 with non-zero probability, the problem can be solved in polynomial time. Specifically, for an instance represented as a graph G, we can consider the corresponding graph G' in which edges with weight 0 are removed. Then, T_1 can beat T_2 with non-zero probability if and only if G' has a matching of size $\lfloor \frac{n}{2} \rfloor + 1$. We state this result below.

Corollary 1. *Given the line-up of T_2, it can be decided in polynomial time whether there exists a line-up of T_1 that beats T_2 with a non-zero probability.*

Second, if the goal is to maximize the probability of winning *all* the matches, the problem reduces to computing a weight maximizing matching, where the weights are the logarithm of the non-zero winning probabilities.

Proposition 2. *Given the line-up of T_2, the line-up of T_1 that maximizes the probability of winning* all *the matches can be computed in polynomial time.*

5 Approximation Algorithm for Team Order

As we have seen, in several cases finding a solution to Team Order is tractable. However, even though it resembles Maximum Weight Matching, its exact solutions are far more nuanced, which suggests its hardness. In this section, we address the practical solvability of our problem by providing a PTAS for Team Order. Assuming the input probabilities are bounded away from 0 and 1 by any arbitrary constant $\varepsilon > 0$, the PTAS computes a solution to Team Order whose winning probability is at most ε less than that of the optimal solution.

5.1 High-Level Ideas

For any perfect matching $M = \{e_1, \ldots, e_n\}$ of $G = (T_1 \cup T_2, E, p)$, let X_M be a random variable counting the number of matches won by T_1. One may observe that X_M follows a Poisson binomial distribution $PB(p_{e_1}, \ldots, p_{e_n})$. Furthermore, TEAM ORDER can be written as the following optimization problem.

$$\min_{M} \quad \mathbf{Pr}\left[X_M \leq \lfloor \frac{n}{2} \rfloor\right]$$
$$\text{subject to:} \quad M \text{ is a perfect matching of } G = (T_1 \cup T_2, E, p)$$

The main idea of our algorithm is as follows. First, we note that the number of matchings M with $\mathbf{Var}\left[X_M\right] < \varepsilon^{-2}$ is bounded from above by a polynomial in n, when ε is a constant. Hence, we can search over all such matchings to find out the optimal one among them. For the other matchings M with a high variance $\mathbf{Var}\left[X_M\right] \geq \varepsilon^{-2}$, we use $\Phi\left(\frac{\lfloor \frac{n}{2} \rfloor - \mathbf{E}[X_M]}{\sqrt{\mathbf{Var}[X_M]}}\right)$ to approximate the objective function, where $\Phi(x) = (\frac{1}{\sqrt{2\pi}}) \int_{-\infty}^{x} e^{\frac{-y^2}{2}} dy$. Since X_M is a Poisson binomial random variable, it holds that if $\mathbf{Var}\left[X_M\right] \geq \varepsilon^{-2}$, then

$$\left| \mathbf{Pr}\left[X_M \leq \lfloor \frac{n}{2} \rfloor\right] - \Phi\left(\frac{\lfloor \frac{n}{2} \rfloor - \mathbf{E}\left[X_M\right]}{\sqrt{\mathbf{Var}\left[X_M\right]}}\right) \right| \leq \varepsilon.$$

Using the fact that $\Phi(x)$ is an increasing and continuous function in x, we get the following optimization problem as an approximation to the original one.

$$\min_{M} \quad \frac{\lfloor \frac{n}{2} \rfloor - \mathbf{E}\left[X_M\right]}{\sqrt{\mathbf{Var}\left[X_M\right]}}$$
$$\text{subject to:} \quad M \text{ is a perfect matching of } G = (T_1 \cup T_2, E, p)$$

The objective function is still non-linear though, but it can be characterized by the mean and variance of X_M. Using the fact that for every matching M we have $0 \leq \mathbf{Var}\left[X_M\right] \leq \frac{n}{4}$ and $\mathbf{E}\left[X_M\right] \leq n$, we can discretize the two dimensional space $\{(x, y) : 0 \leq x \leq \frac{n}{4} \text{ and } 0 \leq y \leq n\}$ and design a search mechanism to eventually hit a matching that is close enough to the optimal matching. The search mechanism is based on an approximation algorithm solving a matching problem that involves both budget and rewards, which we will discuss next.

5.2 Preliminary Results

We introduce necessary preliminary results for designing the PTAS. It has two main ingredients. We apply a normal distribution estimation for a Poisson binomial distribution, and an approximation algorithm for the following BUDGETED/REWARD MATCHING problem. We assume that every $p_e \notin \{0, 1\}$ is bounded away from 0 and 1. Define $\delta = \min_{e \in E, p_e \notin \{0,1\}} \min\{p_e, 1 - p_e\}$. Then, we get that $\frac{1}{\delta} = \Theta(1)$.

Approximation of Poisson Binomial Distribution. We use a normal distribution estimation for a Poisson binomial distribution to approximate $\mathbf{Pr}\left[X_M \leq \lfloor \frac{n}{2} \rfloor\right]$, which is based on the following result.

Theorem 2 ([29, Theorem 3.5]). *Suppose that $X \sim PB(p_1, \ldots, p_n)$ is a Poisson binomial random variable. Then, for every $1 \leq k \leq n$,*

$$\left| \mathbf{Pr}\left[X \leq k\right] - \Phi\left(\frac{k - \mathbf{E}\left[X\right]}{\sqrt{\mathbf{Var}\left[X\right]}}\right) \right| \leq \frac{1}{\sqrt{\mathbf{Var}\left[X\right]}},$$

where $\Phi(x) = (\frac{1}{\sqrt{2\pi}}) \int_{-\infty}^{x} \mathrm{e}^{\frac{-y^2}{2}} dy$.

An immediate application of Theorem 2 results in to the following corollary.

Corollary 2. *Suppose that $X_M \sim PB(p_{e_1}, \ldots, p_{e_n})$ is a Poisson binomial random variable corresponding to a matching $M = \{e_1, \ldots, e_n\}$ with $\mathbf{Var}\left[X_M\right] \geq \varepsilon^{-2}$, for some $\varepsilon > 0$. Then,*

$$\left| \mathbf{Pr}\left[X_M \leq \lfloor \frac{n}{2} \rfloor\right] - \Phi\left(\frac{\lfloor \frac{n}{2} \rfloor - \mathbf{E}\left[X_M\right]}{\sqrt{\mathbf{Var}\left[X_M\right]}}\right) \right| \leq \varepsilon,$$

where $\Phi(x) = (\frac{1}{\sqrt{2\pi}}) \int_{-\infty}^{x} \mathrm{e}^{\frac{-y^2}{2}} dy$.

Budgeted Matching. We will use approximation algorithms for the following Budgeted Matching problem as subroutines in our algorithm.

Budgeted Matching	
Input:	A bipartite graph G, weight $w(e)$ and cost $c(e)$ for each edge e, and a budget B.
Question:	Compute a perfect matching M of G that maximizes $w(M) := \sum_{e \in M} w(e)$, subject to $c(M) := \sum_{e \in M} c(e) \leq B$.

Specifically, we are interested in the following weight and cost functions. For every $e \in E$, we let $w(e) = p_e$ and $c(e) = p_e \cdot (1 - p_e)$. Hence, for every matching M, we have

$$w(M) = \mathbf{E}\left[X_M\right], \quad \text{and} \quad c(M) = \mathbf{Var}\left[X_M\right].$$

We will henceforth stick to the above weight and cost functions, unless otherwise clarified. We use $I_b(G, w, c, B)$ to denote an instance of Budgeted Matching. For convenience, we can also define a "rewarded" variant of Budgeted Matching, where we want the total cost to pass a threshold R, i.e., $c(M) = \sum_{e \in M} c(e) \geq R$, and we denote it by $I_r(G, w, c, B)$. Since $0 \leq c(e) < 1$, we observe that $I_r(G, w, c, B)$ is equivalent to $I_b(G, w, c', n - R)$, where $c'(e) = 1 - c(e)$ for every $e \in E$. Berger et al. [8] designed a PTAS for the Budgeted Matching problem. Using the same idea this PTAS is based on, we get the following.

Lemma 1. *Suppose that $G = (T_1 \cup T_2, E, p)$ is a* Team Order *instance, $w(e) = p_e$ and $c(e) = p_e \cdot (1 - p_e)$ for each $e \in E$. Then, there is a polynomial-time algorithm to compute a feasible solution M to $I_b(G, w, c, B)$ (respectively, $I_r(G, w, c, R)$) such that $w(M) \geq opt - 2$, where opt is the weight of optimal solution of $I_b(G, w, c, B)$ (respectively, $I_r(G, w, c, R)$).*

Small/Large Variance Matchings. We partition the set of edges into edges with fractional and binary weights; let $F = \{e \in E : p_e \notin \{0,1\}\}$ and $\overline{F} = \{e \in E : p_e \in \{0,1\}\}$. Fix an arbitrary constant $\varepsilon \in (0,1]$ and define $\mathcal{M}^+(\varepsilon) = \{N \subset F : N \text{ is a minimum size matching with } c(N) > \varepsilon^{-2}\}$, and

$$\mathcal{M}^-(\varepsilon) = \left\{N \subset F : N \text{ is a matching with } c(N) \leq \varepsilon^{-2}\right\}.$$

Clearly, for every perfect matching M on G, if $c(M) > \varepsilon^{-2}$, then there exists $N \in \mathcal{M}^+(\varepsilon)$ such that $M \cap N = N$. Similarly, if $c(M) \leq \varepsilon^{-2}$, there exists $N \in \mathcal{M}^-(\varepsilon)$ such that $M \cap N = N$.

For every matching $N \subset E$ and every subset of edges $E' \subseteq E$, let $E'_N = \{e \in E' : e \cap N = \emptyset\}$, i.e., E'_N is the set of all edges in E' that do not share all endpoints with N. We now define two families of bipartite graphs as follows. First, $\mathcal{G}^+(\varepsilon) = \{H = (T_1 \cup T_2, N \cup E_N, p) : N \in \mathcal{M}^+(\varepsilon)\}$. Intuitively, we fix the matching N and leave the unmatched part of the graph G free. Then, we define $\mathcal{G}^-(\varepsilon) = \{H = (T_1 \cup T_2, N \cup \overline{F}_N, p) : N \in \mathcal{M}^-_\varepsilon\}$. This differs from $\mathcal{G}^+(\varepsilon)$, as we only consider 0/1-edges in the unmatched part of G. Note that for every perfect matching M of G, if $c(M) > \varepsilon^{-2}$, there is $H \in \mathcal{G}^+(\varepsilon)$ such that $M \subset H$. Similarly, if $c(M) \leq \varepsilon^{-2}$, then there is $H \in \mathcal{G}^-(\varepsilon)$ such that $M \subset H$. Next, we show that the size of these families of graphs is polynomially bounded.

Lemma 2. *It holds that $|\mathcal{G}^+(\varepsilon)| \leq n^{4\delta^{-1}\varepsilon^{-2}}$ and $|\mathcal{G}^-(\varepsilon)| \leq n^{4\delta^{-1}\varepsilon^{-2}}$.*

5.3 The Algorithm

Now we discuss our approximation algorithm, Algorithm 2. Theorem 3 shows that the algorithm produces an ε-approximate solution to Team Order in polynomial time. The proof relies on Lemma 1 and Lemma 2.

Theorem 3. *Algorithm 2 computes an ε-approximate solution to* Team Order *and runs in time $n^{O(\delta^{-1}\varepsilon^{-2})}$, where $\delta = \min_{p_e \notin \{0,1\}} \min_{e \in E}\{p_e, 1 - p_e\}$.*

6 Winning Probability of a Maximum Weight Matching

In this section we investigate the winning probability of a maximum weight matching. Our result provides a lower bound for the winning probability of any maximum weight matching compared with that of the optimal line-up. In particular, the result shows that a sufficiently large/small maximum weight matching performs almost as well as the optimal line-up. In what follows, we view $G = (T_1 \cup T_2, E, p)$ as a weighted bipartite graph where for every $e \in E$, p_e is the weight of e. For every matching M, we use $w(M)$ to denote its weight. In addition, we assume that the size of G is sufficiently large.

ALGORITHM 2: ε-APPROXIMATION ALGORITHM

Input: a TEAM ORDER instance $G = (T_1 \cup T_2, E, p)$.
Output: an ε-approximate solution to TEAM ORDER.

$\varepsilon \leftarrow \frac{\varepsilon}{4}$;
for $H \in \mathcal{G}^-(\varepsilon)$ **do**
  $M^* \leftarrow$ a maximum weight perfect matching of H;
  if *M^* exists and has a higher winning probability than $M_\varepsilon^{(1)}$ (or $M_\varepsilon^{(1)} = null$)* **then**
    $M_\varepsilon^{(1)} \leftarrow M^*$;
  end
end
$x_i \leftarrow \varepsilon^{-2} + \frac{i}{n}$ for each $i = 0, \ldots, \frac{n^2}{4}$;
for $H \in \mathcal{G}^+(\varepsilon)$ **do**
  for $i = 1, \ldots, \frac{n^2}{4}$ **do**
    $M_i^* \leftarrow$ a solution to $I_b(H, w, c, x_i)$ such that $w(M_i^*) > opt - 2$; `// Lemma 1`
    if *M_i^* exists and $w(M_i) < \lfloor \frac{n}{2} \rfloor$* **then**
      $M_i^* \leftarrow$ a solution to $I_r(H, w, c, x_{i-1})$ such that $w(M_i^*) > opt - 2$; `// Lemma 1`
    end
    if *M_i^* exists and $\frac{\lfloor \frac{n}{2} \rfloor - w(M_\varepsilon^{(2)})}{\sqrt{c(M_\varepsilon^{(2)})}} > \frac{\lfloor \frac{n}{2} \rfloor - w(M_i^*)}{\sqrt{c(M_i^*)}}$ (or $M_\varepsilon^{(2)} = null$)* **then**
      $M_\varepsilon^{(2)} \leftarrow M_i^*$;
    end
  end
end
return *$M_\varepsilon^{(1)}$ or $M_\varepsilon^{(2)}$ whichever has the higher winning probability.*

Theorem 4. *Let M^* be a maximum weight matching and let O be an optimal line-up. Then,*

(1) If $w(M^*) = \frac{n}{2} \pm f(n)\sqrt{n}$, where $f(n) \in [1, \frac{\sqrt{n}}{2}]$ is any non-decreasing function in n, then $\mathbf{Pr}[T_1 \text{ wins under } O] \leq \mathbf{Pr}[T_1 \text{ wins under } M^*] + \mathrm{e}^{-2f^2(n)}$.
(2) If $w(M^*) \in [\frac{n}{2} - \sqrt{n \log n},\ \frac{n}{2} + \sqrt{n \log n}]$, then

$$\begin{aligned} &\mathbf{Pr}[T_1 \text{ wins under } O] \\ \leq &\mathbf{Pr}[T_1 \text{ wins under } M^*] + \frac{(4 + o(1))}{n+1} \sum_{e \in M^*} (p_e - \frac{1}{2})^2. \end{aligned}$$

In the proof of Theorem 4 we rely on the following results.

Theorem 5 ([10]). *Suppose that n is a given positive integer and let $X \sim PB(p_1, \ldots, p_n)$ be a Poisson binomial random variable. Then, we have that $\mathbf{Pr}[X \geq \mathbf{E}[X] + \delta] \leq e^{\frac{-2\delta^2}{n}}$, and $\mathbf{Pr}[X \leq \mathbf{E}[X] - \delta] \leq e^{\frac{-2\delta^2}{n}}$.*

Theorem 6 ([29, Theorem 2.1]). *Let $X \sim PB(p_1, \dots, p_n)$ and let $\bar{p} = \sum_{i=1}^{n} \frac{p_i}{n}$. Define $Y \sim Bin(n, \bar{p})$. Then, (1) for every $0 \leq k \leq n\bar{p} - 1$, $\mathbf{Pr}[X \leq k] \leq \mathbf{Pr}[Y \leq k]$, and (2) for every $n\bar{p} \leq k \leq n$, $\mathbf{Pr}[X \leq k] \geq \mathbf{Pr}[Y \leq k]$.*

Theorem 7 ([1, Theorem 1]). *Suppose that $X \sim PB(p_1, \dots, p_n)$, and $\bar{p} = \sum_{i=1}^{n} p_i/n$. Also, let $Y \sim Bin(n, \bar{p})$ is a binomial probability distribution. Then,*

$$\max_{A \subseteq \{0,\dots,n\}} |\mathbf{Pr}[X \in A] - \mathbf{Pr}[Y \in A]| \leq \frac{1 - \bar{p}^n - (1-\bar{p})^n}{(n+1)\bar{p}(1-\bar{p})} \sum_{i=1}^{n} (p_i - \bar{p})^2.$$

We prove the first and the second parts of the theorem separately next.

Part 1. When $w(M^*) = \frac{n}{2} \pm f(n)\sqrt{n}$

Proof. Let $M = \{e_1, \dots, e_n\}$ be an arbitrary matching and X_M be a random variable that counts the number of games won by T_1 under line-up M. Then X_M follows Poisson binomial distribution $PB(p_{e_1}, \dots, p_{e_n})$. , where $M = \{e_1, \dots, e_n\}$. Thus, $\mathbf{E}[X_M] = w(M)$. Let us first assume that $w(M^*) = \frac{n}{2} - f(n)\sqrt{n}$. Then, for every matching M, including the optimal line-up O, we have $w(M) \leq w(M^*)$. Moreover, we have $f(n)\sqrt{n} = \frac{n}{2} - w(M^*) \leq \frac{n}{2} - w(M)$, and

$$\begin{aligned} \mathbf{Pr}[T_1 \text{ wins under } M] &= \mathbf{Pr}\left[X_M \geq \lfloor\frac{n}{2}\rfloor + 1\right] \\ &\leq \mathbf{Pr}\left[X_M \geq w(M) + (\frac{n}{2} - w(M))\right] \\ &= \mathbf{Pr}\left[X_M \geq \mathbf{E}[X_M] + f(n)\sqrt{n}\right] \leq \mathrm{e}^{-2(f(n)\sqrt{n})^2 \frac{2}{n}} = \mathrm{e}^{-2f(n)^2}, \end{aligned}$$

using a concentration bound for Poisson binomial random variables (e.g., see Theorem 5). Following that upper bound, if $w(M^*) = \frac{n}{2} - f(n)\sqrt{n}$, then

$$\mathbf{Pr}[T_1 \text{ wins under } O] \leq \mathrm{e}^{-2f(n)^2} \leq \mathbf{Pr}[T_1 \text{ wins under } M^*] + \mathrm{e}^{-2f(n)^2}. \quad (1)$$

Next, we consider the case where $w(M^*) = \frac{n}{2} + f(n)\sqrt{n}$. Define random variable Y_{M^*} that counts the number of games lost under M^*. Then Y_{M^*} follows Poisson binomial distribution $PB(1 - p_{e_1}, \dots, 1 - p_{e_n})$, where we let $M^* = \{e_1, \dots, e_n\}$. One can check that $\mathbf{E}[Y_{M^*}] = n - w(M^*) = \frac{n}{2} - f(n)\sqrt{n}$.

$$\begin{aligned} \mathbf{Pr}[T_1 \text{ loses under } M^*] &= \mathbf{Pr}\left[Y_{M^*} \geq \lfloor\frac{n}{2}\rfloor + 1\right] \\ &\leq \mathbf{Pr}\left[Y_{M^*} \geq \mathbf{E}[Y_{M^*}] + (\frac{n}{2} - \mathbf{E}[Y_{M^*}])\right] \leq \mathrm{e}^{-2f(n)^2}, \end{aligned}$$

where we have applied the same concentration bound as the previous case. Hence,

$$\mathbf{Pr}[T_1 \text{ wins under } M^*] = 1 - \mathbf{Pr}[T_1 \text{ loses under } M^*] \geq 1 - \mathrm{e}^{-2f(n)^2}.$$

Thus,

$$\mathbf{Pr}[T_1 \text{ wins under } O] \leq 1 \leq \mathbf{Pr}[T_1 \text{ wins under } M^*] + \mathrm{e}^{-2f(n)^2} \quad (2)$$

Hence, combining (1) and (2) gives the first part of Theorem 4.

Part 2. When $w(M^*) \in [\frac{n}{2} - \sqrt{n \log n},\ \frac{n}{2} + \sqrt{n \log n}]$

Proof. Let us first consider the case where $w(M^*) \in [\frac{n}{2} - \sqrt{n \log n}, \frac{n}{2} - 1)$. Define random variables X_O and X_{M^*} that count the number games won by T_1 under O and M^*, respectively. Moreover, define binomial random variables $Z_O \sim Bin(n, \frac{w(O)}{n})$ and $Z_{M^*} \sim Bin(n, \frac{w(M^*)}{n})$. Notice that $w(O) \leq w(M^*)$ and hence Z_{M^*} stochastically dominates Z_O (i.e., $\mathbf{Pr}\left[Z_O \leq \frac{n}{2}\right] \geq \mathbf{Pr}\left[Z_{M^*} \leq \frac{n}{2}\right]$). Since $w(M^*) < \frac{n}{2}$, we apply the stochastic dominance between the Poisson and binomial random variables (e.g., see Theorem 6 (2)) and we have that

$$\mathbf{Pr}\left[T_1 \text{ loses under } O\right] = \mathbf{Pr}\left[X_O \leq \frac{n}{2}\right] \geq \mathbf{Pr}\left[Z_O \leq \frac{n}{2}\right] \geq \mathbf{Pr}\left[Z_{M^*} \leq \frac{n}{2}\right],$$

On the other hand, the optimal line-up O minimizes the losing probability of T_1 and hence, by above inequality we have that

$$\begin{aligned}&\mathbf{Pr}\left[T_1 \text{ loses under } M^*\right]\\ &\quad= \mathbf{Pr}\left[X_{M^*} \leq \frac{n}{2}\right] \geq \mathbf{Pr}\left[T_1 \text{ loses under } O\right] \geq \mathbf{Pr}\left[Z_{M^*} \leq \frac{n}{2}\right].\end{aligned}$$

Applying the above inequality and Theorem 7 results in

$$\begin{aligned}&\mathbf{Pr}\left[T_1 \text{ wins under } O\right] - \mathbf{Pr}\left[T_1 \text{ wins under } M^*\right]\\ &\qquad= (1 - \mathbf{Pr}\left[T_1 \text{ wins under } M^*\right]) - (1 - \mathbf{Pr}\left[T_1 \text{ wins under } O\right])\\ &\qquad= \mathbf{Pr}\left[T_1 \text{ loses under } M^*\right] - \mathbf{Pr}\left[T_1 \text{ loses under } O\right]\\ &\qquad\leq \mathbf{Pr}\left[X_{M^*} \leq \frac{n}{2}\right] - \mathbf{Pr}\left[Z_{M^*} \leq \frac{n}{2}\right]\\ &\qquad\leq \frac{1 - (\bar{p})^n - (1-\bar{p})^n}{(n+1)(1-\bar{p})\bar{p}} \sum_{e \in M^*} (p_e - \bar{p})^2,\end{aligned}$$

where $\bar{p} = \frac{w(M^*)}{n}$. Since we have $n\bar{p} \in (\frac{n}{2} - \sqrt{n \log n}, \frac{\mathbf{n}}{\mathbf{2}})$, and n is an asymptotically large, we have $\bar{p} \approx \frac{1}{2}$ and thus

$$\frac{1 - (\bar{p})^n - (1-\bar{p})^n}{(n+1)(1-\bar{p})\bar{p}} \sum_{e \in M^*} (p_e - \bar{p})^2 \leq \frac{(4 + o(1))}{n+1} \sum_{e \in M^*} (p_e - \frac{1}{2})^2.$$

Therefore, if $w(M^*) \in [\frac{n}{2} - \sqrt{n \log n}, \frac{n}{2} - 1)$, then

$$\mathbf{Pr}\left[T_1 \text{ wins under } O\right] \leq \mathbf{Pr}\left[T_1 \text{ wins under } M^*\right] + \frac{(4 + o(1))}{n+1} \sum_{e \in M^*} (p_e - \frac{1}{2})^2.$$

To derive the same upper bound for the case where $w(M^*) \in [\frac{n}{2}, \frac{n}{2} + \sqrt{n \log n}]$, we define random variables that count the number of games lost by T_1 and the same technique for the above case follows.

7 Conclusion

We proposed the TEAM ORDER problem, which naturally captures several strategic scenarios in information systems and team competitions. We have shown that in the case in which the input probabilities are limited to three values (including 0) it is tractable and have shown that it is possible to efficiently compute a line-up which is close to the optimal in terms of the probability of winning, which is useful when the information about the players' relative strength is limited (e.g., if it is only known when a player is "strong" or "weak" against an opponent). One of our central results is a PTAS for the TEAM ORDER problem. We note that while we focused on the probability of winning against more than a half of opposing players, our results hold for any such threshold.

We conclude by highlighting some important directions for future work. First, the complexity of solving TEAM ORDER exactly is open. We believe that this is a challenging question that also has implications on the related problem of COLORED BIPARTITE MATCHING. It is known that it is NP-complete when d is a variable [16]. However, the complexity of this problem is open if d is a constant larger than 2, or if $d = 2$ but the graph is incomplete (which corresponds to $\{\alpha, \beta, 0\}$) [32]. This motivates further study between the connections of the two discussed problems. It is also not known whether TEAM ORDER admits a fully polynomial-time approximation scheme (FPTAS). Resolving this question would be a strong improvement over our results.

While our result show the complexity of computing a best response to the opponents line-up, it is natural to study the extension in which multiple teams strategize. Regarding sport events, it would also be interesting to see if the results change under other natural assumptions, such as all of the players having an objective level of skill. For example, if a player i has better skill than a player j, then i might always have a better probability of winning against any player k than j's probability of beating k.

Acknowledgments. This work was supported by the NSF-CSIRO grant on "Fair Sequential Collective Decision-Making" (Grant No. RG230833) and by DSTG under the project "Distributed multi-agent coordination for mobile node placement." (Grant No. RG233005). This project has also received funding from the European Research Council (ERC) under the European Union's Horizon 2020 research and innovation programme (grant agreement No. 101002854).

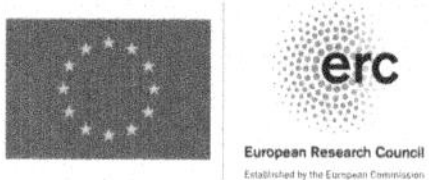

References

1. Ehm, W.: Binomial approximation to the Poisson binomial distribution. Stat. Probab. Lett. **11**(1), 7–16 (1991)
2. Ahani, N., Gölz, P., Procaccia, A.D., Teytelboym, A., Trapp, A.C.: Dynamic placement in refugee resettlement. In: The 22nd ACM Conference on Economics and Computation, p. 5. ACM (2021)
3. Allan, B., Omer, L., Nisarg, S., Tyrone, S.: Primarily about primaries. In: The Thirty-Third AAAI Conference on Artificial Intelligence, pp. 1804–1811 (2019)
4. Aziz, H., Brill, M., Fischer, F., Harrenstein, P., Lang, J., Seedig, H.G.: Possible and necessary winners of partial tournaments. J. Artif. Intell. Res. **54**, 493–534 (2015)
5. Aziz, H., Gaspers, S., Mackenzie, S., Mattei, N., Stursberg, P., Walsh, T.: Fixing a balanced knockout tournament. In: Proceedings of the 28th AAAI Conference on Artificial Intelligence (AAAI), pp. 552–558. AAAI Press (2014)
6. Aziz, H., Biró, P., Gaspers, S., de Haan, R., Mattei, N., Rastegari, B.: Stable matching with uncertain linear preferences. Algorithmica **82**, 1410–1433 (2020)
7. Bansak, K., et al.: Improving refugee integration through data-driven algorithmic assignment. Science **359**(6373), 325–329 (2018)
8. Berger, A., Bonifaci, V., Grandoni, F., Schäfer, G.: Budgeted matching and budgeted matroid intersection via the gasoline puzzle. Math. Program. **128**(1–2), 355–372 (2011)
9. Burkhard, R., Dell'Amico, M., Martello, S.: Assignment Problems. SIAM (2009)
10. Dubhashi, D.P., Panconesi, A.: Concentration of Measure for the Analysis of Randomized Algorithms. Cambridge University Press, Cambridge (2009)
11. Echenique, F., Immorlica, N., Vazirani, V.V.: Online and Matching-Based Market Design. Cambridge University Press, Cambridge (2023)
12. Faliszewski, P., Gourvès, L., Lang, J., Lesca, J., Monnot, J.: How hard is it for a party to nominate an election winner? In: IJCAI (2016)
13. Fu, Q., Lu, J.: On equilibrium player ordering in dynamic team contests. Econ. Inq. **58**(4), 1830–1844 (2020)
14. Fu, Q., Lu, J., Pan, Y.: Team contests with multiple pairwise battles. Am. Econ. Rev. **105**(7), 2120–40 (2015)
15. Gaonkar, A., Raghunathan, D., Weinberg, S.M.: The derby game: an ordering-based Colonel Blotto game. In: EC 2022: The 23rd ACM Conference on Economics and Computation, pp. 184–207. ACM (2022)
16. Geerdes, H.F., Szabó, J.: A unified proof for Karzanov's exact matching theorem. Technical report QP-2011-02, Egerváry Research Group, Budapest (2011). www.cs.elte.hu/egres
17. Kern, W., Paulusma, D.: The new FIFA rules are hard: complexity aspects of sports competitions. Discrete Appl. Math. **108**(3), 317–323 (2001)
18. Konishi, H., Pan, C.Y., Simeonov, D.: Equilibrium player choices in team contests with multiple pairwise battles. Games Econ. Behav. **132**, 274–287 (2022)
19. Kuhn, H.W.: The Hungarian method for the assignment problem. Naval Res. Logist. Q. **2**(1–2), 83–97 (1955)
20. Kuhn, H.W.: The Hungarian method for the assignment problem. In: Jünger, M., et al. (eds.) 50 Years of Integer Programming 1958-2008, pp. 29–47. Springer, Heidelberg (2010). https://doi.org/10.1007/978-3-540-68279-0_2
21. Li, J., Convertino, M.: Inferring ecosystem networks as information flows. Sci. Rep. **11**(1), 1–22 (2021)

22. Lisowski, G.: Strategic nominee selection in tournament solutions. In: Baumeister, D., Rothe, J. (eds.) EUMAS 2022. LNCS, vol. 13442, pp. 239–256. Springer, Cham (2022). https://doi.org/10.1007/978-3-031-20614-6_14
23. Lisowski, G., Ramanujan, M., Turrini, P.: Equilibrium computation for knockout tournaments played by groups. In: International Conference on Autonomous Agents and Multiagent Systems. AAMAS (2022)
24. Lovász, L., Plummer, M.D.: Matching Theory. AMS Chelsea Publishing (2009)
25. Mohamed, M.H., Khafagy, M.H., Ibrahim, M.H.: Recommender systems challenges and solutions survey. In: 2019 International Conference on Innovative Trends in Computer Engineering (ITCE), pp. 149–155. IEEE (2019)
26. Roberson, B.: The Colonel Blotto game. Econ. Theor. **29**(1), 1–24 (2006)
27. Shubik, M., Weber, R.J.: Systems defense games: Colonel Blotto, command and control. Cowles Foundation Discussion Papers 489, Cowles Foundation for Research in Economics, Yale University (1978)
28. Tang, P., Shoham, Y., Lin, F.: Team competition. In: Proceedings of the 8th International Conference on Autonomous Agents and Multiagent Systems (AAMAS), pp. 241–248. IFAAMAS (2009)
29. Tang, W., Tang, F.: The Poisson binomial distribution – old & new (2019)
30. Vassilevska-Williams, V.: Knockout tournaments. In: Brandt, F., Conitzer, V., Endriss, U., Lang, J., Procaccia, A.D. (eds.) Handbook of Computational Social Choice, chap. 19. Cambridge University Press (2016)
31. Vu, T., Altman, A., Shoham, Y.: On the complexity of schedule control problems for knockout tournaments. In: AAMAS (2009)
32. Yi, T., Murty, K.G., Spera, C.: Matchings in colored bipartite networks. Discrete Appl. Math. **121**(1–3), 261–277 (2002)

Fair Division and Resource Allocation

Fair Division of Chores with Budget Constraints

Edith Elkind[1,2], Ayumi Igarashi[3], and Nicholas Teh[1(✉)]

[1] University of Oxford, Oxford, UK
nicholas.teh@cs.ox.ac.uk
[2] Alan Turing Institute, London, UK
[3] University of Tokyo, Bunkyo City, Japan

Abstract. We study fair allocation of indivisible chores to agents under budget constraints, where each chore has an objective size and disutility. This model captures scenarios where a set of chores need to be divided among agents with limited time, and each chore has a specific time needed for completion. We propose a budget-constrained model for allocating indivisible chores, and systematically explore the differences between goods and chores in this setting. We establish the existence of an EFX allocation. We then show that EF2 allocations are polynomial-time computable in general; for many restricted settings, we strengthen this result to EF1.

Keywords: Fair Allocation · Chores · Budget Constraints

1 Introduction

Alice, Bob, and their teenage children Claire and Dan wish to fairly divide a set of household chores. Each chore requires a certain amount of time to complete; for simplicity, assume that this amount, as well as the disutility of the chore, does not depend on who performs the chore (this is approximately true for many chores). Alice works long shifts, so she only has 5 h a week to dedicate to chores. Bob has a more conventional schedule, so he can spend 10 h on chores. Claire and Dan have many extracurricular activities, so they can contribute 7 and 4 h, respectively. How can they divide the chores in a way that is fair and respects time constraints?

In a similar vein, consider a company that needs to allocate several time-consuming tasks to a group of employees, in addition to their regular workload. As employees may have different existing workloads, the amounts of extra work they would be able to take on differ as well.

In both of our examples, it is not immediately clear what it means to be fair, given agents' different time budgets. Thus, we need to adapt the notions of fairness that have been developed in the fair division literature to our setting, and then determine under what conditions fair allocations exist and whether they can be computed in polynomial time. While the budgeted setting has been

G. Schäfer and C. Ventre (Eds.): SAGT 2024, LNCS 15156, pp. 55–71, 2024.
https://doi.org/10.1007/978-3-031-71033-9_4

considered for goods (see Sect. 1.2 for a discussion of related work), extending ideas from prior work to chores poses new challenges.

1.1 Our Contributions

We introduce a framework for allocating chores with objective (i.e., agent-independent) sizes and disutilities under budget constraints. In Sect. 2, we set up our formal model, put forward notions of fairness that are appropriate for this setting, and discuss the challenges that arise when adapting the budget-constrained model from goods to chores.

In Sect. 3, we show the existence of an EFX allocation for indivisible chores. Perhaps surprisingly, an adaptation of the EFX algorithm for goods with objective sizes and utilities in the budgeted setting also works for our scenario; we note that this is not universally true for general restricted instances where EFX is known to exist for goods. This is particularly interesting because EFX for chores is incomparable to EFX for goods, in the sense that the special cases where EFX allocations are known to exist are quite different in these two settings. Moreover, techniques for proving EFX in the goods setting are also known to be very different beyond two agents or identical valuations (unlike for EF1).

In Sect. 4, we provide a polynomial-time algorithm for computing EF2 allocations. Section 5 then looks at five special cases—when chores are identically-valued, identically-sized, identically-dense, agents have identical budgets, and the case of two agents—for which we can compute EF1 allocations efficiently. Most of the results in the above two sections rely on a greedy "densest-item-first" algorithm. In the goods case, the greedy algorithm that achieves similar guarantees is (surprisingly) also a densest-item first greedy algorithm, even though, intuitively, high density is desirable for goods and undesirable for chores. This suggests that the symmetry between chores and goods sometimes presents itself in unexpected ways.

1.2 Related Work

The mathematical framework for fair division has been put forward by Steinhaus [23] over 70 years ago, and this field has seen an explosion of interest in recent years (see, e.g., a survey by Amanatidis et al. [2]). Historically, most works in the field focused on the allocation of *goods*, i.e., items that are valued non-negatively by all agents. While some of the results for fair allocation of goods extend easily to chores, there are many real-world applications for which this is not the case. This observation led to a recent line of work that considers the allocation of *chores*, i.e., items which agents value negatively; see, e.g., [3,8,14] as well as the recent survey by Aziz et al. [4]. Indeed, allocating chores is generally known to be more difficult than allocating goods, with more open problems in chores than goods, and many techniques that work for goods, but do not directly translate to chores; a notable example here is maximization of Nash social welfare. Also, a number of authors have considered the problem of allocating goods to agents under budget constraints [6,15,16,24], as well as rent division with budget constraints [1,22]. However, to the best of our knowledge, we are the first to explore the intersection of these two lines of work, i.e., chore allocation under budget constraints.

Additional motivation for our analysis is provided by the recent work of Igarashi and Yokoyama [19], who have developed an application that helps couples to fairly divide household chores. While their tool captures many aspects of the task, it does not allow the household members to specify budget constraints, and this reduces the usability of the tool. We believe that incorporating such constraints will help many households to come up with a better way of sharing the workload.

We will now discuss prior work on the allocation of goods under budget constraints in more detail. Gan et al. [15,24] assumed identical valuations and size functions, and studied approximation ratios (with respect to EF1) for the maximum Nash welfare rule in budget-constrained scenarios. In particular, they proposed an approximation algorithm for achieving 1/2-EF1, along with special cases where they could guarantee EF1. Barman et al. [6] considered the same model and proposed an algorithm that satisfies EF2 in general, and EF1 in special cases.

Garbea et al. [16] were the first to consider the budget-constrained model with subjective valuation functions (but still identical size functions). However, their results are limited to two- and three-agent cases. They design algorithms that guarantee EFX, while achieving approximation of Nash welfare for these special cases.

Barman et al. [5] studied a more general model of fairly allocating goods under *generalized assignment constraints*, extending the traditional budget-constrained model to one where the sizes and values of the goods can be subjective. They showed the existence (via a pseudopolynomial time algorithm) of EFX allocations for indivisible goods case.

2 Preliminaries

For each positive integer z, let $[z] := \{1, \dots, z\}$. Let $N = [n]$ be a set of n *agents* and $C = \{c_1, \dots, c_m\}$ be a set of m *chores*. Each chore $c \in C$ has an objective *size* $s(c) \in \mathbb{R}_{>0}$ and *disutility* $d(c) \in \mathbb{R}_{\geq 0}$; we write $\rho(c) = \frac{d(c)}{s(c)}$ to denote the *density* of the chore c. Each agent has a *budget* $B_i \in \mathbb{R}_{>0}$; let $\mathbf{B} = (B_1, \dots, B_n)$ be the vector of agents' budgets. For our algorithmic results, we assume that all sizes, disutilities and budgets are rational numbers given in binary.

Unlike in the unconstrained fair allocation model, in our setting it may be impossible to divide all the chores among the agents in N: e.g., the sum of sizes may exceed the sum of the agents' budgets. As we cannot simply discard the chores, this necessitates the introduction of a *housekeeper*, whose role is similar to that of charity in the budget-constrained model for allocating goods. It is assumed that the housekeeper is paid to complete the chores; this payment is exogenous to the model, and an external consideration. Our fairness notions are formulated in such a way that as few chores as possible are allocated to the housekeeper; this is similar to how the allocation to charity is treated in a goods context.

A *bundle* of chores is a subset of C. We assume that sizes and disutilities of chores are additive, so that for each bundle $S \subseteq C$ its size $s(S)$ and disutility $d(S)$ are given by, respectively, $s(S) = \sum_{c \in S} s(c)$ and $d(S) = \sum_{c \in S} d(c)$. This assumption is standard across all works dealing with budget constraints, and is also common for fair division problems in general.

An *allocation* $\mathcal{A} = (A_1, \ldots, A_{n+1})$ is a partition of C into $n+1$ disjoint bundles of chores, where A_i is assigned to agent i, and A_{n+1} is the set of unallocated chores; we will refer to A_{n+1} as the bundle allocated to the housekeeper. We say that an allocation $\mathcal{A}$ is *feasible* if $s(A_i) \leq B_i$ for all $i \in [n]$.

Next, we define several notions of fairness for our setting. Our definitions mirror the respective definitions for allocating goods under budget constraints.

Definition 1 (Envy-freeness). *An allocation $\mathcal{A} = (A_1, \ldots, A_n, A_{n+1})$ is said to be* envy-free (EF) *if for all $i \in [n+1]$, $j \in [n]$ and for every subset $S \subseteq A_i$ with $s(S) \leq B_j$ it holds that $d(S) \leq d(A_j)$.*

Intuitively, an allocation is envy-free if for every agent $i \in [n]$ as well as for the housekeeper it holds that if they consider a subset S of their bundle that could be allocated to an agent $j \in [n]$ (in the sense of having a size that does not exceed B_j), they find S to be at most as unpleasant/objectionable as the actual bundle of j.

Even in the absence of budget constraints, it may be impossible to allocate indivisible items (chores or goods) in an envy-free manner. Clearly, this negative result also applies to the setting with budget constraints. This observation motivates us to adapt several relaxations of EF to our setting. We first consider a popular, relatively strong relaxation of EF, which has been widely studied in the unconstrained setting.

Definition 2 (Envy-freeness up to any chore). *An allocation $\mathcal{A} = (A_1, \ldots, A_n, A_{n+1})$ is said to be* envy-free up to any chore (EFX) *if for all $i \in [n+1]$, $j \in [n]$, for every subset $S \subseteq A_i$ with $s(S) \leq B_j$, and for each $c \in S$ it holds that $d(S \setminus \{c\}) \leq d(A_j)$.*

Next, we consider another class of relaxations of EF.

Definition 3 (Envy-freeness up to k chores). *Given a positive integer k, an allocation $\mathcal{A} = (A_1, \ldots, A_n, A_{n+1})$ is said to be* envy-free up to k chores (EFk) *if for every $i \in [n+1]$, $j \in [n]$, and for every subset $S \subseteq A_i$ with $s(S) \leq B_j$ there exists a subset $S' \subseteq S$ with $|S'| = k$ such that $d(S \setminus S') \leq d(A_j)$.*

The most commonly studied property is EF1 (i.e., EFk with $k = 1$). However, following the analysis of Barman et al. [5] in the budget-constrained goods setting, we will also consider EF2 (i.e., EFk with $k = 2$).

Note that in our definitions of (approximate) envy-freeness i takes values in $[n+1]$ rather than $[n]$, i.e., we want the housekeeper to be (approximately) non-envious towards the agents. This ensures that *sufficiently* many chores are allocated to agents: e.g., an allocation where all chores are allocated to the housekeeper is not envy-free unless no agent can execute any of the chores.

Modeling Assumptions: A Discussion. Our formal model considers only objective (i.e., identical) size and disutility functions. Of course, in practice different agents may assign different disutilities to the same chore: while Alice dislikes dusting more than doing dishes, Bob has the opposite preferences. It may also be the case that the size of the chore varies from one agent to another: while Alice can peel potatoes for dinner in 5 min, Bob will spend 8 min on the same task. However, we chose to leave modeling non–identical disutilities and sizes in the budgeted setting to future work. The reasons for this decision are as follows.

First, as noted by Barman et al. [6], considering budget constraints even under identical valuation functions already constitutes a technically-rich model, due to the additional size (and budget) dimension.

Second, the more general formulation, where agents have subjective size functions (even under identical valuation functions), does not admit a polynomial-time approximation scheme for the value-maximization objective [13].

Third, we have argued that it is necessary to introduce the housekeeper agent, and there is no principled way to define the disutility function for the housekeeper if the agents' disutility functions are non-identical. However, to define (relaxations of) envy-freeness, we would have to reason about the housekeeper's disutility.

We note that most of the existing works studying the allocation of indivisible goods under budget constraints [6,15,24] assume that each good has an objective size, and that agents have an objective valuation function. Models with identical valuations have also been widely studied in the setting of goods without budget constraints [7,20,21]. An important exception is a recent paper by Barman et al. [5], who extend the concepts for the allocation of goods to generalized assignment constraints (as opposed to budget constraints), where sizes are allowed to be agent-specific. They showed the existence (but not polynomial-time computability) of EFX allocations for indivisible goods. However, extending their definitions and results to the setting of chores with non-identical sizes is not straightforward.

3 Existence of EFX Allocations for Indivisible Chores

To begin, we consider the existence of EFX allocations for indivisible chores under budget constraints.

The existence of EFX allocations for more than three agents in the indivisible goods allocation setting is a longstanding open problem in fair division. The setting of chores has been shown to be even more difficult (refer to the surveys of Aziz et al. [4] and Amanatidis et al. [2]). Recently, Barman et al. [5] proved the existence of EFX allocations for indivisible goods under budget constraints, for the case of identical disutility functions. Their algorithm is a close adaptation of the algorithm for finding EFX allocations in restricted settings [12]. In this section, we extend this positive result to the case of chores.

We first introduce the concept of a *manageable set*, which is similar in spirit to the concept of a *minimal envied subset*; the latter is used to prove the existence of EFX allocations for indivisible goods in various restricted settings [5,12,17].

Definition 4 (Manageable set). *A set of chores $T \subseteq C$ is said to be a* manageable set *for an allocation $\mathcal{A} = (A_1, \ldots, A_n, A_{n+1})$ if*

(i) there exists an $i \in [n]$ such that $s(T) \leq B_i$ and $d(T) > d(A_i)$, and
(ii) no strict subset of T satisfies (i), i.e., for each strict subset $T' \subsetneq T$ and each $k \in [n]$, either $s(T') > B_k$ or $s(T') \leq B_k$ and $d(T') \leq d(A_k)$.

Then, consider the following algorithm, which repeatedly finds a manageable set within the housekeeper's bundle and allocates it to one of the agents in N in a feasible way. The algorithm terminates when the housekeeper's bundle no longer contains a manageable set.

Algorithm 1: Computes an EFX allocation

---

1 **Input** disutility function d, size function s, budgets $\mathbf{B}$;
2 Initialize the allocation $\mathcal{A} = (A_1, \ldots, A_n, A_{n+1}) = (\emptyset, \ldots, \emptyset, C)$;
3 **while** *there exists a subset $S \subseteq A_{n+1}$ such that $s(S) \leq B_i$ and $d(S) > d(A_i)$ for some $i \in [n]$* **do**
4   Select a manageable set $T \subseteq A_{n+1}$ and a $k \in [n]$ with $s(T) \leq B_k$, $d(T) > d(A_k)$;
5   Update bundles $A_k \leftarrow T$ and $A_{n+1} \leftarrow C \backslash (\cup_{i \in [n]} A_i)$;
6 **return** *allocation* $\mathcal{A}$;

We will now show that Algorithm 1 returns an EFX allocation, thereby establishing that an EFX allocation is guaranteed to exist.

Theorem 1. *Algorithm 1 returns an* EFX *allocation.*

Proof. First, we note that if the condition of the **while** loop is satisfied, then A_{n+1} contains a manageable set. Indeed, consider a minimum-size set S that satisfies the condition in the **while** loop for some $i \in [n]$. We have $s(S) \leq B_i$, $d(S) > d(A_i)$, so S satisfies condition (i) in the definition of a manageable set. Moreover, by our choice of S no proper subset of S satisfies (i), which means that S satisfies condition (ii) as well. Thus, Algorithm 1 can indeed select a manageable set in line 4.

Next, we observe that Algorithm 1 necessarily terminates. Indeed, if at iteration t we change the bundle of an agent $k \in [n]$ to T in line 5, then k's disutility increases (since before that step we had $d(T) > d(A_k)$) while the disutility of other agents in $[n]$ remains the same. Thus, the sum of disutilities of agents in $[n]$ goes up with each iteration.

Further, once Algorithm 1 terminates, the condition of the **while** loop is no longer satisfied, which means that the housekeeper is not envious towards agents in $[n]$. It remains to argue that the EFX condition is satisfied for all other agents.

Let $\mathcal{A}^{(t)} = (A_1^{(t)}, \dots, A_{n+1}^{(t)})$ denote the allocation maintained by the algorithm just *before* the t-th iteration of the **while** loop (Line 3). We have $\mathcal{A}^{(1)} = (\emptyset, \dots, \emptyset, C)$ and $A_{n+1}^{(t)} = C \backslash (\cup_{i=1}^{n} A_i^{(t)})$ for all $t > 0$. We will write $\mathcal{A}_n^{(t)}$ to denote the vector formed by the first n entries of the vector $\mathcal{A}^{(t)}$ (i.e., excluding the bundle of the housekeeper).

To show that the allocation $\mathcal{A}_n$ is EFX, we use induction. For the base case, note that in the first iteration, EFX trivially holds, as $\mathcal{A}_n^{(1)} = (\emptyset, \dots, \emptyset)$. This allocation is also feasible. Now, consider an iteration $t > 1$. For the inductive step, assume that the allocation $\mathcal{A}_n^{(t)}$ is feasible and satisfies EFX. In the t-th iteration, the algorithm changes the bundle of exactly one agent $k \in [n]$, by replacing it with T. The bundle T satisfies $s(T) \leq B_k$, so this update results in a feasible allocation.

Since the bundles of all agents in $[n] \backslash \{k\}$ remain unchanged, the EFX condition is satisfied for each pair of agents not involving k. Thus, it remains to consider envy by/towards agent k after k has been allocated the bundle T.

We first consider the envy experienced by agent k. Fix a subset $S \subseteq T$ and an agent $k' \in [n] \backslash \{k\}$ such that $s(S) \leq B_{k'}$. For each $c \in S$ the set $S \backslash \{c\}$ is a proper subset of T. Since T is a manageable set and $s(S \backslash \{c\}) \leq s(S) \leq B_{k'}$, condition (ii) of Definition 4 implies that $d(S \backslash \{c\}) \leq d(A_{k'})$, i.e., the envy by agent k towards k' can be eliminated by the removal of any single chore.

Now, consider the envy of agent $k' \in [n] \backslash \{k\}$ towards agent k. By the induction hypothesis, before agent k was allocated the bundle T, the envy by k' towards k could be eliminated by removing a single chore. Moreover, T has a higher disutility than the previous bundle of k. Thus, for every subset $S \subseteq A_{k'}^{(t)} = A_{k'}^{(t+1)}$ with $s(S) \leq B_k$ and every $c \in S$ we have

$$d(S \backslash \{c\}) \leq d(A_k^{(t)}) < d(A_k^{(t+1)}).$$

This implies that the envy by an agent $k' \in [n] \backslash \{k\}$ towards agent k can be eliminated by the removal of any single chore. □

Theorem 1 establishes the existence of an EFX allocation for indivisible chores under budget constraints. While this result is constructive, the running time of Algorithm 1 is pseudopolynomial (to see this, note that checking the condition in Line 3 of the algorithm and finding a manageable set reduces to solving polynomially many instances of Knapsack) rather than polynomial. The existence of a polynomial-time algorithm for computing an EFX allocation for chores remains an open problem (as with the setting of goods).

4 Computing EF2 Allocations for Indivisible Chores

Given that we do not know how to find EFX allocations in polynomial time, a natural follow-up direction is then to look for allocations that satisfy weaker relaxations of EF, but can be computed by algorithms that run in polynomial time. The most popular relaxation of EF after EFX would be EF1. However, even in the setting of allocating goods under budget constraints, the existence of a polynomial-time algorithm for computing EF1 allocations is still a challenging open question [5,15,24]. Consequently, we shift our focus to a property that can be accomplished in polynomial time in the goods case, namely, EF2. Our next result shows that we can replicate this result for chores.

Specifically, it turns out that EF2 allocations can be found by the DENSESTFIRST algorithm (Algorithm 2), which at each iteration picks an agent with minimum disutility and allocates to her the maximally-dense chore. Interestingly, in the goods case a 'densest-item first' greedy algorithm also produces an EF2 allocation [7]. This is surprising, because in the goods setting high-density items are particularly attractive, while in the chores setting high-density items are unattractive. Thus, while one may expect goods and chores to be symmetric, identifying the 'correct' mapping from goods to chores is a non-trivial task.

Algorithm 2: DENSESTFIRST

1 **Input** disutility function d, size function s, budgets $\mathbf{B}$;
2 Initialize the allocation $\mathcal{A} = (A_1, \ldots, A_n, A_{n+1}) = (\emptyset, \ldots, \emptyset, C)$ and the set of live agents $L = [n]$;
3 **while** $L \neq \emptyset$ *and* $A_{n+1} \neq \emptyset$ **do**
4   Let $i := \min \arg\min_{i \in L} d(A_i)$;
5   **if** *for all* $c \in A_{n+1}$ *it holds that* $s(A_{i^*} \cup \{c\}) > B_{i^*}$ **then**
6   Remove i^* from the set of live agents, i.e., $L \leftarrow L \setminus \{i^*\}$;
7   **else**
8   Choose a maximally-dense chore $c^* \in \arg\max_{c \in C: s(A_{i^*} \cup \{c\}) \leq B_{i^*}} \rho(c)$, breaking ties in favor of smaller chores;
9   Update bundles $A_{i^*} \leftarrow A_{i^*} \cup \{c^*\}$ and $A_{n+1} \leftarrow A_{n+1} \setminus \{c^*\}$;
10 **return** *allocation* $\mathcal{A}$;

We first prove that the algorithm runs in polynomial-time.

Theorem 2. *Algorithm 2 runs in time* $\mathcal{O}((n+m)^2)$.

Proof. At each iteration of the **while** loop (Line 3), either an agent is removed from the set of live agents $L \subseteq N$, or one chore is removed from A_{n+1}. Thus, there can be at most $m + n$ iterations. The operation of finding the set of agents in L with minimum disutility (Line 4) takes $\mathcal{O}(n)$ time. Deciding if i^* can be allocated an additional chore (Line 5) and finding a maximally dense chore that is feasible for i^* (Line 8) takes $\mathcal{O}(m)$ time. Together, the algorithm runs in $\mathcal{O}((n+m)^2)$ time. □

Next, we proceed to the main result of this section. Due to space constraints, the proof is deferred to the full version of the paper.

Theorem 3. *Algorithm 2 returns an* EF2 *allocation.*

The above result, coupled with the fact that the complexity of finding EF1 allocations in the budget-constrained goods model is still an open question, leads us to the next natural question: under what circumstances (i.e., special cases) can we compute EF1 allocations in polynomial time? We investigate this in the next section.

5 Computing EF1 Allocations in Special Cases for Indivisible Chores

We consider five special cases where an EF1 allocation can be computed in polynomial time. More specifically, we show that when chores are identically-valued (equivalently, when agents have binary disutility functions), identically-sized, identically-dense, or when agents have identical budgets, the DensestFirst algorithm (Algorithm 2), which guarantees an EF2 allocation in our general model, will, in fact, return an EF1 allocation. For each of these variants, we will only prove correctness, as the polynomial running time has already been established in Theorem 2. We then propose a separate polynomial-time algorithm that returns an EF1 allocation for two agents.

5.1 Binary Disutility Functions or Identically-Valued Chores

The first special case that we will look at is when agents have binary disutility functions (i.e., each chore is valued at either 0 or 1 by all agents). Together with the identical valuation assumption, the case of binary disutilities reduces to that of *identically-valued* chores. This is because we can assume chores valued at 0 is left unallocated (i.e., left in the housekeeper's bundle). In this setting, it suffices to assume that, without loss of generality, each chore has a disutility of 1.

While identical chores are trivial in the traditional fair division model (by simply allocating any $\lfloor m/n \rfloor$ chores to each agent, and then picking an arbitrary set of $m - n \cdot \lfloor m/n \rfloor$ of agents and allocating each of these agents one of the remaining chores), the size dimension of the budget-constrained model leads to EF1 becoming a non-trivial property to prove.

We will now show that executing Algorithm 2 on an instance with identically-valued chores results in an EF1 (in this case, equivalently, EFX) allocation.

Theorem 4. *When agents have binary disutilities or when chores are identically-valued, Algorithm 2 returns an* EF1 *allocation.*

Proof. Consider any two agents $i, j \in [n]$ with $B_i \leq B_j$ and let c_i^t and c_j^t denote the t-th chore added to agent i and j's bundles, respectively. Also, let A_i^t and A_j^t denote the bundles belonging to agents i and j, respectively after the t-th chore was added to their bundles.

We first consider the envy of agent i towards agent j. By the fact that $B_i \leq B_j$ and by construction of the algorithm, we have $|A_i| \leq |A_j| + 1$. Thus, $d(A_i) \leq d(A_j)$ or $d(A_i \backslash \{c\}) \leq d(A_j)$ for some $c \in A_i$, which establishes the EF1 property by i towards j.

Next, we consider the envy by agent j towards agent i. If $|A_j| \leq |A_i| + 1$, then $d(A_j \backslash \{c\}) = |A_j| - 1 \leq |A_i| = d(A_i)$ for any $c \in A_j$ and EF1 is trivially obtained. Thus, we assume $|A_j| > |A_i| + 1$.

Let c_i^α be the last chore that agent i received. Consider the following two cases. Since $|A_j| > |A_i| + 1$, we have $|A_j| > |A_i| = \alpha + 1$.

Case 1: $i < j$. Since $|A_j| > \alpha$, we have that $s(A_i^\alpha \cup \{c_j^{\alpha+1}\}) > B_i$ (otherwise $c_j^{\alpha+1}$ would have been allocated to agent i instead). Then, we get that

$$s(A_j^{\alpha+1}) = s(A_j^\alpha \cup \{c_j^{\alpha+1}\}) \geq s(A_i^\alpha \cup \{c_j^{\alpha+1}\}) > B_i.$$

Consider any subset $S \subseteq A_j$ with $s(S) \leq B_i$. If $|S| > |A_j^\alpha|$, then S contains at least $\alpha + 1$ chores; however, since $A_j^{\alpha+1}$ contains the $\alpha + 1$ smallest chores in A_j, this means that $B_i < s(A_j^{\alpha+1}) \leq s(S)$, a contradiction. Thus, we have $|S| \leq |A_j^\alpha|$, giving us

$$d(S) = |S| \leq |A_j^\alpha| = |A_i^\alpha| = |A_i| = d(A_i).$$

Case 2: $j < i$. Since $|A_j| > \alpha + 1$, we have that $s(A_i^\alpha \cup \{c_j^{\alpha+2}\}) > B_i$ (otherwise $c_j^{\alpha+2}$ would have been allocated to agent i instead). Then, we get that

$$s(A_j^{\alpha+2}) = s(A_j^{\alpha+1} \cup \{c_j^{\alpha+2}\}) \geq s(A_i^\alpha \cup \{c_j^{\alpha+2}\}) > B_i.$$

Together with the fact that $A_j^{\alpha+2}$ contains the $\alpha + 2$ smallest chores in A_j, this means that for any subset $S \subseteq A_j$ with $s(S) \leq B_i$, $|S| \leq |A_j^{\alpha+1}|$, giving us

$$d(S) = |S| \leq |A_j^{\alpha+1}| = |A_i^\alpha| + 1 = |A_i| + 1 = d(A_i) + 1.$$

This is equivalent to $d(S \backslash \{c\}) \leq d(A_i)$ for any chore $c \in S$.

Finally, we consider the envy by the housekeeper towards agents in $[n]$.

Note that every chore $c \in A_{n+1}$ is such that $s(c) \geq s(c')$ for all $c' \in \bigcup_{i=1}^n A_i$. Also, for any $i \in [n]$ and $c \in A_{n+1}$ we have $s(A_i \cup \{c\}) > B_i$.

Suppose for a contradiction there exists a subset $S \subseteq A_{n+1}$ such that $s(S) \leq B_k$ and $|S| - 1 > |A_k|$ for some $k \in [n]$. Since $|S| > |A_k| + 1$, consider the subset $S' \subset S$ such that $|S'| = |A_k|$. Then, $s(S') \geq s(A_k)$. We have that

$$B_k \geq s(S) \geq s(S' \cup \{c\}) \geq s(A_k \cup \{c\}) > B_k$$

for some $c \in S \backslash S'$, a contradiction. Thus, for any subset $S \subseteq A_{n+1}$ such that $s(S) \leq B_k$, we have that

$$d(S \backslash \{c\}) = |S| - 1 \leq |A_k| = d(A_k)$$

for any $c \in S$, as desired. □

5.2 Identically-Sized Chores

In the previous subsection, we showed that when chores are identically-valued but have possibly differing sizes, Algorithm 2 returns an EF1 allocation. Now, we show that when chores are identically-sized but with possibly differing disutilities, the algorithm is also able to compute an EF1 allocation. Note that in this case, the algorithm allocates chores with highest disutility first.

Theorem 5. *When chores are identically-sized, Algorithm 2 returns an* EF1 *allocation.*

Proof. Consider any two agents $i, j \in [n]$ with $B_i \leq B_j$ and let c_i^t and c_j^t denote the t-th chore added to agent i and j's bundle, respectively. Also let A_i^t and A_j^t denote the bundles belonging to agent i and j, respectively, after the t-th chore was added to their bundles. Let c_i^α be the last chore that agent i received.

We first prove the EF1 property by agent i towards agent j. Since $d(c_i^{t+1}) \leq d(c_j^t)$ for all $t = 1, \ldots, \alpha - 1$, by summing over t on both sides, we get

$$d(A_i \backslash \{c_i^1\}) = \sum_{t=1}^{\alpha-1} d(c_i^{t+1}) \leq \sum_{t=1}^{\alpha-1} d(c_j^t) \leq d(A_j).$$

Thus, agent i does not envy j by more than one chore.

Next, we prove the EF1 property by agent j towards agent i. If $|A_j| \leq \alpha + 1 = |A_i| + 1$, by a similar argument as above, we have

$$d(A_j \backslash \{c_i^1\}) = \sum_{t=1}^{|A_j|-1} d(c_j^{t+1}) \leq \sum_{t=1}^{\alpha} d(c_i^t) \leq d(A_i).$$

Thus, consider the case when $|A_j| > \alpha + 1$. Fix any subset $S \subseteq A_j$ such that $s(S) \leq B_i$. Trivially $d(S) \leq d(A_i)$ when $S = \emptyset$, so assume $S \neq \emptyset$. Moreover, we have $|S| \leq \alpha + 1$ as otherwise i would have been allocated $\alpha + 1$ chores instead. Then, since $d(c_j^{t+1}) \leq d(c_j^t)$ for all $t = 1, \ldots, \alpha$, by summing over t on both sides, we get

$$d(A_j^{\alpha+1} \backslash \{c_j^1\}) = \sum_{t=1}^{\alpha} d(c_j^{t+1}) \leq \sum_{t=1}^{\alpha} d(c_i^t) = d(A_i). \tag{1}$$

Since $A_j^{\alpha+1} \backslash \{c_j^1\}$ contains the α chores with highest disutility in $A_j \backslash \{c_j^1\}$,

$$d(S \backslash \{c\}) \leq d(A_j^{\alpha+1} \backslash \{c_j^1\})$$

where $c \in S$ is the chore with highest disutility in S. Together with (1) above, we obtain the EF1 property.

Finally, we prove the EF1 property by the housekeeper towards any agent $k \in [n]$. Fix any $S \subseteq A_{n+1}$ with $s(S) \leq B_k$. Note that every chore $c \in A_{n+1}$ satisfies

$$d(c) \leq d(c') \quad \text{for all } c' \in \bigcup_{i=1}^{n} A_i. \tag{2}$$

Also, for every $c \in A_{n+1}$ we have $s(A_k \cup \{c\}) > B_k$ (otherwise agent k would have been allocated chore c). This means that $s(S) - s(c) < s(A_k)$ for every $c \in A_{n+1}$, implying that $|S| \leq |A_k|$ since all the chores have the same size. Thus, it holds that

$$d(S) \leq \sum_{c' \in A_k} d(c') = d(A_k).$$

where the middle inequality follows from (2). □

Remark 1. We note that for binary disutilities or for identically-sized chores, Algorithm 1 (which computes EFX allocations) runs in polynomial time: indeed, we have argued that the check in Line 3 of that algorithm and computing a feasible set can be reduced to solving Knapsack, and Knapsack is polynomial-time solvable if item sizes or values are polynomially bounded (and if chores have identical sizes, we can assume without loss of generality that each chore has size 1). Consequently, in case of binary disutilities or identical sizes, we can compute an EFX allocation in polynomial time. Since every EFX allocation is also an EF1 allocation, this observation can be seen as a strengthening of Theorems 4 and 5. Nevertheless, Theorems 4 and 5 remain useful, as they provide guarantees on the performance of a specific natural algorithm, namely, DENSESTFIRST.

5.3 Identically-Dense Chores

Next, we consider the case when the chores have identical densities, but with potentially different sizes or disutilities. We will now show that Algorithm 2 is again able to return an EF1 allocation (the tie-breaking in favor of smaller chores in Line 8 is crucial here).

Theorem 6. *When chores are identically-dense, Algorithm 2 returns an* EF1 *allocation.*

Proof. Consider any two agents $i, j \in [n]$ with $B_i \leq B_j$ and let A_i^t and A_j^t denote the bundle belonging to agent i and j, respectively, after the t-th chore was added to their bundle. Since chores are identically-dense, let ρ be the density of each chore. The case when $\rho = 0$ is trivial so we assume that $\rho > 0$.

We first prove the EF1 property by agent i towards agent j. If $d(A_i) \leq d(A_j)$, then we are trivially done. Hence, we assume that $d(A_i) > d(A_j)$. Let there be α chores in A_i, and hence c_i^α is the last chore added to i's bundle. Then, it must be that $d(A_i \backslash \{c_1^\alpha\}) \leq d(A_j)$, otherwise c_1^α would not have been added to agent i's bundle.

Next, we prove the EF1 property by agent j towards agent i. If $d(A_j) \leq d(A_i)$, then we are trivially done. Hence, we assume that $d(A_j) > d(A_i)$. Suppose for a contradiction that for some subset $S \subseteq A_j$ such that $s(S) \leq B_i$, $d(S \backslash \{c\}) > d(A_i)$ for every $c \in S$. Since the density of all the chores is the same, this means that $s(S \backslash \{c\}) > s(A_i)$ for every $c \in S$. Let c be the last chore allocated to A_j among the chores in S (so c is the largest-sized chore in S). Since $s(S \backslash \{c\}) > s(A_i)$,

$$s(A_i \cup \{c\}) < s(S) \leq B_i,$$

and since $d(S\backslash\{c\}) > d(A_i)$, c should have been allocated to agent i instead of agent j, a contradiction. Therefore, $d(S\backslash\{c\}) \leq d(A_i)$, as desired.

Finally, we prove the EF1 property by the housekeeper towards any agent $k \in [n]$, which is similar to the previous case. Suppose for a contradiction that for some subset $S \subseteq A_{n+1}$ such that $s(S) \leq B_k$ it holds that $d(S\backslash\{c\}) > d(A_i)$ for every $c \in S$. Since the density of each chore is the same, this means that $s(S\backslash\{c\}) > s(A_k)$ for every $c \in S$, implying that we have

$$s(A_k \cup \{c\}) < s(S) \leq B_k,$$

for every $c \in S$. Thus, at least one chore $c \in S$ should have been allocated to some agent $k' \in [n]$ (by how the algorithm operates), which is a contradiction. Hence, $d(S\backslash\{c\}) \leq d(A_k)$ as desired. □

5.4 Identical Budgets

In the previous three subsections, we considered the case where chores are identical in some way—be it in value, size, or density. Now, we relax constraints on these three properties, and consider the case where agents' budgets are identical. Again, the same Algorithm 2 is able to return an EF1 allocation.

Theorem 7. *When agents have identical budgets, Algorithm 2 returns an* EF1 *allocation.*

Proof. Let B be the identical budget. We first show the EF1 property between agents. Consider any agent $i \in [n]$. We will show that at each iteration of the **while** loop, the envy agent i has towards any other agent disappears after dropping a new chore from her bundle. Note that since agents have identical budgets, any feasible bundle for agent i will be feasible for any other agent $i' \in [n]$ as well. Now, the claim is clearly true for the initial allocation $\mathcal{A}$ with $A_i = \emptyset$ for all $i \in [n]$. Assume that the claim holds at some iteration, just before agent i^* is allocated a new chore c^*. The new allocation that assigns c^* to i^* is EF1 since i^* does not envy any other agent if we remove the chore c^* from her bundle.

Next, we show the housekeeper is EF1 towards agent i. Consider any subset of chores $S \subseteq A_{n+1}$ with $s(S) \leq B$. Let $c \in S$ be a chore with maximum density among the chores in S, i.e., $c \in \arg\max_{c' \in S} \rho(c')$. Also, let $W_c := \{h \in A_i \mid \rho(h) \geq \rho(c)\}$. Note that $W_c \neq \emptyset$; otherwise, $A_i = \emptyset$ and chore c with $s(c) \leq s(S) \leq B$ would have been added to i's bundle.

By a similar reason, we have that $s(W_c \cup \{c\}) > B$. Since $s(S\backslash\{c\}) \leq B - s(c)$, this means $s(W_c) > B - s(c) \geq s(S\backslash\{c\})$. Thus, we get that $d(A_i) \geq d(W_c) \geq s(W_c) \times \rho(c) > s(S\backslash\{c\}) \times \rho(c) \geq d(S\backslash\{c\})$, as desired. □

5.5 Two Agents

The last special case that we consider is the setting with two agents, which is often studied in the fair allocation literature [9–11,15,16,18]. In fact, the case of two agents is particularly important in the allocation of chores, given that a key application of our results is the domain of household chore division, and many households consist of two adults [19].

We propose an algorithm (Algorithm 3, which is similar to Algorithm 4 in [15]) that returns an EF1 allocation. The algorithm uses Algorithm 2 as a subroutine and runs in polynomial time.

Algorithm 3: Computes an EF1 allocation for two agents

Input disutility function d, size function s, budgets (B_1, B_2) with $B_1 \leq B_2$;
Run Algorithm 2 on both agents with identical budget B_1 and obtain allocation $\mathcal{A}' = (A'_1, A'_2, A'_3)$;
if $d(A'_1) \geq d(A'_2)$ **then**
  $A_1 \leftarrow A'_1$;
else
  $A_1 \leftarrow A'_2$;
Run Algorithm 2 on agent 2 with budget B_2 and set of chores $C \backslash A_1$. Let the output be (A_2, A_3);
return allocation (A_1, A_2, A_3);

Theorem 8. *When there are two agents, Algorithm 3 returns an* EF1 *allocation in polynomial time.*

Proof. The fact that Algorithm 3 runs in polynomial time is easy to observe, given that Algorithm 2 runs in polynomial time (as is proven in Theorem 2), and the other operations take polynomial time as well.

Next, we prove the correctness of the algorithm. Without loss of generality, suppose that $A_1 = A'_1$. Note that $B_1 \leq B_2$. We first show that agent 1 does not envy agent 2 after dropping a chore from his own bundle.

Observe that A'_2 is produced by running Algorithm 2 on a single agent (agent 2) with budget B_2 on items $C \backslash A_1$. If $A_2 \backslash A'_2 = \emptyset$, then $A_2 = A'_2$ (since $B_1 \leq B_2$) and the result follows trivially, so assume that this is not the case. Let c^* be the first chore in $A_2 \backslash A'_2$ allocated to agent 2. Let X be agent 2's bundle right before the algorithm allocates c^*. Then, since Algorithm 2 allocates chores in a densest-first fashion, it must be that $X \subseteq A'_2$. Since $c^* \in A_2$ and $c^* \notin A'_2$, we have

$$s(A'_2) \leq B_1 < s(X \cup \{c^*\}) \leq B_2.$$

Since chores in $A'_2 \backslash X$ have density at most that of c^* and $X \subseteq A'_2$, we get $d(X \cup \{c^*\}) > d(A'_2)$, and

$$d(A_2) \geq d(X \cup \{c^*\}) > d(A'_2).$$

Then, since $d(A'_1 \backslash \{c\}) \leq d(A'_2)$, where c is the last chore added to A'_1, we get that

$$d(A_1 \backslash \{c\}) = d(A'_1 \backslash \{c\}) \leq d(A'_2) < d(A_2),$$

as desired.

Next, we show that agent 2 as well as the housekeeper do not envy agent 1 by more than one chore. To prove this, consider any subset $S \subseteq A_2 \cup A_3$ with $s(S) \leq B_1$. If $S \subseteq A_2'$, then $d(S) \leq d(A_2') \leq d(A_1') = d(A_1)$. Thus, assume that $S \backslash A_2' \neq \emptyset$. Let $c \in S \backslash A_2'$ be a chore with maximum density ρ_c among the chores in $S \backslash A_2'$.

Let $W_c = \{j \in A_2' \mid \rho_j \geq \rho_c\}$. Note that $W_c \neq \emptyset$; otherwise, $A_2' = \emptyset$ and chore c with $s(c) \leq s(S) \leq B_1$ would have been added to A_2'. Since c is not included in A_2', we have $s(W_c \cup \{c\}) > B_1$. Moreover, since $s(S \backslash \{c\}) \leq B_1 - s(c)$, this means that

$$s(W_c) > B_1 - s(c) \geq s(S \backslash \{c\}).$$

Thus, we get

$$d(A_1) \geq d(A_2') \geq d(W_c) \geq s(W_c) \times \rho_c > s(S \backslash \{c\}) \times \rho_c \geq d(S \backslash \{c\}).$$

Finally, the EF1 property by the housekeeper towards agent 2 can be easily verified due to Theorem 7. □

6 Conclusion

In this work, we propose a model of allocating indivisible chores under budget constraints. We prove the existence of EFX allocations. Our proof is constructive and provides a pseudopolynomial time algorithm for finding EFX allocations. Moreover, we put forward a polynomial-time algorithm that returns an EF2 allocation for general instances, and EF1 allocations in five special cases—when chores are identically-valued, identically-sized, identically-dense, when agents have identical budgets, and the case of two agents.

Possible future directions include exploring definitions of envy-freeness when agents have subjective size or disutility functions (i.e., generalized assignment constraints for chores), or considering approximate EF guarantees under non-additive size or disutility functions (while maintaining the identical size/disutility function assumption). Another exciting direction is to extend our formal model and results to mixed manna, i.e., items that are viewed as goods by some agents and as chores by others [3]; anecdotally, this model may be appropriate for some household tasks, such as cooking, gardening, or spending time with children or animals.

References

1. Airiau, S., Gilbert, H., Grandi, U., Lang, J., Wilczynski, A.: Fair rent division on a budget revisited. In: Proceedings of the 26th European Conference on Artificial Intelligence (ECAI), pp. 52–59 (2023)
2. Amanatidis, G., et al.: Fair division of indivisible goods: recent progress and open questions. Artif. Intell. **322**, 103965 (2023)

3. Aziz, H., Caragiannis, I., Igarashi, A., Walsh, T.: Fair allocation of indivisible goods and chores. Auton. Agents Multi-Agent Syst. **36**(1), 3:1–3:21 (2022)
4. Aziz, H., Li, B., Moulin, H., Wu, X.: Algorithmic fair allocation of indivisible items: a survey and new questions. SIGecom Exchanges **20**(1), 24–40 (2022)
5. Barman, S., Khan, A., Shyam, S., Sreenivas, K.V.N.: Guaranteeing envy-freeness under generalized assignment constraints. In: Proceedings of the 24th ACM Conference on Economics and Computation (EC), pp. 242–269 (2023)
6. Barman, S., Khan, A., Shyam, S., Sreenivas, K.: Finding fair allocations under budget constraints. In: Proceedings of the 37th AAAI Conference on Artificial Intelligence (AAAI), pp. 5481–5489 (2023)
7. Barman, S., Sundaram, R.G.: Uniform welfare guarantees under identical subadditive valuations. In: Proceedings of the 29th International Joint Conference on Artificial Intelligence (IJCAI), pp. 46–52 (2020)
8. Bogomolnaia, A., Moulin, H., Sandomirskiy, F., Yanovskaya, E.: Competitive division of a mixed manna. Econometrica **85**(6), 1847–1871 (2017)
9. Brams, S.J., Taylor, A.D.: Fair Division: From cake-cutting to dispute resolution. Cambridge University Press, Cambridge (1996)
10. Budish, E.: The combinatorial assignment problem: approximate competitive equilibrium from equal incomes. J. Polit. Econ. **119**(6), 1061–1103 (2011)
11. Caragiannis, I., Kurokawa, D., Moulin, H., Procaccia, A.D., Shah, N., Wang, J.: The unreasonable fairness of maximum Nash welfare. ACM Trans. Econ. Comput. **7**(3), 12:1–12:32 (2019)
12. Chaudhury, B.R., Kavitha, T., Mehlhorn, K., Sgouritsa, A.: A little charity guarantees almost envy-freeness. SIAM J. Comput. **50**(4), 1336–1358 (2021)
13. Chekuri, C., Khanna, S.: A polynomial time approximation scheme for the multiple knapsack problem. SIAM J. Comput. **35**(3), 713–728 (2005)
14. Dehghani, S., Farhadi, A., HajiAghayi, M., Yami, H.: Envy-free chore division for an arbitrary number of agents. In: Proceedings of the 29th Annual ACM-SIAM Symposium on Discrete Algorithms (SODA), pp. 2564–2583 (2018)
15. Gan, J., Li, B., Wu, X.: Approximation algorithm for computing budget-feasible EF1 allocations. In: Proceedings of the 22nd International Conference on Autonomous Agents and Multiagent Systems (AAMAS), pp. 170–178 (2023)
16. Garbea, M., Gkatzelis, V., Tan, X.: EFx budget-feasible allocations with high Nash welfare. In: Proceedings of the 26th European Conference on Artificial Intelligence (ECAI), pp. 795–802 (2023)
17. Ghosal, P., Vishwa Prakash, H.V., Nimbhorkar, P., Varma, N.: EFX exists for four agents with three types of valuations. arXiv preprint arXiv:2301.10632 (2023)
18. Igarashi, A., Lackner, M., Nardi, O., Novaro, A.: Repeated fair allocation of indivisible items. In: Proceedings of the 38th AAAI Conference on Artificial Intelligence (AAAI), pp. 9781–9789 (2024)
19. Igarashi, A., Yokoyama, T.: Kajibuntan: a house chore division app. In: Proceedings of the 37th AAAI Conference on Artificial Intelligence (AAAI), pp. 16449–16451 (2023)
20. Mutzari, D., Aumann, Y., Kraus, S.: Resilient fair allocation of indivisible goods. In: Proceedings of the 22nd International Conference on Autonomous Agents and Multi-Agent Systems (AAMAS), pp. 2688–2690 (2023)
21. Plaut, B., Roughgarden, T.: Almost envy-freeness with general valuations. In: Proceedings of the 29th Annual ACM-SIAM Symposium on Discrete Algorithms (SODA), pp. 2584–2603 (2018)
22. Procaccia, A., Velez, R., Yu, D.: Fair rent division on a budget. In: Proceedings of the 32nd AAAI Conference on Artificial Intelligence (AAAI), pp. 1177–1184 (2018)

23. Steinhaus, H.: The problem of fair division. Econometrica **16**(1), 101–104 (1948)
24. Wu, X., Li, B., Gan, J.: Budget-feasible maximum Nash social welfare is almost envy-free. In: Proceedings of the 30th International Joint Conference on Artificial Intelligence (IJCAI), pp. 465–471 (2021)

Fair Division with Interdependent Values

Georgios Birmpas[1], Tomer Ezra[2], Stefano Leonardi[3], and Matteo Russo[3](✉)

[1] University of Liverpool, Liverpool, UK
g.birmpas@liverpool.ac.uk
[2] Harvard University, Cambridge, USA
tomer@cmsa.fas.harvard.edu
[3] DIAG "Antonio Ruberti", Sapienza University of Rome, Rome, Italy
{leonardi,mrusso}@diag.uniroma1.it

Abstract. We introduce the study of designing allocation mechanisms for fairly allocating indivisible goods in settings with interdependent valuation functions. In our setting, there is a set of goods that needs to be allocated to a set of agents (without disposal). Each agent is given a private signal, and his valuation function depends on the signals of all agents. Without the use of payments, there are strong impossibility results for designing strategyproof allocation mechanisms even in settings without interdependent values. Therefore, we turn to design mechanisms that always admit equilibria that are fair with respect to their true signals, despite their potentially distorted perception. To do so, we first extend the definitions of pure Nash equilibrium and well-studied fairness notions in literature to the interdependent setting. We devise simple allocation mechanisms that always admit a fair equilibrium with respect to the true signals. We complement this result by showing that, even for very simple cases with binary additive interdependent valuation functions, no allocation mechanism that always admits an equilibrium, can guarantee that all equilibria are fair with respect to the true signals.

Keywords: Fair Division · Interdependent Values · Mechanisms without Money

1 Introduction

The problem of fair division centers around the challenge of allocating a collection of resources to a set of individuals in a fair way. The roots of this problem can be traced back to the early work of Banach, Knaster, and Steinhaus [35], who introduced the notion of *proportionality*, a concept that demands that each person receives at least an equal share of the total value. Another prominent concept in the realm of fairness is *envy-freeness* [23,24,38], which dictates that each individual values their resources as much as anyone else's. However, when resources are indivisible, proportionality and envy-freeness become impossible to attain in general. For example, consider an instance with two agents and one

G. Schäfer and C. Ventre (Eds.): SAGT 2024, LNCS 15156, pp. 72–88, 2024.
https://doi.org/10.1007/978-3-031-71033-9_5

good, being positively valued by both: No allocation exists where no agent envies the other, or every agent gets his proportional share. This motivated the introduction of new (relaxed) fairness notions, such as EF1 (Envy-Freeness up to One Good) [12,28], EFX (Envy-Freeness up to Any Good) [14], and MMS (Maximin Share) [12], to grapple with the division of indivisible resources to agents with equal entitlements, and APS (the AnyPrice Share) [8], WMMS (Weighted Maximin Share) [22], WEF (Weighted Envy-Freeness) [15] or ℓ-out-of-d share [9], to agents with unequal entitlements. For an overview of results on the area, we refer the reader to the survey of [2].

When we demand fair allocations, under agents with incentives, the problem becomes much more challenging. [13,30] and [5] introduced the strategic version of the problem, where the agents are assumed to be selfish, and their goal is to maximize their own utility. In particular, they considered the question of whether it is possible to have truthful allocation mechanisms (without payments) that provide fairness guarantees. This question was later resolved by [1], who showed that truthfulness and fairness are incompatible even for the case of additive valuation functions, as no meaningful fairness notion can be guaranteed by a truthful allocation mechanism. This impossibility, led subsequent works to pursue positive results regarding truthfulness and fairness in more specialized valuation function settings, such as dichotomous additive [6,26], matroid rank functions [7,10,40], or combining fairness with weaker versions of truthfulness, e.g., [33]. In a work that is mostly related to ours, [3] followed a different approach and explored the fairness properties of the pure Nash equilibria (PNE) of non-truthful mechanisms. They focused on the case of additive agents, and showed that there are mechanisms that always admit PNE, all of which induce fair allocations according to the agents' true valuation functions. [4] later expanded the applicability of these results to richer valuation function classes (e.g. cancellable, and submodular).

All the mechanisms that have been devised in the context of strategic fair division assume the agent's valuation functions to be *independent* of each other. In several scenarios, however, such an assumption is either too strong or unrealistic. The concept of agents with interdependent valuation functions was first introduced in the context of auctions by [41] and [31]. It has recently gained a lot of attention in the algorithmic game theory community, for efficient and strategyproof implementation of interdependent value (IDV) auctions [18–21,27,29] and IDV public projects [17] (we defer the interested reader to Sect. 1.2). However, its applicability goes further beyond that. In the case of fair division, consider, for example, a couple of heirs that need to partition (indivisible) inheritance goods between them: Each of them possesses some information (signal) about the goods, but their value depends on the other's signal too. The strategic nature of the situation is apparent, as it becomes clear that both parties have a vested interest in distorting the truth regarding their signals to manipulate the other party's perception and ultimately gain an advantage in the allocation of goods. Whether by exaggerating or downplaying the value of certain goods, each party hopes to influence the other's perception and come out ahead in the final allocation. Another example, is where a company is allocating tasks to its employees, where the perception of the newer employees over the tasks, is affected by the opinion of the employees that work years for the company. Senior employees, possessing a comprehensive understanding of the task difficulties, may sometimes overstate the challenges in order to delegate their workload more effortlessly. Incorporating private signals into the mechanism design process poses a critical challenge now that the goal becomes to ensure fairness that aligns with the true valuation functions of all agents involved, as opposed to the perceived ones. Since allocation mechanisms that do not use monetary transfers cannot compensate the agents for revealing their private signals, having guarantees with respect to the true signal is even more challenging. Our main goal is to design allocation mechanisms that seek to establish equilibria, where the allocation is fair with respect to the agents' *true* values, despite their potentially distorted perception.

1.1 Our Contributions

Before we begin, we highlight that although the aforementioned impossibility results regarding truthfulness and fairness in the independent value model (e.g., [1]) transfer to the setting of interdependent values, the positive results that regard the fair PNE of non-truthful mechanisms (e.g., [3,4]) do not. We demonstrate the latter at Sect. 6 of our paper. Therefore, we initiate the study of mechanisms that admit the existence of fair equilibria with respect to the true signals of the agents, in settings where their values are interdependent.

We extend several notions of fairness, along with the notion of pure Nash equilibrium, to the interdependent value model, and we design a class of mechanisms that, for every signal vector, have at least one pure Nash equilibrium with the following property: the allocation that corresponds to it, is fair with respect to the *true* values (in contrast to what has been reported or perceived by the agents). The common characteristic between the mechanisms of this class is that each agent is not only required to report her own signal (her private information), but her strategy space is larger than the space of the private information she has. In other words, the mechanisms we consider are not *direct revelation* mechanisms.

As the following example demonstrates, such mechanisms are a necessity in the IDV model as there might be cases for which there is no direct revelation mechanism that admits a pure Nash equilibrium that is fair, according to any meaningful fairness notion.

Example 1. Let us consider an instance with two agents and the set of four goods $M = \{a, b, c, d\}$, along with a mechanism $\mathcal{M}$ that always has at least one fair equilibrium for every signal. Agent 1 values good $j \in M$ just according to her signal, i.e., $v_{1j} = s_{1j}$. Agent 2 instead values the same good according to agent 1's signal, i.e., $v_{2j} = s_{1j}$. Since only agent 1 has information regarding the valuations of both agents, it is natural to assume that agent 2 is not able to report something to the mechanism. Consider the following instance where the values of agent 1 (that also defines agent 2's) are $v_1 = (1, 1, 1, 1) = v_2$. By our assumption on $\mathcal{M}$, we know that there exists a report r_1 that agent 1 can declare, which induces a fair PNE according to the true value. This corresponds to allocation (A_1, A_2), where $A_1, A_2 \neq \emptyset$ for the allocation to be fair. The specifics of this allocation depend on the fairness notion that we examine. Now, if we consider a new valuation function for agent 1 that associates $v_{1j} = s_{1j} = 1$ if $j \in A_1$, and 0 otherwise. For this new instance, in any PNE, agent 1 should receive all goods in A_1, as otherwise she could declare r_1 and get them. This means that the goods that agent 2 receives have 0 value for him in any PNE. This shows the absence of fairness with respect to any meaningful fairness notion, a contradiction. This means that allocation mechanisms that don't ask agent 2 for a report cannot guarantee the existence of fair equilibria.

Below we present a roadmap to our paper:

- In Sect. 3, we consider the case of two *general* monotone agents with *equal* entitlements, and we design a mechanism that is based on the *cut and choose protocol.* Our mechanism (Mechanism 1), induces an equilibrium where the corresponding allocation is MMS and EFX for the cutter and EF for the chooser. Although our mechanism is based on the cut-and-choose framework, the implementation of the interdependent version of it, and the proof of its guarantees are very different from the simple implementation and proof of the independent cut-and-choose mechanism.
- In Sect. 4, we consider the case of two *general* monotone agents with *unequal* entitlements, and we devise a mechanism (Mechanism 2) that induces an equilibrium where the corresponding allocation gives at least the APS to one agent, and the proportional share to the other agent as long as her valuation function is XOS. We, in fact, show that achieving the same fairness guarantees is impossible if the valuation functions of both agents are subadditive. As a byproduct, we emphasize that Mechanism 2 for the special case of independent additive values, guarantees that both agents receive their APS in every equilibrium, which to our knowledge, is the first fairness guarantee in a strategic setting for agents with arbitrary entitlement and additive valuations[1].
- In Sect. 5, we consider the case of three or more agents. There, we present a way of transforming, in a black-box manner, any algorithm into a mechanism with at least one PNE that guarantees the same properties of the algorithm with respect to the true signals of the agent.
- In Sect. 6, we complement the above picture with the following negative result: it is impossible to construct allocation mechanisms that always admit a PNE, for which all equilibria are fair with respect to the true signals for either of the fairness notions of MMS and EF1. This impossibility is shown under agents with additive valuation functions and binary signals, and creates a stark separation with the independent valuation model considered in [3].

[1] [36] and [37] showed fairness guarantees in a strategic setting for agents with unequal entitlements in the special cases of binary valuation functions and matroid rank valuation functions respectively.

The missing proofs can be found in the full version of our paper [11].

1.2 Related Work on Interdependent Mechanism Design with Money

As previously underlined, mechanism design (with or without money) has traditionally been conceived for settings where agents' valuations are assumed to be independent. In this setting, a celebrated result by [16,25,39] (the VCG mechanism) resolves the problem of truthful mechanism design optimally. However, the independence assumption falls short of characterizing natural scenarios, such as the illustrative oil drilling auction example of [41]. The auctioned land's value depends on how much oil there is underneath, and different agents might have different pieces of information (signals) about it. Most importantly, unlike traditional settings, each agent's value for that parcel of land depends on *all* such pieces of information. In the context of single-item auctions, [31] formulated the *interdependent value* (IDV) model, which prescribes each agent i has a *private* signal s_i that summarizes the information she possesses about the item being auctioned. Furthermore, the value agent i attributes to the item is a function of all signals, i.e., $v_i(s_i, s_{-i})$.

In the IDV model, it is impossible to design dominant strategy incentive-compatible mechanisms for single-item auctions in that if one agent does not report her signal truthfully, then another agent might not even be aware of its true valuation. More strongly, it is even impossible[2] to design ex-post incentive compatible mechanisms [27], unless a condition called *single-crossing* is met [34]. This signifies that each agent values her own signal above everybody else's. Beyond not being realistic, the single-crossing condition does not generalize well to multi-dimensional signal settings: [18,20,21], and especially [19], circumvent this issue (and later refined by [29]), extending ex-post incentive compatibility results to settings where valuations satisfy *Submodularity over Signals* (SOS), a natural property to impose on valuations. Loosely, this means that the marginal increase in one's value due to her signal's increase is smaller the higher everybody else's signal is. Beyond auctions, [17] have recently studied the problem of interdependent public projects.

2 Model

We study the problem of fairly allocating a set M of m indivisible goods to a set $N = [n]$ of $n \geq 2$ agents with interdependent valuations. Each agent $i \in N$ is characterized by the following three parameters: (1) a private signal s_i from the set of potential signals $S_i \neq \emptyset$, where we denote by S the Cartesian product $\times_{i \in N} S_i$; (2) a publicly known entitlement $\alpha_i \in (0, 1)$; and (3) a publicly[3] known

[2] [27] show this impossibility even for Bayesian settings.

[3] The assumption that the valuations are public (and only the signals are private) is without loss of generality since one can use some dedicated bits of the private signal to encode the private valuation, and these bits will not influence the other agents' valuations.

monotone and normalized[4] combinatorial valuation $v_i : S \times 2^M \rightarrow \mathbb{R}_{\geq 0}$. We assume without loss of generality that the sum of entitlements is 1, and we denote by α the vector of entitlements $(\alpha_1, \ldots, \alpha_n)$. A special case is when all entitlements are $\frac{1}{n}$, (i.e., equal entitlements). An allocation $A = (A_1, \ldots, A_n)$ of the goods is a partition of M where agent i receives the set of goods A_i. We denote by $\mathcal{A}$ the set of all allocations of M (without disposal) to the set of agents N. An instance of our setting is described by $\mathcal{I} = (N, M, \alpha, S, v_1, \ldots, v_n)$.

Fairness Notions. We consider the following natural extensions of the well-studied fairness notions to the interdependent setting. For the case of equal entitlements (i.e., $\alpha_i = \frac{1}{n}$ for all $i \in N$), the *envy*-based fairness notions that we consider are:

Definition 1 (Envy-based Fairness Notions). *An allocation A is EF (respectively, EF1, EFX) for agent $i \in N$ with respect to a signal vector $s = (s_1, \ldots, s_n)$ if, for all agents $i' \in N$ it holds that*

$$v_i(s, A_i) \geq v_i(s, A_{i'}) \tag{EF}$$

$$A_{i'} = \emptyset \text{ or } \exists j \in A_{i'} : \; v_i(s, A_i) \geq v_i(s, A_{i'} \setminus \{j\}) \tag{EF1}$$

$$\forall j \in A_{i'} : \; v_i(s, A_i) \geq v_i(s, A_{i'} \setminus \{j\}) \tag{EFX}$$

An allocation A is EF (respectively, EF1, EFX) with respect to perceived signal vectors $s^{(1)}, \ldots, s^{(n)}$ if for all $i \in N$, A is EF (respectively EF1, EFX) for agent i with respect to a perceived signal vector $s^{(i)}$. If $s^{(1)} = \ldots = s^{(n)} = s$, we say that A is EF (respectively, EF1 or EFX) with respect to s.

The *share*-based fairness notions that we consider are:

Definition 2 (Share-based Fairness Notions). *The PROP (respectively MMS or APS) of agent i with valuation $v_i : S \times 2^M \rightarrow \mathbb{R}_{\geq 0}$ with respect to signal vector $s = (s_1, \ldots, s_n)$ is:*

$$PROP_i(\alpha_i, s) = \alpha_i \cdot v_i(s, M) \tag{PROP}$$

$$MMS_i(\alpha_i = \frac{1}{n}, s) = \max_{(A_1, \ldots, A_n) \in \mathcal{A}} \min_{i' \in N} v_i(s, A_{i'}) \tag{MMS}$$

$$APS_i(\alpha_i, s) = \min_{p \in P} \max_{T \subseteq M} v_i(s, T) \cdot \mathbb{1}\left\{ \sum_{j \in T} p_j \leq \alpha_i \right\} \tag{APS}$$

where P is the set of all non-negative price vectors that sum to 1, and given a price vector $p \in P$, and a subset of goods $T \subseteq M$, we denote by $p(T)$ the sum of prices of goods in T, (i.e., $p(T) = \sum_{j \in T} p_j$). Note that the MMS is only defined for the case of equal entitlements. An allocation A is PROP (respectively, MMS or APS) with respect to signal vectors $s^{(1)}, \ldots, s^{(n)}$ if for every agent $i \in N$, the value of agent i with respect to signal vector $s^{(i)}$ is at least the

[4] A valuation function v is monotone if for every signal vector s and sets $T \subseteq T' \subseteq M$ it holds that $v(s, T) \leq v(s, T')$. A valuation function v is normalized if for every signal vector s it holds $v(s, \emptyset) = 0$.

PROP (respectively, MMS or APS) of agent i with respect to signal vector $s^{(i)}$. If $s^{(1)} = \ldots = s^{(n)} = s$, we say that A is PROP (respectively, MMS or APS) with respect to s.

Allocation Mechanisms.

Definition 3 (Deterministic Mechanisms). *A (deterministic) allocation mechanism $\mathcal{M}$ (without payments) for the interdependent setting is defined by a product set of bids $B = \bigtimes_{i\in N} B_i$ and a mapping $\mathcal{M} : S \times B \to \mathcal{A}$. The allocation mechanism collects from each agent $i \in N$ a report $(r_i, b_i) \in S_i \times B_i$, and allocates the goods according to $\mathcal{M}(r, b)$, where $r = (r_1, \ldots, r_n)$, and $b = (b_1, \ldots, b_n)$. We also denote by $\mathcal{M}_i(r, b)$ as the bundle of goods that agent i receives under reports corresponding to (r, b).*

Definition 4 (Interdependent Pure Nash Equilibria). *For an instance $\mathcal{I}$, and a mechanism $\mathcal{M}$ for signal vector $s \in S$ a report vector $((r_1, b_1), \ldots, (r_n, b_n))$ is a pure Nash equilibrium (PNE) if for every agent $i \in N$ and report (r'_i, b'_i) it holds that:*

$$v_i(r^{(i)}, \mathcal{M}_i(r, b)) \geq v_i(r^{(i)}, \mathcal{M}_i(r', b')),$$

where $b = (b_1, \ldots, b_n), b' = (b'_i, b_{-i}), r = (r_1, \ldots, r_n), r' = (r'_i, r_{-i}), r^{(i)} = (s_i, r_{-i})$. Note that the vector signal that agent i calculates his value with respect to, is the reported signals of the other agents along with his own true signal.

Some of our results apply to the special cases with additive, XOS, and subadditive valuations. We say that an instance is additive if for every agent $i \in N$, every signal vector $s \in S$, and every subset $T \subseteq M$ it holds that $v_i(s, T) = \sum_{j\in T} v_i(s, \{j\})$. We say that an instance is XOS if for every agent $i \in N$, every signal vector $s \in S$, there exist an integer ℓ and ℓ non-negative vectors $a^1, \ldots, a^\ell$ of dimension $|M|$, such that for every subset $T \subseteq M$ it holds that $v_i(s, T) = \max_{\ell' \in \{1,\ldots,\ell\}} \sum_{j\in T} a_j^{\ell'}$. We say that an instance is subadditive if for every agent $i \in N$, every signal vector $s \in S$, and every pair of subsets $T, T' \subseteq M$ it holds that $v_i(s, T) + v_i(s, T') \geq v_i(s, T \cup T')$.

Remark 1. We note that independent private values can be captured by our model by setting all v_i to be independent of s_{-i}. I.e., there exist functions $u_i : S_i \times 2^M \to \mathbb{R}_{\geq 0}$ such that for every agent $i \in N$, for every signal vector $s = (s_1, \ldots, s_n) \in S$, and every subset $T \subseteq M$ it holds that $v_i(s, T) = u_i(s_i, T)$.

Throughout the paper, we use s to denote signals, r to denote reports of signals, and b to report the bidding strategies besides the reported signals.

3 The Case of 2 Equal Entitlement Agents

In this section, we consider the case where there are two agents (i.e., $n = 2$) with equal entitlements (i.e., $\alpha_1 = \alpha_2 = \frac{1}{2}$) and general monotone valuation functions. We devise the IDV Cut-&-Choose mechanism (Mechanism 1), which involves agents reporting their own signal along with a guess of the other agent's signal. The mechanism then divides the goods into two sets in order to maximize the value of the set (according to the cutter's report) the chooser does not select (again, according to the cutter's report). The chooser then selects the better set based on his report. We show that the IDV Cut-&-Choose induces an equilibrium where, once one agent declares her own signal truthfully and guesses correctly the other's signal, the other agent's best response is to also report truthfully his own signal, and repeat the signal of the other agent. The resulting allocation in this equilibrium is MMS and EFX for the cutter and EF for the chooser with respect to the true signals.

3.1 The IDV Cut-&-Choose Mechanism

We consider both shared-based and envy-based notions of fairness and devise a mechanism that has guarantees of both types (EFX and MMS for the cutter, and EF for the chooser). Formally, we show that:

Theorem 1. *For every instance $\mathcal{I} = (N, M, \alpha = (\frac{1}{2}, \frac{1}{2}), S = S_1 \times S_2, v_1, v_2)$ composed of two agents with equal entitlements, there exists an allocation mechanism $\mathcal{M}$ such that for every signal vector $s \in S$ there is a report $((r_1, b_1), (r_2, b_2))$ such that: (1) Report $((r_1, b_1), (r_2, b_2))$ is a PNE; (2) The allocation $\mathcal{M}((r_1, r_2), (b_1, b_2))$ is MMS and EFX for one agent with respect to the true signal vector s; (3) The allocation $\mathcal{M}((r_1, r_2), (b_1, b_2))$ is EF for the other agent with respect to the true signal vector s.*

To prove the above theorem, we devise the IDV Cut-&-Choose mechanism (Mechanism 1). Following Definition 3, in Mechanism 1, the set of bids of agents $1, 2$, are the set of signals of agents $2, 1$ respectively (i.e., $B_1 = S_2$, and $B_2 = S_1$). For ease of understanding, in the statements that follow, we subsume the notation of above.

In our proof and in the mechanism construction, we also use the following theorem that immediately derives from [32, Theorem 4.2], for the case of independent valuations.

Theorem 2 ([32]). *For every monotone valuation function $v : 2^M \to \mathbb{R}_{\geq 0}$, there exists a set T^* such that $v(T^*) \geq v(M \backslash T^*) = MMS(v)$, and for every $j \in T^*$, it holds that $v(T^* \backslash \{j\}) \leq v(M \backslash T^*)$, where $MMS(v) = \max_{T \subseteq M} \min(v(T), v(M \backslash T))$.*

We first observe that since for $T_{\text{MMS}} = \arg\max_{T \subseteq M} \min(v_1(s, T), v_1(s, M \backslash T))$, and since at least one of $T_{\text{MMS}}, M \backslash T_{\text{MMS}}$ is in $\mathcal{T}(s_1, s_2)$, it holds that:

$$\begin{aligned} \xi_1(s_1, s_2) = & \max_{T \in \mathcal{T}(s_1, s_2)} v_1((s_1, s_2), M \backslash T) \\ & \geq \min(v_1((s_1, s_2), T_{\text{MMS}}), v_1((s_1, s_2), M \backslash T_{\text{MMS}})) \\ & = \text{MMS}_1(1/2, s). \end{aligned} \tag{1}$$

Mechanism 1: IDV Cut-&-Choose

Data: Report (r_1, b_1) of agent 1 (cutter) and (r_2, b_2) of agent 2 (chooser).

```
Cutter:
    Let T(r1, b1) := {T ⊆ M | v2((r1, b1), T) ≥ v2((r1, b1), M\T)}
    Let ξ1(r1, b1) := max_{T∈T(r1,b1)} v1((r1, b1), M\T)
    if ξ1(r1, b1) > MMS1(1/2, (r1, b1)) then
        Let T*(r1, b1) be a set in T(r1, b1) for which
          v1((r1, b1), M\T) = ξ1(r1, b1)
    end
    else
        Let T*(r1, b1) be the set of Theorem 2, when applied to v1((r1, b1), ·)
    end
    Partition goods into subsets (T*(r1, b1), M\T*(r1, b1))

Chooser:
    Select A2 = argmax_{T∈{T*(r1,b1), M\T*(r1,b1)}} v2((b2, r2), T), in case of a tie select
      A2 = T*(r1, b1)
```

Result: Return allocation

$$\mathcal{M}((r_1, r_2), (b_1, b_2)) = (A_1, A_2),$$

where $A_1 = M \backslash A_2$.

Lemma 1. *For Mechanism 1 and signal vector $s = (s_1, s_2) \in S$, report vector $((s_1, s_2), (s_2, s_1))$ is a PNE.*

Lemma 2. *For Mechanism 1 and signal vector $s = (s_1, s_2) \in S$, in PNE $((s_1, s_2), (s_2, s_1))$, the cutter receives at least her MMS with respect to signal vector s. Moreover, the allocation $\mathcal{M}((s_1, s_2), (s_2, s_1))$ is EFX for the cutter with respect to signal vector s.*

Since the chooser selects the bundle with higher value according to his report, we can observe the following:

Claim. For Mechanism 1 and signal vector $s = (s_1, s_2) \in S$, in equilibrium $((s_1, s_2), (s_2, s_1))$, the allocation $\mathcal{M}((s_1, s_2), (s_2, s_1))$ is EF for the chooser with respect to s.

The proof of Theorem 1 follows by combining Lemmas 1 and 2 and Sect. 3.1.

3.2 Discussion on Mechanism 1

We next discuss the extent of Theorem 1, and some of its implications. Regarding envy-based notions of fairness, it is clear that there might not exist an allocation that is EF for both agents (e.g., two agents with a single good). Thus, guaranteeing EF for one agent and EFX for the other is the best we can aim for.

For what concerns shared-based notions of fairness, giving the MMS to both agents is impossible for general (subadditive or even XOS) valuations as the next example illustrates.[5] Consider set of goods $M = \{a, b, c, d\}$, and valuations

$$v_1 = \max\left(\mathbb{1}\{a \vee b\} + \mathbb{1}\{a \wedge b\}, \mathbb{1}\{c \vee d\} + \mathbb{1}\{c \wedge d\}\right)$$
$$v_2 = \max\left(\mathbb{1}\{a \vee d\} + \mathbb{1}\{a \wedge d\}, \mathbb{1}\{b \vee c\} + \mathbb{1}\{b \wedge c\}\right),$$

for agent 1 and agent 2, respectively. We observe that (1) these valuations are XOS, and (2) the MMS for both agents is 2. However, no allocation gives the MMS to both agents: If one agent receives one good (and the other three goods), her value is at most 1. Otherwise, both agents receive two goods, and one of them must have a value of at most 1 which is strictly less than the MMS.

In case of subadditive valuations, it holds that EF implies PROP. Therefore, if the chooser's valuation is subadditive then in equilibrium $((s_1, s_2), (s_2, s_1))$, Mechanism 1 guarantees agent 2 his proportional share with respect to signal vector s, and when the chooser's valuation is additive, then it guarantees the MMS with respect to signal vector s to both agents.

In the special case of identical valuations (i.e., $v_1 = v_2$), the MMS can be guaranteed to both agents (for all valuation classes). Indeed, when cutter and chooser have identical valuations, the chooser receives at least his MMS with respect to signal vector s, in equilibrium $((s_1, s_2), (s_2, s_1))$ of Mechanism 1. This is because the chooser gets a value of at least $\xi_1(s_1, s_2)$ which is at least the MMS.

4 The Case of 2 Unequal Entitlement Agents

This section considers the case where two agents (i.e., $n = 2$) have unequal entitlements and general monotone valuation functions. Our main contribution is the design of the IDV Price-&-Choose mechanism (Mechanism 2). Mechanism 2 first prices the goods according to the pricer's report. The pricing tries to maximize the pricer's value of the remaining goods (according to the pricer's report) while assuming that the chooser first picks the set that maximizes his own value (again, according to the pricer's report), and whose total price does not exceed the chooser's entitlement. As for the equal entitlement case, this mechanism guarantees that there exists a PNE of the same form, of reporting your own signal truthfully and guessing the other's signal correctly. The resulting allocation gives at least the PROP share to the pricer as long as his valuation is XOS, and the APS to the chooser. We, in fact, show that achieving the same fairness guarantees is impossible if both valuation functions are subadditive.

4.1 The IDV Price-&-Choose Mechanism

We show the following result:

[5] We omit the dependence on s, since this also holds in non-interdependent settings (and even without incentive considerations).

Theorem 3. *For every instance $\mathcal{I} = (N, M, \alpha = (\alpha_1, \alpha_2), S = S_1 \times S_2, v_1, v_2)$ composed of two agents with unequal entitlements with monotone valuations, there exists an allocation mechanism $\mathcal{M}$ such that for every signal vector $s \in S$ there exists a report $((r_1, b_1), (r_2, b_2))$ such that: (1) Report $((r_1, b_1), (r_2, b_2))$ is a PNE; (2) The allocation $\mathcal{M}((r_1, r_2), (b_1, b_2))$ is APS for one agent with respect to the true signal vector s; (3) If the other agent's valuation is XOS, then the allocation $\mathcal{M}((r_1, r_2), (b_1, b_2))$ is PROP for the other agent with respect to the true signal vector s.*

We devise the IDV Price-&-Choose mechanism (Mechanism 2) to prove the above theorem. Following Definition 3, in Mechanism 2, the set of bids of agents $1, 2$, are the set of signals of agents $2, 1$ respectively (i.e., $B_1 = S_2$, and $B_2 = S_1$).

Mechanism 2: IDV Price-&-Choose

Data: Report (r_1, b_1) of agent 1 (pricer) and (r_2, b_2) of agent 2 (chooser).

`Pricer:`

- Let $\xi_2(r_1, b_1, p) := \max_{T \subseteq M} v_2((r_1, b_1), T) \cdot \mathbb{1}\{p(T) \leq \alpha_2\}$
- Let $\mathcal{T}(r_1, b_1, p) := \{T \subseteq M \mid p(T) \leq \alpha_2 \wedge v_2((r_1, b_1), T) = \xi_2(r_1, b_1, p)\}$
- Let $\mathcal{T}(r_1, b_1) := \bigcup_{p \in P} \mathcal{T}(r_1, b_1, p)$
- Let $T^*(r_1, b_1) := \arg\max_{T \in \mathcal{T}(r_1, b_1)} v_1((r_1, b_1), M \setminus T)$
- Let $p^* \in P$ be a price vector such that $T^*(r_1, b_1) \in \mathcal{T}(r_1, b_1, p^*)$
- Offer goods for prices p^*

`Chooser:`

- **if** $v_2((b_2, r_2), T^*(r_1, b_1)) = \max_{T \subseteq M : p^*(T) \leq \alpha_2} v_2((b_2, r_2), T)$ **then**
  - Select $A_2 = T^*(r_1, b_1)$
- **end**
- **else**
  - Select an arbitrary set A_2 in $\arg\max_{T \subseteq M : p^*(T) \leq \alpha_2} v_2((b_2, r_2), T)$
- **end**

Result: Return allocation $\mathcal{M}((r_1, r_2), (b_1, b_2)) = (A_1, A_2)$, where $A_1 = M \setminus A_2$.

Lemma 3. *For Mechanism 2 and signal vector $s = (s_1, s_2) \in S$, report vector $((s_1, s_2), (s_2, s_1))$ is a PNE.*

Lemma 4. *For Mechanism 2 and signal vector $s = (s_1, s_2) \in S$, in PNE $((s_1, s_2), (s_2, s_1))$, if the pricer's valuation is XOS, then she receives at least her PROP with respect to s.*

Since the chooser selects the bundle with the highest value according to his report subject to his entitlement, we can observe the following:

Claim. For Mechanism 2 and signal vector $s = (s_1, s_2) \in S$, in equilibrium $((s_1, s_2), (s_2, s_1))$, the allocation $\mathcal{M}((s_1, s_2), (s_2, s_1))$ is APS for the chooser with respect to s.

The proof of Theorem 3 follows by combining Lemmas 3 and 4 and Sect. 4.1.

4.2 Discussion on Mechanism 2

We showed that Mechanism 2 under XOS valuations gives in equilibrium $((s_1, s_2), (s_2, s_1))$ the pricer her PROP, and the chooser the APS with respect to the true signals. This implies that in the independent values model, for two XOS agents there is always an allocation that gives one agent her proportional share, and the other his APS. In the special case where the pricer's valuation is additive, since PROP is at least as large as the APS [8], this guarantees the APS for both agents. Moreover, for our Price-&-Choose mechanism when applied to independent additive valuation functions, all equilibria are APS fair with respect to the true values of the agents. Thus, we have the following corollary:

Corollary 1. *When applied to the independent values model (see Remark 1) for additive agents, Mechanism 2 guarantees that in every equilibrium, both agents receive their APS.*

Giving both agents their proportional share is trivially impossible even for the simple case of additive valuations (e.g., when there is a single good). We next show that there are subadditive instances with two agents (even in the non-interdependent case, with no incentive constraints) for which there is no allocation that gives one agent her APS and the other her PROP. This is true even for agents with equal entitlements and same valuation.

Proposition 1. *There exists a subadditive valuation $v : 2^M \to \mathbb{R}_{\geq 0}$, such that for every set $T \subseteq M$, either*

$$v(T) < \frac{v(M)}{2} = PROP$$

or,

$$v(M \backslash T) < \min_{p \in P} \max_{T' \subseteq M} v(T') \cdot \mathbb{1}\left\{\sum_{j \in T'} p_j \leq \frac{1}{2}\right\} = APS.$$

5 The Case of 3 or More Agents

In this section we show how for the case of $n \geq 3$ agents, we can reduce the problem of designing a mechanism that has at least one fair equilibrium with respect to the true signal s, to the purely algorithmic problem. On a high level, we do so as follows: Let each agent report her signal as well as everybody else's.

Since $n \geq 3$, if at least $n-1$ (unanimity up to one vote) agents agree on their reports and guesses, we can execute the algorithm with respect to the reported declarations, excluding the agent that did not agree (if any). This implies that when all the agents bid the same, the produced allocation is identical to the one where all the agents report as before, and just one agent deviates to something different[6]. The latter is crucial to guarantee a PNE with the algorithm's properties.

To formalize this reduction, consider any algorithm $\mathcal{F} : V_1 \times \ldots \times V_n \to \mathcal{A}$ that receives for every agent i an (independent) valuation $v_i : 2^M \to \mathbb{R}_{\geq 0}$ from a set of valuations V_i, and returns an allocation that satisfies some property X (where property X can be EF, EFX, EF1, PROP, MMS, APS, and even non-fairness properties such as maximizing social welfare, or Pareto optimality). Then we prove the following:

Theorem 4. *For every property X, every algorithm $\mathcal{F} : V_1 \times \ldots \times V_n \to \mathcal{A}$ that always satisfies property X, and every instance $\mathcal{I} = (N, M, \alpha, S, v_1, \ldots, v_n)$ with $n \geq 3$ agents, for which for all i, and every signal vector $s \in S$, it holds that $v_i(s, \cdot) \in V_i$, there exists an allocation mechanism $\mathcal{M}$ such that for every signal vector $s \in S$ there is at least one PNE, $((r_1, b_1), \ldots, (r_n, b_n))$, that satisfies that the allocation $\mathcal{M}((r_1, \ldots, r_n), (b_1, \ldots, b_n))$ satisfies property X with respect to s.*

Proof. To prove the theorem, we devise the following mechanism, for which the set of bids of agent i, is the product of the set of signals of all agents, i.e., $B_i = S$, and $B = B_1 \times \ldots \times B_n$. We denote by $b_{i,j}$ the jth coordinate of vector b_i.

Mechanism 3: IDV Black Box Transformation from $\mathcal{F}$ to $\mathcal{M}$

Data: Report vector $((r_1, b_1), (r_2, b_2), \ldots (r_n, b_n))$.
if there exists a set $N' \subseteq N$ of size at least $n-1$, such that 1) for every $i, i' \in N'$, we have $b_i = b_{i'}$, and 2) for every $i \in N'$ it holds that $r_i = b_{i,i}$ **then**
  | Select an arbitrary $i^* \in N'$ Set $\mathcal{M}(r, b) = \mathcal{F}(v_1(b_{i^*}, \cdot), \ldots, v_n(b_{i^*}, \cdot))$
end
else
  | Set $\mathcal{M}(r, b) = A = (A_1, A_2, \ldots, A_n)$, where A is some predefined allocation
end
Result: Return allocation $\mathcal{M}(r, b)$

The mechanism is well defined since because $n \geq 3$, there cannot be more than one value of b_{i^*} that has $n-1$ agents that report it.

Consider the report vector $(s_1, s), \ldots, (s_n, s)$. It is clear that since more than $n-1$ agents agree on the value of b_i (since all of them are s), and for every agent i, $r_i = s_i = b_{i,i}$, then the output of Mechanism 3 is $\mathcal{F}(v_1(s, \cdot), \ldots, v_n(s, \cdot))$, which satisfies all properties that $\mathcal{F}$ satisfies with respect to signal vector s. Therefore,

[6] It is easy to see that this technique cannot be implemented when there are only two agents, something that (surprisingly) makes this case more challenging.

the only thing that remains to show, is that the report vector $(s_1, s), \ldots, (s_n, s)$ is a PNE. This follows directly from the construction of Mechanism 3, as in the case that one agent deviates, the produced allocation remains the same.

Remark 2. It is worth noting that in Mechanism 3, the predefined allocation can be chosen arbitrarily, and the proof applies regardless of the chosen allocation. However, this chosen allocation might not align with the desired properties. The theorem aims to establish the existence of at least one favorable equilibrium that meets property X. This equilibrium occurs when all agents report their true signals. Conversely, a scenario where only $n-1$ agents concur on the signals is not an equilibrium: If an agent deviates, the allocation shifts to the predefined one, and there might be agents that prefer it.

6 Impossibilities

So far, we considered allocation mechanisms that for every signal vector $s \in S$, have at least one PNE, and the allocation that corresponds to one of these PNE is fair (according to some fairness criterion) with respect to the true signals s. The next natural direction is to explore whether there are mechanisms that always admit a PNE for every signal vector s, and *all* equilibria are fair with respect to the true signal vector s.

In the following theorem, we consider the notions of MMS and EF1, and we show that this is impossible. Notice that this creates a separation with the strategic model of non-interdependent values, where it is known that there are allocation mechanisms with this property under agents with additive valuation functions, both for the case of MMS (2 agents), and the case of EF1 (n agents).

Theorem 5. *There is no IDV-allocation mechanism that for every signal vector $s \in S$ has a PNE, and all of its equilibria induce MMS (or EF1) allocations with respect to the true signal vector s. Moreover, this holds even for the special case where all valuation functions are additive dichotomous (where each good has a value of either* 0 *or* 1*).*

Remark 3. Note that our counter-example considers additive valuation functions, with equal entitlements, and binary values, for which in the independent settings, there is a truthful allocation mechanism that is MMS and EF1 fair [6,26]. This creates a separation between the two models.

7 Future Directions

In this paper, we initiated the study of fair division of indivisible goods, under agents with interdependent values. We showed that despite the complexity of this setting, the design of mechanisms that have at least one pure Nash equilibrium, which is fair with respect to the true signals of the agents, is still possible. In addition, the mechanisms that we designed, provide a variety of fairness guarantees, even for agents with valuation functions that go beyond the additive class.

Finally, we also presented a negative result showing that it is not possible to have mechanisms that always admit PNE, and all the PNE of which, are fair (with respect to either MMS or EF1). The latter creates a clear separation between the interdependent and the independent valuation function setting.

Our work leaves several interesting questions open. In particular, for the case of 2 agents, it would be compelling to explore whether it is possible to design mechanisms that, besides having at least one fair PNE for every instance, also have additional *efficiency* properties, e.g., Pareto Optimality (as this is not the case with our mechanisms).

Moreover, moving to the general case of n agents, the fair PNE that our mechanisms have, are of a very specific form, where an agent has to more or less guess the true signals of the other agents. As this may be practically unappealing, it would be nice to examine if we can have mechanisms with fair equilibria of simpler structures.

Finally, another interesting direction would be to see whether we could identify meaningful settings of interdependent values where the impossibility results that we present do not apply, i.e., settings where it is possible to design allocation mechanisms for which all the PNE are fair.

Acknowledgments. Stefano Leonardi and Matteo Russo are partially supported by the ERC Advanced Grant 788893 AMDROMA "Algorithmic and Mechanism Design Research in Online Markets" and MIUR PRIN project ALGADIMAR "Algorithms, Games, and Digital Markets". Tomer Ezra is supported by Harvard University Center of Mathematical Sciences and Applications.

References

1. Amanatidis, G., Birmpas, G., Christodoulou, G., Markakis, E.: Truthful allocation mechanisms without payments: characterization and implications on fairness. In: Proceedings of the 2017 ACM Conference on Economics and Computation, EC 2017, pp. 545–562. ACM (2017)
2. Amanatidis, G., Birmpas, G., Filos-Ratsikas, A., Voudouris, A.A.: Fair division of indivisible goods: a survey. In: IJCAI, pp. 5385–5393. ijcai.org (2022)
3. Amanatidis, G., Birmpas, G., Fusco, F., Lazos, P., Leonardi, S., Reiffenhäuser, R.: Allocating indivisible goods to strategic agents: pure Nash equilibria and fairness. In: Feldman, M., Fu, H., Talgam-Cohen, I. (eds.) WINE 2021. LNCS, vol. 13112, pp. 149–166. Springer, Cham (2022). https://doi.org/10.1007/978-3-030-94676-0_9
4. Amanatidis, G., Birmpas, G., Lazos, P., Leonardi, S., Reiffenhäuser, R.: Round-robin beyond additive agents: existence and fairness of approximate equilibria. In: EC, pp. 67–87. ACM (2023)
5. Amanatidis, G., Birmpas, G., Markakis, E.: On truthful mechanisms for maximin share allocations. In: Proceedings of the 25th International Joint Conference on Artificial Intelligence, IJCAI 2016, pp. 31–37. IJCAI/AAAI Press (2016)
6. Aziz, H.: Simultaneously achieving ex-ante and ex-post fairness. In: Chen, X., Gravin, N., Hoefer, M., Mehta, R. (eds.) WINE 2020. LNCS, vol. 12495, pp. 341–355. Springer, Cham (2020). https://doi.org/10.1007/978-3-030-64946-3_24

7. Babaioff, M., Ezra, T., Feige, U.: Fair and truthful mechanisms for dichotomous valuations. In: Proceedings of the 35th AAAI Conference on Artificial Intelligence, AAAI 2021, pp. 5119–5126. AAAI Press (2021)
8. Babaioff, M., Ezra, T., Feige, U.: Fair-share allocations for agents with arbitrary entitlements. In: EC, p. 127. ACM (2021)
9. Babaioff, M., Nisan, N., Talgam-Cohen, I.: Competitive equilibrium with indivisible goods and generic budgets. Math. Oper. Res. **46**(1), 382–403 (2021)
10. Barman, S., Verma, P.: Truthful and fair mechanisms for matroid-rank valuations. In: AAAI, pp. 4801–4808. AAAI Press (2022)
11. Birmpas, G., Ezra, T., Leonardi, S., Russo, M.: Fair division with interdependent values. arXiv preprint arXiv:2305.14096 (2023)
12. Budish, E.: The combinatorial assignment problem: approximate competitive equilibrium from equal incomes. J. Polit. Econ. **119**(6), 1061–1103 (2011)
13. Caragiannis, I., Kaklamanis, C., Kanellopoulos, P., Kyropoulou, M.: On low-envy truthful allocations. In: Rossi, F., Tsoukias, A. (eds.) ADT 2009. LNCS (LNAI), vol. 5783, pp. 111–119. Springer, Heidelberg (2009). https://doi.org/10.1007/978-3-642-04428-1_10
14. Caragiannis, I., Kurokawa, D., Moulin, H., Procaccia, A.D., Shah, N., Wang, J.: The unreasonable fairness of maximum Nash welfare. ACM Trans. Econ. Comput. **7**(3), 12:1–12:32 (2019)
15. Chakraborty, M., Igarashi, A., Suksompong, W., Zick, Y.: Weighted envy-freeness in indivisible item allocation. ACM Trans. Econ. Comput. **9**(3), 18:1–18:39 (2021)
16. Clarke, E.H.: Multipart pricing of public goods. Public Choice **11**(1), 17–33 (1971)
17. Cohen, A., Feldman, M., Mohan, D., Talgam-Cohen, I.: Interdependent public projects. In: SODA, pp. 416–443. SIAM (2023)
18. Eden, A., Feldman, M., Fiat, A., Goldner, K.: Interdependent values without single-crossing. In: EC, p. 369. ACM (2018)
19. Eden, A., Feldman, M., Fiat, A., Goldner, K., Karlin, A.R.: Combinatorial auctions with interdependent valuations: SOS to the rescue. In: EC, pp. 19–20. ACM (2019)
20. Eden, A., Feldman, M., Talgam-Cohen, I., Zviran, O.: PoA of simple auctions with interdependent values. In: AAAI, pp. 5321–5329. AAAI Press (2021)
21. Eden, A., Goldner, K., Zheng, S.: Private interdependent valuations. In: SODA, pp. 2920–2939. SIAM (2022)
22. Farhadi, A., et al.: Fair allocation of indivisible goods to asymmetric agents. J. Artif. Intell. Res. **64**, 1–20 (2019)
23. Foley, D.K.: Resource allocation and the public sector. Yale Econ. Essays **7**, 45–98 (1967)
24. Gamow, G., Stern, M.: Puzzle-Math. Viking Press (1958)
25. Groves, T.: Incentives in teams. Econometrica **41**(4), 617–631 (1973)
26. Halpern, D., Procaccia, A.D., Psomas, A., Shah, N.: Fair division with binary valuations: one rule to rule them all. In: Chen, X., Gravin, N., Hoefer, M., Mehta, R. (eds.) WINE 2020. LNCS, vol. 12495, pp. 370–383. Springer, Cham (2020). https://doi.org/10.1007/978-3-030-64946-3_26
27. Jehiel, P., Moldovanu, B.: Efficient design with interdependent valuations. Econometrica **69**(5), 1237–1259 (2001)
28. Lipton, R.J., Markakis, E., Mossel, E., Saberi, A.: On approximately fair allocations of indivisible goods. In: Proceedings of the 5th ACM Conference on Electronic Commerce, EC 2004, pp. 125–131. ACM (2004)
29. Lu, P., Sun, E., Zhou, C.: Better approximation for interdependent SOS valuations. In: Hansen, K.A., Liu, T.X., Malekian, A. (eds.) WINE 2022. LNCS, vol. 13778, pp. 219–234. Springer, Cham (2022). https://doi.org/10.1007/978-3-031-22832-2_13

30. Markakis, E., Psomas, C.-A.: On worst-case allocations in the presence of indivisible goods. In: Chen, N., Elkind, E., Koutsoupias, E. (eds.) WINE 2011. LNCS, vol. 7090, pp. 278–289. Springer, Heidelberg (2011). https://doi.org/10.1007/978-3-642-25510-6_24
31. Milgrom, P.R., Weber, R.J.: A theory of auctions and competitive bidding. Econometrica **50**(5), 1089–1122 (1982)
32. Plaut, B., Roughgarden, T.: Almost envy-freeness with general valuations. SIAM J. Discrete Math. **34**(2), 1039–1068 (2020)
33. Psomas, A., Verma, P.: Fair and efficient allocations without obvious manipulations. In: Advances in Neural Information Processing Systems 35: Annual Conference on Neural Information Processing Systems 2022, NeurIPS 2022 (2022)
34. Roughgarden, T., Talgam-Cohen, I.: Optimal and near-optimal mechanism design with interdependent values. In: EC, pp. 767–784. ACM (2013)
35. Steinhaus, H.: Sur la division pragmatique. Econometrica **17**(Supplement), 315–319 (1949)
36. Suksompong, W., Teh, N.: On maximum weighted Nash welfare for binary valuations. Math. Soc. Sci. **117**, 101–108 (2022)
37. Suksompong, W., Teh, N.: Weighted fair division with matroid-rank valuations: monotonicity and strategyproofness. Math. Soc. Sci. **126**, 48–59 (2023)
38. Varian, H.R.: Equity, envy and efficiency. J. Econ. Theory **9**, 63–91 (1974)
39. Vickrey, W.: Counterspeculation, auctions, and competitive sealed tenders. J. Finance **16**(1), 8–37 (1961)
40. Viswanathan, V., Zick, Y.: Yankee swap: a fast and simple fair allocation mechanism for matroid rank valuations. In: AAMAS, pp. 179–187. ACM (2023)
41. Wilson, R.: Communications to the editor-competitive bidding with disparate information. Manag. Sci. **15**(7), 446–452 (1969)

Fair Division with Bounded Sharing: Binary and Non-degenerate Valuations

Samuel Bismuth[1(✉)], Ivan Bliznets[2], and Erel Segal-Halevi[1]

[1] Ariel University, 40700 Ariel, Israel
samuelbismuth101@gmail.com
[2] University of Groningen, 9747 AG Groningen, The Netherlands

Abstract. A set of objects is to be divided fairly among agents with different tastes, modeled by additive utility-functions. An agent is allowed to share a bounded number of objects between two or more agents in order to attain fairness.

The paper studies various notions of fairness, such as proportionality, envy-freeness, equitability, and consensus. We analyze the run-time complexity of finding a fair allocation with a given number of sharings under several restrictions on the agents' valuations, such as: binary generalized-binary and non-degenerate.

—

NOTE: due to space constraints, we had to move several parts that appeared on the submitted version to appendices. All material can be found in the full version at https://arxiv.org/abs/1912.00459 [2].

Keywords: Fair Division · Indivisible Goods · Allocation of Indivisible and Divisible Goods

1 Introduction

Consider several siblings who have inherited some assets and need to decide how to allocate them, or several parties who form a coalitional government and need to allocate the government ministries, or several faculty members who have moved to a new building and need to allocate the offices. In all these cases, a set of valuable objects has to be allocated among several agents, who may have different preferences over the objects, and it is important that all agents view the allocation as *fair*.

In many cases, fairness can only be attained by giving fractions of the same object to different agents. For example, if three siblings inherit four identical houses, then fairness requires that each sibling receives one house plus 1/3 of the fourth house.

Fractional allocation means that some objects must be *shared* among two or more agents. Sharing can be implemented in various ways. For example, sharing a house can be implemented by renovating it in a way that will enable all three siblings to live in it simultaneously; sharing a cabinet ministry is often done by

G. Schäfer and C. Ventre (Eds.): SAGT 2024, LNCS 15156, pp. 89–107, 2024.
https://doi.org/10.1007/978-3-031-71033-9_6

a rotational agreement, in which each party controls the ministry for a fraction of the time. Still, sharing an object is inconvenient, so it is desirable to share as few objects as possible. In the above example, a fair allocation could also be attained by sharing all four houses, giving each of the three siblings 1/3 of each house, but this allocation is clearly less desirable than the allocation in which only one house is shared.

This motivates the following generic computational problem, which is at the heart of the present research:

> (*) *Given m objects, n agents with different valuations over the objects, a fairness notion, and an integer $s \geq 0$, find a fair allocation in which at most s objects are shared, if such an allocation exists.*

We also consider a variant in which s is the number of *sharings* rather than the number of shared objects (e.g. a single object shared between 10 agents counts as 9 sharings). We assume that agents have linear additive valuations, and consider four common fairness notions: *proportionality* (each agent values his share as at least $1/n$ of the total value, where n is the number of agents), *envy-freeness* (each agent values his share as at least the share of any other agent), *equitability* (the subjective value of all agents is the same), and *consensus* (all n agents value all n bundles exactly the same); see Sect. 2 for the formal definitions.

Throughout the paper we focus on the allocation of goods—objects with a non-negative utility.

1.1 Related Work

Fair allocation with bounded sharing was first studied by Brams and Taylor [5,6]. Their *Adjusted Winner (AW)* procedure finds an allocation for $n = 2$ agents with at most $s = 1$ shared object, that is simultaneously proportional, envy-free, equitable, and *fractionally Pareto-optimal* (fPO: no other fractional allocation is at least as good for all agents and strictly better for some agent). They also show an example with three agents [5] where no allocation is simultaneously fPO, envy-free and equitable. This does not rule out the option of satisfying each of these properties on its own.

For $n \geq 3$ agents, the number of required sharings was studied in an unpublished manuscript of Wilson [19]. He proved the existence of an *egalitarian* allocation of goods—an allocation in which all agents have the largest possible equal utility [16]—with $n-1$ sharings. Egalitarian allocations are proportional but not necessarily envy-free.

Several more recent works have proved a polynomial upper bound on the required number of sharings for other fairness notions ([2][Appendix B] provides complete proofs and references):

- For proportionality, envy-freeness and equitability, there always exists a fair allocation with $s = n - 1$ (sharings or shared objects), and there may not exist a fair allocation with smaller s.
- There always exists a consensus allocation with $s = (n - 1)n$ (sharings or shared objects), and there may not exist a fair allocation with smaller s.

In all cases, an allocation satisfying the worst-case upper bound can be computed in polynomial time (see [2][Appendix B]).

Although the worst-case upper-bounds on sharing are well-understood, experiments on real-life and simulated instances show that most instances admit a fair allocation with fewer sharings [3,17]. This motivates the question of deciding whether a specific instance admits a fair allocation with s shared objects or sharings, where s is smaller than the upper bound. In our opening example, as there are $n = 3$ siblings, the worst-case upper bound is $n-1 = 2$, but there exists an allocation in which only $s = 1$ house is shared, which is more convenient than having to share two houses.

In general, the problem (*) is strongly NP-hard, as even for $s = 0$, it can be used to solve the strongly NP-hard problem 3-PARTITION (Given a 3-PARTITION instance with $3p$ items, we construct an instance of the problem (*) with $m = 3p$ objects and $n = p$ agents with identical valuations, and $s = 0$). But, as many real-life problems (e.g. inheritance allocation or cabinet ministries) involve a small number of agents, there is value in studying the case when n is a small, fixed constant, and the run-time should be polynomial in m. Even for fixed n the problem is NP-hard in general, as the special case of $s = 0$ and $n = 2$ agents with identical valuations is equivalent to the NP-hard PARTITION problem. However, several recent works provide polynomial-time algorithms for some other special cases:

1. Bismuth, Makarov, Segal-Halevi and Shapira [3] consider agents with *identical* valuations. With identical valuations, all fairness notions coincide, and are equivalent to finding a perfect scheduling of m jobs on n identical machines.[1] They develop a polytime algorithm for deciding if there exists a fair allocation with $s = n - 2$ shared objects, that is, one fewer than the worst-case upper bound, for any fixed $n \geq 3$. They prove that the $n-2$ is tight, as the problem is NP-hard for any fixed $n \geq 3$ and $s \leq n - 3$.
2. Sandomirskiy and Segal-Halevi [17] go to the other extreme and consider agents with *non-degenerate* valuations (for every two agents, their value ratios for the m objects are all different). They also require the allocation to be fPO in addition to being fair. They prove that, with non-degenerate valuations, the number of fPO allocations with s shared objects is polynomial in m (for every fixed n), so it is possible to enumerate all such allocations in polynomial time and check whether one of them satisfies any desired fairness notion. Therefore, the problem (*) is in P for every fixed n. Misra and Sethia [13] complement this result by proving that, when n is not fixed, the problem is NP-hard even for non-degenerate valuations and $s = 0$.
3. Goldberg, Hollender, Igarashi, Manurangsi and Suksompong [9] study *consensus splitting*—a partition of objects into k subsets each of which has a value of exactly $1/k$ for all agents. They show that computing a partition with at most $(k - 1)n$ sharings can be done in polynomial time. However, computing a partition with fewer sharings is hard even for $k = 2$: for any

[1] In fact, their result pertains also to the more general case of *uniform machines*, which is equivalent to agents having identical valuations but different entitlements.

fixed n and any $s < n$, it is NP-hard to decide whether a partition with at most s sharings exists.

Further related work is surveyed in [2][Appendix D].

1.2 Our Contribution

Our goal is to better understand the computational aspects of fair allocation with bounded sharing. We provide results for various special cases of the problem (*).

We start with the case of *binary valuations* (every agent values every object at 0 or 1). This case was not handled before, but it is easy to show that, for any constant n and s, the existence of a fair allocation for n agents with s sharings/shared objects can be decided in polynomial time by a mixed integer linear program with a fixed number of variables (Sect. 3).

Second, we consider agents with *generalized binary valuations* [7], also known as *cost valuations* [4]—for every object o there is a price p_o such that each agent i values o at either p_o or 0. In particular, we focus on a subset of the generalized binary valuations in which the sum of the utilities is equal for each agent. We call these *equal-sum generalized binary* utilities. These valuations generalize identical valuations that were studied in [3]; the results there imply that deciding existence of fair allocation with $s \leq n-3$ shared objects, or with $s \leq n-2$ sharings, are both NP-hard for any fixed $n \geq 3$. The remaining unsolved cases for equal-sum generalized binary valuations are the cases with $s = n-2$ shared objects. We present a polynomial-time algorithm for deciding the existence of a proportional allocation for $n = 3$ agents and $s = 1$ shared object. We note that, in fair item allocation, even the case of three agents is often interesting and non-trivial. For example, an important recent paper [8] is devoted to finding an EFX allocation for three agents, whereas the case of four agents is still open. Our algorithm is presented in Sect. 4.

Our third set of results involves agents with *non-degenerate valuations*. Non-degeneracy is arguably a weak requirement, as, informally, almost all valuations are non-degenerate [17] (see formal statement in Proposition 1 in Sect. 5). Polynomial-time solvability for non-degenerate valuations means that almost all instances are "easy"; this result is in the spirit of smoothed analysis [14,18]. But fPO—the other assumption made by [17]—is a strong requirement, as, informally, "almost all" allocations are not fPO.[2] Dropping the fPO may enable allocation with fewer sharings, which the agents may prefer to an fPO allocation with many sharings. This raises the question of whether a fair allocation (not necessarily fPO) can be found efficiently for agents with non-degenerate valuations. Our findings are mostly negative: for most fairness notions, we prove NP-hardness even for non-degenerate valuations.

Our proofs use an unusual reduction technique, which may be of independent interest. Usually, NP-hardness of a problem P_2 is shown by reduction from a

[2] Formally, even without sharing, the number of fPO allocations is in $O(m^{\binom{n}{2}+2})$ [17], whereas the total number of allocations is at least n^m, so the fraction of fPO allocations goes to 0 as $m \to \infty$, when n is fixed.

known NP-hard problem P_1, where each instance of P_1 is transformed to a *single* instance of P_2 with the same answer. We define *multi-reductions*, in which, for each instance of P_1, we construct a large set of instances of P_2 with the same answer. We use these multi-reduction to prove that (unless P = NP) it is *not* true that "almost all instances of P_2 can be solved in polynomial time". These results are presented in Sect. 5.

Finally, we study *truthful mechanisms* for fair allocation. A truthful mechanism is an algorithm which incentivizes the agents to reveal their true valuations. Truthfulness is another reason to drop the fPO requirement: it is known that no truthful mechanism can guarantee both fairness and Pareto-efficiency; it may be desired to give up efficiency to get truthfulness. In [2][Appendix C] we survey several truthful fair allocation algorithms, and check whether they can be adapted to construct an allocation with bounded sharing.

The results from this and previous works are summarized and compared in Table 1.

2 Preliminaries

2.1 Agents, Objects, and Allocations

There is a set $[n] = \{1, \dots, n\}$ of agents and a set $[m] = \{1, \dots, m\}$ of objects. For each agent $i \in [n]$ and object $o \in [m]$, the value $v_{i,o} \in \mathbb{Q}$ represents agent i's utility of receiving the object $o \in [m]$ in its entirety. The set of all instances is $\mathbb{Q}^{nm}$, representing the set of all $n \times m$ matrices. The total value of agent i to all objects is denoted by $V_i := \sum_{o \in [m]} v_{i,o}$. In general, the matrix $\mathbf{v}$ may contain values of mixed signs; but in this paper, we focus on allocation of goods and assume that all elements of $\mathbf{v}$ are non-negative.

A *bundle* $\mathbf{x}$ of objects is a vector $(x_o)_{o \in [m]} \in [0, 1]^m$, where the component x_o represents the fraction of object o in the bundle. Each agent $i \in [n]$ has a *utility function* u_i, assigning a numeric utility to each bundle. The utility functions are assumed to be *linear* and *additive*, which means that $u_i(\mathbf{x}) = \sum_{o \in [m]} v_{i,o} \cdot x_o$.

An *allocation* $\mathbf{z}$ is a collection of bundles $(\mathbf{z_i})_{i \in [n]}$, one for each agent, with the condition that all the objects are fully allocated. An allocation can be identified with the matrix $\mathbf{z} := (z_{i,o})_{i \in [n], o \in [m]}$ such that all $z_{i,o} \geq 0$ and $\sum_{i \in [n]} z_{i,o} = 1$ for each $o \in [m]$.

2.2 Fairness and Efficiency Concepts

We focus on four common fairness concepts. An allocation $\mathbf{z}$ is called:

- *Proportional (PROP)*—if every agent prefers her bundle to the equal division. Formally, for all $i \in [n]$: $u_i(\mathbf{z}_i) \geq V_i/n$.
- *Envy-free (EF)*—if every agent prefers her bundle to the bundles of others. Formally, for all $i, j \in [n]$: $u_i(\mathbf{z}_i) \geq u_i(\mathbf{z}_j)$. Every envy-free allocation is also proportional; with $n = 2$ agents, envy-freeness and proportionality are equivalent.

Table 1. Run-time complexity of allocating m objects among agents, with a bound on the number of sharings/shared objects.

Valuations	Allocation	Num of agents	Measure	bound	Run-time complexity
Identical	Fair[a]	Unbounded	sharing	s (any)	**Strong NP-hard [[3]]**
			shared ob.	s (any)	**Strong NP-hard [[3]]**
		Const. n	sharing	$s \leq n-2$	**NP-hard [[3]]**
				$s \geq n-1$	$O(m+n)$ *[cut-the-line]*
			shared	$s \leq n-3$	**NP-hard [[3]]**
				$s = n-2$	$O(poly(m, \log(V_1)))$ [3]
				$s \geq n-1$	$O(m+n)$ *[cut-the-line]*
Binary	EF	Unbounded	both	$s = 0$	**NP-complete [Section 3.1]**
	PROP			$s = 0$	$O(poly(m, n))$ *[Section 3.1]*
	EQ			$s = 0$	$O(poly(m, n))$ *[Section 3.1]*
	All	Const. n	both	s (any)	*MILP [Theorem 1]*
Equal-sum-generalized	Fair	Const. n	shared ob.	$s \leq n-3$	**NP-hard [see Identical]**
				$s \geq n-1$	*Weak poly [see Arbitrary]*
	PROP	3	shared ob.	$s = 1$	$O(poly(m, \log(V_1))$ *[Theorem 2]*
	All	Const. n	sharing	$s \leq n-2$	**NP-hard [See Identical]**
				$s \geq n-1$	*Weak poly [see Arbitrary]*
Non-degenerate	PROP, EF	Unbounded	sharing	s (any)	**Strong NP-hard [[2]Thm 11]**
			shared ob.	s (any)	**Strong NP-hard [[2]Thm 12]**
	PROP, EF	Const. n	sharing	$n \geq 2, s = 0$ or $n \geq 3$,	**NP-complete [Theorem 3]**
			shared ob.	$s \leq n-3$	**NP-complete [Theorem 4]**
	PROP+ dPO		both	$s = 0$	**NP-complete [Theorem 5]**
	EF+ dPO				
	Fair+ fPO			$s \geq n-1$	$O(m^{poly(n)})$ [17]
Arbitrary	PROP	Const. n	both	$s \geq n-1$	$O(mn\log(n))$ *[*[2]*[App. B]]*
	CONS		both	$s \geq n(n-1)$	*Strong poly* [10] *LP [*[2]*[Appendix C]]*
	EQ		both	$s \geq n-1$	*LP [*[2]*[Appendix B]]*
	EF		both	$s \geq n-1$	$O((n+m)^4 \log(n+m))$ [15] *LP [*[2]*[Appendix B]]*

[a] With identical valuations, all fairness concepts coincide.

- *Equitable (EQ)*—if it gives each agent exactly the same relative value, defined as the value of the bundle divided by the total value. Formally, for all $i, j \in [n]$: $u_i(\mathbf{z}_i)/V_i = u_j(\mathbf{z}_j)/V_j$.
- *Consensus (CONS)*—if every agent attributes exactly the same value to every bundle: for all $i, j \in [n]$: $u_i(\mathbf{z}_j) = V_i/n$. A consensus allocation is proportional, envy-free and equitable. Often a more general notion is considered, in which

the number of parts may be different than n; a *consensus k-partition* is a partition of the objects into k bundles such that for all $i \in [n], j \in [k]$: $u_i(\mathbf{z}_j) = V_i/k$.

We also consider two common efficiency concepts. We say that an allocation $\mathbf{z}$ is *Pareto-dominated* by an allocation $\mathbf{y}$ if $\mathbf{y}$ gives at least the same utility to all agents and strictly more to at least one of them. An allocation $\mathbf{z}$ is called:

- *Fractionally Pareto-optimal (fPO)*: if it is not dominated by any allocation $\mathbf{y}$.
- *Discretely Pareto-optimal (dPO)*: if it is not dominated by any allocation $\mathbf{y}$ with no sharing.

2.3 Measures of Sharing

If for some $i \in [n]$, $z_{i,o} = 1$, then the object o is not shared—it is fully allocated to agent i. Otherwise, object o is shared between two or more agents. Throughout the paper, we consider two measures quantifying the amount of sharing in a given allocation $\mathbf{z}$.

The simplest one is *the number of shared objects*:

$$\#\text{shared}(\mathbf{z}) := \Big| \{o \in [m] \,:\, z_{i,o} \in (0,1) \text{ for some } i \in [n]\} \Big|.$$

Alternatively, one can take into account the number of times each object is shared. This is captured by *the number of sharings*

$$\#\text{sharings}(\mathbf{z}) := \sum_{o \in [m]} \Big(\big|\{i \in [n] \,:\, z_{i,o} > 0\}\big| - 1 \Big)$$

Both measures are zero for discrete allocations. They differ, for example, if only one object o is shared but each agent consumes a bit of o: the number of shared objects in this case is 1 while the number of sharings is $n-1$. Clearly, $\#\text{shared}(\mathbf{z}) \leq \#\text{sharings}(\mathbf{z})$ for every allocation $\mathbf{z}$.

2.4 Types of Utilities

Recall that for each agent $i \in [n]$ and object $o \in [m]$, the value $v_{i,o} \in \mathbb{Q}$ represents agent i's utility of receiving the object $o \in [m]$ in its entirety. In general, $v_{i,o}$ can be any value in $\mathbb{Q}$ with no relation with other agent values. We relate to this as *arbitrary valuations*. We consider several special classes:

- *Identical valuations*—for each object $o \in [m]$, there is a rational number $p_o > 0$ such that $v_{i,o} = p_o$ for every agent $i \in [n]$.
- *Binary valuations*—Each agent $i \in [n]$ values every object $o \in [m]$ as either $v_{i,o} = 0$ or $v_{i,o} = 1$.

- *Generalized binary valuations*—for each object $o \in [m]$, there is a rational number $p_o > 0$ such that, for every agent $i \in [n]$, either $v_{i,o} = p_o$ or $v_{i,o} = 0$.[3] In the special case *Equal-sum generalized binary valuations*, the sum of values of all objects is the same for all agents.
- *Non-degenerate valuations*—for every two agents $i, j \in [n]$, there are no two objects $o_1, o_2 \in [m]$ such that the value-ratios are equal ($v_{i,o_1} \cdot v_{j,o_2} = v_{i,o_2} \cdot v_{j,o_1}$).

2.5 Computational Problems

In this paper we consider computational problems of the following kinds.

Definition 1. *Let F be a fairness criterion (proportionality, envy-freeness, etc.).*

(a) For any fixed integers $n \geq 2$ and $s \geq 0$, F-Sharings(n, s) is the problem of deciding if a given instance with n agents admits an F allocation with at most s sharings. F-SharedObj(n, s) is the same problem where s is the maximum number of shared objects.

(b) For any fixed integer $s \geq 0$, F-Sharings(s) is the problem of deciding if a given instance admits an F allocation among n agents (where n is part of the input) with at most s sharings; F-SharedObj(s) is the same problem where s is the maximum number of shared objects.

All those problems are obviously in NP, so proving NP-hardness is enough to prove that those problems are NP-complete.

3 Binary Utilities

3.1 Unbounded n

When all agents have binary valuations, the following results are known (or easy to derive) for the setting without sharing, when n is unbounded:

- EF-Sharings$(s = 0)$ is NP-complete, by reduction from Exact 3-Cover or from Equitable Coloring [1,11].
- In contrast, Prop-Sharings$(s = 0)$ can be solved in polynomial time by a standard reduction to network flow: there is an arc from the source to each agent i with capacity $\lceil V_i/n \rceil$; from each agent i to each object he values at 1 with capacity ∞; and from each object to the sink with a capacity 1. An maximum integral network flow can be found in polynomial time. The network has an integral flow in which all arcs from source to agents are saturated, if and only if the instance has a PROP allocation with no sharing.

[3] A similar utility function is used in [4,7].

- Eq-Sharings($s = 0$) can be solved in polynomial time by checking, for each $k \in 0, \ldots, m$, whether there exists an equitable allocation with common value k. This can be done by reduction to maximum-weight matching: we construct a weighted bipartite agent-object graph in which each agent has m copies. k copies are connected only to objects that the agent values at 1, and their weight is 1; $m - k$ copies are connected only to objects the agent values at 0, and their weight is ϵ (sufficiently small). An EQ allocation with common value k corresponds to a matching with weight $n \cdot k + \epsilon \cdot (m - nk)$. Assume first that we have an EQ allocation with common value k. In the graph, connect each object received by an agent with one of the agent's copies. The weight is at least $n \cdot k$. All the remaining objects are connected randomly, adding $\epsilon \cdot (m - nk)$ to the total weight. That is, we reach a matching with weight $n \cdot k + \epsilon \cdot (m - nk)$. Assume second that we have a matching with weight $n \cdot k + \epsilon \cdot (m - nk)$. This means that with ϵ sufficiently small, in the matching, there are exactly $n \cdot k$ edges with weight 1 and $(m - nk)$ edges with weight ϵ. For each edge with weight 1, give the object to the corresponding agent. Obviously, every agent gets a bundle with a value of exactly k, corresponding to an EQ allocation.

All results are not immediately applicable to the setting with $s \geq 1$. In particular, to extend the network-flow algorithm to solve Prop-Sharings(s) for $s \geq 1$, we would need to make the capacity of each edge from the source to agent i equal V_i/n, and apply a variant of the integer network flow which finds the maximum network flow with at most s non-integral edges. Currently, we do not know of any polynomial-time algorithm for this problem.

Open Problem 1. *(a) What is the running time of* Prop-Sharings(s), EF-Sharings(s) *and* Eq-Sharings(s) *for any $s \geq 1$, for agents with binary valuations?*
(b) What is the running time of Cons-Sharings(s) *for any $s \geq 0$, for agents with binary valuations?*

3.2 Fixed n

When the number of agents n is fixed, finding a PROP, EF, EQ or CONS allocation becomes polynomial, for every fixed number of agents n and number of sharings/shared objects s.

Theorem 1. *For every fixed number of agents n and number of sharings s:*

(a) Prop-Sharings(n, s), EF-Sharings(n, s), Eq-Sharings(n, s), Cons-Sharings(n, s) *with binary utilities can be solved in polynomial time.*
(b) Prop-SharedObj(n, s), EF-SharedObj(n, s), Eq-SharedObj(n, s), *and* Cons-SharedObj(n, s) *can be solved in polynomial time.*

Proof.[4] If $s \geq n - 1$ then the theorem follows from the results in [2][Appendix B]. Therefore we assume $s \leq n - 2$.

[4] We are grateful to Rohit Vaish for the proof idea for $s = 0$.

For each subset N of agents, let M_N be the set of objects that are valued at 1 by all and only the agents in N. Denote by $\mathcal{N}$ the set of all these subsets, there are at most 2^n subsets in $\mathcal{N}$.

We brute-force all possibilities of objects to share. We call the ℓ shared objects $o_1, o_2, \ldots, o_\ell$ (note that $\ell \leq s$, for both sharings and shared objects). Every shared object can belong to any one of the 2^n subsets in $\mathcal{N}$, and all objects in the same subset are equivalent, so it is enough to consider at most $(2^n)^\ell \leq 2^{ns}$ cases.

In addition, for each shared object $k \in [\ell]$, we denote by H_k the nonempty subset of agents who get a non-zero fraction from o_k. Overall there are $2^n - 1$ options for each H_k, so at most $(2^n-1)^\ell < 2^{ns}$ cases to check. Overall, for fixed constants n, s we consider only a constant number of cases.

For part (a), where s is the number of *sharings*, in each subcase we check that

- $\sum_{k\in[\ell]}(|H_k| - 1) \leq s$.

If this is not true, we simply discard this case. For part (b), where s is the number of *shared objects*, this check is not required, it is sufficient that $\ell \leq s$.

For each case, we construct a mixed integer linear program with the following variables:

- For every subset $N \in \mathcal{N}$ and agent $i \in N$, create an integer variable $x_{i,N}$ representing how many indivisible objects in M_N agent i obtains.
- For each $k \in [\ell]$ and each agent $i \in H_k$ create a real variable $y_{i,k}$ representing the fraction agent i gets from object o_k.

Now, we describe the constraints for variables $x_{i,N}, y_{i,k}$ that are true if and only if the corresponding allocation distributes all items among agents:

- $x_{i,N} \geq 0, \forall i \in [n], N \in \mathcal{N}$, and $x_{i,N}$ is an integer number;
- $\sum_{i\in[n]} x_{i,N} = |M_N \backslash \{o_1, \ldots, o_\ell\}|$, $\forall N \in \mathcal{N}$—we distribute all non-shared objects from M_N among all agents;
- $y_{i,k} \geq 0$, $\forall i \in H_k, k \in [\ell]$—each agent in H_k gets a non-negative fraction from o_k;
- $\sum_{i\in H_k} y_{i,k} = 1$, $\forall k \in [\ell]$—the divisible object o_k is completely distributed among agents from the set H_k.

Next, we add equations for computing the value each agent i attributes to each bundle j

$$u_{i,j} = \sum_{N:N\ni i}\left(x_{j,N} + \sum_{k:\ H_k\ni j,\ o_k\in M_N} y_{j,k}\right) \qquad \forall i,j \in [n].$$

The value of agent i is determined only by objects in sets M_N for which N contains i. For each such N, we add the number of objects given completely to j ($x_{j,N}$), and the fractions of divisible objects o_k given to j. Based on these equations, it is easy to write constraints for any desired fairness notion, according to the definitions in Sect. 2.2.

An MILP can be solved in polynomial time for any constant number of integer variables [12][Section 5]. The number of variables in our MILPs bounded by some function of n, s, as n and s are fixed constants. So we have that all our MILP-s is solvable in polynomial time.

4 Generalized Binary Utilities with Equal Sum

We recall the definition of the generalized binary utilities—assume that for each object $o \in [m]$, there is a rational number $p_o > 0$, for every agent $i \in [n]$ and object $o \in [m]$, $v_{i,o} \in \{0, p_o\}$. We assume that the sum of the utilities is equal for every agent, so for every two agents, $V_i = V_j$.

Our main positive result in this section is for the case $n = 3, s = 1$.

Theorem 2. Prop-SharedObj$(3, 1)$ *with equal-sum generalized binary valuations can be solved in polynomial time.*

To design our algorithm, we use another algorithm for the max-min variant of the n-way partition problem in which $s = 1$ object is allowed to be shared. The problem is defined as follows: given m objects and $n = 3$ agents with identical valuations, return an allocation with at most $s = 1$ shared object, for which the smallest bundle value is as large as possible. We call this problem MaxMinIdentical$(3, 1)$. A polynomial-time algorithm for MaxMinIdentical$(3, 1)$ is given in [3]. Their proof uses two structure Lemmas, that we will use later:

Lemma 1. *In any instance of* MaxMinIdentical$(3, 1)$ *either the output is perfect (all the bundle values are equal), or, the shared object is shared only between the two smallest bundles, say bundle 1 and bundle 2, and their values are equal.*

Lemma 2. *With identical valuations, for every allocation with $s = 1$ shared object and bundle sums b_1, b_2, b_3, there exists an allocation with the same bundle sums b_1, b_2, b_3 in which only the* highest-valued *object is shared.*

When there are three agents, it is easy to visualize the utilities of the agents using a table that we introduce next. Consider Table 2, that divide the set of objects into seven different categories, from $\mathcal{X}_1$ to $\mathcal{X}_7$.

Table 2. equal-sum generalized binary utilities for three agents

$\mathcal{X}_l$	$\sum_{j \in \mathcal{X}_l} u_1(x_j)$	$\sum_{j \in \mathcal{X}_l} u_2(x_j)$	$\sum_{j \in \mathcal{X}_l} u_3(x_j)$
$\mathcal{X}_1$	$U_1 = \sum_{j \in \mathcal{X}_1} p_j$	$U_1 = \sum_{j \in \mathcal{X}_1} p_j$	$U_1 = \sum_{j \in \mathcal{X}_1} p_j$
$\mathcal{X}_2$	$U_2 = \sum_{j \in \mathcal{X}_2} p_j$	$U_2 = \sum_{j \in \mathcal{X}_2} p_j$	0
$\mathcal{X}_3$	$U_3 = \sum_{j \in \mathcal{X}_3} p_j$	0	$U_3 = \sum_{j \in \mathcal{X}_3} p_j$
$\mathcal{X}_4$	0	$U_4 = \sum_{j \in \mathcal{X}_4} p_j$	$U_4 = \sum_{j \in \mathcal{X}_4} p_j$
$\mathcal{X}_5$	$U_5 = \sum_{j \in \mathcal{X}_5} p_j*$	0	0
$\mathcal{X}_6$	0	$U_6 = \sum_{j \in \mathcal{X}_6} p_j*$	0
$\mathcal{X}_7$	0	0	$U_7 = \sum_{j \in \mathcal{X}_7} p_j*$

There is no interest in assigning an object to an agent that values it at zero, so we automatically assign objects in $\mathcal{X}_5$ to agent 1, objects in $\mathcal{X}_6$ to agent 2, and objects in $\mathcal{X}_7$ to agent 3. To notify that the entire set of objects is assigned to an agent, we add a star to the corresponding cell. Denote by V the sum of the utilities. Note that:

$$V = U_1 + U_2 + U_3 + U_5 = U_1 + U_2 + U_4 + U_6 = U_1 + U_3 + U_4 + U_7.$$

It is easy to see that:

$$U_3 + U_5 = U_4 + U_6; \qquad U_2 + U_6 = U_3 + U_7; \qquad U_2 + U_5 = U_4 + U_7.$$

Assume w.l.o.g. that $U_5 \geq U_6 \geq U_7$, so $U_2 \leq U_3 \leq U_4$.

Using this notation, we present the algorithm for proving Theorem 2. The algorithm is based on a detailed case analysis, which we provide in [2][Appendix A]. In general, there are three main cases:

- $U_1 \leq U_4$—there is always a PROP allocation. The intuitive meaning of this case is that some set of objects (namely $\mathcal{X}_4$) is worth a lot to agents 2 and 3, but worth little to agent 1. We can allocate these objects among agents 2 and 3 in any way we like with a single shared object, and use the other objects (which are less valuable) to compensate agent 1 without additional splits (see Case 1 in [2][Appendix A]);
- $U_4 < U_1 \leq \frac{2}{3} \cdot V$—there is always a PROP allocation. The intuitive meaning of this case is that there is a set of objects (namely $\mathcal{X}_1$) that are worth a lot to all three agents, but the sum of these objects is at most $2/3$ of the total value V. Therefore, it is enough to allocate this set among two of the three agents (giving each of them at least $V/3$), and use the other objects to compensate the third agent (see Case 2 in [2][Appendix A]);
- $U_1 > \frac{2}{3} \cdot V$. The intuitive meaning of this case is that there is a set of objects (namely $\mathcal{X}_1$) that are worth a lot to all three agents, and their sum is larger than $2/3$ the total value. Therefore, we must partition these objects among all three agents. To do this, we use $\mathsf{MaxMinIdentical}(3, 1)$. If the largest bundle sum is at most $\frac{U_1+U_2+2\cdot(U_3+U_4+U_5+U_6)}{3}$, then a PROP allocation exists and we answer "yes"; otherwise, we prove that no PROP allocation can exist, so we answer "no" (see Case 3 in [2][Appendix A]).

This is a polynomial time algorithm for $\mathsf{Prop\text{-}SharedObj}(3, 1)$ with equal-sum generalized binary utilities, proving Theorem 2.

Open Problem 2. *Given some fixed $n \geq 3$, what is the run-time complexity of the problems* $\mathsf{Prop\text{-}SharedObj}(n, n-2)$ *and of* $\mathsf{EF\text{-}SharedObj}(n, n-2)$ *for generalized binary and arbitrary utilities?*

In particular, what is the run-time complexity of $\mathsf{Prop\text{-}SharedObj}(3, 1)$ *for arbitrary utilities and* $\mathsf{EF\text{-}SharedObj}(3, 1)$ *for generalized binary utilities?*

The same questions are interesting for CONS and EQ.

5 Non-degenerate Valuations

NOTE: following reviewers' comments, this section was rewritten in a substantially more formal way. Most of the proofs have been moved to appendices and can be found in the full version [2].

As mentioned in the introduction, Sandomirskiy and Segal-Halevi [17] prove that problems PropFpo-Sharings(n, s), EFFpo-Sharings(n, s), PropFpo-SharedObj(n, s), EFFpo-SharedObj(n, s) are solvable in time $O(poly(m))$ for any fixed n and s, whenever the valuations are *non-degenerate*, and argue—somewhat informally—that "almost all" valuations are non-degenerate; this means that almost all instances of the above problems are easy. In this section we would like to show that the requirement of fractional Pareto-optimality (fPO) is essential for this result: when this requirement is dropped, or even just relaxed to dPO, computational hardness strikes even for non-degenerate valuations, and it is no longer true that almost all instances are easy.

To prove this statement, we first have to formally define the notion of "almost all valuations are easy".

5.1 Definitions

We consider a decision problem P, whose input is a vector of some t non-negative integers. As the input size is measured by t, the binary encoding length of the numbers should be polynomial in t. For simplicity, we assume that all input integers are in the range $[0, 2^{ct}]$, for some constant c that may depend on the problem. We denote the set of possible inputs of size t (that is, t-sized vectors of integers in $[0, 2^{ct}]$) by $\mathcal{I}(t, c)$.

For any input vector $x \in \mathcal{I}(t, c)$, we denote its infinity norm by $\|x\|$. For any x and integer r, we denote by $\mathcal{B}(x, r)$ the *ball of radius* r around x: $\mathcal{B}(x, r) := \{x' \in \mathcal{I}(t, c) : \|x - x'\| \leq r\}$.

The notion of "almost all instances are easy" is formalized by the following definition of "generically polynomial-time algorithm".

Definition 2. *Given an algorithm A for a problem P, we say that A* runs generically in polynomial time *if there exists a polynomial function f_p such that, for every size t and input $x \in \mathcal{I}(t, c)$, there is a subset of "Good inputs" $G(x) \subseteq \mathcal{B}(x, f_p(t))$, such that the following holds:*

(a) Algorithm A runs in time poly(t) on all inputs in $G(x)$;
(b) The fraction of good inputs approaches 1, that is,

$$\lim_{t\to\infty} \min_{x\in\mathcal{I}(t,c)} \frac{|G(x)|}{|\mathcal{B}(x, f_p(t))|} = 1,$$

(c) Given x, it is possible to compute in time poly(t) a vector in $G(x)$.

The results in [17] imply:

Proposition 1. *For every fixed* n, s*, decision problems* PropFpo-Sharings(n, s), EFFpo-Sharings(n, s),PropFpo-SharedObj(n, s),EFFpo-SharedObj(n, s) *have algorithms that run in generically-polynomial time.*

Proof. In these problems, the input is $x \equiv v =$ a valuation matrix, and the number of integers in the input is $t = mn =$ the number of values in the valuation matrix. We choose the polynomial $f_p(t) := t^3$. We define the set $G(v)$ as the subset of matrices in $\mathcal{B}(v, t^3)$ that define non-degenerate valutions. We show that this set satisfies Definition 2.

(a) Is satisfied by the algorithms in [17].
(b) For each input (valuation matrix) $v' \in \mathcal{B}(v, t^3)$, for each value $v'_{i,o}$ in the matrix there are some $r_{i,o}$ options, where $t^3 + 1 \leq r_{i,o} \leq 2t^3 + 1$.[5] The inputs in $G(v)$ are the matrices v' in which each value $v'_{i,o}$ (the value of agent i to object o) satisfies inequalities of the form: $v'_{i,o}/v'_{j,o} \neq v'_{i,p}/v'_{j,p}$ for other agents j and objects p; the number of such inequalities is smaller than $mn = t$. Therefore, the number of options to choose $v'_{i,o}$ is at least $r_{i,o} - t$. Overall,

$$\frac{|G(v)|}{|\mathcal{B}(v, t^3)|} \geq \frac{\prod_{i,o}(r_{i,o} - t)}{\prod_{i,o}(r_{i,o})}$$
$$= \prod_{i,o}(1 - t/r_{i,o}) \geq (1 - t/(t^3 + 1))^t,$$

which approaches 1 as $t \to \infty$; this formalizes the claim that "almost all inputs are non-degenerate".
(c) To compute an input in $G(v)$, it is sufficient to check at most t options for changing each coefficient $v_{i,o}$; this can be done using polynomial in t time. □

Our goal is to prove that some problems do *not* have generically-polynomial-time algorithms. To this end, we first define a *multi-reduction.* We present a simplified version first, and then the full version.

Definition 3 (multi-reduction – simplified). *Given two decision problems* P_1 *and* P_2*, both defined on inputs in* $\mathcal{I}(t, c)$ *for some* $c \geq 1$*, a* polynomial-time multi-reduction *from* P_1 *to* P_2 *is a family of functions,* $h_t : \mathcal{I}(t, c) \to \mathcal{I}(t, c)$*, which maps an input for* P_1 *to an input for* P_2*, and satisfies the following:*

(a) h_t *runs in time poly*(t)*;*
(b) There exists a super-polynomial function f_e *such that, for all* t *and all* $x_1 \in \mathcal{I}(t, c)$*, when* $x_2 := h_t(x_1)$*,*

$$P_2(x'_2) = P_1(x_1) \qquad \textit{for all } x'_2 \in \mathcal{B}(x_2, f_e(t)),$$

[5] $r_{i,o} = t^3 + 1$ when the original value $v_{i,o}$ is at the bottom or top end of the allowed interval, and $r_{i,o} = 2t^3 + 1$ when $v_{i,o}$ is at the middle of the allowed interval.

A multi-reduction is stronger than a usual reduction in that each input to P_1 is transformed simultaneously to an super-polynomially-large set of inputs to P_2, all of which have the same output.

Definition 3 is "simplified" since it assumes that the size and encoding length of the inputs to P_1 is equal to the size and encoding length of the inputs to P_2. In fact, reductions often add some inputs or increase the encoding length. We handle this technical issue by assuming that the input x_1 is in $\mathcal{I}(t_1, c_1)$ and the transformed input x_2 is in $\mathcal{I}(t_2, c_2)$, where t_2 depends polynomially on t_1 but they do not have to be equal. We also allow the constants c_1 and c_2 to differ.

Definition 4 (multi-reduction – full). *Given two decision problems, P_1 defined on inputs in $\mathcal{I}(t_1, c_1)$ for some constant $c_1 \geq 1$ and P_2 defined on inputs in $\mathcal{I}(t_2, c_2)$ for some constant $c_2 \geq 1$, a* multi-reduction *from P_1 to P_2 consists of a polynomial-time-computable function $t_2 : \mathbb{N} \to \mathbb{N}$ and a family of functions, $h_{t_1} : \mathcal{I}(t_1, c_1) \to \mathcal{I}(t_2(t_1), c_2)$, which maps an input for P_1 to an input for P_2, and satisfies the following:*

(a) h_{t_1} runs in time poly(t_1);
(b) There exists a super-polynomial function f_e such that, for all t_1 and all $x_1 \in \mathcal{I}(t_1, c_1)$, when $x_2 := h_{t_1}(x_1)$,

$$P_2(x_2') = P_1(x_1) \qquad \text{for all } x_2' \in \mathcal{B}(x_2, f_e(t_2)),$$

Definition 5. *A decision problem P_2 is called* generically NP hard *if there exists a multi-reduction from some NP-hard problem P_1 to P_2.*

In [2][Appendix B] we prove that the relation between "generically-polynomial" and "generically-NP-hard" is analogous to the relation between "polynomial" and "NP-hard":

Proposition 2. *If a problem is generically-NP-hard, then it does not have a generically-polynomial algorithm unless P = NP.*

Now we are ready for our main results: showing that fair division problems without the fPO requirement are generically-NP-hard, and therefore probably do not have generically-polynomial-time algorithms (in particular, they are NP-hard even for non-degenerate valuations).

5.2 EF and PROP Allocations—Fixed n

Theorem 3. *The decision problems* Prop-Sharings(n, s) *and* EF-Sharings(n, s) *are generically NP-hard in the following cases:*

(a) For any fixed $n \geq 2$ and $s = 0$;
(b) For any fixed $n \geq 3$ and $s \in \{0, \dots, n-3\}$.

The proof is by showing a multi-reduction from k-WAY PARTITION, where $k = n$ in part (a) and $k = n - s - 1$ in part (b). Due to space constraints, we moved all proofs to [2][Appendix B].

We could not adapt the proof of Theorem 3 to *shared objects*. We can still prove similar results for shared objects, but with a different multi-reduction.

Theorem 4. *Problems* Prop-SharedObj(n, s) *and* EF-SharedObj(n, s) *are generically NP-hard in the following cases:*

(a) For any fixed $n \geq 2$ *and* $s = 0$*;*
(b) For any fixed $n \geq 3$ *and* $s \in \{0, \ldots, n-3\}$.

The hardness of Theorem 3(a) and Theorem 4(a) remains even if we add the requirement of *discrete* PO (in contrast to fractional PO). We consider only part (a) as the requirement of discrete PO makes sense only for allocations without sharing.

Theorem 5. *For any fixed integer* $n \geq 2$*, the decision problems* PropDpo-Sharings$(n, 0)$ *(*$\equiv$ PropDpo-SharedObj$(n, 0)$*) and* EFDpo-Sharings$(n, 0)$ *(*$\equiv$ EFDpo-SharedObj$(n, 0)$*) are generically NP-hard.*

Theorem 5 is interesting as it shows a crucial difference between the apparently-similar concepts fPO and dPO: whilst the fPO allocations can be enumerated in polynomial time (as their number is polynomial in m when the valuations are non-degenerate [17]), the dPO allocations cannot.

When $s = n - 1$, an EF and PROP allocation with s sharings always exists (see [2][Appendix B]). Therefore, only the case $s = n - 2$ remains open. With identical valuations, this case is NP-hard with sharings and polynomial with shared objects [3]. We do not know if the same is true with non-degenerate valuations.

Open Problem 3. *For any* $n \geq 3$ *and* $s = n - 2$*, do the problems* Prop-Sharings(n, s), EF-Sharings(n, s), Prop-SharedObj(n, s), EF-SharedObj(n, s) *have a generically-polynomial-time algorithm?*

For equitability and consensus allocation, we do not yet have any generic results:

Open Problem 4. *(a) For any* $n \geq 2$ *and* $s \in [0, n - 1)$*, do the problems* Eq-Sharings(n, s), Eq-SharedObj(n, s) *have a generically-polynomial-time algorithm?*
(b) For any $n \geq 2$ *and* $s \in [0, n(n - 1))$*, do the problems* Cons-Sharings(n, s), Cons-SharedObj(n, s) *have a generically-polynomial-time algorithm?*

5.3 EF and PROP Allocations—Unbounded n

So far we assumed that n is fixed. If n is unbounded (part of the input), then we can prove *strong* NP-hardness for both sharings and shared objects. This requires to adapt the notions of multi-reduction and generic NP-hardness to strong NP-hardness. All definitions and proofs are in [2][Appendix B].

6 Conclusion

Our work presented several results related to fairly allocating objects among agents, with a mixture of divisible and indivisible objects. In our work all objects are divisible, however, divisibility is highly discouraged. We covered many fairness and efficiency concepts, with a bound on the number of sharings or shared objects. We tackle the difficulty of finding an allocation for agents with arbitrary valuations by restricting the agent valuations to some well-studied domains, especially the *binary*, *generalized-binary*, and the *non-degenerate* valuations. In addition, our work shows that sometimes a polynomial algorithm can be designed if we allow a bounded number of objects to be shared among agents. Such a behavior is already analyzed in [3], and our work can be seen as a confirmation—when searching for a fair allocation is too hard, considering the objects as divisible, but still constraining the number of shared objects, might be reasonable for the agents, and may significantly decrease the runtime of the search. Thus, we highlighted a new way to relax the difficult fair-division problem, as it is common in the literature using for example Envy-freeness up to one good (EF1).

Our main result considers 3 agents under equal-sum generalized binary utilities. Despite the fact that we consider only 3 agents, the algorithm is complicated, and involves an exhaustive cases analysis. That is why, as a next challenge, one can try to generalize our algorithm and see if it is possible to extend it for any fixed number n of agents. Also, it is interesting to see if the same result holds for agents under general additive utilities.

We dedicated one section to non-degenerate valuations. Despite the surprising polynomial time algorithm designed in [17], several results in our paper point up the hardness of this problem, even when allowing shared objects.

Along the paper, we left many open problems. This paper is an invitation for researchers interested in the fair-division field, to find new results, with the goal to find theoretic and practical solutions, to numerous fair-division problems.

Acknowledgements. This work was inspired by Achikam Bitan, Steven Brams and Shahar Dobzinski, who expressed their dissatisfaction with the current trend of approximate-fairness (SCADA conference, Weizmann Institute, Israel 2018). We are grateful to participants of De Aequa Divisione Workshop on Fair Division (LUISS, Rome, 2019), Workshop on Theoretical Aspects of Fairness (WTAF, Patras, 2019) and the rationality center game theory seminar (HUJI, Jerusalem, 2019) for their helpful comments. Suggestions of Herve Moulin, Antonio Nicolo, Nisarg Shah and Rohit Vaish were especially useful.

Several members of the stack-exchange network provided helpful answers, in particular: D.W, Peter Taylor, Gamow, Sasho Nikolov, Chao Xu, Mikhail Rudoy, xskxzr, Philipp Christophel, Kevin Dalmeijer, Bodo Manthey, and Niklas Rieken.

We are grateful to the anonymous referees of WTAF 2019 and SAGT 2024 for their helpful comments.

This work was partly funded by Israel Science Foundation grant no. 712/20.

References

1. Aziz, H., Gaspers, S., Mackenzie, S., Walsh, T.: Fair assignment of indivisible objects under ordinal preferences. Artif. Intell. **227**, 71–92 (2015). https://doi.org/10.1016/J.ARTINT.2015.06.002
2. Bismuth, S., Bliznets, I., Segal-Halevi, E.: Fair division with bounded sharing: binary and non-degenerate valuations (2024). https://arxiv.org/abs/1912.00459
3. Bismuth, S., Makarov, V., Segal-Halevi, E., Shapira, D.: Uniform machines scheduling with bounded splittable jobs. CoRR abs/2204.11753 (2022). https://doi.org/10.48550/ARXIV.2204.11753
4. Botan, S., Ritossa, A., Suzuki, M., Walsh, T.: Maximin fair allocation of indivisible items under cost utilities. In: Deligkas, A., Filos-Ratsikas, A. (eds.) SAGT 2023. LNCS, vol. 14238, pp. 221–238. Springer, Cham (2023). https://doi.org/10.1007/978-3-031-43254-5_13
5. Brams, S.J., Taylor, A.D.: Fair Division: From Cake Cutting to Dispute Resolution. Cambridge University Press, Cambridge (1996)
6. Brams, S.J., Taylor, A.D.: The Win-Win Solution: Guaranteeing Fair Shares to Everybody (Norton Paperback). W. W. Norton & Company, reprint edn. (2000)
7. Camacho, F., Fonseca-Delgado, R., Pérez, R.P., Tapia, G.: Generalized binary utility functions and fair allocations. Math. Soc. Sci. **121**, 50–60 (2023). https://doi.org/10.1016/J.MATHSOCSCI.2022.10.003
8. Chaudhury, B.R., Garg, J., Mehlhorn, K.: EFX exists for three agents. J. ACM **71**(1), 1–27 (2024)
9. Goldberg, P.W., Hollender, A., Igarashi, A., Manurangsi, P., Suksompong, W.: Consensus halving for sets of items. Math. Oper. Res. **47**(4), 3357–3379 (2022)
10. Goldberg, P.W., Hollender, A., Igarashi, A., Manurangsi, P., Suksompong, W.: Consensus halving for sets of items. Math. Oper. Res. **47**(4), 3357–3379 (2022). https://doi.org/10.1287/moor.2021.1249
11. Hosseini, H., Sikdar, S., Vaish, R., Wang, H., Xia, L.: Fair division through information withholding. In: The Thirty-Fourth AAAI Conference on Artificial Intelligence, AAAI 2020, The Thirty-Second Innovative Applications of Artificial Intelligence Conference, IAAI 2020, The Tenth AAAI Symposium on Educational Advances in Artificial Intelligence, EAAI 2020, New York, NY, USA, 7–12 February 2020, pp. 2014–2021. AAAI Press (2020). https://doi.org/10.1609/AAAI.V34I02.5573
12. Lenstra, H.W., Jr.: Integer programming with a fixed number of variables. Math. Oper. Res. **8**(4), 538–548 (1983). https://doi.org/10.1287/MOOR.8.4.538
13. Misra, N., Sethia, A.: Fair division is hard even for amicable agents. In: Bureš, T., et al. (eds.) SOFSEM 2021. LNCS, vol. 12607, pp. 421–430. Springer, Cham (2021). https://doi.org/10.1007/978-3-030-67731-2_31
14. Moitra, A., O'Donnell, R.: Pareto optimal solutions for smoothed analysts. In: Proceedings of the 43rd Annual ACM Symposium on Theory of Computing - STOC 2011 (2011). https://doi.org/10.1145/1993636.1993667
15. Orlin, J.B.: Improved algorithms for computing fisher's market clearing prices: computing fisher's market clearing prices. In: Proceedings of the Forty-Second ACM Symposium on Theory of Computing, pp. 291–300. ACM (2010)
16. Pazner, E.A., Schmeidler, D.: Egalitarian equivalent allocations: a new concept of economic equity. Quart. J. Econ. **92**(4), 671–687 (1978)
17. Sandomirskiy, F., Segal-Halevi, E.: Efficient fair division with minimal sharing. Oper. Res. **70**(3), 1762–1782 (2022)

18. Spielman, D.A.: The smoothed analysis of algorithms. In: Liśkiewicz, M., Reischuk, R. (eds.) FCT 2005. LNCS, vol. 3623, pp. 17–18. Springer, Heidelberg (2005). https://doi.org/10.1007/11537311_2
19. Wilson, S.J.: Fair division using linear programming. Preprint, Departement of Mathematics, Iowa State University (1998)

Incentives in Dominant Resource Fair Allocation Under Dynamic Demands

Giannis Fikioris(✉), Rachit Agarwal, and Éva Tardos

Cornell University, Ithaca, NY 14853, USA
{gfikioris,ragarwal}@cs.cornell.edu, eva.tardos@cornell.edu

Abstract. Every computer system performs resource allocation across system users. The defacto allocation policies used in most of these systems are max-min fairness for single resource settings and dominant resource fairness for multiple resources. These allocation schemes guarantee desirable properties like incentive compatibility, envy-freeness, and Pareto efficiency. Assuming that user demands are static (time-independent) the allocation is also fair. However, in modern real-world production systems, user demands are dynamic, that is, vary over time. As a result, there is now a fundamental mismatch between the resource allocation goals of computer systems and the properties enabled by classical resource allocation policies. This paper aims to bridge this mismatch. When demands are dynamic, instant-by-instant max-min fairness can be extremely unfair over a longer period of time, i.e., lead to unbalanced user allocations, as previous large allocations have no effect in the current time step. We consider a natural generalization of the classic algorithm for max-min fair allocation and dominant resource fairness for multiple resources when users have dynamic demands. This algorithm guarantees Pareto optimality while ensuring that resources allocated to users are as max-min fair as possible *up to* any time instant, given the allocation in previous periods. While this dynamic allocation scheme remains Pareto optimal and envy-free, unfortunately, it is not incentive compatible. We study the strength of the incentive to misreport; our results show that the possible increase in utility by misreporting demand is bounded and, since this misreporting can lead to a significant decrease in overall useful allocation, this suggests that it is not a useful strategy.

Keywords: Dominant Resource Fairness · Resource Allocation

1 Introduction

Resource allocation is a fundamental problem in computer systems. Companies like Google [45] and Microsoft [20] use schedulers in private clouds to allocate a limited and divisible amount of resources (e.g., CPU, memory, servers, etc.) among many selfish and possibly strategic users that want to maximize their allocation; the goal of the scheduler is to maximize resource utilization while achieving fairness in resource allocation. The defacto allocation policies used in many

G. Schäfer and C. Ventre (Eds.): SAGT 2024, LNCS 15156, pp. 108–125, 2024.
https://doi.org/10.1007/978-3-031-71033-9_7

of these systems are the classic *max-min fairness* (MMF) and its generalization, *dominant resource fairness* (DRF) [19] policies, for single and multiple resource settings, respectively. For instance, these policies are used in most schedulers in private clouds [8,19–22,25,42,43,45]; they are deeply entrenched in congestion control protocols like TCP and its variants [2,11]; and are the default policies for resource allocation in most operating systems and hypervisors [9,33]. DRF has also attracted a lot of attention in the economics and computing community, starting with [38] and with followup work [17,18,27,30,34].

The strong prevalence of MMF and DRF is rooted in their guarantees: Pareto-efficiency, sharing incentives (users are not better off by getting their fair share of resources every round), incentive compatibility, envy-freeness (no user envies the allocation of another user), and fairness. However, to guarantee these properties, both MMF and DRF policies assume that user demands do not change over time. This assumption is far from realistic in modern real-world deployments: several recent studies in the systems community have shown that user demands have become highly dynamic, that is, vary over time [10,40,46,47,49]. For such dynamic user demands, naively using these policies (e.g., to perform a new instantaneously max-min fair allocation every round) can result in vastly disparate user allocations over time—intuitively, since MMF does not take past allocations into account, dynamic user demands can result in increasingly unfair user allocations over time. We have also studied this issue in [46], where we implement a dynamic version of MMF that aims to equalize users' total allocations and show that it results in reasonably fair allocation on practical user data.

Motivated by the realistic case of dynamic user demands over divisible resources with multiple resources, we study Dynamic DRF[1], a mechanism that generalizes DRF for dynamic demands; just like DRF generalizes MMF for multi-resource allocations, Dynamic DRF generalizes Dynamic MMF [16] for multi-resource allocations over dynamic demands by taking past allocations into account. Our model is the same as the original DRF paper [19]: every round, each user specifies a vector of *ratios* (the proportions according to which the user uses the different resources, e.g., for her application) and a *demand* (the maximum allocation of resources that would be useful to her). Users have *Leontief preferences*—as they are known in economics—which capture the idea that the user's workload requires fixed proportions of the resources. Formally, the utility in each round is equal to the minimum over the amount of every resource received divided by their ratio for it, up to their demand. However, in contrast to [19], we consider scenarios where the ratios and the demands can vary over time and users want to maximize the sum of their utilities across rounds.

In every round, Dynamic DRF allocates resources while being as fair as possible given the past allocations: first the minimum total utility of any user is maximized, then the second minimum, etc. Besides being fair, Dynamic DRF is also Pareto-efficient by construction: every round, either every user's demand

[1] The name Dynamic DRF has also been used by [30] for the extension of DRF when users arrive and depart sequentially. See more about the difference in our Relate work.

is satisfied or for every user a resource she wants to use is saturated. However, neither Dynamic MMF [16], nor Dynamic DRF are *incentive compatible*, i.e., it is possible that a user can misreport her demand or her ratios on one round to increase her total useful allocation in the future. The lack of incentive compatibility was already true for Dynamic MMF [16], but our study leads to improved lower bounds on the incentive compatibility violation of Dynamic MMF; see Theorems 4 and 8.

Despite Dynamic DRF not being fully incentive compatible, studying it is both important and interesting. First, similar to the widely-used classic MMF and DRF (also referred to as static MMF and DRF), Dynamic DRF is simple and easy to understand; thus, it has the potential for real-world adoption (similar to many other non-incentive compatible mechanisms used in practice, e.g., non-incentive compatible auctions used by U.S. Treasury to sell treasury bills since 1929 and by the U.K to sell electricity [23,32,39]). Second, our results show that Dynamic DRF is approximately incentive compatible, that is, strategic users can increase their allocation by misreporting their demands but this increase is bounded by a relatively small constant factor, independent of the number of users and the number of resources. Moreover, effective misreporting not only requires knowledge of future demands but can significantly decrease overall useful allocation, suggesting that misreporting is unlikely to be a useful strategy for any user. Approximate incentive compatibility has been extensively used to analyze various systems in other areas, with the idea that closeness to incentive compatibility (and the danger of reward loss by misreporting) will make truthful reporting a likely strategy in practice. We review this literature at the end of this section.

Our Contribution. Our goal is to study Dynamic DRF and the incentive to misreport in it. A popular relaxation of incentive compatibility is *γ-incentive compatibility* [3–5,12,13,31,36], which requires that the possible increase in utility by untruthful reporting must be bounded by a factor of $\gamma \geq 1$ (γ is referred to as *incentive compatibility ratio*). Using this notion we show that users have limited incentive to be untruthful:

Our main results are presented in Sect. 3, where we focus on the setting of multiple resources. In the case that every user demands every resource when using the system, we show that in Dynamic DRF user i cannot increase her utility more than a factor of $(1 + \rho_i)$ (Theorem 3) and give a matching lower bound (Theorem 4) where ρ_i is a parameter that quantifies the relative importance of every resource between user i and the other users. We also show that in this case users have no incentive to over-report their demand or misreport their ratios (Theorem 2); these guarantee that resources allocated to the users are always in use. The assumption that every user demands every resource when using the system, even if in different ratios, is quite natural; in computer systems where the resources shared are CPU, memory, storage, etc. the users run applications that use every resource. This assumption is also used by [30], where they extend DRF to the case of users arriving and leaving sequentially resulting in dynamically changing the allocations in the system. Additionally, we show that Dynamic

DRF is envy-free (Theorem 5), and the variant where every user is guaranteed an α fraction of her fair share satisfies α-sharing incentives while retaining every one of the aforementioned properties.

In Sect. 4 we consider the simpler problem of single resource environments. We study Dynamic weighted MMF which generalizes Dynamic MMF for the case when every user i has a positive weight w_i that indicates her priority. Our guarantees for Dynamic DRF (that are tight in that setting) carry over to the single resource setting but prove to be too weak: we offer better incentive compatibility bounds under more adverse conditions. More specifically, we consider the setting where players can create coalitions to increase their total utility. In this case, Dynamic weighted MMF is 2-incentive compatible (Theorem 7) and demand over-reporting does not increase utility (Theorem 6). The former of these results strikes a big contrast between single and multiple resource settings: assuming that users do not collude, we can directly apply our results from Dynamic DRF to the Dynamic weighted MMF mechanism. In this case, we have that $\rho_i = \max_{k \neq i}\{w_i/w_k\}$, giving an upper bound of $1+\rho_i = 1+\max_{k \neq i}\{w_i/w_k\}$ for the incentive compatibility ratio, that can possibly be unbounded. In contrast, we prove a 2-upper bound for Dynamic weighted MMF. In addition, we prove a lower bound: in Dynamic MMF a user can increase her utility by a factor of $\sqrt{2}$ by under-reporting her demand (Theorem 8).

Related Work. The simplest algorithm for resource allocation is *strict partitioning* [44,48], which allocates a fixed amount of resources to each user independent of their demands. While incentive compatible, strict partitioning can have arbitrarily bad efficiency. Static MMF and DRF [16,19,20,22,38,42] are Pareto-efficient, incentive compatible, envy-free, and satisfy sharing incentives, but are fair only when user demands are static.

[16] prove that Dynamic MMF is not incentive compatible under the same utility model as ours. The papers [16,26] study resource allocation for single resource settings with dynamic demands focusing on the case when users have utility for resources above their demand, only at a lower rate. They offer alternate mechanisms where past allocations have some effect on the current ones (unlike static MMF) while maintaining incentive compatibility, but the mechanisms they consider are closer to MMF separately in each round, and aim less to be fair overall. Under this model, they present two mechanisms that are incentive compatible but either satisfy sharing incentives and have no Pareto-efficiency guarantees or approximately satisfy sharing incentives and are approximately Pareto-efficient under strong assumptions (user demands being i.i.d. random variables and number of rounds growing large).

[41] present minor improvements in fairness over static DRF for dynamic demands while maintaining incentive compatibility. Their mechanism allocates resources in an incentive compatible way according to DRF while marginally penalizing users with larger past allocations using a parameter $\delta \in [0, 1)$. Specifically, if t rounds ago a user received an allocation of r, that allocation penalizes the user in the current round by $r(1-\delta)\delta^t$. This means that the penalty of $(1-\delta)\delta^t \leq 0.25$ reduces exponentially fast with time for any fixed $\delta < 1$ and

as $\delta \to 1$ the penalty tends to 0. Thus, for every δ (and, especially for $\delta = 0$ and $\delta \to 1$), their mechanism suffers from similar problems as static DRF: past allocations are barely taken into account.

In [46] we implement a version of Dynamic MMF and analyze its performance under practical user data. Our results show that Dynamic MMF yields much fairer outcomes for users with dynamic demands compared to other static mechanisms like classic MMF. In [46] we also prove that Dynamic MMF is 3/2-incentive compatible. Here we generalize this result, obtaining the aforementioned one as one of the corollaries of Theorem 7 (were we study Dynamic weighted MMF, the generalization of Dynamic MMF).

Several other papers study resource allocation when user demands are dynamic, but with significantly different settings than ours. [1,50] examine the setting where indivisible items arrive over time and are allocated to users whose utilities are random; however [1] study a very weak version of incentive compatibility in which a mechanism is incentive compatible if misreporting cannot increase a user's utility in the current round and [50] do not consider strategic users. [28] study single resource allocation and assume the users do not know their exact demands every round, needing to provide feedback after each round of allocation to allow the mechanism to learn. The goal of the paper is to offer a version of MMF that approximately satisfies incentive compatibility, sharing incentives, and Pareto-efficiency, despite the lack of information, but is not considering the long-term fairness that is the focus of our mechanisms.

Another series of work study users arriving and leaving consecutively after some period of time: [17,34] focus on single resource settings, while [18,27,30] also study the allocation of multiple resources. Even though their setting is dynamic, users have constant demands and cannot re-arrive after leaving, making the user demands static. After every arrival or departure of a user, resources need to be re-distributed while maintaining some sort of fairness, e.g., the users' utilities need to be approximately similar. [30] focus on never decreasing a user's allocation when re-distributing resources, thus creating a mechanism that allocates at most k/n fraction of every resource when k out of n users are present in the system. They offer a mechanism for this setting that they also call dynamic DRF, however, in their setting the dynamic nature of the problem comes from churn in users, as well as the corresponding changes in total resources, and not from dynamic demands. The papers [17,18,27,34] consider resource allocation, but incentive compatibility is not taken into account and the focus is to maximize fairness while bounding the "disruptions" of the system every time a user enters or leaves, which is how many users' allocations are altered.

There are many reasons why mechanisms used in practice are often not incentive compatible, including the relative simplicity of these mechanisms that makes it easier for users to understand and use them and the fact that mechanisms claimed to be incentive compatible may not turn out to actually be incentive compatible in practice, depending on the information structure, e.g., when participants collude (see [5] for more examples and a nice discussion of a long list of other reasons). In most settings, even if the mechanism is manipulable, finding a profitable manipulation is hard. In our setting, finding such manipulation requires knowledge of all users' future demands, and while under-reporting demand has the potential to increased future utility, it seems more likely to lead to decreased utility. When using a manipulable mechanism, it is important to understand how large is the incentive to manipulate. γ-incentive compatibility has been considered in many settings. [6] show that there is a gap of $O(\sqrt{\gamma})$ in the optimal achievable revenue between γ-approximate incentive compatible mechanisms and incentive compatible ones in the context of auctions. [7] design mechanisms that have a constant incentive compatibility ratio when the auctioneer has an estimate of constant error for the users' private information. [14] show that when designing contracts there is a provable trade-off between the approximate incentive compatibility ratio and the approximation of the optimal payoff of the contract. As we mentioned before, [28] study a setting similar to ours, but where long-term fairness is not their focus, instead they try to learn users' demands; to achieve their results they propose several mechanisms most of which are only approximately incentive compatible. [3,13,31,36] study combinatorial auctions that are almost incentive compatible. [12] study approximate incentive compatibility in machine learning, when users are asked to label data. [4] examines approximate incentive compatibility in large markets, where the number of users grows to infinity. [5] develop algorithms that can estimate how incentive compatible are various mechanisms for buying, matching, and voting. [29] propose a mechanism for repeated second price auctions where the mechanics of one auction depend one the previous one, making the overall mechanism approximately incentive compatible. [35] study different concepts of approximate incentive compatibility that can be used to design mechanisms that have better guarantees (e.g., computational efficiency, bypassing impossibility results, etc.) than incentive compatible ones in the context of mechanism design with or without money. [37] reviews the limitations imposed by incentive compatibility, e.g., in some settings multi-dimensional user information makes incentive compatibility imply a constant outcome—one independent of user types. [24] develop auctions that use samples from past non-incentive compatible auctions to improve social welfare or revenue guarantees.

2 Preliminaries

We first make some definitions that apply to all sections. We use $[n]$ to denote the set $\{1, 2, \dots, n\}$ for any $n \in \mathbb{N}$. Additionally, we define $x^+ = \max\{x, 0\}$ and denote with $\mathbb{1}[\cdot]$ the indicator function.

There are n users, where $n \geq 2$. The set of users is denoted with $[n]$. In some settings, every user i is associated with a weight, $w_i > 0$ which indicates each user's priority and in this case we view the allocation fair, if user i has (approximately) w_i/w_j more utility than user j. Additionally, we sometimes assume that user i has initial endowment or *fair share* a $w_i/\sum_{j\in[n]} w_j$ fraction of the total resources. The game is divided into *rounds* $1, 2, \dots, t, \dots$

A mechanism is called *envy-free* if for any users i and j, where i is truthful, she would not have gained utility if she had been allocated the resources user j was allocated. Similarly, we define *weighted envy-freeness*: if every user i is associated with a weight $w_i > 0$, a mechanism is envy-free if for users i and j, user i would not have gained utility if she had been allocated the resources user j was allocated, scaled by w_i/w_j (note that this scaling sometimes results in comparing to usage of more resources than what is available).

A mechanism satisfies *sharing incentives* if every truthful user's utility is not less than her utility if she had been allocated her fair share every round, i.e., a $\frac{1}{n}$ fraction of the total resources. For weighted users, a mechanism satisfies sharing incentives if the utility of user i is not less than her utility if she had been allocated her fair share every round, i.e., a $w_i/\sum_{j\in[n]} w_j$ fraction of the total resources. For $\alpha \in [0, 1]$, there is also the notion of *α-sharing incentives*, in which user i's utility must be at least α times her utility if she had been allocated her fair share every round.

3 Multiple Resources Setting

In this section we analyze Dynamic Dominant Resource Fairness (Dynamic DRF), the generalization of DRF for dynamic demands.

Notation and User Utilities. We consider users that have varying demand for a set of $m \geq 1$ different resources over time. We use $q \in [m]$ to denote the m resources, and w.l.o.g., we assume that for every resource the amount available is the same, $\mathcal{R}$. A typical example of such a system may focus on users running applications that use resources such as CPU, memory, storage, etc.

Every round, each user demands an amount of every resource which they report to the mechanism. With the multidimensional nature of demand, users have very complex ways to misreport their demand. Throughout this paper we will assume that users have *Leontief preferences*, which we define next. Leontief preferences have been considered by much of the previous work in resource sharing, e.g., [19,30,38].

Formally, with Leontief preferences a user i's demand in a round t is characterized by the vector of m *ratios* $a_i^t = (a_{i1}^t, \ldots, a_{im}^t)$ she needs for the resources and a *demand* d_i^t. The ratios indicate the proportions according to which the user demands the resources: for some $\xi \geq 0$, user i's application in that round is going to use ξa_{iq}^t amount of every resource $q \in [m]$. The demand d_i^t of user i in round t represents the maximum fraction ξ (possibly, $\xi > 1$) of the ratios the user can take advantage of. Specifically, user i demands (or is asking for) $d_i^t a_{iq}^t$ amount of every resource q. The resource q with the maximum ratio a_{iq}^t is called the *dominant resource* of user i in round t. W.l.o.g. we assume that for every round t, $\max_q a_{iq}^t = 1$ for all users. If a user i receives $x_{i1}^t, \ldots, x_{im}^t$ of every resource, respectively, her utility that round is equal to $u_i^t = \min\left\{d_i^t, \min_{q:a_{iq}>0}\left\{x_{iq}^t/a_{iq}^t\right\}\right\}$. The total utility of user i by round t is $U_i^t = \sum_{\tau=1}^{t} u_i^\tau$.

In each round, users will be asked to report both their ratios for the round as well as their demand. Note that users can misreport their type in two different ways: they can either request less or more from all the resources or they can demand the resources in different proportions (or they can do both).

Dynamic Dominant Resource Fairness. DRF is the generalization of MMF for multiple resources, where the fairness criterion is applied to each user's dominant resource and the rest of the resources are distributed according to their ratios. If r_i^t is the amount that user i receives of her dominant resource in round t, then she receives $r_i^t a_{iq}^t$ of every resource q (recall $\max_q a_{iq}^t = 1$). We call r_i^t the *allocation* of user i in round t.

Dynamic DRF also extends to a weighted version, the case when users have different weights representing their priorities. In this version the fairness criterion is applied to the users' allocation normalized by the weights. More specifically, if every user i is associated with a weight $w_i > 0$, then the mechanism tries to give each user i an allocation proportional to her weight w_i.

We also consider Dynamic DRF with a parameter $\alpha \in [0, 1]$ that guarantees a fraction of her fair share of each resource to each user in every round, independent of previous allocations. User i's fair share of a resource q is $\mathcal{R}w_i/\sum_j w_j$. When $\alpha = 1$, we guarantee at least the fair share of their dominant resource (assuming a big enough demand to use it), when $\alpha < 1$ we guarantee a smaller share. Beyond this guarantee, the goal of the mechanism in round t is to be as fair as possible to the cumulative allocation of every user, normalized by their weights. We use $R_i^t = \sum_{\tau=1}^t r_i^\tau$ for the sum of allocations till time t. Using this notation, Dynamic DRF is easy to describe. For a given round t, assuming that every user i has cumulative allocation R_i^{t-1} on round $t-1$:

$$\begin{aligned} &\text{choose} && r_1^t, r_2^t, \ldots, r_n^t \\ &\text{applying MMF on} && \frac{R_1^{t-1} + r_1^t}{w_1}, \frac{R_2^{t-1} + r_2^t}{w_2}, \ldots, \frac{R_n^{t-1} + r_n^t}{w_n} \\ &\text{subject to} && \forall i \in [n], \min\left\{d_i^t, \alpha\frac{\mathcal{R}w_i}{\sum_{k\in[n]} w_k}\right\} \le r_i^t \le d_i^t, \\ & && \forall q \in [m], \sum_{i\in[n]} r_i^t a_{iq}^t \le \mathcal{R} \end{aligned} \tag{1}$$

We define $g_i(d_i^t) = \min\left\{d_i^t, \alpha\frac{\mathcal{R}w_i}{\sum_k w_k}\right\}$ to be the guaranteed amount that every user receives every round.

We note a few properties that follow from the description. If all users share the same dominant resource q^t in each round t and have equal weights, then the mechanism will become identical to Dynamic MMF, as at each round the allocation of the shared dominant resource is the bottleneck for all users. Second, if all the users share their dominant resource q^t, $\alpha = 1$, and demands are high, then the minimum guarantee becomes $g(d_i^t) = \mathcal{R}w_i/\sum_k w_k$ which will become user i's allocation, as this saturates resource q^t (the guaranteed total use of q^t is $\sum_i g(d_i^t)a_{iq^t}^t = \sum_i g(d_i^t) = \mathcal{R}$). However, even with large demands each iteration and $\alpha = 1$, the dynamic fair sharing nature of our allocation will play an important role when applications (users) do not always share their dominant resource. Third, because of the guarantee of every user, Dynamic DRF satisfies α-sharing incentives.

Theorem 1. *Dynamic DRF with a guarantee of α satisfies α-sharing incentives.*

User's Utility when Misreporting Ratios. In defining the Dynamic DRF mechanism, we have not considered the difference of truthful reporting and misreporting demands or ratios to the mechanism. The main topic of this section is explaining how a user's utility behaves in these two scenarios.

When user i truthfully reports her demand d_i^t and ratios a_i^t, and gets an allocation of r_i^t, her utility in that round is $u_i^t = \min\{d_i^t, \min_{q:a_{iq}^t>0} r_i^t a_{iq}^t / a_{iq}^t\} = r_i^t$, since Dynamic DRF guarantees $r_i^t \le d_i^t$.

When user i misreports $\hat{a}_i^t$ and $\hat{d}_i^t$, and Dynamic DRF gives her an allocation $\hat{r}_i^t$ based on the reported values, let $\hat{u}_i^t$ denote the user's true utility in round t under that reporting. In this case she receives $x_{iq}^t = \hat{a}_{iq}^t \hat{r}_i^t$ of every resource q and thus she gets true utility $\hat{u}_i^t = \min\{d_i^t, \min_{q:a_{iq}{}^t>0} x_{iq}^t / a_{iq}^t\} = \min\{d_i^t, \hat{r}_i^t \min_{q:a_{iq}{}^t>0} \hat{a}_{iq}^t / a_{iq}^t\}$ We define $\hat{\lambda}_i^t = \min_{q:a_{iq}{}^t>0}\{\hat{a}_{iq}^t / a_{iq}^t\}$ making the above expression equal to $\hat{u}_i^t = \min\{d_i^t, \hat{r}_i^t \hat{\lambda}_i^t\}$. We note that if the user reports ratios truthfully ($\hat{a}_i^t = a_i^t$), then $\hat{\lambda}_i^t = 1$. Additionally, because each user is constrained to declare $\max_q \hat{a}_{iq}^t = \max_q a_{iq}^t = 1$, it holds $\hat{\lambda}_i^t \le 1$ for any $\hat{a}_i^t$.

3.1 Incentives Assuming Positive Ratios for All Resources

Our main results in this paper consider resource allocation with multiple resources in Dynamic DRF under the assumption that users always demand at least some of each of the resources, i.e., $a_{iq}^t > 0$ for all i, q, t. With typical system resources, such as CPU, RAM, etc., it is indeed the likely scenario. While different applications have different dominant resources (e.g., some have heavy use of compute power, while in others the main bottleneck is bandwidth), each uses at least some of each one of these basic resources.

As observed by [38], zero ratios for some resources significantly changes the problem from having a tiny $\epsilon > 0$ ratio. In the full version of the paper [15] we show that with zero ratios users can have an incentive to over-report their demand, which we show is not true with positive ratios. Further, the benefit of such over-reporting can increase the user's utility by a factor of $\Omega(m)$, increasing with the number of resources in the system. With positive ratios, users are bottlenecked by the same resource being saturated. In contrast, when ratios are zero, different users are bottlenecked by different resources, resulting in users' allocations being almost independent from one another.

In contrast, the main results of this section are that, assuming users have positive ratios, misreporting them, as well as over-reporting demand, is not beneficial (Theorem 2). Further, the approximate incentive compatibility ratio for user i is bounded by $1 + \rho_i$, where $\rho_i = \max_{k \ne i, q, t} \{w_i a_{iq}^t / w_k a_{kq}^t\}$ (Theorem 3), and this bound is tight (Theorem 4) even with just two resources and ratios that do not change over time.

Upper Bound on Incentive Compatibility Ratio. We now focus on upper bounds on how much utility a user can gain by deviating. For ease of presentation, we are going to focus on how much utility user 1 can get from deviating.

The assumption that all users are using each resource allows us to prove a lemma that offers a simple condition on which pair of users can gain overall allocations from one another. When users' demands are not satisfied, for a user to get more resources someone else needs to get less. The lemma will allow us to reason about how a deviation by user 1 can lead to a user i getting more resources and another user j getting less (possibly $i = 1$ or $j = 1$).

Lemma 1. *Fix a round t and the total allocations up to round $t-1$ of any two outcomes $\{\hat{R}_k^{t-1}\}_{k\in[n]}$ and $\{\bar{R}_k^{t-1}\}_{k\in[n]}$. Let i, j be two different users. Assume that: (i) for i, $\bar{r}_i^t < \hat{r}_i^t$, $\hat{d}_i^t \leq \bar{d}_i^t$, and $\hat{a}_{iq}^t > 0$ for all q, and (ii) for j, $\bar{r}_j^t > \hat{r}_j^t$, $\bar{d}_j^t \leq \hat{d}_j^t$, and $\bar{a}_{jq}^t > 0$ for all q. Then, for any $\alpha \in [0,1]$ used by Dynamic DRF, it holds that $\bar{R}_i^t/w_i \geq \bar{R}_j^t/w_j$ and $\hat{R}_i^t/w_i \leq \hat{R}_j^t/w_j$, implying $\frac{\hat{R}_i^t - \bar{R}_i^t}{w_i} \leq \frac{\hat{R}_j^t - \bar{R}_j^t}{w_j}$.*

The intuition behind this lemma is due to the conditions in the bullets: it is feasible to trade resources between users i and j in each of the two outcomes. This proves the bound for their total allocations due to the definition of Dynamic DRF, which tries to equalize the users' normalized total allocations. We defer the full proof to the full version [15].

The main technical tool in our work is the following lemma bounding the total amount a user can "win" because of user 1 deviating, i.e., $\max_k\{\hat{R}_k^t - R_k^t\}$. Rather than just bounding the deviating user 1's gain directly, it is better to consider the maximum increase of any user. This is because user 1 can increase her utility by using the increase in utility of some other user.

Lemma 2. *Fix a round t and two outcomes $\{\hat{R}_k^{t-1}\}_{k\in[n]}$ and $\{\bar{R}_k^{t-1}\}_{k\in[n]}$ which, for some $X \geq 0$, satisfy $\max_{k\in[n]}\left\{\frac{\hat{R}_k^{t-1} - \bar{R}_k^{t-1}}{w_k}\right\} \leq X$. If in round t users have positive ratios, user 1 reports her ratios truthfully (i.e., $\hat{a}_1^t = \bar{a}_1^t = a_1^t$) and $\hat{d}_1^t \leq \bar{d}_1^t$ then, for all $\alpha \in [0,1]$ used by Dynamic DRF it holds that $\max_{k\in[n]}\left\{\frac{\hat{R}_k^t - \bar{R}_k^t}{w_k}\right\} \leq X + \mathbb{1}\left[\hat{d}_1^t < \bar{d}_1^t\right] \rho_1^t \frac{\bar{r}_1^t}{w_1}$, where $\rho_1^t = \max_{k\neq 1,\, q\in[m]} \frac{w_1 a_{1q}^t}{w_k a_{kq}^t}$.*

The above lemma, by assuming truthful reporting of ratios and that user 1 does not over-report her demand ($\hat{d}_1^t \leq d_1^t$), inductively proves a $\left(1 + \max_t\{\rho_1^t\}\right)$ incentive compatibility ratio.

In order to prove this lemma we need to consider two cases. If $\hat{d}_1^t = \bar{d}_1^t$, then users with larger $\hat{R}_k^{t-1} - \bar{R}_k^{t-1}/w_k$ will not gain additional resources because they are less favored by Dynamic DRF. If $\hat{d}_1^t < \bar{d}_1^t$, user 1 can increase the allocation of some other user k by under-reporting her demand, but by at most $O(\bar{r}_1^t)$. The full proof can be found in the full version of the paper [15].

Before stating the upper bound on the incentive compatibility ratio, we first show that users have no incentive to over-report their demand or misreport their ratios. Aside from being useful in proving the upper bound on incentive compatibility, this property also proves that the resources allocated are always in use: any misreporting that aims to increase utility only under-reports demand, leading to full utilization of the resources allocated.

The immediate effect of over-reporting or misreporting ratios for user 1 is getting allocated resources that do not contribute to her utility. Intuitively, this suggests that user 1 is put into a disadvantageous position: other users may get less resources which makes them be favored by the allocation algorithm in the future, while user 1 becomes less favored. However, a small change in the users' resources causes a cascading change in future allocations making the proof of this theorem not trivial. As we show in the full version [15] that this is no longer the case when users only use a subset of them. The full proof can be found in the full version of our paper [15].

Theorem 2. *Assume that the users' true ratios are positive, i.e., for all users $i \in [n]$, resources $q \in [m]$, and rounds t, it holds that $a_{iq}^t > 0$. Then, for any $\alpha \in [0, 1]$ used by Dynamic DRF, the users have nothing to gain by declaring a demand higher than their actual demand, and any gain achievable by misreporting ratios can also be obtained by under-reporting demand.*

We now prove the desired upper bound on the incentive to deviate. We do this by combining Lemma 2 and Theorem 2.

Theorem 3. *Assume that all users have positive ratios: $a_{iq}^t > 0$ for all users i, resources q, and rounds t. Then for any user i and $\alpha \in [0, 1]$ used by Dynamic DRF, user i cannot misreport her demand or ratios to increase her utility by a factor larger than $(1 + \rho_i)$, where $\rho_i = \max_{k \neq i, q, t} \left\{ {}^{w_i a_{iq}^t} / {}_{w_k a_{kq}^t} \right\}$.*

Proof. W.l.o.g. we are going to prove the theorem for $i = 1$. Because of Theorem 2 we can assume that user 1 does not over-report her demand or misreport her ratios and thus we can bound $\hat{R}_1^t$ instead of $\hat{U}_1^t$. Because of this condition, we can use Lemma 2 which inductively implies that for any t, $\hat{R}_1^t - R_1^t \le \rho_1 R_1^t$. This proves the theorem.

We now prove Theorem 2. First, we prove the following lemma, with which Theorem 2 is easily proven using induction. The lemma says that if in rounds $T_0 + 1$ to T (for any $T_0 \le T$) user 1 does not over-report her demand and reports her ratios truthfully, she cannot increase her utility in round T by over-reporting demand or misreporting ratios in T_0.

Lemma 3. *Fix a round T_0 and the allocations of an outcome $\{\hat{R}_k^{T_0 - 1}\}_{k \in [n]}$. Fix another round $T \ge T_0$ and assume that in rounds $T_0 + 1, T_0 + 2, \ldots, T$ user 1 reports her ratios truthfully and does not over-report her demand. Then, for any $\alpha \in [0, 1]$ used by Dynamic DRF, any increase in user 1's utility in round T gained by over-reporting demand or misreporting ratios in T_0 can be achieved with truthful ratio reporting and no demand over-reporting in round T_0.*

By over-reporting her demand or misreporting ratios in T_0, user 1 (potentially) gains some resources that do not contribute to her utility, while other users get less resources. This puts user 1 in a disadvantage entailing that in the rounds after T_0 she cannot increase her total allocation further (even though it is possible that her allocation in a single round can increase). Since she doesn't over-report her demand in rounds after T_0, her utility in those rounds is the same as her allocation, hence her total utility also does not increase. To prove the lemma we consider an alternative reporting that both reports ratios truthfully and guarantees no demand over-reporting. With this reporting user 1 is guaranteed the same utility without increasing her allocation or decreasing the other user's allocations; this entails a more advantageous position for her in the following rounds. We defer the proof of Lemma 3 to the full version in [15].

Using Lemma 2 we can now easily prove Theorem 2.

Proof (of Theorem 2). Fix a round T and let $T_0 \leq T$ be the last round where user 1 over-reported or misreported her ratios. Lemma 3 allows us to change user 1's reporting in T_0 to one that does not over-report demand and reports ratios truthfully, without decreasing her total utility in T. Doing this inductively for every such T_0 creates a demand profile that does not decrease user 1's total utility in T.

Lower Bound on Incentive Compatibility Ratio. In Theorem 3 we proved that the incentive compatibility ratio of user 1 is at most $(1 + \rho_1)$. We now show that if the only constraints on users' ratios and weights are that they are positive and ρ_1 is fixed, then it is possible for the incentive compatibility ratio of user 1 to be $(1 + \rho_1)$. We prove this even if the users' ratios do not change over time. We defer the full proof to the full version of the paper [15].

Theorem 4. *For any $\epsilon \in (0, 1)$, $w_1, w_2 > 0$, and $\alpha \in [0, 1]$ used by Dynamic DRF, there is an instance where the users' ratios are constant every round, user 1 has weight w_1 and another user has weight w_2, $\rho_1 = \frac{w_1}{w_2 \epsilon}$, and user 1 can under-report her demand to increase her total utility by a factor of $1 + \rho_1$.*

3.2 Envy-Freeness in Dynamic DRF

In this section we examine the envy-freeness of Dynamic DRF. More specifically, we show that in Dynamic DRF a user is envy-free if they are truthful.

Theorem 5. *For every $\alpha \in [0, 1]$, Dynamic DRF is envy-free according to the weights $w_1, \ldots, w_n$, i.e., for every round t, no user i envies the total allocation of user j scaled by w_i/w_j: $U_i^t \geq \sum_{\tau=1}^{t} \min\left\{d_i^\tau, \frac{w_i}{w_j} r_j^\tau \min_{q: a_{iq}^\tau > 0} \frac{a_{jq}^\tau}{a_{iq}^\tau}\right\}$.*

The key to this proof is to study the (potential) first round t user i envies user j. In the simple case that $w_i = w_j$, that proves that $r_j^t > r_i^t$, which leads to a reasoning similar to Lemma 1 proving that $R_j^t \leq R_i^t$. Therefore, since user j has an allocation smaller than i's it is impossible for user i to envy her. We defer the full proof of the theorem to the full version [15].

4 Improved Guarantees in Single Resource Settings

In this section we examine the simpler single-resource setting and more specifically Dynamic weighted MMF, version of Dynamic DRF for only one resource. The results of Sect. 3 apply to the single resource setting, making Dynamic weighted MMF envy-free, satisfy α-sharing incentives when each user is guaranteed a α fraction of her fair share, and have bounded incentive compatible ratio. However, as we show, one can prove much stronger incentive compatibility guarantees for this setting under more adverse conditions.

Dynamic Weighted Max-Min Fairness. Dynamic weighted MMF is a special case of Dynamic DRF where there is only one resource, $m = 1$. This means that every round t, given the allocations of the previous rounds $\{R_i^{t-1}\}$, the following problem is solved

$$\begin{aligned} &\text{choose} \quad r_1^t, r_2^t, \ldots, r_n^t \\ &\text{applying MFF on} \quad \frac{R_1^{t-1} + r_1^t}{w_1}, \frac{R_2^{t-1} + r_2^t}{w_2}, \ldots, \frac{R_n^{t-1} + r_n^t}{w_n} \\ &\text{subject to} \quad \sum_{i \in [n]} r_i^t \leq \mathcal{R} \quad \text{and} \quad \min\left\{d_i^t, \alpha\frac{\mathcal{R}w_i}{\sum_k w_k}\right\} \leq r_i^t \leq d_i^t, \forall i \end{aligned} \tag{2}$$

where $\alpha \in [0,1]$ and $w_1, \ldots, w_n$ are positive numbers, similar to the definition of Dynamic DRF. Similar to the previous section, for a given α, we denote $g_i(d_i^t) = \min\{d_i^t, \alpha\mathcal{R}w_i / \sum_k w_k\}$ the *guarantee* of user i in round t.

Coalitions. In this section we are going to also consider that a deviating user might not be acting alone. More specifically, we consider that users form coalitions to increase the sum of their utilities by each member of the coalition deviating. We bound that increase, i.e., if the set $I \subset [n]$ of users forms a coalition and report demands $\{\hat{d}_i^t\}_{i \in I, t}$ instead of $\{d_i^t\}_{i \in I, t}$, then for some $\gamma \geq 1$ and for all t we want to prove that $\sum_{i \in I} \hat{U}_i^t \leq \gamma \sum_{i \in I} U_i^t$.

Incentive Compatibility Upper Bound. We first present the analogue of Theorem 2 in this setting: even if users can form coalitions, they have nothing to gain in Dynamic weighted MMF by over-reporting.

Theorem 6. *Let $I \subset [n]$ be a set of users that form a coalition. Then, for any value of $\alpha \in [0,1]$ used by Dynamic weighted MMF, the users in I have nothing to gain by over-reporting their demand.*

Similar to Theorem 2, the proof of this theorem makes intuitive sense: if one user of the coalition over-reports her demand she puts herself into disadvantage by gaining unnecessary resources which harms the entire coalition. We defer the proof of this theorem to the full version [15].

Next we upper bound the incentive of the users in the coalition to deviate. If there is no coalition ($I = \{i\}$), Theorem 3 yields a bound. More specifically, using the notation of that section, we have that $\rho_i = \max_{j \neq i}\{w_i/w_j\}$, making that theorem provide an upper bound of $1 + \rho_i$. Even though this is possibly unbounded, the incentive compatibility ratio we prove here is at most 2.

Theorem 7. *Let $I \subset [n]$ be a set of users that form a coalition and $w_1, \dots, w_n$ be any weights, according to which Dynamic weighted MMF allocates resources. Then, for any $\alpha \in [0,1]$, any deviation of the users in I, and any round t it holds that $\sum_{i \in I} \hat{U}_i^t \leq 2 \sum_{i \in I} U_i^t$. Additionally, when $I = \{i\}$ for any user i, it holds $\hat{U}_i^t \leq \left(1 + \max_{j \neq i} \frac{w_i}{w_i + w_j}\right) U_i^t$.*

Note that if $I = \{i\}$ and users have the same weights the above theorem recovers the bound of 3/2 of [46] for Dynamic MMF without coalitions.

To prove Theorems 6 and 7 we prove the following lemma, which is similar to Lemma 2 but stronger. In Lemma 2 we bounded the increase of $\max_k \{\hat{R}_k^t - R_k^t\}$ across rounds using the allocation of the deviating user. In the lemma that follows we bound the increase of $\sum_k \{\hat{R}_k^t - R_k^t\}$ again using the allocation of the deviating user, leading to a strictly stronger result.

We note that unlike Lemma 2, here we do not have to normalize the users' allocations by their weights. Because there is only one resource users' allocations are more easily comparable: r_i^t is not only the amount of user i' dominant resource received by her but rather the amount of the only available resource. This intuition could likely prove that if all users have the same dominant resource we can prove stronger guarantees of Dynamic DRF, but this is not a realistic assumption that we did not explore.

Lemma 4. *Fix a round t and the allocations of two different outcomes $\{\hat{R}_k^{t-1}\}_{k \in [n]}$ and $\{\bar{R}_k^{t-1}\}_{k \in [n]}$. Assume that $\{\bar{d}_i^t\}_{i \in [n]}$ are some users' demands and that $\{\hat{d}_i^t\}_{i \in [n]}$ are the same demands except users in I, who deviate but not by over-reporting, i.e., $\hat{d}_i^t \leq \bar{d}_i^t$ for $i \in I$. Then, for any $\alpha \in [0,1]$, it holds that*

$$\sum_{k \in [n]} \left(\hat{R}_k^t - \bar{R}_k^t\right)^+ - \sum_{k \in [n]} \left(\hat{R}_k^{t-1} - \bar{R}_k^{t-1}\right)^+ \leq \mathbb{1}\left[\sum_{k \in I} \hat{d}_k^t < \sum_{k \in I} \bar{d}_k^t\right] \sum_{k \in I} \bar{r}_k^t \quad (3)$$

When the users in the coalition are truthful in round t, then the l.h.s. of (3) is at most 0: Dynamic weighted MMF allocates resources such that the large $\hat{R}_k^t - \bar{R}_k^t$ are decreased and the small $\hat{R}_k^t - \bar{R}_k^t$ are increased. Finally, if the users in I report lower demands, then the (at most) $\sum_{k \in I} \bar{r}_1^t$ resources these users do not get might increase the total over-allocation by the same amount.

Using this lemma and Theorem 6 it is not hard to prove the incentive compatibility ratio of 2 in Theorem 7 by summing (3) across rounds.

Inventive Compatibility Lower Bound for Dynamic MMF. Finally, we provide a lower bound for the single resource setting. We provide one for the simpler case when users' weights are equal and there is no coalition.

We prove in Dynamic MMF a user can increase her utility by a factor of $\sqrt{2}$, which is close to the 3/2 upper bound for the same setting that Theorem 7 proves. We defer the proof of this theorem to the full version [15].

Theorem 8. *For any value of $\alpha \in [0,1]$, there is an instance with n users, where in Dynamic MMF a user can under-report her demand to increase her utility by a factor of $\sqrt{2}$ as $n \to \infty$.*

5 Getting Increased Utility Multiple Times

In this section, we study what happens in Dynamic MMF when user 1 deviates and gets more utility over an extended period, or multiple times. Specifically, we study for how many rounds user 1 can get $\gamma > 1$ times more utility by deviating and for how many rounds she needs to have reduced utility because of deviating before having γ times more again. We first make the following definitions:

- For $\ell = 0, 1, 2, \ldots$ let s_ℓ be distinct and ordered times (i.e., $s_{\ell-1} < s_\ell$) when user 1 *begins* having more resources by misreporting, i.e., $\hat{R}_1^{s_\ell - 1} \le R_1^{s_\ell - 1}$ and $\hat{R}_1^{s_\ell} > R_1^{s_\ell}$.
- For $\ell = 0, 1, 2, \ldots$ let e_ℓ be the first time after round s_ℓ when user 1 *begins* having less resources by misreporting, i.e., $\hat{R}_1^{e_\ell - 1} \ge R_1^{e_\ell - 1}$ and $\hat{R}_1^{e_\ell} < R_1^{e_\ell}$.

Note that $0 < s_0 < e_0 < s_1 < e_1 < \ldots$ by definition. Using the above notation we prove that if during every interval $[s_\ell, e_\ell]$ user 1 got γ more resources in some round $t_\ell \in [s_\ell, e_\ell]$ by misreporting, then t_ℓ cannot be much larger than s_ℓ, implying that user i cannot keep having γ times more utility for a long period of time. We also prove that t_ℓ scales exponentially with ℓ, implying that the ℓ-th time user i gets increased resources must happen exponentially far away. The proof of the theorem is presented in the full version of the paper [15].

Theorem 9. *Assume that for every t, $R_1^t \in \Theta(t)$ and for every $\ell = 0, 1, \ldots$ there exists a round $t_\ell \in [s_\ell, e_\ell)$ for which $\hat{R}_1^{t_\ell} \ge \gamma R_1^{t_\ell}$, for some $\gamma > 1$. Then, in Dynamic MMF for any $\alpha \in [0,1]$, any $\ell = 0, 1, \ldots$, and any $t_\ell \in [s_\ell, e_\ell)$ such that $\hat{R}_1^{t_\ell} \ge \gamma R_1^{t_\ell}$, it holds that $t_\ell = O(s_\ell)$ and $t_\ell = \left(\frac{2-\gamma}{3-2\gamma}\right)^\ell \Omega(t_0)$.*

References

1. Aleksandrov, M., Walsh, T.: Strategy-proofness, envy-freeness and pareto efficiency in online fair division with additive utilities. In: Nayak, A.C., Sharma, A. (eds.) PRICAI 2019. LNCS (LNAI), vol. 11670, pp. 527–541. Springer, Cham (2019). https://doi.org/10.1007/978-3-030-29908-8_42
2. Alizadeh, M., et al.: Data center TCP (DCTCP). In: Proceedings of the ACM SIGCOMM 2010 Conference (2010)
3. Archer, A., Papadimitriou, C.H., Talwar, K., Tardos, É.: An approximate truthful mechanism for combinatorial auctions with single parameter agents. In: Proceedings of the Fourteenth Annual ACM-SIAM Symposium on Discrete Algorithms, Baltimore, Maryland, USA, 12–14 January 2003 (2003)
4. Azevedo, E.M., Budish, E.: Strategy-proofness in the large. Rev. Econ. Stud. **86**(1), 81–116 (2018)

5. Balcan, M., Sandholm, T., Vitercik, E.: Estimating approximate incentive compatibility. In: Proceedings of the 2019 ACM Conference on Economics and Computation, EC 2019, Phoenix, AZ, USA, 24–28 June 2019 (2019)
6. Balseiro, S.R., Besbes, O., Castro, F.: Mechanism design under approximate incentive compatibility. Oper. Res. **72**(1), 355–372 (2024)
7. Bei, X., Huang, Z.: Bayesian incentive compatibility via fractional assignments. In: Randall, D. (ed.) Proceedings of the Twenty-Second Annual ACM-SIAM Symposium on Discrete Algorithms, SODA 2011, San Francisco, California, USA, 23–25 January 2011. SIAM (2011)
8. Boutin, E., et al.: Apollo: scalable and coordinated scheduling for cloud-scale computing. In: Flinn, J., Levy, H. (eds.) 11th USENIX Symposium on Operating Systems Design and Implementation, OSDI 2014, Broomfield, CO, USA, 6–8 October 2014. USENIX Association (2014)
9. Chaubal, C.: The architecture of VMWare ESXi. VMware White Paper (7) (2008)
10. Cheng, Y., Anwar, A., Duan, X.: Analyzing Alibaba's co-located datacenter workloads. In: 2018 IEEE International Conference on Big Data (Big Data). IEEE (2018)
11. Chiu, D.M., Jain, R.: Analysis of the increase and decrease algorithms for congestion avoidance in computer networks. Comput. Netw. ISDN Syst. **17**(1), 1–14 (1989)
12. Dekel, O., Fischer, F.A., Procaccia, A.D.: Incentive compatible regression learning. In: Proceedings of the Nineteenth Annual ACM-SIAM Symposium on Discrete Algorithms, SODA 2008, San Francisco, California, USA, 20–22 January 2008 (2008)
13. Dütting, P., Fischer, F.A., Jirapinyo, P., Lai, J.K., Lubin, B., Parkes, D.C.: Payment rules through discriminant-based classifiers. In: Proceedings of the 13th ACM Conference on Electronic Commerce, EC 2012, Valencia, Spain, 4–8 June 2012 (2012)
14. Dütting, P., Roughgarden, T., Talgam-Cohen, I.: The complexity of contracts. In: Chawla, S. (ed.) Proceedings of the 2020 ACM-SIAM Symposium on Discrete Algorithms, SODA 2020, Salt Lake City, UT, USA, 5–8 January 2020. SIAM (2020)
15. Fikioris, G., Agarwal, R., Tardos, É.: Incentives in dominant resource fair allocation under dynamic demands (2022). https://arxiv.org/abs/2109.12401
16. Freeman, R., Zahedi, S.M., Conitzer, V., Lee, B.C.: Dynamic proportional sharing: a game-theoretic approach. In: Proceedings of the ACM on Measurement and Analysis of Computing Systems (2018)
17. Friedman, E.J., Psomas, C., Vardi, S.: Dynamic fair division with minimal disruptions. In: Proceedings of the Sixteenth ACM Conference on Economics and Computation, EC 2015, Portland, OR, USA, 15–19 June 2015. ACM (2015)
18. Friedman, E.J., Psomas, C., Vardi, S.: Controlled dynamic fair division. In: Proceedings of the 2017 ACM Conference on Economics and Computation, EC 2017, Cambridge, MA, USA, 26–30 June 2017. ACM (2017)
19. Ghodsi, A., Zaharia, M., Hindman, B., Konwinski, A., Shenker, S., Stoica, I.: Dominant resource fairness: fair allocation of multiple resource types. In: Andersen, D.G., Ratnasamy, S. (eds.) Proceedings of the 8th USENIX Symposium on Networked Systems Design and Implementation, NSDI 2011, Boston, MA, USA, 30 March–1 April 2011 (2011)
20. Grandl, R., Ananthanarayanan, G., Kandula, S., Rao, S., Akella, A.: Multi-resource packing for cluster schedulers. In: ACM SIGCOMM 2014 Conference, SIGCOMM 2014, Chicago, IL, USA, 17–22 August 2014 (2014)

21. Grandl, R., Chowdhury, M., Akella, A., Ananthanarayanan, G.: Altruistic scheduling in multi-resource clusters. In: 12th USENIX Symposium on Operating Systems Design and Implementation, OSDI 2016, Savannah, GA, USA, 2–4 November 2016 (2016)
22. Grandl, R., Kandula, S., Rao, S., Akella, A., Kulkarni, J.: GRAPHENE: packing and dependency-aware scheduling for data-parallel clusters. In: 12th USENIX Symposium on Operating Systems Design and Implementation, OSDI 2016, Savannah, GA, USA, 2–4 November 2016 (2016)
23. Harada, M.: The ad exchanges place in a first-price world (2018)
24. Hartline, J.D., Taggart, S.: Sample complexity for non-truthful mechanisms. In: Karlin, A., Immorlica, N., Johari, R. (eds.) Proceedings of the 2019 ACM Conference on Economics and Computation, EC 2019, Phoenix, AZ, USA, 24–28 June 2019. ACM (2019)
25. Hindman, B., et al.: Mesos: a platform for fine-grained resource sharing in the data center. In: Proceedings of the 8th USENIX Symposium on Networked Systems Design and Implementation, NSDI 2011, Boston, MA, USA, 30 March–1 April 2011 (2011)
26. Hossain, R.: Sharing is caring: dynamic mechanism for shared resource ownership. In: Elkind, E., Veloso, M., Agmon, N., Taylor, M.E. (eds.) Proceedings of the 18th International Conference on Autonomous Agents and MultiAgent Systems, AAMAS 2019, Montreal, QC, Canada, 13–17 May 2019. International Foundation for Autonomous Agents and Multiagent Systems (2019)
27. Im, S., Moseley, B., Munagala, K., Pruhs, K.: Dynamic weighted fairness with minimal disruptions. In: Abstracts of the 2020 SIGMETRICS/Performance Joint International Conference on Measurement and Modeling of Computer Systems, Boston, MA, USA, 8–12 June 2020. ACM (2020)
28. Kandasamy, K., Sela, G., Gonzalez, J.E., Jordan, M.I., Stoica, I.: Online learning demands in max-min fairness. CoRR (2020)
29. Kanoria, Y., Nazerzadeh, H.: Incentive-compatible learning of reserve prices for repeated auctions. In: Companion of The 2019 World Wide Web Conference, WWW 2019, San Francisco, CA, USA, 13–17 May 2019. ACM (2019)
30. Kash, I.A., Procaccia, A.D., Shah, N.: No agent left behind: dynamic fair division of multiple resources. In: Gini, M.L., Shehory, O., Ito, T., Jonker, C.M. (eds.) International Conference on Autonomous Agents and Multi-Agent Systems, AAMAS 2013, Saint Paul, MN, USA, 6–10 May 2013. IFAAMAS (2013)
31. Kothari, A., Parkes, D.C., Suri, S.: Approximately-strategyproof and tractable multi-unit auctions. In: Proceedings 4th ACM Conference on Electronic Commerce (EC-2003), San Diego, California, USA, 9–12 June 2003 (2003)
32. Krishna, V.: Auction Theory. Academic Press (2009)
33. KVM: Main page—KVM (2016)
34. Li, B., Li, W., Li, Y.: Dynamic fair division problem with general valuations. In: Lang, J. (ed.) Proceedings of the Twenty-Seventh International Joint Conference on Artificial Intelligence, IJCAI 2018, Stockholm, Sweden, 13–19 July 2018 (2018)
35. Lubin, B., Parkes, D.C.: Approximate strategyproofness. Curr. Sci. **103**(9), 1021–1032 (2012)
36. Mennle, T., Seuken, S.: An axiomatic approach to characterizing and relaxing strategyproofness of one-sided matching mechanisms. In: ACM Conference on Economics and Computation, EC 2014, Stanford , CA, USA, 8–12 June 2014 (2014)
37. Milgrom, P.: Critical issues in the practice of market design. Econ. Inq. **49**(2), 311–320 (2011)

38. Parkes, D.C., Procaccia, A.D., Shah, N.: Beyond dominant resource fairness: extensions, limitations, and indivisibilities. In: Proceedings of the 13th ACM Conference on Electronic Commerce, EC 2012, Valencia, Spain, 4–8 June 2012. ACM (2012)
39. Parkin, R.: A year in first-price (2018)
40. Reiss, C., Tumanov, A., Ganger, G.R., Katz, R.H., Kozuch, M.A.: Heterogeneity and dynamicity of clouds at scale: Google trace analysis. In: Proceedings of the Third ACM Symposium on Cloud Computing (2012)
41. Sadok, H., Campista, M.E.M., Costa, L.H.M.K.: Stateful DRF: considering the past in a multi-resource allocation. IEEE Trans. Comput. **70**(7), 1094–1105 (2021)
42. Shue, D., Freedman, M.J., Shaikh, A.: Performance isolation and fairness for multi-tenant cloud storage. In: Thekkath, C., Vahdat, A. (eds.) 10th USENIX Symposium on Operating Systems Design and Implementation, OSDI 2012, Hollywood, CA, USA, 8–10 October 2012 (2012)
43. Vavilapalli, V.K., et al.: Apache hadoop yarn: yet another resource negotiator. In: Proceedings of the 4th Annual Symposium on Cloud Computing (2013)
44. Verbitski, A., et al.: Amazon aurora: design considerations for high throughput cloud-native relational databases. In: Proceedings of the 2017 ACM International Conference on Management of Data, SIGMOD Conference 2017, Chicago, IL, USA, 14–19 May 2017. ACM (2017)
45. Verma, A., Pedrosa, L., Korupolu, M.R., Oppenheimer, D., Tune, E., Wilkes, J.: Large-scale cluster management at Google with Borg. In: Proceedings of the European Conference on Computer Systems (EuroSys), Bordeaux, France (2015)
46. Vuppalapati, M., Fikioris, G., Agarwal, R., Cidon, A., Khandelwal, A., Tardos, É.: Karma: resource allocation for dynamic demands. In: 17th USENIX Symposium on Operating Systems Design and Implementation (OSDI 2023) (2023)
47. Vuppalapati, M., Miron, J., Agarwal, R., Truong, D., Motivala, A., Cruanes, T.: Building an elastic query engine on disaggregated storage. In: 17th USENIX Symposium on Networked Systems Design and Implementation, NSDI 2020, Santa Clara, CA, USA, 25–27 February 2020. USENIX Association (2020)
48. Vuppalapati, M., Miron, J., Agarwal, R., Truong, D., Motivala, A., Cruanes, T.: Building an elastic query engine on disaggregated storage. In: Bhagwan, R., Porter, G. (eds.) 17th USENIX Symposium on Networked Systems Design and Implementation, NSDI 2020, Santa Clara, CA, USA, 25–27 February 2020 (2020)
49. Yang, J., Yue, Y., Rashmi, K.V.: A large scale analysis of hundreds of in-memory cache clusters at Twitter. In: 14th USENIX Symposium on Operating Systems Design and Implementation, OSDI 2020, Virtual Event, 4–6 November 2020. USENIX Association (2020)
50. Zeng, D., Psomas, A.: Fairness-efficiency tradeoffs in dynamic fair division. In: EC 2020: The 21st ACM Conference on Economics and Computation, Virtual Event, Hungary, 13–17 July 2020. ACM (2020)

Mechanism Design

Agent-Constrained Truthful Facility Location Games

Argyrios Deligkas[1], Mohammad Lotfi[2], and Alexandros A. Voudouris[3](✉)

[1] Royal Holloway University of London, Egham, UK
argyrios.deligkas@rhul.ac.uk
[2] Sharif University of Technology, Tehran, Iran
mohammad.lot@sharif.edu
[3] University of Essex, Colchester, UK
alexandros.voudouris@essex.ac.uk

Abstract. We consider a truthful facility location problem in which there is a set of agents with private locations on the line of real numbers, and the goal is to place a number of facilities at different locations chosen from the set of those reported by the agents. Given a feasible solution, each agent suffers an individual cost that is either its total distance to all facilities (sum-variant) or its distance to the farthest facility (max-variant). For both variants, we show tight bounds on the approximation ratio of strategyproof mechanisms in terms of the social cost, the total individual cost of the agents.

Keywords: Mechanism design · Facility location · Approximation ratio

1 Introduction

Suppose you are the mayor of a small town and your task is to decide where to build a park and a library on a very busy street to accommodate the needs of the citizens. One way to make this decision is to simply place the facilities arbitrarily. Even though this is easy to implement, the chosen locations might not be very accessible and the citizens most probably will end up complaining and not vote for you in the next election. Instead, you could ask the citizens to suggest the possible locations where the facilities could be built and choose one that collectively satisfies them. While this now seems sufficient enough to get you re-elected, you also need to make sure that the citizens are incentivized to truthfully suggest their real ideal locations and not lie in order to minimize the distance they have to walk. This is known as the *truthful facility location problem.*

Since the seminal work of Procaccia and Tennenholtz [18] on *approximate mechanism design without money*, many different variants of the problem have been studied under assumptions about the number of facilities to be placed, the preferences of the agents for the facilities, and the feasible locations where the

G. Schäfer and C. Ventre (Eds.): SAGT 2024, LNCS 15156, pp. 129–146, 2024.
https://doi.org/10.1007/978-3-031-71033-9_8

facilities can be built; we refer the reader to the survey of Chan *et al.* [2] for an overview, and to our discussion of related work below. In this work we consider a previously unexplored, yet fundamental model where the facilities can be built at locations that are dynamically proposed by the agents, in contrast to previously studied models where the facilities could be placed either at any location on the line or only at a predetermined set of fixed candidate locations.

1.1 Our Model

We consider the following agent-constrained truthful facility location problem. An instance I consists of a set of $n \geq 2$ *agents* with *private locations* on the line of real numbers, and $k \geq 2$ *facilities* that can be placed at different locations chosen from the (multi-)set of locations reported by the agents. Given a feasible *solution* $\mathbf{x}$ which determines the agent locations where the k facilities are placed, each agent i suffers an *individual cost*. We consider two different models that differ on the cost function of the agents. In the *sum-variant*, the cost of i in instance I is its *total* distance from the facilities:

$$\text{cost}_i^{\text{sum}}(\mathbf{x}|I) = \sum_{x \in \mathbf{x}} d(i, x),$$

where $d(i, x) = |i - x|$ is the distance between the location of agent i and point x on the line. In the *max-variant*, the cost of i in instance I is its distance to the *farthest* facility:

$$\text{cost}_i^{\text{max}}(\mathbf{x}|I) = \max_{x \in \mathbf{x}} \{d(i, x)\}.$$

Whenever the variant we study is clear from context, we will drop the sum and max from notation, and simply write $\text{cost}_i(\mathbf{x})$ for the individual cost of i when solution $\mathbf{x}$ is chosen; similarly, we will drop I from notation when the instance is clear from context. We are interested in choosing solutions that have a small effect in the overall cost of the agents, which is captured by the *social cost* objective function, defined as:

$$\text{SC}(\mathbf{x}|I) = \sum_i \text{cost}_i(\mathbf{x}|I).$$

A solution can also be *randomized* in the sense that it is a probability distribution $\mathbf{p} = (p_\mathbf{x})_\mathbf{x}$ over all feasible solutions; the *expected social cost* of such a randomized solution is defined appropriately as

$$\mathbb{E}[\text{SC}(\mathbf{p}|I)] = \sum_{\mathbf{x}} p_\mathbf{x} \cdot \text{SC}(\mathbf{x}|I).$$

The solution is decided by a *mechanism* based on the locations reported by the agents; let $M(I)$ be the solution computed by a mechanism M when given as input an instance I. A mechanism M is said to be *strategyproof* if no agent i can misreport its true location and decrease its individual cost; that is,

$$\text{cost}_i(M(I)|I) \leq \text{cost}_i(M(J)|I)$$

for every pair of instances I and J that differ only on the location reported by agent i. In case the mechanism is randomized, then it is said to be *strategyproof-in-expectation* if no agent i cannot misreport its true location and decrease its *expected* individual cost.

The *approximation ratio* of a mechanism is the worst-case ratio (over all possible instances) of the (expected) social cost of the chosen solution over the minimum possible social cost:

$$\sup_I \frac{\mathbb{E}[\text{SC}(M(I)|I)]}{\min_{\mathbf{x}} \text{SC}(\mathbf{x}|I)}.$$

Our goal is to design mechanisms that are strategyproof and achieve an as small approximation ratio as possible.

1.2 Our Contribution

For both individual cost variants, we show tight bounds on the best possible approximation ratio that can be achieved by strategyproof mechanisms. We start with the case of $k = 2$ facilities for which we study both deterministic and randomized mechanisms. For the sum-variant, in Sect. 2, we show a tight bound of $3/2$ for deterministic mechanisms and a bound of $10 - 4\sqrt{5} \approx 1.0557$ for randomized ones. For the max-variant, in Sect. 3, we show bounds of 3 and 2 on the approximation ratio of deterministic and randomized mechanisms, respectively. In Sect. 4, we switch to the general case of k facilities and focus exclusively on deterministic mechanisms. For the sum-variant, we show that the approximation ratio is between $2-1/k$ and 2, while for the max-variant, we show a tight bound of $k+1$. Due to space constraints, the proofs of some statements are ommitted.

Our upper bounds follow by appropriately defined *statistic-type* mechanisms that choose the agent locations where the facilities will be placed according to the ordering of the agents on the line from left to right. In particular, for $k = 2$, our mechanisms locate one facility at the median agent m and the other either at the agent ℓ that is directly to the left of m or the agent r that is directly to the right of m. To be even more specific, our deterministic mechanism always chooses the solution (m, r), while our randomized mechanisms choose the solutions (ℓ, m) and (m, r) according to some probability distribution. Interestingly, for the sum-variant, it turns out that the probabilities are functions of the distances $d(\ell, m)$ and $d(m, r)$; to the best of our knowledge, this is one of few settings in which the best possible randomized strategyproof mechanism is not required to assign fixed, constant probabilities. For the general case of k facilities, our (deterministic) upper bounds for both variants follow by a mechanism that is a natural generalization of the one for $k = 2$; in particular, the mechanism places the facilities around the median agent(s) within a radius of about $k/2$.

1.3 Related Work

Truthful facility location problems have a long history within the literature of *approximate mechanism design without money*, starting with the paper of

Procaccia and Tennenholtz [18]. Various different models have been studied depending on parameters such as the number of facilities whose location needs to be determined [11,17,18], whether the facilities are obnoxious [5], whether the agents have different types of preferences over the facilities (for example, optional [4,13,15,19], fractional [10], or hybrid [7]), and whether there are other limitations or features (for example, the facilities might only be possible to be built at specific fixed locations [8,12,14,22], there might be limited resources that can be used to build some of the available facilities rather than all [6], there might be limited available information during the decision process [3,9], or there might be even more information in the form of predictions about the optimal facility locations which can be leveraged [1,21]). We refer the reader to the survey of Chan *et al.* [2] for more details on the different dimensions along which facility location problems have been studied over the years.

When there are multiple facilities to locate, the typical assumption about the individual behavior of the agents is that they aim to minimize their distance to the closest facility [11,17,18,20,21]; such a cost model essentially assumes that the facilities are homogeneous (in the sense that they offer the same service) and thus each agent is satisfied if it is close enough to one of them. In contrast, both variants (sum and max) we consider here model different cases in which the facilities are heterogeneous (in the sense that they offer different services) and each agent aims to minimize either the total or the maximum distance to the facilities. These variants have also been considered in previous work under different assumptions than us; in particular, the sum-variant has been studied in [12,14,19,22], while the max-variant has been studied in [4,16,23].

The main differences between our work and the aforementioned ones are the following: In most of these papers, the agents have optional preferences over the facilities; that is, some agents approve one facility and are indifferent to the other, while some agents approve both facilities. Here, we focus exclusively on the fundamental case where all agents approve both facilities. In addition, some of these papers study a constrained model according to which the facilities can only be built at different locations chosen from a set of fixed, predetermined candidate ones. In our model, the facilities can also only be built at different locations, which, however, are chosen from the set of locations that are reported by the agents; this is a more dynamic setting in the sense that the candidate locations can change if agents misreport. We remark that, in continuous facility location settings (where the facilities can be placed anywhere on the line) such as those studied in the original paper of Procaccia and Tennenholtz [18] and follow-up work, the class of strategyproof mechanisms mainly consists of mechanisms that place the facilities at agent locations (according to an ordering). However, to the best of our knowledge, there has not been any previous work that has studied the model where the candidate locations are restricted to the ones reported by the agents, an assumption that also affects the optimal solution in terms of social cost.

2 Sum-Variant for Two Facilities

We start the presentation of our technical results with the case of $k = 2$ facilities and the sum-variant. Recall that in this variant the individual cost of any agent is its distance from both facilities. We will first argue about the structure of the optimal solution; this will be extremely helpful in bounding the approximation ratio of our strategyproof mechanisms later on. We start with the case where the number of agents n is an even number, for which the optimal solution is well-defined and actually leads to an optimal strategyproof mechanism.

Lemma 1. *For any even $n \geq 2$, an optimal solution is to place the facilities at the two median agents.*

Proof. Let m_1 and m_2 be two median agents. Suppose that there is an optimal solution (o_1, o_2) with $o_1 \leq o_2$. Since any point $x \in [m_1, m_2]$ minimizes the total distance of all agents from any other point of the line, we have

$$\begin{aligned}\mathrm{SC}(m_1, m_2) &= \sum_i d(i, m_1) + \sum_i d(i, m_2) \\ &\leq \sum_i d(i, o_1) + \sum_i d(i, o_2) = \mathrm{SC}(o_1, o_2),\end{aligned}$$

and thus (m_1, m_2) is also an optimal solution. □

Before we continue, we remark that the TWO-MEDIANS mechanism, which is implied by Lemma 1, is indeed strategyproof: To change the solution of the mechanism, an agent i would have to report a location $x > m_1$ in case $i \leq m_1$ or a location $x < m_2$ in case $i \geq m_2$; such a misreport leads to an individual cost of at least $\min\{d(i,x), d(i,m_2)\} + d(i,m_2)$ in the first case and of at least $d(i,m_1) + \min\{d(i,x), d(i,m_1)\}$ in the second case, which is at least the true individual cost $d(i,m_1) + d(i,m_2)$ of i. Hence, agent i has no incentive to deviate and the mechanism is strategyproof.

For the case where the number of agents $n \geq 3$ is an odd number, it will be useful to calculate the social cost of the solutions (ℓ, m) and (m, r), where ℓ and r are the agents directly to the left and right of the median agent m, respectively.

Lemma 2. *For any $x \in \{\ell, r\}$, the social cost of the solution (x, m) is*

$$\mathrm{SC}(x, m) = 2 \cdot \sum_i d(i, m) + d(m, x).$$

Using this, we can argue about the structure of the optimal solution.

Lemma 3. *For any odd $n \geq 3$, an optimal solution is to place the facilities at the median agent and the agent that is closest to it.*

Proof. Clearly, one of ℓ or r is the closest agent to m, say ℓ; hence, $d(\ell, m) \leq d(m, r)$. To simplify our notation, for any x let $f(x) = \sum_i d(i, x)$ denote the total distance of all agents from x. It is well-known that f is monotone such that $f(i) \geq f(\ell) \geq f(m)$ for every $i \leq \ell \leq m$, and $f(i) \geq f(r) \geq f(m)$ for every $i \geq r \geq m$. Consequently, the optimal solution is either (ℓ, m) or (m, r). By Lemma 2 with $x = \ell$ and $x = r$, we get

$$\mathrm{SC}(\ell, m) - \mathrm{SC}(m, r) = d(\ell, m) - d(m, r).$$

Since $d(\ell, m) \leq d(m, r)$, we conclude that $\mathrm{SC}(\ell, m) \leq \mathrm{SC}(m, r)$ and the solution (ℓ, m) is indeed the optimal one. □

It is not hard to observe that when n is odd, computing the optimal solution is not strategyproof; the second-closest agent to the median might have incentive to misreport a location slightly closer to the median to move the second facility there. However, we do know that one of the solutions (ℓ, m) and (m, r) must be optimal. Based on this, we consider the following Median-Right mechanism: Place one facility at the position the median agent m and the other at the position of the agent r directly to the right of m.[1] One can verify that this mechanism is strategyproof using an argument similar to the one we presented above for the Two-Medians mechanism in the case of even n. So, we continue by bounding its approximation ratio.

Theorem 1. *For any odd $n \geq 3$, the approximation ratio of the* Median-Right *mechanism is at most* $3/2$.

Proof. The solution of the mechanism is $\mathbf{w} = (m, r)$. If r is the closest agent to m, then $\mathbf{w}$ is optimal by Lemma 3. So, assume that this is not the case and the optimal solution is $\mathbf{o} = (\ell, m)$. By Lemma 2 with $x = r$, we get

$$\mathrm{SC}(\mathbf{w}) = 2 \cdot \sum_i d(i, m) + d(m, r).$$

Similarly, for $x = \ell$, we get

$$\begin{aligned} \mathrm{SC}(\mathbf{o}) &= 2 \cdot \sum_i d(i, m) + d(\ell, m) \\ &\geq 2 \cdot \sum_i d(i, m) \\ &\geq 2 \cdot |\{i \geq r\}| \cdot d(m, r) = (n-1) \cdot d(m, r). \end{aligned}$$

Using these two lower bounds on the optimal social cost, we can now upper-bound the social cost of $\mathbf{w}$ as follows:

$$\mathrm{SC}(\mathbf{w}) \leq \left(1 + \frac{1}{n-1}\right) \cdot \mathrm{SC}(\mathbf{o}) = \frac{n}{n-1} \cdot \mathrm{SC}(\mathbf{o}).$$

Therefore, the approximation ratio is at most $n/(n-1) \leq 3/2$ for any $n \geq 3$. □

[1] Clearly, since we are dealing with the case of odd n, instead of this mechanism, one could also consider the Median-Left mechanism which places the second facility to the agent ℓ that is directly to the left of m; both mechanisms are symmetric and achieve the same approximation ratio.

The approximation ratio of 3/2 is in fact the best possible that can be achieved by any deterministic strategyproof mechanism; this follows directly by Theorem 10 for $k = 2$.

Theorem 2. *The approximation ratio of any deterministic strategyproof mechanism is at least* 3/2.

Since the optimal solution is either (ℓ, m) or (m, r), it is reasonable to think that randomizing over these two solutions, rather than blindly choosing one of them, can lead to an improved approximation ratio. Indeed, we can show a significantly smaller tight bound of $10 - 4\sqrt{5} \approx 1.0557$ for randomized strategyproof mechanisms when $n \geq 3$ is an odd number; recall that, for even $n \geq 2$, we can always compute the optimal solution. For the upper bound, we consider the following Reverse-Proportional randomized mechanism: With probability $p_\ell = \frac{d(m,r)}{d(\ell,r)}$ choose the solution (ℓ, m), and with probability $p_r = \frac{d(\ell,m)}{d(\ell,r)}$ choose the solution (m, r).

Theorem 3. *For any odd $n \geq 3$, the* Reverse-Proportional *mechanism is strategyproof-in-expectation and achieves an approximation ratio of at most* $10 - 4\sqrt{5} \approx 1.0557$.

Proof. We here focus on bounding the approximation ratio of the mechanism. Without loss of generality, suppose that $d(\ell, m) \leq d(m, r)$ and thus the optimal solution is $\mathbf{o} = (\ell, m)$. By the definition of the mechanism, the solutions $d(\ell, m)$ and $d(m, r)$ are chosen with probability $p_\ell = d(m, r)/d(\ell, r)$ and $p_r = d(\ell, m)/d(\ell, r)$, respectively; observe that $p_\ell \geq p_r$. By Lemma 2 with $x = \ell$ and using the fact that that $d(\ell, m) + d(m, r) = d(\ell, r)$, we can lower-bound the optimal social cost as follows:

$$\text{SC}(\mathbf{o}) = 2 \cdot \sum_i d(i, m) + d(\ell, m) \geq 2 \cdot d(\ell, r) + d(\ell, m).$$

Again using Lemma 2 with $x = \ell$ and $x = r$, as well as the fact that $p_\ell = 1 - p_r$, we can write the expected social cost of the randomized solution $\mathbf{w}$ chosen by the mechanism as

$$\begin{aligned}
\mathbb{E}[\text{SC}(\mathbf{w})] &= p_\ell \cdot \left(2 \cdot \sum_i d(i, m) + d(\ell, m)\right) + p_r \cdot \left(2 \cdot \sum_i d(i, m) + d(m, r)\right) \\
&= 2 \cdot \sum_i d(i, m) + (1 - p_r) \cdot d(\ell, m) + p_r \cdot d(m, r) \\
&= 2 \cdot \sum_i d(i, m) + d(\ell, m) + p_r \cdot \left(d(m, r) - d(\ell, m)\right) \\
&= \text{SC}(\mathbf{o}) + p_r \cdot \left(d(m, r) - d(\ell, m)\right).
\end{aligned}$$

Consequently, the approximation ratio is

$$\frac{\mathbb{E}[\mathrm{SC}(\mathbf{w})]}{\mathrm{SC}(\mathbf{o})} \leq 1 + p_r \cdot \frac{d(m,r) - d(\ell,m)}{2 \cdot d(\ell,r) + d(\ell,m)}$$

$$= 1 + p_r \cdot \frac{\frac{d(m,r)}{d(\ell,r)} - \frac{d(\ell,m)}{d(\ell,r)}}{2 + \frac{d(\ell,m)}{d(\ell,r)}}$$

$$= 1 + p_r \cdot \frac{p_\ell - p_r}{2 + p_r}$$

Using the fact that $p_\ell = 1 - p_r$, we finally have that

$$\frac{\mathbb{E}[\mathrm{SC}(\mathbf{w})]}{\mathrm{SC}(\mathbf{o})} \leq 1 + p_r \cdot \frac{1 - 2 \cdot p_r}{2 + p_r}.$$

The last expression attains its maximum value of $10 - 4\sqrt{5} \approx 1.0557$ for $p_r = \sqrt{5} - 2$. □

Next, we will argue that the Reverse-Proportional mechanism is the best possible by showing a matching lower bound on the approximation ratio of any randomized strategyproof-in-expectation mechanism. To do this, we will use instances with three agents for which we first show the following technical lemma that reduces the class of mechanisms to consider.

Lemma 4. *Consider any instance with three agents located at $x < y < z$. For any randomized mechanism M that assigns positive probability to the solution (x, z), there exists a randomized mechanism M_0 that assigns 0 probability to that solution and achieves at most as much expected social cost as M.*

Using the above lemma, we can now show the desired lower bound.

Theorem 4. *For the sum-variant, the approximation ratio of any randomized strategyproof-in-expectation mechanism is at least* $10 - 4\sqrt{5} \approx 1.0557$.

Proof. Consider any randomized strategyproof mechanism and an instance I with three agents located at 0, 1 and 2. Let $p_0(I)$ and $p_1(I)$ be the probabilities assigned to solutions $(0, 1)$ and $(1, 2)$, respectively. By Lemma 4, we can assume that $p_0(I) + p_1(I) = 1$, and thus suppose that $p_0(I) \geq 1/2$ without loss of generality. The expected individual cost of the agent i that is located at 2 is then

$$3 \cdot p_0(I) + 1 \cdot p_1(I) \cdot 1 = 3 \cdot p_0(I) + 1 - p_0(I) = 2 \cdot p_0(I) + 1 \geq 2.$$

Now consider an instance J with three agents located at 0, 1 and $x = 1/q \in (1, 2)$, where $q = 3 - \sqrt{5} \approx 0.764$; hence, the only different between I and J is that agent i is now located at x rather than 2. Let $p_0(J)$ and $p_x(J)$ be the probabilities assigned to solutions $(0, 1)$ and $(1, x)$, respectively. Again, using Lemma 4 we can assume that $p_0(J) + p_x(J) = 1$; any other case would achieve worse approximation ratio. Suppose that $p_x(J) > q$. Then, the expected cost of agent i when misreporting its position as $1/q$ rather than 2 would be

$$3 \cdot p_0(J) + \left(1 + 2 - \frac{1}{q}\right) \cdot p_x(J) = 3 \cdot \left(1 - p_x(J)\right) + \left(3 - \frac{1}{q}\right) \cdot p_x(J)$$
$$= 3 - \frac{1}{q} \cdot p_x(J) < 2$$

and agent i would manipulate the mechanism. Therefore, for the mechanism to be strategyproof, it has to be the case that $p_x(J) \leq q$, and thus $p_0(J) \geq 1 - q$.

In instance J, the optimal solution is $(1, x)$ with social cost $1+1/q+2(1/q-1) = 3/q-1$. Since the social cost of the solution $(0, 1)$ is $2+1/q+1/q-1 = 2/q+1$, the approximation ratio is

$$\frac{p_0(J) \cdot \mathrm{SC}(0,1) + p_x(J) \cdot \mathrm{SC}(1,x)}{\mathrm{SC}(1,x)} = p_x(J) + p_0(J) \cdot \frac{2/q+1}{3/q-1}$$
$$= 1 - p_0(J) + p_0(J) \cdot \frac{2+q}{3-q}$$
$$= 1 + p_0(J) \cdot \frac{2q-1}{3-q}$$
$$\geq 1 + (1-q) \cdot \frac{1-2(1-q)}{2+(1-q)} = 10 - 4\sqrt{5}.$$

Hence, the approximation ratio is at least $10 - 4\sqrt{5} \approx 1.0557$. □

3 Max-Variant for Two Facilities

We now turn our attention to the max-variant in which the individual cost of any agent is its distance from the farthest facility. One might be tempted to assume that the optimal solution has the same structure as in the sum-variant, which trivially holds for the case of $n = 2$ agents. However, this is not true as the following example demonstrates: Consider an instance with $n = 4$ agents with locations $-1/2$, 0, 1, and 2. The optimal solution is $(-1/2, 0)$ with a social cost of 5; note that the two-medians solution $(0, 1)$, which is optimal for the sum-variant according to Lemma 1, has social cost $11/2$.

In spite of this, we do not require the exact structure of the optimal solution to identify the best possible strategyproof mechanisms. For the class of deterministic mechanisms, we once again consider the Median-Right mechanism; recall that this mechanism places one facility at the (leftmost) median agent m and the other at agent r that is directly to the right of m. This mechanism is strategyproof for the max-variant as well: The true individual cost of any agent $i \geq r$ is $d(i, m)$, and any misreport $x \geq m$ of does not change it, while any misreport $x < m$ can only lead to a larger cost; the case of $i < m$ is similar. We next show that this mechanism always achieves an approximation ratio of at most 3, and it can achieve an improved approximation ratio of at most 2 when the number of agents is even. The bound can be derived by setting $k = 2$ to the more general bound of $k + 1$ that we show for the case of multiple facilities in Theorem 11.

Theorem 5. *The approximation ratio of the* Median-Right *mechanism is at most* 2 *for any even* $n \geq 4$ *and at most* 3 *for any odd* $n \geq 3$.

We next show that the Median-Right mechanism is the best possible by showing a matching lower bound of 3 on the worst-case (over all possible instances) approximation ratio of any deterministic strategyproof mechanism; this bound follows directly by Theorem 12 for $k = 2$.

Theorem 6. *The approximation ratio of any deterministic strategyproof mechanism is at least* 3.

While no deterministic strategyproof mechanism can achieve an approximation ratio better than 3 in general, as we have already seen in Theorem 5, the Median-Right mechanism actually has an approximation ratio of at most 2 when n is an even number. We next show that when the number of agents $n \geq 3$ is odd (which is the worst class of instances for deterministic mechanisms), it is possible to design a randomized strategyproof mechanism with improved approximation ratio of at most 2. In particular, we consider the following Uniform mechanism: With probability 1/2 choose the solution (ℓ, m), and with probability 1/2 choose the solution (m, r). This mechanism is clearly strategyproof-in-expectation as it is defines a constant probability distribution over two deterministic strategyproof mechanisms (the Median-Left and the Median-Right).

Theorem 7. *For any odd* $n \geq 3$*, the approximation ratio of the* Uniform *mechanism is at most* 2.

Proof. Since there is an odd number $n \geq 3$ of agents, by the definition of m, we have that $|\{i \geq m\}| = |\{i \leq m\}| = (n+1)/2$. Hence, we can write the expected social cost of the randomized solution $\mathbf{w}$ chosen by the mechanism as follows:

$$\begin{aligned}
\mathbb{E}[\text{SC}(\mathbf{w})] &= \frac{1}{2}\Big(\sum_{i \leq \ell} d(i,m) + \sum_{i \geq m} d(i,\ell)\Big) + \frac{1}{2}\Big(\sum_{i \leq m} d(i,r) + \sum_{i \geq r} d(i,m)\Big) \\
&= \sum_i d(i,m) + \frac{1}{2}|\{i \geq m\}| \cdot d(\ell,m) + \frac{1}{2}|\{i \leq m\}| \cdot d(m,r) \\
&= \sum_i d(i,m) + \frac{1}{2} \cdot \frac{n+1}{2} \cdot d(\ell,r).
\end{aligned}$$

For the optimal solution $\mathbf{o}$, since the position of the median agent is the point that minimizes the total distance from all agents, we have that

$$\text{SC}(\mathbf{o}) \geq \sum_i d(i,m).$$

Since there are two facilities to be placed, in $\mathbf{o}$ one facility must be placed at the position of some agent $o \le \ell$ or $o \ge r$. In the former case, we have that

$$\forall i \ge r : \text{cost}_i(\mathbf{o}) \ge d(i,o) = d(i,r) + d(r,m) + d(m,\ell) + d(\ell,o) \ge d(\ell,r).$$

In the latter case, we have that

$$\forall i \le \ell : \text{cost}_i(\mathbf{o}) \ge d(i,o) = d(i,\ell) + d(\ell,m) + d(m,r) + d(r,o) \ge d(\ell,r).$$

Since $|\{i \ge r\}| = |\{i \le \ell\}| = (n-1)/2$ by the definition of ℓ and r, we have established that

$$\text{SC}(\mathbf{o}) \ge \frac{n-1}{2} \cdot d(\ell,r).$$

Using these two lower bounds on the optimal social cost, we can upper-bound the social cost of $\mathbf{w}$ as follows:

$$\mathbb{E}[\text{SC}(\mathbf{w})] \le \left(1 + \frac{1}{2} \cdot \frac{n+1}{2} \cdot \frac{2}{n-1}\right) \cdot \text{SC}(\mathbf{o}) = \frac{3n-1}{2n-2} \cdot \text{SC}(\mathbf{o}).$$

Hence, the approximation ratio is at most $(3n-1)/(2n-2) \le 2$ for $n \ge 3$. □

Finally, we show 2 is the best possible approximation ratio for any randomized strategyproof-in-expectation mechanism.

Theorem 8. *The approximation ratio of any randomized strategyproof-in-expectation mechanism is at least* 2.

Proof. We again consider the same instance I with three agents that are located at 0, 1, and 2. Since there are three possible locations for two facilities, there is probability $p \ge 1/2$ that one of the facilities will be placed at 0 or 2, say 0. Then, the expected cost of the agent at position 2 is equal to $2p + 1 - p = p + 1$.

Now consider the instance J in which this agent moves to 1. If there is probability $q < p$ that a facility is placed at 0 in J, then the agent would have decreased her expected cost from $p+1$ to $q+1$, which contradicts that the mechanism is strategyproof-in-expectation. Hence, one facility must be placed at 0 with probability at least $p \ge 1/2$ in J, which means that the expected social cost is

$$p \cdot \text{SC}(0,1) + (1-p) \cdot \text{SC}(1,1) = 3p + 1 - p = 2p + 1 \ge 2.$$

However, the optimal social cost is $\text{SC}(1,1) = 1$, leading to an approximation ratio of at least 2. □

4 Deterministic Mechanisms for Multiple Facilities

Having completely resolved the case of $k = 2$ facilities in the previous sections, we now consider the general case of k facilities for which we present (asymptotically) tight bounds on the approximation ratio of *deterministic* strategyproof mechanisms.

4.1 Sum-Variant

We again start with the sum-variant and first argue about the structure of the optimal solution when there are k facilities to be placed.

Lemma 5. *For the sum-variant, an optimal solution is to place the facilities at a set of consecutive agents that includes the median agent(s).*

We now show our upper bound by considering a generalization of the MEDIAN-RIGHT mechanism that we used for $k = 2$. If $k \geq 2$ is even, our mechanism places the facilities at the (leftmost) median agent m, at the $k/2 - 1$ agents at the left of m, and at the $k/2$ agents at the right of m (which might include the second median agent in case of an even overall number of agents). If $k \geq 3$ is odd, the mechanism places the facilities at the (leftmost) median agent m, at the $(k-1)/2$ agents at the left of m, and at the $(k-1)/2$ agents at the right of m. We will refer to this mechanism as MEDIAN-BALL (given that it places the facilities around the median agent within a radius of about $k/2$ in each direction).

Since the mechanism bases its decision only on the ordering of the agents on the line, it is clearly strategyproof for the same reason that MEDIAN-RIGHT is strategyproof when $k = 2$, so in the following we focus on bounding its approximation ratio.

Theorem 9. *For the sum-variant, the approximation ratio of the* MEDIAN-BALL *mechanism is at most* 2.

Proof. We present the proof for an odd number $k \geq 3$ of facilities; the proof is similar for even k. Let $\mathbf{w} = (x_{(k-1)/2}, \ldots, x_1, m, y_1, \ldots, y_{(k-1)/2})$ be the solution computed by the mechanism. To compute the social cost of $\mathbf{w}$, we first consider the agents that are not part of the solution. Let $S_<$ and $S_>$ be the sets of agents that are to the left of agent $x_{(k-1)/2}$ and to the right of agent $y_{(k-1)/2}$, respectively. Also, let X be the indicator variable that is 1 if n is even and 0 otherwise. By definition, we have that $|S_<| = |S_>| - 1$ if $X = 1$, and $|S_<| = |S_>|$ otherwise. In any case, since $|S_<| \leq |S_>|$, we can match every agent $i \in S_<$ to an agent $\mu(i) \in S_>$ and observe that, for any $w \in \mathbf{w}$,

$$d(i, w) + d(\mu(i), w) = d(i, m) + d(\mu(i), m).$$

Clearly, if the number of agents is even, there will be an agent $R \in S_>$ that is left unmatched;[2] for this agent R, if it exists, we use the fact that $d(x_\ell, R) = d(x_\ell, y_\ell) + d(y_\ell, R)$. Given this, we have

[2] Note that, if k is even, there might be an agent in $S_<$ that is left unmatched instead of an agent in $S_>$.

$$\begin{aligned}
\sum_{i \notin \mathbf{w}} \text{cost}_i(\mathbf{w}) &= \sum_{i \in S_<} \Big(\text{cost}_i(\mathbf{w}) + \text{cost}_{\mu(i)}(\mathbf{w}) \Big) + X \cdot \text{cost}_R(\mathbf{w}) \\
&= \sum_{i \in S_<} \sum_{w \in \mathbf{w}} \Big(d(i,m) + d(\mu(i),m) \Big) + X \cdot \sum_{w \in \mathbf{w}} d(R,w) \\
&= k \cdot \sum_{i \notin \mathbf{w} \cup \{R\}} d(i,m) \\
&\quad + X \cdot \left(\sum_{\ell=1}^{(k-1)/2} d(x_\ell, y_\ell) + 2 \sum_{\ell=1}^{(k-1)/2} d(R, y_\ell) + d(R,m) \right).
\end{aligned}$$

Next, we consider the agents that are part of the solution $\mathbf{w}$ and the distances between them. Consider any two agents $x, y \in \mathbf{w}$ between which there are t different agents. For each such agent $i \in (x, y)$, we need to take into account the distance of x to i, the distance of i to x, the distance of y to i, and the distance of i to y. All together, these distances are exactly

$$2\,(d(x,i) + d(i,y)) = 2 \cdot d(x,y).$$

Accounting for the agents x and y as well, we have that the contribution of the distances of all agents in $[x, y]$ to the social cost is

$$(2t + 2) \cdot d(x,y).$$

We can now use this observation for all pairs of agents (x_ℓ, y_ℓ) for $\ell \in [(k-1)/2]$ (note that by doing this we will have calculated the distances of all agents in $\mathbf{w}$ from all agents in $\mathbf{w}$, including m). Since there are $2\ell - 1$ agents between x_ℓ and y_ℓ, the distance $d(x_\ell, y_\ell)$ has a coefficient of 4ℓ in the social cost.[3] Hence,

$$\sum_{i \in \mathbf{w}} \text{cost}_i(\mathbf{w}) = \sum_{\ell=1}^{(k-1)/2} 4\ell \cdot d(x_\ell, y_\ell) \le 2(k-1) \sum_{\ell=1}^{(k-1)/2} d(x_\ell, y_\ell).$$

Putting everything together, we have

$$\begin{aligned}
\text{SC}(\mathbf{w}) \le k \cdot \sum_{i \notin \mathbf{w} \cup \{R\}} d(i,m) + X \cdot \left(2 \sum_{\ell=1}^{(k-1)/2} d(R, y_\ell) + d(R,m) \right) \\
+ (2k - 2 + X) \sum_{\ell=1}^{(k-1)/2} d(x_\ell, y_\ell). \qquad (1)
\end{aligned}$$

We now focus on bounding the optimal social cost. By Lemma 4.1, the optimal solution $\mathbf{o}$ can be thought of as a shift of $\mathbf{w}$ towards the left or the right. We will only consider the case where the shift is towards the right; the other

[3] If k is even, for any (x_ℓ, y_ℓ) for $\ell \in [k/2]$ with $x_1 = m$, there are $2\ell - 2$ agents between x_ℓ and y_ℓ, leading to a coefficient of $4\ell - 2$ for the distance $d(x_\ell, y_\ell)$.

case can be handled similarly and is simpler since the agent R, if it exists, will have larger cost in the optimal solution, thus leading to a smaller bound on the approximation ratio. We again start by considering the agents that are not part of the solution $\mathbf{w}$. As before, consider the same matching μ of the agents in $S_<$ to the agents in $S_>$. Let $o \in \mathbf{o}$ be some agent that is part of the optimal solution. For any agent $i \in S_<$ such that $o \le \mu(i)$, we have that

$$d(i,o) + d(\mu(i),o) = d(i,m) + d(\mu(i),m).$$

On the other hand, for any agent $i \in S_<$ such that $\mu(i) < o$,

$$d(i,o) = d(i,\mu(i)) + d(\mu(i),o) \ge d(i,m) + d(\mu(i),m).$$

Therefore,

$$\begin{aligned}
\sum_{i \notin \mathbf{w}} \text{cost}_i(\mathbf{o}) &= \sum_{i \in S_<} \Big(\text{cost}_i(\mathbf{o}) + \text{cost}_{\mu(i)}(\mathbf{o}) \Big) + X \cdot \text{cost}_R(\mathbf{o}) \\
&= \sum_{i \in S_<} \sum_{o \in \mathbf{o}} \Big(d(i,m) + d(\mu(i),m) \Big) + X \cdot \sum_{o \in \mathbf{o}} d(R,o) \\
&\ge k \cdot \sum_{i \notin \mathbf{w} \cup \{R\}} d(i,m) + X \cdot \left(\sum_{\ell=1}^{(k-1)/2} d(R,y_\ell) + d(R,m) \right).
\end{aligned}$$

Next, consider agent x_ℓ for $\ell \in [(k-1)/2]$ and let $o \in \mathbf{o}$. If $o \le y_\ell$, then

$$d(x_\ell,o) + d(o,y_\ell) = d(x_\ell,y_\ell),$$

Otherwise, if $o > y_\ell$, then

$$d(x_\ell,o) = d(x_\ell,y_\ell) + d(y_\ell,o) > d(x_\ell,y_\ell).$$

Hence, we overall have that

$$\sum_{i \in \mathbf{w}} \text{cost}_i(\mathbf{o}) \ge \sum_{\ell=1}^{(k-1)/2} \sum_{o \in \mathbf{o}} \Big(d(x_\ell,o) + d(y_\ell,o) \Big) \ge k \cdot \sum_{\ell=1}^{(k-1)/2} d(x_\ell,y_\ell).$$

Putting everything together, we have

$$\begin{aligned}
\text{SC}(\mathbf{o}) \ge k \cdot \sum_{i \notin \mathbf{w} \cup \{R\}} d(i,m) + X \cdot \left(\sum_{\ell=1}^{(k-1)/2} d(R,y_\ell) + d(R,m) \right) \\
+ k \cdot \sum_{\ell=1}^{(k-1)/2} d(x_\ell,y_\ell) \\
\ge X \cdot \sum_{\ell=1}^{(k-1)/2} d(R,y_\ell) + k \cdot \sum_{\ell=1}^{(k-1)/2} d(x_\ell,y_\ell).
\end{aligned}$$

By (1), we now obtain

$$\begin{aligned}
\mathrm{SC}(\mathbf{w}) &\leq \mathrm{SC}(\mathbf{o}) + X \cdot \sum_{\ell=1}^{(k-1)/2} d(R, y_\ell) + (k-2+X) \cdot \sum_{\ell=1}^{(k-1)/2} d(x_\ell, y_\ell) \\
&\leq \mathrm{SC}(\mathbf{o}) + X \cdot \sum_{\ell=1}^{(k-1)/2} d(R, y_\ell) + k \cdot \sum_{\ell=1}^{(k-1)/2} d(x_\ell, y_\ell) \\
&\leq 2 \cdot \mathrm{SC}(\mathbf{o}).
\end{aligned}$$

This completes the proof. □

We next provide an asymptotically tight lower bound of 2–$1/k$.

Theorem 10. *For the sum-variant and k facilities, the approximation ratio of any deterministic strategyproof mechanism is at least* $2 - 1/k$.

Proof. Consider an instance with $n = k+1$ agents with one agent at 0, $k-1$ agents at 1 (or very close to 1) and one agent at 2. Since not all facilities can be placed at 1, at least one of them has to be placed 0 or 2, say 0. Then, the cost of the agent i that is located 2 is at least k (in particular, the cost of i is $2+k-1 = k+1$ if no facility is placed at 2, and $2+k-2 = k$ if a facility is placed at 2).

Now consider a new instance in which i has moved to $1+\varepsilon$ for some infinitesimal $\varepsilon > 0$. Due to strategyproofness, the mechanism must place one of the facilities at 0 as well. Otherwise, agent i would have cost $k - \varepsilon$ according to its position in the original instance, and would thus prefer to misreport its position as $1+\varepsilon$ instead of 2. So, in the new instance, the social cost of any possible solution that is restricted to having a facility at 0 is approximately $k-1+k = 2k-1$, while the social cost of the remaining solution is only k, leading to an approximation ratio of $2 - 1/k$. □

4.2 Max-Variant

For the max-variant, we will show a tight bound of $k+1$ on the approximation ratio of deterministic strategyproof mechanisms. The upper bound again follows by the MEDIAN-BALL mechanism; note the upper bound of 2 on the approximation ratio of MEDIAN-BALL for the sum-variant immediately implies an upper bound of $2k$ for the max-variant, which however is not the best possible we can show.

Theorem 11. *For the max-variant, the approximation ratio of the* MEDIAN-BALL *mechanism is at most* $k+1$.

Proof. Let ℓ and r be the leftmost and rightmost agents in the solution $\mathbf{w}$ computed by the mechanism. By the definition of $\mathbf{w}$, we have that $||\{i \leq \ell\}| - |\{i \geq$

$r\}|| \leq 1$. Since the individual cost of any agent i is the distance to its farthest facility, we have

$$\text{cost}_i(\mathbf{w}) = \begin{cases} d(i,r) & \text{if } i \leq \ell \\ \max\{d(i,\ell), d(i,r)\} & \text{if } i \in \mathbf{w} \setminus \{\ell, r\} \\ d(i,\ell) & \text{if } i \geq r. \end{cases}$$

Given this, and using the fact that $d(i,x) \leq d(i,m) + d(m,x)$ for any $x \in \{\ell, r\}$, we can bound the social cost of $\mathbf{w}$ as

$$\begin{aligned} \text{SC}(\mathbf{w}) &= \sum_{i \leq \ell} d(i,r) + \sum_{i \in \mathbf{w} \setminus \{\ell,r\}} \max\{d(i,\ell), d(i,r)\} + \sum_{i \geq r} d(i,\ell) \\ &\leq \sum_i d(i,m) + |\{i \leq \ell\}| \cdot d(m,r) + (k-2) \cdot \max\{d(\ell,m), d(m,r)\} \\ &\quad + |\{i \geq r\}| \cdot d(\ell,m) \\ &\leq \sum_i d(i,m) + \Big(\max\big\{|\{i \leq \ell\}|, |\{i \geq r\}|\big\} + k - 2 \Big) \cdot d(\ell,r). \end{aligned}$$

We now bound the social cost of an optimal solution $\mathbf{o}$. Since the location of the median agent m minimizes the total distance of all agents, if we were allowed to place the facilities at the same location, we would place all k facilities at m to minimize the social cost. Since this is not allowed in our model, the optimal social cost is larger than that, and we obtain

$$\text{SC}(\mathbf{o}) \geq \sum_i d(i,m).$$

In addition, since $\mathbf{w}$ is not optimal (as otherwise the approximation ratio would be 1), at least one facility must be placed at an agent o that is weakly to the left of ℓ or weakly to right of r. Let S be the set of agents that are not part of the solution $\mathbf{w}$ and are on the opposite side of o; that is, $S = \{i \geq r\}$ if $o \leq \ell$ and $S = \{r \leq \ell\}$ if $o \geq r$. For each agent $i \in S$, we have that $\text{cost}_i(\mathbf{o}) \geq d(i,o) \geq d(\ell,r)$, which implies

$$\text{SC}(\mathbf{o}) \geq |S| \cdot d(\ell,r) \geq \min\big\{|\{i \leq \ell\}|, |\{i \geq r\}|\big\} \cdot d(\ell,r).$$

Putting everything together, we have that

$$\text{SC}(\mathbf{w}) \leq \left(1 + \frac{\max\big\{|\{i \leq \ell\}|, |\{i \geq r\}|\big\} + k - 2}{\min\big\{|\{i \leq \ell\}|, |\{i \geq r\}|\big\}} \right) \cdot \text{SC}(\mathbf{o}).$$

Since $\max\big\{|\{i \leq \ell\}|, |\{i \geq r\}|\big\} \leq \min\big\{|\{i \leq \ell\}|, |\{i \geq r\}|\big\} + 1$ and $\min\big\{|\{i \leq \ell\}|, |\{i \geq r\}|\big\} \geq 1$, we obtain an upper bound of $k+1$ on the approximation ratio. □

We conclude the presentation of our technical results with a matching lower bound of $k+1$ on the approximation ratio of deterministic mechanisms for the max-variant, thus completely resolving this setting.

Theorem 12. *For the max-variant and k facilities, the approximation ratio of any deterministic strategyproof mechanism is at least $k + 1$.*

Proof. Consider an instance with $n = k + 1$ agents with one agent at 0, $k - 1$ agents at 1 (or very close to 1) and one agent at 2. Since not all facilities can be placed at 1, at least one of them has to be placed 0 or 2, say 0. Then, the cost of the agent i that is located 2 is 2.

Now consider a new instance in which i has moved to 1. Due to strategyproofness, the mechanism must place one of the facilities at 0 as well. Otherwise, if all facilities are placed at 1, agent i would have cost 1 according to its position in the original instance, and would thus prefer to misreport its position as 1 instead of 2. So, in the new instance, the social cost of the solution chosen by the mechanism is $k + 1$, while the social cost of solution that places all facilities at 1 is just 1, leading to an approximation ratio of $k + 1$. □

5 Conclusion and Open Problems

In this work, we showed tight bounds on the best possible approximation ratio of deterministic and randomized strategyproof mechanisms for the two-facility location problem where the facilities can be placed at the reported agent locations and the individual cost of an agent is either its distance from both facilities or its distance to the farthest facility. We believe there are many directions for future work. In terms of our results, it would be interesting to close the gap between $2 - 1/k$ and 2 for the sum-variant and multiple facilities, as well as consider randomized mechanisms. One can also generalize our model in multiple dimensions, for example, by considering agents that might have different preferences over the facilities (such as optional or fractional preferences), and the efficiency of mechanisms is measured by objective functions beyond the social cost (such as the egalitarian cost, or the more general family of ℓ-centrum objectives).

Acknowledgement. Argyrios Deligkas is supported by the UKRI EPSRC grant EP/X039862/1.

References

1. Agrawal, P., Balkanski, E., Gkatzelis, V., Ou, T., Tan, X.: Learning-augmented mechanism design: leveraging predictions for facility location. In: Proceedings of the 23rd ACM Conference on Economics and Computation (EC), pp. 497–528 (2022)
2. Chan, H., Filos-Ratsikas, A., Li, B., Li, M., Wang, C.: Mechanism design for facility location problems: a survey. In: Proceedings of the Thirtieth International Joint Conference on Artificial Intelligence (IJCAI), pp. 4356–4365 (2021)
3. Chan, H., Gong, Z., Li, M., Wang, C., Zhao, Y.: Facility location games with ordinal preferences. Theor. Comput. Sci. **979**, 114208 (2023)

4. Chen, Z., Fong, K.C.K., Li, M., Wang, K., Yuan, H., Zhang, Y.: Facility location games with optional preference. Theor. Comput. Sci. **847**, 185–197 (2020)
5. Cheng, Y., Yua, W., Zhang, G.: Strategy-proof approximation mechanisms for an obnoxious facility game on networks. Theor. Comput. Sci. **497**, 154–163 (2013)
6. Deligkas, A., Filos-Ratsikas, A., Voudouris, A.A.: Heterogeneous facility location with limited resources. Games Econom. Behav. **139**, 200–215 (2023)
7. Feigenbaum, I., Sethuraman, J.: Strategyproof mechanisms for one-dimensional hybrid and obnoxious facility location models. In: AAAI Workshop on Incentive and Trust in E-Communities, vol. WS-15-08 (2015)
8. Feldman, M., Fiat, A., Golomb, I.: On voting and facility location. In: Proceedings of the 2016 ACM Conference on Economics and Computation (EC), pp. 269–286 (2016)
9. Filos-Ratsikas, A., Kanellopoulos, P., Voudouris, A.A., Zhang, R.: The distortion of distributed facility location. Artif. Intell. **328**, 104066 (2024)
10. Fong, C.K.K., Li, M., Lu, P., Todo, T., Yokoo, M.: Facility location games with fractional preferences. In: Proceedings of the 32nd AAAI Conference on Artificial Intelligence (AAAI), pp. 1039–1046 (2018)
11. Fotakis, D., Tzamos, C.: On the power of deterministic mechanisms for facility location games. ACM Trans. Econ. Comput. **2**(4), 15:1–15:37 (2014)
12. Gai, L., Liang, M., Wang, C.: Two-facility-location games with mixed types of agents. Appl. Math. Comput. **466**, 128479 (2024)
13. Kanellopoulos, P., Voudouris, A.A., Zhang, R.: On discrete truthful heterogeneous two-facility location. SIAM J. Discret. Math. **37**, 779–799 (2023)
14. Kanellopoulos, P., Voudouris, A.A., Zhang, R.: Truthful two-facility location with candidate locations. In: Proceedings of the 16th International Symposium on Algorithmic Game Theory (SAGT) (2023)
15. Li, M., Lu, P., Yao, Y., Zhang, J.: Strategyproof mechanism for two heterogeneous facilities with constant approximation ratio. In: Proceedings of the 29th International Joint Conference on Artificial Intelligence (IJCAI), pp. 238–245 (2020)
16. Lotfi, M., Voudouris, A.A.: On truthful constrained heterogeneous facility location with max-variant cost. Oper. Res. Lett. **52**, 107060 (2024)
17. Lu, P., Sun, X., Wang, Y., Zhu, Z.A.: Asymptotically optimal strategy-proof mechanisms for two-facility games. In: Proceedings of the 11th ACM Conference on Electronic Commerce (EC), pp. 315–324 (2010)
18. Procaccia, A.D., Tennenholtz, M.: Approximate mechanism design without money. ACM Trans. Econ. Comput. **1**(4), 18:1–18:26 (2013)
19. Serafino, P., Ventre, C.: Heterogeneous facility location without money. Theor. Comput. Sci. **636**, 27–46 (2016)
20. Tang, Z., Wang, C., Zhang, M., Zhao, Y.: Mechanism design for facility location games with candidate locations. In: Proceedings of the 14th International Conference on Combinatorial Optimization and Applications (COCOA), pp. 440–452 (2020)
21. Xu, C., Lu, P.: Mechanism design with predictions. In: Proceedings of the 31st International Joint Conference on Artificial Intelligence (IJCAI), pp. 571–577 (2022)
22. Xu, X., Li, B., Li, M., Duan, L.: Two-facility location games with minimum distance requirement. J. Artif. Intell. Res. **70**, 719–756 (2021)
23. Zhao, Q., Liu, W., Nong, Q., Fang, Q.: Constrained heterogeneous facility location games with max-variant cost. J. Comb. Optim. **45**(3), 90 (2023)

The k-Facility Location Problem via Optimal Transport: A Bayesian Study of the Percentile Mechanisms

Gennaro Auricchio[1(✉)] and Jie Zhang[2]

[1] Università degli Studi di Padova, Padua, Italy
gennaro.auricchio@unipd.it
[2] University of Bath, Bath, UK
jz2558@bath.ac.uk

Abstract. In this paper, we investigate the k-Facility Location Problem (k-FLP) within the Bayesian Mechanism Design framework, in which agents' preferences are samples of a probability distributed on a line. Our primary contribution is characterising the asymptotic behavior of percentile mechanisms, which varies according to the distribution governing the agents' types. To achieve this, we connect the k-FLP and projection problems in the Wasserstein space. Owing to this relation, we show that the ratio between the expected cost of a percentile mechanism and the expected optimal cost is asymptotically bounded. Furthermore, we characterize the limit of this ratio and analyze its convergence speed. Our asymptotic study is complemented by deriving an upper bound on the Bayesian approximation ratio, applicable when the number of agents n exceeds the number of facilities k. We also characterize the optimal percentile mechanism for a given agent's distribution through a system of k equations. Finally, we estimate the optimality loss incurred when the optimal percentile mechanism is derived using an approximation of the agents' distribution rather than the actual distribution.

Keywords: Bayesian Mechanism Design · Facility Location Problem · Optimal Transport

1 Introduction

The scope of Mechanism Design is defining procedures that aggregate a group of agents' private information for optimizing a social objective. Nevertheless, merely optimizing the social objective based on the reported preferences often leads to undesired manipulation due to the agents' self-interested behaviour. For this reason, one of the most important properties a mechanism should possess is *truthfulness*, which guarantees that no agent benefits from misreporting its private information. This stringent property is often incompatible with the optimization of the social objective, so we have to compromise on a sub-optimal solution. To quantify the efficiency loss, Nisan and Ronen introduced the notion

G. Schäfer and C. Ventre (Eds.): SAGT 2024, LNCS 15156, pp. 147–164, 2024.
https://doi.org/10.1007/978-3-031-71033-9_9

of *approximation ratio*, which is the highest ratio between the social objective achieved by a truthful mechanism and the optimal social objective achievable over all the possible agents' reports [39].

One of the most famous examples of these problems is the k-Facility Location Problem (k-FLP), where a central authority has to locate k facilities amongst n self-interested agents. Every agent needs to access the facility, so they would prefer to have one of the facilities placed as close as possible to their position. Despite its simplicity, this problem and its variants have found a wide range of applications in fields such as disaster relief [12], supply chain management [38], healthcare [1], and public facilities accessibility [14]. The study of the k-FLP from an algorithmic mechanism design viewpoint was initiated by Procaccia and Tennenholtz. In their seminal work [41], they considered the problem of locating one facility amongst a group of agents situated in a line. They were the first to design an allocation process that places the facility while keeping in mind that every agent is self-interested, i.e. that agents would manipulate the process in its favour if able. Subsequently, a variety of methods with fixed approximation ratios for positioning one or two facilities on different types of structures such as double-peaked [24], trees [23], circles [33,34], graphs [2,21], and metric spaces [37,46] were introduced. These positive outcomes, however, pertain to scenarios with a limited number of agents or when the facilities to place are at most 2. The approximation ratio results are much more negative when we move to three or more facilities. Fotakis and Tzamos [26] showed that for every $k \geq 3$, there does not exist any deterministic, anonymous, and truthful mechanisms with a bounded approximation ratio for the k-FLP on the line, even for instances with $k+2$ agents. Nonetheless, it is possible to define truthful mechanisms with bounded approximation ratio when the number of agents is equal to the number of facilities plus one [22] or by considering randomized mechanisms [25].

Our study concerns a class of truthful mechanisms for the generic k-FLP, the *percentile mechanisms*, introduced in [45]. Although every percentile mechanism has an unbounded approximation ratio, we prove that this is not the case if the agents' type is sampled from a probability distribution. This framework is also known as Bayesian Mechanism Design [18,29]. Our main contribution shows that it is possible to select a percentile mechanism that asymptotically behaves optimally, i.e. it minimizes the expected social objective.

1.1 Our Contribution

In this paper, we conduct a comprehensive investigation of the k-Facility Location Problem (k-FLP) from a Bayesian Mechanism Design perspective, where we assume that agents' positions on the line follow a distribution μ [18,29]. We focus specifically on the class of percentile mechanisms [45] and explore the conditions under which the Bayesian approximation ratio of these mechanisms – defined as the ratio between the expected cost induced by a mechanism and the expected optimal cost – is bounded. We establish that each percentile mechanism exhibits different performances depending on the measure μ, and we identify the optimal percentile mechanism tailored to a distribution μ. Our study establishes a connection between the k-FLP and a projection problem in the Wasserstein space. Through this connection, we import tools and techniques from Optimal Transport theory to approach the k-FLP. In particular, we demonstrate that when the number of agents on the line tends to infinity, the ratio between the expected cost induced by the mechanism and the expected optimal cost converges to a bounded value. Moreover, we characterize both the limit value of the ratio and the speed of convergence. To retrieve these results, we make massive use of Bahadur's representation formula, which relates the j-th ordered statistic of a random variable to a suitable quantile of the probability distribution associated with the random variable Finally, leveraging the characterization of the limit and its convergence rate, we derive a bound on the performances of percentile mechanisms for any finite number of agents.

We then tackle the problem of retrieving the best percentile mechanism tailored to a distribution μ and the number of facilities k. We show that there always exists a percentile vector, namely $\boldsymbol{v}_\mu \in (0,1)^k$, that induces the optimal percentile mechanism, *i.e.* a mechanism whose expected social cost is asymptotic to the optimal expected cost when the number of agents increases. We characterize this vector as the solution to a system of k equations and employ it to compute the optimal percentile vector associated with common probability measures, such as the Uniform and Gaussian distributions. Lastly, we show that the optimal percentile vector is invariant under positive affine transformations of the probability measures describing agents. In particular, $\boldsymbol{v}_\mu$ does not depend on the specific mean and variance of the distribution μ.

To conclude the paper, we present a study on the stability of the optimal percentile vector. Specifically, let $\tilde{\mu}$ be an approximation of the true agents' distribution μ. Additionally, let $\boldsymbol{v}_{\tilde{\mu}}$ and $\boldsymbol{v}_\mu$ represent the optimal percentile vectors associated with $\tilde{\mu}$ and μ, respectively. We demonstrate that when the agents are distributed according to μ, the Bayesian approximation ratio limit of the percentile mechanism induced by $\boldsymbol{v}_{\tilde{\mu}}$ deviates from 1 by an amount proportional to the infinity Wasserstein distance between μ and $\tilde{\mu}$. The more precise the approximation of μ, the better the asymptotic performance of the optimal percentile mechanism induced by $\boldsymbol{v}_\mu$ when the agents are distributed according to μ. Due to space limits, proofs and additional results are deferred to the full version of the paper [10].

1.2 Related Work

The study of k-FLP research from an algorithmic mechanism design viewpoint was initiated by Procaccia and Tennenholtz in [41]. When $k = 1, 2$ there are several truthful mechanisms, such as the median mechanism [15] and its generalizations [13], that achieve small constant worst-case approximation ratios. When $k > 2$, however, these efficiency guarantees are much more negative: there are no truthful, deterministic and anonymous mechanisms with bounded approximation ratio [26]. It is worthy of notice however, that this impossibility result does not apply to randomized mechanisms [25], to instances where the number of agents is precisely equal to the number of facilities plus one, as shown in [22], and to problems in which facilities have capacity limits [7,9].

The *Percentile Mechanisms* are a class of mechanisms for the k-FLP that, similarly to the median mechanism, places the facilities at the locations of k agents depending on their order on the line, [45]. Due to the dictatorial-like nature of these mechanisms, it is easy to build an *ad hoc* instance of the k-FLP on which the optimal social cost is as small as we like, but the cost attained by the mechanisms is greater than a positive constant. These instances depend on the percentile mechanism and carry little practical sense in applied contexts.

Bayesian Mechanism Design is an alternative paradigm to evaluate the performances of a mechanism in which every agent's type is drawn from a known probability distribution, [18,29]. Consequentially, this defines a distribution over the set of inputs over which the mechanism is defined, allowing us to introduce the notion of expected cost of a mechanism. To the best of our knowledge, the only other two papers studying the k-FLP in a Bayesian Mechanism Design framework, are [11], where the k-Capacitated Facility Location Problem is considered, and [48], in which the authors study how to use the Lugosi-Mendelson median [36] to define approximately truthful mechanisms for the 1-FLP. Other fields in which Bayesian Mechanism Design framework has been applied are: routing games [28], combinatorial mechanisms based on ϵ-greedy mechanisms [35], and auction mechanism design problems [17,30].

Over the past few decades, Optimal Transport (OT) methods have gradually found their application within the broad landscape of Theoretical Computer Science. Notable examples include Computer Vision [8,40,42], Computational Statistics [32], Clustering [6], and Machine Learning in general [27,43,44]. However, there has been limited advancement in applying OT theory to the field of mechanism design. To the best of our knowledge, the only field related to mechanism design that has been explored using OT theory is auction design [19]. In their work, the authors demonstrated that the optimal auction mechanism for independently distributed items can be characterized by the Dual Formulation of an OT problem. Moreover, they utilized this relationship to derive the optimal mechanism for various item classes, thereby establishing a fruitful application of OT theory in the context of mechanism design.

2 Preliminaries

In this section, we fix the necessary notations on the k-Facility Location Problem (k-FLP), Bayesian Mechanism Design, and Optimal Transport (OT). Furthermore, we recall the definition of the percentile mechanisms.

The k-Facility Location Problem. Given a set of self-interested agents $\mathcal{N} = [n] := \{1, 2, \dots, n\}$, we denote with $\mathcal{X} := \{x_i\}_{i\in[n]}$ the set of their positions over $\mathbb{R}$. Without loss of generality, assume that the agents are indexed such that the positions x_i's are in non-decreasing order. We denote with $\boldsymbol{x} := (x_1, x_2, \dots, x_n) \in \mathbb{R}^n$ the vector containing the elements of $\mathcal{X}$. In this setting, if the k facilities are located at the entries of the vector $\boldsymbol{y} := (y_1, y_2, \dots, y_k) \in \mathbb{R}^k$, an agent positioned in x_i incurs a cost of $c_i(x_i, \boldsymbol{y}) = \min_{j\in[k]} |x_i - y_j|$ to access a facility. In what follows, we will use $\boldsymbol{y} = (y_1, \dots, y_k)$ and the set of points $\{y_j\}_{j\in[k]}$, interchangeably. Given a vector $\boldsymbol{x} \in \mathbb{R}^n$ containing the agents' positions, the *Social Cost* (*SC*) of $\boldsymbol{y}$ is the sum of all the agents' utilities, that is $SC(\boldsymbol{x}, \boldsymbol{y}) = \sum_{i\in[n]} c_i(x_i, \boldsymbol{y})$. The *k-Facility Location Problem*, consists in finding the locations for k facilities that minimize the function $\boldsymbol{y} \to SC(\boldsymbol{x}, \boldsymbol{y})$. Given that multiplying the cost function by a constant does not alter the approximation ratio of the mechanisms, we rescaled the Social Cost as $SC(\boldsymbol{x}, \boldsymbol{y}) = \frac{1}{n}\sum_{i\in[n]} c_i(x_i, \boldsymbol{y})$.

Mechanism Design and the Worst-Case analysis. A *k-facility location mechanism* is a function $f : \mathbb{R}^n \to \mathbb{R}^k$ that takes the agents' reports $\boldsymbol{x}$ in input and returns a set of k locations $\boldsymbol{y}$ for the facilities. In general, an agent may misreport its position if it this results in a set of facility locations such that the agent's incurred cost is smaller than reporting truthfully. A mechanism f is said to be *truthful* (or *strategyproof*) if, for every agent, its cost is minimized when it reports its true position. That is, $c_i(x_i, f(\boldsymbol{x})) \le c_i(x_i, f(\boldsymbol{x}_{-i}, x_i'))$ for any misreport $x_i' \in \mathbb{R}$, where $\boldsymbol{x}_{-i}$ is the vector $\boldsymbol{x}$ without its i-th component. Albeit deploying a truthful mechanism instead of computing the optimal location prevents agents from misreporting their positions, it comes with a loss in terms of efficiency. To evaluate this efficiency loss, Nisan and Ronen introduced the notion of approximation ratio of a truthful mechanism [39]. Given a truthful mechanism f, its approximation ratio is defined as

$$ar(f) := \sup_{\boldsymbol{x}\in\mathbb{R}^n} \frac{SC_f(\boldsymbol{x})}{SC_{opt}(\boldsymbol{x})}, \tag{1}$$

where $SC_f(\boldsymbol{x})$ is the Social Cost of placing the facilities at $f(\boldsymbol{x})$ and $SC_{opt}(\boldsymbol{x})$ is the optimal Social Cost achievable when the agents' report is $\boldsymbol{x}$. In what follows, we will refer to the worst-case approximation ratio defined in (1) as the approximation ratio. Evaluating a mechanism f from its approximation ratio is also known as the worst-case analysis of f.

Bayesian Analysis. In Bayesian Mechanism Design, we assume that the agents' types follow a probability distribution and study the performance of mechanisms from a probabilistic viewpoint. Every agent's type is then described by a random variable X_i. In what follows, we assume that every X_i is identically distributed according to a law μ and independent from the other random variables. A mechanism is said to be truthful if, for every agent i, it holds

$$\mathbb{E}_{\boldsymbol{X}_{-i}}[c_i(x_i, f(x_i, \boldsymbol{X}_{-i}))] \le \mathbb{E}_{\boldsymbol{X}_{-i}}[c_i(x_i, f(x_i', \boldsymbol{X}_{-i}))] \qquad \forall x_i \in \mathbb{R}, \tag{2}$$

where x_i agent i's true type, $\boldsymbol{X}_{-i}$ is the $(n-1)$-dimensional random vector that describes the other agents' type, and $\mathbb{E}_{\boldsymbol{X}_{-i}}$ is the expectation with respect to the joint distribution of $\boldsymbol{X}_{-i}$. Given $\beta \in \mathbb{R}$, a mechanism f is a β-approximation if $\mathbb{E}[SC_f(\boldsymbol{X}_n)] \le \beta\,\mathbb{E}[SC_{opt}(\boldsymbol{X}_n)]$ holds, so that the lower β is, the better the mechanism is. To unify the notation, we define the Bayesian approximation ratio for the Social Cost as the ratio between the expected Social Cost of a mechanism and the expected Social Cost of the optimal solution. More formally, given a mechanism f, its Bayesian approximation ratio is defined as follows

$$B_{ar}^{(n)}(f) := \frac{\mathbb{E}[SC_f(\boldsymbol{X}_n)]}{\mathbb{E}[SC_{opt}(\boldsymbol{X}_n)]}, \tag{3}$$

where the expected value is taken over the joint distribution of the vector $\boldsymbol{X}_n := (X_1, \dots, X_n)$. Notice that, if $B_{ar}^{(n)}(f) < +\infty$, then f is a $B_{ar}^{(n)}(f)$-approximation. Since we consider only truthful mechanisms, in what follows we use $\boldsymbol{x}$ to denote the vector containing the agents' reports and the agents' real position interchangeably. Moreover, we use the capital letter $\boldsymbol{X}_n$ to denote the random vector describing the agents' types.

The Percentile Mechanisms. The class of *percentile mechanisms* has been introduced in [45]. Given a vector $\boldsymbol{v} = (v_1, v_2, \dots, v_k)$, such that $0 \le v_1 \le v_2 \le \dots \le v_k \le 1$, the percentile mechanism induced by $\boldsymbol{v}$, namely $\mathcal{PM}_v$, proceeds as follows: (i) The mechanism designer collects all the reports of the agents, namely $\{x_1, \dots, x_n\}$ and reorders them non-decreasingly. Without loss of generality, let us assume that the reports are already ordered in non-decreasing order, i.e. $x_i \le x_{i+1}$. (ii) The designer places the k facilities at the positions $y_j = x_{i_j}$, where $i_j = \lfloor (n-1)v_j \rfloor + 1$. Notice that, if the values x_i are sampled from a distribution, the output of any percentile mechanism is composed by the $(\lfloor (n-1)v_j \rfloor + 1)$-th order statistics of the sample. Percentile mechanisms are truthful whenever the cost of an agent placed at x_i is $c_i = \min_{j \in [k]} |x_i - y_j|$, where y_j are the position of the facilities. Thus, when $k > 2$, the approximation ratio of any percentile mechanism becomes unbounded since the percentile mechanisms are also anonymous and deterministic, that is $ar(\mathcal{PM}_v) = +\infty$ for every percentile vector $\boldsymbol{v}$. Moreover, it is worth noting that since percentile mechanisms are truthful in the classic setting, they also retain their truthfulness within the Bayesian framework [29].

Basic Notions on Optimal Transport. In the following, we denote with $\mathcal{P}(\mathbb{R})$ the set of probability measures over $\mathbb{R}$. Given a measure $\gamma \in \mathcal{P}(\mathbb{R})$, we denote with $spt(\gamma) \subset \mathbb{R}$ the support of γ, that is, the smallest closed set $C \subset \mathbb{R}$ such that $\gamma(C) = 1$. Furthermore, we denote with $\mathcal{P}_k(\mathbb{R})$ the set of probability measures over $\mathbb{R}$ whose support consists of k points. That is, $\nu \in \mathcal{P}_k(\mathbb{R})$ if and only if $\nu = \sum_{j=1}^k \nu_j \delta_{x_j}$, where $x_j \in \mathbb{R}$ for every $j \in [k]$, $\nu_j \ge 0$ are real values such that $\sum_{j=1}^k \nu_j = 1$, and δ_{x_j} is the Dirac's delta centered in x_j. Given two measures $\alpha, \beta \in \mathcal{P}(\mathbb{R})$, the Wasserstein distance between α and β is defined as

$$W_1(\alpha, \beta) = \min_{\pi \in \Pi(\alpha,\beta)} \int_{\mathbb{R}\times\mathbb{R}} |x-y| d\pi, \tag{4}$$

where $\Pi(\alpha, \beta)$ is the set of probability measures over $\mathbb{R} \times \mathbb{R}$ whose first marginal is equal to α and the second marginal is equal to β, [31]. Lastly, the infinity Wasserstein distance is defined as $W_\infty(\alpha, \beta) = \min_{\pi \in \Pi(\alpha,\beta)} \max_{(x,y)\in spt(\pi)} |x - y|$. It is well-known that both W_1 and W_∞ are metrics over $\mathcal{P}(\mathbb{R})$. For a complete introduction to the Optimal Transport theory, we refer the reader to [47].

Basic Assumptions. In the remainder of the paper, we tacitly assume that the underlying distribution μ satisfies the following properties: (i) The measure μ is absolutely continuous. We denote with ρ_μ its density. (ii) The support of μ is an interval, which can be bounded or not, and that ρ_μ is strictly positive on the interior of the support. (iii) The density function ρ_μ is differentiable on the support of μ. Notice that the cumulative distribution function (c.d.f.) F_μ of a probability measure μ satisfying these properties is locally bijective. Thus the pseudo-inverse function of F_μ, namely $F_\mu^{[-1]}$, is well-defined on $(0,1)$.

3 The Bayesian Analysis of the Percentile Mechanism

In this section, we study the percentile mechanisms in the Bayesian Mechanism Design framework. Specifically, we consider a scenario where the agents' reports are drawn from a shared distribution μ, which satisfies the basic assumptions outlined in Sect. 2. First, we establish a connection between the k-FLP and the Wasserstein distance and use it to investigate the convergence behaviour of the Bayesian approximation ratio as the number of agents tends to infinity.

3.1 The k-FLP as a Wasserstein Projection problem

Given a vector $\boldsymbol{x} := (x_1, x_2, \dots, x_n)$ containing the reports of n agents, we define the measure $\mu_{\boldsymbol{x}} := \frac{1}{n}\sum_{i=1}^n \delta_{x_i}$. Using the map $\boldsymbol{x} \to \mu_{\boldsymbol{x}}$, we are able to associate any agents' profile to a probability measure in $\mathcal{P}_n(\mathbb{R}) \subset \mathcal{P}(\mathbb{R})$. Let us now consider the following minimization problem

$$\min_{\lambda \in \mathcal{P}_k(\mathbb{R})} W_1(\mu_{\boldsymbol{x}}, \lambda). \tag{5}$$

Due to the metric properties of W_1, problem (5) is also known as the Wasserstein projection problem on $\mathcal{P}_k(\mathbb{R})$. Since $\mathcal{P}_k(\mathbb{R})$ is closed with respect to any W_1 metric, any Wasserstein projection problem admits at least a solution [3]. When $\mu_{\boldsymbol{x}}$ is clear from the context, we denote with $\nu^{(k,n)}$ the solution to problem (5). In general, given a measure ζ, we say that ν is the projection of ζ over $\mathcal{S} \subset \mathcal{P}(\mathbb{R})$ with respect to W_1 if $\nu \in \mathcal{S}$ and $W_1(\zeta,\nu) \leq W_1(\zeta,\rho)$ for every $\rho \in \mathcal{S}$. In particular, $\nu^{(k,n)}$ is the projection of $\mu_{\boldsymbol{x}}$ over $\mathcal{P}_k(\mathbb{R})$ with respect to W_1.

The starting point of our Bayesian analysis of the percentile mechanisms connects the k-FLP to a Wasserstein projection problem. In particular, the objective value of problem (5) is the same as the objective value of the k-FLP.

Theorem 1. *Let $\boldsymbol{x}$ be the reports of n agents. Let $\boldsymbol{y}$ be the solution to the k-FLP, i.e. the facility locations that minimize the Social Cost. Then the set $\{y_j\}_{j\in[k]}$ is the support of a measure $\nu^{(k,n)}$ that solves problem (5). Moreover, we have that*

$$SC_{opt}(\boldsymbol{x}) = W_1(\mu_{\boldsymbol{x}}, \nu^{(k,n)}) = \min_{\lambda \in \mathcal{P}_k(\mathbb{R})} W_1(\mu_{\boldsymbol{x}}, \lambda).$$

Vice-versa, if $\nu \in \mathcal{P}_k(\mathbb{R})$ is a solution to problem (5), then its support $\{y_j\}_{j\in[k]}$ is a solution to the k-FLP.

Proof. Let $\boldsymbol{x}$ be the vector containing the reports of n agents, and let $\boldsymbol{y}$ be the vector containing the optimal location for k facilities when the agents are located according to $\boldsymbol{x}$. Without loss of generality, we assume that the closest facility to each agent x_i is unique so that the sets A_j, defined as $A_j := \Big\{x_i : \min_{l\in[k]}|x_i - y_l| = |x_i - y_j|\Big\}$, are well-defined and disjoint. First, we show that, given an optimal facility location $\boldsymbol{y}$, it is possible to retrieve a measure $\nu \in \mathcal{P}_k(\mathbb{R})$ that solves the projection problem (5) and whose support is $\{y_j\}_{j\in[k]}$.

For every y_j, let us set $\nu_j = \frac{\ell_j}{n}$, where $\ell_j := |A_j|$ is the number of agents whose closest facility is located at y_j. We then set $\nu = \sum_{j\in[k]} \nu_j \delta_{y_j}$. Since A_j are disjoint sets, we have $\nu \in \mathcal{P}_k(\mathbb{R})$. Let us now consider the transportation plan, namely π, between $\mu_{\boldsymbol{x}}$ and ν defined as

$$\pi_{i,j} := \pi_{x_i,y_j} = \begin{cases} \frac{1}{n} & \text{if} \quad x_i \in A_j \\ 0 & \text{otherwise.} \end{cases}$$

Since according to π every agent goes to its closest facility, π is optimal, thus we have $W_1(\mu_{\boldsymbol{x}}, \nu) = \sum_{i\in[n],j\in[k]} |x_i - y_j|\pi_{i,j} = \frac{1}{n}\sum_{j\in[k]}\sum_{x_i\in A_j}|x_i - y_j|$. We now show that ν solves problem (5). Toward a contradiction, let us assume that $\tilde{\nu} = \sum_{j=1}^{k} \tilde{\nu}_j \delta_{\tilde{y}_j} \in \mathcal{P}_k(\mathbb{R})$ is such that $W_1(\mu_{\boldsymbol{x}}, \tilde{\nu}) < W_1(\mu_{\boldsymbol{x}}, \nu)$. Let us define the partition of agents A'_j related to the set of points $\{y'_j\}_{j\in[k]}$.[1] Then we have

$$\frac{1}{n}\sum_{j\in[k]}\sum_{x_i\in A'_j}|x_i - y'_j| = W_1(\mu_{\boldsymbol{x}}, \tilde{\nu}_j) < W_1(\mu_{\boldsymbol{x}}, \nu) = \frac{1}{n}\sum_{j\in[k]}\sum_{x_i\in A_j}|x_i - y_j|, \tag{6}$$

which contradicts the optimality of $\boldsymbol{y}$, proving the first part of the Theorem.

For the inverse implication, it suffices to repeat the same argument backwards. Indeed, let ν' be a solution to the W_1 Projection problem. Toward a contradiction, let us assume that the support of ν' is not a solution to the k-FLP. Then, given a solution to the k-FLP problem, we can use the argument used in the first part of the proof to build a new measure that has a lower cost than ν', which would contradict the optimality of the initial solution. □

By restricting the set on which the projection problem is defined, we retrieve a similar characterization for the cost of any k-facility location mechanism.

Theorem 2. *Let $f : \mathbb{R}^n \to \mathbb{R}^k$ be a k-facility location mechanism. Then, the following identity holds*

$$SC_f(\boldsymbol{x}) = \min_{\{\lambda_j\}_{j\in[k]}\subset\mathbb{R}} W_1(\mu_{\boldsymbol{x}}, \lambda), \tag{7}$$

where $\lambda = \sum_{j\in[k]} \lambda_j \delta_{y_j}$ and $\boldsymbol{y} = (y_1, y_2, \ldots, y_k) = f(\boldsymbol{x})$.

[1] Again, without loss of generality, we can assume that the facility that is closest to a given agent is unique.

Notice that the projection problem (7) is a further restricted version of the projection problem (5). Indeed, in (5), the support of the solution can be any subset of $\mathbb{R}$ containing k elements, while in (7), the support of the solution is fixed by the mechanism f.

3.2 The Bayesian Analysis of the Percentile Mechanisms

In this section, we use the results presented in Theorem 1 and Theorem 2 to study the limiting behaviour of the Bayesian approximation ratio of any percentile mechanism $\mathcal{PM}_{\boldsymbol{v}}$ with $\boldsymbol{v} \in (0,1)^k$. From Theorem 1, the k-FLP is equivalent to a projection problem in the space of probability distributions with respect to W_1. It is well-known that, in order to ensure that the W_1 distance between two measures is finite, both the measures must have a finite first moment [47]. We recall that a measure μ has a finite first moment if

$$\int_{\mathbb{R}} |x| d\mu < +\infty. \tag{8}$$

Lemma 1. *Let $\boldsymbol{X}_n := (X_1, X_2, \dots, X_n)$ be the random vector describing the reports of n i.i.d. agents distributed as μ. If μ satisfies (8), then, for every $k \in \mathbb{N}$, we have that $\mathbb{E}[SC_{opt}(\boldsymbol{X}_n)]$ converges to $W_1(\mu, \nu^{(k)})$ as $n \to \infty$, where $\nu^{(k)}$ is the solution to the following projection problem*

$$\min_{\lambda \in \mathcal{P}_k(\mathbb{R})} W_1(\mu, \lambda). \tag{9}$$

In particular, we have that $\mathbb{E}[SC_{opt}(\boldsymbol{X}_n)]$ is strictly positive for n large enough.

Proof. Let $\nu^{(k,n)}$ be the solution to problem (5) and let $\nu^{(k)}$ be the solution to problem (9). Owing to the triangular inequality and to the properties of the projection problem, we have

$$W_1(\mu_{\boldsymbol{x}}, \nu^{(k,n)}) \leq W_1(\mu_{\boldsymbol{x}}, \nu^{(k)}) \leq W_1(\mu_{\boldsymbol{x}}, \mu) + W_1(\mu, \nu^{(k)})$$

and, similarly

$$W_1(\mu, \nu^{(k)}) \leq W_1(\mu, \nu^{(k,n)}) \leq W_1(\mu, \mu_{\boldsymbol{x}}) + W_1(\mu_{\boldsymbol{x}}, \nu^{(k,n)}),$$

which implies $|W_1(\mu, \nu^{(k)}) - W_1(\mu_{\boldsymbol{x}}, \nu^{(k,n)})| \leq W_1(\mu, \mu_{\boldsymbol{x}})$ and thus

$$\mathbb{E}[|W_1(\mu, \nu^{(k)}) - W_1(\mu_{\boldsymbol{x}}, \nu^{(k,n)})|] \leq \mathbb{E}[W_1(\mu, \mu_{\boldsymbol{x}})].$$

Since $\lim_{n \to \infty} \mathbb{E}[W_1(\mu, \mu_{\boldsymbol{x}})] = 0$, see [16], we infer that $\mathbb{E}[W_1(\mu_{\boldsymbol{x}}, \nu^{(k,n)})]$ converges to $W_1(\mu, \nu^{(k)})$ as $n \to \infty$. Finally, since $W_1(\mu, \nu^{(k)})$ is strictly positive, for n large enough, we have $\mathbb{E}[W_1(\mu_{\boldsymbol{x}}, \nu^{(k,n)})]$ is strictly positive as well. □

It is worthy of notice that in the proof of Lemma 1, we have shown a slightly stronger result: the random variable $SC_{opt}(\boldsymbol{X}_n)$ converges with respect to the L^1 norm to the constant value $W_1(\mu, \nu^{(k)})$. Moving on to the limit cost of the mechanism, we observe that the set characterizing the projection problem (7) is dependent on the output of the percentile mechanism. Hence, the argument used to prove Lemma 1 cannot be directly applied in this case. However, by employing a more sophisticated construction and leveraging the convergence properties of the k-th order statistics, it is possible to identify the limit of $\mathbb{E}[SC_{\boldsymbol{v}}(\boldsymbol{X}_n)]$ and ensure convergence by imposing mild assumptions on the percentile vector $\boldsymbol{v}$.

Lemma 2. *Let μ be a measure that satisfies (8). Given $k \in \mathbb{N}$, let $\boldsymbol{v} \in (0,1)^k$ be a percentile vector. Then, $\mathbb{E}[SC_{\boldsymbol{v}}(\boldsymbol{X}_n)]$ converges to $W_1(\mu, \nu_{Q_{\boldsymbol{v}}})$, where $\nu_{Q_{\boldsymbol{v}}}$ is defined as*

$$\nu_{Q_{\boldsymbol{v}}} := \sum_{i=1}^{k} (F_\mu(z_i) - F_\mu(z_{i-1}))\delta_{F_\mu^{[-1]}(v_i)}, \tag{10}$$

where $z_i = \frac{(F_\mu^{[-1]}(v_i)+F_\mu^{[-1]}(v_{i+1}))}{2}$ for $i = 1, \dots, k-1$, $z_0 = \inf_{x\in I} x$, and $z_k = \sup_{x\in I} x$, F_μ is the cumulative distribution function of μ, and $F_\mu^{[-1]}$ is the pseudo-inverse function related to μ.

Proof. First, we notice that the measure (10) is well-defined since there exists a j such that $v_j \neq 0, 1$. Let $\boldsymbol{v} = (v_1, \dots, v_k)$ be a percentile vector, $\boldsymbol{x}$ be the vector containing the reports of the agents, $\nu^{(k,n)}$ be the solution to problem (7), and let $\boldsymbol{y}$ be the vector containing the facility positions returned by the percentile mechanisms, so that $\nu^{(k,n)} = \sum_{j\in[k]}(\nu^{(k,n)})_j\delta_{y_j}$. To lighten-up the notation, we set $\nu_{Q_{\boldsymbol{v}}} := \nu_Q$, thus $\nu_Q := \sum_{j\in[k]}(\nu_Q)_j\delta_{F_\mu^{[-1]}(v_j)}$, where $(\nu_Q)_j := (F_\mu(z_j) - F_\mu(z_{j-1}))$, where $z_0 = -\infty$, $z_k = +\infty$, and $z_i = \frac{y_i+y_{i+1}}{2}$ for every $i = 2, \dots, k-1$. We now show that ν_Q is the solution to the following minimization problem

$$\min_{\{\lambda_j\}_{j\in[k]}} W_1(\lambda, \mu), \tag{11}$$

where $\lambda = \sum_{j=1}^k \lambda_j \delta_{F^{[-1]}(v_j)}$. We can rewrite the W_1 distance between μ and ν_Q as it follows

$$W_1(\mu, \nu_Q) = \sum_{j=1}^{k} \int_{F^{[-1]}(\sum_{i=1}^{k}(\nu_Q)_i)}^{F^{[-1]}(\sum_{i=1}^{k+1}(\nu_Q)_i)} |x - F^{[-1]}(v_j)| d\mu.$$

By definition of ν_Q we have that $\sum_{i=1}^j (\nu_{Q_{\boldsymbol{v}}})_i = F_\mu(z_j)$, thus

$$W_1(\mu, \nu_Q) = \sum_{j=0}^{k} \int_{z_j}^{z_{j+1}} |x - F^{[-1]}(v_j)| d\mu = \int_{-\infty}^{+\infty} \min_{j\in[k]} |x - F^{[-1]}(v_j)| d\mu, \tag{12}$$

where we used the fact that $z_0 = -\infty$, $z_k = +\infty$, and $z_i = \frac{F^{[-1]}(v_i)+F^{[-1]}(v_{i+1})}{2}$ for every $i = 2, \dots, k-1$. Thus every point in the support of μ is assigned to its closest facility, thus ν_Q is a solution to (11).

We are now ready to study the convergence of $\mathbb{E}[SC_{\boldsymbol{v}}(\boldsymbol{X}_n)]$. For every $n \in \mathbb{N}$, let us define $\gamma_n = \sum_{j\in[k]}(\nu_Q)_j\delta_{y_j}$, where y_j is the j-th point in the support of $\nu^{(k,n)}$. By a similar argument to the one used to prove Lemma 1, we have

$$W_1(\mu_{\boldsymbol{x}}, \nu^{(k,n)}) \leq W_1(\mu_{\boldsymbol{x}}, \gamma_n) \leq W_1(\mu_{\boldsymbol{x}}, \mu) + W_1(\mu, \nu_Q) + W_1(\nu_Q, \gamma_n).$$

For every $n \in \mathbb{N}$, let us now define $\eta_n := \sum_{j\in[k]}(\nu^{(k,n)})_j\delta_{F_\mu^{[-1]}(v_j)}$. We then have

$$W_1(\mu, \nu_Q) \leq W_1(\mu, \eta_n) \leq W_1(\mu, \mu_{\boldsymbol{x}}) + W_1(\mu_{\boldsymbol{x}}, \nu^{(k,n)}) + W_1(\nu^{(k,n)}, \eta_n).$$

Since $W_1(\nu_Q, \gamma_n), W_1(\nu^{(k,n)}, \eta_n) \geq 0$, we infer that

$$\mathbb{E}[|W_1(\mu, \nu_Q) - W_1(\mu_{\boldsymbol{x}}, \nu^{(k,n)})|] \leq \mathbb{E}[W_1(\mu, \mu_{\boldsymbol{x}})] + \mathbb{E}[W_1(\nu_Q, \gamma_n)] + \mathbb{E}[W_1(\nu^{(k,n)}, \eta_n)]. \tag{13}$$

From [16], we have that $\lim_{n\to\infty}\mathbb{E}[W_1(\mu, \mu_{\boldsymbol{x}})] = 0$. To conclude the thesis, we need to prove that both $\mathbb{E}[W_1(\nu_Q, \gamma_n)]$ and $\mathbb{E}[W_1(\nu^{(k,n)}, \eta_n)]$ go to zero as $n \to \infty$. If we express $\nu^{(k,n)}$ and η_n explicitly, we infer that $\mathbb{E}[W_1(\nu^{(k,n)}, \eta_n)]$ converges to zero if the $(\lfloor v_j(n-1)\rfloor + 1)$-th quantile converges to $F_\mu^{[-1]}(v_j)$ with respect to the l_1 norm, which can be done using Bahadur's representation formula [20] and leveraging our hypothesis on the regularity of μ. A similar argument allows us to handle $\mathbb{E}[W_1(\nu_Q, \gamma_n)]$ and conclude the proof. □

By combining the convergence results shown in Lemma 1 and 2, we infer that the Bayesian approximation ratio of any $\mathcal{PM}_{\boldsymbol{v}}$ converges to a bounded quantity.

Theorem 3. *Let $\boldsymbol{X}_n$ be a random vector of n i.i.d. variables distributed as μ and let $\boldsymbol{v} \in (0,1)^k$ be a percentile vector. If μ satisfies* (8)*, we have*

$$\lim_{n\to\infty}\frac{\mathbb{E}[SC_{\boldsymbol{v}}(\boldsymbol{X}_n)]}{\mathbb{E}[SC_{opt}(\boldsymbol{X}_n)]} = \frac{W_1(\mu, \nu_{Q_{\boldsymbol{v}}})}{W_1(\mu, \nu^{(k)})}.$$

Theorem 3 ensures that the limit of the Bayesian approximation ratio of any percentile mechanisms is equal to a quantity that depends only on μ, k, and $\boldsymbol{v}$. For an illustration, we compute this quantity for a generic k and $\boldsymbol{v}$, and μ is the uniform distribution over $[0,1]$.

Example 1. Let $\boldsymbol{v} = (v_1, \dots, v_k)$ be a percentile vector and let the underlying distribution μ be the uniform distribution over $[0,1]$. The measure $\nu_{Q_{\boldsymbol{v}}}$ is then defined as $\nu_{Q_{\boldsymbol{v}}} := \sum_{i=1}^k \frac{v_{i+1}-v_{i-1}}{2}\delta_{v_i}$, where $v_0 = 0$ and $v_{k+1} = 1$. It is easy to see that the projection of μ over $\mathcal{P}_k(\mathbb{R})$ is $\nu^{(k)} := \frac{1}{k}\sum_{j=1}^k \delta_{\frac{2j-1}{2k}}$. From a simple computation, we infer that $W_1(\mu, \nu_{Q_{\boldsymbol{v}}}) = \sum_{i=1}^k\left[\frac{(v_{i+1}-v_i)^2+(v_i-v_{i-1})^2}{2}\right]$ and $W_1(\mu, \nu^{(k)}) = \frac{1}{4k}$. Moreover, since $v_i \leq v_{i+1}$ and $v_j \in [0,1]$, we have that $(v_{i+1}-v_i)^2 \leq v_{i+1} - v_i$, and obtain

$$\lim_{n\to\infty} B_{ar}^{(n)}(\mathcal{PM}_{\boldsymbol{v}}) \leq 4k\sum_{i=1}^k\left[\frac{(v_{i+1}-v_{i-1})}{2}\right] \leq 4k.$$

That is, when the agents are distributed according to an uniform distribution, the Bayesian approximation ratio of any percentile mechanism for the k-FLP is upper bounded by $4k$.

We now characterize the convergence rate of the Bayesian approximation ratio. To do so, μ must have compact support or there exists $\delta > 0$ such that

$$\int_{\mathbb{R}} |x|^{2+\delta} d\mu < +\infty. \tag{14}$$

In both cases, we have that the convergence rate is at most of the order $n^{-\frac{1}{2}}$.

Theorem 4. *Under the hypothesis of Theorem 3, let us further assume that either μ is supported on a compact set or satisfies* (14)*. Then, we have that*

$$\left| \frac{\mathbb{E}[SC_v(\boldsymbol{X}_n)]}{\mathbb{E}[SC_{opt}(\boldsymbol{X}_n)]} - \frac{W_1(\mu, \nu_{Q_v})}{W_1(\mu, \nu^{(k)})} \right| \leq O(n^{-\frac{1}{2}}). \tag{15}$$

Thus the convergence rate of the Bayesian approximation ratio of $\mathcal{PM}_v$ is $O(n^{-\frac{1}{2}})$. Moreover, for every $\boldsymbol{v} \in (0,1)^k$, there exists $C > 0$ such that

$$B_{ar}^{(n)}(\mathcal{PM}_v) \leq \frac{W_1(\mu, \nu_{Q_v})}{W_1(\mu, \nu^{(k)})} + \frac{C}{\sqrt{n}} \qquad \forall n > k, \tag{16}$$

where ν_{Q_v} is defined in (10) *and $\nu^{(k)}$ is a minimizer of* (9).

Notice that the term $\frac{W_1(\mu,\nu_{Q_v})}{W_1(\mu,\nu^{(k)})}$ in (16), is a constant that does not depend on n, but depends only on the specifics of the problem, that is μ, k, and $\boldsymbol{v}$.

Remark 1. To conclude the section, we discuss the case in which $v_j \in \{0,1\}$ for at least one index $j \in [m]$. For the sake of argument, let us consider a percentile mechanism induced by a percentile vector $\boldsymbol{v}$ such that $v_1 = 0$. In this case, the mechanism places a facility at the position of the leftmost agent. The asymptotic behaviour of the mechanism then depends on whether the support of μ is bounded from left or not. Indeed, if $-\infty < a := \inf_{x \in spt(\mu)} x$, we have that the position of the leftmost agent converges to a. In this case, we can study the limit Bayesian approximation ratio of the percentile mechanism, but we will not be able to retrieve any guarantee on the convergence speed. If $\inf_{x \in spt(\mu)} x = -\infty$, the position of the leftmost agent does not converge, thus we cannot adapt Theorem 3 to suit this case.

4 The Optimal Percentile Mechanism

Owing to Theorem 3, if $W_1(\mu, \nu_{Q_v}) = W_1(\mu, \nu^{(k)})$ the Bayesian approximation ratio of $\mathcal{PM}_v$ converges to 1 when $n \to \infty$. We now show that, for any $k \in \mathbb{N}$ and any underlying distribution μ, there exists a percentile vector whose associated mechanism asymptotically behaves optimally, *i.e.* the limit of the Bayesian approximation ratio of the induced mechanism is equal to 1. Given an underlying distribution μ, we denote with $\boldsymbol{v}_\mu$ its related optimal percentile vector.

Theorem 5. *Let μ be the underlying distribution and $\{y_j\}_{j\in[k]}$ be the support of the solution to problem (9). Then, the vector $\boldsymbol{v}_\mu$ defined as*

$$(v_\mu)_j = F_\mu(y_j), \tag{17}$$

is an optimal percentile vector.

Remark 2. It is also worth of notice that $\nu_{Q_{\boldsymbol{v}_\mu}} = \nu^{(k)}$ holds. Indeed, toward a contradiction, let us assume that $\nu_{Q_{\boldsymbol{v}_\mu}} \neq \nu^{(k)}$. Then there exists a $\bar{j} \in [k]$ such that $(\nu_{Q_{\boldsymbol{v}_\mu}})_i = \nu_i^{(k)}$ for every $i = 1, \ldots, \bar{j}-1$ and $(\nu_{Q_{\boldsymbol{v}_\mu}})_{\bar{j}} \neq \nu_{\bar{j}}^{(k)}$. Since the optimal transportation plan between two measures supported over a line is monotone, we have that $W_1(\mu, \nu^{(k)}) = \sum_{j=0}^{k} \int_{l_j}^{l_{j+1}} |x - y_j| d\mu$, where $l_0 = -\infty$ and $l_r = F_\mu^{[-1]}(\sum_{i=1}^{r} \nu_i^{(k)})$ for every $r \in [k]$. Since $(\nu_{Q_{\boldsymbol{v}_\mu}})_{\bar{j}} \neq \nu_{\bar{j}}^{(k)}$, we have that $l_{\bar{j}} \neq \frac{y_{\bar{j}}+y_{\bar{j}+1}}{2}$. Thus we have $W_1(\mu, \nu^{(k)}) \neq \int_{-\infty}^{+\infty} \min_{j\in[k]} |x - y_j| d\mu$, which contradicts the definition of $\nu^{(k)}$ and (12), thus $\nu_{Q_{\boldsymbol{v}_\mu}} = \nu^{(k)}$.

Given $k \in \mathbb{N}$ and a probability measure μ, it is possible to retrieve a system of k equations that characterizes the optimal percentile mechanism. Indeed, let us denote with $y_1, \ldots, y_k$ the support of the solution to $\min_{\lambda\in\mathcal{P}_k(\mathbb{R})} W_1(\mu, \lambda)$ and let $z_i = \frac{y_i+y_{i+1}}{2}$ for $i = 1, \ldots, k-1$, $z_0 = -\infty$, and $z_k = +\infty$. Since every agent's cost is defined by its distance to the closest facility, we know that every agent in (z_0, z_1) will access the facility located in y_1. Due to the optimality of the solution, we infer that y_1 is locally optimal over the set (z_0, z_1). Otherwise, we could reduce the cost of the solution by replacing y_1 with the optimal facility location for the problem restricted to (z_0, z_1). Since we are considering the Social Cost, the local optimality of y_1 is expressed by the identity $2(F_\mu(y_1) - F_\mu(z_0)) = F(z_1) - F_\mu(z_0)$, since y_1 has to be the median of μ when the measure is restricted to (z_0, z_1).

Theorem 6. *Given $k \in \mathbb{N}$ and $\mu \in \mathcal{P}(\mathbb{R})$, let ν be a solution to Problem (9). Then the optimal percentile vector is $\boldsymbol{v}_\mu = (F_\mu(y_1), \ldots, F_\mu(y_m)) \in (0,1)^k$, where $y_1 \leq y_2 \leq \cdots \leq y_k$ satisfy the following system of k equations*

$$\begin{cases} 2F_\mu(y_1) = F_\mu\left(\frac{y_1+y_2}{2}\right) \\ 2\left(F_\mu(y_{j-1}) - F_\mu\left(\frac{y_{j-2}+y_{j-1}}{2}\right)\right) = F_\mu\left(\frac{y_j+y_{j-1}}{2}\right) - F_\mu\left(\frac{y_{j-1}+y_{j-2}}{2}\right), \quad j \in [k-1]. \\ 2\left(F_\mu(y_k) - F_\mu\left(\frac{y_k+y_{k-1}}{2}\right)\right) = 1 - F_\mu\left(\frac{y_k+y_{k-1}}{2}\right) \end{cases} \tag{18}$$

Since the projection problem (9) admits a solution, system (18) admits at least a solution. In Table 1, we report the optimal percentile vectors associated with the uniform, normal, and exponential distributions. The details on how to apply Theorem 6 to compute them are deferred to the full version of the paper.

Theorem 7. *Given a probability distribution μ such that condition (8) is satisfied, let $\boldsymbol{v}_\mu$ be the optimal percentile vector associated with μ. Then, there exists a constant C such that, for every $n > k$, we have $B_{ar}^{(n)}(\mathcal{PM}_{\boldsymbol{v}_\mu}) \leq 1 + \frac{C}{\sqrt{n}}$.*

Table 1. The asymptotically optimal percentile vectors for the Normal ($\mathcal{N}$), Exponential ($\mathcal{E}$), and Uniform distribution ($\mathcal{U}$). Every row contains the optimal percentile vectors of a distribution for 1, 2, and 3 facilities and with respects to the Social Cost.

	$k=1$	$k=2$	$k=3$
$\mathcal{N}$	(0.5)	(0.25, 0.75)	(0.15, 0.5, 0.85)
$\mathcal{E}$	(0.5)	(0.33, 0.83)	(0.25, 0.67, 0.92)
$\mathcal{U}$	(0.5)	(0.25, 0.75)	(0.16, 0.5, 0.83)

To conclude, we show that the limit of the Bayesian approximation ratio of the percentile mechanisms and the percentile vector $\boldsymbol{v}_\mu$ defined in (17) are immune to scale changes. This is particularly useful when the mechanism designer only knows the class of distribution to which the agents' distribution belongs. For example, the designer might know that the agents' type follows a Gaussian distribution but is unaware of its mean and/or its standard deviation. In the following, we show that the optimal percentile and the limit of the Bayesian approximation ratio of the percentile mechanisms are the same regardless of the mean or standard deviation of the distribution.

Theorem 8. *Let X be the random variable that describes the agents' type distribution. If $\boldsymbol{v}_\mu$ is the optimal percentile vector associated with X, then $\boldsymbol{v}_\mu$ is also the optimal percentile vector for any random variable of the form $X' := \sigma X + m$, where $m \in \mathbb{R}$ and $\sigma > 0$.*

Theorem 8 formalizes the following observation: the optimal facility locations and the output of any percentile mechanism do not depend on the scale. Indeed, given a percentile vector $\boldsymbol{v}$ and the number of agents n, if the agents' positions are sampled from a random variable X, the output of $\mathcal{PM}_v$ is the vector containing the $(\lfloor (n-1)v_j \rfloor + 1)$-th order statistics of the sample. Since the ordering of the values is unaffected by positive affine transformations, scaling any sample just magnifies (or shrinks) the cost of the output according to σ. Similarly, if we scale the agents' positions, the optimal facility locations will scale accordingly. Hence the ratio of the two costs is immune to scale changes.

5 Computing the Optimal Percentile Mechanism from an Approximation of μ

In the previous section, we have shown that if the agents' types are sampled from a common probability distribution μ, the mechanism designer is able to detect a percentile mechanism whose cost is asymptotically optimal when the number of agents increases. To detect the optimal percentile vector, however, it is necessary to have access to μ. In many cases, this is not feasible since the designer has only access to an approximation or a prediction of agents' distribution, namely $\tilde{\mu}$. Thus, the designer is able to compute $\boldsymbol{v}_{\tilde{\mu}}$ rather than the real optimal percentile vector $\boldsymbol{v}_\mu$. We show that it is possible to estimate the difference between the limit of the Bayesian approximation ratio of $\mathcal{PM}_{\boldsymbol{v}_{\tilde{\mu}}}$ and 1, i.e. the limit of the Bayesian approximation ratio of the optimal percentile mechanism. In particular, we show that the closer μ and $\tilde{\mu}$ are with respect to W_∞, the closer the asymptotic cost of $\mathcal{PM}_{\boldsymbol{v}_{\tilde{\mu}}}$ gets to the optimal cost.

Theorem 9. *Let $\tilde{\mu}$ and μ be two probability measures supported over a compact interval I. Let $\boldsymbol{v}_{\tilde{\mu}}$ be the percentile vector obtained by solving the system* (17) *by using $\tilde{\mu}$ instead of μ. Then, we have*

$$\lim_{n\to\infty}\left|\frac{\mathbb{E}[SC_{\boldsymbol{v}_{\tilde{\mu}}}(\boldsymbol{X}_n)]}{\mathbb{E}[SC_{opt}(\boldsymbol{X}_n)]}-1\right|\leq\frac{W_\infty(\mu,\tilde{\mu})+2W_1(\mu,\tilde{\mu})}{W_1(\mu,\nu^{(k)})}\leq 3\frac{W_\infty(\mu,\tilde{\mu})}{W_1(\mu,\nu^{(k)})},\tag{19}$$

where $\mathbb{E}$ is the expected value with respect to the real agents' distribution μ. In particular, for every $n>k$, there exists a constant $C>0$ such that

$$|B_{ar}^{(n)}(\mathcal{PM}_{\boldsymbol{v}_\mu})-B_{ar}^{(n)}(\mathcal{PM}_{\boldsymbol{v}_{\tilde{\mu}}})|\leq 3\frac{W_\infty(\mu,\tilde{\mu})}{W_1(\mu,\nu^{(k)})}+\frac{C}{\sqrt{n}},$$

where $\boldsymbol{v}_\mu$ is the optimal percentile vector associated with μ.

Proof. Let us denote with $\tilde{\nu}^{(k)}$ the projection of $\tilde{\mu}$ over $\mathcal{P}_k(\mathbb{R})$ and with $\nu^{(k)}$ the projection of μ over $\mathcal{P}_k(\mathbb{R})$. We denote with $\{y_j\}_{j\in[k]}$ the support of $\nu^{(k)}$ and with $\{\tilde{y}_j\}_{j\in[k]}$ the support of $\tilde{\nu}^{(k)}$. Accordingly, we denote with $\boldsymbol{v}_{\tilde{\mu}}$ and $\boldsymbol{v}_\mu$ the optimal percentile vectors associated with $\tilde{\mu}$ and μ, respectively. By Lemma 2, we have that the numerator of the Bayesian approximation ratio converges to $W_1(\mu,\nu_{Q_{\boldsymbol{v}_{\tilde{\mu}}}})$, where $\nu_{Q_{\boldsymbol{v}_{\tilde{\mu}}}}$ is defined as in (10). Let us now consider, $\beta_{\boldsymbol{v}_{\tilde{\mu}}}$ defined as $\beta_{\boldsymbol{v}_{\tilde{\mu}}}:=\sum_{j\in[k]}(\tilde{\nu}^{(k)})_j\delta_{z_j}$, where $\boldsymbol{z}=(z_1,\dots,z_k)$ is the support of $\nu_{Q_{\boldsymbol{v}_{\tilde{\mu}}}}$, then

$$\begin{aligned}W_1(\mu,\nu_{Q_{\boldsymbol{v}_{\tilde{\mu}}}})&\leq W_1(\mu,\beta_{\boldsymbol{v}_{\tilde{\mu}}})\leq W_1(\mu,\tilde{\mu})+W_1(\tilde{\mu},\beta_{\boldsymbol{v}_{\tilde{\mu}}})\\&\leq W_1(\mu,\tilde{\mu})+W_1(\tilde{\mu},\tilde{\nu}^{(k)})+W_1(\tilde{\nu}^{(k)},\beta_{\boldsymbol{v}_{\tilde{\mu}}})\\&\leq W_1(\mu,\tilde{\mu})+W_1(\tilde{\mu},\nu^{(k)})+W_1(\tilde{\nu}^{(k)},\beta_{\boldsymbol{v}_{\tilde{\mu}}})\\&\leq 2W_1(\mu,\tilde{\mu})+W_1(\mu,\nu^{(k)})+W_1(\tilde{\nu}^{(k)},\beta_{\boldsymbol{v}_{\tilde{\mu}}}).\end{aligned}\tag{20}$$

By definition of $\beta_{\boldsymbol{v}_{\tilde{\mu}}}$ and $\tilde{\nu}^{(k)}$, we have $W_1(\tilde{\nu}^{(k)},\beta_{\boldsymbol{v}_{\tilde{\mu}}})\leq\sum_{j\in[k]}(\tilde{\nu}^{(k)})_j|F_{\tilde{\mu}}^{[-1]}(\tilde{p}_j)-F_\mu^{[-1]}(\tilde{p}_j)|$. Since $W_1(\mu,\tilde{\mu})\leq W_\infty(\mu,\tilde{\mu})$ and $W_\infty(\mu,\tilde{\mu})=\max_{\ell\in[0,1]}|F_\mu^{[-1]}(\ell)-F_{\tilde{\mu}}^{[-1]}(\ell)|$, we infer $W_1(\tilde{\nu}^{(k)},\beta_{\boldsymbol{v}_{\tilde{\mu}}})\leq W_\infty(\mu,\tilde{\mu})$. To conclude, we notice that,

$$\begin{aligned}&|B_{ar}^{(n)}(\mathcal{PM}_{\boldsymbol{v}_\mu})-B_{ar}^{(n)}(\mathcal{PM}_{\boldsymbol{v}_{\tilde{\mu}}})|\\&\quad\leq|1-B_{ar}(\mathcal{PM}_{\boldsymbol{v}_{\tilde{\mu}}})|+|B_{ar}^{(n)}(\mathcal{PM}_{\boldsymbol{v}_\mu})-1|+|B_{ar}(\mathcal{PM}_{\boldsymbol{v}_{\tilde{\mu}}})-B_{ar}^{(n)}(\mathcal{PM}_{\boldsymbol{v}_{\tilde{\mu}}})|\\&\quad\leq 3\frac{W_\infty(\mu,\tilde{\mu})}{W_1(\mu,\nu^{(k)})}+|B_{ar}^{(n)}(\mathcal{PM}_{\boldsymbol{v}_\mu})-1|+|B_{ar}(\mathcal{PM}_{\boldsymbol{v}_{\tilde{\mu}}})-B_{ar}^{(n)}(\mathcal{PM}_{\boldsymbol{v}_{\tilde{\mu}}})|,\end{aligned}$$

where $B_{ar}(\mathcal{PM}_{\boldsymbol{v}_{\tilde{\mu}}})=\lim_{n\to\infty}B_{ar}^{(n)}(\mathcal{PM}_{\boldsymbol{v}_{\tilde{\mu}}})$. Since μ has compact support and $\boldsymbol{v}_\mu,\boldsymbol{v}_{\tilde{\mu}}\in(0,1)^k$, we infer that $|B_{ar}^{(n)}(\mathcal{PM}_{\boldsymbol{v}_\mu})-1|\leq O(n^{-\frac{1}{2}})$ and $|B_{ar}(\mathcal{PM}_{\boldsymbol{v}_{\tilde{\mu}}})-B_{ar}^{(n)}(\mathcal{PM}_{\boldsymbol{v}_{\tilde{\mu}}})|\leq O(n^{-\frac{1}{2}})$, which concludes the proof. □

6 Conclusion and Future Works

In this paper, we studied the percentile mechanisms in the Bayesian Mechanism Design framework. We have shown that the ratio between the expected cost of the mechanisms and the expected optimal cost converges to a constant as the number of agents goes to infinity. We have characterized both the limit value and the convergence speed. We then showed that for every underlying distribution μ, there exists an optimal percentile vector $\boldsymbol{v}_\mu$ that does not depend on the mean or the variance of the distribution. The scale invariance property allows us to compute the optimal percentile vector when the designer only knows the class to which the probability measure belongs. Lastly, we have shown that determining the optimal percentile mechanism from an approximation of the underlying distribution leads to a mechanism whose performance is quasi-optimal as long as the approximation is close to the real distribution with respect to W_∞.

An open question is whether our formalism could be adopted to higher dimensional cases. In [45], the percentile mechanisms are generalized to higher dimensions by dealing with each dimension separately. This suggests that our approach can be extended to handle higher-dimensional problems since the Wasserstein Distance can be separated along each cardinal direction [4,5]. Moreover, our framework can be extended beyond the classic k-FLP. In particular, it is foreseeable to use our results to tackle the case in which agents have fractional preferences. Anotherdirection is to adapt our reformulation of the problem through OT theory to design and study randomized mechanisms for the k-FLP.

Acknowledgments. Jie Zhang was partially supported by a Leverhulme Trust Research Project Grant (2021–2024) and the EPSRC grant (EP/W014912/1).

References

1. Ahmadi-Javid, A., Seyedi, P., Syam, S.S.: A survey of healthcare facility location. Comput. Oper. Res. **79**, 223–263 (2017)
2. Alon, N., Feldman, M., Procaccia, A.D., Tennenholtz, M.: Strategyproof approximation of the minimax on networks. Math. Oper. Res. **35**(3), 513–526 (2010)
3. Ambrosio, L., Gigli, N., Savaré, G.: Gradient Flows: In Metric Spaces and in the Space of Probability Measures. Springer Science & Business Media, Berlin (2005). https://doi.org/10.1007/978-3-7643-8722-8
4. Auricchio, G.: On the pythagorean structure of the optimal transport for separable cost functions. Rendiconti Lincei **34**(4), 745–771 (2024)
5. Auricchio, G., Bassetti, F., Gualandi, S., Veneroni, M.: Computing kantorovich-wasserstein distances on d-dimensional histograms using $(d+1)$-partite graphs. In: Advances in Neural Information Processing Systems, vol. 31, pp. 5798–5808 (2018)
6. Auricchio, G., Bassetti, F., Gualandi, S., Veneroni, M.: Computing Wasserstein Barycenters via linear programming. In: Rousseau, L.-M., Stergiou, K. (eds.) CPAIOR 2019. LNCS, vol. 11494, pp. 355–363. Springer, Cham (2019). https://doi.org/10.1007/978-3-030-19212-9_23
7. Auricchio, G., Clough, H.J., Zhang, J.: On the capacitated facility location problem with scarce resources. In: The 40th Conference on Uncertainty in Artificial Intelligence (2024)

8. Auricchio, G., Codegoni, A., Gualandi, S., Toscani, G., Veneroni, M.: The equivalence of fourier-based and wasserstein metrics on imaging problems. Rendiconti Lincei **31**(3), 627–649 (2020)
9. Auricchio, G., Wang, Z., Zhang, J.: Facility location problems with capacity constraints: two facilities and beyond. arXiv preprint arXiv:2404.13566 (2024)
10. Auricchio, G., Zhang, J.: The k-facility location problem via optimal transport: a Bayesian study of the percentile mechanisms (2024). https://arxiv.org/abs/2407.06398
11. Auricchio, G., Zhang, J., Zhang, M.: Extended ranking mechanisms for the m-capacitated facility location problem in Bayesian mechanism design. In: Proceedings of the 23rd International Conference on Autonomous Agents and Multiagent Systems, pp. 87–95. AAMAS '24 (2024)
12. Balcik, B., Beamon, B.M.: Facility location in humanitarian relief. Int. J. Log. Res. Appl. **11**(2), 101–121 (2008)
13. Barbera, S., Dutta, B., Sen, A.: Strategy-proof social choice correspondences. J. Econ. Theory **101**(2), 374–394 (2001)
14. Barda, O.H., Dupuis, J., Lencioni, P.: Multicriteria location of thermal power plants. Eur. J. Oper. Res. **45**(2–3), 332–346 (1990)
15. Black, D.: On the rationale of group decision-making. J. Polit. Econ. **56**(1), 23–34 (1948)
16. Bobkov, S., Ledoux, M.: One-dimensional empirical measures, order statistics, and Kantorovich transport distances, vol. 261. American Mathematical Society, Providence, Rhode Island (2019)
17. Chawla, S., Hartline, J.D., Kleinberg, R.: Algorithmic pricing via virtual valuations. In: Proceedings of the 8th ACM Conference on Electronic Commerce, pp. 243–251. ACM, New York, NY, USA (2007)
18. Chawla, S., Sivan, B.: Bayesian algorithmic mechanism design. ACM SIGecom Exchanges **13**(1), 5–49 (2014)
19. Daskalakis, C., Deckelbaum, A., Tzamos, C.: Mechanism design via optimal transport. In: Proceedings of the Fourteenth ACM Conference on Electronic Commerce, pp. 269–286. ACM, New York, NY, USA (2013)
20. De Haan, L., Taconis-Haantjes, E.: On bahadur's representation of sample quantiles. Ann. Inst. Statist. Math **31**(Part A), 299–308 (1979)
21. Dokow, E., Feldman, M., Meir, R., Nehama, I.: Mechanism design on discrete lines and cycles. In: EC, pp. 423–440. ACM, New York, NY, USA (2012)
22. Escoffier, B., Gourvès, L., Kim Thang, N., Pascual, F., Spanjaard, O.: Strategy-proof mechanisms for facility location games with many facilities. In: Brafman, R.I., Roberts, F.S., Tsoukiàs, A. (eds.) ADT 2011. LNCS (LNAI), vol. 6992, pp. 67–81. Springer, Heidelberg (2011). https://doi.org/10.1007/978-3-642-24873-3_6
23. Filimonov, A., Meir, R.: Strategyproof facility location mechanisms on discrete trees. In: AAMAS, pp. 510–518. ACM, New York, NY, USA (2021)
24. Filos-Ratsikas, A., Li, M., Zhang, J., Zhang, Q.: Facility location with double-peaked preferences. Auton. Agents Multi Agent Syst. **31**(6), 1209–1235 (2017)
25. Fotakis, D., Tzamos, C.: Strategyproof facility location for concave cost functions. In: Proceedings of the Fourteenth ACM Conference on Electronic Commerce, pp. 435–452. ACM, New York, NY, USA (2013)
26. Fotakis, D., Tzamos, C.: On the power of deterministic mechanisms for facility location games. ACM Trans. Econ. Comput. 1–37 (2014)
27. Frogner, C., Zhang, C., Mobahi, H., Araya, M., Poggio, T.A.: Learning with a Wasserstein loss. Adv. Neural Inf. Process. Syst. **28** (2015)

28. Gairing, M., Monien, B., Tiemann, K.: Selfish routing with incomplete information. In: Proceedings of the Seventeenth Annual ACM Symposium on Parallelism in Algorithms and Architectures, pp. 203–212. ACM, New York, NY, USA (2005)
29. Hartline, J.D.: Bayesian mechanism design. Found. Trends® Theor. Comput. Sci. **8**(3), 143–263 (2013)
30. Hartline, J.D., Roughgarden, T.: Simple versus optimal mechanisms. In: Proceedings of the 10th ACM Conference on Electronic Commerce. ACM (2009)
31. Kantorovich, L.V.: On the translocation of masses. J. Math. Sci. **133**(4), 1381–1382 (2006)
32. Levina, E., Bickel, P.: The earth mover's distance is the Mallows distance: some insights from statistics. In: Proceedings of the IEEE International Conference on Computer Vision, vol. 2, pp. 251–256, February 2001
33. Lu, P., Sun, X., Wang, Y., Zhu, Z.A.: Asymptotically optimal strategy-proof mechanisms for two-facility games. In: EC, pp. 315–324. ACM (2010)
34. Lu, P., Wang, Y., Zhou, Y.: Tighter bounds for facility games. In: Leonardi, S. (ed.) WINE 2009. LNCS, vol. 5929, pp. 137–148. Springer, Heidelberg (2009). https://doi.org/10.1007/978-3-642-10841-9_14
35. Lucier, B., Borodin, A.: Price of anarchy for greedy auctions. In: Proceedings of the Twenty-First Annual ACM-SIAM Symposium on Discrete Algorithms, pp. 537–553. SIAM, New York, NY, USA (2010)
36. Lugosi, G., Mendelson, S.: Sub-Gaussian estimators of the mean of a random vector. Ann. Stat. **47**(2), 783–794 (2019)
37. Meir, R.: Strategyproof facility location for three agents on a circle. In: Fotakis, D., Markakis, E. (eds.) SAGT 2019. LNCS, vol. 11801, pp. 18–33. Springer, Cham (2019). https://doi.org/10.1007/978-3-030-30473-7_2
38. Melo, M., Nickel, S., da Gama, F.S.: Facility location and supply chain management - a review. Eur. J. Oper. Res. **196**(2), 401–412 (2009)
39. Nisan, N., Ronen, A.: Algorithmic mechanism design. In: Proceedings of the Thirty-First Annual ACM Symposium on Theory of Computing, pp. 129–140. ACM, New York, NY, USA (1999)
40. Pele, O., Werman, M.: Fast and robust Earth Mover's Distances. In: Computer Vision, 2009 IEEE 12th International Conference on, pp. 460–467. IEEE (2009)
41. Procaccia, A.D., Tennenholtz, M.: Approximate mechanism design without money. ACM Trans. Econ. Comput. (TEAC) **1**(4), 1–26 (2013)
42. Rubner, Y., Tomasi, C., Guibas, L.J.: The earth mover's distance as a metric for image retrieval. Int. J. Comput. Vis. **40**(2), 99–121 (2000)
43. Scagliotti, A.: Deep learning approximation of diffeomorphisms via linear-control systems. Math. Control Related Fields **13**(3), 1226–1257 (2023)
44. Scagliotti, A., Farinelli, S.: Normalizing flows as approximations of optimal transport maps via linear-control neural odes (2023)
45. Sui, X., Boutilier, C., Sandholm, T.: Analysis and optimization of multi-dimensional percentile mechanisms. In: IJCAI, pp. 367–374. Citeseer (2013)
46. Tang, P., Yu, D., Zhao, S.: Characterization of group-strategyproof mechanisms for facility location in strictly convex space. In: EC, pp. 133–157. ACM (2020)
47. Villani, C.: Optimal Transport: Old and New, vol. 338. Springer, Berlin (2009). https://doi.org/10.1007/978-3-540-71050-9
48. Zampetakis, E., Zhang, F.: Bayesian strategy-proof facility location via robust estimation. In: International Conference on Artificial Intelligence and Statistics, pp. 4196–4208. PMLR (2023)

Discrete Single-Parameter Optimal Auction Design

Yiannis Giannakopoulos[1] and Johannes Hahn[2(✉)]

[1] University of Glasgow, Glasgow, UK
yiannis.giannakopoulos@glasgow.ac.uk
[2] University of Technology Nuremberg, Nuremberg, Germany
johannes.hahn@utn.de

Abstract. We study the classic single-item auction setting of Myerson, but under the assumption that the buyers' values for the item are distributed over *finite* supports. Using strong LP duality and polyhedral theory, we rederive various key results regarding the revenue-maximizing auction, including the characterization through virtual welfare maximization and the optimality of deterministic mechanisms, as well as a novel, generic equivalence between dominant-strategy and Bayesian incentive compatibility. Inspired by this, we abstract our approach to handle more general auction settings, where the feasibility space can be given by arbitrary convex constraints, and the objective is a linear combination of revenue and social welfare. We characterize the optimal auctions of such systems as generalized virtual welfare maximizers, by making use of their KKT conditions, and we present an analogue of Myerson's payment formula for general discrete single-parameter auction settings. Additionally, we prove that total unimodularity of the feasibility space is a sufficient condition to guarantee the optimality of auctions with integral allocation rules. Finally, in the full version of our paper, we demonstrate this KKT approach by applying it to a setting where bidders are interested in buying feasible flows on trees with capacity constraints, and provide a combinatorial description of the (randomized, in general) optimal auction.

Keywords: Optimal auction design · Revenue maximization · LP duality · KKT conditions · Discrete auctions · Network auctions · Ironing · Virtual valuations

1 Introduction

The design of optimal auctions [12,15] that maximize the seller's revenue is a cornerstone of the field of mechanism design (see, e.g., [11, Ch. 9] and [9]), established into prominence by the highly-influential work of Myerson [16], and traced back to the seminal work of Vickrey [17].

A full version of this paper can be found at https://arxiv.org/abs/2406.08125 [7].

G. Schäfer and C. Ventre (Eds.): SAGT 2024, LNCS 15156, pp. 165–183, 2024.
https://doi.org/10.1007/978-3-031-71033-9_10

In its most classical form [16], which is the basis for the setting we are studying in our paper as well, there is a single item to sell and the problem is modelled as a Bayesian game. The seller has only incomplete information about the bidders' true valuations of the item, in the form of independent (but not necessarily identical) probability distributions; these distributions are assumed to be public knowledge across all participants in the auction. The players/bidders submit bids to the auctioneer/seller and the seller decides (a) who gets the item, and with what probability (since lotteries are allowed), and (b) how much the winning bidders are charged for this transaction.

In this game formulation, the strategies of the players are the different bids they can submit, and it could well be the case that bidders misreport their true valuations, if this can result in maximizing their own personal utility. Therefore, a desirable feature of mechanism design in such settings is the implementation of auctions which provably guarantee that truth-telling is an *equilibrium* of the game; such auctions are called *truthful* (or *incentive compatible (IC)*). Perhaps surprisingly, the celebrated Revelation Principle of Myerson [16] ensures that restricting our attention within the class of such well-behaved selling mechanisms is without loss for our purposes.

The seminal work of Myerson [16] provides a complete and mathematically satisfying characterization of revenue-maximizing truthful auctions in the aforementioned single-item setting, under the assumption that valuation/bidding spaces are *continuous*. It explicitly constructs an optimal auction that (a) is deterministic, i.e. the item is allocated to a single bidder (with full probability), or not sold at all, (b) satisfies truthfulness in a very strong sense, namely under dominant-strategy equilibrium, and not just in-expectation (see Sect. 2.2 for more details), and (c) has a very elegant description, enabled via the well-known *virtual valuation* "trick" (see (1)); this casts the problem into the domain of welfare-maximization, simplifying it significantly by stripping away the game-theoretic incentives components, and transforming it to a "purely algorithmic" optimization problem—resembling the familiar, to any computer scientist, notion of a reduction (a formalization of this connection, even for more general environments, can be found in the work of Cai, Daskalakis, and Weinberg [2,3]).

Still, the assumption of continuity may be considered as too strong for many practical, and theoretical, purposes. Any conceivable instantiation of an auction on a computing system will require some kind of discretization; not only as a trivial, unavoidable consequence of the fundamentally discrete nature of computation (i.e., "bits"), but also for practical reasons: bids are usually expected to be submitted as increments of some common denomination (e.g., "cents"). And any implementation of optimal auction design as an optimization problem, would need to be determined by finitely many parameters and variables, to be passed, e.g., to some solver. Furthermore, although many of the key properties and results for the continuous setting can be derived as a limiting case of a sequence of discrete ones, in general the opposite is not true: most of the techniques used in traditional auction theory rely on real analysis and continuous probability, thus breaking down when called to be applied to discrete spaces.

The above reasons highlight the importance of deriving a clear and robust theory of optimal auction design, under the assumption of *finite* value spaces. In other words, a discrete analogue of Myerson's [16] theory. During the last couple of decades, various papers within the field of algorithmic game theory have dealt with this task; see Sect. 1.1 for a more detailed overview. Our goal in this paper is to first rederive existing key results, in a unified way, with an emphasis on clarity, simplicity, and rigorousness; and, do this via purely discrete optimization tools (namely, LP duality and polyhedral combinatorics), "agnostically", rather than trying to mimic and discretize Myerson's [16] approach for the continuous setting. Secondly, this comprehensiveness and transparency allows us to lift our approach up to handle quite general single-parameter mechanism design environments, by concisely formulating our problem as an elegant KKT system.

1.1 Related Work

To the best of our knowledge, the first to explicitly study optimal auction design at a discrete setting were Bergmann and Pesendorfer [1] and Elkind [6]; the latter offers a more complete treatment, providing a natural discretization of Myerson's [16] techniques, including "ironing" of non-regular distributions. A limitation of [6] is that it establishes that the discrete analogue of Myerson's auction is optimal within the more restrictive class of dominant-strategy incentive compatible (DSIC) mechanisms, instead of using the standard, weaker notion of Bayesian incentive compatibility (BIC).

In a discussion paper, Malakhov and Vohra [14] study discrete auction environments with identical bidders under BIC, providing a simpler, equivalent characterization of truthfulness, through a set of local constraints. We will make critical use of this characterization, appropriately adapted to our general, non-symmetric setting of our paper (see Sect. 2.3). The treatment of [14] puts emphasis on linear programming (LP) formulations, and derive an interesting, flow-based description of optimality for general, multi-dimensional mechanism design settings; the monograph of Vorah [18] provides a comprehensive treatment of this approach.

All aforementioned approaches work, essentially, by adapting the key steps of Myerson's derivations, from the continuous to the discrete setting. Cai, Devanur, and Weinberg [4] provide a totally different, and very powerful, approach based on Lagrangian duality. Conceptually, their paper is clearly the closest to ours. [4] followed a line of work, where duality proved very useful in designing optimal multiple-item auctions in the continuous case (see, e.g., [5,8]). Although the duality framework of [4] is fundamentally discrete, it was also designed for multi-dimensional revenue-maximization, a notoriously difficult and complex problem. Therefore, its instantiation for a single-parameter Myersonian setting (see [4, Sec. 4]) results, arguably, in a rather involved presentation. One of the goals of our paper is exactly to demystify duality for single-item domains, by making use of classical LP duality, particularly tailored for our problem, instead of the more obscure Lagrangian flows interpretation in [4], resulting in greater transparency and a wider spectrum of questions that we can attack (see Sect. 3).

1.2 Our Results

We begin our presentation by introducing our *single-parameter* auction design setting, and fixing some overarching notation, in Sect. 2. Our model formulation is deliberately general, allowing for arbitrary feasibility domains $\mathcal{A}$ for the auction's allocation; we will specialize this to the standard distributional simplex when studying the classical Myersonian single-item setting in Sect. 3, however we want to be able to capture the abstract convex environments we study later in Sect. 4. Importantly, in Sect. 2.2 we discuss in detail the two different notions of truthfulness used for our problem, and in Sect. 2.3 we provide a local characterization of truthfulness, essentially taken from Malakhov and Vohra [14], which we will extensively use in our optimization formulation throughout our paper.

Section 3 includes our rederivation of the key components of Myerson's [16] theory for single-item revenue-maximization, but for finite-support distributions, as well as some novel results. They all arise, in a unified way, through a chain of traditional LP duality, presented in Sect. 3.1 (see Fig. 1 for a concise pictorial view). The resulting revenue-maximizing auction, together with some key results characterizing optimality, are given in the "master" Theorem 1: in a nutshell, the optimal auction first transforms the submitted bids to *virtual* bids and then *irons* them, finally allocating the item to the highest nonnegative (virtual, ironed) bidder. Similar to the classical results of [16] for continuous domains, this auction turns out to be deterministic and truthful in the strongest DSIC sense, "for free", although we are optimizing within the much wider space of lotteries under BIC. To the best of our knowledge, Point 2 where we formalize the equivalence of DSIC and BIC, under revenue-maximization, as a more fundamental and general consequence of the polyhedral structure of our feasibility space, rather than just a feature of the particular optimal auction solution format, is novel. The remaining Sects. 3.2, and 3.3, are dedicated in elaborating and formally proving the various components of Theorem 1. A point worth noting is that our virtual value (12) and ironing (14) transformations are not "guessed" and the proven to impose optimality, as is the case with prior work in the area, but rather arise organically as a necessity of our strong LP duality technique.

Inspired by the transparency of our duality framework in Sect. 3, we try to generalize our approach to a more general single-parameter mechanism design setting, where the feasibility space $\mathcal{A}$ is given by arbitrary convex constraints, and the optimization objective is a linear combination of revenue and social welfare; see Sect. 4.1. Our results are summarized in master Theorem 2, which is essentially the analogue of Theorem 1. Given the generality of our model in this section, we have to depart from our basic LP duality tools of Sect. 3, and make use of the more general KKT conditions framework, including duality and complementary slackness; our KKT formulation is discussed in the full version of our paper [7, Sec 4.2]. The abstraction of our model allows for a very concise and elegant description of the optimal auction's allocation and payment rules (see Sect. 4.2). Similarly to the single-item setting of Sect. 3, we can again show that optimizing under the more restrictive notion of DSIC truthfulness is without loss for our optimization objective. Furthermore, we investigate under what structural conditions of our underlying feasibility space we can "generically" guarantee that there exists an optimal auction that does not need to allocate fractionally/randomly, i.e. it is integral; it turns out, that *total unimodularity* is such a sufficient condition (see Definition 2, and the full version of our paper [7, Sec. 4.4], for more details and definitions).

Due to space constraints, all omitted proofs and missing content, including an application of our framework to a combinatorial problem of buying flows on a capacitated tree network, can be found in the full version of our paper [7].

2 Preliminaries

2.1 Model and Notation

Single-Parameter Settings. In a (Bayesian) single-parameter auction design setting there are $n \geq 1$ bidders, and each bidder $i \in [n]$ has a value $v_i \in \mathbb{R}_{\geq 0}$ for being allocated a single "unit" of some "service". Each value v_i is drawn independently from a distribution (with cdf) F_i with support $V_i \subseteq \mathbb{R}_{\geq 0}$, called the *prior* of bidder i. We will use f_i to denote the probability mass function (pmf) of F_i. These distributions are public knowledge, however the realization v_i is private information of bidder i only. In this paper we only study *discrete* auction settings, where the prior supports V_i are *finite*. For notational convenience we denote the corresponding product distribution of the *value profiles* $\boldsymbol{v} = (v_1, v_2, \dots, v_n) \in \boldsymbol{V} := \times_{i=1}^{n} V_i$ by $\boldsymbol{F} := \times_{i=1}^{n} F_i$, and we also use $\boldsymbol{V}_{-i} := \times_{j \in [n] \setminus i} V_j$ and $\boldsymbol{F}_{-i} := \times_{j \in [n] \setminus i} F_j$.

There is also a set of feasible outcomes $\mathcal{A} \subseteq \mathbb{R}_{\geq 0}^{n}$, each outcome $\boldsymbol{a} = (a_1, a_2, \dots, a_n) \in \mathcal{A}$ corresponding to bidder i being allocated a "quantity" a_i. Throughout this paper we assume that $\mathcal{A}$ is *convex*. A canonical example is the classical single-item auction setting (which we study in Sect. 3), where a_i can be interpreted as the probability of a lottery assigning the item to bidder i, in which case the feasibility set $\mathcal{A}$ is the n-dimensional simplex $\mathcal{S}_n := \left\{\boldsymbol{a} \in \mathbb{R}_{+}^{n} \mid \sum_{i=1}^{n} a_i \leq 1\right\}$.

Auctions. An auction $M = (\boldsymbol{a}, \boldsymbol{p})$ consists of an allocation rule $\boldsymbol{a} : \boldsymbol{V} \longrightarrow \mathcal{A}$ and a payment rule $\boldsymbol{p} : \boldsymbol{V} \longrightarrow \mathbb{R}^n$ that, given as input a vector of bids $\boldsymbol{b} \in \boldsymbol{V}$, dictates that each bidder i should get allocated quantity $a_i(\boldsymbol{b})$ and submit a payment of $p_i(\boldsymbol{b})$ to the auctioneer.

Given such an auction M, the (ex-post) utility of a bidder i, when their true value is $v_i \in V_i$ and bidders submit bids $\boldsymbol{b} \in \boldsymbol{V}$, is

$$u_i^M(\boldsymbol{b}; v_i) = u_i(\boldsymbol{b}; v_i) := a_i(\boldsymbol{b}) \cdot v_i - p_i(\boldsymbol{b}). \tag{1}$$

Using the distributional priors F_i to capture the uncertainty about other bidders' behaviour, we can also define the *interim* utility of a bidder, when having true value $v_i \in V_i$ and bidding $b_i \in V_i$ as

$$U_i(b_i; v_i) := \mathbb{E}_{\boldsymbol{b}_{-i} \sim \boldsymbol{F}_{-i}} \left[u_i(b_i, \boldsymbol{b}_{-i}; v_i)\right] = A_i(b_i) \cdot v_i - P_i(b_i),$$

where

$$A_i(b_i) := \mathbb{E}_{\boldsymbol{b}_{-i}\sim \boldsymbol{F}_{-i}}\left[a_i(b_i, \boldsymbol{b}_{-i})\right] \qquad \text{and} \qquad P_i(b_i) := \mathbb{E}_{\boldsymbol{b}_{-i}\sim \boldsymbol{F}_{-i}}\left[p_i(b_i, \boldsymbol{b}_{-i})\right]$$

are the interim versions of the allocation and payment rules of the mechanism, respectively.

An auction whose allocations lie in the n-simplex, i.e. $\boldsymbol{a}(\boldsymbol{v}) \in \mathcal{S}_n$ for all $\boldsymbol{v} \in \boldsymbol{V}$, will be called a *lottery*, since its fractional allocations $a_i \in [0, 1]$ can be equivalently interpreted as the probability of assigning 1 unit of service to bidder i, given the linearity of the utilities (1). In particular, lotteries with only integral 0-1 allocations, i.e. $\boldsymbol{a} \in \mathcal{S}_n \cap \{0, 1\}^n$ will be called *deterministic auctions*. More generally, any auction with allocation rule $\boldsymbol{a} \in \mathbb{N}^n$ will be called *integral*.

2.2 Incentive Compatibility

From the perspective of each bidder i, the goal is to bid so that they can maximize their own utility. In particular, this means that bidders can lie and misreport $b_i \neq v_i$. Therefore, one of the goals of mechanism design is to construct auctions that avoid this pitfall, and which *provably* guarantee that truthful participation is to each bidder's best interest. From a game-theoretic perspective this can be formalized by demanding that truthful bidding $b_i = v_i$ is an equilibrium of the induced Bayesian game.

This gives rise to the following constraints, known as *dominant-strategy incentive compatibility (DSIC)*: for any bidder i, any true value $v_i \in V_i$, and any bidding profile $\boldsymbol{b} \in \boldsymbol{V}$, it holds that

$$u_i(v_i, \boldsymbol{b}_{-i}; v_i) \geq u_i(b_i, \boldsymbol{b}_{-i}; v_i), \tag{DSIC}$$

and its more relaxed version of *Bayesian incentive compatibility (BIC)*, involving the interim utilities:

$$U_i(v_i; v_i) \geq U_i(b_i; v_i), \tag{BIC}$$

for any bidder i, true value $v_i \in V_i$ and bid $b_i \in V_i$.

Individual Rationality. Another desired property of our mechanisms is that no bidder should harm themselves by truthfully participating in our auction, known as *individual rationality (IR)*. Similarly to the truthfulness conditions (DSIC) and (BIC), this can be formalized both in an ex-post and interim way: $u_i(v_i, \boldsymbol{b}_{-i}) \geq 0$ and $U_i(v_i; v_i) \geq 0$, respectively, for all bidders i, true values $v_i \in V_i$ and other bidders' bid profile $\boldsymbol{b}_{-i} \in \boldsymbol{V}_{-i}$, respectively.

One elegant way to merge the (IR) constraints into truthfulness, is to extend the bidding space of bidder i in (DSIC) and (BIC) from V_i to $\bar{V}_i := V_i \cup \{\varnothing\}$ and define

$$a_i(\varnothing, \boldsymbol{b}_{-i}) = p_i(\varnothing, \boldsymbol{b}_{-i}) = 0 \tag{2}$$

for all bidders i and other bidders' bids $\boldsymbol{b}_{-i} \in \boldsymbol{V}_{-i}$. Then, bidding $\varnothing$ can be interpreted as an option to "abstain" from the auction for a utility of

$u_i(\varnothing, \boldsymbol{b}_{-i}; v_i) = U_i(\varnothing; v_i) = 0$. From now on we will assume that our truthfulness conditions (DSIC) and (BIC) are indeed extended in that way to $\bar{V}_i$, thus including the (IR) constraints. An auction will be called DSIC (resp. BIC) if it satisfies those (extended) (DSIC) (resp. (BIC)) constraints. Observe that, since (DSIC) $\subseteq$ (BIC), any DSIC auction is also BIC.

Optimal Auctions. The main focus of our paper is the design of *optimal auctions*, for discrete value domains. That is, maximize the seller's *revenue* within the space of all feasible *truthful* auctions. Formally, if for a given auction $M = (\boldsymbol{a}, \boldsymbol{p})$ we denote its expected revenue, with respect to the value priors $\boldsymbol{F}$, by

$$\mathsf{Rev}(M) := \mathbb{E}_{\boldsymbol{v} \sim \boldsymbol{F}} \left[\sum_{i=1}^{n} p_i(\boldsymbol{v}) \right], \tag{3}$$

then our optimization problem can be stated as $\sup_{M:\mathcal{A}\wedge(\mathrm{DSIC})} \mathsf{Rev}(M)$, or $\sup_{M:\mathcal{A}\wedge(\mathrm{BIC})} \mathsf{Rev}(M)$, depending on whether we choose the notion dominant-strategy, or Bayesian truthfulness. An optimal solution to the former problem will be called *optimal DSIC* auction, and to the latter, *optimal BIC* auction. Following the standard convention in the field (see, e.g., [12] and [16]), the term *optimal auction* that does not explicitly specify the underlying truthfulness notion, will refer to the optimal BIC auction. Notice that, since (DSIC) $\subseteq$ (BIC), for an optimal DSIC auction M and an optimal BIC auction M' it must be that $\mathsf{Rev}(M) \leq \mathsf{Rev}(M')$.

Nevertheless, as we demonstrate in Sect. 4, our general duality approach provides for greater flexibility with respect to the optimization objective. For example, this will allow us to instantiate our framework for a convex combination or revenue and another important objective in auction theory, that of *social welfare*:

$$\mathsf{SW}(M) := \mathbb{E}_{\boldsymbol{v} \sim \boldsymbol{F}} \left[\sum_{i=1}^{n} a_i(\boldsymbol{v}) v_i \right]. \tag{4}$$

2.3 Locality of Truthfulness

It turns out our truthfulness constraints can be simplified, and expressed through a set of constraints that are "local" in nature, in the sense that they only involve deviations between adjacent values. To formalize this, recall that our value spaces V_i are finite, so we can define the notion of *predecessor* and *successor* values for a given bidder i and a value $v_i \in V_i$:

$$v_i^+ := \min \{v \in V_i \mid v > v_i\} \quad \text{and} \quad v_i^- := \max \{v \in V_i \mid v < v_i\},$$

if the above sets are nonempty, otherwise we define $v_i^+ := \varnothing$ for $v_i = \max V_i$ and $v_i^- := \varnothing$ for $v_i = \min V_i$.

Now we can state the local characterization of truthfulness, first for (DSIC), but a totally analogous lemma holds for (BIC) as well – see the full version of our paper [7, Appendix A]. This result is essentially proven in [14, Theorem 2].

Lemma 1 (Malakhov and Vohra [14]). *For any discrete, single-dimensional auction* $(\boldsymbol{a}, \boldsymbol{p})$*, the (DSIC) condition is* equivalent *to the following set of constraints:*

$$u_i(\boldsymbol{v}; v_i) \geq u_i(v_i^-, \boldsymbol{v}_{-i}; v_i) \tag{5}$$

$$u_i(\boldsymbol{v}; v_i) \geq u_i(v_i^+, \boldsymbol{v}_{-i}; v_i) \tag{6}$$

$$a_i(\boldsymbol{v}) \geq a_i(v_i^-, \boldsymbol{v}_{-i}), \tag{7}$$

for all bidders $i \in [n]$ *and any value profile* $\boldsymbol{v} \in \boldsymbol{V}$.

Conditions (5) and (6) are called *downwards* and *upwards* DSIC constraints, respectively, and (7) are called *monotonicity* constraints.

3 The Discrete Myerson Auction: An LP Duality Approach

In this section we begin our study of optimal single-parameter auctions, by considering the canonical single-item setting of Myerson [16], but under discrete values. That is, the feasibility set for our allocations is the simplex $\mathcal{S}_n$ (see Sect. 2.1), giving rise to the following *feasibility constraints*:

$$\sum_{i=1}^{n} a_i(\boldsymbol{v}) \leq 1, \qquad \text{for all } \boldsymbol{v} \in \boldsymbol{V}. \tag{8}$$

Our results of this section are summarized in the following main theorem:

Theorem 1 (Optimal Discrete Single-Item Auction). *For any discrete, single-item auction setting, the following hold for revenue maximization:*

1. *There always exists an optimal auction which is deterministic.*
2. Any *optimal DSIC auction is an optimal BIC auction.*
3. *The following deterministic DSIC auction is optimal (even within the class of randomized BIC auctions):*
   - *Allocate (fully) the item to the bidder with the highest nonnegative ironed virtual value* (14)*, breaking ties arbitrarily.*[1]
   - *Collect from the winning bidder a payment equal to their critical bid* (13).

Point 3 of Theorem 1 is essentially a discrete analogue of Myerson's optimal auction for the continuous case. As we mentioned in our introduction (see Sects. 1.1 and 1.2), this result can be already derived by readily combining prior work on discrete auctions (see, e.g., [2,6]); our contribution here is not the result itself, but the proof technique, which makes use of classical LP

[1] In order to maintain determinism, this can be any fixed deterministic tie-breaking rule; e.g., allocating the bidder with the smallest index i. Fractionally splitting the item among bidders that tie would still ensure revenue optimality (and DSIC), but the mechanism would be randomized.

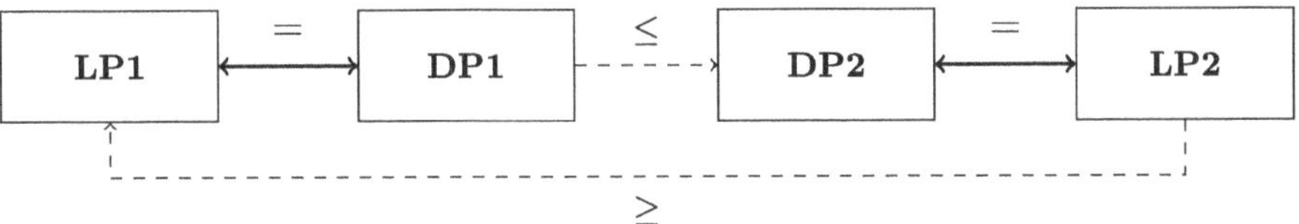

Fig. 1. An overview of the linear programs used in our derivation throughout Sect. 3 and the relations between their optimal values.

duality theory. This allows us to make use of powerful and transparent results from polyhedral combinatorics, to structurally characterize optimal auctions. In particular, we establish the optimality of DISC mechanisms, in a very general sense (see 2), which to the best of our knowledge was not known before. This is also enabled by our discrete optimization view of the problem, through the use of polyhedral properties (see Sect. 3.3). Finally, observe that Point 1 can be derived directly as a corollary of Point 3; nevertheless, we choose to state it independently, in order to reflect the logical progression of our derivation in this paper, which actually allows us to establish Point 1 more generally, as a result of the polyhedral structure of our problem (see Sect. 3.2), *before* we determine the actual optimal solution in Point 3.

We start our presentation by considering the revenue-maximization problem under the more restricted DSIC truthfulness notion. We do this for reasons of clarity of exposition, and then in Sect. 3.3 we carefully discuss how our formulations adapt for the more relaxed (BIC) constraints, and the relation between the two notions with respect to optimality, completing the picture for Theorem 1.

3.1 A Chain of Dual Linear Programs

In this section we develop the skeleton of our approach for proving Theorem 1. It consists of a sequence of LPs, as summarized in Fig. 1. We begin by formulating our single-item, revenue-maximization problem as an LP in (**LP1**). Next, we dualize it in (**DP1**), and then restrict its constraints to derive (**DP2**) that can only have a worse (i.e., higher) optimal objective. Then, we dualize again, deriving a maximization program in (**LP2**). Finally, we prove (see Lemma 3) that our original maximization program (**LP1**) is a relaxation of (**LP2**), thus establishing a collapse of the entire duality chain, and the equivalence of all involved LPs. This closure of the chain is exactly from where virtual values (12), virtual welfare maximization (**LP2**), optimality of determinism (see Lemma 4), and the optimal payment rule (**LP2**) naturally emerge.

Before we formally present and start working within the LPs, we need to fix some notation.

LP Notation. Since our value sets are finite, for each player i we can enumerate their support as $V_i = \{v_{i,1}, v_{i,2}, \dots, v_{i,K_i}\}$, for some positive integer K_i. For notational convenience we denote $\boldsymbol{K} := [K_1] \times [K_2] \times \cdots \times [K_n]$ and $\boldsymbol{K}_{-i} :=$

$[K_1] \times \cdots \times [K_{i-1}] \times [K_{i-1}] \times \cdots \times [K_n]$. To keep our LP formulations below as clean as possible, we will feel free to abuse notation and use the support indices $k \in [K_i]$ instead of the actual values $v_{i,k} \in V_i$, as arguments for the allocations a_i, payments p_i, and prior cdf's F_i and pmf's f_i. That is, e.g., we will denote $a_i(k, \boldsymbol{k}_{-i})$, $p_i(k, \boldsymbol{k}_{-i})$, $f_i(k)$, and $F_i(k)$, instead of $a_i(v_i, \boldsymbol{v}_{-i})$, $p_i(v_i, \boldsymbol{v}_{-i})$, $f_i(v_i)$, and $F_i(v_i)$, respectively, when the valuation profile $\boldsymbol{v}$ is such that $v_i = v_{i,k_i}$ for $i \in [n]$. As all values are independently drawn from distributions F_i, the probability of a bid profile $\boldsymbol{v} \in \boldsymbol{V}$ being realized is given by the pmf of their product distribution $\boldsymbol{F}$, denoted by $f(\boldsymbol{k}) = f(\boldsymbol{v}) = \prod_{i\in[n]} f_i(v_{i,k})$. Analogously, we denote $f(\boldsymbol{k}_{-i}) = f(\boldsymbol{v}_{-i}) = \prod_{j\in[n]\setminus\{i\}} f_j(v_{j,k_j})$. Finally, given that we make heavy use of duality, we choose to label each constraint of our LPs with the name of its corresponding dual variable, using square brackets (see, e.g., (**LP1**)).

For our starting (**LP1**), we want to formulate an LP maximizing expected revenue (3), under the single-item allocation constraints (8) of our current section, and DSIC truthfulness, through its equivalent formulation via Lemma 1. Since we want to optimize over the space of all feasible auctions, the real-valued variables of our LP are the allocation and payment rules of the auction, over all possible bidding profiles, namely $\{a_i(\boldsymbol{v}), p_i(\boldsymbol{v})\}_{\boldsymbol{v}\in\boldsymbol{V}}$. Putting everything together, we derive the following LP:

$$
\begin{aligned}
\max \quad & \sum_{\boldsymbol{v}\in\boldsymbol{V}} \sum_{i=1}^{n} p_i(\boldsymbol{v}) f(\boldsymbol{v}) && (\textbf{LP1})\\
\text{s.t.} \quad & v_{i,k} a_i(k, \boldsymbol{k}_{-i}) - p_i(k, \boldsymbol{k}_{-i}) \geq v_{i,k} a_i(k-1, \boldsymbol{k}_{-i}) - p_i(k-1, \boldsymbol{k}_{-i}), \\
& \qquad \text{for } i \in [n], k \in [K_i], \boldsymbol{k}_{-i} \in \boldsymbol{K}_{-i}, && [\lambda_i(k, k-1, \boldsymbol{k}_{-i})]\\
& v_{i,k} a_i(k, \boldsymbol{k}_{-i}) - p_i(k, \boldsymbol{k}_{-i}) \geq v_{i,k} a_i(k+1, \boldsymbol{k}_{-i}) - p_i(k+1, \boldsymbol{k}_{-i}), \\
& \qquad \text{for } i \in [n], k \in [K_i], \boldsymbol{k}_{-i} \in \boldsymbol{K}_{-i}, && [\lambda_i(k, k+1, \boldsymbol{k}_{-i})]\\
& a_i(k, \boldsymbol{k}_{-i}) \geq a_i(k-1, \boldsymbol{k}_{-i}), \\
& \qquad \text{for } i \in [n], k \in [K_i], \boldsymbol{k}_{-i} \in \boldsymbol{K}_{-i}, && [\tau_i(k, k-1, \boldsymbol{k}_{-i})]\\
& \sum_{i=1}^{n} a_i(\boldsymbol{v}) \leq 1, && [\psi(\boldsymbol{v})]\\
& \qquad \text{for } \boldsymbol{v} \in \boldsymbol{V}.
\end{aligned}
$$

Notice how our LP can readily incorporate the no-participation IR constraints (2), by fixing the under-/overflowing corner cases as constants

$$
a_i(0, \boldsymbol{k}_{-i}) = p_i(0, \boldsymbol{k}_{-i}) = a_i(K_i+1, \boldsymbol{k}_{-i}) = p_i(K_i+1, \boldsymbol{k}_{-i}) = 0 \tag{9}
$$

for all bidders i, on any bidding profile $\boldsymbol{k}_{-i}$ of the other bidders.

According to this we formulate the dual LP (**DP1**). Similar to the borderline cases (9) in the primal LP some restrictions on the dual variables are necessary to obtain a correct dual problem formulation. There we have

$$
\lambda_i(K_i, K_i+1, \boldsymbol{k}_{-i}) = \lambda_i(K_i+1, K_i, \boldsymbol{k}_{-i}) = \lambda_i(0, 1, \boldsymbol{k}_{-i}) = \tau_i(K_i+1, K_i, \boldsymbol{k}_{-i}) = 0 \tag{10}
$$

for all bidders i, on any bidding profile $\boldsymbol{k}_{-i}$ of the other bidders, for constraints that do not exist in (**LP1**). To ensure dual feasibility, all dual variables corresponding to inequality constraints in the primal have to be non-negative, thus all $\lambda, \psi, \tau \geq 0$. It is worth pointing out that $\lambda_i(1, 0, \boldsymbol{k}_{-i})$ and $\tau_i(1, 0, \boldsymbol{k}_{-i})$ are explicitly not fixed to zero as the corresponding constraints, the local downward DSIC constraint that ensures IR, $v_{i,1}a_i(1, \boldsymbol{k}_{-i}) - p_i(1, \boldsymbol{k}_{-i}) \geq 0$, as well as the monotonicity constraint that ensures the non-negativity of the allocation variables, $a_i(1, \boldsymbol{k}_{-i}) \geq 0$, are crucial for the problem. By that we write the dual LP as

$$
\begin{aligned}
\min \quad & \sum_{\boldsymbol{v} \in V} \psi(\boldsymbol{v}) && (\textbf{DP1}) \\
\text{s.t.} \quad & \psi(k, \boldsymbol{k}_{-i}) \geq v_{i,k}\lambda_i(k, k-1, \boldsymbol{k}_{-i}) + v_{i,k}\lambda_i(k, k+1, \boldsymbol{k}_{-i}) \\
& \qquad - v_{i,k+1}\lambda_i(k+1, k, \boldsymbol{k}_{-i}) - v_{i,k-1}\lambda_i(k-1, k, \boldsymbol{k}_{-i}) \\
& \qquad + \tau_i(k, k-1, \boldsymbol{k}_{-i}) - \tau_i(k+1, k, \boldsymbol{k}_{-i}), && [a_i(k, \boldsymbol{k}_{-i})] \\
& \quad \text{for } \ i \in [n], k \in [K_i], \boldsymbol{k}_{-i} \in \boldsymbol{K}_{-i}, \\
& \lambda_i(k, k-1, \boldsymbol{k}_{-i}) + \lambda_i(k, k+1, \boldsymbol{k}_{-i}) \\
& \qquad - \lambda_i(k+1, k, \boldsymbol{k}_{-i}) - \lambda_i(k-1, k, \boldsymbol{k}_{-i}) = f(\boldsymbol{v}), && [p_i(k, \boldsymbol{k}_{-i})] \\
& \quad \text{for } \ i \in [n], k \in [K_i], \boldsymbol{k}_{-i} \in \boldsymbol{K}_{-i}.
\end{aligned}
$$

In the same spirit as denoting the local DSIC constraints, that consider a deviation to the lower value, as *downwards constraint* (5), we call the corresponding dual variables $\lambda_i(k, k-1, \boldsymbol{k}_{-i})$ where the index in the first argument is greater than in the second *downward λ variables*. The dual variables $\lambda_i(k, k+1, \boldsymbol{k}_{-i})$ corresponding to the upwards DSIC constraints (6) are the *upward λ variables.* Putting together the dual borderline variables (10) and the set of equations in (**DP1**) we can state the following Lemma.

Lemma 2. *In any feasible solution of (**DP1**) all downward λ variables are strictly positive, i.e.,*

$$\lambda_i(k, k-1, \boldsymbol{k}_{-i}) > 0,$$

for all $i \in [n], k \in [K_i], \boldsymbol{k}_{-i} \in \boldsymbol{K}_{-i}$.

This motivates us to reformulate the dual program in a certain way. Recall, that any dual solution has to satisfy the set of equations

$$\lambda_i(k, k-1, \boldsymbol{k}_{-i}) + \lambda_i(k, k+1, \boldsymbol{k}_{-i}) = f(\boldsymbol{v}) + \lambda_i(k+1, k, \boldsymbol{k}_{-i}) + \lambda_i(k-1, k, \boldsymbol{k}_{-i}). \tag{11}$$

Using this we reformulate the dual inequality constraints

$$
\begin{aligned}
\psi(\boldsymbol{v}) \overset{(11)}{\geq} \; & v_{i,k}f(\boldsymbol{v}) - (v_{i,k+1} - v_{i,k})\lambda_i(k+1, k, \boldsymbol{k}_{-i}) \\
& + (v_{i,k} - v_{i,k-1})\lambda_i(k-1, k, \boldsymbol{k}_{-i}) + \tau_i(k, k-1, \boldsymbol{k}_{-i}) - \tau_i(k+1, k, \boldsymbol{k}_{-i})
\end{aligned}
$$

Note, that by the use of (11), i.e. exclusively equations, this is only a reformulation and does not affect the set of feasible dual solutions of (**DP1**). Now we unconventionally fix specific values of the λ variables. As the dual's objective aims to minimize the sum of the ψ variables, according to the reformulated inequality constraints it seems convenient to choose all upward λ as small and all downward λ as large as possible. To do so we set $\lambda_i(k, k+1, \boldsymbol{k}_{-i}) = 0$, for all $k \in [K_i], i \in [n]$ and $\boldsymbol{k}_{-i} \in \boldsymbol{K}_{-i}$. Fixing variables, essentially adding equality constraints, can only increase the optimal value of (**DP1**) in terms of minimization. As a next critical step we introduce *free* variables ρ and substitute the expression

$$\rho_i(k, \boldsymbol{k}_{-i}) := \lambda_i(k, k-1, \boldsymbol{k}_{-i}) - \lambda_i(k+1, k, \boldsymbol{k}_{-i})$$

for all bidders i with value index $k \in [K_i]$, and any bidding profile $\boldsymbol{k}_{-i}$ of the other bidders. These variables are all bound to fixed values and by dropping the λ variables from the problem formulation we do not lose any information about feasible dual solutions as by $\lambda_i(K+1, K, \boldsymbol{k}_{-i}) = 0$ we keep track of all fixed values. The reformulated dual LP then is

$$\min \quad \sum_{\boldsymbol{v} \in \boldsymbol{V}} \psi(\boldsymbol{v}) \tag{DP2}$$

$$\begin{aligned}
\text{s.t.} \quad & \psi(k, \boldsymbol{k}_{-i}) \geq v_{i,k}\rho_i(k, \boldsymbol{k}_{-i}) - (v_{i,k+1} - v_{i,k}) \sum_{l=k+1}^{K_i} \rho_i(l, \boldsymbol{k}_{-i}) \\
& \qquad + \tau_i(k, k-1, \boldsymbol{k}_{-i}) - \tau_i(k+1, k, \boldsymbol{k}_{-i}), && [a_i(k, \boldsymbol{k}_{-i})] \\
& \qquad \text{for } \ i \in [n], k \in [K_i], \boldsymbol{k}_{-i} \in \boldsymbol{K}_{-i}, \\
& \rho_i(k, \boldsymbol{k}_{-i}) = f(\boldsymbol{v}), && [p_i(k, \boldsymbol{k}_{-i})] \\
& \qquad \text{for } \ i \in [n], k \in [K_i], \boldsymbol{k}_{-i} \in \boldsymbol{K}_{-i}.
\end{aligned}$$

The inequality constraints now can also be written with all explicit values of ρ inserted. By that we obtain for a fixed bidder i and bids $\boldsymbol{v}_{-i}$

$$\begin{aligned}
\psi(k, \boldsymbol{k}_{-i}) \geq f(\boldsymbol{v}) \Big[v_{i,k} - (v_{i,k+1} - v_{i,k}) \frac{1 - F_i(k)}{f_i(k)} \\
+ \frac{\tau_i(k, k-1, \boldsymbol{k}_{-i})}{f(\boldsymbol{v})} - \frac{\tau_i(k+1, k, \boldsymbol{k}_{-i})}{f(\boldsymbol{v})} \Big].
\end{aligned}$$

This gives rise to the well known definition of a sequence of values for player i which is independent of all other bidders' values $\boldsymbol{v}_{-i}$.

Definition 1 (Virtual values). *The* virtual values *of bidder $i \in [n]$ are defined as*

$$\varphi_i(k) = \varphi_i(v_{i,k}) := v_{i,k} - (v_{i,k+1} - v_{i,k}) \frac{1 - F_i(v_{i,k})}{f_i(v_{i,k})} \qquad \text{for } k \in [K_i]. \tag{12}$$

We return to the primal setting of allocation and payment variables by now taking *the dual of the dual.* To get the full transparency of the gained insights within the reformulation to (**DP2**) we do two things at the same time: We insert the true values of all ρ in the inequalities and obtain the virtual values as the coefficients of the allocation variables in the new primal objective. At the same time we stick with ρ as free variables in the dual inequalities and obtain the payment formula in (**LP2**) as the coefficients of ρ in the dual become the coefficients of the allocation variables in the primal payment formula. Note, that equivalently we could still maximize the expected payments in the new primal LP without using the explicit values for ρ.

$$\begin{aligned}
\max \quad & \sum_{v \in V} \sum_{i=1}^{n} a_i(\boldsymbol{v}) \varphi_i(k) f(\boldsymbol{v}) && \textbf{(LP2)} \\
\text{s.t.} \quad & p_i(k, \boldsymbol{k}_{-i}) = v_{i,k} a_i(k, \boldsymbol{k}_{-i}) - \sum_{l=1}^{k-1} (v_{i,l+1} - v_{i,l}) a_i(l, \boldsymbol{k}_{-i}), && [\rho_i(k, \boldsymbol{k}_{-i})] \\
& \quad \text{for } i \in [n], k \in [K_i], \boldsymbol{k}_{-i} \in \boldsymbol{K}_{-i}, \\
& a_i(k, \boldsymbol{k}_{-i}) \geq a_i(k-1, \boldsymbol{k}_{-i}), && [\tau_i(k, k-1, \boldsymbol{k}_{-i})] \\
& \quad \text{for } i \in [n], k \in [K_i], \boldsymbol{k}_{-i} \in \boldsymbol{K}_{-i}, \\
& \sum_{i=1}^{n} a_i(\boldsymbol{v}) \leq 1, && [\psi(\boldsymbol{v})] \\
& \quad \text{for } \boldsymbol{v} \in \boldsymbol{V}.
\end{aligned}$$

As our interest lies in optimal auctions, we *close the chain* of LPs using Lemma 2 and strong LP duality to verify that the set of optimal solutions of (**LP1**) and of (**LP2**) are equivalent.

Lemma 3. *Any optimal solution of (**LP2**) represents an optimal (DSIC) auction, i.e. an optimal solution of (**LP1**) and vice versa.*

The immediate result is that the problem of finding an optimal (DSIC) auction reduces to finding an optimal solution of (**LP2**), i.e. a feasible, virtual welfare maximizing, monotone allocation rule $\boldsymbol{a}$. The optimal payments are computed afterwards as a linear function of the allocations according to the payment rule

$$p_i(k, \boldsymbol{k}_{-i}) = v_{i,k} a_i(k, \boldsymbol{k}_{-i}) - \sum_{l=1}^{k-1} (v_{i,l+1} - v_{i,l}) a_i(l, \boldsymbol{k}_{-i}). \tag{13}$$

3.2 Deterministic vs Randomized Auctions

In this section we essentially establish the foundation for Point 1 of Theorem 1. We are using the property of *total unimodularity* [10] of the constraint matrix of (**LP2**). This is enough to show that the optimal allocations of (**LP1**) and (**LP2**) are the convex hull of optimal binary solutions.

Lemma 4 (Optimality of determinism). *The vertices of the polyhedron of feasible allocations for (**LP2**) are integral, hence, binary.*

Thus, determinism of optimal DSIC auctions is without loss. Also, since the set of optimal solutions is convex, any fractional optimal solution is only a convex combination of multiple integer solutions and for given $\boldsymbol{v} \in \boldsymbol{V}$ represents a probability distribution.

3.3 Dominant-Strategy Vs Bayesian Truthfulness

The optimal auction problem typically is considered in a setting where truthfulness constraints are a relaxed version of (DSIC) and a bidder's truthfulness only has to hold in expectation over all other bidders' distributions, i.e., in the (BIC) sense. Essentially the same steps as in Sect. 3.1 where we considered DSIC truthfulness can be performed for the BIC formulation. A detailed analysis can be found in the full version of our paper [7].

It appears that in the BIC setting the virtual values not only arise in the same manner but also assume the exact same values. Furthermore, in an optimal dual solution the τ variables uniquely can be chosen such that in combination with the virtual values they form a non-decreasing sequence $\tilde{\varphi}_i(k)$, the *ironed virtual values*. This can be done for the DSIC as well as for the BIC setting, and even though the values for τ may be different in the two settings, ultimately the ironed virtual values are the same and optimality transfers from DSIC to BIC.

Lemma 5 (DSIC optimality). *Let* $(\boldsymbol{a}, \boldsymbol{p})$ *be an optimal DSIC auction, i.e. an optimal solution of (**LP2**). Then* $(\boldsymbol{a}, \boldsymbol{p})$ *is an optimal BIC auction.*

As we know by Lemma 4 about the existence of integral solutions for DSIC auctions case and by Lemma 5 that DSIC is without loss we directly proceed to the following.

Lemma 6 (Optimality of Determinism). *The set of optimal BIC auctions always contains a deterministic auction, and it can be computed by solving the linear program (**LP2**).*

Beyond formally ensuring the existence of a deterministic optimal solution, we want to derive the explicit auction when we are given a bid profile $\boldsymbol{v} \in \boldsymbol{V}$. By the complementary slackness condition

$$a_i(k, \boldsymbol{k}_{-i}) > 0 \quad \Longrightarrow \quad \psi(\boldsymbol{v}) = f(\boldsymbol{v})\tilde{\varphi}_i(k) = f(\boldsymbol{v}) \max_{i \in [n]} \tilde{\varphi}_i(k), \tag{14}$$

and the existence of an integral solution, we essentially have shown Point 3 of Theorem 1, i.e., receiving a bid profile $\boldsymbol{v} \in \boldsymbol{V}$ the item is allocated fully to the highest non-negative ironed virtual bidder, breaking ties by a fixed deterministic rule. The corresponding payments are computed via (13) which by determinism reduces to the *critical bid*, i.e., the threshold value of such that the player still wins.

4 General Single-Parameter Auction Design: A KKT Approach

In general, single-parameter auctions go far beyond the single-item case. In this section we generalize our formulation from the previous section and present a framework for a wider range of feasibility spaces. In fact, the specialization on the single-item setting emerges solely from the feasibility constraints (8). In a more general single-parameter setting we want to relax feasibility while still holding on to truthfulness, i.e., that the players have no incentive to misreport their true values. We maintain the linearity of the truthfulness constraints that arises from the definition of a player's utility (1), which is natural for the single-parameter auction design. Our framework which unites the techniques from the single-item setting, i.e., the duality approach connected by complementary slackness, is a KKT system formulation [13].

Unfortunately, due to space constraints, we need to defer most of our presentation and results of this section to the full version of our paper [7, Sec. 4]. Below, we summarize our results in a main theorem and provide a quick overview of our approach and some key notions:

Theorem 2 (Optimal Single-Parameter Auction). *For any discrete convex single-parameter auction setting, under the objective of maximizing a linear combination of revenue and social welfare (see* (15)*), the following hold:*

1. *If our setting is TU, then there exists an optimal auction which is integral.*[2]
2. Any *optimal DSIC auction is an optimal BIC auction.*
3. *The following DSIC auction is optimal (even within the class of BIC auctions):*
   - *Choose an allocation that maximizes the expected generalized virtual social welfare (**GM2**).*
   - *Collect from the allocated bidders a payment equal to their critical bids* (16).

The framework we present in [7, Sec. 4.2] allows us to assume that any feasible solution of the KKT system is also an optimal solution. Within this rather abstract formulation we are free to leave the ambiguity whether to interpret the truthfulness constraint as DSIC or BIC. This not only shows us how similar the two interpretations are, but the framework allows us great clearness when drawing the connection establish Point 2. Motivated by this, in [7, Sec. 4.4] we establish a setting where we can guarantee that the optimal auction is integral and randomization or fractional allocation is not necessary. Even in the very general case of Sect. 4, Point 3 gives a description of the optimal auction. We not only are able to maintain the transition to welfare maximization (see [7, Sec. 4.3]), but also derive the identical payment rule as in the single-item setting. Although complementary slackness cannot guarantee such a clear optimal

[2] Recall the definition of an integral auction from Sect. 2.1, Page 6. The definition of a totally unimodular (single-parameter) auction setting can be found in Definition 2.

auction as in Theorem 1, in the full version of our paper [7, Sec. 5] we give an application to show that the auction can still be nicely described.

4.1 Notation

For the general model formulation we want to use a notation that provides simplicity while at the same time allows to model very general settings. Still, we frequently draw the connection to the single-item LP formulation such that the reader can always recall this as a special case. In the following we will use a unified notation: In both settings of truthfulness, DSIC and BIC, each allocation and payment variable represents an outcome per given bid profile $\boldsymbol{v} \in \boldsymbol{V}$ and per player $i \in [n]$. We write the allocations $\boldsymbol{a}$ and payments $\boldsymbol{p}$ as vectors of dimension $\mathcal{N} := n|\boldsymbol{V}| = n \cdot K_1 \cdots K_n$. One entry is a single variable, e.g., $a_i(k, \boldsymbol{k}_{-i})$. We further define $\boldsymbol{f}$ as a vector of the same dimension. Each entry is the probability that a specific bid profile $\boldsymbol{v}$ is realized, i.e., $f(\boldsymbol{v})$ corresponding to the respective allocation or payment variable $a_i(k, \boldsymbol{k}_{-i})$ or $p_i(k, \boldsymbol{k}_{-i})$ for all players $i \in [n]$. To remain accurate with the dimensions of the objects that represent social and virtual welfare, we also define $\boldsymbol{\nu}$ as the quadratic $\mathcal{N} \times \mathcal{N}$ matrix with all values and similarly $\boldsymbol{\varphi}$ with all virtual values corresponding to player i's value of the respective allocation on the diagonal and zero elsewhere.

Objective Function. With this notation we write the generalized objective, a linear combination of expected revenue and expected social welfare, as

$$\alpha \, \mathsf{Rev}(M) + \beta \, \mathsf{SW}(M) = \alpha \boldsymbol{f}^\top \boldsymbol{p} + \beta \boldsymbol{f}^\top \boldsymbol{\nu} \boldsymbol{a}, \tag{15}$$

with $\boldsymbol{f}, \boldsymbol{a} \in \mathbb{R}^{\mathcal{N}}_{\geq 0}, \boldsymbol{p} \in \mathbb{R}^{\mathcal{N}}, \boldsymbol{\nu} \in \mathbb{R}^{\mathcal{N} \times \mathcal{N}}_{\geq 0}$ and $\alpha, \beta \in \mathbb{R}_{\geq 0}$.

Truthfulness. Independent of a more general feasibility space the locality of the linear truthfulness constraints is maintained. They can be expressed by matrix vector notation: Matrix A contains the coefficients of the allocation variables $\boldsymbol{a}$ and B the coefficients of the payment variables $\boldsymbol{p}$ of the upward and downward truthfulness constraints. Matrix M contains the coefficients required to model the monotonicity constraints. Whether we consider DSIC or BIC truthfulness then depends on the coefficients and dimensions of the matrices A, B and M, and we do not restrict ourselves to only one of the settings.

Feasibility Space. Besides the truthfulness conditions the allocations' feasibility space $\mathcal{A}$ is represented by a finite set of convex and continuously differentiable constraints. We assume that for each bid profile $\boldsymbol{v} \in \boldsymbol{V}$, there are $m \in \mathbb{N}$ constraints. Each constraint $g_j(\boldsymbol{a}) : \boldsymbol{V} \longrightarrow \mathbb{R}_{\geq 0}$ involves *only* allocation variables corresponding to this very bid profile. That is, $g_j(\boldsymbol{a}) = g(a_1(\boldsymbol{v}), a_2(\boldsymbol{v}), \dots, a_n(\boldsymbol{v}))$ for $j \in [m]$ and some $\boldsymbol{v} \in \boldsymbol{V}$. To maintain ex-post feasibility the constraints are copied for each bid profile varying over the $\boldsymbol{v} \in \boldsymbol{V}$ such that the total number of constraints then is $\mathcal{M} := m|V|$. E.g., in the single-item case $\mathcal{M} = |V|$ and each $g_j(\boldsymbol{a})$ represents *the one* feasibility constraint per fixed bid profile, see (8).

Hence, an allocation $\boldsymbol{a}$ is feasible, i.e. $\boldsymbol{a} \in \mathcal{A}$, if and only if $g_j(\boldsymbol{a}) \leq 0$ for all $j \in [\mathcal{M}]$. In our framework we use the notion G which can be seen as a vector of the g_j functions,

$$G(\boldsymbol{a}) = \begin{pmatrix} g_1(\boldsymbol{a}) \\ \vdots \\ g_{\mathcal{M}}(\boldsymbol{a}) \end{pmatrix}, \quad (G(\boldsymbol{a}))^\top \psi = 0 \iff g_1(\boldsymbol{a})\psi_1 = 0, \ldots, g_{\mathcal{M}}(\boldsymbol{a})\psi_{\mathcal{M}} = 0.$$

$\nabla_{\boldsymbol{a}} G(\boldsymbol{a})$ is the corresponding Jacobian matrix of $G(\boldsymbol{a})$ where column i contains all functions' derivatives with respect to player i for a given bid profile $\boldsymbol{v}$. Note, that we can always hide the non-negativity of the allocations within these constraints.

We now give a definition of an auction setting with sufficient conditions for obtaining an integral auction (see Point 1 of Theorem 2).

Definition 2 (TU setting). *A (single-parameter) auction setting will be called* totally unimodular (TU), *if the allocation feasibility constraints are given by a TU matrix. More precisely, if there exists a TU matrix G and an integral vector b such that*

$$\mathcal{A} = \{\boldsymbol{a} \mid G\boldsymbol{a} \leq b\}.$$

Notice how this implies that several single-parameter auction settings have integral solutions also in the BIC setting. Examples are, of course, the single-item auction, but also the k-unit auction, the digital good auction, and in general combinatorial auctions where all constraints can be described via a totally unimodular matrix G. We dive deeper in such a combinatorial auction in the application presented in the full version of our paper [7, Sec. 5].

4.2 General Virtual Welfare Maximization

A key contribution in the full version of our paper (see [7, Sec. 4] for an appropriate presentation and all details) is showing that the problem of finding an optimal single-parameter auction, in the general setting of the current Sect. 4, essentially reduces to solving the following optimization problem

$$\begin{aligned} \max \quad & \boldsymbol{f}^\top(\alpha\tilde{\boldsymbol{\varphi}} + \beta\tilde{\boldsymbol{\nu}})\boldsymbol{a} \\ \text{s.t.} \quad & \boldsymbol{p} = C\boldsymbol{a}, \\ & G(\boldsymbol{a}) \leq 0, \end{aligned} \tag{GM2}$$

where

$$(\alpha\tilde{\boldsymbol{\varphi}} + \beta\tilde{\boldsymbol{\nu}})\boldsymbol{f} := \alpha\boldsymbol{\varphi}\boldsymbol{f} + \beta\boldsymbol{\nu}\boldsymbol{f} - M^\top\tau$$

are the *ironed generalized virtual values* (see the full version of our paper [7, Sec. 4.3]), and matrix C is such that (see [7, Lemma 9])

$$\boldsymbol{p} = C\boldsymbol{a}. \tag{16}$$

Acknowledgement. We would like to thank Deutsche Forschungsgemeinschaft (DFG) for their support within project B07 of the Sonderforschungsbereich/Transregio 154 "Mathematical Modelling, Simulation and Optimization using the Example of Gas Networks".

Disclosure of Interests. The authors have no competing interests to declare that are relevant to the content of this article.

References

1. Bergemann, D., Pesendorfer, M.: Information structures in optimal auctions. J. Econ. Theory **137**(1), 580–609 (2007). https://doi.org/10.1016/j.jet.2007.02.001
2. Cai, Y., Daskalakis, C., Weinberg, S.M.: Optimal multi-dimensional mechanism design: Reducing revenue to welfare maximization. In: Proceedings of the 53rd Annual Symposium on Foundations of Computer Science (FOCS), pp. 130–139 (2012). https://doi.org/10.1109/FOCS.2012.88
3. Cai, Y., Daskalakis, C., Weinberg, S.M.: Understanding incentives: mechanism design becomes algorithm design. In: Proceedings of the 54th Annual Symposium on Foundations of Computer Science (FOCS), pp. 618–627 (2013). https://doi.org/10.1109/FOCS.2013.72
4. Cai, Y., Devanur, N.R., Weinberg, S.M.: A duality-based unified approach to Bayesian mechanism design. SIAM J. Comput. **50**(3) (2019). https://doi.org/10.1137/16M1100113
5. Daskalakis, C., Deckelbaum, A., Tzamos, C.: Strong duality for a multiple-good monopolist. Econometrica **85**(3), 735–767 (2017). https://doi.org/10.3982/ECTA12618
6. Elkind, E.: Designing and learning optimal finite support auctions. In: Proceedings of the 18th Annual ACM-SIAM Symposium on Discrete Algorithms (SODA), pp. 736–745 (2007). https://dl.acm.org/doi/10.5555/1283383.1283462
7. Giannakopoulos, Y., Hahn, J.: Discrete single-parameter optimal auction design. CoRR **abs/2406.08125**, June 2024. https://arxiv.org/abs/2406.08125
8. Giannakopoulos, Y., Koutsoupias, E.: Duality and optimality of auctions for uniform distributions. SIAM J. Comput. **47**(1), 121–165 (2018). https://doi.org/10.1137/16M1072218
9. Hartline, J.D., Karlin, A.R.: Profit maximization in mechanism design. In: Nisan, N., Roughgarden, T., Tardos, É., Vazirani, V. (eds.) Algorithmic Game Theory, chap. 13, pp. 331–362. Cambridge University Press (2007). https://doi.org/10.1017/CBO9780511800481.015
10. Hoffman, A.J., Kruskal, J.B.: Integral boundary points of convex polyhedra. 50 Years of Integer Programming 1958–2008: From the Early Years to the State-of-the-Art, pp. 49–76 (2010). https://doi.org/10.1007/978-3-540-68279-0_3
11. Jehle, G.A., Reny, P.J.: Advanced Microeconomic Theory. Prentice Hall, Hoboken (2011)
12. Krishna, V.: Auction Theory, second edn. Academic Press, Cambridge (2009)
13. Kuhn, H.W., Tucker, A.W.: Nonlinear programming. In: Proceedings of the 2nd Berkeley Symposium on Mathematical Statistics and Probability (1951)
14. Malakhov, A., Vohra, R.V.: Single and multi-dimensional optimal auctions: a network approach. Discussion paper (2004). https://hdl.handle.net/10419/31166

15. Milgrom, P.: Putting Auction Theory to Work. Cambridge University Press, Cambridge (2004). https://doi.org/10.1017/CBO9780511813825.009
16. Myerson, R.B.: Optimal auction design. Math. Oper. Res. **6**(1), 58–73 (1981). https://doi.org/10.1287/moor.6.1.58
17. Vickrey, W.: Counterspeculation, auctions and competitive sealed tenders. Journal of Finance **16**(1), 8–37 (1961). https://doi.org/10.1111/j.1540-6261.1961.tb02789.x
18. Vohra, R.V.: Mechanism Design: A Linear Programming Approach. Cambridge University Press, Cambridge (2011). https://doi.org/10.1017/CBO9780511835216

Estimating the Expected Social Welfare and Cost of Random Serial Dictatorship

Ioannis Caragiannis(✉) and Sebastian Homrighausen

Department of Computer Science, Aarhus University, Åbogade 34, 8200 Aarhus N, Denmark
{iannis,homrighausen}@cs.au.dk

Abstract. We consider the assignment problem, where n agents have to be matched to n items. Each agent has a preference order over the items. In the serial dictatorship (SD) mechanism the agents act in a particular order and pick their most preferred available item when it is their turn to act. Applying SD using a uniformly random permutation as agent ordering results in the well-known random serial dictatorship (RSD) mechanism. Accurate estimates of the (expected) efficiency of its outcome can be used to assess whether RSD is attractive compared to other mechanisms. In this paper, we explore whether such estimates are possible by sampling a (hopefully) small number of agent orderings and applying SD using them. We consider a value setting in which agents have values for the items as well as a metric cost setting where agents and items are assumed to be points in a metric space, and the cost of an agent for an item is equal to the distance of the corresponding points. We show that a (relatively) small number of samples is enough to approximate the expected social welfare of RSD in the value setting and its expected social cost in the metric cost setting despite the #P-hardness of the corresponding exact computation problems.

1 Introduction

We consider assignment problems in which a set of n agents must be assigned (matched) to a set of n items. In an assignment instance each agent has a preference ranking over the items. According to the most straightforward mechanism known as *serial dictatorship* (SD), the agents are asked to act in a predefined order, and when it is their turn to act, they are assigned to their favourite item that has not been selected by other agents in previous steps.

SD achieves several desirable properties. For example, the interaction with the mechanism is minimal and intuitive from an agent's perspective. Also, SD is strategyproof and probably the most natural representative in the field of mechanism design without money [21,23]. Furthermore, it produces Pareto-efficient assignments when the agent preferences are expressed via strict rankings [1]. Still, it may produce unfair outcomes as agents who act early have a clear advantage over agents who act later. The obvious way to fix the fairness issue without

G. Schäfer and C. Ventre (Eds.): SAGT 2024, LNCS 15156, pp. 184–201, 2024.
https://doi.org/10.1007/978-3-031-71033-9_11

harming the other two properties is to have the agents act in a uniformly random order. This gives us the well-known *random serial dictatorship* (RSD) or *random priority* mechanism [1,9].

In addition to the abstract setting of assignment problems in which agents have *ordinal* preferences for the items, we consider two settings which use additional *cardinal* information. In the first one (the *value setting*), we assume that agents have values for the items and the items of higher value for an agent appear higher in their preference ranking. In the second one (the *metric cost setting*), agents and items are assumed to be points in a metric space, and each agent has cost for an item equal to the distance of the corresponding points. Now items of lower cost for an agent are those which appear higher in the preference ranking. In the two settings, we would like to compute matchings with high *social welfare* and low *social cost*, respectively, defined as the sum of the values in the former case and costs of the agents for their assigned items in the latter.

Even though RSD can neither optimise the social welfare nor the social cost (e.g. see [10,15]), it could be attractive compared to alternatives that do not have its other favourable properties. To explore whether this is the case, we need to be able to assess the outcome of RSD in terms of efficiency. In our two settings, this translates to computing the expected social welfare or the expected social cost of RSD, which in turn requires knowledge of the *RSD lottery* matrix. This matrix consists of the probabilities for all agent-item pairs that an agent is assigned to an item by RSD. Unfortunately, computing even a single entry of this matrix is a #P-complete problem [3,22]. Still, approximations of the expected social welfare and cost would be sufficient to compare RSD with other mechanisms.

So, can we compute —in reasonable time— accurate estimates of the expected social welfare and expected social cost of RSD when applied to assignment instances in the value and metric cost settings, respectively? This is the question we study in the current paper.

1.1 Our Contribution

We first give a formal argument (in Sect. 3) explaining how the previous #P-hardness results of Aziz et al. [3] and Sabán and Sethuraman [22] for computing the RSD lottery matrix for a given assignment instance in the abstract setting imply the #P-hardness of computing the expected social welfare and expected social cost of the RSD outcome in the value and metric cost setting, respectively. Specifically, given an assignment instance in the abstract setting, we show how to construct equivalent instances in the value and metric cost setting so that the binary representations of the expected social welfare and expected social cost have polynomial size (in terms of n), and furthermore contain the binary representation of the probabilities in the RSD lottery matrix. Then, the existence of a polynomial-time algorithm for computing the expected social welfare/cost in the value/metric cost setting would imply the existence of a polynomial-time algorithm for computing the RSD lottery.

We then consider a simple algorithm which randomly samples a number of agent orderings, applies the serial dictatorship using each of them, and returns the average social welfare (or average social cost) in the computed matchings. In spite of our #P-hardness result, we show that $\Theta\left(\frac{n}{\varepsilon^2}\ln\frac{1}{\delta}\right)$ samples are sufficient and necessary so that the value returned by the algorithm when applied to assignment instances in the value setting with n agents/items approximates the expected social welfare of RSD within a factor of $1 \pm \varepsilon$ with probability at least $1-\delta$. These results are presented in Sect. 4. To prove the upper bound, we use Bernstein's inequality, which allows us to use a simple bound on the variance that depends on the relation of the expected social welfare of RSD to the optimal social welfare. The lower bound follows by a reverse Chernoff bound.

Unfortunately, in the metric cost setting, the same algorithm needs a much higher number of samples to obtain similar guarantees. Specifically, we show that in order to approximate the expected social cost of RSD within a factor of $1 \pm \varepsilon$ with probability at least $1-\delta$ for all values of parameters n, ε, and δ, the number of samples this algorithm uses should depend exponentially on either n or $\ln\frac{1}{\delta}$. To bypass this barrier, we prove a non-trivial bound on the variance of the social cost of RSD. This is the most technically interesting among our results and yields that, using only $O\left(\frac{n^3}{\varepsilon^2}\right)$ samples, the algorithm approximates the expected social cost within $1 \pm \varepsilon$ with (sufficiently high) constant probability. Then, using an idea from approximate counting and taking the median of values returned by $O\left(\ln\frac{1}{\delta}\right)$ executions of the averaging algorithm, we obtain the desired approximation guarantee using at most $O\left(\frac{n^3}{\varepsilon^2}\ln\frac{1}{\delta}\right)$ samples in total. These results are presented in Sect. 5.

We continue with a discussion on the literature. A quick overview of the sharp concentration bounds we use in our proofs is presented in Sect. 2, together with our notation and definitions. We conclude with Sect. 6.

1.2 Further Related Work

House allocation has been the generic assignment problem; Abdulkadiroğlu and Sönmez [1], Bogomolnaia and Moulin [9], Crès and Moulin [13], and Sönmez and Ünver [25] discuss further applications. Besides its simplicity, the SD mechanism has already received considerable attention. For example, in the economic literature, Svensson [24] characterized it as the only deterministic assignment mechanism that is strategy-proof, non-bossy and neutral. Several authors (e.g. Abdulkadiroğlu and Sönmez [1], Abraham et al. [2]) have observed that a solution to the assignment problem is Pareto-optimal if and only if it can be produced by SD with an appropriate agent ordering. In the computer science literature, SD has been studied in the metric cost setting, where it has been proved to be highly inefficient in the worst-case but very efficient under resource augmentation assumptions [10,18]. Recently, Caragiannis and Rathi [12] consider the problem of optimizing the agent ordering so that SD yields good results not only in assignment problems but in combinatorial optimization more generally.

The properties of RSD are discussed extensively by Abdulkadiroğlu and Sönmez [1] and Bogomolnaia and Moulin [9], where it is also compared to other mechanisms such as the probabilistic serial mechanism and the mechanism of

Hylland and Zeckhauser [17]. The complexity of computing the RSD lottery matrix is studied by Aziz et al. [3] and Sabán and Sethuraman [22], who prove that it is #P-complete; see [16] for an introduction to the complexity class #P. On the positive side, Aziz and Mestre [5] present fixed-parameter tractable and polynomial-time algorithms that compute the RSD lottery for restricted assignment instances. In the metric cost model, RSD has been proved to be highly superior to SD, approximating the optimal social cost within a factor that is at most n and at least $n^{0.29}$ [10]. We remark that we use some of the results from [10] and [18] in our proofs for the metric cost setting.

Sampling techniques have found important applications in social choice. Indicative works include their use in deciding the winning alternative according to voting rules or estimates of notions like the distortion [11] or margin of victory [8]. More related in spirit to our work are papers aiming at estimating the Shapley value in cooperative games [4] or the Banzhaf index in voting [6].

2 Preliminaries

An assignment instance consists of n agents and n items. We will use the set $[n] := \{1, 2, ..., n\}$ to represent both the set of agents and the set of items.

In the *abstract setting* usually considered in the literature, each agent $i \in [n]$ has a strict preference ranking of all items. In this paper, we consider two more settings. In the first one, called the *value setting*, the agents have *values* for the items. For an agent $i \in [n]$ and item $g \in [n]$, we denote by $v_i(g)$ the (non-negative) value agent i has for item g. A (perfect) matching $M = (M_1, M_2, ..., M_n)$ is an assignment of the items to the agents so that each item is assigned to one agent, and each agent gets one item. The *social welfare* of a matching is the total value the agents have for the items they are assigned, i.e. $\mathrm{SW}(M) = \sum_{i \in [n]} v_i(M_i)$. For an assignment instance $\mathcal{I}$, we denote by $\mathrm{OPT}(\mathcal{I})$ the maximum social welfare among all possible matchings in $\mathcal{I}$.

In the second setting, called the *metric cost setting*, the agents have *costs* for the items. For an agent $i \in [n]$ and item $g \in [n]$, we denote by $c_i(g)$ the (non-negative) cost agent i has for item g. We assume that the agents and items correspond to points in a metric space, and the cost $c_i(g)$ is the distance between the points corresponding to agent i and item g. Thus, the costs satisfy the triangle inequality, e.g. for agents i_1 and i_2 and items g_1 and g_2, the triangle inequality implies that $c_{i_1}(g_1) \leq c_{i_1}(g_2) + c_{i_2}(g_2) + c_{i_2}(g_1)$. The *social cost* of a matching is the total cost the agents have for their allocated items, i.e. $\mathrm{SC}(M) = \sum_{i \in [n]} c_i(M_i)$. For an assignment instance $\mathcal{I}$, we slightly abuse notation and also use $\mathrm{OPT}(\mathcal{I})$ to denote the minimum social cost among all possible matchings in $\mathcal{I}$.

The *serial dictatorship* mechanism (or SD for short) takes an assignment instance $\mathcal{I}$ and an ordering π of the agents as input and computes a matching of items and agents as follows. The mechanism considers the agents one by one according to the ordering π. Whenever an agent is considered, they select their most preferable item that has not been selected by an agent until that point.

This will be the agent's highest-ranked item in their preference ranking in the abstract setting, their most valuable item in the value setting, or their least costly item in the metric cost setting. We denote by $\mathrm{SD}(\mathcal{I}, \pi)$ the matching computed by the SD mechanism when applied on instance $\mathcal{I}$ using the agent ordering π.

The *random serial dictatorship* mechanism (or RSD for short) applies SD using an ordering π that has been selected uniformly at random among all agent orderings. The RSD *lottery* is an $n \times n$ matrix $P(\mathcal{I})$ (or simply P) in which the entry $P_{i,g}$ denotes the probability that item g is assigned to agent i when RSD is applied on instance $\mathcal{I}$. We use $\mathrm{RSD}(\mathcal{I})$ to denote both the expected social welfare in the value setting and the expected social cost in the metric cost setting of the matching returned by RSD, when applied on instance $\mathcal{I}$. For $\varepsilon > 0$, we say that a quantity Q is an ε-approximation of $\mathrm{RSD}(\mathcal{I})$ if $|Q - \mathrm{RSD}(\mathcal{I})| < \varepsilon \cdot \mathrm{RSD}(\mathcal{I})$. We sometimes use the terms *over* and *under* ε-approximation to refer to a quantity Q satisfying $Q < (1+\varepsilon) \cdot \mathrm{RSD}(\mathcal{I})$ and $Q > (1-\varepsilon) \cdot \mathrm{RSD}(\mathcal{I})$, respectively.

In the following, we present some inequalities and bounds that we use later on. The Bernstein inequality is the first one.

Lemma 1 (Bernstein inequality, e.g. see [14], page 9). *Let X_1, X_2, ..., X_k be independent random variables satisfying $|X_i| \leq \alpha$ for $i \in [k]$, with mean 0 and variance $\sigma^2(X_i)$. Then*

$$\Pr\left[\left|\sum_{i=1}^{k} X_i\right| \geq t\right] \leq 2\exp\left(-\frac{3t^2}{6\sum_{i=1}^{k}\sigma^2(X_i) + 2\alpha t}\right)$$

Knowing the mean as well as upper and lower bounds for a random variable, we can use the Bhatia-Davis inequality to bound its variance.

Lemma 2 (Bhatia-Davis inequality [7]). *Consider a random variable X that takes values from the interval $[\alpha, \beta]$ and has expectation μ. Then, its variance is*

$$\sigma^2 \leq (\beta - \mu)(\mu - \alpha).$$

In addition to the Bernstein inequality, we will use the Chebyshev-Cantelli inequality and the Chernoff bound for proving (one-sided) concentration bounds.

Lemma 3 (Chebyshev-Cantelli inequality, e.g. see [20], page 64). *Let X be a random variable with expectation μ and variance σ^2. Then, for $t > 0$, it holds*

$$\Pr[X - \mu \geq t\sigma] \leq \frac{1}{1+t^2}.$$

Lemma 4 (Chernoff bound, e.g. see [20], page 68). *Let $X_1, X_2, \ldots, X_k$ be independent random variables taking values in $\{0,1\}$ and $X = \sum_{i=1}^{k} X_i$ be their sum with expectation $\mathbb{E}[X] = \mu$. Then, for $\eta > 0$, it holds*

$$\Pr(X \geq (1+\eta)\mu) \leq \left(\frac{e^{\eta}}{(1+\eta)^{1+\eta}}\right)^{\mu}.$$

Moreover, we will use an anti-concentration bound, also known as the reverse Chernoff bound.

Lemma 5 (Reverse Chernoff bound, Klein and Young [19]). *Let $X_1, X_2, \ldots, X_k$ be independent and identically distributed Bernoulli random variables with expectation $p \in (0, 1/2]$. Then, for every $\eta \in (0, 1/2]$ so that $\eta^2 pk \geq 3$, it holds*

$$\Pr\left[\frac{1}{k}\sum_{i=1}^{k} X_i \geq (1+\eta)p\right] \geq \exp\left(-9\eta^2 pk\right).$$

3 #P-Hardness of Random Serial Dictatorship

Before presenting our estimation results, let us explain how the #P-hardness of computing the RSD lottery [3,22] for a given assignment instance implies the #P-hardness of computing the expected social welfare or the expected social cost in the value or metric cost setting, respectively.

The above-mentioned papers deal with assignment instances in abstract form. For $j \in [n]$, let $r_i(j)$ denote the j-th most preferred item of agent $i \in [n]$. Given such an instance in abstract form, we show how to construct equivalent instances in the value and metric cost settings, in the sense that the outcome of RSD applied to the three instances, and consequently their RSD lotteries, coincide.

Given the preference orderings r_i of each agent $i \in [n]$, we define consistent agent values v_i and metric costs c_i as follows:

- In the value setting, we define the value of agent i for item $r_i(j)$ to be $2^{(in-j)\lceil \log(n!+1)\rceil}$.
- In the metric cost setting, we define the cost of agent i for item $r_i(j)$ to be $2^{n^2\lceil \log(n!+1)\rceil} + 2^{((i-1)n+j-1)\lceil \log(n!+1)\rceil}$. Notice that all agent costs differ by less than a multiplicative factor of 2 and, thus, they define a metric.

Also, notice that the number of bits in the representation of values and costs is at most $O(n^3 \log n)$. Hence, the construction of the instances in the value and metric cost setting takes only polynomial time.

Let P be the RSD lottery and L be the $n \times n$ matrix with entry L_{ij} being the number of different agent orderings π so that $\mathrm{SD}(\mathcal{I}, \pi)$ assigns item $r_i(j)$ to agent i. Clearly, $L_{ij} = n! \cdot P_{i,r_i(j)}$ for every $i, j \in [n]$.

In the value setting, we have

$$\sum_{j\in[n]} P_{ij} \cdot v_i(j) = \sum_{j\in[n]} P_{i,r_i(j)} \cdot v_i(r_i(j)) = \frac{1}{n!} \cdot \sum_{j\in[n]} L_{ij} \cdot v_i(r_i(j))$$

for agent $i \in [n]$. Thus, we have

$$n! \cdot \mathbb{E}[\mathrm{SW}(\mathrm{SD}(\mathcal{I}, \pi))] = n! \cdot \sum_{i\in[n]}\sum_{j\in[n]} P_{ij} \cdot v_i(j) = \sum_{i\in[n]}\sum_{i\in[n]} L_{ij} \cdot v_i(r_i(j)),$$

i.e. the quantity $n! \cdot \mathbb{E}[\text{SW}(\text{SD}(\mathcal{I}, \pi))]$ is a non-negative integer. Now, recall that $v_i(r_i(j)) = 2^{(in-j)\lceil \log(n!+1) \rceil}$, and thus the binary representation of the integer $L_{ij} \cdot v_i(r_i(j))$ has the binary representation of L_{ij} in bit positions[1] from $(in - j)\lceil \log(n!+1) \rceil$ to $(in-j+1)\lceil \log(n!+1) \rceil - 1$ and 0s everywhere else. Notice that these bit positions are disjoint for different pairs of i and j in $[n]$; indeed, $\lceil \log(n!+1) \rceil$ bits are enough to encode L_{ij}, which can take (integer) values between 0 and $n!$. Thus, the binary representation of $n! \cdot \mathbb{E}[\text{SW}(\text{SD}(\mathcal{I}, \pi))]$ has the binary representation of L_{ij} in bit positions from $(in - j)\lceil \log(n!+1) \rceil$ to $(in-j+1)\lceil \log(n!+1) \rceil - 1$ for every $i, j \in [n]$.

Similarly, in the metric cost setting, we have

$$\sum_{j \in [n]} P_{ij} \cdot c_i(j) = \frac{1}{n!} \cdot \sum_{i \in [n]} L_{ij} \cdot c_i(r_i(j)),$$

and, hence,

$$n! \cdot \mathbb{E}[\text{SC}(\text{SD}(\mathcal{I}, \pi))] = n! \cdot \sum_{i \in [n]} \sum_{j \in [n]} P_{ij} \cdot c_i(j) = \sum_{i \in [n]} \sum_{i \in [n]} L_{ij} \cdot c_i(r_i(j)),$$

i.e. the quantity $n! \cdot \mathbb{E}[\text{SC}(\text{SD}(\mathcal{I}, \pi))]$ is again a non-negative integer. Now, recall that $c_i(r_i(j)) = 2^{n^2 \lceil \log(n!+1) \rceil} + 2^{((i-1)n+j-1)\lceil \log(n!+1) \rceil}$, and thus the binary representation of the integer $L_{ij} \cdot c_i(r_i(j))$ has the binary representation of L_{ij} in bit positions from $((i-1)n+j-1)\lceil \log(n!+1) \rceil$ to $((i-1)n+j)\lceil \log(n!+1) \rceil - 1$, 1 in bit position $n^2 \lceil \log(n!+1) \rceil$, and 0s everywhere else. Thus, the binary representation of $n! \cdot \mathbb{E}[\text{SC}(\text{SD}(\mathcal{I}, \pi))]$ has the binary representation of n^2 in the $\lceil \log(n^2) \rceil$ bit positions from $n^2 \lceil \log(n!+1) \rceil$ to $n^2 \lceil \log(n!+1) \rceil + \lceil \log(n^2) \rceil - 1$, and the binary representation of L_{ij} in the bit positions from $((i-1)n+j-1)\lceil \log(n!+1) \rceil$ to $((i-1)n+j)\lceil \log(n!+1) \rceil - 1$ for every $i, j \in [n]$.

From the discussion above we conclude that any polynomial-time algorithm that computes the expected social welfare in the value setting or the expected social cost in the metric cost setting can be used to compute the entries of the matrix L in polynomial time, and thus the RSD lottery of the assignment instance. The following statement summarises the discussion above.

Theorem 1. *Given an assignment instance $\mathcal{I}$ in the value or metric cost setting, computing the expected social welfare or expected social cost of the outcome of RSD when applied on $\mathcal{I}$ is #P-hard.*

4 Approximating the Expected Social Welfare

We now consider a simple algorithm (Algorithm 1) that estimates (approximates) the expected social welfare of RSD. Given an assignment instance $\mathcal{I}$ in the value setting, Algorithm 1 samples k agent orderings (uniformly at random and with

[1] We number the bit positions by assuming that the least significant bit is at position 0.

replacement), computes k matchings by applying SD on $\mathcal{I}$ using each of the sampled agent orderings, and returns the average of the social welfare of these matchings as output. We will present upper and lower bounds on k so that Algorithm 1 returns an ε-approximation of RSD($\mathcal{I}$) with probability at least $1 - \delta$.

Algorithm 1. Approximating the expected social welfare of Random Serial Dictatorship

Input: An assignment instance $\mathcal{I}$ with n agents/items and an integer $k \geq 1$
Output: A non-negative number
1: Select independently k uniformly random orderings $\pi^1, \pi^2, \ldots, \pi^k$
2: **return** $\frac{1}{k} \sum_{i=1}^{k} \mathrm{SW}(\mathrm{SD}(\mathcal{I}, \pi^i))$

We start with a simple lemma that relates the maximum social welfare and the expected social welfare of the RSD outcome. This will come in handy in the proof of Theorem 2 below.

Lemma 6. *For every assignment instance $\mathcal{I}$ with n agents/items in the value setting, it holds $OPT(\mathcal{I}) \leq n \cdot RSD(\mathcal{I})$.*

Proof. Let $M = (M_1, M_2, \ldots, M_n)$ be the matching of maximum social welfare in the assignment instance $\mathcal{I}$. For $i \in [n]$, denote by g_i the item of maximum value for agent i. Notice that the RSD mechanism allocates item g_i to agent i with a probability of at least $1/n$. Then,

$$\mathrm{RSD}(\mathcal{I}) \geq \frac{1}{n} \cdot \sum_{i=1}^{n} v_i(g_i) \geq \frac{1}{n} \cdot \sum_{i=1}^{n} v_i(M_i) = \mathrm{OPT}(\mathcal{I})/n,$$

as desired. □

We are now ready to prove our upper bound. The key ideas in the proof are the use of the Bhatia-Davis inequality (Lemma 2) to bound the variance of the social welfare of the matching returned by RSD and the application of Bernstein inequality (Lemma 1).[2]

Theorem 2. *Let $n \geq 1$ be an integer and $\delta, \varepsilon \in (0, 1]$. For $k \geq \frac{8n}{3\varepsilon^2} \ln \frac{2}{\delta}$, the output of Algorithm 1, when applied on the assignment instance $\mathcal{I}$ with n agents/items in the value setting, is an ε-approximation to the expected social welfare of the RSD mechanism with probability at least $1 - \delta$.*

[2] Readers familiar with probabilistic analysis may wonder why we do not use the simpler Hoeffding inequality to prove (a statement similar to) Theorem 2. We have verified that such an analysis yields a weaker bound of $O\left(\frac{n^2}{\varepsilon^2} \ln \frac{1}{\delta}\right)$ on k. The main reason for this weaker result is the lack of information about the variance, something that the Bernstein inequality exploits.

Proof. For $i \in [k]$ and a uniformly random ordering π^i of the agents in $[n]$, notice that the random variable $\mathrm{SW}(\mathrm{SD}(\mathcal{I}, \pi^i))$ takes values in $[0, \mathrm{OPT}(\mathcal{I})]$ and has expectation $\mathrm{RSD}(\mathcal{I})$. By the Bhatia-Davis inequality (Lemma 2), we have that the variance of the random variable $\mathrm{SW}(\mathrm{SD}(\mathcal{I}, \pi^i))$ is

$$\begin{aligned}\sigma^2(\mathrm{SW}(\mathrm{SD}(\mathcal{I}, \pi^i))) &\leq (\mathrm{OPT}(\mathcal{I}) - \mathrm{RSD}(\mathcal{I})) \cdot \mathrm{RSD}(\mathcal{I}) \\ &\leq \mathrm{OPT}(\mathcal{I}) \cdot \mathrm{RSD}(\mathcal{I}). \end{aligned} \tag{1}$$

Now, for $i \in [k]$, define the random variable Z_i as $Z_i := \mathrm{SW}(\mathrm{SD}(\mathcal{I}, \pi^i)) - \mathrm{RSD}(\mathcal{I})$. We have that the probability that the output of Algorithm 1 is not an ε-approximation of RSD is

$$\begin{aligned}&\Pr\left[\left|\frac{1}{k}\sum_{i=1}^{k} \mathrm{SW}(\mathrm{SD}(\mathcal{I}, \pi^i)) - \mathrm{RSD}(\mathcal{I})\right| \geq \varepsilon \cdot \mathrm{RSD}(\mathcal{I})\right] \\ &= \Pr\left[\left|\sum_{i=1}^{k} Z_i\right| \geq \varepsilon k \cdot \mathrm{RSD}(\mathcal{I})\right]. \end{aligned} \tag{2}$$

To complete the proof, we will bound the RHS of Eq. (2). By the definition of the random variable Z_i, its variance is equal to the variance of the random variable $\mathrm{SW}(\mathrm{SD}(\mathcal{I}, \pi^i))$, i.e. by Eq. (1), it holds that

$$\sigma^2(Z_i) \leq \mathrm{OPT}(\mathcal{I}) \cdot \mathrm{RSD}(\mathcal{I}).$$

Furthermore, Z_i takes values in $[-\mathrm{OPT}(\mathcal{I}), \mathrm{OPT}(\mathcal{I})]$ and has expectation 0.

We now apply Bernstein inequality (Lemma 1) for the random variable $\sum_{i=1}^{k} Z_i$ using $t = \varepsilon k \cdot \mathrm{RSD}(\mathcal{I})$ and $\alpha = \mathrm{OPT}(\mathcal{I})$. By the discussion above, we have $\sum_{i=1}^{k} \sigma^2(Z_i) \leq k \cdot \mathrm{OPT}(\mathcal{I}) \cdot \mathrm{RSD}(\mathcal{I})$. Thus,

$$\begin{aligned}&\Pr\left[\left|\sum_{i=1}^{k} Z_i\right| \geq \varepsilon k \cdot \mathrm{RSD}(\mathcal{I})\right] \\ &\leq 2\exp\left(-\frac{3\varepsilon^2 k^2 \cdot \mathrm{RSD}(\mathcal{I})^2}{6k \cdot \mathrm{OPT}(\mathcal{I}) \cdot \mathrm{RSD}(\mathcal{I}) + 2\varepsilon k \cdot \mathrm{OPT}(\mathcal{I}) \cdot \mathrm{RSD}(\mathcal{I})}\right) \\ &\leq 2\exp\left(-\frac{3\varepsilon^2 \cdot k \cdot \mathrm{RSD}(\mathcal{I})}{8 \cdot \mathrm{OPT}(\mathcal{I})}\right) \leq 2\exp\left(-\frac{3\varepsilon^2 \cdot k}{8n}\right). \end{aligned} \tag{3}$$

The second inequality follows since $\varepsilon \leq 1$ and the third one by Lemma 6. By Eq. (2) and Eq. (3), we conclude that for $k \geq \frac{8n}{3\varepsilon^2} \cdot \ln \frac{2}{\delta}$, we have

$$\Pr\left[\left|\frac{1}{k}\sum_{i=1}^{k} \mathrm{SW}(\mathrm{SD}(\mathcal{I}, \pi^i)) - \mathrm{RSD}(\mathcal{I})\right| \geq \varepsilon \cdot \mathrm{RSD}(\mathcal{I})\right] \leq \delta,$$

as desired. □

We now prove our lower bound for the value setting, by applying the reverse Chernoff bound (Lemma 5).

Theorem 3. *Let $n \geq 2$ be an integer, $\varepsilon \in (0,1]$, $\delta \in (0, e^{-27})$, and k be such that $\frac{3n}{\varepsilon^2} \leq k < \frac{n}{9\varepsilon^2} \ln \frac{1}{\delta}$. Then, there exists an assignment instance $\mathcal{I}$ with n agents/items in the value setting, so that the output of Algorithm 1 when applied on $\mathcal{I}$ is an ε-approximation to the expected social welfare of the RSD mechanism with probability smaller than $1-\delta$.*

Proof. Let $\mathcal{I}$ be the instance in which agent 1 has value 1 for item 1 and 0 for any other items. All other agents have a valuation of 0 for all items. All ties regarding the item an agent picks under SD are resolved in favour of the minimum-index item. Thus, the serial dictatorship returns a matching of social welfare 1 when applied with an ordering that has agent 1 first and social welfare 0 otherwise. So, for a uniformly random ordering π of the agents, $\text{SW}(\text{SD}(\mathcal{I},\pi))$ is a Bernoulli random variable with expectation $1/n$. Thus, $\text{RSD}(\mathcal{I}) = 1/n$. We will use the reverse Chernoff bound (Lemma 5) to bound the probability that Algorithm 1 computes an ε-approximation of $\text{RSD}(\mathcal{I})$ from below.

We apply Lemma 5 to the random variables $X_1, X_2, \ldots, X_k$ denoting the k independent copies of the random variable $\text{SW}(\text{SD}(\mathcal{I},\pi))$ used by Algorithm 1. Notice that we have $p = 1/n$, meaning that the lower bound on k in the statement of the theorem guarantees that $\varepsilon^2 pk \geq 3$, and hence the conditions of Lemma 5 are satisfied. We obtain that the probability that the quantity returned by Algorithm 1 is not an ε-approximation is

$$\Pr\left[\left|\frac{1}{k}\sum_{i=1}^{k} \text{SW}(\text{SD}(\mathcal{I},\pi^i)) - \text{RSD}(\mathcal{I})\right| \geq \varepsilon \cdot \text{RSD}(\mathcal{I})\right]$$
$$\geq \Pr\left[\frac{1}{k}\sum_{i=1}^{k} X_i \geq (1+\varepsilon)\text{RSD}(\mathcal{I})\right] \geq \exp\left(-\frac{9\varepsilon^2 k}{n}\right) > \delta,$$

implying that the probability that the output of Algorithm 1 is an ε-approximation to the expected social welfare of random serial dictatorship is less than $1-\delta$. □

5 Approximating the Expected Social Cost

To approximate the expected social cost of RSD in the metric cost setting, we can modify Algorithm 1 by changing SW with SC in Line 2, i.e. the algorithm now returns the average social cost of the k matchings returned by executing SD with each of the k random agent orderings. We will refer to this modification as Algorithm 1 as well.

Unfortunately, as we show in Sect. 5.1, to return an ε-approximation with probability at least $1-\delta$ for all assignment instances and all values of parameters, Algorithm 1 must use a value for k that depends exponentially on either n or $\ln \frac{1}{\delta}$. To bypass this issue, we use a technique from the literature on approximate counting (see [26, Chapter 28]) by executing Algorithm 1 several times and taking the median value returned in these executions. This is implemented in Algorithm 2 below.

Algorithm 2. Approximating the expected social cost of Random Serial Dictatorship

Input: An assignment instance $\mathcal{I}$ with n agents/items and integers $k, \lambda \geq 1$
Output: A non-negative number

```
for j ← 1, 2, ... λ do
    ξ_j ← Algorithm1(I, k)
end for
return median(ξ)
```

For the analysis of Algorithm 2, we will need upper bounds on the variance of RSD. Unfortunately, while 0 and $\text{OPT}(\mathcal{I})$ are natural bounds on the social welfare of a matching returned by RSD when applied on an assignment instance $\mathcal{I}$ in the value setting, the corresponding bounds for the social cost in the metric cost setting are much further apart, and the Bhatia-Davis inequality is not useful anymore in bounding the variance. Instead, we prove a new bound on the variance of the social cost returned by RSD, which we present in Sect. 5.2. Finally, in Sect. 5.3, we prove bounds on the parameters k and λ used by Algorithm 2 so that it computes an ε-approximation of RSD.

5.1 A Lower Bound for Algorithm 1

We begin our study of the metric cost setting by showing an exponential lower bound on the number of samples needed by Algorithm 1.

Theorem 4. *If Algorithm 1 returns an ε-approximation to the expected social cost of the RSD mechanism with probability at least $1-\delta$ on input any assignment instance in the metric cost setting with n agents/items and for every $\delta, \varepsilon \in (0, 1)$, then k should depend exponentially on either n or $\ln \frac{1}{\delta}$.*

Proof. We use a family of assignment instances that are very similar to the worst-case instances used by Caragiannis et al. [10] (see also [18]) to prove lower bounds on the performance of serial dictatorship in the metric cost setting.

For $n \geq 1$, the assignment instance $\mathcal{I}_n$ is defined as follows. There are n agents at locations $1, 2, 4, \ldots, 2^{n-1}$ and n items at locations $-1, 2, 4, \ldots, 2^{n-1}$ on the real line. Notice that the assignment which matches the agent at location 1 to the item at location -1 and, for $i = 1, ..., n-1$, the agent at location 2^i to the item at the same location has a social cost of 2. Thus, $\text{OPT}(\mathcal{I}_n) \leq 2$.

Now, consider the agent ordering $\pi_n^* = \langle 1, 2, ..., n \rangle$ and observe that the execution of serial dictatorship on instance $\mathcal{I}_n$ using this agent ordering returns the assignment in which, for $i = 1, 2, ..., n-1$, the agent at location 2^{i-1} is matched to the item at location 2^i, and the agent at location 2^{n-1} is matched to the item at location -1. Thus, $\text{SC}(\text{SD}(\mathcal{I}_n, \pi_n^*)) = \sum_{i=1}^{n-1} 2^{i-1} + 2^{n-1} + 1 = 2^n$.

For $n \geq 1$, consider the instance $\mathcal{I}_n$ of the above family with n agents/items. Let $\delta = \frac{1}{2n!}$ and $\varepsilon \in (1/2, 1)$, and assume, for the sake of contradiction, that

$k \le \frac{2^n}{4n}$. The probability that Algorithm 1 does not return an ε-approximation on input instance $\mathcal{I}_n$ is

$$\begin{aligned}
&\Pr\left[\left|\frac{1}{k}\sum_{i=1}^{k}\mathrm{SC}(\mathrm{SD}(\mathcal{I}_n,\pi^i)) - \mathrm{RSD}(\mathcal{I}_n)\right| \ge \varepsilon\cdot\mathrm{RSD}(\mathcal{I}_n)\right] \\
&\ge \Pr\left[\frac{1}{k}\sum_{i=1}^{k}\mathrm{SC}(\mathrm{SD}(\mathcal{I}_n,\pi^i)) \ge 2\cdot\mathrm{RSD}(\mathcal{I}_n)\right] \\
&\ge \Pr\left[\sum_{i=1}^{k}\mathrm{SC}(\mathrm{SD}(\mathcal{I}_n,\pi^i)) \ge 4kn\right] \ge \Pr\left[\sum_{i=1}^{k}\mathrm{SC}(\mathrm{SD}(\mathcal{I}_n,\pi^i)) \ge 2^n\right]. \qquad (4)
\end{aligned}$$

The first inequality follows since $\varepsilon < 1$, the second one from the result of Caragiannis et al. [10], stating that $\mathrm{RSD}(\mathcal{I}_n) \le n\cdot\mathrm{OPT}(\mathcal{I}_n) \le 2n$, and the third inequality follows since $k \le \frac{2^n}{4n}$.

Now, recall that $\mathrm{SC}(\mathrm{SD}(\mathcal{I}_n,\pi_n^*)) \ge 2^n$; thus, the probability in the last line of derivation (4) is lower-bounded by the probability that the ordering π_n^* is selected as one of the k orderings used by Algorithm 1 which in turn is at least $1/n! > \delta$. Thus,

$$\Pr\left[\left|\frac{1}{k}\sum_{i=1}^{k}\mathrm{SC}(\mathrm{SD}(\mathcal{I}_n,\pi^i)) - \mathrm{RSD}(\mathcal{I}_n)\right| \ge \varepsilon\cdot\mathrm{RSD}(\mathcal{I}_n)\right] > \delta,$$

which implies that Algorithm 1 returns an ε-approximation of $\mathrm{RSD}(\mathcal{I}_n)$ with probability at least $1-\delta$ only when $k > \frac{2^n}{4n}$. In this case, k depends exponentially on either n or $\ln 1/\delta < n^2$. The theorem follows. □

5.2 Bounding the Variance of Social Cost

We will shortly turn our attention to Algorithm 2. For its analysis, we will prove an upper bound on the variance of the social cost of RSD, or more precisely, on the expectation of the square of its social cost; we do so in the following lemma.

Lemma 7. *For every assignment instance $\mathcal{I}$ with n agents/items in the metric cost setting, it holds that*

$$\mathbb{E}[SC(SD(\mathcal{I},\pi))^2] \le n^3\cdot OPT(\mathcal{I})^2,$$

where π is a uniformly random ordering of the agents.

Proof. We will prove the statement using induction on n. Observe that the statement holds trivially if $n = 1$ since there is a single perfect matching in this case. Assuming the statement holds for all assignment instances with $n-1$ agents/items in the metric cost setting, we show that this is true for instances with n agents/items as well.

We make use of some additional notation throughout the proof. For any $i \in [n]$, we denote by r_i the item that agent i prefers the most (breaking ties arbitrarily). Also, for any $i \in [n]$, given an assignment instance $\mathcal{I}$ with n agents/items, we denote by $\mathcal{I}_{-i}$ the assignment instance obtained by $\mathcal{I}$ after removing agent i and item r_i. We also use π_{-i} to denote a uniformly random ordering of the agents in $[n] \setminus \{i\}$. We can view the execution of RSD as a uniformly random selection of the first agent i who picks their most preferred item r_i followed by running the RSD mechanism on the reduced instance $\mathcal{I}_{-i}$. Therefore, we have

$$\begin{aligned}\mathbb{E}\left[\mathrm{SC}(\mathrm{SD}(\mathcal{I},\pi))^2\right] &= \frac{1}{n}\sum_{i=1}^{n}\mathbb{E}\left[\left(c_i(r_i)+\mathrm{SC}(\mathrm{SD}(\mathcal{I}_{-i},\pi_{-i}))\right)^2\right]\\ &= \frac{1}{n}\sum_{i=1}^{n}c_i(r_i)^2+\frac{2}{n}\sum_{i=1}^{n}c_i(r_i)\cdot\mathbb{E}\left[\mathrm{SC}(\mathrm{SD}(\mathcal{I}_{-i},\pi_{-i}))\right]\\ &\quad+\frac{1}{n}\sum_{i=1}^{n}\mathbb{E}\left[\mathrm{SC}\left(\mathrm{SD}(\mathcal{I}_{-i},\pi_{-i})\right)^2\right]. \end{aligned} \tag{5}$$

We proceed by presenting two lemmas which will be useful to upper-bound the two final terms of the RHS in Eq. (5).

Lemma 8. *For every assignment instance $\mathcal{I}$ with n agents/items in the metric cost setting, it holds that $\sum_{i=1}^{n} c_i(r_i) \le OPT(\mathcal{I})$ and $\sum_{i=1}^{n} c_i\left(r_i\right)^2 \le OPT(\mathcal{I})^2$.*

Proof. Consider any perfect matching M for the assignment instance $\mathcal{I}$. Clearly, for every $i \in [n]$, the cost of agent i for the item she is matched to in M is at least $c_i(r_i)$. Thus, $\mathrm{OPT}(\mathcal{I}) \ge \sum_{i=1}^{n} c_i(r_i)$ and $\mathrm{OPT}(\mathcal{I})^2 \ge \left(\sum_{i=1}^{n} c_i(r_i)\right)^2 \ge \sum_{i=1}^{n} c_i\left(r_i\right)^2$, as desired. □

The next lemma follows from Caragiannis et al. [10], who proved that the expected social cost of the outcome of RSD on any assignment instance with n agents/items in the metric cost setting is at most n times the optimal social cost.

Lemma 9. *For every assignment instance $\mathcal{I}$ with n agents/items in the metric cost setting and every agent $i \in [n]$, it holds that $\mathbb{E}\left[SC(\mathcal{I}_{-i},\pi_{-i})\right] \le (n-1)\cdot OPT(\mathcal{I}_{-i})$.*

By the induction hypothesis, we have

$$\mathbb{E}\left[\mathrm{SC}\left(\mathrm{SD}(\mathcal{I}_{-i},\pi_{-i})\right)^2\right] \le (n-1)^3\cdot\mathrm{OPT}(\mathcal{I}_{-i})^2. \tag{6}$$

Using Lemma 8 and Lemma 9 as well as Eq. (6), Eq. (5) yields

$$\begin{aligned}\mathbb{E}\left[\mathrm{SC}\left(\mathrm{SD}(\mathcal{I},\pi)\right)^2\right] &\le \frac{1}{n}\cdot\mathrm{OPT}(\mathcal{I})^2+\left(2-\frac{2}{n}\right)\sum_{i=1}^{n}c_i(r_i)\cdot\mathrm{OPT}(\mathcal{I}_{-i})\\ &\quad+\frac{(n-1)^3}{n}\sum_{i=1}^{n}\mathrm{OPT}(\mathcal{I}_{-i})^2. \end{aligned} \tag{7}$$

We now use a lemma that has also been used by Caragiannis et al. [10]. The proof has been included here for the sake of completeness.

Lemma 10. *For every assignment instance $\mathcal{I}$ with n agents/items in the metric cost setting and every agent $i \in [n]$, it holds that $OPT(\mathcal{I}_{-i}) \le OPT(\mathcal{I}) + c_i(r_i)$.*

Proof. Let M be a matching of minimum social cost on instance $\mathcal{I}$. If agent i is matched to item r_i in M, then the restriction of M that does not include the pair (i, r_i) is a matching of $\mathcal{I}_{-i}$ of social cost at most $\mathrm{OPT}(\mathcal{I})$, and the statement follows. Assume now that agent i is matched to some item g different from r_i in M, while some agent $j \neq i$ is matched to item r_i. The set of agent-item pairs consisting of pair (j, g) and the restriction of M not including the pairs (i, g) and (j, r_i) is a matching for instance $\mathcal{I}_{-i}$ of social cost

$$\mathrm{OPT}(\mathcal{I}) - c_i(g) - c_j(r_i) + c_j(g) \le \mathrm{OPT}(\mathcal{I}) + c_i(r_i)\,.$$

The inequality follows from applying the triangle inequality which states that $c_j(g) \le c_j(r_i) + c_i(r_i) + c_i(g)$. □

Using Lemma 8 and Lemma 10, the sum in the second term of the RHS of Eq. (7) becomes

$$\begin{aligned}\sum_{i=1}^{n} c_i(r_i) \cdot \mathrm{OPT}(\mathcal{I}_{-i}) &\le \sum_{i=1}^{n} c_i(r_i) \cdot (\mathrm{OPT}(\mathcal{I}) + c_i(r_i)) \\ &= \mathrm{OPT}(\mathcal{I}) \cdot \sum_{i=1}^{n} c_i(r_i) + \sum_{i=1}^{n} c_i(r_i)^2 \le 2 \cdot \mathrm{OPT}(\mathcal{I})^2\,. \end{aligned} \tag{8}$$

Similarly, making use of Lemma 8 and Lemma 10 once more, the sum in the third term of the RHS of Eq. (7) results in

$$\begin{aligned}\sum_{i=1}^{n} \mathrm{OPT}(\mathcal{I}_{-i})^2 &\le \sum_{i=1}^{n} (\mathrm{OPT}(\mathcal{I}) + c_i(r_i))^2 \\ &= n \cdot \mathrm{OPT}(\mathcal{I})^2 + 2\mathrm{OPT}(\mathcal{I}) \cdot \sum_{i=1}^{n} c_i(r_i) + \sum_{i=1}^{n} c_i(r_i)^2 \\ &\le (n+3) \cdot \mathrm{OPT}(\mathcal{I})^2\,. \end{aligned} \tag{9}$$

Finally, using Eq. (8) and Eq. (9), Eq. (7) yields

$$\begin{aligned}\mathbb{E}\left[\mathrm{SC}\left(\mathrm{SD}(\mathcal{I}, \pi)\right)^2\right] &\le \left(\frac{1}{n} + 2\left(2 - \frac{2}{n}\right) + \frac{(n-1)^3}{n}(n+3)\right) \cdot \mathrm{OPT}(\mathcal{I})^2 \\ &= \left(n^3 - \frac{6}{n}(n-1)^2\right) \mathrm{OPT}(\mathcal{I})^2 \le n^3 \cdot \mathrm{OPT}(\mathcal{I})^2\,,\end{aligned}$$

as desired. This completes the proof of Lemma 7. □

5.3 The Upper Bound for Algorithm 2

We are now ready to present bounds on the parameters k and λ so that Algorithm 2 computes an ε-approximation of RSD with probability at least $1 - \delta$. This requires sampling only $O\left(\frac{n^3}{\varepsilon^2}\ln\frac{1}{\delta}\right)$ agent orderings and running RSD according to them. The proof has two parts. First, in Lemma 11, we exploit our bound on the variance of the social cost of RSD from Lemma 7 to prove that the probability that an execution of Algorithm 1 does not return an over or an under ε-approximation is at most 1/4. This is enough to conclude (using a Chernoff bound argument in the proof of Theorem 5) that the median of the social cost values returned by all executions of RSD provides the desired ε-approximation with high probability.

Lemma 11. *Let $n \geq 1$ be an integer and $\varepsilon \in (0, 1]$. For $k \geq \frac{3n^3}{\varepsilon^2}$, the probability that the output of Algorithm 1, when applied on the assignment instance $\mathcal{I}$ with n agents/items in the metric cost setting, is not an over (respectively, not an under) ε-approximation to the expected social cost of RSD is at most $\frac{1}{4}$.*

Proof. By Lemma 7, for $i \in [k]$, the random variable $\mathrm{SC}(\mathrm{SD}(\mathcal{I}, \pi^i))$ has variance at most $n^3 \cdot \mathrm{OPT}(\mathcal{I})^2$. Thus, the random variable $\frac{1}{k}\sum_{i=1}^{k}\mathrm{SC}(\mathrm{SD}(\mathcal{I}, \pi^i))$ has expectation $\mathrm{RSD}(\mathcal{I})$ and variance $\sigma^2 \leq \frac{n^3}{k} \cdot \mathrm{OPT}(\mathcal{I})^2$, since the random orderings π^i are independent for $i \in [k]$. We will now apply the Chebyshev-Cantelli's inequality (Lemma 3) to the random variable $\frac{1}{k}\sum_{i=1}^{k}\mathrm{SC}(\mathrm{SD}(\mathcal{I}, \pi^i))$ with $t = \frac{\sqrt{k}\cdot\varepsilon\cdot\mathrm{RSD}(\mathcal{I})}{\sigma}$. Notice that using the condition on k as well as the fact that $\mathrm{OPT}(\mathcal{I}) \leq \mathrm{RSD}(\mathcal{I})$, we get

$$t^2 = \frac{k\cdot\varepsilon^2\cdot\mathrm{RSD}(\mathcal{I})^2}{\sigma^2} \geq \frac{k\cdot\varepsilon^2\cdot\mathrm{RSD}(\mathcal{I})}{n^3\cdot\mathrm{OPT}(\mathcal{I})} \geq \frac{k\cdot\varepsilon^2}{n^3} \geq 3.$$

Lemma 3 then yields

$$\Pr\left[\frac{1}{k}\sum_{i=1}^{k}\mathrm{SC}(\mathrm{SD}(\mathcal{I}, \pi^i)) \geq (1+\varepsilon)\cdot\mathrm{RSD}(\mathcal{I})\right] \leq \frac{1}{4}.$$

Hence, the probability that the output of Algorithm 1 is an over ε-approximation to $\mathrm{RSD}(\mathcal{I})$ is at least $\frac{1}{4}$, as desired. The proof for it being an under ε-approximation follows along the same lines by considering the random variable $-\frac{1}{k}\sum_{i=1}^{k}\mathrm{SC}(\mathrm{SD}(\mathcal{I}, \pi^i))$ instead of $\frac{1}{k}\sum_{i=1}^{k}\mathrm{SC}(\mathrm{SD}(\mathcal{I}, \pi^i))$. □

We are now ready to prove our main statement for Algorithm 2.

Theorem 5. *Let $n \geq 1$ be an integer and $\delta, \varepsilon \in (0, 1]$. For $k \geq \frac{3n^3}{\varepsilon^2}$ and $\lambda \geq \frac{4}{\ln 4/e}\ln\frac{2}{\delta}$, the output of Algorithm 2 when applied to the assignment instance $\mathcal{I}$ with n agents/items in the metric cost setting is an ε-approximation to the expected social cost of RSD with probability at least $1 - \delta$.*

Proof. We will show that the probability that $\mathrm{median}(\boldsymbol{\xi})$ returned by Algorithm 2 is not an upper ε-approximation (respectively, a lower ε-approximation) of RSD is at most $\delta/2$. Theorem 5 then follows by applying a simple union bound.

We first bound the probability that $\mathrm{median}(\boldsymbol{\xi})$ is not an upper ε-approximation of RSD. Notice that the values $\xi_1, \ldots, \xi_\lambda$ computed in line 2 of Algorithm 2 are independent random variables with expectation $\mathrm{RSD}(\mathcal{I})$. For $j = 1, 2, \ldots, \lambda$, we define Z_j as the Bernoulli random variable that indicates whether ξ_j as computed in line 2 of Algorithm 2 is not an upper ε-approximation of $\mathrm{RSD}(\mathcal{I})$, i.e.

$$Z_j = \begin{cases} 1 & \text{if } \xi_j \geq (1+\varepsilon) \cdot \mathrm{RSD}(\mathcal{I}) \\ 0 & \text{otherwise} \end{cases}.$$

By Lemma 11, we have that $\Pr[Z_j = 1] \leq \frac{1}{4}$, and hence the random variable $Z = \sum_{j=1}^{\lambda} Z_j$ indicating the number of ξ_j values that are not upper ε-approximations of $\mathrm{RSD}(\mathcal{I})$ has expectation at most $\frac{\lambda}{4}$. Then, the median($\boldsymbol{\xi}$) exceeds $(1+\varepsilon) \cdot \mathrm{RSD}(\mathcal{I})$ when the random variable Z has value at least $\frac{\lambda}{2}$. We obtain that

$$\Pr[\mathrm{median}(\boldsymbol{\xi}) \geq (1+\varepsilon) \cdot \mathrm{RSD}(\mathcal{I})] = \Pr[Z \geq 2\mathbb{E}[Z]] \leq \left(\frac{e}{4}\right)^{\frac{\lambda}{4}} \leq \frac{\delta}{2},$$

as desired. The first inequality follows by applying the Chernoff bound (Lemma 4) for the random variable Z (recall that the random variables $Z_1, \ldots, Z_\lambda$ are independent) with $\eta = 1$ and the last inequality follows due to the choice of λ.

To bound the probability that median($\boldsymbol{\xi}$) is not a lower ε-approximation of RSD is almost identical; the only change required is in the definition of the random variable Z_j which should use $\xi_j \leq (1-\varepsilon) \cdot \mathrm{RSD}(\mathcal{I})$ instead. □

6 Conclusion

We have presented a formal statement showing that earlier #P-hardness results on computing the RSD lottery matrix in the abstract setting imply the #P-hardness of computing both the expected social welfare and the expected social cost of assignment instances in the value and metric cost settings, respectively. Furthermore, we have presented bounds on the number of samples sufficient and necessary to approximate these expectations with simple algorithms. Even though our analysis of Algorithm 1 for the expected social welfare is asymptotically tight, for Algorithm 2 and the expected social cost there seems to be some room for improvement. We believe that such improvements can benefit from better bounds on the expectation and variance of the social cost in terms of n and $\mathrm{OPT}(\mathcal{I})$, compared to the linear bound of Caragiannis et al. [10] (see also Lemma 9) and our polynomial bound in Lemma 7, respectively. Regarding extensions of the techniques, it would be interesting to consider scenarios with more items than agents and the round-robin algorithm. We believe that our analysis for the value setting carries over to this scenario but more detailed arguments are needed for the metric cost setting.

Acknowledgments. This work was supported by the Independent Research Fund Denmark (DFF) under grant 2032-00185B.

References

1. Abdulkadiroğlu, A., Sönmez, T.: Random serial dictatorship and the core from random endowments in house allocation problems. Econometrica **66**(3), 689–702 (1998)
2. Abraham, D.J., Cechlárová, K., Manlove, D.F., Mehlhorn, K.: Pareto optimality in house allocation problems. In: Proceedings of the 16th International Symposium on Algorithms and Computation (ISAAC), pp. 3–15 (2005)
3. Aziz, H., Brandt, F., Brill, M.: The computational complexity of random serial dictatorship. Econ. Lett. **121**(3), 341–345 (2013)
4. Aziz, H., de Keijzer, B.: Shapley meets shapley. In: Proceedings of the 31st International Symposium on Theoretical Aspects of Computer Science (STACS), pp. 99–111 (2014)
5. Aziz, H., Mestre, J.: Parametrized algorithms for random serial dictatorship. Math. Soc. Sci. **72**, 1–6 (2014)
6. Bachrach, Y., Markakis, E., Resnick, E., Procaccia, A.D., Rosenschein, J.S., Saberi, A.: Approximating power indices: theoretical and empirical analysis. Auton. Agents Multi Agent Syst. **20**(2), 105–122 (2010)
7. Bhatia, R., Davis, C.: A better bound on the variance. Am. Math. Mon. **107**(4), 353–357 (2000)
8. Bhattacharyya, A., Dey, P.: Predicting winner and estimating margin of victory in elections using sampling. Artif. Intell. **296**, 103476 (2021)
9. Bogomolnaia, A., Moulin, H.: A new solution to the random assignment problem. J. Econ. Theory **100**(2), 295–328 (2001)
10. Caragiannis, I., Filos-Ratsikas, A., Frederiksen, S.K.S., Hansen, K.A., Tan, Z.: Truthful facility assignment with resource augmentation: an exact analysis of serial dictatorship. Math. Program. **203**(1), 901–930 (2024)
11. Caragiannis, I., Micha, E., Peters, J.: Can a few decide for many? The metric distortion of sortition. In: Proceedings of the 41st International Conference on Machine Learning (ICML). To appear (2024)
12. Caragiannis, I., Rathi, N.: Optimizing over serial dictatorships. In: Proceedings of the 16th International Symposium on Algorithmic Game Theory (SAGT), pp. 329–346 (2023)
13. Crès, H., Moulin, H.: Scheduling with opting out: improving upon random priority. Oper. Res. **49**(4), 565–577 (2001)
14. Dubhashi, D.P., Panconesi, A.: Concentration of Measure for the Analysis of Randomized Algorithms. Cambridge University Press, Cambridge (2009)
15. Filos-Ratsikas, A., Frederiksen, S.K.S., Zhang, J.: Social welfare in one-sided matchings: random priority and beyond. In: Proceedings of the 7th International Symposium on Algorithmic Game Theory (SAGT), pp. 1–12 (2014)
16. Fortnow, L.: Counting complexity. In: Hemaspaandra, L., Selman, A. (eds.) Complexity Theory Retrospective II, pp. 81–107. Springer (1997)
17. Hylland, A., Zeckhauser, R.: The efficient allocation of individuals to positions. J. Polit. Econ. **87**(2), 293–314 (1979)
18. Kalyanasundaram, B., Pruhs, K.: Online weighted matching. J. Algorithms **14**(3), 478–488 (1993)
19. Klein, P., Young, N.E.: On the number of iterations for Dantzig-Wolfe optimization and packing-covering approximation algorithms. SIAM J. Comput. **44**(4), 1154–1172 (2015)

20. Motwani, R., Raghavan, P.: Randomized Algorithms. Cambridge University Press, Cambridge (1995)
21. Procaccia, A.D., Tennenholtz, M.: Approximate mechanism design without money. ACM Trans. Econ. Comput. **1**(4), 18:1–18:26 (2013)
22. Sabán, D., Sethuraman, J.: The complexity of computing the random priority allocation matrix. Math. Oper. Res. **40**(4), 1005–1014 (2015)
23. Schummer, J., Vohra, R.: Mechanism design without money. In: Nisan, N., Roughgarden, T., Tardos, E., Vazirani, V.V. (eds.) Algorithmic Game Theory. Cambridge University Press (2007)
24. Svensson, L.: Strategy-proof allocation of indivisible goods. Soc. Choice Welf. **16**(4), 557–567 (1999)
25. Sönmez, T., Ünver, M.: Matching, allocation, and exchange of discrete resources. In: Benhabib, J., Bisin, A., Jackson, M.O. (eds.) Handbook of Social Economics, vol. 1, pp. 781–852. North-Holland (2011)
26. Vazirani, V.V.: Approximation Algorithms. Springer, Berlin Heidelberg (2001). https://doi.org/10.1007/978-3-662-04565-7

Game Theory and Repeated Games

Swim till You Sink: Computing the Limit of a Game

Rashida Hakim[1(✉)], Jason Milionis[1], Christos Papadimitriou[1], and Georgios Piliouras[2]

[1] Columbia University, New York, USA
{rashida.hakim,christos}@columbia.edu, jm@cs.columbia.edu
[2] Google DeepMind, London, UK
gpil@deepmind.com

Abstract. During 2023, two interesting results were proven about the limit behavior of game dynamics: First, it was shown that there is a game for which no dynamics converges to the Nash equilibria. Second, it was shown that the sink equilibria of a game adequately capture the limit behavior of natural game dynamics. These two results have created a need and opportunity to articulate a principled computational theory of the meaning of the game that is based on game dynamics. Given any game in normal form, and any prior distribution of play, we study the problem of computing the asymptotic behavior of a class of natural dynamics called the noisy replicator dynamics as a limit distribution over the sink equilibria of the game. When the prior distribution has pure strategy support, we prove this distribution can be computed efficiently, in near-linear time to the size of the best-response graph. When the distribution can be sampled—for example, if it is the uniform distribution over all mixed strategy profiles—we show through experiments that the limit distribution of reasonably large games can be estimated quite accurately through sampling and simulation.

Keywords: Replicator Dynamics · Sink Equilibria

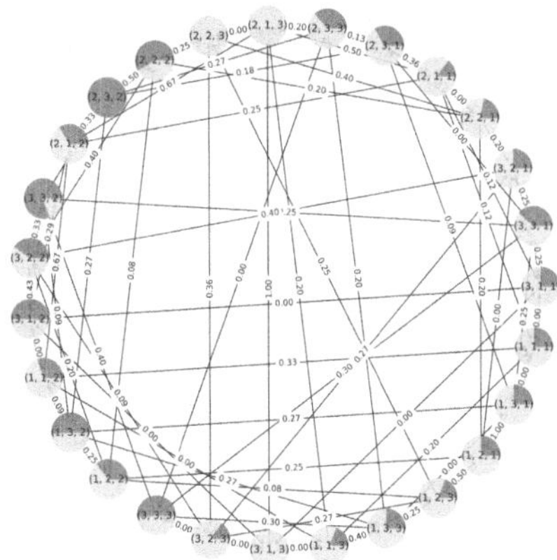

Fig. 1. The better-response graph of the $3 \times 3 \times 3$ game depicting the hitting probabilities of the pure profiles as pie charts.

G. Schäfer and C. Ventre (Eds.): SAGT 2024, LNCS 15156, pp. 205–222, 2024.
https://doi.org/10.1007/978-3-031-71033-9_12

1 Introduction

The Nash equilibrium has been the quintessence of Game Theory. The field started its modern existence in 1950 with Nash's definition and existence theorem, and the Nash equilibrium remained for three quarters of a century its paramount solution concept—the others exist as refinements, generalizations, or contradistinctions. During the past three decades, during which Game Theory came under intense computational scrutiny, the Nash equilibrium has lost some of its appeal, as its fundamental incompatibility with computation became apparent. The Nash equilibrium has been shown intractable to compute or approximate in normal-form games [10,16,17], while its other well-known deficiency of computational nature—the ambiguity of non-uniqueness and the quagmire of equilibrium selection [23]—had already been known.

A long list of *game dynamics*—that is, dynamical systems, continuous- or discrete-time, defined on mixed strategy profiles—proposed by economists over the decades are all known to *fail* to converge consistently to the Nash equilibrium. This included Nash's own discrete-time dynamics used in his proof, the well-known replicator dynamics treated in this paper, and many others. Given this, the following question acquired some importance:

> *Question 1: We know that every game has a Nash equilibrium. But does every game have a dynamics that converges to the Nash equilibria of the game?*

A negative answer would be another serious setback for the Nash equilibrium, and impetus would be added to efforts (see for example [34,39]) to elevate the limit behavior of natural game dynamics as a proposed "meaning of the game," an alternative to the Nash equilibrium. An important obstacle to these efforts was that the nature of the limit behavior of natural dynamics in general games had been lacking the required clarity. It had been known for 40 years since Conley's seminal work [14] that the right concept of limit behavior in a general dynamical system is a system of topological objects known as its *chain recurrent components*. However, this concept is mathematically intractable for general dynamical systems: there can be infinitely many such sets, of unbounded complexity.

> *Question 2: Is there a concrete characterization, in terms of familiar game-theoretic concepts, of the chain recurrent sets in the special case of natural game dynamics in normal-form games?*

During this past year, there was important progress on both questions.

1. It was proven in [31] that the answer of Question 1 above is negative: there is a game for which no game dynamics can converge to the Nash equilibria—that is, there is no dynamics such that the fate of all initial strategy profiles are the Nash equilibria, and the Nash equilibria are themselves fixed points of the dynamical system. Thus, the Nash equilibrium is fundamentally incapable of capturing asymptotic player behavior.

2. Biggar and Shames establish a useful characterization of the chain recurrent sets of natural recurrent dynamics [8]: it was shown that each chain recurrent component of a game under the replicator dynamics contains the union of one or more *sink equilibria* of the game. Sink equilibria, first defined by Goemans et al. [21] in the context of the price of anarchy, are the sink strongly connected components of the better response graph of the game.

We believe that these two results open an important opportunity to articulate a new approach to understanding a normal-form game. Instead of considering it, as game theorists have been doing so far, as the specification of an intractable equilibrium selection problem, we propose to see it as a specification of the limit behavior of the players. According to this point of view [39], a game is a mapping from a prior distribution over mixed strategy profiles (MSPs) to the resulting limit distribution if the players engage in an intuitive and well accepted natural behavior called *noisy replicator dynamics*, which is related to multiplicative weight updates and will be defined soon. This is the quest we are pursuing and advancing in this paper.

Our Contributions

- We propose a concrete, unambiguous, and computationally tractable conception of a game as a mapping from any prior distribution over MSPs to the sink equilibria of the game, namely the limit distribution of the noisy replicator dynamics when initialized at the prior.
- We initiate the study of the efficiency of its computation. As a baby step, in the next section we show that the sink equilibria can be computed in time linear in the description of the game. We also point out that they are intractable for various families of implicit games.
- We prove that the mapping from a prior to a distribution over sink equilibria can be calculated explicitly and efficiently (near linear in the size of the game description) when the prior has pure strategy support. This is highly nontrivial because the better response graph of the game may contain many directed cycles of length two with infinitesimal transition probability ϵ, corresponding to tie edges; the analysis must be carried out at the $\epsilon \to 0$ limit. The algorithm involves a number of novel graph-theoretic concepts and techniques relating to Markov chains, and the deployment of near-linear algorithms for directed Laplacian system solving as well as a dynamic algorithm for incrementally maintaining the strongly connected components (SCCs) of a graph.
- We also show through extensive experimentation that the general case (arbitrary prior) can be solved efficiently for quite large games.

Related Work

Non-convergence of Learning Dynamics in Games. The difficulty of learning dynamics to converge to Nash equilibria in games is punctuated by a plethora of diverse negative results spanning numerous disciplines such as game theory, economics, computer science and control theory [1,2,4,5,11,15,20,24,26,27,30, 39,48,49]. Recently, [31] capped off this stream of negative results with a general impossibility result showing that there is no game dynamics that achieve global convergence to Nash for all games, a result that is independent of any complexity theoretic or uncoupledness assumptions on the dynamics. Besides such worst case theoretical results, detailed experimental studies suggest that chaos is commonplace in game dynamics, and emerges even in low dimensional systems across a variety of game theoretic applications [6,12,29,36,37,41,44,45].

Dynamical Systems for Learning in Games. This extensive list of non-equilibrating results has inspired a program for linking game theory to dynamical systems [38,39] and Conley's fundamental theorem of dynamical systems [14]. These tools have since been applied in multi-agent ML settings such as developing novel rankings as well as training methodologies for agents playing games [33–35,43]. Finally, Peyton Young's paper on conventions [40] is an important precursor of our point of view in the Economics literature, focusing in the special case of games in which the sink SCCs are pure strategy equilibria.

Sink Equilibria. The notion of sink equilibrium, a strongly connected component with no outgoing arcs in the strategy profile graph associated with a game, was introduced in [21]. They also defined an analogue to the notion of Price of Anarchy [28], the Price of Sinking, the ratio between the worst case value of a sink equilibrium and the value of the socially optimal solution. The value of a sink equilibrium is defined as the expected social value of the steady state distribution induced by a random walk on that sink. Later work established further connections between Price of Sinking and Price of Anarchy via the (λ, μ)-robustness framework [42]. A number of negative, PSPACE-hard complexity results for analyzing and approximating sink equilibria have been established in different families of succinct games [19,32].

Finally, [25] compute the limiting stationary distribution of an irreducible MC with vanishing edges; their technique can be used in our framework to solve for the time averaged long run behavior within a sink SCC.

2 Preliminaries

We assume the standard definitions of a normal-form game G with p players with pure strategy sets $\{S_i\}$, and their utilities U_i. We denote by $|G|$ the size of the description of G. The *better-or-equal response graph* $B(G)$ has the pure strategy profiles as nodes, and an edge from u to v if u and v differ in the strategy of only one player i, and $U_i(v) \geq U_i(u)$. Let E be the set of edges of $B(G)$. Notice that, because of the tie edges and the transitive edges, the number of edges in the response graph can be much larger than the size of the description of the game, i.e., $|E| = \Omega(|G|)$. The *sink equilibria* of G are the sink strongly connected components (sink SCCs) of $B(G)$, that is, maximal sets of nodes with paths between all pairs, such that there is no edge leaving this set. We shall define many other novel graph-theoretic concepts for this graph in the next section. Our first theorem delineates the complexity of finding the sink equilibria of a game:

Theorem 1. *The sink equilibria can be computed in time near-linear in the description of the game presented in normal form, whereas computing them in a graphical game is PSPACE-complete.*

Proof. The first claim follows from the fact that, even though (as pointed out in the preceding paragraph) $B(G)$ has more edges than the size of the description of the game, there is an equivalent graph of linear size with the same transitive closure (and therefore strongly connected components) obtained as follows: For each player and each pure strategy profile for other players, consider the subgraph of only the nodes that correspond to the strategy profile for the other players together with some action for the player. Sort the nodes by their outcome for the player. For each node in order from lowest to highest outcome, create an edge to the next node in increasing order as well as the last node of the same outcome if the next node has a higher outcome (only if this would not be a self-loop). This preserves transitive closure as compared to $B(G)$ and each node has at most outgoing 2 edges (it sums to $\leq 3/2$ edges per node on average in the worst case). Besides sorting this can be done in linear time.

The second claim follows from known results [18]; the result holds for other forms of succinct descriptions of games, such as Bayesian or extensive form games.

Next we define the *noisy replicator dynamics* on G [39], a noisy generalization of the classical replicator dynamics [46]. It is a function mapping the set of MSPs of G to itself as follows:

- $\phi(x) = \partial G(x + \eta \cdot \mathrm{BR}_{\mathrm{x}} + \mathcal{N}_{\mathrm{x}}(0, \delta))$, where
- BR_x is the unit best response vector at x *projected to the subspace of* x—that is, containing zeros at all coordinates in which x is zero; this ensures that the support of x never increases;
- $\mathcal{N}(0, \delta)_x$ is Gaussian noise, also projected;
- the function ∂G maps $(x + \eta \cdot \mathrm{BR}_{\mathrm{x}} + \mathcal{N}(0, \delta))_{\mathrm{x}}$ either to itself, if it is inside the domain of the game's MSPs, or to the closest point in the support's boundary, otherwise;
- and $\delta, \eta > 0$ are important small parameters.

Justification. The replicator dynamics [46,47] has been for four decades the standard model for the evolution of strategic behavior. In connection with Economics and Game Theory, it has the important advantage of invariance under positive affine changes in the players' utilities. For our purposes, it is approximated via the noisy version of Multiplicative Weights Update (MWU) [3]. Projecting the noise to the support of the current MSP x is motivated by evolution and extinction, and is instrumental for fast convergence. This precise dynamics has been used extensively in reinforcement learning for game play, see for example [34,35].

Finally, we define a dynamics on the *pure* strategy profiles (that is, a Markov chain), called the *Conley-Markov Chain* of G, or CMC(G) [21,39]. If (u, v) is an edge of $B(G)$ corresponding to a defection of player i, its probability in CMC(G) is proportional to $U_i(v) - U_i(u)$, with the edges out of each node u normalized to one. It is not hard to see that this is the limit of the noisy replicator dynamics as the noise goes to zero and the MSP goes to u. Importantly, however, CMC(G) also has an infinitesimal probability ϵ for each tie edge. (Note that this probability is used symbolically as it descends to zero, and, in the interest of clarity, it does not affect the normalization at u.) This treatment of tie edges reflects two things: First, it was shown in [7] that tie edges must be included in the calculation of the sink equilibria for their theorem to hold; and second, to incorporate tie edges in a way compatible with Conley's theorem [14] is to think of them as conduits of a *balanced random walk* on the undirected edge between the two nodes in which the MSP is changing via tiny steps of σ at a time, so that it will take $\Theta(1/\sigma^2)$ steps for the transition to be completed, justifying its infinitesimal transition probability.

3 A Combinatorial Algorithm for the Hitting Probabilities

We start by collapsing all sink SCCs of CMC(G) to single absorbing nodes. Our main goal is to compute the hitting probabilities from each node i of $CMC(G)$ to each of the absorbing nodes—that is, the probability that a path starting from i will end up in the node—albeit *in the limit as* $\epsilon \to 0$. The hitting probabilities can be defined in two equivalent ways, both of which we will use in our proofs. Define h_{iS} to be the hitting probability of node i to sink S, that is, the probability that a path from i will eventually be absorbed by S. Let p_{ij} be the transition probability of node i to node j (the weight of the edge (i, j) or 0 if there is no edge). Then the hitting probabilities are the smallest non-negative numbers that satisfy the following system of equations.

$$h_{iS} = \begin{cases} \sum_{j \in \text{nodes}} p_{ij} h_{jS}, & \text{if } i \notin S. \\ 1, & \text{if } i \in S \end{cases} \tag{1}$$

If we define Ψ_{iS} as the (potentially infinite) set of paths that start at i and end at some node in S, then we equivalently have the following set of equations.

$$h_{iS} = \sum_{p \in \Psi_{iS}} \mathbf{Pr}[p] \tag{2}$$

In this section we prove the following:

Theorem 2. *The limit hitting probabilities of CMC(G) can be computed in time $O(|E|^{1+\delta})$, where E is the set of edges of CMC(G) and $\delta > 0$.*

Significant progress has been made in solving linear systems associated with weighted directed graphs such as Eq. 1 faster than the time required to solve arbitrary linear systems. Two problems can now be solved in almost-linear time in the size of the graph: computing the stationary distribution of an irreducible Markov chain (hence abbreviated MC); and computing the escape probabilities [13]. The computation of escape probabilities in a random walk maps directly to the problem of computing hitting probabilities in a MC. So we have fast algorithms for our problem in the case of no tie edges. However, the introduction of tie edges creates an ill-conditioned problem, and we are interested in its solution as $\epsilon \to 0$. A possible approach would be to solve the system of equations given in Eq. 1 symbolically and then take limits as $\epsilon \to 0$; however, solving large systems of equations symbolically is intractable. Instead, we take a combinatorial approach to transform any given CMC(G) into a simpler MC that preserves the limit hitting probabilities of the original CMC(G) but eventually has no tie-edges. The hitting probabilities of this simplified MC can be computed in almost linear time as mentioned above.

3.1 Outline of the Algorithm

1. The input to the algorithm is CMC(G)—actually, it could be any ϵ-MC M with absorbing nodes. The output is the list of hitting probabilities $\{h_{iS}\}$, the probabilities that the sink SCCs of the graph will be reached by each of the nodes in the rest of the graph.
2. We start by collapsing the sink SCCs of M.
3. We calculate the SCC's of M *without* the ϵ-edges, called *rSCC's*. This makes sense since ϵ edges are traversed at a far slower rate than the rest.
4. Next we must handle a phenomenon called a *pseudosink*, an rSCC that only has ϵ-edges outgoing. Within a pseudosink, the MC achieves convergence to a steady state before exiting, and therefore all of its nodes have the same hitting probabilities. The pseudosinks are identified one by one and collapsed, with their outgoing ϵ-edges replaced by regular edges in accordance with Definition 6. A simple disjoint set data structure can track the original vertices through the collapses.
5. A complication is that the collapsed pseudosinks acquire new regular edges to the rest of the graph, and as a result the rSCCs of the graph must be recalculated. This procedure may also create new pseudosinks, so steps 3 and 4 are repeated until no more pseudosinks exist.
6. Once all pseudosinks have been removed this way, any remaining ϵ-edges do not affect the hitting probabilities and can therefore be deleted. At this point, the hitting probabilities can be computed in almost linear time.

3.2 Definitions

Definition 1. *ϵ-Markov Chain:* *A ϵ-Markov Chain (ϵ-MC) is a Markov chain that has two types of edges: regular edges which have weights $c_r - c_{re}\epsilon$ for constants $c_r > 0$ and $c_{re} \geq 0$ and ϵ edges which have weights $c_e\epsilon$ for constant*

$c_e > 0$. Thus, the CMC(G) is an ϵ-MC. As the values of the coefficients c_{re} in the regular edges do not affect the limiting hitting probabilities we shall ignore them.

Definition 2. ***Sink SCC:*** *A sink SCC S is a maximal set of nodes that are strongly connected (including connected via ϵ-edges) that has no outgoing edges.*

Definition 3. ***rSCC:*** *A rSCC is a maximal set of nodes that is strongly connected via regular edges. An rSCC may contain ϵ-edges between nodes within the rSCC, but every node is reachable from every other node without requiring ϵ-edges.*

Definition 4. ***Pseudosink:*** *A pseudosink P is a rSCC that has at least one outgoing ϵ-edge and no outgoing regular edges.*

Definition 5. ***Order:*** *The order of a node i, $\text{ORDER}(i)$, is the minimum number of tie edges that exist on a path from i to any sink SCC. The maximum order of the current MC, $\text{MAXORDER}(M)$, is our gauge of progress in the algorithm.*

Definition 6. *The* **weight** *of a new regular edge from a collapsed pseudosink to node y is as follows (here O is the set of outgoing ϵ-edges from the pseudosink P).*

$$W(P, y) = \frac{\sum_{e \in O : e=(x,y)} c_e \pi_P[x]}{\sum_{e'=(x',y') \in O} c_{e'} \pi_P[x']},$$

where $\pi_P[x]$ is the steady-state probability of x within P, computed using only regular edges.

3.3 Algorithm Correctness

Throughout the algorithm, we maintain a MC that we denote M, initially the MCM(G) with all sink SCCs collapsed. There are two aspects to validate. The first is that the algorithm progresses until M has no remaining ϵ-edges. The second is that M maintains the property that at all stages the limit hitting probabilities of the original nodes are maintained through collapsing pseudosinks (step 2) and deleting ϵ-edges (step 6).

Algorithm Progression

Lemma 1. *If $\text{MAXORDER}(M) \geq 1$ then M contains a pseudosink.*

Proof. Consider the set of all nodes i that achieve $\text{ORDER}(i) = \text{MAXORDER}(M)$. By definition, from this set of nodes, the MC cannot reach any other nodes without using ϵ-edges. Consider the rSCC decomposition of this set of nodes. The regular edges between rSCCs induce a directed acyclic graph on this set of nodes. Every finite DAG has at least one leaf, defined as a vertex that has no outgoing edges. Each leaf is a rSCC with no outgoing regular edges and therefore is a pseudosink.

Lemma 2. *Collapsing all pseudosinks in M reduces the maximum order of M by at least 1.*

Proof. Again, consider the set of all nodes i that have ORDER(i) equal to MAXORDER(M) and the DAG representing the rSCC structure of this set. Each leaf of the DAG is a pseudosink. From each pseudosink, there exists a path to some sink SCC that achieves the order MAXORDER(M). Collapsing the pseudosink replaces all outgoing ϵ-edges with regular edges and therefore this same path must now achieve the order MAXORDER(M) − 1. Since every rSCC in the DAG is either a leaf or has a regular path to a leaf, all nodes that previously achieved i will now have ORDER(i) $\leq$ MAXORDER(M) − 1. So the maximum order of M is reduced by at least 1.

Combining these two lemmas means that at each stage of our algorithm we will find one or more pseudosinks and collapse them, decreasing the maximum order by at least 1. So the algorithm will progress until the maximum order reaches 0, at which point we delete all remaining ϵ-edges and are left with a Markov Chain with only regular edges.

Pseudosink Collapse

Lemma 3. *Let P be a pseudosink and N_P be the node set of P. Let O be the set of outgoing ϵ-edges from P. Finally, let L_e be the event that a Markov chain started at any $i \in N_P$ will take $(e = (x, y)) \in O$, where $W_M(e) = c_e\epsilon$ is the weight in M of edge e. Then*

$$\lim_{\epsilon \to 0} \mathbf{Pr}[L_e] = \frac{c_e \pi_P[x]}{\sum_{e' \in O} c_{e'} \pi_P}$$

This immediately implies that $W(P, y)$ from Definition 6 is the probability that y is the first node outside of P that a chain started at any $i \in N_P$ will travel to.

Proof Sketch. The key idea behind the proof is that, with high probability, the chain converges to its stationary distribution within the pseudosink P before taking it's first outgoing epsilon edge. Therefore, the probabilities are dependent on the stationary distribution on P as well as the weights on the ϵ-edges.

Lemma 4. *Let P be a pseudosink, and let the stationary distribution on P without ϵ-edges be denoted by π_P. The hitting probabilities to sink SCCs of the overall graph are not affected by collapsing P to a single node A_P with outgoing edges corresponding to $W(A_P, y)$.*

Proof Sketch. We analyze the probability of all paths between nodes in M by dividing up the set of paths by the number of times each path enters pseudosink P. We then use Lemma tie edge coefficient to match up these paths with paths in M' (where M' is the MC after collapsing P) that have equal probability.

The proofs of Lemmas 3 and 4 are in the full version of the paper [22].

Deletion of ϵ-Edges at the Final Step

Lemma 5. *If for all nodes $i \in N_M$ have a regular path (only using regular edges) to an absorbing state, then $\forall i$, $\lim_{\epsilon \to 0} P[V_i] = 1$ where V_i is the event that the MC started at i is absorbed before taking any ϵ-edges.*

Proof. Consider the set of regular edges of M which have weights of the form $c_r - c_{re}\epsilon$. Set $\epsilon \leq \frac{\min(c_r)}{2 \max c_{re}}$, so that every regular edge has weight at least $w_{min} = \min(c_r)/2$. Now set $C_{min} = w_{min}^{|N_M|}$. Note that C_{min} is a constant independent of ϵ. For every node i there exists a regular path p_{iS} of length at most $|N_M|$ to an absorbing state, which must have probability at least C_{min}. In addition, recall that L_{max} is a constant defined in Lemma 3, such that $L_{max}\epsilon$ upper bounds the probability of taking an ϵ-edge at any timestep.

We will calculate $P[V_i]$ by analyzing a new MC M' that depends on M. This new MC begins in the "start" state, with the original MC at some node i' which is not an absorbing state (initially this node is i). One step of M' is as follows: It evolves the M from i' for len(p) steps, where p is the path from i' to an absorbing state that has probability $\geq C_{min}$. If M ends up at some absorbing state (one way to do this is to take path p) then M' moves to the absorbing "success" state. If M takes any ϵ transitions during this evolution, then M' moves to the absorbing "failure" state. If neither of these happen, M will be at some non-absorbing state i' and M' will stay in the start state.

Observe that the probability that M' reaches the success state is exactly $P[V_i]$, since M' reaches the success state if and only if M reaches some absorbing state before taking any ϵ-edges. Denote the event that M' reaches the success state by V_S'. The edge from the start state to the success state has probability at least C_{min}. The edge from the start state to the failure state has probability at most $F(\epsilon) = 1 - (1 - L_{min}\epsilon)^{|N_M|}$. This is because there are $len(p) \leq |N_M|$ steps of M taken by a single step of M' and the chance of taking an ϵ transition at each of those steps is upper bounded by $L_{max}\epsilon$.

We can use these bounds on the probabilities of the edges of M' to compute that:

$$\begin{aligned} \mathbf{Pr}[V_S'] &\geq (1 \cdot C_{min}) + (0 \cdot F) + (\mathbf{Pr}[V_S'] \cdot (1 - C_{min} - F)) \\ &\geq \frac{C_{min}}{C_{min} + F(\epsilon)} \end{aligned}$$

Since $\lim_{\epsilon \to 0} F(\epsilon) = 0$, $\lim_{\epsilon \to 0} P[V_S'] = 1$ and therefore $\lim_{\epsilon \to 0} P[V_i] = 1$.

Lemma 6. *If* MaxOrder$(M) = 0$, *deleting all ϵ-edges (that are not within a sink SCC) does not affect the hitting probabilities.*

Proof. By definition of order, if MaxOrder$(M) = 0$, every node in M has a regular path to an absorbing state, so we can apply Lemma 5 to get that for all states i, $\lim_{\epsilon \to 0} P[V_i] = 1$ where V_i is the event that M transitions from i to some absorbing state without taking any ϵ-edges.

Let M' be the graph with all ϵ-edges removed and all edge weights of the form $c_r + c_{re}\epsilon$ set to c_r. Let h'_{iS} be the hitting probability from i to absorbing state S in M'. Define Ψ_r to be the set of paths in M that only use regular edges, and Ψ'_r to be the analogous set for M'. Then $\Psi_{iS} \cap \Psi_r$ is the set of paths in M that go from i to S only using regular edges. Define P as the set of all paths in M, we have that $\Psi_{iS} \cap (\Psi \backslash \Psi_r)$ is the set of paths in M that go from i to S using one or more ϵ-edges.

$$h_{iS}(\epsilon) = \sum_{p \in (\Psi_{iS} \cap \Psi_r)} \mathbf{Pr}[p] + \sum_{p \in (\Psi_{iS} \cap (\Psi \backslash \Psi_r))} \mathbf{Pr}[p]$$

The probability of taking a path from i to S that has one or more ϵ-edges must be less than the probability of taking a ϵ-edge before absorption (since the first event is a subset of the second).

$$\sum_{p \in (\Psi_{iS} \cap (\Psi \backslash \Psi_r))} \mathbf{Pr}[p] \leq 1 - \mathbf{Pr}[V_{iJ}]$$

$$\lim_{\epsilon \to 0} \sum_{p \in (\Psi_{iS} \cap (\Psi \backslash \Psi_r))} \mathbf{Pr}[p] \leq \lim_{\epsilon \to 0} (1 - \mathbf{Pr}[V_i]) = 0$$

We can plug this limit into the hitting probability equation and use the property that, since the only structural difference between M and M' is the ϵ-edges, $P_{iS} \cap P_r = P'_{iS} \cap P'_r$. In addition, for all regular edges $e \in E_M$ we have that $\lim_{\epsilon \to 0} W_M(e) = W_{M'}(e)$ due to our renormalization. Note that we can interchange limits because each weight on an regular edge converges to a positive constant.

$$\begin{aligned}
\lim_{\epsilon \to 0} h_{iS}(\epsilon) &= \lim_{\epsilon \to 0} \sum_{p \in (\Psi_{iS} \cap \Psi_r)} \mathbf{Pr}[p] = \sum_{p \in (\Psi_{iS} \cap \Psi_r)} \prod_{e \in p} \lim_{\epsilon \to 0} W_M(e) \\
&= \sum_{p \in (\Psi'_{iS} \cap \Psi'_r)} \prod_{e \in p} W_{M'}(e) = h'_{iS}
\end{aligned}$$

Deleting the ϵ edges and re-normalizing the regular edges therefore has no effect on the limiting hitting probabilities.

3.4 Running Time of the Algorithm

Only steps 4, 5, and 6 of the algorithm have superlinear complexity. For 4 we need to calculate the steady-state probabilities of the nodes of the pseudosink, because they are needed in the calculation of the weights of the edges leaving the collapsed pseudosink. For 6, we need to compute the hitting probability in an ordinary graph (no ϵ-edges). Both of these problems can be solved in time $O(|E|^{1+\delta})$ for all $\delta > 0$ [13]. In Step 5, we do incremental maintenance of (r)SCCs. The fastest known algorithms for incremental SCC maintenance take amortized time $O(E^{1+\delta})$ [9].

4 Experiments

We have implemented our algorithm and experimented with random games with various values of the parameters p (players) and s (strategies per player) ranging for both parameters from 2 to 12. In the next subsection, we present certain examples that exhibit interesting behavior viz. our algorithm. Since our main message is a new way to view a game as a algorithmic map from a prior to a posterior distribution, in the second subsection we demonstrate how this works for various reasonably large games. Given a prior distribution (typically the uniform distribution over all MSPs), we sample from this distribution and then simulate the noisy replicator. We repeat until our convergence criteria are satisfied, and output the posterior distribution. This accomplishes our overarching goal, the empirical computation of the meaning of the game. We repeat this experiment for larger and larger games, taking this simulation to its practical laptop limits. The code used to generate the entire section is available at https://jasonmili.github.io/files/gd_hittingprobabilities_code.zip.

4.1 Some Interesting Games and Their Better-Response Graphs

We use the following plotting conventions: in each better-response graph, every node of every sink SCC will be colored with a unique color. Other pure profile nodes of the graph will be depicted as a "pie" graph with colored areas that indicate the hitting probabilities towards each of the sink SCCs it reaches, as identified by the former colors (of the sink SCCs). Tie edges (ϵ edges) appear in the graph as bidirectional "0.00" edges; this is only for plotting convenience.

3 × 3 Game. We start with a modified version of a game presented by [39] that exhibits two sink SCCs: a directed cycle of length four (corresponding to a periodic orbit in the replicator space) and another that is a single pure profile (corresponding to a strict pure NE); see Fig. 2.

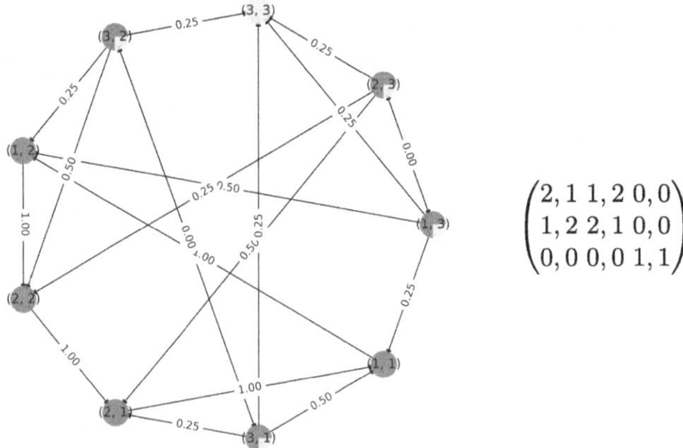

Fig. 2. 3 × 3 game. Left: the better-response graph. Right: the game utilities.

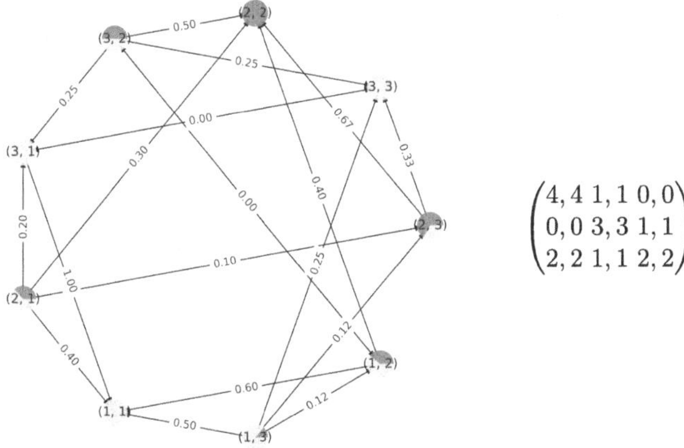

$$\begin{pmatrix} 4,4 & 1,1 & 0,0 \\ 0,0 & 3,3 & 1,1 \\ 2,2 & 1,1 & 2,2 \end{pmatrix}$$

Fig. 3. Tie game. The profile (3, 3) is order 1. Left: the better-response graph. Right: the game utilities.

Game with Order 1 Profile. We construct a game, depicted in Fig. 3 with two sink SCCs, and a pure profile of order 1 (which is also a pseudosink SCC) that needs exactly one tie edge to reach any sink SCC. Notice that the presence of this pure profile affects the hitting probabilities towards the sink SCCs, as described in Sect. 3. This example shows that there are cases where a pure NE may not be a sink SCC, or as a matter of fact, not even inside *any* sink equilibrium. That is, this NE is not *stochastically stable* in the terminology of [40].

$3 \times 3 \times 3$ Game. The utilities for this game can be found in our code. See Figs. 1 and 4.

4.2 Convergence Statistics

Methodology. We generate games of various sizes and random utilities (see the figures below), and we carry out a number of independently-randomized experiments of running noisy replicator dynamics (RD) on each. For each sampled point of the prior distribution (typically uniform), we run multiple independently-randomized instances of the noisy RD to obtain an empirical distribution. We consider the outcome of the game as the *empirical last-iterate distribution*, i.e., the average of all obtained distributions after run T. We keep track of the total variation (TV) distance between the running average distribution (e.g., at time $t < T$) and the ex-post empirical last-iterate (average at time T). We consider that a distribution has achieved good enough convergence when the TV distance is less than 1%—we found that this is roughly the accuracy that is feasible in a laptop-like experimental setup. All calculations in this section were performed on

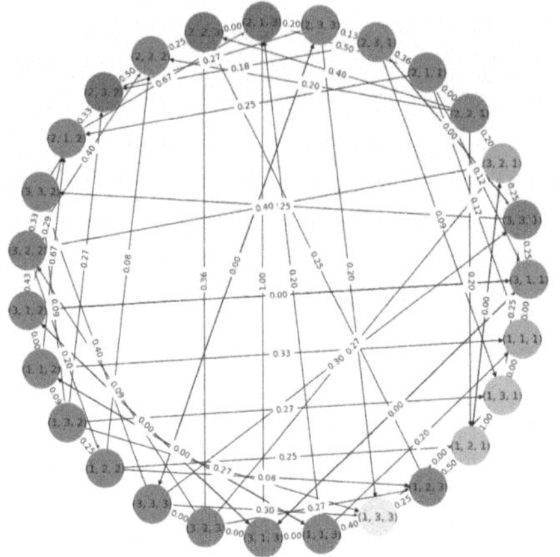

(a) The better-response graph of the game; nodes of sink SCCs are depicted in red.

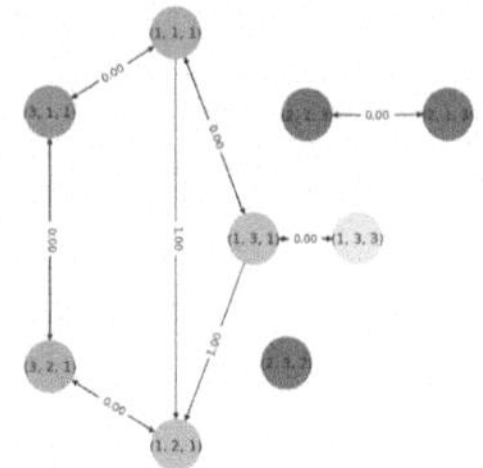

(b) For added clarity, we show a subgraph of (a) of all pure profiles of order ≥ 1, along with the sink SCCs.

(c) The color coding that depicts the orders of the various pure-profile nodes.

Fig. 4. The $3 \times 3 \times 3$ game. (Color figure online)

an Apple M2 processor with the use of multi-threading with 8 parallel threads. All graphs can be found in the full version of the paper [22]; here, we present only the final aggregate in Fig. 5.

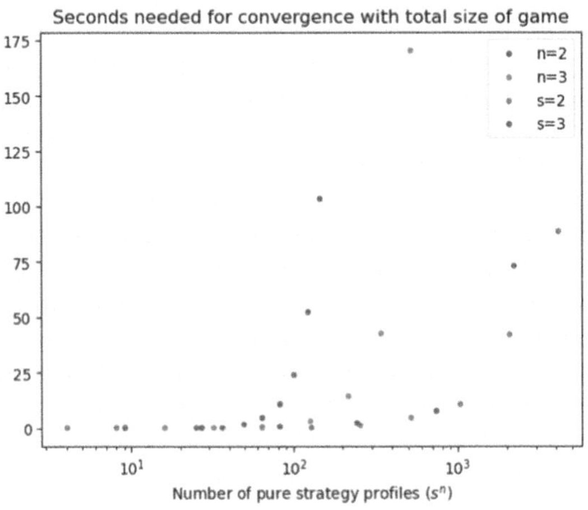

Fig. 5. Convergence of our algorithm with total size of the game.

5 Discussion and Open Questions

We have proposed that a useful way of understanding a game in normal form is as a map between a prior distribution over mixed strategy profiles to a distribution over sink equilibria; namely, the distribution induced by the noisy replicator dynamics if started at the prior. We showed that this distribution can be computed quite efficiently starting from any pure strategy profile, through a novel algorithm that handles the infinitesimal transitions associated with tie edges. By implementing this algorithm and dynamical system we conducted experiments which we believe demonstrate the feasibility of this approach to understanding the meaning of a game.

There are many problems left open by this work.

- In our simulations we approximated the meaning of the game for quite large games. We believe that more sophisticated statistical methods can yield more informative results for larger games. Another possible front of improvement in our simulations would be a better theoretical understanding of the trade-off between the parameters δ and η of the dynamics—the length of the jump and the intensity of the noise.
- Under which assumptions do the sink equilibria coincide with the chain recurrent components of the replicator dynamics (the solution concept suggested by the topological theory of dynamical systems)? Sharpening the result of Biggar and Shames in this way is an important open problem. On the other hand, a counterexample showing that it cannot be sharpened would also be an important advance; we note that experiments such as the ones in this paper are a fine way of generating examples of systems of sink equilibria which could eventually point the way to a counterexample. Another question in the interface with the topological theory is, does the time average behavior within a sink SCC correspond to the behavior within a chain component of the replicator?
- It would be very interesting to try—defying PSPACE-completeness—to compute sink equilibria and simulate the noisy replicator on succinct games such as extensive form, Bayesian, or graphical.

References

1. Andrade, G.P., Frongillo, R., Piliouras, G.: Learning in matrix games can be arbitrarily complex. In: Conference on Learning Theory, pp. 159–185. PMLR (2021)
2. Andrade, G.P., Frongillo, R., Piliouras, G.: No-regret learning in games is Turing complete. In: ACM Conference on Economics and Computation (EC) (2023)
3. Arora, S., Hazan, E., Kale, S.: The multiplicative weights update method: a meta-algorithm and applications. Theory Comput. **8**(1), 121–164 (2012)
4. Babichenko, Y.: Completely uncoupled dynamics and Nash equilibria. Games Econ. Behav. **76**(1), 1–14 (2012)
5. Bailey, J.P., Piliouras, G.: Multiplicative weights update in zero-sum games. In: ACM Conference on Economics and Computation (2018)

6. Bielawski, J., Chotibut, T., Falniowski, F., Kosiorowski, G., Misiurewicz, M., Piliouras, G.: Follow-the-regularized-leader routes to chaos in routing games. In: International Conference on Machine Learning, pp. 925–935. PMLR (2021)
7. Biggar, O., Shames, I.: The attractor of the replicator dynamic in zero-sum games. arXiv preprint arXiv:2302.00253 (2023)
8. Biggar, O., Shames, I.: The replicator dynamic, chain components and the response graph. In: Agrawal, S., Orabona, F. (eds.) Proceedings of the 34th International Conference on Algorithmic Learning Theory. Proceedings of Machine Learning Research, vol. 201, pp. 237–258. PMLR (2023). https://proceedings.mlr.press/v201/biggar23a.html
9. Chen, L., Kyng, R., Liu, Y.P., Meierhans, S., Gutenberg, M.P.: Almost-linear time algorithms for incremental graphs: cycle detection, SCCs, s-t shortest path, and minimum-cost flow. arXiv preprint arXiv:2311.18295 (2023)
10. Chen, X., Deng, X., Teng, S.H.: Computing Nash equilibria: approximation and smoothed complexity. In: FOCS 2006, pp. 603–612. IEEE Computer Society (2006)
11. Cheung, Y.K., Piliouras, G.: Online optimization in games via control theory: connecting regret, passivity and poincaré recurrence. In: International Conference on Machine Learning, pp. 1855–1865. PMLR (2021)
12. Chotibut, T., Falniowski, F., Misiurewicz, M., Piliouras, G.: The route to chaos in routing games: when is price of anarchy too optimistic? In: Advances in Neural Information Processing Systems, vol. 33, pp. 766–777 (2020)
13. Cohen, M.B., et al.: Almost-linear-time algorithms for Markov chains and new spectral primitives for directed graphs. In: Proceedings of the 49th Annual ACM SIGACT Symposium on Theory of Computing, pp. 410–419 (2017)
14. Conley, C.: Isolated Invariant Sets and the Morse Index. No. 38 in Regional Conference Series in Mathematics. American Mathematical Society, Providence (1978)
15. Daskalakis, C., Frongillo, R., Papadimitriou, C.H., Pierrakos, G., Valiant, G.: On learning algorithms for Nash equilibria. In: Kontogiannis, S., Koutsoupias, E., Spirakis, P.G. (eds.) SAGT 2010. LNCS, vol. 6386, pp. 114–125. Springer, Heidelberg (2010). https://doi.org/10.1007/978-3-642-16170-4_11
16. Daskalakis, C., Goldberg, P.W., Papadimitriou, C.H.: The complexity of computing a Nash equilibrium. In: STOC 2006, pp. 71–78. ACM (2006)
17. Etessami, K., Yannakakis, M.: On the complexity of Nash equilibria and other fixed points. In: FOCS 2007, pp. 113–123. IEEE Computer Society (2007)
18. Fabrikant, A., Papadimitriou, C.: The complexity of game dynamics: BGP oscillations, sink equlibria, and beyond. In: SODA (2008). http://www.cs.berkeley.edu/~alexf/papers/fp08.pdf
19. Fabrikant, A., Papadimitriou, C.H.: The complexity of game dynamics: BGP oscillations, sink equilibria, and beyond. In: SODA, vol. 8, pp. 844–853. Citeseer (2008)
20. Galla, T., Farmer, J.D.: Complex dynamics in learning complicated games. Proc. Natl. Acad. Sci. **110**(4), 1232–1236 (2013)
21. Goemans, M., Mirrokni, V., Vetta, A.: Sink equilibria and convergence. In: 46th Annual IEEE Symposium on Foundations of Computer Science (FOCS 2005), pp. 142–151 (2005). https://doi.org/10.1109/SFCS.2005.68
22. Hakim, R., Milionis, J., Papadimitriou, C., Piliouras, G.: Swim till you sink: computing the limit of a game (2024). https://jasonmili.github.io/files/gd_hittingprobabilities.pdf
23. Harsanyi, J.C., Selten, R.: A General Theory of Equilibrium Selection in Games, 2nd edn. MIT Press, Cambridge (1992)
24. Hart, S., Mas-Colell, A.: Uncoupled dynamics do not lead to Nash equilibrium. Am. Econ. Rev. **93**(5), 1830–1836 (2003)

25. Hassin, R., Haviv, M.: Mean passage times and nearly uncoupled Markov chains. SIAM J. Discrete Math. **5**(3), 386–397 (1992)
26. Hsieh, Y.P., Mertikopoulos, P., Cevher, V.: The limits of min-max optimization algorithms: convergence to spurious non-critical sets. arXiv preprint arXiv:2006.09065 (2020)
27. Kleinberg, R., Ligett, K., Piliouras, G., Tardos, É.: Beyond the Nash equilibrium barrier. In: Symposium on Innovations in Computer Science (ICS) (2011)
28. Koutsoupias, E., Papadimitriou, C.: Worst-case equilibria. In: Meinel, C., Tison, S. (eds.) STACS 1999. LNCS, vol. 1563, pp. 404–413. Springer, Heidelberg (1999). https://doi.org/10.1007/3-540-49116-3_38
29. Leonardos, S., Reijsbergen, D., Monnot, B., Piliouras, G.: Optimality despite chaos in fee markets. In: Baldimtsi, F., Cachin, C. (eds.) FC 2023. LNCS, vol. 13951, pp. 346–362. Springer, Cham (2023). https://doi.org/10.1007/978-3-031-47751-5_20
30. Mertikopoulos, P., Papadimitriou, C., Piliouras, G.: Cycles in adversarial regularized learning. In: Proceedings of the Twenty-Ninth Annual ACM-SIAM Symposium on Discrete Algorithms, pp. 2703–2717. SIAM (2018)
31. Milionis, J., Papadimitriou, C., Piliouras, G., Spendlove, K.: An impossibility theorem in game dynamics. Proc. Natl. Acad. Sci. **120**(41), e2305349120 (2023). https://doi.org/10.1073/pnas.2305349120. https://www.pnas.org/doi/abs/10.1073/pnas.2305349120
32. Mirrokni, V.S., Skopalik, A.: On the complexity of Nash dynamics and sink equilibria. In: Proceedings of the 10th ACM Conference on Electronic Commerce, pp. 1–10 (2009)
33. Muller, P., et al.: A generalized training approach for multiagent learning. arXiv preprint arXiv:1909.12823 (2019)
34. Omidshafiei, S., et al.: α-rank: multi-agent evaluation by evolution. Sci. Rep. **9**(1), 9937 (2019)
35. Omidshafiei, S., et al.: Navigating the landscape of multiplayer games. Nat. Commun. **11**(1), 5603 (2020)
36. Palaiopanos, G., Panageas, I., Piliouras, G.: Multiplicative weights update with constant step-size in congestion games: convergence, limit cycles and chaos. In: Advances in Neural Information Processing Systems, pp. 5872–5882 (2017)
37. Pangallo, M., Sanders, J., Galla, T., Farmer, D.: A taxonomy of learning dynamics in 2×2 games. arXiv e-prints arXiv:1701.09043 (2017)
38. Papadimitriou, C., Piliouras, G.: From Nash equilibria to chain recurrent sets: an algorithmic solution concept for game theory. Entropy **20**(10), 782 (2018)
39. Papadimitriou, C., Piliouras, G.: Game dynamics as the meaning of a game. ACM SIGecom Exchanges **16**(2), 53–63 (2019)
40. Peyton Young, H.: The evolution of conventions. Econometrica **61**(1), 57–84 (1993). http://www.jstor.org/stable/2951778
41. Piliouras, G., Yu, F.Y.: Multi-agent performative prediction: from global stability and optimality to chaos. In: Proceedings of the 24th ACM Conference on Economics and Computation, pp. 1047–1074 (2023)
42. Roughgarden, T.: Intrinsic robustness of the price of anarchy. In: ACM Symposium on Theory of Computing (STOC), pp. 513–522. ACM (2009)
43. Rowland, M., et al.: Multiagent evaluation under incomplete information. arXiv preprint arXiv:1909.09849 (2019)
44. Sanders, J.B., Farmer, J.D., Galla, T.: The prevalence of chaotic dynamics in games with many players. Sci. Rep. **8**(1), 1–13 (2018)

45. Sato, Y., Akiyama, E., Farmer, J.D.: Chaos in learning a simple two-person game. Proc. Natl. Acad. Sci. **99**(7), 4748–4751 (2002). https://doi.org/10.1073/pnas.032086299. https://www.pnas.org/content/99/7/4748
46. Schuster, P., Sigmund, K.: Replicator dynamics. J. Theor. Biol. **100**(3), 533–538 (1983). https://doi.org/10.1016/0022-5193(83)90445-9. http://www.sciencedirect.com/science/article/pii/0022519383904459
47. Taylor, P.D., Jonker, L.B.: Evolutionary stable strategies and game dynamics. Math. Biosci. **40**(1), 145–156 (1978). https://doi.org/10.1016/0025-5564(78)90077-9. http://www.sciencedirect.com/science/article/pii/0025556478900779
48. Vlatakis-Gkaragkounis, E.V., Flokas, L., Lianeas, T., Mertikopoulos, P., Piliouras, G.: No-regret learning and mixed Nash equilibria: they do not mix. In: Advances in Neural Information Processing Systems, vol. 33, pp. 1380–1391 (2020)
49. Young, H.P.: The possible and the impossible in multi-agent learning. Artif. Intell. **171**(7), 429–433 (2007)

The Investment Management Game: Extending the Scope of the Notion of Core

Vijay V. Vazirani(✉)

University of California, Irvine, CA 92697, USA
vazirani@ics.uci.edu
https://ics.uci.edu/~vazirani/

Abstract. The core is a dominant solution concept in economics and cooperative game theory; it is used for profit—equivalently, cost or utility—sharing. Starting with the classic work of Shapley and Shubik [25] on the assignment game, the cores of several natural games have been characterized using total unimodularity. The purpose of our paper is two-fold:

1. We give the first game for which total unimodularity does not hold and *total dual integrality* is needed for characterizing its core.
2. We demonstrate the *versatility* of the notion of core by proposing a completely different use: in a so-called investment management game, which is a *game against nature* rather than a cooperative game.

Our game has only one agent, whose strategy set is all possible ways of distributing her money among investment firms. The agent wants to pick a strategy such that in each of exponentially many future "scenarios", sufficient money is available in the "right" firms so she can buy an "optimal investment" for that scenario. Such a strategy constitutes a core imputation under a broad interpretation—though traditional mathematical framework—of the core.

Our game is defined on *perfect graphs*, since the maximum stable set problem (which plays a key role in this game) can be solved in polynomial time for such graphs. We completely characterize the core of this game, analogous to Shapley and Shubik's characterization of the core of the assignment game. A key difference is that whereas their characterization follows from total unimodularity, ours follows from total dual integrality.

Keywords: Core of a game · Total unimodularity · Total dual integrality · Perfect graphs · Games against nature

1 Introduction

The core is a dominant solution concept in economics and game theory. Its origins lie in the 19th century book of Edgeworth [6] in which it was referred to as the *contract curve* and was first used in general equilibrium theory. In 1959,

G. Schäfer and C. Ventre (Eds.): SAGT 2024, LNCS 15156, pp. 223–239, 2024.
https://doi.org/10.1007/978-3-031-71033-9_13

Gillies [10] gave an improved definition which started being used in cooperative game theory for "fair" profit sharing. Since then, the stature of this solution concept grew considerably and today it is considered the gold standard for profit—equivalently, cost or utility—sharing. Indeed, it is considered much more desirable than other notions, e.g., least core and nucleolus[1].

The assignment game forms a paradigmatic setting for studying the notion of core of a transferable utility (TU) market game, in large part because of the work of Shapley and Shubik [25]; in particular, they completely characterized the core of this game, see Sect. 3. The key to this characterization is that the polytope defined by the constraints of LP (1) is integral; in turn, this follows from the fact that the constraint matrix of this LP is totally unimodular (TUM), see Definition 7. As observed in [4,5,30], TUM underlies the characterization of the cores of several natural games which involve finding optimal integral solutions, e.g., matchings and flows. As is well known, there is a more general condition than TUM which leads to integrality, namely total dual integrality (TDI), see Definition 8. These facts raise the following questions.

1. Are there natural games whose constraint matrix is not TUM and the integrality of its polytope follows from TDI? If so, can TDI lead to a characterization of the core of this game?
2. How versatile is the solution concept of core, i.e., can it find applications outside of profit—equivalently, cost or utility—sharing?

This paper provides positive answers to both these questions. In order to give a context for the second question, let us consider Nash equilibrium, which is perhaps the most important solution concept in game theory. Even for the special case of bimatrix games, it provides deep insights in a rich milieu of situations, each having its own special character, e.g., Prisoner's Dilemma, Matching Pennies, Battle of the Sexes, and Rock-Paper-Scissor; the last game is in fact zero-sum. Indeed, most "big" solution concepts tend to be similarly multi-faceted. In contrast, within game theory, the core has been used for profit—equivalently, cost or utility—sharing only.

This paper shows that the notion of core is more versatile than previously believed. We will do so in the context of the investment management game, in which the notion of core is not used for profit sharing. Obviously, in studying the core of this game, we must leave the mathematical formulation of core, given in Sect. 3, unchanged. However, we will need to reinterpret the *meaning* of the mathematical terms in order to move them away from a profit-sharing setting to the setting at hand and we will also provide more appropriate names for these terms. The correspondence with the traditional names is given in Remark 2 in Sect. 4.

Definition 1 introduces the *investment management game*[2], whose core we will characterize. Interestingly, it is not a cooperative game; in fact, it has *only*

[1] Some drawbacks of these two notions are mentioned in [28].

[2] The game given in Definition 1 may not seem realistic. However, we note that the purpose of our paper is not immediate applicability—its purpose is to make a conceptual advance on a fundamental solution concept.

one agent. Our game is best viewed as a *game against nature*, i.e., a game against a player whose pay-offs, as well as probability distribution over strategies, are unknown. The origin of this name comes from the experience of farmers from a time when weather prediction and irrigation systems were not well developed, making it difficult for them to predict the best choice of crops and forcing them to play in complete ignorance [1]. Our game has exactly this character, see Definition 1. The saving grace is that in our game, a core imputation helps salvage the situation by enabling the agent to invest in such a way she is able to respond to *every strategy* of nature optimally!

After defining our game, in Remark 1 we draw a clear distinction between the manner in which the core is applied in cooperative game theory and in our game. In Sect. 4, we define our game on a graph. The main computational problem that arises in it is to find a maximum weight independent set, which is $\mathcal{NP}$-hard in arbitrary graphs. However, the restriction of this problem to *perfect graphs* (Sect. 5) is in $\mathcal{P}$ [11]; this will be the setting for our game[3].

In Sect. 5 we characterize the core of this game. A good way of describing this result is by drawing an analogy with the classic paper of Shapley and Shubik [25] which characterized the core of the assignment game. They showed that the core of this game is precisely the set of optimal solutions to the dual of the LP-relaxation of the maximum weight matching problem in the underlying bipartite graph, see Sect. 3. As observed in [4,5,30], at the heart of the Shapley-Shubik proof lies the fact that the polytope defined by the constraints of this LP is integral, see Definition 6. In turn, integrality follows from the fact that the constraint matrix of this LP is *totally unimodular (TUM)*, see Definition 7. The underlying reason for this requirement of integrality is an inherent indivisibility in the game, e.g., in a cooperative game, such as the assignment game, agents are indivisible.

Integrality holds for the maximum weight independent set problem for perfect graphs as well. However, the underlying reason is not TUM, but the more general condition of *total dual integrality (TDI)*, see Definition 8. Building on this fact, we show that the set of core imputations of our game is precisely the set of optimal solutions to the dual of the LP-relaxation of the maximum weight independent set problem for perfect graphs. Ours appears to be the first work which uses TDI for characterizing the core of a game; previous characterizations were based on TUM, see Sect. 2.

Another novelty of our work is the following: In cooperative games, the profit of a sub-coalition under an imputation was defined via a *bottom-up process* in which the worth of the game was distributed among agents and the profit of a sub-coalition was simply the sum of profits of agents in it. In our game there is only one agent and the natural process is *top-down*, as described in Sect. 4 and summarized in Remark 3.

Definition 1. *The* investment management game *involves* one agent, *a set V of* assets *with a* cost function $w : V \to \mathbb{Q}_+$, *and a set $\mathcal{M}$ of* investment firms. *For*

[3] Many classes of perfect graphs are known by now, including bipartite graphs, line graphs and chordal graphs [11].

concreteness, assume that assets are shares of specific companies, each worth a specific dollar amount; obviously, the number of shares in an asset changes according to the going price. Assume further that, much like a mutual fund, each investment firm specializes in holding shares of specific types of companies, e.g., Internet companies, software companies, computer hardware companies, automobile companies, etc. The difference is that whereas each share of a mutual fund represents a collection of assets in some predetermined proportions, in our setting, the agent can buy individual assets *from a firm.*

The set of assets sold by a specific firm $f \in \mathcal{M}$ *is denoted by* $Q_f \subseteq V$. *The same asset, such as "five thousand dollars worth of Microsoft", may be sold by more than one firm, e.g., this asset may be sold by a firm specializing in Internet companies as well as a firm specializing in software companies. Each subset* $S \subseteq V$ *of assets is called a* scenario*; hence there are exponentially many, namely* $2^{|V|}$*, scenarios.*

Two or more assets which are sold by the same firm have obvious correlations and therefore do not constitute a low-risk investment. On the other hand, a set of assets which picks at most one asset from any investment firm constitutes a diversified portfolio*—it avoids correlations and is therefore considered a "healthy" investment. An* optimal investment *for a scenario* S *is a largest possible diversified portfolio in it; formally, it is defined to be a maximum cost set* $S' \subseteq S$ *which picks at most one asset from any investment firm.*

The total money, $T \in \mathbb{Q}_+$ *of the agent is just sufficient to buy an optimal investment in scenario* V. *The* strategy set *of the agent is all possible ways of distributing money* T *among the firms; each strategy will also be called an* imputation. *The rules of this game dictate that the money allocated to a firm* $f \in \mathcal{M}$ *is available for use at* any asset $v \in Q_f$. *Note that this money need not be used for buying* v*; it is simply available at* v. *Therefore, the* money available in scenario S *is the sum of money allocated to all firms which have at least one asset in* S. *Clearly, our game is a transferable utility (TU) game.*

The game *is the following: Find an imputation such that when, at a certain time in the future, nature picks a strategy, i.e., a scenario* $S \subseteq V$, *the money available in* S *is sufficient to buy an optimal investment in* S. *Such an imputation is said to be in the* core *of the game. Thus a core imputation enables the agent to invest* T *money in firms in such a way she is able to respond to* every strategy *of nature optimally.*

Remark 1. In light of Definition 1, a clear distinction can be given between the manner in which the core is applied in cooperative game theory and in our game. Whereas in the former, a core imputation defines payoffs of all players that results in the stability of the grand coalition, in the latter, a core imputation captures an investment strategy of the unique player, that encapsulates risk aversion (for the stated definition of "risk") under every future scenario.

2 Related Works

Results characterizing cores of natural Transferable Utility (TU) games are given below. First, we mention the stable matching game, defined by Gale and Shapley [13], in which preferences are ordinal, i.e., it is an NTU game. The only coalitions that matter in this game are ones formed by one agent from each side of the bipartition. A stable matching ensures that no such coalition has the incentive to secede and the set of such matchings constitute the core of this game. Vande Vate [27] and Rothblum [22] gave linear programming formulations for stable matchings; the vertices of their underlying polytopes are integral and are stable matchings. Interestingly enough, Kiraly and Pap [16] showed that the linear system of Rothblum is in fact TDI.

A core imputation has to ensure that *each* of the exponentially many sub-coalitions is "satisfied"—clearly, that is a lot of constraints. As a result, the core is known to be non-empty only for a handful of games, some of which are mentioned below; total unimodularity plays a key role in these results.

Deng et al. [5] observed the role of integrality in the Shapley-Shubik Theorem and used this insight to distilled its underlying ideas to obtain a general framework which helps characterize the cores of several games that are based on fundamental combinatorial optimization problems, including maximum flow in unit capacity networks both directed and undirected, maximum number of edge-disjoint s-t paths, maximum number of vertex-disjoint s-t paths and maximum number of disjoint arborescences rooted at a vertex r.

The survey of Demange and Deng [4] on the notion of balancedness of Bondareva and Shapley [3,24] explored the role of integrality in depth. Towards the end of their paper, they note that TDI of a linear system is a very general condition that leads to integrality and therefore non-emptyness of the core and balancedness. However, they did not give a specific game for which TUM does not hold and TDI had to be invoked for establishing integrality.

A natural generalization of the assignment game is the b-matching game in bipartite graphs. Biro et al. [2] showed that the core non-emptiness and core membership problems for the b-matching game are solvable in polynomial time if $b \leq 2$ and are co-NP-hard even for $b = 3$. More recently, Vazirani [29] showed that if b is the constant function, then core imputations are precisely optimal solutions to the dual LP; this is analogous to the Shapley-Shubik theorem. Furthermore, Vazirani [29] showed that if b is arbitrary, then every optimal solutions to the dual LP is a core imputations; however, there are core imputations that are not optimal solutions to the dual LP.

We next describe results for the facility location game. First, Kolen [15] showed that for the unconstrained facility location problem, each optimal solution to the dual of the classical LP-relaxation is a core imputation if and only if this relaxation has no integrality gap. Later, Goemans and Skutella [14] showed a similar result for any kind of constrained facility location game. They also proved that in general, for facility locations games, deciding whether the core is non-empty and whether a given allocation is in the core is NP-complete.

Samet and Zemel [26] study games which are generated by linear programming optimization problems; these are called LP-games. For such games, It is well known that the set of optimal dual solutions is contained in the core and [26] gives sufficient conditions under which equality holds. These games do not ask for integral solutions and are therefore different in character from the ones studied in this paper.

Granot and Huberman [8,9] showed that the core of the minimum cost spanning tree game is non-empty and gave an algorithm for finding an imputation in it. Koh and Sanita [17] settle the question of efficiently determining if a spanning tree game is submodular; the core of such games is always non-empty. Nagamochi et al. [21] characterize non-emptyness of core for the minimum base game in a matroid; the minimum spanning tree game is a special case of this game.

3 Definitions and Preliminary Facts

In this section, we will give the standard definition of core in the setting of a cooperative game. For completeness, we will also formally state the characterization of the core of the assignment game given by Shapley and Shubik. In Sect. 4 we will modify some of these definitions, and appropriately rename others, so that the notion of core can be used to study the investment management game.

We will study *transferable utility (TU)* games, i.e., a games in which utilities of the agents are stated in monetary terms and side payments are allowed. For an extensive coverage of these notions, see the book by Moulin [20].

Definition 2. *A* cooperative game *consists of a pair* (N, c) *where* N *is a set of* n *agents and* c *is the* characteristic function*;* $c : 2^N \rightarrow \mathcal{R}_+$*, where for* $S \subseteq N$*,* $c(S)$ *is the* worth *that the sub-coalition* S *can generate by itself.* N *is also called the* grand coalition.

Definition 3. *An* imputation *gives a way of dividing the worth of the game,* $v(N)$*, among the agents. It can be viewed as a function* $x : N \rightarrow \mathbb{Q}_+$*. For each sub-coalition* $S \subseteq N$*, we will define its* profit *as* $\text{profit}(S) = \sum_{i \in S} x(i)$.

Definition 4. *Let* x *be an imputation and* $S \subseteq N$ *a sub-coalition. We will say that* x satisfies S *if its profit is at least as large as its worth, i.e.,* $\text{profit}(S) \geq \text{worth}(S)$.

Definition 5. *An imputation* x *is said to be in the* core of the game *if it satisfies each sub-coalition* $S \subseteq N$.

The *assignment game*, consists of a bipartite graph $G = (U, V, E)$ and a weight function $w : E \rightarrow \mathbb{Q}_+$. The agents of this game are $U \cup V$ and for each sub-coalition $(S_u \cup S_v)$, with $S_u \subseteq U$ and $S_v \subseteq V$, its worth is defined to be the weight of a maximum weight matching in $G(S_u \cup S_v)$, where the latter is the subgraph of G induced on the vertices $(S_u \cup S_v)$.

Linear program (1) gives the LP-relaxation of the problem of finding such a matching. In this program, variable x_{ij} indicates the extent to which edge (i, j) is picked in the solution.

$$
\begin{aligned}
\max \quad & \sum_{(i,j)\in E} w_{ij}x_{ij} \\
\text{s.t.} \quad & \sum_{(i,j)\in E} x_{ij} \leq 1 \quad \forall i \in U \\
& \sum_{(i,j)\in E} x_{ij} \leq 1 \quad \forall j \in V \\
& x_{ij} \geq 0 \quad \forall (i,j) \in E
\end{aligned}
\tag{1}
$$

Taking u_i and v_j to be the dual variables for the first and second constraints of (1), we obtain the dual LP:

$$
\begin{aligned}
\max \quad & \sum_{i\in U} u_i + \sum_{j\in V} v_j \\
\text{s.t.} \quad & u_i + v_j \geq w_{ij} \quad \forall (i,j) \in E \\
& u_i \geq 0 \quad \forall i \in U \\
& v_j \geq 0 \quad \forall j \in V
\end{aligned}
\tag{2}
$$

The constraint matrix of LP (1) is totally unimodular (TUM), see Definition 7, and therefore the polytope defined by its constraints is integral, see Definition 6 and [19]. This integrality is the key to proving Theorem 1.

Theorem 1. *(Shapley and Shubik [25]) The imputation (u, v) is in the core of the assignment game if and only if it is an optimal solution to the dual LP, (2).*

Definition 6. *We will say that a* polytope is integral *if its vertices have all integral coordinates.*

Definition 7. *Let $Ax \leq b$ be a linear system where A is an $m \times n$ matrix and b an m-dimensional vector, both with integral entries. A is said to be* totally unimodular (TUM) *if every submatrix of A has determinant $0, 1$ or -1. If so, the polytope of this linear system is integral.*

Definition 8. *Let $Ax \leq b$ be a linear system where A is an $m \times n$ matrix and b an m-dimensional vector, both with rational entries. We will say that this linear system is* totally dual integral (TDI) *if for any integer-valued vector c^T such that the linear program*

$$\max\{cx : Ax \leq b\}$$

has an optimum solution, the corresponding dual linear program has an integer *optimal solution. If so, the polytope of this linear system is integral.*

Note that TDI is a more general condition than TUM for integrality of polyhedra. If A is TUM then the polyhedron of the linear system $Ax \leq b$ is integral for every integral vector b. However, even if A is not TUM, for specific choices of an integral vector b, the polytope of the linear system $Ax \leq b$ may be integral, and TDI may apply in this situation. It is important to remark that TDI is not a property of the polytope but of the particular linear system chosen to define it. See [11,23] for further details.

4 The Investment Management Game Defined on a Graph

As stated in the Introduction, the investment management game, which is described at a high level in Definition 1, is a game against nature and it is not a cooperative game; in fact it has only one agent. A natural setting for this game is a graph, as described below. This is a TU game: the money available at a firm can be used to buy assets from other firms, as specified below.

In studying the core of the investment management game, we will leave the mathematical formulation of core, given in Sect. 3, unchanged. However, we will need to reinterpret the *meaning* of the mathematical terms in order to move them away from a profit-sharing setting to the setting being studied. In the process we will also provide more appropriate names for these terms; the correspondence with the traditional names will always be explicitly stated, see Remark 2.

Let $G = (V, E)$ be a graph whose vertices are *assets*; let $|V| = n$. The function $w : V \rightarrow \mathbb{Q}_+$ defines the *cost* of each asset. Every *maximal clique*[4] in G is an *investment firm*; let $\mathcal{M}$ denote the set of all firms. For each firm $f \in \mathcal{M}$, the assets sold by f are represented by the vertices in its clique, $Q_f \subseteq V$. As stated in Definition 1, each investment firm specializes in holding shares of a specific type of companies. As a result, two or more assets held by the same firm will have obvious correlations.

We now explain why investment firms are defined to be cliques, via the following analogy. Consider all the ATMs in the US of a certain bank. Since the money deposited in this bank, or in an ATM of this bank, can be withdrawn from any of its ATMs, the set of all ATMs of this bank can be viewed as a clique, interconnected via a network. Similarly, since the money allocated to an investment firm is available for use at any of its assets, as stated in Definition 1, we have defined the investment firm to be a clique—over its assets.

Every set $S \subseteq V$ is called a *scenario*, i.e., scenario plays the same role as subcoalition in a cooperative game. Let $G(S)$ denote the subgraph of G induced on vertices in S. As required by Definition 1, an *optimal investment* in scenario S is defined to be any maximum cost independent set[5] in $G(S)$; clearly, this is a

[4] It is easy to see that our result will hold even if we had defined every clique to be an investment firm. However, under that formulation, there would be numerous pairs of firms f, f' with $Q_{f'} \subset Q_f$, making firm f' redundant. Restricting to maximal cliques avoids this deficiency in the formulation.

[5] An independent set is also called a stable set, see Definition 11.

diversified investment since it picks at most one asset from any investment firm. Let $O_S \subseteq S$ denote such an investment and define

$$\mathrm{cost}(S) := \sum_{v \in O_S} w_v.$$

Note that cost plays the same role as worth in Definition 2 and the function cost : $2^V \rightarrow \mathbb{Q}_+$ plays the same role as the characteristic function. The *total money* of the agent is defined to be $T = \mathrm{cost}(V)$, i.e., just sufficient to buy an optimal investment in G.

A function $y : \mathcal{M} \rightarrow \mathbb{Q}_+$ where $\sum_{Q \in \mathcal{M}} y_Q = T$ is called an *imputation*, i.e., it is a way of distributing T money among the investment firms. The set of all such functions y can also be viewed as the *strategy set* of the unique agent in the game. For any scenario S, the *money available* for buying assets in S is defined to be the sum of money in all investment firms which contain at least one asset from S, i.e.,

$$\mathrm{money}(S) := \sum_{Q \in G(S):\ Q \cap S \neq \emptyset} y_Q,$$

where "$Q \in G(S)$" is short for "clique Q in $G(S)$". Strictly speaking, the summation should be over maximal cliques, but since for a non-maximal clique Q, y_Q can be assumed to be zero, this distinction can be dropped. Notice that money(S) plays the role of *profit* of S in a cooperative game.

Remark 2. So far we have renamed three important terms. The correspondence between the traditional names and the new names is:

- sub-coalition $\equiv$ scenario
- worth $\equiv$ cost
- profit $\equiv$ money

We now use the correspondence stated in Remark 2 to reinterpret Definitions 4 and 5. By Definition 4, scenario S is satisfied by imputation y if money$(S) \geq$ cost(S) and by Definition 5, imputation y is said to be in the *core* of this game if it satisfies every scenario, i.e.,

$$\forall S \subseteq V, \quad \mathrm{money}(S) \geq \mathrm{cost}(S).$$

Remark 3 and points out an important difference between our setting and the setting of cooperative games.

Remark 3. In a cooperative game, an *imputation* distributes the total worth of the game among agents and the *profit of a sub-coalition* is defined to be the sum of the profits of its agents; the latter can be viewed as a *bottom-up process*. Clearly these processes do not apply to our game, since it has a unique agent and the notion of a "sub-coalition" is replaced by that of a scenario.

The natural way of defining an imputation and the money available in a scenario can be summarized as follows: an imputation distributes the total money of the game among "large" objects—the firms, which are maximal cliques—and the money available in a scenario is defined via a *top-down process*, by summing the money of all cliques which intersect this scenario.

4.1 Limitations of This Model, and Desired Properties

The problem of computing a maximum cost independent set in an arbitrary graph is NP-hard, even if all vertex costs are unit. This NP-hardness also indicates that the game defined above lacks structural properties that could lead to an understanding of its core. In sharp contrast, the assignment game is in $\mathcal{P}$ and its LP-relaxation supports integrality of the underlying polytope, therefore leading to a characterization of its core. Hence, the model defined above, on an arbitrary graph, is too general to be useful.

To be useful, the model should allow for:

1. A characterization of the core; in particular, determine if the core is non-empty.
2. Efficient computation of T, and $\text{cost}(S)$ for any scenario S.
3. An efficient algorithm for computing an imputation in the core.

In the next section, we show that restricting the game to perfect graphs gives all these properties.

5 The Investment Management Game on Perfect Graphs

In this section, we will study a restriction of the investment management game to perfect graphs. In Sect. 5.1 we give the required definitions and facts from the (extensive) theory of perfect graphs. We will not credit individual papers for these facts; instead, we refer the reader to Chap. 9 of the book [11] as well as the remarkably clear and concise exposition of this theory, presented as an "appetizer" by Groetschel [12]. In Sect. 5.2 we will use these facts to characterize the core of this game.

5.1 Definitions and Preliminaries

Definition 9. *Given a graph $G = (V, E)$, $\omega(G)$ denotes its* clique number, *i.e., the size of the largest clique in it and $\chi(G)$ denotes its* chromatic number, *i.e., the minimum number of colors needed for its vertices so that the two endpoints of any edge get different colors.*

Definition 10. *A graph $G = (V, E)$ is said to be* perfect *if and only if the clique number and chromatic number are equal for each vertex-induced subgraph of G, i.e.,*

$$\forall S \subseteq V, \ \omega(G(S)) = \chi(G(S)).$$

Let $\overline{G}$ denote the *complement* of G, i.e., $\overline{G} = (V, \overline{E})$, where $\overline{E}$ is the complement of E, with $\forall\ u, v \in V, (u, v) \in E$ if and only if $(u, v) \notin \overline{E}$. A central fact about perfect graphs is that G is perfect if and only if $\overline{G}$ is perfect.

Definition 11. *Set $S \subseteq V$ is said to be a* stable set *in G, also sometimes called an* independent set, *if no two vertices of S are connected by an edge, i.e., $\forall\ u, v \in S,\ (u, v) \notin E$. Let $w : V \to \mathbb{Q}_+$ be a weight function on the vertices of G.*

Let G be an arbitrary graph. Clearly any clique in G can intersect a stable set in at most one vertex, and therefore the constraint in LP (3) is satisfied by every stable set; note that variable x_v indicates the extent to which v is picked in a fractional stable set. LP (3) contains such a constraint for each clique in G and is an LP-relaxation of the maximum weight stable set problem in G. However, LP (3) has exponentially many constraints, one corresponding to each clique in G; moreover, it is NP-hard to solve in general [11].

$$\begin{aligned} \max \quad & \sum_{v \in V} w_v x_v \\ \text{s.t.} \quad & x(Q) \leq 1 \quad \forall \text{ clique } Q \text{ in } G \\ & x_v \geq 0 \quad \forall v \in V \end{aligned} \tag{3}$$

The situation is salvaged in case G is a perfect graph: with the help of the Lovasz theta function, one can can show that LP (3) can be solved in polynomial time using the ellipsoid algorithm [11,12].

Below is the dual LP, which is obtained by taking y_Q to be the dual variable for the constraint of LP (3). The dual LP is solving a clique covering problem.

$$\begin{aligned} \max \quad & \sum_{\text{clique } Q \text{ in } G} y_Q \\ \text{s.t.} \quad & \sum_{Q \ni v} y_Q \geq w_v \quad \forall v \in V \\ & y_Q \geq 0 \quad \forall \text{ clique } Q \text{ in } G \end{aligned} \tag{4}$$

A key fact for our purpose is that the linear system of LP (3) is totally dual integral (TDI), see Definition 8, for perfect graphs [11,12]. We provide a proof sketch below.

By Definition 9, if G is perfect, $\omega(G) = \chi(G)$. Next, consider the complement of G, namely $\overline{G}$; two vertices are adjacent in G if and only if they are not adjacent in $\overline{G}$. Denote by $\alpha(G)$ and $\overline{\chi}(G)$ the size of the largest stable set and the minimum number of disjoint cliques needed to cover all vertices of G, respectively. Clearly, $\omega(G) = \alpha(\overline{G})$ and $\chi(G) = \overline{\chi}(\overline{G})$. Furthermore, since the complement of a perfect graph is also perfect, we get that for a perfect graph G, $\alpha(G) = \overline{\chi}(G)$.

Consider LP (3) and its integer programming formulation for the case that the weight function $w \in \{0, 1\}^n$, and let $L_p(w)$ and $I_p(w)$ denote their optimal objective function values. For the same restriction on w, let $L_d(w)$ and $I_d(w)$ denote the optimal objective function values of LP (4) and its integer programming formulation. Now,

$$I_p(w) \leq L_p(w) = L_d(w) \leq I_d(w),$$

where the equality follows from the LP-duality theorem, and the inequalities follow from the relation between integral and fractional solutions.

For $w \in \{0, 1\}^n$, let G' be the subgraph of G induced on vertices v for which $w_v = 1$. Since G' is also perfect, $I_p(w) = \alpha(G')$ and $I_d(w) = \overline{\chi}(G')$. Since $\alpha(G') = \overline{\chi}(G')$, we get that $I_p(w) = I_d(w)$ and hence equality holds for all four programs defined above.

To show that LP (3) is TDI we must show that for every weight function $w \in \boldsymbol{Z}_+^n$, the optimal objective function value of the dual is integral. Since $L_d(w) = I_d(w)$, this is the case if $w \in \{0, 1\}^n$. Finally, via Fulkerson [7] and Lovasz [18] we get that this integrality implies integrality of the optimal dual even if $w \in \boldsymbol{Z}_+^n$. Therefore the linear system of LP (3) is TDI and hence the polytope defined by it is integral; the vertices of this polytope are stable sets.

5.2 Characterizing the Core of the Game

Since the firms in the investment management game are *maximal* cliques, we first restrict the constraint in LP(3) to maximal cliques only to obtain LP (5). Observe that if Q is a clique in G and Q' is a sub-clique of Q, then the constraint in LP (3) corresponding to Q' is redundant and can be removed, since it is implied by the constraint corresponding to Q. Continuing in this manner, we are left with constraints corresponding to maximal cliques only. Therefore, the two LPs are equivalent.

$$
\begin{array}{lll}
\max & \sum_{v \in V} w_v x_v & \\
\text{s.t.} & x(Q) \leq 1 & \forall \text{ maximal clique } Q \text{ in } G \\
& x_v \geq 0 & \forall v \in V
\end{array}
\tag{5}
$$

The dual of LP (5) is given below.

$$
\begin{array}{lll}
\max & \sum_{\text{clique } Q \text{ in } G} y_Q & \\
\text{s.t.} & \sum_{Q \ni v} y_Q \geq w_v & \forall v \in V \\
& y_Q \geq 0 & \forall \text{ maximal clique } Q \text{ in } G
\end{array}
\tag{6}
$$

We next show that the linear system of LP (5) is also TDI. This is not immediate, since as stated after Definition 8, TDI is not a property of the polytope but of the particular linear system chosen to define it.

Lemma 1. *The linear system of LP (5) is TDI.*

Proof. Let Q be a maximal clique in G, Q' be its sub-clique and x be a vector of variables x_v for each vertex $v \in V$. For any x whose coordinates have been set to non-negative numbers, if $x(Q) \leq 1$ then $x(Q') \leq 1$, i.e., the latter constraint is redundant. Therefore the linear system of (5) is obtained by removing redundant constraints from the linear system of (3). Hence the two primal LPs (3) and (5) are equivalent.

Since the linear system of LP (3) is TDI, for any integer-valued cost vector w such that the linear program (3) has an optimum solution, the dual linear program (4) has an *integer* optimal solution. We will use this fact to show that the same holds for LPs (5) and (6) as well.

Let w be an integer-valued cost vector such that the linear program (3) has an optimum solution and let y be the corresponding integer optimal solution to LP (4). Since the two primal LPs given above are equivalent, LP (5) also has an optimum solution for w.

Use y to construct a dual solution y' for LP (6) repeating the following operation: Suppose $y_{Q'} > 0$, where Q' is not a maximal clique. Let Q be any maximal clique of which Q' is a sub-clique. Let $y_{Q'} = \alpha$; now decrease $y_{Q'}$ by α and increase y_Q by α. Since y is integral, the new dual, y', is also integral; moreover, its objective value remains unchanged and it is feasible for LP (4) since for each vertex v, $\sum_{Q \ni v} y_Q$ remains unchanged.

Repeat this operation until $y_Q > 0$ only if Q is a maximal clique. Call the resulting dual y'. Clearly, y' is an integral optimal dual for LP (6). Therefore LP (5) is TDI.

In our setting, an optimal dual distributes the worth of the game among the maximal cliques of G. However, dividing the money y_Q, given to clique Q, among the vertices in the clique in not very meaningful, see Example 1 for a detailed explanation. As described in Theorem 2, in our game, the money available to a *any scenario* is defined via a different process, see also Remark 3.

Example 1. The 3×3 Paley graph, which is a perfect graph, is shown in Fig. 1. It has three disjoint maximum stable sets of size three each, one of each color, and three disjoint maximal cliques, one of which is shown in bold. Under unit cost for each asset, the worth of the investment management game on this graph is 3 and the optimal dual assigns 1 to each of the three cliques. Consider three scenarios, each consisting of a maximum stable set. The optimal investment in each scenario is to buy all three of its assets, requiring 3 units of money. If the cliques were to distribute their unit money to the assets, then at most one of these scenarios can be satisfied: by each clique allocating its money to a different colored asset of the *same* scenario. Therefore, the dual on a clique cannot be distributed among its vertices.

Theorem 2. *An imputation y is in the core of the stable set game over a perfect graph if and only if it is an optimal solution to the dual LP(6).*

Proof. ($\Leftarrow$) Let y be an optimal solution to the dual LP(6). By Lemma 1, the linear system of LP(5) is TDI and therefore a maximum cost stable set in G is

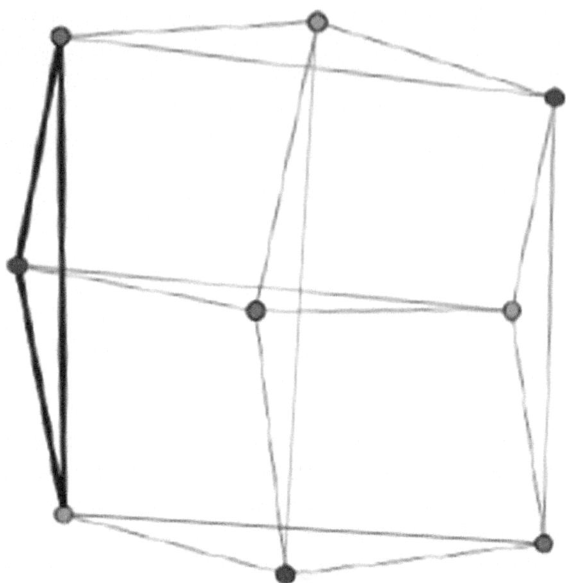

Fig. 1. The 3×3 Paley graph for Example 1

an optimal solution to this LP. This fact together with the LP-duality theorem give:

$$T = \text{cost}(V) = \sum_{Q \in G} y_Q = \text{money}(V),$$

Where cost(V) is the optimal value of LP (5). Therefore y is an imputation.

Consider a scenario $S \subseteq V$. By Definition 3,

$$\text{money}(S) := \sum_{Q \in G:\ Q \cap S \neq \emptyset} y_Q.$$

Clearly $Q \cap S$ is a clique in $G(S)$. Next, we will define a function, z, on cliques in $G(S)$ as follows: For a clique Q' in $G(S)$, define

$$z_{Q'} := \sum_{Q \in G:\ Q \cap S = Q'} y_Q.$$

Since each clique Q of G which has a non-empty intersection with S will contribute to exactly one clique in $G(S)$, namely $Q \cap S$, we get

$$\text{money}(S) = \sum_{Q' \in G(S)} z_{Q'}.$$

Next, we observe that z is a feasible solution to the restriction of LP (4) to $G(S)$ because

$$\forall v \in S: \quad \sum_{Q' \in G(S):\ Q' \ni v} z_{Q'} = \sum_{Q \in G:\ Q \ni v} y_Q \geq w_v,$$

where the last inequality holds because y is a feasible solution to LP 6. Since the subgraph of a perfect graph is also perfect, $G(S)$ is a perfect graph and the restriction of LP (3) to $G(S)$ satisfies TDI. Therefore a maximum cost stable set in $G(S)$ is an optimal solution to the latter LP. This fact together with weak duality give us

$$\text{cost}(S) \leq \sum_{Q \in G(S)} z_Q = \text{money}(S),$$

i.e., imputation y satisfies scenario S. Therefore y is a core imputation.

($\Rightarrow$) Next, assume that y is a core imputation of the investment management game over a perfect graph G.

For $v \in V$, consider the scenario $S = \{v\}$. Since y is in the core, by Definition 5,

$$\text{money}(S) = \sum_{Q \in G:\ Q \cap S \neq \emptyset} y_Q \geq \text{cost}(S) = w_v.$$

Therefore, y is a feasible solution for LP (6).

By Definition 3,

$$\text{money}(V) = \sum_{Q \in G} y_Q = T = \text{cost}(V).$$

Therefore by the TDI of LP (5), the objective function value of y is the same as that of the optimal value of the primal. This together with the feasibility of y establishes that y is an optimal solution to the dual LP (6).

Corollary 1. *The core of the stable set game over a perfect graph is non-empty.*

6 Discussion

Remark 1 draws a clear contrast between the way the core is used in cooperative game theory and in our game. Are there other ways of interpreting and using the notion of core, without taking liberties with its formal framework?

A simple way of defining the game presented in this paper would use the graph-theoretic language of stable sets and cliques. Definition 1 moves a step towards modeling an economic situation. There is no doubt that our model falls short of providing a solution concept for a realistic economic situation; however, that was not the intent of this paper. We simply wanted to provide evidence that the beautiful solution concept of core is more versatile than previously envisaged. We hope other researchers will be able to build on this viewpoint to find realistic applications of the notion of core outside of cooperative game theory.

Acknowledgments. I wish to thank Martin Bullinger for insightful comments on the write up, and Gerard Cornuejols, Federico Echenique, Naveen Garg, Martin Groetschel, Ruta Mehta, Joseph Root, Thorben Trobst and Richard Zeckhauser for valuable discussions.

This study was supported in part by NSF grant CCF-2230414.

Disclosure of Interests. The author has no competing interests to declare that are relevant to the content of this article.

References

1. Biswas, T.: Games against nature and the role of information in decision-making under uncertainty. In: Decision-Making under Uncertainty, pp. 185–196. Palgrave, London (1997). https://doi.org/10.1007/978-1-349-25817-8_14
2. Biró, P., Kern, W., Paulusma, D.: Computing solutions for matching games. Int. J. Game Theory **41**(1), 75–90 (2012)
3. Bondareva, O.N.: Some applications of linear programming methods to the theory of cooperative games. Problemy Kibernetiki **10**(119), 139 (1963)
4. Demange, G., Deng, X.: Universally balanced combinatorial optimization games. Games **1**(3), 299–316 (2010)
5. Deng, X., Ibaraki, T., Nagamochi, H.: Algorithmic aspects of the core of combinatorial optimization games. Math. Oper. Res. **24**(3), 751–766 (1999)
6. Edgeworth, F.Y.: Mathematical Psychics: An Essay on the Application of Mathematics to the Moral Sciences, vol. 10. CK Paul, Karukapally (1881)
7. Fulkerson, D.R.: Anti-blocking polyhedra. J. Comb. Theory Ser. B **12**(1), 50–71 (1972)
8. Granot, D., Huberman, G.: Minimum cost spanning tree games. Math. Program. **21**(1) (1981)
9. Granot, D., Huberman, G.: On the core and nucleolus of minimum cost spanning tree games. Math. Program. **29**(3), 323–347 (1984)
10. Gillies, D.B.: Solutions to general non-zero-sum games. Contrib. Theory Games **4**(40), 47–85 (1959)
11. Grotschel, M., Lovasz, L., Schirjver, A.: Geometric Algorithms and Combinatorial Optimization. Springer-Verlag, Berlin, Heidelberg (1988). https://doi.org/10.1007/978-3-642-78240-4
12. Grötschel, M.: My favorite theorem: characterizations of perfect graphs (1999)
13. Gale, D., Shapley, L.S.: College admissions and the stability of marriage. Am. Math. Mon. **69**(1), 9–15 (1962)
14. Goemans, M.X., Skutella, M.: Cooperative facility location games. J. Algorithms **50**(2), 194–214 (2004)
15. Kolen, A.: Solving covering problems and the uncapacitated plant location problem on trees. Eur. J. Oper. Res. **12**(3), 266–278 (1983)
16. Király, T., Pap, J.: Total dual integrality of rothblum's description of the stable-marriage polyhedron. Math. Oper. Res. **33**(2), 283–290 (2008)
17. Koh, Z.K., Sanità, L.: An efficient characterization of submodular spanning tree games. Math. Program. **183**(1), 359–377 (2020)
18. Lovász, L.: Normal hypergraphs and the perfect graph conjecture. Discret. Math. **2**(3), 253–267 (1972)
19. Lovász, L., Plummer, M.D.: Matching Theory. North-Holland, Amsterdam, New York (1986)
20. Moulin, H.: Cooperative Microeconomics: A Game-Theoretic Introduction, vol. 313. Princeton University Press, Princeton (2014)
21. Nagamochi, H., Zeng, D.-Z., Kabutoya, N., Ibaraki, T.: Complexity of the minimum base game on matroids. Math. Oper. Res. **22**(1), 146–164 (1997)

22. Rothblum, U.G.: Characterization of stable matchings as extreme points of a polytope. Math. Program. **54**(1), 57–67 (1992)
23. Schrijver, A.: Theory of Linear and Integer Programming. John Wiley & Sons, New York, NY (1986)
24. Shapley, L.S.: On balanced sets and cores. Technical report, RAND Corp Santa Monica, California (1965)
25. Shapley, L.S., Shubik, M.: The assignment game i: the core. Int. J. Game Theory **1**(1), 111–130 (1971)
26. Samet, D., Zemel, E.: On the core and dual set of linear programming games. Math. Oper. Res. **9**(2), 309–316 (1984)
27. Vate, J.H.V.: Linear programming brings marital bliss. Oper. Res. Lett. **8**(3), 147–153 (1989)
28. Vazirani, V.V.: The general graph matching game: approximate core. Games Econ. Behav. **132** (2022)
29. Vazirani, V.V.: New characterizations of core imputations of matching and *b*-matching games. Found. Softw. Technol. Theor. Comput. Sci. (2022)
30. Vazirani, V.V.: LP-duality theory and the cores of games. arXiv preprint arXiv:2302.07627, 2023

Edge-Dominance Games on Graphs

Farid Arthaud(✉), Edan Orzech, and Martin Rinard

MIT CSAIL, Cambridge, MA 02139, USA
{farto,iorzech,rinard}@csail.mit.edu

Abstract. We consider zero-sum games in which players move between adjacent states, where in each pair of adjacent states one state dominates the other. The states in our game can represent positional advantages in physical conflict such as high ground or camouflage, or product characteristics that lend an advantage over competing sellers in a duopoly. We study the equilibria of the game as a function of the topological and geometric properties of the underlying graph. Our main result characterizes the expected payoff of both players starting from any initial position, under the assumption that the graph does not contain certain types of small cycles. This characterization leverages the block-cut tree of the graph, a construction that describes the topology of the biconnected components of the graph. We identify three natural types of (on-path) pure equilibria, and characterize when these equilibria exist under the above assumptions. On the geometric side, we show that strongly connected outerplanar graphs with undirected girth at least 4 always support some of these types of on-path pure equilibria. Finally, we show that a data structure describing all pure equilibria can be efficiently computed for these games.

Keywords: Zero-sum repeated game · Hotelling model · Graph movement game

1 Introduction

Consider two players moving among the vertices of a graph G, where at each timestep both players simultaneously move to a neighbor of their current vertex, or remain. The game ends with some constant probability $(1-\delta) \in]0;1[$ at each timestep. The edges of the graph are undirected for movement but directed for payoff: at the conclusion of the game, each player receives a payoff of 1 if they are at a parent of the other player, -1 if the other player is at one of their parents, and 0 if their vertices are not neighbors (or if they are at the same vertex).

This setting is a general way of constraining how players change actions between rounds in a repeated symmetric zero-sum game: each vertex of the graph G is a possible action, the edges of the graph describe the payoffs of the game (the symmetric zero-sum game matrix is seen as an adjacency matrix) and players can only move between *similar* actions between rounds (i.e. actions that have an edge between them).

G. Schäfer and C. Ventre (Eds.): SAGT 2024, LNCS 15156, pp. 240–257, 2024.
https://doi.org/10.1007/978-3-031-71033-9_14

These games model aspects of physical conflict in which adjacent vertices represent positions where adversaries can engage. For instance, terrain features such as high ground or easy camouflage may yield an advantage to one of the adjacent positions. These games also model aspects of duopoly markets in which each vertex of the graph represents a set of features of a product manufactured by two competing firms. The firms can only incrementally change their product, by moving along the edges of the graph. Adjacent sets of features compete for the same consumers, with the direction of the edge indicating which product is more profitable to sell when these two products are on the market (as a function of consumer preference, production prices and selling prices). On the other hand, products further away from each other in the graph have features that appeal to sufficiently different segments of consumers that they have no discernible advantage over one another. In turn, this can be seen as a dynamic variant of discrete HOTELLING models [9] (recall location in HOTELLING models is often interpreted as product differentiation), an area of interest in economics and algorithmic game theory [4,5,15].

We focus on characterizing the payoff that each player can obtain from any given initial position in a NASH equilibrium of the game. We study the conditions under which one player has a *winning strategy*, namely a strategy that gives the player strictly positive expected payoff, as well as how much payoff this player can obtain in expectation. We also study the conditions under which both players have *safe strategies*, namely strategies where they do not incur strictly negative payoff.

We identify three types of on-path pure NASH equilibria (meaning players do not mix unless a player has deviated) in safe strategies that are extremal equilibria in cycle graphs C_n (for $n \geq 4$). The first type is the *k-chase* equilibrium, in which one player follows k steps behind the other player, without ever reaching them. The second type is the *walking together* equilibrium, in which both players start at the same vertex and always deterministically move together to a same vertex at each round, ensuring a stalemate. The third type is the *static* equilibrium, where both players remain static at a distance from each other. We provide characterizations under which k-chase, walking together and static equilibria exist on certain classes of graphs. These characterizations imply sufficient conditions for the existence of on-path pure NASH equilibria in safe strategies in general.

We also consider the existence of these types of equilibria in *planar graphs*. These graphs are important as possible models of maps of geographical locations, and because as our results will imply in Sects. 6 and 7, nonplanarity plays a role in ensuring the existence of the above equilibria. We provide sufficient conditions on *outerplanar graphs*, a particular class of planar graphs, such that walking together and 2-chase equilibria exist.

1.1 Results and Techniques

Main Result. Our main result characterizes the value of the game for any given initial position, under the assumption that the graph does not contain certain types of small cycles as a subgraph.

Theorem 1 (Informal version of Theorem 4). *In any graph of undirected girth at least* 4 *that does not contain small* unbalanced *cycles, a player is in a losing position if and only if there is a unique way of cutting the graph halfway between the two players, and the resulting connected component of this player is an (out)-directed rooted tree, rooted at the vertex where the graph was cut.*

An *unbalanced* cycle is a cycle with edges oriented such that there is exactly one maximal vertex (with no incoming edges) and one minimal vertex (with no outgoing edges). The excluded cycles in this theorem are all 3-cycles, and the four unbalanced cycles of length 4 or 5. In particular, our characterization holds for trees and more generally all graphs of (undirected) girth at least 6.

The key concept in this characterization's precise formulation is the position of the players in the *block-cut tree* of the (undirected) graph. The block-cut tree of a graph is the tree whose vertices are the maximal biconnected components and the cut vertices of the graph, with edges between each biconnected component and all of its cut vertices. The block-cut tree is important because we establish that players have a safe strategy as long as they find themselves in a biconnected component. Intuitively, if cutting the graph halfway between the players leaves a player with biconnected components (or other types of safe positions) in their half, they can reach this safe position before the other player reaches them.

Static and Cycle-Based Equilibria. We then characterize the existence of walking together, 2-chase, and static equilibria. Under the same topological assumptions as above we show that walking together and 2-chase equilibria exist if and only if the graph contains a directed cycle – the weakest possible condition we could hope for. We also provide an exact characterization of graphs with static equilibria under these assumptions (Proposition 2). We then show the importance of small unbalanced cycles for these characterizations by presenting constructions that contain some small unbalanced cycles, verify our characterization above, and yet do not admit these equilibria – despite being strongly connected and of girth 5 (see Sect. 6 for the constructions).

We provide another sufficient condition using graph geometry rather than topology, showing that any strongly connected *outerplanar* graph with undirected girth at least 4 supports a 2-chase and a walking together equilibrium (Theorem 8). An outerplanar graph is a planar graph in which all of the vertices belong to the outer face of the graph (for some planar embedding of the graph). Intuitively, this result holds for two reasons. One, such a graph has a face whose edges form a directed cycle, which gives a player some safety from the other player's deviations when they remain on it. Two, because of outerplanarity, a player who exits that directed cycle from some vertex v can only reenter it at a vertex that is either v or its neighbors on the cycle. Therefore, the player who does not deviate can remain at a safe distance from the deviator at all times. This result falls into the larger context of the existence of static and cycle-based equilibria in *planar graphs*. We conjecture in Sect. 7 that all strongly connected planar graphs with girth at least 4 have a static or cycle-based equilibrium, given our result on outerplanar graphs and that our constructions without cycle-based equilibria are nonplanar.

Computational Aspects. We show that a data structure describing all pure equilibria can be computed efficiently, despite the number of pure equilibria potentially being exponential. Computing all mixed equilibria in the game (for all starting positions) results from a straightforward application of BELLMAN's equations and value iteration algorithms.

Roadmap. The remainder of this section contains related work. In Sect. 2 we define our game formally, and characterize equilibria in cycle graphs to introduce important definitions and lemmas. Section 3 characterizes winning positions in trees and Sect. 4 characterizes all equilibria in graphs of girth at least 6. The results from these two sections are combined to obtain our main results in Sect. 5 for graphs with no small unbalanced cycles. Section 6 contains our strongly connected constructions without cycle-based or static equilibria. Section 7 contains our results for outerplanar graphs, and finally Sect. 8 presents our algorithm to efficiently compute pure equilibria.

1.2 Related Work

Pursuit-Evasion Games. The branch of research conceptually closest to ours is pursuit-evasion games on graphs, also known as cops and robbers problems, originally introduced by Quilliot [14]. In this game, a robber player chooses a vertex in a finite graph, after which k cops choose positions in the graph. Then, turn by turn, the robber and cops move along the edges in the graph, until either a cop reaches the same vertex as the robber or until a configuration repeats twice. In the latter case, the robber manages to evade the cops and wins, whereas if a cop reaches the robber, the robber loses. In the original paper, Quilliot [14] characterizes the graphs in which one cop is sufficient to capture the robber, and a long line of work followed on finding bounds on the *cop number* of a graph (the number of cops necessary to capture the robber) [2,3,11].

Bonato and Nowakowski [3] provide a general survey of variants of cops and robbers. Several variants have some similarities to our setting. Konstantinidis and Kehagias [11] study simultaneous-move cops and robbers and show that the (appropriately defined) cop number of a graph is unchanged relative to the classical cops and robbers. Hamidoune [7] introduces a variant on directed graphs, however the direction of edges has a different meaning than in our game: players are constrained to follow edges only in one direction, and the goal of the cops is still to reach the same vertex as the robber. Bonato, Chiniforooshan and

Prałat [2] introduce capture from a distance, where cops win if any one of them comes within a certain distance of the robber. As cops and robbers has traditionally been studied from a graph-theoretical or combinatorial point of view, some works investigate game-theoretic formulations, such as the recent work of Kehagias and Konstantinidis [10], and the survey by Luckraz [13] of game-theoretic formulations.

There are several important differences between the game of cops and robbers and our model: (i) in our model the players are symmetric whereas in cops and robbers there is a pursuer and an evader; (ii) the winning condition in our model is to reach a parent vertex of the other player, whereas in standard cops and robbers it is to reach the same vertex (and edges are undirected); (iii) our model is simultaneous move whereas standard cops and robbers take turns. This means that results from cops and robbers do not apply in our setting. In particular, players can find themselves mixing between situations where they have strictly positive, zero and strictly negative payoff (whereas in usual cops and robbers there are no draws – one player can force a win by ZERMELO's theorem – see Remark 1 for a simple example). Moreover, no player can win with probability 1 in our model and players randomize to evade capture (see Remark 3 for a simple example and Proposition 3 for a proof). In contrast, even in simultaneous cops and robbers, the cop number is defined as the minimum number of cops such that the robber is captured with probability 1 [11], significantly altering the analysis. The main idea in simultaneous cops and robbers is for the cop to guess the next move of the robber, and play as if their guess is correct – with probability 1, they will eventually guess correctly for sufficiently long that they will capture the robber. Such a strategy is not viable in our game since it requires that for every pair of positions, the cop can win against the robber in the usual turn-based cops and robbers. Since our players are symmetric, if one player has a superior position to the other then the converse cannot be true, and a guessing strategy can lead the player in a superior position to end up in an inferior position by misguessing. The optimal strategies in our setting can therefore ensure positive expected payoff at best, but never capture with probability 1 as in cops and robbers and its variants – and they very well can lead a player with strictly positive expected payoff to obtain strictly negative payoff with non-zero probability.

More importantly, beyond our characterization of winning positions in the game, many of our results concern properties of 0-payoff equilibria the players can be in, characterizing player behavior when neither player has an advantage over the other. To the best of our knowledge this has not been explored for cops and robbers, where most work only focuses on characterizing when one player has an advantage over the other (in particular through the cop number) [13].

Stochastic Games. Our game is an instance of a stochastic game [16], i.e. an extensive-form game with a state that is affected by the actions of both players (and potentially also external randomness, but not in our case) and which affects the players' payoffs. However, it has much more added structure which makes our analysis possible: the state space is a cartesian product (of the set of the graph's vertices with itself), each player affects only one component of the state, and the payoff is related to the allowed transitions (through the graph edges). To our knowledge, there are no results in this area concerning games with this structure.

Discrete HOTELLING Models. Finally, as mentioned earlier, our model has similarities with discrete HOTELLING models, such as those presented in Serra and Revelle [15]. These games are sometimes also called VORONOI games on graphs or *competitive facility location* games when the players can only locate at vertices (similarly to our model). In these models, two players choose a vertex of a graph (or sometimes a position along an edge of a graph) and then are rewarded as a function of the quantity of vertices or edges (sometimes weighted) that are closer to them than to the other player [5]. The fundamental difference between these models and ours is that only adjacent vertices in the graph have an advantage over one another in our model, whereas in VORONOI games it is likely that most pairs of distinct vertices have unequal payoff. Moreover, to the best of our knowledge, these models are static and do not model dynamics of relocation like ours. Some works analyze best response dynamics in these games [4], but with no restrictions on where players can move.

2 Preliminaries

Let $G = (V, E)$ be a connected oriented graph. An oriented graph is a directed graph with no loops or parallel edges, they are the graphs obtained by assigning an orientation to each edge in an undirected graph. If $(u, v) \in E$, we often write $(u \to v) \in E$ or simply $u \to v$ when E is clear from context. We write $u - v$ to say that there is an edge $u \to v$ or $v \to u$ in the graph. For convenience, we denote $\hat{E} = \{(v, u) \mid (u, v) \in E\}$ the reversed set of edges.

Definition 1. *A* **path** *in G is a list of at least 2 distinct vertices $u_0 \to u_1 \to \cdots \to u_k$ with directed edges, whereas an* **undirected path** *is a list of at least 2 distinct vertices $u_0 - u_1 - \cdots - u_k$ with edges in either direction. We denote $\tilde{G}$ the* undirected graph *underlying G. An* **undirected cycle** *in G is a cycle in $\tilde{G}$, i.e. an undirected path of length at least 3 such that the last vertex is a neighbor of the first. The* **ball** *of radius r centered at vertex u is,*

$$\mathcal{B}_r(u) = \{v \in V \mid \exists k \in [\![0; r]\!],\ \exists v_0, \ldots, v_k,\ (u = v_0) - v_1 - \cdots - v_{k-1} - (v_k = v)\}.$$

$g(\tilde{G})$ denotes the **undirected girth** *of G, i.e. the length of a shortest cycle in $\tilde{G}$. We refer to it as g when G is clear from context.*

The game is defined by the graph G and initial positions for both players $(x_0, y_0) \in V^2$. At each timestep $t \in \mathbb{N}$, we denote (x_t, y_t) the positions of the players in the graph. The strategies of the players are mappings from their current positions (x_t, y_t) to a distribution over their neighborhoods $\mathcal{B}_1(x_t)$ and $\mathcal{B}_1(y_t)$. We denote $\varphi_x : V^2 \to \Delta(V)$ the strategy of player x and φ_y for player y, where $\Delta(V)$ is the simplex over the vertices of G. For a given initial state (x_0, y_0), a distribution over histories of play $(h_t)_{t\in\mathbb{N}}$ is naturally induced by φ_x and φ_y: we write $h \sim (\varphi_x, \varphi_y)$ for a random variable h following this distribution when (x_0, y_0) is clear from context. Note the game is defined in such a way that the strategies are memoryless: they only depend on the current state and not on the history of play.

The game ends with probability $(1-\delta)$ at the end of each round, for some fixed parameter $\delta \in]0;1[$. At that point, the payoff of each player is 1 if they are at a parent of the other player, -1 if they are at a child of the other player, and 0 otherwise (in particular, if both players are at the same vertex). The game is a zero-sum game, and the expected payoff of player x can be written,

$$u_x(\varphi_x, \varphi_y) = (1-\delta)\,\mathbb{E}_{h\sim(\varphi_x,\varphi_y)}\left[\sum_{t\in\mathbb{N}} \delta^t \left(\mathbb{1}_{h_t\in E} - \mathbb{1}_{h_t\in\hat{E}}\right)\right]. \tag{1}$$

As is often done in the repeated games literature and in order to simplify analysis, we often interpret Eq. (1) as the payoff of a discounted game: the game is then always infinite, and payoff at round t is multiplied by a factor δ^t. We also refer to the sum starting at $t = 1$ in Eq. (1) as the **continuation payoff** of player x.

Definition 2. *A pair of strategies is called a* NASH *equilibrium if neither player can increase their expected payoff by changing strategies. Note that by the minmax theorem, there always exists a* NASH *equilibrium. We call the* ***value*** *of a vertex u over a vertex v the minmax equilibrium value of player x when the players start at (u, v). A strategy is called* ***safe*** *for a player if its expected payoff is (weakly) positive, and* ***winning*** *if its expected payoff is strictly positive.*

2.1 Cycle-Based and Static Equilibria

We begin by considering cycle graphs, in which we characterize all equilibria and identify three particular types of extremal pure equilibria. This leads us to define the three types of equilibria we investigate in general graphs in the following sections. We also introduce important definitions and lemmas for the rest of the paper by studying the directed 3-path.

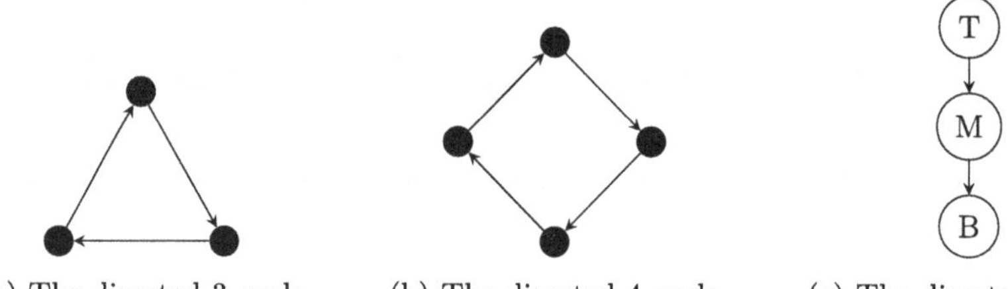

(a) The directed 3-cycle (b) The directed 4-cycle (c) The directed 3-path

Fig. 1. Example graphs

3-Cycle. In Fig. 1a the directed 3-cycle is shown: on this graph, our game reduces to (repeated) rock-paper-scissors game, as the players are unconstrained in which actions they can choose at each round. Therefore, there exists a unique NASH equilibrium where both players uniformly mix over all three possible movements (stay or move to one of their two neighbors).

Remark 1. Note this example highlights a fundamental difference with cops and robber games: both players are mixing between outcomes that have strictly positive, zero, or strictly negative value for them, meaning players need to take a chance of capturing and a risk of being captured. This cannot occur in cops and robbers for several reasons: the moves are not simultaneous and the roles are asymmetric (the cop cannot be captured). In simultaneous cops and robbers [11], it would be a cop-win graph since the cop will eventually collide with the robber with probability 1 regardless of starting positions.

4-Cycle. The directed 4-cycle in Fig. 1b has more equilibria. If both players start opposite from each other, their optimal strategy is to randomize between their two neighbors with any distribution that puts at most 1/2 probability on their parent. Indeed, if one player moves counterclockwise to their parent with probability strictly more than 1/2, the other player can ensure strictly positive payoff by remaining at their current vertex; if one player puts any probability on staying at their vertex then the other player can ensure strictly positive payoff by moving to their child (clockwise).

If both players start at the same vertex, the optimal strategies are similar: any mixing between their two neighbors that puts at most 1/2 probability on their child is optimal. In both starting positions the minmax one-round expected payoff is 0 for both players, and by induction the overall minmax expected payoff is also 0. The set of equilibria consists of all distributions that satisfy the above conditions, therefore the players will find themselves either at the same vertex or at opposite vertices at every round if they start in one of these positions. Note two particular extremal equilibria in this graph are: (i) both players are at the same vertex and move deterministically counterclockwise to the parent (together) and (ii) both players are opposite from each other and move deterministically clockwise to their child at each round.

In a longer cycle, more strategies exist: when players are far from each other, all actions are equivalent, whereas when the players are at distance 2 or 3 from each other, one player will have to avoid the other (by moving in the opposite direction). In particular, another type of on-path pure equilibrium arises when the cycle is of length at least 6: both players can remain static at vertices that are distance at least 3 from one another. The following definition generalizes the three extremal equilibria we have seen in cycles so far.

Definition 3. *A* ***walking together equilibrium*** *(WT) is an equilibrium such that for every $t \in \mathbb{N}$, $x_t = y_t$ and $x_{t+1} \neq x_t$ with probability 1. A* ***k-chase equilibrium****, for $k \in [\![2; +\infty[\![$, verifies $x_{t+k} = y_t$ for all $t \in \mathbb{N}$. A* ***static equilibrium*** *is such that with probability 1, there is a t_0 and vertices $x_\infty, y_\infty \in V$ such that for every $t \geq t_0$, $(x_t, y_t) = (x_\infty, y_\infty)$.*

Before our final example, we define an important property of certain graphs that simplifies analysis of equilibria. In all generality, having negative payoff at a given round could still lead to compensations later on, for instance a player could accept one immediate round of negative payoff to ensure many rounds of positive payoff later on. We define edges and graphs for which this does not have an effect on winning strategies.

Definition 4. *For a given value of δ, an edge $u \to v$ is called* ***decisive*** *if the value of a player at u over a player at v is strictly positive. A graph G is called* ***edge-decisive*** *for a given δ if all of its edges are decisive. δ will be omitted when clear from context.*

Remark 2. In an edge-decisive graph, a walking together equilibrium always steps towards parents: $\forall t,\ (x_{t+1} \to x_t) \in E$. This is because when walking together to a child, either player could make a profitable deviation by not moving. Inversely, a 2-chase equilibrium in an edge-decisive graph always steps towards children: $\forall t,\ (y_t \to y_{t+1}) \in E$ (otherwise the chased player can stop after having taken an edge in the opposite direction). In particular, both types of equilibria correspond to a directed cycle in the graph G.

3-Path. To illustrate cases where one player has an advantage over the other, we look at the directed 3-path example illustrated in Fig. 1c, when one player lies at T and the other at B. The one-step game is equivalent to matching pennies (where the two sides of the pennies are 'move' or 'stay'): the top player wins if exactly one of them moves to M whereas the bottom player wins if either both or neither of them move. However, the two outcomes where the bottom player wins the one-shot game are not equivalent: if they both move, the continuation payoff is 0, since both players will simply move to T in the next round. If neither moves, the game repeats and the top player has some chance of winning again. Similarly, if just the top player moves then they gain payoff 1 and the game repeats (since the players have the same two actions each, the top player going to B is dominated). If just the bottom player moves, the game ends at the next round with both players reaching T. Regardless, the top player has strictly positive expected payoff starting from the initial condition. The proof of the following lemma is deferred to the full version of the paper [1, Lemma 2.5].

Lemma 1. *In the 3-path illustrated in Fig. 1c, a player at T has strictly positive payoff over a player at B.*

Remark 3. Note that unlike cops and robbers, the bottom player always has some probability of avoiding capture, by randomizing between moving to the middle vertex and staying. This is true of any position in the graph: if a player randomizes between all of their parents and staying at their current vertex, the other player cannot ensure capture in one round (and ensuring capture in one round has higher payoff than any strategy that does not, hence even in minmax play capture is never ensured).

3 Trees

A simple generalization of the ideas behind the case of a single path is a tree, where we find a characterization of positions that have positive minmax value for a player. We first define some useful notions to express our results.

Definition 5. *For a tree T rooted at r and a vertex v, let T_v be the subtree of T rooted at v. In a rooted tree, a directed edge $u \to v$ is said to be pointing* ***upwards*** *if $u \neq r$ and v is on the (undirected) path from u to r – otherwise, it is a* ***downwards*** *edge. A rooted directed tree is called* ***outgoing*** *if all of its edges are downwards edges.*

The main intuition behind the characterization is to root the tree at the midpoint of the path between the two players' positions. If either player can reach an upwards edge in the tree without going through the root, they are safe – it ensures the other player would have to go through their child to reach them. Otherwise, the other player can reach the root and then start chasing them down (all edges go downwards) until they reach a leaf, and Lemma 1 for Fig. 1c shows they can obtain strictly positive payoff. Figure 2 shows the two situations where player y has a winning strategy over player x. The proof of the following theorem is deferred to the full version [1, Theorem 3.2].

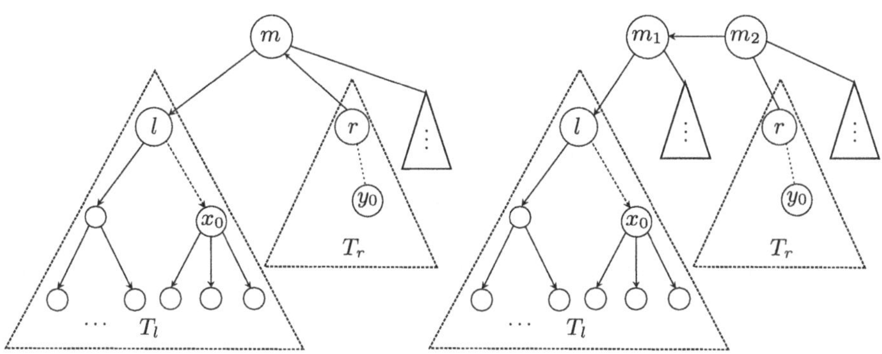

Fig. 2. The two winning configurations for player y in a tree

Theorem 2. *For a given initial position (x_0, y_0) in a tree, player y has strictly positive payoff over player x if and only if,*

1. *The path between x and y is of even length, has the form $x_0 - \cdots - l \leftarrow m \leftarrow r - \cdots - y_0$, where $d(x_0, m) = d(y_0, m)$ and the subtree T_l (containing player x) of T rooted at m has no upwards edges (left tree in Fig. 2); or*
2. *The path between x and y is of odd length, has the form $x_0 - \cdots - l \leftarrow m_1 \leftarrow m_2 - r - \cdots - y_0$, where $d(x_0, m_1) = d(y_0, m_2)$ and the subtree T_l (containing player x) of T rooted at m_2 has no upwards edges (right tree in Fig. 2).*

4 Girth at Least 6

The first extension of our results on trees (acyclic connected graphs) are graphs with high girth (only big cycles). We show that with strong connectivity, players are essentially always safe from one another – unless one player is at a parent of the other player in the initial position. The proof of the following theorem is deferred to the full version of the paper [1, Theorem 4.1]. The intuition is that both players are always on a long cycle, therefore they can maintain distance from one another by moving on such a cycle away from the other player.

Theorem 3. *If G is strongly connected and $g \geq 6$, the minmax value of a pair of vertices is 0 if and only if they are not neighbors. For neighboring vertices, the player at the child of the other has payoff in the range $\left[-\frac{4(1-\delta)}{4-\delta}; -(1-\delta)\right]$. In particular, there is a static equilibrium at any pair of vertices at distance 3 from each other, a 2-chase equilibrium for any starting vertices with a (directed) 2-path from one to the other, and a WT equilibrium starting at every vertex.*

Corollary 1. *In a strongly connected graph with $g \geq 6$ and initial positions (x_0, y_0) with no edge between them, the set of* Nash *equilibria is the set of mixed strategies (φ_x, φ_y) such that at each round, x puts 0 probability on going to a child of $\mathcal{B}_1(y_t)$ and vice versa.*

Proof. We first note such distributions always exist: if $x_0 = y_0$ then any parent of the vertex verifies the condition, and otherwise $d(x_0, y_0) \geq 2$ and Theorem 3 proves the existence of such a move.

If at round t player x puts non-zero probability on such a vertex v, then a possible move of player y is to go to a parent u of v deterministically. This gives y strictly positive expected payoff, as we showed in Theorem 3 the graph is edge-decisive. Therefore, every Nash equilibrium has the property above. Conversely, any pair of distributions that verifies this property is clearly a Nash equilibrium since if a player deviates, they can never reach a parent of the other player by definition of their strategy. □

5 Graphs with No Unbalanced Small Cycles

For our main result, we combine the results of the two previous section on trees and graphs with girth at least 6. We remove the strong connectivity assumption and replace it with an analysis of the *block-cut tree* of the graph: biconnected components will be analogous to the strongly connected graphs of Theorem 3 (though they are not always strongly connected), whereas cut vertices will behave more like tree vertices seen in Theorem 2. This results in weakening the assumptions from the previous section in two ways: strong connectivity is no longer required, and we replace the assumption $g \geq 6$ with the assumption $g \geq 4$ and the absence of *small unbalanced cycles* as subgraphs of the graph. Let us begin by defining these concepts.

Definition 6. *An* (x, y)-***cut vertex*** *of* G *for* $x, y \in V$ *is a cut vertex of* G *such that removing it separates* x *and* y *into two distinct connected components.*

Definition 7 ([6,8])**.** *For an undirected graph* $G = (V, E)$ *define its* ***block-cut tree*** *as the tree containing a vertex for each maximal biconnected component of* G*, a vertex for each of its cut vertices, and an edge connecting each cut vertex to the biconnected components it belongs to. We call the* ***thinned block-cut tree*** *of* G*, denoted* $T(G)$*, the following transformation of its block-cut tree: for each maximal biconnected component of size* 2 *containing two cut vertices, remove its vertex from the tree and add an edge between its two cut vertices; for all other biconnected components of size* 2*, remove its vertex and replace it with a vertex labeled by its non-cut vertex. A biconnected component remaining in the thinned block-cut tree (equivalently, a biconnected component with strictly more than* 2 *vertices) is called a* ***nontrivial biconnected component****.*

Notice each vertex in the thinned block-cut tree is labeled either by a maximal nontrivial biconnected component or a vertex of the graph, and all vertices labeled by a vertex of the graph are either cut vertices or leaves of the block-cut tree. The proof of the following lemma is deferred to the full version of the paper [1, Lemma 5.3].

Lemma 2. *Along any shortest undirected path between* x *and* y*, the indices containing* (x, y)*-cut vertices are always the same, and each index always contains the same cut vertex. Moreover, in all shortest undirected paths the indices that do not contain* (x, y)*-cut vertices correspond to a vertex in the common biconnected component of the previous and next* (x, y)*-cut vertices in the path.*

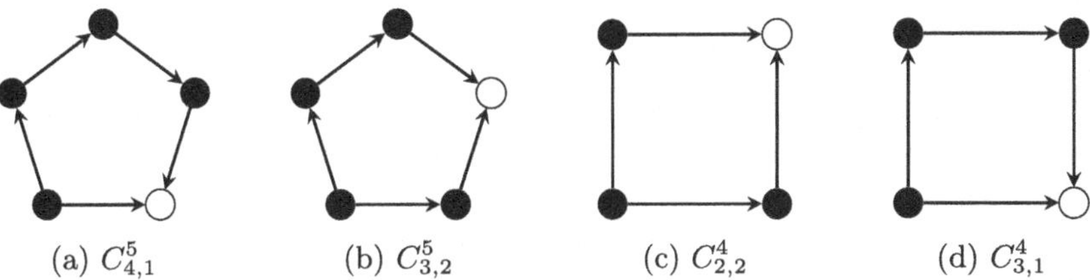

(a) $C^5_{4,1}$ (b) $C^5_{3,2}$ (c) $C^4_{2,2}$ (d) $C^4_{3,1}$

Fig. 3. The four small unbalanced cycles, with their minimal vertex highlighted

Definition 8. *Let* $C^k_{a,b}$ *be the length* k *undirected cycle with* a *consecutive edges in one direction and the remaining* b *edges in the opposite direction. We call the* ***small unbalanced cycles*** *the four cycles* $C^5_{4,1}, C^5_{3,2}, C^4_{2,2}$ *and* $C^4_{3,1}$*, illustrated in Fig. 3.*

In the full version of the paper [1, Appendix A], we show the remaining 4-cycles and 5-cycles are not unbalanced.

We now characterize winning positions in graphs with no small unbalanced cycles. The proof of the following theorem is deferred to the full version [1, Theorem B.1], we offer a proof sketch here.

Theorem 4. *Suppose G satisfies $g \geq 4$ and does not contain any of the small unbalanced cycles as subgraphs (in particular, this is verified when $g \geq 6$). Consider the thinned block-cut tree $T(\tilde{G})$ of G. For a given initial position (x_0, y_0) in G, player y has strictly positive payoff over player x if and only if there exists a shortest path between x and y which is either,*

1. *of even length, of the form $x_0 - \cdots - l \leftarrow m \leftarrow r - \cdots - y_0$, where $d(x_0, m) = d(y_0, m)$, the midway vertex m is an (x, y)-cut vertex of G and the subtree T_l (containing player x) of $T(\tilde{G})$ rooted at m has no nontrivial biconnected components or upwards edges; or*
2. *of odd length, of the form $x_0 - \cdots - l \leftarrow m_1 \leftarrow m_2 - r - \cdots - y_0$, where $d(x_0, m_1) = d(y_0, m_2)$, the vertex m_1 is an (x, y)-cut vertex of G and the subtree T_l (containing player x) of $T(\tilde{G})$ rooted at m_1 has no nontrivial biconnected components or upwards edges.*

Proof (sketch). First note that by Lemma 2 the criteria are well-defined, i.e. they do not depend on the chosen shortest path. The main idea is to show that both players have a safe strategy when they are both in a nontrivial biconnected component and at distance at least 2 from one another.

Indeed, suppose player x is at distance exactly 2 from player y. For staying at their current node or moving towards y not to be safe strategies, there must be a directed 2-path from y to x. Moreover, x must have some other neighbor in the biconnected component than the one between x and y: moving to this neighbor not being safe must mean y is a parent of that neighbor or is adjacent to a parent of that neighbor. The constructed edges so far create a 4 or 5 cycle with some orientations fixed: one can check that no orientations for the remaining edges avoid creating a small unbalanced cycle. The reasoning when y is outside of the biconnected component is similar, because y can only enter the component through a unique cut vertex.

The rest of the characterization is similar to Theorem 2, with the added subtlety of the case where the root is part of a nontrivial biconnected component. In these cases, we show that before reaching the midpoint, both players will enter its biconnected component, and be distance at least 2 from one another. Since we have shown these are both safe positions, we deduce that for a player to have a winning strategy, the root must be a cut vertex. From there, we have shown that nontrivial biconnected components are safe, therefore the connected component containing the losing player must be composed of only cut vertices, which makes it a tree. By similar reasoning to Theorem 2 once more, we show this tree must be outdirected from the midway point. □

We now characterize the presence of cycle-based equilibria under the assumptions of Theorem 4. We show the weakest necessary condition one could hope for (under edge decisiveness, which we show holds here) is necessary and sufficient: cycle-based equilibria exist if and only if a directed cycle is present. The proof of the following proposition is deferred to the full version [1, Proposition 5.6].

Proposition 1. *If G satisfies $g \geq 4$ and does not contain any of the small unbalanced cycles as subgraphs, G has a WT equilibrium and a 2-chase equilibrium if* *and only if there is a directed cycle in G. In particular, G always has either a cycle-based or a static equilibrium.*

We state a necessary and sufficient condition under the previous assumptions for there to be a static equilibrium, albeit for concision we state it in negative form. The proof of the following proposition is deferred to the full version of the paper [1, Proposition B.2].

Proposition 2. *If G satisfies $g \geq 4$ and contains no unbalanced cycles, G has no static equilibria if and only if the following are all true,*

1. *G has exactly one nontrivial biconnected component B;*
2. *The thinned BC-tree is an outdirected tree rooted at B (all edges go downwards);*
3. *B is of diameter exactly 2 and all pairs of distance-2 vertices have a common neighbor that is a parent of one of the two;*
4. *Every vertex in B has a parent.*

We finally show that up to added outgoing branches, the only graph with no static equilibria with no small unbalanced cycles and $g \geq 5$ is the directed 5-cycle. The proof of the following proposition is deferred to the full version [1, Corollary B.3].

Corollary 2. *If $g \geq 5$ and small unbalanced cycles are forbidden, the only graphs with no static equilibria are composed of a directed 5-cycle with outgoing edges from its nodes forming a directed outgoing tree rooted at the cycle. In particular, all graphs with $g \geq 5$ and no small unbalanced cycles either have a static equilibrium or both a 2-chase and a WT equilibrium.*

6 Constructions with No Cycle-Based or Static Equilibria

We now argue that unbalanced cycles play an important role in the existence of cycle-based equilibria by exposing constructions without cycle-based or static equilibria. We show that even under strong connectivity (a much stronger assumption than before, meaning all vertices are part of a nontrivial biconnected component and of a directed cycle) and the absence of *any* 4-cycles, these equilibria do not always exist. Exhibiting such counter-examples is subtle since ensuring strong connectivity often creates many new directed cycles, which creates opportunities for cycle-based equilibria.

We first show that a constant upper bound on δ along with a girth assumption ensures the edge-decisiveness of a graph, which will be used in the proofs in this section. The proof of the following proposition is deferred to the full version [1, Proposition C.1]. The intuition is to lower bound the number of rounds necessary for the losing player to reach a parent of the winning player, which leads to an upper bound on the discount factor for all future positive payoffs.

Lemma 3. *Let γ_a be the unique positive root of $\gamma^{a-2} + \gamma - 1 = 0$ for $a \geq 4$. If $g(G) \geq a$, then G is edge-decisive for all $\delta < \gamma_a$.*

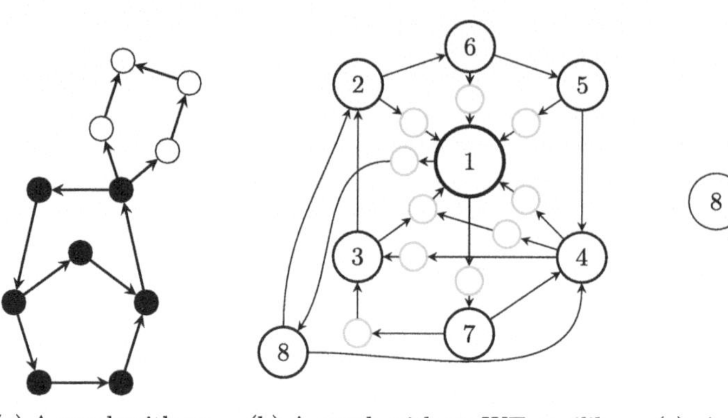

(a) A graph with no static equilibria

(b) A graph with no WT equilibria

(c) A graph with no 2-chase equilibria

Fig. 4. Constructions without static or cycle-based equilibria

In particular, when $g = 5$, G is edge-decisive for all $\delta < \gamma_5 \approx 0.68233$.

We begin with a construction (Fig. 4a) that supports no static equilibria, whilst only containing one type of small unbalanced cycles ($C^5_{3,2}$) and no cycles of length 4. To be relevant, this construction must violate at least one item of Proposition 2: we show it violates all of them but the last, which cannot be violated by any such construction (a vertex with no parents always has a static equilibrium). We note that unlike the other constructions, this construction is not strongly connected: it can easily be made to be by removing the white vertices while preserving its properties (except it would no longer violate item 1 of Proposition 2). The proof of the following theorem is deferred to the full version [1, Theorem C.2].

Theorem 5. *The graph in Fig. 4a has $g \geq 5$, only contains $C^5_{3,2}$ of the small unbalanced cycles and has no static equilibria for every $\delta \in]0; 1[$. Moreover, it violates every condition of Proposition 2 except for item 4 (which clearly is a necessary condition).*

We now show a construction (Fig. 4b) that supports no WT equilibria for all $\delta < \gamma_5$ (in this case the construction is edge-decisive). This construction is more involved as it contains many directed cycles and deviations take more rounds to be profitable. Intuitively, it supports no WT equilibria because it has a central vertex (vertex 1 in the figure) that has many parents. No matter which parent the WT equilibrium prescribes to go to, there is always a parent to which a deviator can profitably deviate. We then show that every directed cycle that does not go through 1 verifies a similar property with vertex 2. The proof of the following theorem is deferred to the full version [1, Theorem C.3].

Theorem 6. *The graph in Fig. 4b is strongly connected, satisfies $g \geq 5$, and has no WT equilibria for every $\delta < \gamma_5 \approx 0.68233$.*

We conclude this section with a construction (Fig. 4c) that supports no 2-chase equilibria. It is based on two $C^5_{4,1}$ cycles offering profitable deviations in all directed cycles, at vertices 6 and 3. The proof of the following theorem is deferred to the full version of this paper [1, Theorem 6.4].

Theorem 7. *The graph in Fig. 4c is strongly connected, satisfies $g \geq 5$, only contains $C^5_{4,1}$ of the small unbalanced cycles and has no 2-chase equilibria for every $\delta < \gamma_5 \approx 0.68233$.*

Observe that the graphs in Figs. 4b and 4c are nonplanar by KURATOWSKI's theorem [12]: Fig. 4b contains a subdivision of $K_{3,3}$ (the complete bipartite graph with 3 vertices on each side) with the vertices $1, 2, 4$ and $3, 6, 8$ on the two sides of $K_{3,3}$; Fig. 4c contains a subdivision of $K_{3,3}$ with the vertices $1, 4, 7$ and $3, 6, 9$ on the two sides of $K_{3,3}$.

7 Outerplanar Graphs

Say that G is *outerplanar* if it is planar and all of its vertices are part of the unbounded face of G. In this section we prove that all strongly connected outerplanar graphs with $g \geq 4$ have both a WT and a 2-chase equilibria. We note that they do not necessarily have static equilibria, for example the 4-cycle. The proof is deferred to the full version [1, Lemma 7.1, Theorem 7.2].

Suppose that G is outerplanar and strongly connected. Fix some outerplanar embedding of G. As a planar graph, G consists of bounded faces $C_1, \ldots, C_k$ (which we also call *minimal (undirected) cycles*). These cycles are minimal in the sense that each C_i bounds exactly one face.

We now prove that G contains a well-directed minimal cycle. The proof of this property works for any planar graph. The idea is to start from some well-directed cycle, and carve out parts of it, while keeping the cycle at hand well-directed, until reaching a well-directed face.

Lemma 4. *All strongly connected planar graphs have a well-directed face.*

To prove the existence result, we pick a well-directed minimal cycle C that corresponds to a well-directed face, which exists by Lemma 4, and we define both the WT and 2-chase strategies of the players on C.

Theorem 8. *If G is outerplanar and strongly connected then it supports both a WT and a 2-chase equilibria.*

In light of this result and given the constructions of Sect. 6, we conjecture the following,

Conjecture 1. All strongly connected planar graphs with $g \geq 4$ have either a static, a 2-chase or a walking together equilibrium.

8 Computing Equilibria

For every pair of vertices, the value of minmax play when the players are situated at these vertices must verify a system of Bellman equations:

$$V_x(u,v) = \max_{s_x \in \Delta(\mathcal{B}_1(u))} \min_{s_y \in \Delta(\mathcal{B}_1(v))} \mathbb{E}_{(u',v') \sim s_x \times s_y} \left[(1-\delta) r_x(u',v') + \delta V_x(u',v')\right], \tag{2}$$

where we define $r_x(u,v) = \mathbb{1}_{(u,v)\in E} - \mathbb{1}_{(u,v)\in \hat{E}}$. This can be computed using value iteration [16], thus leading to efficient computation of optimal player strategies. Once V_x is computed for all pairs of vertices, all mixed equilibria can be deduced by computing the set of all minmaxes satisfying Eq. (2) for each state.

We now show that the same can be said of pure equilibria, using Algorithm 1. Recall the **strong product graph** $\tilde{G} \boxtimes \tilde{G}$ is defined as the graph with vertices V^2 and edges E' such that $((u,v),(u',v')) \in E'$ if and only if $u = v$ and $(u',v') \in \tilde{G}$ or $(u,v) \in \tilde{G}$ and $u' = v'$ or $(u,v) \in \tilde{G}$ and $(u',v') \in \tilde{G}$.

Algorithm 1. Algorithm computing pure equilibria

Require: The strong product graph $\tilde{G} \boxtimes \tilde{G}$.
Ensure: A subgraph $\mathcal{F}_\infty$ of $\tilde{G} \boxtimes \tilde{G}$ indicating possible states in a pure equilibrium.
Ensure: A labeling $\ell : V^2 \to V$ for each state of possible moves.

```
t ← 0
ℓ(u, v) ← B₁(u) for (u, v) ∈ V²
F₀ ← Ê
while F_t ≠ F_{t−1} do
    F_{t+1} ← F_t
    for (u, v) ∈ V² \ F_{t+1} do
        for u′ ∈ ℓ(u, v) do
            if ∃v′ ∈ B₁(v) | (u′, v′) ∈ F_t then
                ℓ(u, v) ← ℓ(u, v) \ {u′}
            end if
        end for
        if ℓ(u, v) = ∅ then
            F_{t+1} ← F_{t+1} ∪ {(u, v)}
        end if
    end for
    t ← t + 1
end while
return F_∞ = F_{t−1} ∪ {(v, u) | (u, v) ∈ F_{t−1}}
```

Algorithm 1 computes a subgraph $\mathcal{F}_\infty$ of the strong product graph $\tilde{G} \boxtimes \tilde{G}$, and pure equilibria are exactly the trajectories in the complementary of $\mathcal{F}_\infty$.[1] The proof of the following proposition is deferred to the full version [1, Proposition 8.1].

[1] A *trajectory* is a list of pairwise distinct nodes except for the last node. It is a path that ends with a cycle.

Proposition 3. *Pure* NASH *equilibria all have payoff* 0 *at every round after the initial condition, and can be computed using Algorithm 1.*

Acknowledgement. The authors would like to thank Laurel Britt and Stephen Morris for valuable insights.

Disclosure of Interests. The authors have no competing interests to declare that are relevant to the content of this article.

References

1. Arthaud, F., Orzech, E., Rinard, M.: Edge-dominance games on graphs. CoRR abs/2407.07785 (2024). https://doi.org/10.48550/ARXIV.2407.07785
2. Bonato, A., Chiniforooshan, E., Prałat, P.: Cops and robbers from a distance. Theor. Comput. Sci. **411**(43), 3834–3844 (2010). https://doi.org/10.1016/J.TCS.2010.07.003
3. Bonato, A., Nowakowski, R.J.: The Game of Cops and Robbers on Graphs. Student Mathematical Library, vol. 61. American Mathematical Society (2011)
4. Dürr, C., Thang, N.K.: Nash equilibria in Voronoi games on graphs. In: Arge, L., Hoffmann, M., Welzl, E. (eds.) ESA 2007. LNCS, vol. 4698, pp. 17–28. Springer, Heidelberg (2007). https://doi.org/10.1007/978-3-540-75520-3_4
5. Fournier, G.: General distribution of consumers in pure hotelling games. Int. J. Game Theory **48**(1), 33–59 (2019). https://doi.org/10.1007/S00182-018-0648-4
6. Gallai, T.: Elementare relationen bezüglich der glieder und trennenden punkte von graphen. A Magyar Tudományos Akadémia Matematikai Kutató Intézetének közleményei **9**(1–2), 235–236 (1964)
7. Hamidoune, Y.O.: On a pursuit game on Cayley digraphs. Eur. J. Comb. **8**(3), 289–295 (1987). https://doi.org/10.1016/S0195-6698(87)80034-3
8. Harary, F., Prins, G.: The block-cutpoint-tree of a graph. Publ. Math. Debrecen **13**, 103–107 (1966)
9. Hotelling, H.: Stability in competition. Econ. J. **39**(153), 41–57 (1929). https://doi.org/10.2307/2224214
10. Kehagias, A., Konstantinidis, G.: Some game-theoretic remarks on two-player generalized cops and robbers games. Dyn. Games Appl. **11**(4), 785–802 (2021). https://doi.org/10.1007/S13235-021-00385-0
11. Konstantinidis, G., Kehagias, A.: Simultaneously moving cops and robbers. Theor. Comput. Sci. **645**, 48–59 (2016). https://doi.org/10.1016/J.TCS.2016.06.039
12. Kuratowski, C.: Sur le problème des courbes gauches en topologie. Fundam. Math. **15**(1), 271–283 (1930)
13. Luckraz, S.: A survey on the relationship between the game of cops and robbers and other game representations. Dyn. Games Appl. **9**(2), 506–520 (2019). https://doi.org/10.1007/S13235-018-0275-5
14. Quilliot, A.: Jeux et pointes fixes sur les graphes. Thèse de 3ème cycle, Université de Paris VI (1978)
15. Serra, D., Revelle, C.: Competitive location in discrete space. Economics Working Papers 96, Department of Economics and Business, Universitat Pompeu Fabra (1994). https://ideas.repec.org/p/upf/upfgen/96.html
16. Shapley, L.S.: Stochastic games. Proc. Natl. Acad. Sci. **39**(10), 1095–1100 (1953). https://doi.org/10.1073/pnas.39.10.1095

Playing Repeated Games with Sublinear Randomness

Farid Arthaud(✉)

MIT CSAIL, Cambridge, MA 02139, USA
farto@csail.mit.edu

Abstract. The seminal result of NASH in game theory states that any normal-form game has a NASH equilibrium if each player can randomize their strategy. The assumption that players can randomize arbitrarily is non-trivial, as true randomness might be scarce or costly and humans are known to have difficulty generating truly random sequences. In a repeated game, the assumption that players are unconstrained in their capability to randomize their strategies is particularly strong if the amount of random bits required to play the repeated game scales linearly with the number of repetitions.

We identify conditions on a normal-form game under which, if players have a limited capability to randomize, certain NASH equilibria of its finitely repeated version cannot be played. We provide a complete characterization of normal-form games for which there exists NASH equilibria of its finitely repeated version using $O(1)$ randomness, closing an open question posed by Budinich and Fortnow [3] (EC '11) and Hubáček, Naor and Ullman [8] (SAGT '15, TCSys '16). Moreover, we prove a 0–1 law for randomness in repeated games, showing that any repeated game either has $O(1)$-randomness NASH equilibria, or all of its NASH equilibria require $\Omega(n)$ randomness. Our techniques are general and naturally characterize the payoff space of sublinear-entropy equilibria, and could be of independent interest to the study of players with other bounded capabilities in repeated games.

Keywords: Randomness · Repeated games · Bounded entropy

1 Introduction

Randomization is a fundamental concept in computer science, and the necessity of randomness is an important question in several areas of the field (for instance, $\mathsf{P} \stackrel{?}{=} \mathsf{BPP}$). The seminal result of NASH in game theory states that any normal-form game has a NASH equilibrium if each player can randomize their strategy. As noted in previous work [3,7,8], the assumption that players can randomize arbitrarily is non-trivial, as true randomness might be scarce or costly and humans are known to have difficulty generating truly random sequences.

Some NASH equilibria of a repeated game require large amounts of entropy from the players to execute. For instance, a straightforward NASH equilibrium

G. Schäfer and C. Ventre (Eds.): SAGT 2024, LNCS 15156, pp. 258–276, 2024.
https://doi.org/10.1007/978-3-031-71033-9_15

of any repeated game is to repeat a NASH equilibrium of the stage game at each round, irrespective of the outcomes of previous rounds. When the chosen NASH equilibrium of the stage game requires a player to randomize, this equilibrium requires this player to use an amount of entropy that grows linearly with the number of repetitions. However, not all equilibria of the repeated game are of this form: in some cases, the repeated game has equilibria in which all players use only constant randomness. From here, several natural questions arise,

Question 1. When do there exist NASH equilibria of the repeated game in which all players use sublinear entropy?

Question 2. When do all equilibria of the repeated game require all players to use linear entropy?

Question 3. What are the answers to the two previous questions when only restricting the entropy used by a subset of the players?

Question 4. How does restricting randomization affect the payoffs of the NASH equilibria of the repeated game?

Budinich and Fortnow [3] first studied the question of equilibria with bounded randomness in repeated games in the case of the two-player matching pennies game. They find that each player needs n independent random bits to play any equilibrium of the repeated game, where n is the number of repetitions. This further motivates Questions 1 and 2 above: it shows that matching pennies fits in the case specified by Question 2 and shows there are simple games where the answer to Question 1 is negative.

Hubáček, Naor and Ullman [8] further extend this study to general repeated games with any number of players. They identify a sufficient condition on the stage game for there to exist an equilibrium of the repeated game requiring $O(1)$ entropy for all players, therefore providing a sufficient condition for Question 1. They also identify a sufficient condition under which all NASH equilibria of the repeated game require $\Omega(n)$ entropy for each player, therefore providing a sufficient condition for Question 2. This latter condition is satisfied by all two-player zero-sum games where all NASH equilibria require all players to randomize, in particular the matching pennies game.

These results show that there are at least two regimes for the required entropy to play a NASH equilibrium of a repeated game (constant or linear). Moreover, both Budinich and Fortnow [3] and Hubáček, Naor and Ullman [8] posed the open question of whether a characterization of games satisfying Question 1 can be found.

1.1 Our Results

We close an open question posed by Budinich and Fortnow [3] and Hubáček, Naor and Ullman [8] by completely characterizing normal-form games for which there are NASH equilibria of the repeated game in which all players use $O(1)$ randomness as a function of the number of repetitions (Theorem 6).

As an important consequence of our proof of this characterization, we show that when $O(1)$ randomness is not sufficient, then in all equilibria at least one player uses $\Omega(n)$ entropy. As such, the existence of sublinear-randomness NASH equilibria of a repeated game implies the existence of $O(1)$-randomness NASH equilibria, meaning that Question 1 and its analog for constant-randomness equilibria always have the same answer. The only previously known result on sublinear-entropy NASH equilibria in general repeated games assumes that one-way functions do not exist and that players are computationally bounded [8].

This closes Question 1 above. While precise statements require more definitions, we informally summarize the stated results so far.

Theorem 1 (Informal version of Theorem 6). *For a normal-form game G, there are constant-randomness* NASH *equilibria (or equivalently, sublinear-randomness* NASH *equilibria) of its repeated games if and only if there exists a subset S' of the pure strategy profiles of G and a convex combination γ_s for $s \in S'$ such that,*

1. *the payoff vector $\sum_{s \in S'} \gamma_s u(s)$ is at least the pure-strategy minmax of each player;*
2. *and for each player $j \in A$, either*
   (a) j is best responding in every strategy profile in S';
   (b) or j can be punished by a finite number of repetitions of G.

We also close Question 1 in the case of repeated games with observable distributions (Theorem 5), a setting in which the distribution chosen by each player to randomize their actions is revealed after reach round (Definition 9). Regarding Question 3, for any subset B of the players, we provide a sufficient condition and a necessary condition for there to exist a NASH equilibrium of the repeated game in which all players in B use $O(1)$ entropy (Theorem 7). Another consequence of our results is that they naturally characterize the set of payoffs to which the average payoff of a sublinear-entropy NASH equilibrium can converge, closing Question 4.

Two-Player Games. For two-player games, our characterization of Question 1 (Theorem 3) only requires testing the feasibility of a polynomial-size linear program over the stage game G and testing for the existence of a NASH equilibrium of G satisfying certain linear inequalities. This means the condition is decidable and is even in NP (this is also true for the observable distributions case with any number of players mentioned above).

Theorem 2 (Informal version of Theorem 3). *For a two-player game G, there are constant-randomness* NASH *equilibria of G's repeated game (or equivalently, sublinear-randomness* NASH *equilibria) if and only if G has a pure* NASH *equilibrium or there exists a convex combination γ_s over pure strategy profiles s of G such that,*

1. *the payoff vector $\sum_s \gamma_s u(s)$ is at least the pure-strategy minmax of each player;*

2. *and G has a* Nash *equilibrium that is strictly better than the minmax of one of the players.*

Still in the case of two-player games, we also close Questions 2, 3 and 4: we provide a complete characterization of two-player games in which both players each require $\Omega(n)$ entropy to play any equilibrium (Theorem 4), a complete characterization of two-player games in which one player only needs sublinear entropy (in the full version [1, Theorem D.2]) and a characterization of asymptotically achievable payoffs (Corollary 2). All of these conditions are in NP.

Note above that Budinich and Fortnow [3] deal with random bits whereas Hubáček, Naor and Ullman [8] bound entropy. We work with entropy (Definition 5) as in Hubáček, Naor and Ullman [8]. In Subsect. 2.1 we introduce a computational model that precisely relates the expected random bits used by an algorithm implementing a player's strategy and the entropy as defined by Hubáček, Naor and Ullman [8]. A difference between Hubáček, Naor and Ullman [8] and our results is the distinction between *effective entropy* and *total entropy* of an equilibrium, which is explained in Subsect. 1.3. We extend our main result to effective entropy in the full version of the paper [1, Theorem F.4].

Roadmap. The remainder of this section contains an overview of our techniques and further related work. Section 2 contains the main definitions and introduces our computational model under which the entropy characterization relates to exact play of equilibria using balanced random bits. Section 3 contains proofs of the main technical tools we use throughout the paper, both for sufficient and necessary conditions. We progressively build up towards the most general case by presenting our results on sublinear-entropy Nash equilibria of repeated games in Sect. 4 for two-player games, Sect. 5 for m-player games with observable distributions, and Sect. 6 for general m-player games.

1.2 Our Techniques: Playing with Sublinear Randomness

Our starting point is the following sufficient condition, which we then weaken into a necessary (and sufficient) condition.

Proposition 1 (Informal version of Proposition 5). *If there exists a convex combination $(\gamma_s)_{s \in S}$ over pure strategies s of G such that the payoff vector $\sum_s \gamma_s u(s)$ is at least the pure-strategy minmax of each player, and every player either is best responding in every strategy s where $\gamma_s > 0$ or has a* Nash *equilibrium that has a payoff strictly better than their minmax value, then there exists $O(1)$-entropy* Nash *equilibria of G's repeated game.*

Note this proposition generalizes [8, Theorem 2] in two ways: (i) it does not require all players to have a strictly individually rational NE and (ii) it does not require the coefficients γ_s to be rationals. When $\gamma_s \in \mathbb{Q}$ for all s (as assumed by Hubáček, Naor and Ullman [8]), a construction similar to the one from folk theorems such as Benoît and Krishna [2] is possible. These constructions fail otherwise, which requires us to construct a different type of equilibria, we explain this distinction in more detail in Sect. 3.

We use two main tools to weaken this statement into a necessary condition. The first shows that the last assumption in the proposition above (that each player has a NASH equilibrium strictly better than their minmax) must be true, not of all players, but only of at least one player.

Proposition 2 (Informal version of Proposition 6). *Assume G does not have a pure* NASH *equilibrium, and has* NASH *equilibria of its repeated game using $o(n)$ entropy. Then G has a* NASH *equilibrium in which at least one player's payoff is strictly greater than their minmax.*

The proof uses the compactness of NASH equilibria of G to find a round where a player is not best responding. For this player to not be incentivized to deviate to a better response at that round, there must be a future round that rewards/punishes that player depending on whether they deviate or not. By taking the latest such round, we show it must be a NASH equilibrium that is strictly better than that player's minmax.

Let v_i be the pure-strategy minmax of player i, i.e. the worst punishment that can be imposed on player i without randomizing (see Definition 1). The second tool to weaken the above Proposition 1 is a lemma that bounds the long-term punishments that can be imposed on a player for deviating (see Definition 8 for a formal definition), when all players are bounded in their entropy.

Lemma 1 (Informal version of Lemma 5). *For any small enough $\varepsilon > 0$ and any n-round punishment of player i in which all players use at most $O(f(n))$ entropy, the average payoff of player i is at least $v_i - O\left(\frac{f(n)}{\varepsilon n}\right) - O(\varepsilon)$.*

This is important because it means that in a low-entropy equilibrium, the on-path payoffs must converge to a value higher than the pure minmax of each player. This ties back to our sufficient conditions, where the only payoff profiles we were able to approximate were those that were better than each player's pure minmax. The idea behind the proof is that rounds in the punishment either use a low amount of entropy or a high amount of entropy: low-entropy rounds cannot have payoffs too far off from the pure minmax, whereas there cannot be too many high-entropy rounds.

This bound allows us to prove the following lemma, which shows that the average payoff of any sublinear-entropy equilibrium must be at least the pure-strategy minmax. Intuitively, if the worst punishment possible approximately imposes the payoff v_i on player i, then their average payoff on-path must be at least v_i in the long run, otherwise player i could gain a linear advantage from deviating as the punishment would be better than on-path payoffs.

Lemma 2 (Informal version of Lemma 6). *If G has* NASH *equilibria of its repeated game using $o(n)$ randomness, then there exists a convex combination γ_s over pure strategies s of G such that the payoff vector $\sum_s \gamma_s u(s)$ is at least the pure minmax of each player.*

Note that this recovers the first assumption made in Proposition 1 above, by showing it is true as soon as there are sublinear-entropy Nash equilibria. In the two-player case, we show in Theorem 3 that this condition and the one in Proposition 2 are together sufficient.

The proofs for repeated games with observable distributions (Sect. 5) and for general repeated games (Sect. 6) rely on similar ideas, but require stronger tools to identify punishments in the necessary condition. In particular, Proposition 2 is no longer sufficient on its own, and the proofs of Theorems 5 and 6 must instead produce strategy profiles where a subset of players are best responding and a particular player's payoff is strictly better than their minmax. This results in an ordering of the players in their rewards/punishments: after a player has been punished, they must not have any other opportunities for deviation during the punishment of other players.

1.3 Further Related Work

There is one difference between the setting in which Hubáček, Naor and Ullman [8]'s results hold and our setting, namely the definition used for entropy of an equilibrium. The sufficient condition for $O(1)$ randomness equilibria in [8, Theorem 2] bounds *effective entropy* rather than total entropy: instead of measuring the amount of entropy along any path of the game, they only measure entropy along paths which are sampled with non-zero probability under equilibrium play (see Definition 5 for formal definitions). Effective entropy could model a situation in which linear entropy is costly yet achievable (players agree that they could hurt each other using linear entropy so they won't deviate), but it does not model situations where players are unable to produce linear amounts of entropy. Our sufficient conditions are therefore stronger in this regard, as bounding total entropy implies the same bound for effective entropy, and Proposition 3 does so under similar assumptions to [8, Theorem 2]. Our techniques are sufficiently general that our characterizations extend to the case of effective entropy naturally: the full version of the paper [1, Appendix F.2] contains an analog of our most general result for effective entropy.

Costly Randomness. Halpern and Pass [7] consider adding a cost of computation to games in a setting where players are modeled by Turing machines. They show that Nash equilibria do not necessarily exist when players pay a cost for randomness, but that whenever randomness is free (but computation can be costly) Nash equilibria always exist.

Off-Equilibrium Play with Bounded Randomness. There is a line of research on the maxmin payoff in repeated games where $\Omega(n)$ entropy is needed but only sublinear entropy is available to one player, and specifically in two-player zero-sum games. Budinich and Fortnow [3] had already shown that in repeated matching pennies, if one player's strategy uses $(1-\delta)n$ random bits there exists a deterministic strategy resulting in a payoff of δn against them. This results in an exact characterization of approximate Nash equilibria and required randomness to play them in matching pennies. Both of these results were later shown [8] to be consequences of Neyman and Okada [12]'s results, which characterize the maxmin payoff of a player with bounded entropy in a repeated two-player zero-sum game (against an opponent with access to unbounded entropy). Gossner and Vieille [6] in turn generalize this result by assuming that the player with bounded entropy only has access to realizations of random variables $X_t \sim \mathcal{L}(X)$ of entropy $h = \mathcal{H}(\mathcal{L}(X))$ at each round t, whose *arbitrary distribution* $\mathcal{L}(X)$ is also publicly known. Follow-up work by Valizadeh and Gohari [13] further extends this work to a setting where the source of entropy X can also be leaked to the adversary, and study the non-asymptotic behavior of the maxmin value [14]. Kalyanaraman and Umans [9] undertake this problem with a learning flavor: in two-player zero-sum games with payoffs in $\{0, 1\}$, they provide an algorithm using $O(\log \log n + \log(1/\varepsilon))$ random bits (with high probability) yielding a $O\left(1/\sqrt{n} + \varepsilon\right)$ additive regret term against an adaptive adversary. All of this work differs from our results in that it characterizes off-equilibrium outcomes in games where players are bounded by their randomness, whereas we characterize the amount of randomness players require to play equilibria.

Computational Nash Equilibria with Bounded Randomness. Both Budinich and Fortnow [3] and Hubáček, Naor and Ullman [8] also study *computational* Nash *equilibria* in repeated games, the former for matching pennies and the latter in two-player zero-sum games. A computational Nash equilibrium is an approximate Nash equilibrium that can be implemented using polynomial-size circuits. Hubáček, Naor and Ullman [8] show that if one-way functions do not exist, then there are no computational Nash equilibria using sublinear entropy in two-player zero-sum games with no weakly dominant pure strategies (recall that they show there are no constant-entropy equilibria in the computationally unbounded case).

2 Preliminaries

Let G be a normal-form game, with a set $A = [\![1; m]\!]$ of $m \geq 2$ players, and denote S_i the set of strategies of player i and $S = S_1 \times \cdots \times S_m$ the entire strategy space. We assume that $|S_i| \in \mathbb{N}^* \setminus \{1\}$ for all i. σ denotes a (potentially mixed) strategy profile $(\sigma_1, \ldots, \sigma_m)$, where σ_i is the distribution with which player i samples over S_i. When referring to a strategy profile σ_n that already has a subscript, we denote $(\sigma_n)_i$ the strategy profile of player i in σ_n. When strategies are known to be pure (i.e. the actions are deterministic) we use the letter $s \in S$. σ_{-i} designates the strategy profile of all players except player i in σ, i.e. $(\sigma_1, \ldots, \sigma_{i-1}, \sigma_{i+1}, \ldots, \sigma_m)$. The utility of player i is $u_i : S \to [0; 1]$ (we assume all payoffs are normalized without loss of generality). This is then extended to mixed strategies in the natural way, $u_i(\sigma) = \sum_{s \in S} \left(\prod_{j=1}^{m} \sigma_j(s) \right) u_i(s)$. We write $u_i(\sigma_j, \sigma_{-j}) = u_i(\sigma)$, and $u(s) = (u_1(s), \ldots, u_m(s))$.

Definition 1. *The* ***minmax*** *of player i is the best payoff player i can ensure under any play by the other players,*

$$\tilde{v}_i = \min_{\sigma_{-i}} \max_{\sigma_i} u_i(\sigma) = \min_{\sigma_{-i}} \max_{s_i \in S_i} u_i(s_i, \sigma_{-i}). \tag{1}$$

The ***pure minmax*** *is the best payoff player i can ensure assuming other players only play pure strategies,*

$$v_i = \min_{s_{-i} \in \prod_{j \neq i} S_j} \max_{\sigma_i} u_i(\sigma_i, s_{-i}) = \min_{s_{-i} \in \prod_{j \neq i} S_j} \max_{s_i \in S_i} u_i(s). \tag{2}$$

Note the pure minmax of a player is always larger than their minmax.

Definition 2. *The set of* ***feasible payoff profiles*** *(or feasible payoffs) is the convex hull of $\{(u_1(s), \ldots, u_m(s)),\ s \in S\} \subset [0;1]^m$. We say that a particular feasible payoff profile p is* ***supported by*** *strategies $(s_1, \ldots, s_r) \in S^r$ and coefficients $(\gamma_1, \ldots, \gamma_r) \in \left(\mathbb{R}_+^*\right)^r$ if $\sum_{k=1}^r \gamma_k u(s_k) = p$ and $\sum_{k=1}^r \gamma_k = 1$. We will also say that p is* ***supportable by*** *$\{s_1, \ldots, s_r\} \subseteq S$ in this case.*

A payoff being feasible does not necessarily imply there exists a mixed strategy σ achieving it, since mixed strategies only span product distributions. Note the difference with [8, Definition 4]'s definition of feasible payoff profiles, where they only consider convex combinations with coefficients in $\mathbb{Q}$.

Definition 3. *A feasible payoff profile is called* ***$\mathbb{Q}$-feasible*** *if it is supported by strategies and coefficients such that $\gamma_k \in \mathbb{Q}$ for all k (when such an assumption is not made, we sometimes refer to the payoff profile as $\mathbb{R}$-feasible).*

A payoff profile p is ***individually rational for player*** *i if its payoff is larger than its minmax, $p_i \geq \tilde{v}_i$, and* ***individually rational*** *if $p \geq \tilde{v}$. Strictly individually rational means $p > \tilde{v}$.*

p is ***purely individually rational for player*** *i if its payoff is larger than its pure minmax, $p_i \geq v_i$ (and the previous variants extend).*

The extensive-form game where G is repeated $n \in \mathbb{N}^*$ times is denoted G^n and is called the n**-repeated game** of G, and G is called the **stage game** of G^n. A **history of play** during the first $k \in [\![1; n-1]\!]$ rounds is a k-tuple $h \in S^k$. When $h \in S^{n-1}$ is a history of play for all first $n-1$ rounds, subscripts denote its truncation to its first elements: h_k is the history of play for the first k rounds of h. A strategy profile for G^n is a mapping from histories of play to mixed strategy profiles of the stage game G for all players. They are also denoted by the symbol σ, and $\sigma(h_k)$ refers to the strategy profile of G played at round $k+1$ if the history of play is h_k in the first k rounds. The symbol $\emptyset$ denotes the empty history, and $\sigma(\emptyset)$ is therefore the strategy profile played at the first round in σ. The concatenation of two histories h and h' is denoted $h \cdot h'$.

Definition 4. *For a given strategy profile σ of a repeated game, we call the* ***on-path tree*** *$T(\sigma)$ the tree of all histories that are sampled with non-zero probability by σ: its root is $\emptyset$ and the children of some history $h \in S^k$ are the $h \cdot s \in S^{k+1}$ such that s is played with non-zero probability in $\sigma(h)$. We denote $H(\sigma)$ the set of leaves of $T(\sigma)$, and $\mathbb{P}_\sigma(h)$ the probability of history $h \in S^k$ when all players play according to σ.*

Recall that for a discrete distribution p over a set S, its SHANNON base-2 entropy is defined as $\mathcal{H}(p) = -\sum_{s\in S} p(s)\log_2(p(s))$.

Definition 5 ([8, Definition 10]). *For a given strategy σ of the repeated game G^n, we define its **entropy** (or total entropy) for player i as,*

$$\mathcal{H}_i(\sigma) = \mathcal{H}_i(\sigma_i) = \max_{h\in S^{n-1}} \left[\mathcal{H}(\sigma_i(\emptyset)) + \sum_{k=1}^{n-1} \mathcal{H}(\sigma_i(h_k))\right].$$

*Its **effective entropy** only considers histories that happen with non-zero probability during play,*

$$\mathcal{H}_i^{\mathsf{eff}}(\sigma) = \max_{h\in H(\sigma)} \left[\mathcal{H}(\sigma_i(\emptyset)) + \sum_{k=1}^{n-1} \mathcal{H}(\sigma_i(h_k))\right].$$

Definition 6. *The **amount of randomness** or **amount of entropy** of an equilibrium σ of a repeated game is $\sum_{i=1}^{m} \mathcal{H}_i(\sigma)$.*

2.1 Computational Model

There are three immediate issues when the framework above is applied to computationally bounded players (such as TURING-equivalent players):

(i) All NASH equilibria of the stage game G can have irrational coefficients in all generality (even when payoff matrices have coefficients in $\mathbb{Q}$) and cannot immediately be efficiently represented;[1]
(ii) As a second consequence, it is not clear that sampling from a player's distribution for such an equilibrium can be done exactly (non-exact play could be exploited by other players) and;
(iii) Rather than sampling from one distribution with entropy $\mathcal{H}_i^{\mathsf{eff}}(\sigma)$, a player might be sampling many times from conditionally independent very low-entropy distributions (at each round), which naïvely requires a much larger number of expected bits (because of overhead and uncertainty on what distributions it will have to sample from in the future).

We therefore have to specify a computational model that addresses all of these issues. To represent NASH equilibria or more generally any mixed strategy profile, we assume that players have access to an oracle that can specify the probabilities of their strategy up to any finite precision at each round. A particular case of this model could be that agents have all approximately computed the strategy profile themselves, and are able to compute approximations of the distributions up to any precision efficiently.[2]

[1] However, since the set of NEs is a semialgebraic variety, the TARSKI-SEIDENBERG principle ensures the existence of an efficiently representable NE satisfying any linear conditions provided there exists such a NASH equilibrium [11, Theorem 1].

[2] As per the previous footnote, if all played NASH equilibria (and strategy profiles) are algebraic this would simply mean computing approximations to exactly-represented numbers.

For point (ii), to sample from a NASH equilibrium represented in this way, the players can use inversion sampling and request the oracle for more bits as required. The full version of the paper [1, Appendix B] contains an inversion sampling algorithm for this oracle model, and shows it requires a finite number of bits in expectation (with a geometrically-decaying tail), a finite number of requests to the oracle in expectation, and produces an exact sample.

Regarding point (iii), throughout the paper we will ensure that whenever there exists an equilibrium of a repeated game using $O(1)$ entropy, it will also use mixing in $O(1)$ rounds along each path. This ensures that it can be played with a constant expected amount of random bits using the inversion sampling algorithm from point (ii) at each round. Note that without this condition, using inversion sampling could require $\Omega(n)$ expected random bits to play some $O(1)$-entropy equilibria of the repeated game due to overhead.

Definition 7. *Analogously to the definition of entropy of a repeated* NASH *equilibrium* $\mathcal{H}(\sigma)$, *we define its* ***worst-case expected random bits***[3] *for player* i,

$$\mathsf{Bits}_i(\sigma) = \max_{h \in S^{n-1}} \min_{N} \mathbb{E}(N(\sigma_i(\emptyset), \sigma_i(h_1), \ldots, \sigma_i(h_n))),$$

where N *spans over the number of expected bits used by algorithms terminating with probability* 1 *and producing an exact sample of the distribution.*

Theorem 2.2 of Knuth and Yao [10] and its corollary yield that $\mathcal{H}_i(\sigma) \leq \mathsf{Bits}_i(\sigma) < \mathcal{H}_i(\sigma) + 2$. This completes our computational model as it means that NASH equilibria of the repeated game requiring $\Omega(n)$ entropy on their worst-case path will also induce an algorithm using $\Omega(n)$ bits in expectation at least along their worst-case path.

3 Playing with Sublinear Randomness

3.1 Tools for Sufficient Conditions

This first proposition shows how a constant-entropy equilibrium can be built under conditions similar to folk theorems, when the payoff profile is $\mathbb{Q}$-feasible. As the punishment phase is the only place where randomness is used, the built equilibrium uses constant entropy for all players. In the interest of space, the proof is deferred to the full version of the paper [1, Proposition C.1].

Proposition 3. *If* G *has a* $\mathbb{Q}$*-feasible payoff profile* p *that is purely individually rational for all players and every player has a* NASH *equilibrium that is strictly individually rational for them, then there exists a* NASH *equilibrium of the repeated game requiring* $O(1)$ *entropy.*

[3] Note that this concept is distinct from the expected random bits conditioned on a worst-case path being taken for several reasons: (i) it could be ill-defined, as the worst-case path for player i could involve deviations by player i and (ii) the definitions of total and effective entropy do not condition on the path being taken, and would be very different if they did.

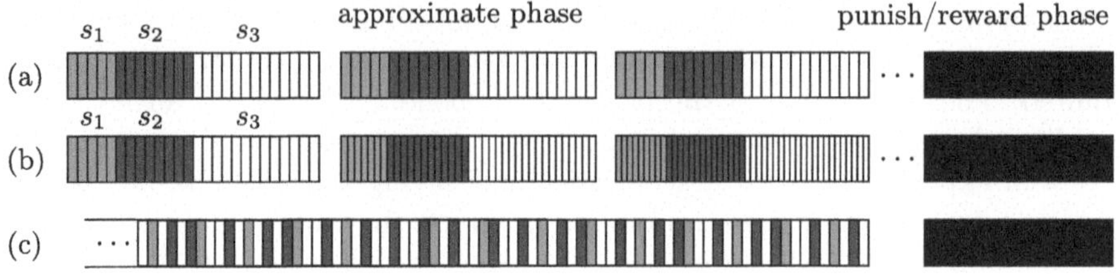

Fig. 1. Possible approximation schemes for a payoff profile consisting of three strategies s_1, s_2 and s_3

The construction is illustrated in Fig. 1 (a): a cycle is repeated, in which each strategy is played proportionally to its weight in the payoff profile (which is possible as it is $\mathbb{Q}$-feasible). This is followed by a punishment/reward phase, constructed similarly to folk theorems using the strictly individually rational NASH equilibrium of each player.

Now, if the payoff profile is $\mathbb{R}$-feasible instead of $\mathbb{Q}$-feasible, such an approximation scheme can no longer be used: the fixed sequence being repeated can only approximate $\mathbb{Q}$-feasible payoff profiles. The equilibrium from Proposition 3 is illustrated in Fig. 1 in line (a); a possible adaptation to $\mathbb{R}$-feasible payoff profiles is illustrated in line (b): by using finer and finer approximations of the coefficients of the $\mathbb{R}$-feasible payoff profile and concatenating them as previously, one obtains a first phase that approximates the payoff profile arbitrarily well. However, an important property is lost: if s_3 in Fig. 1 is strictly worse than a particular player's pure minmax, that player could profit a lot by deviating in the last (and finest) approximation (since they will avoid many rounds of s_3). As the approximations grow in length arbitrarily, the reward/punishment phase would need to be arbitrarily long to avoid these types of deviations and the equilibrium would no longer be $O(1)$-entropy.

To remedy this, we show the following lemma which constructs an approximation phase without any fixed repetitions of cycles. The construction, illustrated in line (c) of Fig. 1, starts from the end, and ensures that any suffix of the first phase is not too far from the pure minmax of each player: this prevents any player from having a profitable deviation (beyond a constant profit, compensated by the punishment phase).

Lemma 3. *If p is a $\mathbb{R}$-feasible payoff profile of G, then for all $n \in \mathbb{N}^*$ there exists a series of strategy profiles $(s_1, \dots, s_n)$ such that for all $k \in [\![1; n]\!]$, the average payoff of each player i over the last k rounds is at least $p_i - |S|/k$.*

Proof. The following lemma is shown in the full version [1, Lemma A.3].

Lemma 4. *For any vector $(x_1, \dots, x_r) \in \left(\mathbb{R}_+^*\right)^r$ such that $\sum_{i=1}^r x_i = 1$, there exists a sequence $(a_k)_{k \in \mathbb{N}}$ such that $a_k \in \mathbb{N}^r$ and $\sum_{j=1}^r a_{k,j} = k$ and $a_{k+1} \geq a_k$ and $a_{k,j} \geq \lfloor k x_j \rfloor$ for all $k \in \mathbb{N}$ and $j \in [\![1; r]\!]$.*

Let $\sum_{j=1}^r x_j u(y_j) = p$ be a support for p, where $y_j \in S$ for all j. Build the series of integer vectors $(a_k)_{k \in \mathbb{N}}$ according to the construction of Lemma 4. For a fixed

$n \in \mathbb{N}^*$, define for $k \in [\![1; n]\!]$ the series of strategies $s_k = y_j$ where j is such that $\mathbb{1}_j = a_{n-k+1} - a_{n-k}$.

The strategies played in the last k rounds are $s_{n-k+1}, \ldots, s_n$. The number of times each of the strategies $y_1, \ldots, y_r$ appear in these rounds is the vector,

$$(a_k - a_{k-1}) + (a_{k-1} - a_{k-2}) + \cdots + (a_1 - a_0) = a_k.$$

Therefore, the average payoff for the last k rounds is the matrix product,

$$\frac{a_k}{k} \cdot \begin{pmatrix} u(y_1) \\ \vdots \\ u(y_r) \end{pmatrix} \geq \frac{\lfloor kx \rfloor}{k} \cdot \begin{pmatrix} u(y_1) \\ \vdots \\ u(y_r) \end{pmatrix} \geq \frac{kx - \mathbb{1}}{k} \cdot \begin{pmatrix} u(y_1) \\ \vdots \\ u(y_r) \end{pmatrix} \geq p - \frac{r}{k}\mathbb{1}. \qquad \square$$

We now use this result to extend Proposition 3 to $\mathbb{R}$-feasible payoff profiles.

Proposition 4. *If G has an $\mathbb{R}$-feasible payoff profile p that is purely individually rational for all players and every player has a* Nash *equilibrium that is strictly individually rational for them, then there exists a* Nash *equilibrium of the repeated game requiring $O(1)$ entropy.*

Proof. The equilibrium used follows the same structure as the one from Proposition 3, except the first phase is replaced by the series from Lemma 3. The second phase consists of $\left\lceil \frac{|S|+1}{\delta_i} \right\rceil$ repetitions of each player's strictly individually rational NE σ_i, similarly to Proposition 3. The total entropy is still constant independently from the length of the first phase.

It is left to show that it is indeed a Nash equilibrium. We claim that any player deviating during the first phase can gain at most $|S|$ in expected payoff compared to their pure minmax over the entire first phase of play. Indeed, if player l deviates k rounds before the end of the first phase, their maximum benefit over the rest of the phase is $kv_l - k\left(p_l - |S|/k\right) \leq |S|$ by Lemma 3. Adding to this the benefit from deviating at the round where they deviate, their benefit is indeed upper bounded by $|S| + 1$. Following the same proof as in Proposition 3, it follows that the second phase contains sufficient incentives to compensate this benefit, and therefore this is indeed a Nash equilibrium. $\square$

We finally weaken the assumption that each player can be punished with a Nash equilibrium, remarking that some players may not need to be punished. The proof is deferred to the full version of the paper [1, Proposition C.4].

Proposition 5. *If G has an $\mathbb{R}$-feasible purely individually rational payoff profile p supported by a convex combination of strategy profiles $\sum_{k=1}^{r} \gamma_k s_k$ such that for each player i, either they have a strictly individually rational* Nash *equilibrium σ_i or they are best responding in all of the s_k, then the repeated games G^n have* Nash *equilibria requiring $O(1)$ entropy.*

3.2 Tools for Necessary Conditions

We now prove a first necessary condition, showing there must exist a strictly individually rational NASH equilibrium for at least one player as soon as there are NASH equilibria of the repeated game using sublinear entropy. This proposition is similar to the contraposition of [8, Theorem 1], with the two differences that it only requires no pure NASH equilibria rather than all NASH equilibria being mixed for all players, and it requires $o(n)$ randomness for all players rather than for at least one player. Our proof technique is different from theirs, and is reused in various necessary conditions throughout the paper. The intuition is to find a round where a player is not best responding on-path: at this round, the player must be incentivized not to deviate to a better response. We prove the only way this can happen is when a strictly individually rational NASH equilibrium exists for that player (and is played in a future round). The proof is deferred to the full version of the paper [1, Proposition C.5].

Proposition 6. *Assume G does not have a pure* NASH *equilibrium, and has* NASH *equilibria of its repeated game using $o(n)$ entropy. Then G has a* NASH *equilibrium that is strictly individually rational for at least one player.*

Our second necessary condition concerns the existence a purely individually rational payoff profile. We begin by defining punishments.

Definition 8. *An n-round punishment σ of player i in G is a strategy profile of the repeated game G^n such that player i cannot benefit from deviating to any other σ'_i, i.e. $\forall \sigma'_i$, $u_i(\sigma) \geq u_i(\sigma'_i, \sigma_{-i})$.*

The first step is to show a lower bound on the worst punishment that can be inflicted on a player with limited entropy. This is important because it means that in a low-entropy equilibrium, the on-path payoffs must converge to a value higher than this lower bound – otherwise, a player could deviate at the beginning of the game, and the low-entropy punishment would be higher than the on-path payoff! This ties back to our sufficient conditions, where the only approximable payoff profiles were those that were purely individually rational.

Lemma 5. *For any $\varepsilon \in]0; 1 - 1/e]$ and any n-round punishment of player i using at most $O(f(n))$ entropy the average payoff of player i is at least*

$$v_i - O\left(\frac{f(n)}{\varepsilon n}\right) - O(\varepsilon). \tag{3}$$

The proof is deferred to the full version of the paper [1, Lemma C.7]. The main idea of the proof is to divide rounds (or rather, nodes of the game tree) into high-entropy rounds and low-entropy rounds. The high-entropy rounds can have any feasible payoff, therefore payoff arbitrarily close to the minmax of the player being punished. However, the low-entropy rounds must have a payoff somewhat close to the pure minmax of the punished player. Since the entropy of the entire tree is bounded, the number of high-entropy rounds is also limited, yielding the desired lower bound on payoffs.

We finally use this lower bound to deduce that the average play on-path must converge to a purely individually rational payoff profile. The proof is deferred to the full version of the paper [1, Lemma C.8].

Lemma 6. *If G has a* NASH *equilibrium of its repeated game for all large enough n using $o(n)$ randomness, then G has a $\mathbb{R}$-feasible payoff profile that is* purely *individually rational for all players.*

4 Two-Player Games

In this section, we show that the necessary condition from Proposition 2, along with Lemma 2 are together sufficient in the two-player case, providing a complete characterization of the existence of constant-entropy NASH equilibria of the repeated game. The complete proof is in the full version of the paper [1, Theorem D.1], and a proof sketch is offered here.

Theorem 3. *A two-player game G has $O(1)$-randomness* NASH *equilibria of its repeated game if and only if G has a pure* NASH *equilibrium or if it has a purely individually rational feasible payoff profile and has a* NASH *equilibrium that is strictly individually rational for a player.*

Proof (sketch). Begin by the sufficient condition, and assume G has no pure NASH equilibria. We build an equilibrium in two phases, much like Proposition 4, except we need to distinguish between two cases. Up to renaming, assume player 2 is the one that has a strictly individually rational NASH equilibrium σ_2, and denote p the purely individually rational payoff profile.

The first case is when the maximum payoff of player 1 is exactly their minmax $\tilde{v}_1$. In this case, one can directly apply Proposition 6 since player 1 already achieves their maximum payoff at every round (and is therefore best responding).

In the second case, we know the maximum payoff of player 1 is strictly greater than their minmax $\tilde{v}_1$: let $s \in S$ be a pure strategy profile achieving player 1's maximum payoff. The first phase is the same as in the proof of Proposition 4.

Let $\delta_1 = u_1(s) - \tilde{v}_1 > 0$ by assumption. The second phase begins with $k_1 = \left\lceil \frac{|S|+1}{\delta_1} \right\rceil$ repetitions of s, followed by $k_2 = \left\lceil \frac{|S|+1+k_1}{\delta_2} \right\rceil$ repetitions of σ_2. As in Proposition 4, any deviation in the first phase is punished by that player's pure minmax for the rest of the phase and that player's minmax during the entire second phase. Moreover, any deviation by player 2 during the play of s is punished by playing the minmax of player 2 for the rest of the game. Deviations during the play of σ_2 are ignored, since it is already a NASH equilibrium, and deviations by player 1 during s are also ignored since it is its maximum payoff.

This is a constant-entropy strategy profile of the repeated game, since randomness is only used in at most $k_1 + k_2 = O(1)$ rounds along any path. It is also a NASH equilibrium since Lemma 3 bounds the benefit from deviating in the first phase, and the number of repetitions of the reward (or punishment) for each player is fixed appropriately as in Proposition 4.

The converse is a direct application of Proposition 6 and Lemma 6. $\square$

Sublinear Randomness. Notice that the proof of the necessary condition only relies on Proposition 6 and Lemma 6, which both only require sublinear-entropy equilibria. We deduce that the same condition holds for sublinear-entropy NASH equilibria, yielding the following 0–1 law for entropy in two-player repeated games.

Corollary 1. *For any two-player game, there are either $O(1)$-randomness equilibria of its repeated game or all equilibria require $\Omega(n)$ randomness.*

Asymptotically Achievable Payoffs. Another consequence of the proof is that by applying the necessary condition and then the sufficient condition to a series of sublinear-entropy NASH equilibria, we find constant-entropy NASH equilibria with the same asymptotic average payoffs. This provides a folk theorem-like result for constant-entropy NASH equilibria, which can be compared to folk theorems such as Benoît and Krishna [2].

Corollary 2. *When the condition of Theorem 3 is satisfied, the set of payoff profiles achievable asymptotically by sublinear and constant-entropy equilibria are both exactly the set of purely individually rational payoff profiles. When the condition is not satisfied, there are no sublinear-entropy* NASH *equilibria and the set of achievable payoffs with sublinear entropy is empty.*

Both Players Requiring $\Omega(n)$ Entropy Each. The full version of the paper [1, Theorem D.2] provides a characterization for the existence of NASH equilibria in which one player uses $O(1)$ entropy. This yields the following characterization of two-player games in which all NASH equilibria of the repeated game require $\Omega(n)$ entropy *from both players*. Recall that Hubáček, Naor and Ullman [8] proved that a sufficient condition is that all NASH equilibria of G have payoffs exactly $\tilde{v}$ (the minmax of all players) and all NASH equilibria of G are mixed for all players (for effective entropy). We complete this into a necessary and sufficient condition for two-player games using total entropy.

Theorem 4. *Suppose G is a two-player game. All* NASH *equilibria of the repeated game G^n require $\Omega(n)$ randomness from both players if and only if all* NASH *equilibria of G are mixed for all players and either all* NASH *equilibria of G have payoffs exactly $\tilde{v}$ or there is no individually rational feasible payoff profile p that is also purely individually rational for one of the two players.*

5 Observable Distributions

In this section, we prove a similar result for games with more than two players in a setting where players are able to observe the distributions that other players used to select their action at the end of each round. This setting serves as an intermediate between the two-player case and the general case, and the condition we obtain is a natural extension of the one from Theorem 3. The assumption of observable distributions is standard in the repeated games literature [4,5].

Definition 9. *For a given game G and $n \in \mathbb{N}^*$, $\tilde{G}^n$ is the* n***-repeated game with observable distributions*** *of G. It is the extensive-form game where strategy profiles are mappings from histories of actions and distributions for all players to strategy profiles of G,*

$$\sigma : \bigcup_{k=0}^{n-1} \left(S \times \prod_{j \in A} \Delta(S_j) \right)^k \to \prod_{j \in A} \Delta(S_j),$$

where $\Delta(X)$ is the simplex over X.

The main advantage of observable distributions is that it records any deviation in the history, allowing for deviations in mixed rounds to be punished, which cannot be done in the general case. The proof of the following theorem is deferred to the full version of the paper [1, Theorem E.1].

Theorem 5. *An m-player game G has $O(1)$-randomness* Nash *equilibria of its repeated game with observable distributions $\tilde{G}^n$ (or equivalently sublinear-randomness* Nash *equilibria) if and only if it has a purely individually rational feasible payoff profile p supportable by $S' \subseteq S$ and there exists a mapping $f : A \to \{-\infty\} \cup \mathbb{N}$ such that,*

- *If $f(i) = -\infty$ then player i best responds in all strategy profiles in S';*
- *If $f(i) \in \mathbb{N}$ then there exists a strategy profile σ of G such that $u_i(\sigma) > \tilde{v}_i$ and in which the players $\{j,\ f(j) \leq f(i)\}$ are all best responding.*

6 General Case

We finally extend our results to general repeated games (without observable distributions). The proof of the following theorem is deferred to the full version of the paper [1, Theorem F.1].

Theorem 6. *An m-player game G has $O(1)$-randomness* Nash *equilibria of its repeated game G^n if and only if G has a feasible purely individually rational payoff profile p supportable by $S' \subseteq S$, there exists some constant $n_0 \in \mathbb{N}$, a* Nash *equilibrium σ_{n_0} of the repeated game G^{n_0}, and a partition $A = A_0 \cup A_1$ of the players such that,*

- *Every player in A_0 is best responding in every strategy profile in S',*
- *Every player in A_1 has an average payoff in σ_{n_0} that is strictly better than their minmax.*

Moreover, if G does not satisfy this condition then all equilibria of its repeated game require $\Omega(n)$ entropy.

As in Corollary 2, note the set of achievable payoffs by constant or sublinear-entropy equilibria are the same, and are exactly the payoff profiles p that satisfy the condition in the theorem above for some support S', integer n_0, equilibrium σ_{n_0} and partition $A_0 \cup A_1$.

The structure of the conditions from Theorems 3 and 5 is partially lost, as there is no ordering but simply a partition: A_0 can be seen as $f^{-1}(\{-\infty\})$ and A_1 as $f^{-1}(\mathbb{N})$. We now give a sufficient condition and a necessary condition that are closer to our earlier characterizations. The proof of the following proposition is deferred to the full version of the paper [1, Proposition F.2].

Proposition 7 (Sufficient condition). *If G has a feasible purely individually rational payoff profile p supportable by $S' \subseteq S$ and there exists a mapping $f : A \rightarrow \{-\infty\} \cup \mathbb{N}$ such that,*

- *If $f(i) = -\infty$ then player i is best responding in every strategy profile in S'*
- *If $f(i) \in \mathbb{N}$ then there exists a strategy profile σ_i such that $u_i(s) > \tilde{v}_i$ and in which the players $\{j,\ f(j) \leq f(i)\}$ are best responding and the players $\{j,\ f(j) > f(i)\}$ have the same payoff for all the strategies they are mixing over,*

then G has $O(1)$-randomness Nash *equilibria of its repeated game.*

Corollary 3 (Necessary condition). *If G has $o(n)$-entropy* Nash *equilibria of its repeated game, then the condition of Theorem 5 is verified by G.*

Proof. An $o(n)$-randomness equilibrium of the repeated game G^n is immediately an $o(n)$-randomness equilibrium of the repeated game with observable distributions $\tilde{G}^n$. By Theorem 5, it satisfies all of its conditions. □

Subset of Players Using Bounded Entropy. Finally, we show that our techniques can be adapted to provide conditions on games where only a subset of players use sublinear randomness.

Definition 10. *For $T \subseteq A$, a payoff profile p is T-**individually rational** if,*

$$\forall i \in A, \qquad p_i \geq \min_{\substack{s_j \in S_j \\ j \in T \setminus \{i\}}} \min_{\substack{\sigma_j \\ j \in A \setminus (T \cup \{i\})}} \max_{s_i \in S_i} u_i(s_i, s_{T \setminus \{i\}}, \sigma_{A \setminus (T \cup \{i\})}) = v_{T,i}.$$

*The term on the right $v_{T,i}$ is called the T-**pure minmax** of player i.*

In short, it must ensure the player's minmax assuming players from T do not use mixing and those from $A \setminus T$ do. The proof of the following theorem is deferred to the full version of the paper [1, Theorem F.7].

Theorem 7. *Suppose G is an m-player game, and $T \subseteq A$. If G has a feasible T-individually rational payoff profile p supportable by $S' \subseteq S$, there exists some constant $n_0 \in \mathbb{N}$, a* Nash *equilibrium σ_{n_0} of the repeated game G^{n_0}, and a partition $A = A_0 \cup A_1$ of the players such that,*

- *Every player in A_0 best responds in every strategy profile in S',*
- *Every player in A_1 has an average payoff in σ_{n_0} that is strictly better than their minmax,*

then G has NASH *equilibria of its repeated game such that all players in T use O(1) randomness. Conversely, if G has* NASH *equilibria of its repeated game such that all players in T use O(1) randomness, then it has a feasible T-individually rational payoff profile p.*

Acknowledgement. The author would like to thank Edan Orzech and Kai Jia for their feedback on the computational model. Laurel Britt, Martin Rinard, and Sandeep Silwal are also thanked for their feedback on early versions of the introduction.

Disclosure of Interests. The author has no competing interests to declare that are relevant to the content of this article.

References

1. Arthaud, F.: Playing repeated games with sublinear randomness. CoRR abs/2312.13453 (2023). https://doi.org/10.48550/ARXIV.2312.13453
2. Benoît, J.P., Krishna, V.: Nash equilibria of finitely repeated games. Int. J. Game Theory **16**, 197–204 (1987). https://doi.org/10.1007/BF01756291
3. Budinich, M., Fortnow, L.: Repeated matching pennies with limited randomness. In: Shoham, Y., Chen, Y., Roughgarden, T. (eds.) Proceedings 12th ACM Conference on Electronic Commerce (EC 2011), San Jose, CA, USA, 5–9 June 2011, pp. 111–118. ACM (2011). https://doi.org/10.1145/1993574.1993592
4. Fudenberg, D., Kreps, D.M., Maskin, E.S.: Repeated games with long-run and short-run players. Rev. Econ. Stud. **57**(4), 555–573 (1990). https://doi.org/10.2307/2298086
5. Fudenberg, D., Maskin, E.: On the dispensability of public randomization in discounted repeated games. J. Econ. Theory **53**(2), 428–438 (1991). https://doi.org/10.1016/0022-0531(91)90163-X
6. Gossner, O., Vieille, N.: How to play with a biased coin? Games Econ. Behav. **41**(2), 206–226 (2002). https://doi.org/10.1016/S0899-8256(02)00507-9
7. Halpern, J.Y., Pass, R.: Algorithmic rationality: game theory with costly computation. J. Econ. Theory **156**, 246–268 (2015). https://doi.org/10.1016/J.JET.2014.04.007
8. Hubáček, P., Naor, M., Ullman, J.R.: When can limited randomness be used in repeated games? Theory Comput. Syst. **59**(4), 722–746 (2016). https://doi.org/10.1007/s00224-016-9690-4
9. Kalyanaraman, S., Umans, C.: Algorithms for playing games with limited randomness. In: Arge, L., Hoffmann, M., Welzl, E. (eds.) ESA 2007. LNCS, vol. 4698, pp. 323–334. Springer, Heidelberg (2007). https://doi.org/10.1007/978-3-540-75520-3_30
10. Knuth, D.E., Yao, A.C.C.: The complexity of nonuniform random number generation. In: Traub, J.F. (ed.) Algorithms and Complexity: New Directions and Recent Results, pp. 357–428. Academic Press Inc. (1976)
11. Lipton, R.J., Markakis, E.: Nash equilibria via polynomial equations. In: Farach-Colton, M. (ed.) LATIN 2004. LNCS, vol. 2976, pp. 413–422. Springer, Heidelberg (2004). https://doi.org/10.1007/978-3-540-24698-5_45
12. Neyman, A., Okada, D.: Repeated games with bounded entropy. Games Econ. Behav. **30**(2), 228–247 (2000). https://doi.org/10.1006/GAME.1999.0725

13. Valizadeh, M., Gohari, A.: Playing games with bounded entropy. Games Econ. Behav. **115**, 363–380 (2019). https://doi.org/10.1016/J.GEB.2019.03.013
14. Valizadeh, M., Gohari, A.: Playing games with bounded entropy: convergence rate and approximate equilibria. CoRR abs/1902.03676 (2019). https://doi.org/10.48550/arXiv.1902.03676

Pricing, Revenue, and Regulation

Mind the Revenue Gap: On the Performance of Approximation Mechanisms Under Budget Constraints

Ahuva Mu'alem[1](✉) and Juan Carlos Carbajal[2](✉)

[1] Department of Computer Science, Holon Institute of Technology, Holon, Israel
ahumu@yahoo.com

[2] School of Economics, UNSW Sydney, Sydney, Australia
jc.carbajal@unsw.edu.au

Abstract. We consider a buyer-seller interaction where the revenue-maximizer seller has one object to allocate and the buyer has private valuation and private budget. The presence of private budgets is one among the several triggers of complexity in mechanism design models. Che and Gale [8] show that the optimal mechanism for this setting may require a continuum of menu entries. We focus on a restricted class of simple mechanisms, consisting of those whose associated menus have a small number of entries. We show that for distributions supported on $[1, \overline{v}] \times [1, \overline{w}]$, an arbitrarily high fraction of the optimal revenue can be obtained using a simple mechanism with poly-logarithmic menu size. This result applies even if the valuation and budget are arbitrarily correlated. However, if the distribution has unbounded support, then any selling mechanism that contains a fixed number of menu entries cannot guarantee any positive fraction of the optimal revenue. In fact, we are able to strengthen this negative result and show that, for some family of finite distributions, any mechanism that contains an asymptotically sub-linear number of menu entries (in the size of the finite support) cannot guarantee a positive fraction of the optimal revenue.

Keywords: Approximation Mechanisms · Menu Complexity · Budget Constraints

1 Introduction

We study a buyer–seller interaction where the seller has a single object to allocate and the buyer has private valuations and private budgets. In this setting, we are interested in measuring the performance, in terms of expected revenue for the seller, of approximation mechanisms against optimal mechanisms.

The presence of the private budget, which acts as a hard constraint on the purchasing ability of the buyer, is the only complication we consider. Remove it, and we are back in the classic "single-product monopoly" setting with incomplete information. Previous work in the literature that addresses variations of

G. Schäfer and C. Ventre (Eds.): SAGT 2024, LNCS 15156, pp. 279–296, 2024.
https://doi.org/10.1007/978-3-031-71033-9_16

this problem has provided strong justifications for the inclusion of budget constraints.[1] Importantly, this points to the limits of our understanding of mechanism design outside single-item, single-type models. For example, Che and Gale [8] consider revenue maximization by a seller who faces a buyer with private valuations and private budgets distributed on a bounded rectangle. Even when the joint distribution is well-behaved, the optimal mechanism is extremely complex to obtain and to understand. It relies on minor details of the distribution and, when translated into a menu of lotteries for implementation, it contains a continuum of such options.

The presence of private budgets is one among the several triggers of complexity in mechanism design models. Indeed, the literature recognized early on that when the seller has many items to allocate (the so called "multi-product monopoly" case), searching for optimal mechanisms or even characterizing incentive compatibility is challenging.[2] As a result, a branch of the literature adopted an alternative route and studied approximation mechanisms in the multi-product monopoly case without considering budget constraints.[3] The performance of an approximation mechanism is measured in terms of its expected revenue compared against the expected revenue of the optimal mechanism. For instance, Hart and Nisan [13,14] consider revenue maximization with k goods and valuations given by random vectors of the form $(V_1, \ldots, V_k)$. They show that if that $V_1, \ldots, V_k$ are independent random variables, a nontrivial fraction of the optimal revenue can always be approximated by simple mechanisms. Simplicity here refers to the size of the menu associated with a mechanism. Surprisingly, Hart and Nisan [14] show that no such "good approximation" is possible if the components of the random vector $(V_1, \ldots, V_k)$ are correlated.[4]

The dramatic effect that different assumptions have on the revenue gap between approximation mechanisms and optimal mechanisms is cause for further study. A significant contribution of our paper is to point out how sensitive this revenue gap is to some modeling assumptions, even in single-item settings, as long as the buyer has private budgets and private valuations.

The preliminaries are presented in Sect. 2. We consider a seller with a single item to allocate (and zero opportunity cost). The buyer is characterized by a random vector (V, W), where V refers to the valuation for the item and W to the budget. We formalize the seller's problem in terms of direct mechanisms that satisfy incentive compatibility, (ex-post) individual rationality, and (ex-post) budget feasibility. The set of these ex-post feasible and incentive compatible mechanisms is denoted by $\mathcal{M}$. The seller's problem is thus to choose a mechanism in $\mathcal{M}$ that maximizes expected revenue. Approximation mechanisms are obtained from $\mathcal{M}$ in two different ways: (i) by imposing additional constraints, thus restricting $\mathcal{M}$ to a subclass of mechanisms, say $\mathcal{N}$; or (ii) by relaxing some original constraints, thus expanding $\mathcal{M}$ to a superclass of mechanisms, say $\mathcal{N}'$.

[1] See for example [8,9,17,22,23] and [4].

[2] See the pioneering contributions in [20] and [24].

[3] [25] present a recent and enlightening survey.

[4] For related results, see [2,16,18] and [5].

We measure the performance of a restricted class of mechanisms $\mathcal{N} \subseteq \mathcal{M}$, in terms of expected revenue, using the *guaranteed fraction of optimal revenue* ratio introduced by [13]. We measure the performance of a relaxation class of mechanisms $\mathcal{N}' \supseteq \mathcal{M}$ using a related ratio, which we call the *maximal value of relaxation*. Both ratios perform robust revenue comparisons, i.e., over families of (V, W)–buyers. We conclude this section with a *revenue monotonicity* result that expands a similar result of Hart and Reny [15] to single-item settings where, in addition to the valuation, the buyer has a private budget, which may be of independent interest.[5]

In Sect. 3 we consider different classes of approximation mechanisms.

Subsections 3.1 and 3.2 focus on *simple* approximation mechanisms. Following [14], we measure the complexity of a mechanism by the cardinality of its associated menu. There are several reasons to focus on simple mechanisms, notably that the menu size of a mechanism is related to its communication complexity [1]. In settings without budget constraints, Myerson [21] shows that the optimal mechanism contains one non-trivial option. When the buyer's budget is publicly known, Chawla et al. [6] demonstrate that the optimal mechanism has a menu that contains at most two non-trivial options. In contrast, we show that when the budget is private, simple mechanisms (i.e., mechanisms with poly-logarithmic many options) generate an arbitrarily small revenue gap with respect to optimal mechanisms as long as one considers a family of random vectors (V, W) whose support is bounded from above and bounded away from zero from below (Proposition 2). However, if the family of (V, W)–buyers has unbounded support, or includes random vectors whose support is a subset of the unit square, then any selling mechanism that contains a fixed number of options cannot guarantee any positive fraction of the optimal revenue (see Proposition 3 and Proposition 4). In fact, we are able to strengthen these negative results and show that, for some family of distributions, any mechanism that contains an asymptotically sub-linear number of options cannot guarantee a positive fraction of the optimal revenue (Proposition 5). Our positive result on the revenue gap of simple mechanisms is derived using the revenue monotonicity result mentioned above. The proofs of our negative results are constructive.

We switch gears in Subsect. 3.3 and pay attention to a relaxation of the seller's problem, where the seller can costlessly prevent the buyer from over-reporting her budget.[6] Our analysis here is motivated by both computational and economic considerations, which we discuss in detail later on.[7] Che and Gale [8] show that if valuations and budgets are positively correlated, then this relaxation to mechanisms where the seller ignores deviations to budget over-reporting has no revenue advantage for the seller. In contrast, we show that

---

[5] For reasons of space, we only provide proofs of Propositions 2 and 3. All other proofs can be found in the working paper version accessible at https://sites.google.com/site/ahuvamualem/ or https://sites.google.com/site/carbajaleconomics/.

[6] There are several ways the seller can prevent over-reporting. For instance, the buyer could be required to post a cash-bond prior to the interaction with the seller.

[7] Previous work considering this relaxation include [11] and [10].

when valuations and budgets are negatively correlated, the value of relaxation can be unboundedly large (Proposition 6). We obtain this result using a family of (V, W)–buyers for whom the budget constraint is never binding under truthful reporting. Finally, in Subsect. 3.4 we consider a restricted problem in which the seller considers deviations from truthful reporting even when these deviations may lead to unaffordable choices for the buyer. The imposition of these stronger incentive constraints has received attention in the literature because it provides computationally feasible solutions. We show that the optimal restricted mechanisms perform badly in comparison with the optimal mechanisms when the family of (V, W)–buyers has unbounded support (Proposition 7) or has support restricted to the unit square (Proposition 8).

Overall, our research highlights the limitations of using approximation mechanisms in settings with private valuations and private budgets. In only one of the six cases we consider do approximation mechanisms yield a negligible revenue gap. Specifically, we show that for distributions with bounded support (above and below away from zero), an arbitrarily high fraction of the optimal revenue can be obtained using a simple mechanism with poly-logarithmic menu size. In all others, the revenue gap between the optimal mechanism and the optimal approximation mechanism is large. Thus, despite some clear advantages (including computational) in dealing with restricted or relaxed classes of mechanisms, one has to be aware of large potential revenue loses. Our research also stresses the crucial role some of our assumptions play in the performance of our theoretical constructs. Our results, positive and negative, apply to a single-item setting. While we expect most of the negative results to hold in multi-item settings, it is hard to forecast a priori which of the positive results will survive. We leave this important question for future research.

2 Setting

We study a buyer–seller interaction where the seller has one object to sell and the buyer has a private valuation and a private budget. This last element is the only complication we consider. Without the presence of a hard financial constraint on the buyer's side, we are back in the canonical single-product monopoly setting in which the revenue maximization problem is thoroughly understood. Our focus on the single-product case allows us to highlight the limits of using approximation mechanisms in the presence of budget constraints.

The seller's objective is to maximize expected revenue—the seller's opportunity cost is zero. The buyer has a private valuation $v \geq 0$ for the object and a private budget $w \geq 0$ constraining her ability to pay. Valuations and budgets are given by a random vector (V, W) that takes values in $\mathbb{R}^2_+$. We do not exclude the possibility of V and W being correlated. The realization of this random vector is private information of the buyer. The seller only knows the distribution of the (V, W)–buyer involved in the exchange.[8]

[8] We use the expressions (V, W)*–buyer* and *random vector* (V, W) interchangeably.

The seller offers *lotteries* of the form (q, P_e). Here $q \in [0,1]$ denotes the probability that the buyer gets the object, and $P_e \geq 0$ denotes the price the buyer pays in case the good exchanges hands—if there is no exchange, the buyer pays nothing. The expected payment to the seller generated by lottery (q, P_e) is thus $P_e\, q$. A *menu* is a collection (finite or infinite) of lotteries offered by the seller.

2.1 Buyer's Behavior

We consider a buyer who, due to her hard budget constraint, employs a two-stage approach in deciding which lottery to choose from those available in a given menu M. In the first stage, the buyer makes a shortlist composed of all lotteries in M that are *ex-post individually rational* and *ex-post budget feasible* given her budget. In the second stage the buyer selects, from this shortlist, a lottery that maximizes her expected utility.

Formally, given menu M, a (V, W)–buyer with type realization $(v, w) \in \mathbb{R}^2_+$ sorts out the collection of lotteries

$$M(v, w) \;=\; \big\{(q, P_e) \in M :\; P_e \;\leq\; \min(v, w)\big\}.$$

Any lottery in $M(v, w)$ is ex-post budget feasible for the buyer—she is always able to afford the price associated to either outcome, trade or no trade—and ex-post individually rational—she weakly prefers purchasing any lottery in $M(v, w)$ to not interacting with the seller, trade or no trade. Note that the value of the buyer's outside option is normalized to zero. From the shortlist $M(v, w)$, the buyer selects a lottery (q^*, P_e^*) that maximizes her expected utility; i.e.,

$$(q^*, P_e^*) \;\in\; \arg\max\, \big\{(v - P_e)\, q :\; (q, P_e) \;\in\; M(v, w)\big\}.$$

It is convenient to let $p = P_e\, q$ denote the *expected payment* to the seller. We now write lotteries simply as pairs (q, p), where $q \in [0,1]$ and $p \geq 0$. The *actual payment* from the buyer to the seller, conditional on trade, is of course p/q as long as $q > 0$, and zero otherwise.

Remark 1. We present the buyer's behavior as a two-stage decision process to highlight the fact that the buyer is an expected utility maximizer and has quasi-linear preferences. This two-stage process is reminiscent of certain sequential elimination processes employed in behavioral economic theory to model choices by a boundedly rational agent (e.g., Manzini and Mariotti [19]). The difference is that in our case the buyer is not boundedly rational, merely budget constrained: the shortlist $M(v, w)$ expresses only financial constraints.

2.2 Seller's Problem

By the Revelation Principle [21], we formalize the seller's problem in terms of direct mechanisms. A *direct (selling) mechanism* $\mu = (x, s)$ is composed of a pair of Borel measurable functions, where $x\colon \mathbb{R}^2_+ \to [0,1]$ describes the probability of the good being allocated to the buyer, and $s \colon \mathbb{R}_+^2 \to \mathbb{R}_+$ describes the expected payment to the seller. To be more explicit, given a report $(\tilde{v}, \tilde{w}) \in \mathbb{R}_+^2$ from the buyer, the mechanism $\mu = (x, s)$ assigns the object with probability $x(\tilde{v}, \tilde{w}) \in [0, 1]$ in exchange for an expected payment of $s(\tilde{v}, \tilde{w}) \geq 0$ to the seller.

Given the buyer's behavior, the seller takes into account the following restrictions in the design of direct mechanisms. First, $\mu = (x, s)$ must satisfy the *ex-post budget feasibility* constraint; i.e., for all $(v, w) \in \mathbb{R}_+^2$,

$$s(v, w) \leq w\, x(v, w). \tag{BF}$$

(Recall the buyer's actual payment upon receiving the object is $s(v, w)/x(v, w)$ as long as $x(v, w) > 0$, and zero otherwise.) Second, the mechanism $\mu = (x, s)$ must be *ex-post individually rational*; i.e., for all $(v, w) \in \mathbb{R}_+^2$,

$$v\, x(v, w) - s(v, w) \geq 0. \tag{IR}$$

(Recall that the buyer pays zero if the object doesn't exchange hands.) Finally, the mechanism $\mu = (x, s)$ must also be *incentive compatible*: for all $(v, w) \in \mathbb{R}_+^2$ and all $(\tilde{v}, \tilde{w}) \in \mathbb{R}_+^2$ such that $s(\tilde{v}, \tilde{w}) \leq w\, x(\tilde{v}, \tilde{w})$, it must be that

$$v\, x(v, w) - s(v, w) \geq v\, x(\tilde{v}, \tilde{w}) - s(\tilde{v}, \tilde{w}). \tag{IC}$$

In words, the buyer participates in the mechanism and truthfully reveals her type if (i) the ex-post payment upon receiving the object is less than her budget, and the ex-post payment when trade does not occur is zero; (ii) the ex-post utility and, thus the expected utility, associated to participating in the mechanism is weakly greater than the value of her outside option, which is normalized to zero; and (iii) the expected utility from truthfully revealing her type to the mechanism is weakly greater than the expected utility associated with any other report, as long as this deviation is ex-post affordable for the buyer, given her true type.[9]

Our focus on ex-post budget feasibility and ex-post individual rationality is not uncommon in the literature on auctions and mechanisms with budget constrained agents. The analysis of first-price and second-price auctions under budget constraints is usually performed under these two assumptions—see for instance [6,7], and [17]. Some work on efficient and revenue-maximizing resource allocation mechanisms with many agents also focuses on ex-post constraints, e.g., [9] and [23]. On the other hand, [8] and [22] focus on interim constraints in their analyses.

The ex-post constraints make more sense in our setting, where we focus on mechanisms that offer lotteries to a single buyer. Che et al. [9] argue that, while intellectually interesting, lotteries with positive entry fees are not common in the real world because they are susceptible to manipulation by the seller. Focusing on ex-post instead of interim constraints is not without consequences, as expected revenue generally will be lower under the ex-post constraints.[10]

[9] Because of this budget affordability requirement, the incentive constraints in this case do not allow the seller's problem to be expressed as a linear program.

[10] Example 1 provides a concrete illustration—see also [6].

Remark 2. We define a direct mechanism $\mu = (x, s)$ on $\mathbb{R}^2_+$ even when the random vector (V, W) may not have full support. This is without loss of generality. As we show in the working paper version, any ex-post budget feasible and incentive compatible mechanism defined on the support of (V, W) can be extended to an ex-post budget feasible and incentive compatible mechanism defined on $\mathbb{R}^2_+$. From there, it is immediate to show that if the original mechanism defined on the support of (V, W) is also ex-post individual rational, then so is its extension to $\mathbb{R}^2_+$. Hart and Reny [15] show a similar *Extension Lemma* in a setting without budget constraints, but with multiple goods. To the best of our knowledge, we are the first to point out its validity under hard budget constraints.

The expected revenue raised by $\mu = (x, s)$ when the seller interacts with a (V, W)–buyer is $R(\mu; V, W) := \mathbb{E}[s(V, W)]$. Let $\mathcal{M}$ denote the class of all selling mechanisms defined on $\mathbb{R}^2_+$ that satisfy the (BF), (IR) and (IC) constraints. The *seller's problem* is to find a direct mechanism $\mu \in \mathcal{M}$ that maximizes expected revenue. The *optimal revenue from the* (V, W)*–buyer* is the value of the solution to the seller's problem at (V, W), if one exists; i.e.,

$$\mathrm{Rev}(V, W) \; := \; \sup_{\mu \in \mathcal{M}} R(\mu; V, W). \tag{1}$$

An *optimal mechanism for the* (V, W)*–buyer* is any mechanism μ^* in $\mathcal{M}$ that generates expected revenue $\mathrm{Rev}(V, W)$.

2.3 Estimating the Revenue Gaps

Several papers in the literature have studied variations of the seller's problem. This body of work points to the complexity of optimal mechanisms under budget constraints. As a result, important contributions in the literature focus on solving the revenue maximization problem over a more restricted set of (simpler) mechanisms or a more relaxed set of mechanisms. In this paper, we refer to feasible solutions of these related problems as *approximation mechanisms* and explore the limits of using approximation mechanisms to solve the seller's revenue maximization problem. When focusing on a different class of mechanisms $\mathcal{N} \neq \mathcal{M}$, we write

$$\mathrm{Rev}(V, W|\mathcal{N}) \; := \; \sup_{\nu \in \mathcal{N}} R(\nu; V, W)$$

for the optimal revenue that can be raised by a seller who interacts with a (V, W)–buyer and uses mechanisms in the class $\mathcal{N}$. An *optimal approximation mechanisms in the class* $\mathcal{N}$ is any mechanism $\nu^* \in \mathcal{N}$ that generates expected revenue $\mathrm{Rev}(V, W|\mathcal{N})$.

A branch of the literature considers restricted classes of mechanisms that are 'simpler' than $\mathcal{M}$ and thus have lower communication complexity or computational complexity. This branch focuses on the expected revenue that is lost by using simple mechanisms. To measure any potential losses, we follow Hart and Nisan [13] and focus on worst-case scenarios. In what follows, let $\mathscr{B}$ denote a given family of random vectors (V, W) that take values on $\mathbb{R}^2_+$. When specific properties of random vectors in $\mathscr{B}$ are required (e.g., bounded support), we will be explicit about it.

Definition 1. *Let $\mathscr{B}$ be a given family of (V, W)–buyers and $\mathcal{N} \subseteq \mathcal{M}$ be a non-empty subclass of mechanisms. The* Guaranteed Fraction of Optimal Revenue *for $\mathcal{N}$, denoted by GFOR($\mathcal{N}; \mathscr{B}$), is defined as*

$$GFOR(\mathcal{N}; \mathscr{B}) \ := \ \inf_{(V,W)\in\mathscr{B}} \frac{\text{Rev}(V, W|\mathcal{N})}{\text{Rev}(V, W)}.$$

In words, GFOR($\mathcal{N}; \mathscr{B}$) is the maximal number $0 \leq \alpha \leq 1$ such that, for every (V, W)–buyer in $\mathscr{B}$, there is a mechanism ν in the class $\mathcal{N}$ that generates at least α times the optimal revenue $\text{Rev}(V, W)$. Hart and Nisan [13] discuss the importance of this concept in the multiple-goods monopoly model and its relationship with the *competitive ratio* concept in the computer science literature. When α is close to 1, the revenue gap between an optimal approximation mechanism in the subclass $\mathcal{N}$ and an optimal mechanism in $\mathcal{M}$ is negligible, and thus $\mathcal{N}$ can serve as a desirable replacement for $\mathcal{M}$.

A related branch of the literature considers relaxing some of the constraints embedded in the class $\mathcal{M}$ of (IC), (IR) and (BF) mechanisms. Sometimes this relaxation is motivated by computational requirements—e.g., weakening some of the constraints transforms the seller's problem into a computationally feasible linear program. In other instances, a relaxation is motivated by economic or institutional considerations. Che and Gale [8] explore revenue maximization when the seller can force the buyer to set up a cash bond, thus preventing the buyer from over-reporting her budget. We consider best-case scenarios in measuring any potential gains from considering a relaxed environment.

Definition 2. *Let $\mathscr{B}$ be a given family of (V, W)–buyers and $\mathcal{N}' \supseteq \mathcal{M}$ be a non-empty superclass of mechanisms. The* Maximal Value of Relaxation *for $\mathcal{N}'$, denoted by MVR($\mathcal{N}'; \mathscr{B}$), is*

$$\text{MVR}(\mathcal{N}'; \mathscr{B}) \ := \ \sup_{(V,W)\in\mathscr{B}} \frac{\text{Rev}(V, W|\mathcal{N}')}{\text{Rev}(V, W)}.$$

Thus, MVR($\mathcal{N}'; \mathscr{B}$) is the minimal number $\beta \geq 1$ such that, for every (V, W)–buyer in the family of random vectors $\mathscr{B}$, every mechanism ν' in the class $\mathcal{N}'$ generates at most a multiple β of the optimal revenue $\text{Rev}(V, W)$. When β is close to 1, the revenue gap between an optimal mechanism in $\mathcal{M}$ and an optimal approximation mechanism in the superclass $\mathcal{N}'$ is negligible, and so is the value of relaxing the seller's problem.

An additional reason to focus on GFOR and MVR is that they provide a measure of robustness under higher orders of uncertainty. The seller knows that a buyer is characterized by the random vector (V, W) in $\mathscr{B}$, and thus is aware of the fact that facing a (V, W)–buyer is different from facing a (V', W')–buyer. However, the seller may not know which buyer is present at the time of exchange and thus may be interested in prior-free measurements of the revenue gap. When considering the use of simpler mechanisms in $\mathcal{N}$, the seller focuses on a 'worst-case' measure of lost revenue, which is precisely what $\text{GFOR}(\mathcal{N};\mathscr{B})$ provides. When considering the value of relaxing some of the constraints in the revenue maximization problem, the seller focuses on a 'best-case' measure of gained revenue, which is what $\text{MVR}(\mathcal{N}';\mathscr{B})$ represents.

2.4 Revenue Monotonicity

Before we present our contributions on the revenue gaps, it is convenient to show a revenue monotonicity result: under a reasonable tie-breaking rule, expected revenue increases whenever the valuation and/or the budget of the buyer increase. This holds for any mechanism satisfying (IC), (BF) and (IR), not just revenue-maximizing mechanisms. We follow Hart and Reny [15] and define *seller-favorable mechanisms* as those that break ties in favor of the seller. Since we are interested in the revenue gaps between optimal mechanisms and optimal approximation mechanisms, focusing on seller-favorable mechanisms does not entail any loss of generality.

Recall that a random vector (V',W') first order stochastically dominates (V,W) if and only if $\mathbb{E}\left[u(V',W')\right] \geq \mathbb{E}\left[u(V,W)\right]$ for any non decreasing function $u\colon \mathbb{R}^2 \to \mathbb{R}$.

Proposition 1. *Suppose that $\mu=(x,s)\in\mathcal{M}$ is a seller-favorable mechanism. For any two buyers (V,W) and (V',W') such that (V',W') first order stochastically dominates (V,W), one has*

$$R(\mu;V,W) \;\leq\; R(\mu;V',W').$$

3 Revenue Gaps

In this section, we explore the gap between the revenue generated by optimal mechanisms in $\mathcal{M}$ and the maximal revenue generated by approximation mechanisms (that is, mechanisms that lie in some class $\mathcal{N}\neq\mathcal{M}$).

3.1 Good Approximations with Simple Mechanisms

For expositional purposes, it is convenient to define the *menu* generated by a mechanism $\mu=(x,s)$ as

$$\text{Menu}(\mu) \;:=\; \left\{\big(x(v,w),s(v,w)\big) \,\in\, [0,1]\times\mathbb{R}_+ \,:\, (v,w)\in\mathbb{R}_+^2\right\}\setminus\{(0,0)\}.$$

In other words, the menu of $\mu=(x,s)$ is given by its image, excluding the trivial lottery.[11] The *menu size* of the mechanism μ is the cardinality of its menu. As before, we refer to the different components of $\text{Menu}(\mu)$ as lotteries or menu entries. When the mechanism is fixed and no confusion arises, we write generic

[11] Any ex-post individually rational mechanism can include the trivial lottery $(0,0)$.

lotteries as pairs (q, p), where $q \in [0, 1]$ is the probability that the buyer gets the object and $p \geq 0$ is the expected price associated with this probability. A lottery (q, p) belongs to Menu(μ) if there exists some type (v, w) for which $(x(v, w), s(v, w)) = (q, p)$.

For any positive integer m, let $\mathcal{S}_m \subseteq \mathcal{M}$ denote the subclass of mechanisms that have a menu size of (at most) m. In a setting where the buyer faces no financial constraints, Myerson [21] shows that for any random valuation V, the menu size of the optimal mechanism is $m = 1$. Thus, without budgets, there is no revenue gap between the subclass $\mathcal{S}_1$ and $\mathcal{M}$ (where these two sets are defined ignoring the (BF) constraint).

Unfortunately this result doesn't extend to a setting with financial constraints. Indeed, [8] characterize the optimal selling mechanism for a single item in a seller–buyer setting with private valuation and private budget, where valuation and budget may be correlated. They show that the revenue maximizing mechanism may require a continuum of lotteries. This is true even in cases where the buyer has a publicly known valuation. As a result, any mechanism with finite menu size is suboptimal. We adapt their example to our setting with ex-post budget feasibility and participation constraints.

Example 1. The buyer has a publicly known valuation of $\hat{v} > 1$ and a private budget that lies in the unit interval. Budgets are uniformly distributed on $[0, 1]$, which is common knowledge. Thus, the support of the random vector (V, W) is $\{\hat{v}\} \times [0, 1]$.

It is not difficult to show that the optimal mechanism $\mu^* = (x^*, s^*)$, which we define only on the support of the (V, W)–buyer for simplicity, is associated with a critical budget level $0 < w_c < 1$ such that

$$\begin{aligned} x^*(\hat{v}, w) &= \frac{\hat{v} - w_c}{\hat{v} - w} \quad \text{and} \quad s^*(\hat{v}, w) = w\, x^*(\hat{v}, w), \quad \text{for } 0 \leq w < w_c, \\ x^*(\hat{v}, w) &= 1 \quad \text{and} \quad s^*(\hat{v}, w) = w_c, \quad \text{for } w_c \leq w \leq 1. \end{aligned}$$

Thus, the seller pools all types with budgets above w_c and offers them the degenerate lottery $(1, w_c)$. Upon receiving the good (which happens with probability one), any buyer that buys this lottery pays a price of w_c. Types with budgets below w_c are offered different lotteries, where the probability of trade is increasing in the budget level.

Standard arguments can be used to verify that $\mu^* = (x^*, s^*)$ is implementable. Crucially, one can also show that the budget level w_c above, which depends on $\hat{v}$, is bounded away from zero. Thus, the menu size of the optimal mechanism in this example is infinite, since Menu(μ^*) contains a continuum of entries. △

Feng et al. [12] investigate the power of posted prices as a selling mechanism in the presence of private budgets. They showed that the revenue gap between any single take-it-or-leave-it offer and the exact optimal mechanism can be infinite, even for a single buyer setting where the budget distribution and the valuation distribution are independent from each other. Therefore, the optimal revenue in their setting cannot be well approximated using mechanisms in the subclass $\mathcal{S}_1$. In this subsection, we show that if the family of random vectors contains only those with bounded support in terms of both the valuation and the budget realizations, then simple, finite menu size mechanisms provide a good approximation to the optimal revenue. In the next subsection we show how things change dramatically when one considers random variables with unbounded support. In these two cases, we do not assume independence between the valuation and the budget.

Our first result on the revenue gap between optimal mechanisms and optimal approximation mechanisms states that if valuations and budgets are bounded above and below (away from zero), then an arbitrarily high fraction of the optimal revenue can be extracted using a simple mechanism with finite (poly-logarithmic) menu size. This result holds even if the family of buyers considered by the seller includes random vectors where valuation and budget are correlated.

Proposition 2. *Fix $1 < \overline{v} < \infty$ and $1 < \overline{w} < \infty$. Let $\mathscr{B}_b$ be a family of (V, W)–buyers that contains all distributions whose supports are subsets of $[1, \overline{v}] \times [1, \overline{w}]$. Then, for every $\epsilon > 0$ there exists a positive integer $m \leq (1 + \lfloor \log_{1+\epsilon} \overline{v} \rfloor)(1 + \lfloor \log_{1+\epsilon} \overline{w} \rfloor)$ such that*

$$GFOR(\mathcal{S}_m; \mathscr{B}_b) \;\geq\; \frac{1}{1+\epsilon}.$$

Proof. Let (V, W) be a random vector in $\mathscr{B}_b$, so that any realization (v, w) of (V, W) lies in the bounded rectangle $[1, \overline{v}] \times [1, \overline{w}]$. Fix any $\epsilon > 0$, and consider a (V', W')–buyer such that

$$(V', W') \;=\; \frac{1}{1+\epsilon}(V, W).$$

In words, type $(v', w') = (1 + \epsilon)^{-1}(v, w)$ belongs to the support of (V', W') if and only if type (v, w) is in the support of (V, W). Clearly,

$$\text{Rev}(V', W') \;=\; \frac{1}{1+\epsilon}\,\text{Rev}(V, W), \tag{2}$$

since $(x^*, s^*) \in \mathcal{M}$ is a revenue maximizing mechanism for the (V, W)–buyer if and only if $(x^*, s^*/(1 + \epsilon)) \in \mathcal{M}$ is a revenue maximizing mechanism for the scaled down (V', W')–buyer.

We construct a new random vector $(\hat{V}, \hat{W})$ from (V, W) as follows. Replace every realization (v, w) in the support of (V, W) with $(\hat{v}, \hat{w})$, where $\hat{v}$ is the highest non-negative integer power of $1 + \epsilon$ that is smaller than or equal to v, and likewise $\hat{w}$ is the highest non-negative integer power of $1 + \epsilon$ that is smaller than or equal to w. Observe that different types in the support of (V, W) may 'collapse' to the same type in the rounded down random vector $(\hat{V}, \hat{W})$, in which case the probability of type $(\hat{v}, \hat{w})$ is the sum of probabilities of the collapsing types in (V, W). By construction, the $(\hat{V}, \hat{W})$–buyer contains at most

$$m^* \;=\; (1 + \lfloor \log_{1+\epsilon} \overline{v} \rfloor)(1 + \lfloor \log_{1+\epsilon} \overline{w} \rfloor) \tag{3}$$

distinct types.[12] Thus, the optimal mechanism in $\mathcal{M}$ for the $(\hat{V}, \hat{W})$–buyer has menu size $m \leq m^*$.

Notice that for every type (v, w) in the support of (V, W), we have $v/(1+\epsilon) \leq \hat{v} \leq v$ and $w/(1+\epsilon) \leq \hat{w} \leq w$, where $(\hat{v}, \hat{w})$ is the corresponding 'collapsed' type in the support of $(\hat{V}, \hat{W})$. Since $(V', W') = 1/(1+\epsilon)(V, W)$, using Proposition 1, we conclude that $\text{Rev}(V', W') \leq \text{Rev}(\hat{V}, \hat{W})$. A further application of Proposition 1 shows that any mechanism μ with menu size m generates a larger expected revenue from the (V, W)–buyer than from the $(\hat{V}, \hat{W})$–buyer. But this implies $\text{Rev}(\hat{V}, \hat{W}) \leq \text{Rev}(V, W|\mathcal{S}_m)$, since the optimal mechanism for the $(\hat{V}, \hat{W})$–buyer has menu size of at most m. Using Eq. 2 obtains

$$\frac{1}{1+\epsilon} \leq \frac{\text{Rev}(V, W|\mathcal{S}_m)}{\text{Rev}(V, W)}.$$

Our argument holds for any (V, W)–buyer in the class of distributions $\mathscr{B}_b$ with support inside the bounded rectangle $[1, \overline{v}] \times [1, \overline{w}]$. Thus, we conclude that

$$\frac{1}{1+\epsilon} \leq \text{GFOR}(\mathcal{S}_m; \mathscr{B}_b),$$

as desired. □

Remark 3. Notice that we can bound m^*, which itself is an upper bound for the menu size complexity needed in Proposition 2, by the following:

$$\begin{aligned} m^* &= (1 + \lfloor \log_{1+\epsilon} \overline{v} \rfloor)(1 + \lfloor \log_{1+\epsilon} \overline{w} \rfloor) \leq 4 \log_{1+\epsilon} \overline{v} \log_{1+\epsilon} \overline{w} \\ &\leq 4 \left(1 + \tfrac{1}{\epsilon}\right)^2 \log_2 \overline{v} \log_2 \overline{w}, \end{aligned}$$

where the last inequality is obtained by changing the base of the logarithm to 2 and using the fact that $1/\log_2(1+\epsilon) \leq 1 + 1/\epsilon$, for every $\epsilon > 0$. Consequently, the upper bound on menu size complexity for the family of buyers $\mathscr{B}_b$ depends on ϵ, $\overline{v}$ and $\overline{w}$ in a poly-logarithmic way. Thus, it has a reasonable size even for large upper bounds for the valuation and/or the budget. For fixed bounds on the support of the distributions considered, a lower $\epsilon > 0$ (equivalently, a better approximation to the optimal revenue) demands a larger menu size complexity. For instance when $\overline{v} = \overline{w} = 2$, Eq. 3 shows for $\epsilon = 1$ (hence, to approximate half the optimal revenue) it suffices to consider mechanisms with menu size of at most 4. But the upper bound on menu size complexity increases to 8 if $\epsilon = 1/2$.

3.2 Bad Approximations with Simple Mechanisms

A conjecture that seems to be suggested from Proposition 2 is that, as the upper bounds $\overline{v}$ and $\overline{w}$ of the support of the random vector (V, W) increase without bound, the menu size complexity required to guarantee an arbitrarily

[12] To see this, notice that $\hat{v} \in \left\{(1+\epsilon)^0, (1+\epsilon)^1, (1+\epsilon)^2, \ldots, (1+\epsilon)^{\lfloor \log_{1+\epsilon} \overline{v} \rfloor}\right\}$, and likewise for $\hat{w}$.

high fraction of the optimal revenue tends towards infinity. The validity of this conjecture is not immediately verified for three reasons. First, Proposition 2 provides a poly-logarithmic upper bound on the menu size complexity, but it does not provide a matching lower bound. Thus, we cannot be sure of how tight the upper bound is. Second, the right-hand side of Eq. 3 grows in a sub-linear way. So, it is possible that, as $\overline{v}$ and $\overline{w}$ grow, letting the size of the menu grow faster (i.e., linearly) would keep the good approximation result in place. Finally, it is not clear what happens when the support of the marginal distribution of V is bounded and that of the marginal distribution of W is unbounded (or vice versa).

However, our next result shows that despite these considerations, the conjecture holds. If the menu size of the approximation mechanisms grows sub-linearly when the upper bounds for the valuation and budget, $\overline{v}$ and $\overline{w}$ respectively, tend to infinity, the revenue loss can become unboundedly large.

Proposition 3. *Let $\mathscr{B}_u$ be a family of (V, W)–buyers that contains all distributions with supports that are unbounded from above. Then for any fixed positive integer m,*

$$GFOR(\mathcal{S}_m; \mathscr{B}_u) \;=\; 0.$$

Proof. Fix $k \geq 2$ and let $B = 2k$. Consider a (V, W)–buyer with a public valuation and k private budgets. More specifically, the support for the (V, W)–buyer is given by

$$\left\{(v_i, w_i) : w_i \;=\; B^i, \text{ for all } i = 1, \ldots, k, \text{ and } v_i = \overline{v} \geq B^{k+1}\right\}. \tag{4}$$

Further, assume that the probability mass function f for the random vector (V, W) is

$$f(v_i, w_i) \;\equiv\; f_i \;=\; \begin{cases} (1 - B^{-1})\, B^{-(i-1)} & : i \;=\; 1, \ldots, k-1, \\ 1 - \sum\limits_{j=1}^{k-1} f_j & : i = k. \end{cases} \tag{5}$$

The proof of Proposition 3 follows from two lemmas that we make in regard to the revenues that can be raised from the (V, W)–buyer.

Lemma 1. *A lower bound for the optimal revenue from the (V, W)–buyer is $B^2/8$; i.e.,*

$$\text{Rev}(V, W) \;\geq\; B^2/8.$$

Lemma 2. *For any $1 \leq m < k$, an upper bound for the optimal revenue from the (V, W)–buyer in the class of mechanisms with menu size m is $(m+1)B$; i.e.,*

$$(m \,+\, 1)\, B \;\geq\; \text{Rev}(V, W | \mathcal{S}_m).$$

Since $B = 2k$, the following ratio obtains from Lemmas 1 and 2

$$\frac{\text{Rev}(V, W | \mathcal{S}_m)}{\text{Rev}(V, W)} \;\leq\; \frac{8(m+1)}{B} \;=\; \frac{4(m+1)}{k}.$$

We are looking at the family of buyer's distributions with unbounded support for W, letting $k \to \infty$ shows that $\text{GFOR}(\mathcal{S}_m; \mathscr{B}_u) = 0$, as desired. □

It would appear that the fact that $\mathscr{B}_u$ contains random vectors (V, W) with unbounded support in both V and W is crucial for our bad approximation result in Proposition 3. Indeed, if either the marginal distribution of V or W has bounded support, then $\text{Rev}(V, W) \leq \mathbb{E}\big[\min\{V, W\}\big] < +\infty$. But this is not correct. The key to the above result is the observation that, while both $\text{Rev}(V, W)$ and $\text{Rev}(V, W|\mathcal{S}_m)$ may increase without bound, the first one does so at a faster rate. This can happen as well when the lower bounds for the valuation and budget become arbitrarily close to zero.

In particular, we can show the following.

Proposition 4. *Let $\mathscr{B}_0$ be a family of (V, W)–buyers that contains all distributions whose supports are subsets of $[0, 1] \times [0, 1]$. Then for any fixed positive integer m,*

$$GFOR(\mathcal{S}_m; \mathscr{B}_0) = 0.$$

Technically, the proof of Proposition 3 shows a stronger result. Indeed, the revenue gap between the optimal mechanism and the optimal approximation mechanism can be exceedingly large as long as the menu size complexity of the class of approximation mechanisms we consider grows in a sub-linear way. Recall that a positive real valued function ϕ defined on the set of positive integers is called *sub-linear* if $\lim_{k \to \infty} \phi(k)/k = 0$. Thus, we can show that any mechanism with asymptotically sub-linear menu size complexity cannot guarantee a positive fraction of the optimal revenue, if the buyer's family of random-vectors is either $\mathscr{B}_u$ or $\mathscr{B}_0$.

Proposition 5. *Let $\{(V_k, W_k) \colon k = 1, 2, \ldots\}$ be a sequence of random vectors, where for each k the distribution of (V_k, W_k) is given in Eq. 5. Then, for any sub-linear function ϕ,*

$$\lim_{k \to \infty} \frac{\text{Rev}(V_k, W_k|\mathcal{S}_{\phi(k)})}{\text{Rev}(V_k, W_k)} = 0.$$

Remark 4. While the marginal distribution of the budgets in the proofs of Proposition 3 and Proposition 4 is important in our construction, this distribution shares similar properties with the standard geometric distribution. In particular, it gives diminishing weight to types with higher budgets.

3.3 Cash Bond Relaxations

In this part of the paper we consider a relaxation of the seller's revenue maximization problem. Following Che and Gale [8], we study a setting where the seller can prevent the buyer from over-reporting her budget at no cost.[13] In practice, ruling out budget over-reporting can be accomplished by additional institutional

[13] See Section 4 of their paper.

arrangements. For example, the seller can require the buyer to post a cash bond prior to choosing a lottery, or by some information disclosure rules (e.g., financial reports, bank statements, etc.). In this setting, the incentive compatibility constraints are required to hold only for under-reporting the private budget. As a result, higher budgeted types might be offered better deals (discounts).

To formalize this possibility, let $\mathcal{N}_{cb}$ be the class of ex-post budget feasible and ex-post individually rational mechanisms that require a (V, W)–buyer to post a cash bond to prevent her from over-reporting her budget. More explicitly, a mechanism $\nu = (x, s)$ belongs to $\mathcal{N}_{cb}$ if it satisfies (BF), (IR) and the following constraint, which replaces (IC): for all (v, w) and all $(\tilde{v}, \tilde{w})$ with $\tilde{w} \leq w$, it must be that

$$v\, x(v,w) \;-\; s(v,w) \;\geq\; v\, x(\tilde{v},\tilde{w}) \;-\; s(\tilde{v},\tilde{w}). \tag{CB}$$

In words, because the seller exacts a cash bond from the buyer, the mechanism $\nu \in \mathcal{N}_{cb}$ only needs to prevent a higher-budget type from mimicking a lower-budget type. It is immediate to realize that $\mathcal{M} \subseteq \mathcal{N}_{cb}$. Thus, for any family $\mathscr{B}$ of buyers, we have that

$$\mathrm{MVR}(\mathcal{N}_{cb}; \mathscr{B}) \;\geq\; 1,$$

since for any (V, W)-buyer in $\mathscr{B}$, it must be $\mathrm{Rev}(V, W) \leq \mathrm{Rev}(V, W|\mathcal{N}_{cb})$.

From a computational point of view, the importance of focusing on this relaxed problem comes from the fact that it admits a natural linear programming representation in discrete type spaces, which has been studied before (see Bhattacharya et al. [3] for example). In particular, Devanur and Weinberg [11] compute the optimal mechanism in this relaxed setting and show that it has a menu with exponentially-many non-trivial options (in the number of possible budgets).

In principle, then, the optimal revenue obtained in the superclass of mechanisms $\mathcal{N}_{cb}$ can be used to approximate the optimal revenue in $\mathcal{M}$. Of course, this provides a good approximation as long as $\mathrm{MVR}(\mathcal{N}_{cb}; \mathscr{B})$ has a reasonable bound. Che and Gale [8] show that if the random variables V and W are positively affiliated, then $\mathrm{Rev}(V, W) = \mathrm{Rev}(V, W|\mathcal{N}_{cb})$.[14] Thus, when one focuses on the family of (V, W)–buyers where the valuation and the budget are affiliated random variables, there is no revenue gap between optimal mechanisms in $\mathcal{M}$ and optimal approximation mechanisms in $\mathcal{N}_{cb}$.

Unfortunately, positive affiliation is crucial for this good approximation result to hold. Che and Gale [8] provide an example showing that a higher revenue can be extracted (compared to the optimal auction) when the seller uses cash-bonds to prevent over-reporting of private budgets. By adjusting this example, we show that the optimal revenue in the class $\mathcal{N}_{cb}$ can be unboundedly large compared to the optimal revenue in $\mathcal{M}$. This shows that optimal approximation mechanism in the class $\mathcal{N}_{cb}$ cannot serve as proxies to the revenue maximization problem in the class $\mathcal{M}$.

---

[14] This result uses an additional *declining marginal revenue* assumption—see Assumptions 1 and 2 in [8].

Proposition 6. *Let $\mathscr{B}_{na}$ be a family of (V, W)–buyers that contains all distributions where V and W are negatively affiliated. Then one has*

$$\mathrm{MVR}(\mathcal{N}_{cb}; \mathscr{B}_{na}) \;=\; +\infty.$$

What do we make of the result stated in Proposition 6? In other words, should one take it as a positive or a negative approximation result? From a computational perspective, it is certainly not positive. Using the class of cash-bond mechanisms to approximate the optimal revenue in $\mathcal{M}$, or equivalently ignoring the incentive constraints that consider over-reporting, can yield an exceedingly large overestimation of the optimal revenue generated by a fully incentive compatible mechanism. At the same time, the fact that the maximal value of relaxation in $\mathcal{N}_{cb}$ is potentially large may provide a justification for the seller to look for institutional fixes; i.e., lobby financial or other regulatory authorities to make the use of cash bonds legal.

3.4 Restricted Mechanisms Under Strong Incentive Constraints

Daskalakis et al. [10] studied selling mechanisms when the incentive compatibility constraints hold even if the deviation produces a non-affordable outcome. In other words, [10] increase the number of incentive constraints, by considering deviations to lotteries that are not ex-post affordable for a buyer given her budget. Formally, they consider the following incentive constraint on the mechanism $\mu = (x, s)$: for all $(v, w) \in \mathbb{R}^2_+$ and all $(\tilde{v}, \tilde{w}) \in \mathbb{R}^2_+$, it must be that

$$v\,x(v, w) \;-\; s(v, w) \;\geq\; v\,x(\tilde{v}, \tilde{w}) \;-\; s(\tilde{v}, \tilde{w}). \tag{SIC}$$

Comparing it with (IC), this constraint omits the requirement that the deviation of the buyer's type (v, w) to purchasing a lottery for the type $(\tilde{v}, \tilde{w})$ must be within (v, w)'s budget.

Let $\mathcal{N}_{sic} \subseteq \mathcal{M}$ denote the class of mechanisms that satisfy (BF), (IR) and (SIC). Of course, the imposition of these 'extra' incentive constraints means that, for any (V, W)–buyer, the optimal revenue in the class $\mathcal{N}_{sic}$ will be weakly lower than the optimal revenue in the class $\mathcal{M}$. The advantage of imposing these constraints is that the resulting setting has a natural linear programming formulation. So, in principle, it can be used to approximate the revenue generated by the optimal mechanisms in $\mathcal{M}$.

Our two final results on revenue gaps in the presence of budget constraints show that, sometimes, the expected revenue of the optimal approximation mechanism in the subclass $\mathcal{N}_{sic}$ can be arbitrarily small compared to the expected revenue of the optimal mechanism. Thus, despite its computational advantages, the subclass $\mathcal{N}_{sic}$ of mechanisms with the strong incentive constraint cannot serve as a good proxy to the class $\mathcal{M}$ of incentive compatible, ex-post budget feasible and ex-post individually rational mechanisms. Below we argue that this holds at least in two important cases: when one considers the family of (V, W)–buyers with unbounded support (Proposition 7), or the family of (V, W)–buyers with support in the unit square (Proposition 8).

Proposition 7. *Let $\mathscr{B}_u$ be the family of (V, W)–buyers that contains all distributions with unbounded support. Then one has that*

$$GFOR(\mathcal{N}_{sic}; \mathscr{B}_u) \;=\; 0.$$

Proposition 8. *Let $\mathscr{B}_0$ be a family of (V, W)–buyers that contains all distributions whose supports are subsets of $[0, 1] \times [0, 1]$. One has that*

$$GFOR(\mathcal{N}_{sic}; \mathscr{B}_0) \;=\; 0.$$

These bad approximation results in terms of the revenue gap between computationally feasible mechanisms in $\mathcal{N}_{sic}$ and optimal mechanisms in $\mathcal{M}$ mimic the bad approximation results for simple mechanisms previously obtained. Unfortunately, we have not been able to show either a good approximation result or a bad approximation result in the subclass $\mathcal{N}_{sic}$ when the family of (V, W)–buyers consists of random vectors with support bounded from above and from below (away from zero). We leave this for future work.

Acknowledgments and Disclosure of Interest. We would like to thank comments and discussions with Murali Agastya, Isa Hafalir, Daniel Lehmann, Simon Loertscher, Idione Meneghel, Noam Nisan and the audience at the 2023 Markets, Contracts and Organizations Conference organized by the Australian National University. Financial support from the Australian Research Council under grant DP190102064 is gratefully acknowledged. Declaration of interest: none.

References

1. Babaioff, M., Gonczarowski, Y.A., Nisan, N.: The menu-size complexity of revenue approximation. Games Econom. Behav. **134**, 281–307 (2022). https://doi.org/10.1016/j.geb.2021.03.001
2. Babaioff, M., Nisan, N., Rubinstein, A.: Optimal deterministic mechanisms for an additive buyer. In: Proceedings of the 2018 ACM Conference on Economics and Computation, EC 2018, p. 429. ACM (2018)
3. Bhattacharya, S., Conitzer, V., Munagala, K., Xia, L.: Incentive compatible budget elicitation in multi-unit auctions. In: Proceedings of the Twenty-First Annual ACM-SIAM Symposium on Discrete Algorithms, pp. 554–572 (2010)
4. Boulatov, A., Severinov, S.: Optimal and efficient mechanisms with asymmetrically budget constrained buyers. Games Econom. Behav. **127**, 155–178 (2021). https://doi.org/10.1016/j.geb.2021.02.001
5. Briest, P., Chawla, S., Kleinberg, R., Weinberg, S.M.: Pricing lotteries. J. Econ. Theory **156**, 144–174 (2015). https://doi.org/10.1016/j.jet.2014.04.011
6. Chawla, S., Malec, D., Malekian, A.: Bayesian mechanism design for budget-constrained agents. In: Proceedings of the 12th ACM Conference on Electronic Commerce, EC 2011, pp. 253–262. ACM (2011)
7. Che, Y.K., Gale, I.: Standard auctions with financially constrained bidders. Rev. Econ. Stud. **65**(1), 1–21 (1998)
8. Che, Y.K., Gale, I.: The optimal mechanism for selling to a budget-constrained buyer. J. Econ. Theory **92**, 198–233 (2000)

9. Che, Y.K., Gale, I., Kim, J.: Assigning resources to budget-constrained agents. Rev. Econ. Stud. **80**(1), 73–107 (2013)
10. Daskalakis, C., Devanur, N.R., Weinberg, S.M.: Revenue maximization and ex-post budget constraints. ACM Trans. Econ. Comput. **6**(3–4), 1–19 (2018)
11. Devanur, N.R., Weinberg, S.M.: The optimal mechanism for selling to a budget-constrained buyer: the general case. In: Proceedings of the 2017 ACM Conference on Economics and Computation, EC 2017, pp. 39–40. Microsoft Research, ACM Press, New York (2017)
12. Feng, Y., Hartline, J.D., Li, Y.: Simple mechanisms for agents with non-linear utilities (2022). arXiv – https://arxiv.org/abs/2003.00545
13. Hart, S., Nisan, N.: Approximate revenue maximization with multiple items. J. Econ. Theory **172**, 313–347 (2017). https://doi.org/10.1016/j.jet.2017.09.001
14. Hart, S., Nisan, N.: Selling multiple correlated goods: revenue maximization and menu-size complexity. J. Econ. Theory **183**, 991–1029 (2019)
15. Hart, S., Reny, P.J.: Maximal revenue with multiple goods: nonmonotonicity and other observations. Theor. Econ. **10**(3), 893–922 (2015). https://doi.org/10.3982/te1517
16. Hart, S., Reny, P.J.: The better half of selling separately. ACM Trans. Econ. Comput. **7**(4), 18 (2019). https://doi.org/10.1145/3369927
17. Kotowski, M.H.: First-price auctions with budget constraints. Theor. Econ. **15**(1), 199–237 (2020). https://doi.org/10.3982/te2982
18. Li, X., Yao, A.C.C.: On revenue maximization for selling multiple independently distributed items. Proc. Natl. Acad. Sci. U.S.A. **110**(28), 11232–11237 (2013). http://www.jstor.org/stable/42712717
19. Manzini, P., Mariotti, M.: Sequentially rationalizable choice. Am. Econ. Rev. **97**(5), 1824–1839 (2007)
20. McAfee, R., McMillan, J.: Multidimensional incentive compatibility and mechanism design. J. Econ. Theory **46**(2), 335–354 (1988). https://doi.org/10.1016/0022-0531(88)90135-4
21. Myerson, R.B.: Optimal auction design. Math. Oper. Res. **6**(1), 58–73 (1981)
22. Pai, M.M., Vohra, R.V.: Optimal auctions with financially constrained buyers. J. Econ. Theory **150**, 383–425 (2014)
23. Richter, M.: Mechanism design with budget constraints and a population of agents. Games Econ. Behav. **115**, 30–47 (2019)
24. Rochet, J.C., Chone, P.: Ironing, sweeping, and multidimensional screening. Econometrica **66**(4), 783–826 (1998). https://doi.org/10.2307/2999574
25. Roughgarden, T., Talgam-Cohen, I.: Approximately optimal mechanism design. Ann. Rev. Econ. **11**, 355–381 (2019). https://doi.org/10.1146/annurev-economics-080218-025607

Sublogarithmic Approximation for Tollbooth Pricing on a Cactus

Andrzej Turko^(✉) and Jarosław Byrka

University of Wrocław, Wrocław, Poland
andrzej.turko@gmail.com, jby@cs.uni.wroc.pl

Abstract. We study an envy-free pricing problem, in which each buyer wishes to buy a shortest path connecting her individual pair of vertices in a network owned by a single vendor. The vendor sets the prices of individual edges with the aim of maximizing the total revenue generated by all buyers. Each customer buys a path as long as its cost does not exceed her individual budget. In this case, the revenue generated by her equals the sum of prices of edges along this path. We consider the unlimited supply setting, where each edge can be sold to arbitrarily many customers. The problem is to find a price assignment which maximizes vendor's revenue. A special case in which the network is a tree is known under the name of the tollbooth problem. Gamzu and Segev proposed a $\mathcal{O}\left(\frac{\log m}{\log\log m}\right)$-approximation algorithm for revenue maximization in that setting. Note that paths in a tree network are unique, and hence the tollbooth problem falls under the category of single-minded bidders, i.e., each buyer is interested in a single fixed set of goods.

In this work we step out of the single-minded setting and consider more general networks that may contain cycles. We obtain an algorithm for pricing cactus shaped networks, namely networks in which each edge can belong to at most one simple cycle. Our result is a polynomial time $\mathcal{O}\left(\frac{\log m}{\log\log m}\right)$-approximation algorithm for revenue maximization in tollbooth pricing on a cactus graph. It builds upon the framework of Gamzu and Segev, but requires substantially extending its main ideas: the recursive decomposition of the graph, the dynamic programming for rooted instances and rounding the prices.

Keywords: envy-free pricing · tollbooth problem · cactus graphs

1 Introduction

The problem of maximizing revenue by setting optimal prices has been widely studied in various settings (see, e.g., [1,2,12]). This work discusses the problem of envy-free pricing for revenue maximization. In general, this problem can be

Supported by NCN grant number 2020/39/B/ST6/01641.
Full version available at https://arxiv.org/abs/2305.05405.

G. Schäfer and C. Ventre (Eds.): SAGT 2024, LNCS 15156, pp. 297–314, 2024.
https://doi.org/10.1007/978-3-031-71033-9_17

modeled as a two phase game. In the first step, vendor assigns prices to the offered goods. Then, each buyer purchases her most preferred subset of goods based on given prices and her own preferences. Every buyer aims to maximize her utility, and the seller aims to maximize the total price paid by customers. The problem is to find an optimal strategy for the vendor.

More precisely, an instance of the envy-free pricing problem consists of m goods and n buyers. Each buyer is defined by a function which assigns a non-negative valuation to every subset of the goods. It is assumed that the valuation of an empty set for each customer equals zero. A solution to the problem is formed by non-negative prices of goods and an envy-free allocation of goods to the buyers. Utility of a buyer from a set of goods equals her valuation of this set minus the total price of its elements. An allocation is envy-free when no buyer would like to change her assigned set of goods. In other words, the set assigned to her must maximize her utility.

In this work we focus on the unlimited supply setting, where each one of the m goods can be sold to arbitrarily many buyers. Such goods may be thought of as intellectual property or access to infrastructure. Sometimes the limited supply setting is also considered, where each good is available only in a certain number of copies. In that case, the solution must not only satisfy the envy-freeness constraints, but also the number of buyers any good is allocated to must not exceed its supply.

We study a natural case of the envy-free pricing with unlimited supply, where the goods can be modeled by edges in a graph and buyers wish to purchase cheapest paths. More precisely, each buyer has equal positive valuations for paths connecting a certain pair of vertices and zero valuation for all the other sets of goods. Such a problem may be used to model a situation where the vendor is an owner of a road network and buyers are drivers wishing to travel from one city to another.

Guruswami et al. [11] have defined and studied two subcases of this scenario called: the tollbooth and the highway problems. In the former the underlying graph is a tree and in the latter it is a path. We extend this collection by adding cactus graphs that allow edge-disjoint cycles and hence allow more than one path being attractive for a client. To the best of our knowledge this is the first work that addresses envy-free tollbooth pricing of networks where clients have alternative routes (are not single-minded).

1.1 Related Work

The problem of envy-free pricing for revenue maximization has been studied in various settings. We are going to survey mostly the results for single-minded buyers, a model where each buyer has positive valuation for exactly one set of goods.

Guruswami et al. [11] defined the single-minded buyers setting and presented a polynomial $\mathcal{O}(\log m + \log n)$-approximation algorithm for the variant with unlimited supply. Also for the unlimited supply setting, Balcan, Blum and Mansour [3] have shown that a logarithmic guarantee on expected revenue can be achieved by randomly setting a single price to all the goods. Notably, this result holds for buyers with arbitrary valuations. By taking it as a reference point, a natural question is: For what valuation classes sublogarithmic approximation of revenue is possible?

Indeed, for special cases of the unlimited supply setting with single-minded buyers, such results were obtained. For the tollbooth problem, Gamzu and Segev [8] achieved a $\mathcal{O}\left(\frac{\log m}{\log\log m}\right)$-approximation of revenue with a polynomial algorithm. For the highway problem, Grandoni and Rothvoß [9] have designed a polynomial time approximation scheme (PTAS).

Already for these two problems hardness results for envy-free pricing with single-minded buyers are known. Guruswami et al. [11] have proven that the tollbooth problem is NP-hard. This was followed by a result from Briest and Krysta [4], who showed the same for the highway problem.

For the general envy-free pricing Demaine et al. [6] showed several inapproximability results under various complexity assumptions. They proved a lower bound of $\Omega(\log n)$, under a hardness hypothesis regarding the balanced bipartite independent set problem. In this context, the result of Gamzu and Segev [8] shows that pricing is strictly simpler to approximate on trees. We extend it to show that sublogarithmic approximation of revenue is also possible on cactus graphs.

Of course, the mentioned impossibility results hold for limited supply as well. In that setting there also are several approximation results. Cheung and Swamy [5] have designed a $\mathcal{O}\left(\sqrt{m}\log u_{max}\right)$-approximation algorithm for the general envy-free pricing problem with single-minded buyers (u_{max} denotes the maximal number of copies of a single good). In the tollbooth and highway problems they have obtained approximation ratio of $\mathcal{O}\left(\log u_{max}\right)$. Elbassioni, Fouz and Swamy [7] have obtained a matching approximation guarantee for the non-envy-free tollbooth problem without the single-mindedness constraint. More recently, Grandoni and Wiese [10] obtained a PTAS for the limited supply version of the highway problem.

1.2 Our Result

We consider the tollbooth problem on cactus graphs, a natural generalization of the original tollbooth problem (on trees) with unlimited supply. Instead of requiring the graph to be a tree, we only require that the underlying graph is a cactus, i.e. its every edge belongs to at most one simple cycle. The main difference between the two models is that, unlike in a tree, in a cactus there can be multiple simple paths connecting a single pair of vertices. Thus, each buyer can be interested in purchasing multiple sets of goods, i.e. is not single-minded. We obtain the following result:

Theorem 1. *There exists a polynomial time approximation algorithm for the tollbooth problem on cactus graphs with unlimited supply which achieves an approximation guarantee for revenue of* $\mathcal{O}\left(\frac{\log m}{\log\log m}\right)$, *where* m *is the number of edges of the graph.*

Our approximation algorithm utilizes a similar framework as the algorithm by Iftah Gamzu and Danny Segev [8] for the classical tollbooth problem (on trees). However, various parts of the algorithmic construction are carefully adapted to handle cycles and the freedom of clients to choose one of two routes on each cycle they have on their way.

To the best of our knowledge, this is the first such result for graphs more general than trees, and hence the first one not restricted to the single-minded bidder case.

1.3 Model and Preliminaries

Let us consider an instance of the tollbooth problem on cactus graphs with m goods and n buyers. Its description consists of a simple graph G with m edges such that no edge lies on two simple cycles and a set B of buyers. Each buyer $i \in B$ is described by a pair of vertices u_i and v_i, and her budget $b_i > 0$. For each subset of edges S, her valuation is defined in the following way:

$$f_i(S) = \begin{cases} b_i, & \text{if } S \text{ consists of edges along a } u_i\text{-}v_i \text{ path} \\ 0, & \text{otherwise} \end{cases} \tag{1}$$

A solution is a real vector p assigning non-negative prices to the edges of G. Let us treat the prices as lengths of edges and let d_i denote the distance between v_i and u_i. If $b_i \geq d_i$, i-th buyer purchases all edges along a shortest u_i-v_i path. Otherwise, she buys nothing. Such an allocation is envy-free. Note that if there are many shortest u_i-v_i paths, choosing either one does not change the revenue.

In this work we present an algorithm for finding such prices that the above-mentioned way of allocating goods to buyers results in revenue $\mathcal{O}\left(\frac{\log m}{\log\log m}\right)$ times smaller than optimal.

1.4 Overview of Techniques

Our algorithm follows the classify-and-select paradigm of Gamzu and Segev. Buyers are split into $\mathcal{O}\left(\frac{\log m}{\log\log m}\right)$ subsets, which define separate instances of the problem. Those subproblems are processed independently and constant factor approximations for each of them are computed. The final solution is obtained by choosing the one yielding the biggest revenue. Supply of the goods is unlimited and, thus, the revenue of a solution to a subproblem does not decrease when applied to the initial instance with all the buyers. This gives an $\mathcal{O}\left(\frac{\log m}{\log\log m}\right)$ approximation of revenue, because the total revenue is at most the sum of the revenues of the subproblems.

For each subproblem the algorithm constructs a subgraph of G, called the skeleton, such that all paths desired by a buyer in the given instance enter and leave the skeleton exactly once and in the same vertices. This way, each such path is split into three parts, one of which is in the skeleton and the other two are not. Note that, for a constant factor approximation, it suffices to collect revenue either only on the skeleton or only outside the skeleton. In the former case the revenue is achieved by setting appropriate prices of the skeleton's edges. In the latter it suffices to focus on groups entering or leaving the skeleton through the same vertex leading to rooted instances. Due to cycles in the underlying graph, several significant challenges arise in both subproblems.

Rooted Instances: The algorithm by Gamzu and Segev solved rooted instances with a black-box dynamic programming algorithm from [11]. In our case the subgraphs forming rooted instances can contain cycles. Thus, that dynamic programming, which has been designed for trees, could not be applied verbatim. In order to define its subproblems, we have generalized the notion of a subtree using the tree-like structure of biconnected components. Another key element of our solution is a technique which effectively transforms cycles into paths. It is based on the observation, that, as far as shortest paths from vertices to the root are concerned, one edge of a cycle is always redundant. By guessing this edge, one can tackle the problem on a cycle as if it was a path. For an optimization problem it is enough to iterate over all possible choices of this edge and calculate the solutions independently using the dynamic programming for a tree. We believe that this technique has a wide range of applications in generalizing algorithms for trees to cyclic graphs whose biconnected components have simple structure. Its usage for our problem is described in Sect. 4.1.

Dependent Subgraphs of the Skeleton: The classification of buyers is based on a recursive decomposition of the input graph. The algorithm for the original tollbooth problem in each step splits the tree into several connected subgraphs and processes buyers who wish to buy paths with endpoints belonging to different ones. Those connected subgraphs can be then processed completely independently because all paths relevant to the next instances are fully contained in the individual subgraphs. This is not the case in cactus graphs, namely for subgraphs which contain edges lying on the same cycle. As the cycles can be arbitrarily long, our algorithm may need to divide them when splitting the skeleton into smaller parts. It turns out that the dependencies between resulting subgraphs regard the cost of paths in the cycle shared between them. By making assumptions about their costs, the algorithm can isolate the subgraphs and process them independently. This approach, however, results in multiple solutions for each subgraph based on different assumed costs of individual parts of shared cycles. While merging those solutions into an approximately optimal global one, the procedure from Sect. 5 controls the cost of each cycle using dynamic programming inspired by the knapsack problem. In order to make this possible, we have extended the price rounding techniques, which allowed to relax the assumptions about the costs of shared cycles and to calculate approximate revenue. Those techniques are described in detail in the full version of the paper [13].

Decomposing the Graph: In the original tollbooth problem it was sufficient to split each subtree at a given level of decomposition into connected subtrees of the right size, which formed the next step of the decomposition. Our solution for handling the dependencies between subgraphs of the skeleton has been made possible by additional properties ensured by the decomposition. For example, we ensure that two subgraphs forming the decomposition can share at most one cycle. Another way of limiting the dependencies between subgraphs, to which G is split, is limiting the number of vertices each subgraph shares with the other ones. The decomposition used by our algorithm is characterized in full detail by Lemma 1.

Pricing the Segments: Segments are edge-disjoint subgraphs of the skeleton. In each of them there are two vertices, called endpoints, and only they can be shared with other segments. Thus, one can think of them as generalization of edges. In one of the subproblems the algorithm fixes the lengths of whole segments and sets the prices of the edges inside a given segment so that the revenue from selling the paths starting inside and ending outside it is maximized. In the original tollbooth problem the buyers are single-minded, so for each such path the endpoint through which it will leave the segment is fixed. For a cactus graph it is not the case, a buyer may choose a path passing though any of the two endpoints depending on the prices.

We handle that additional complexity in the following way. With fixed lengths of segments, it is possible to calculate the maximal amount of money a buyer is able to spend on edges inside the segment where her path begins. Our algorithm uses this to split the buyers into two categories. Each buyer who cannot afford to pay half of the segment's length, for fixed prices of its edges, can only purchase paths passing through a single fixed endpoint. As for the remaining buyers, the vendor can charge half of the segment's cost for edges incident to any of the endpoints and they will be able to pay this much. The procedure based on this idea is described in full detail in the long version of this paper [13]. It is a part of the algorithm for the skeleton subproblem, which is summarized in Sect. 5.

2 Graph Decomposition

Using a recursive decomposition of the cactus graph our algorithm splits the buyer set B into disjoint subsets, which are later processed independently. Here we define this partition.

2.1 The Tree of Biconnected Components

We begin by discussing the structure of biconnected components of the cactus graph. Let us fix an arbitrary vertex of G, denoted r_G, as the root of G for the duration of the whole algorithm. A vertex or edge is said to be above another one if it is closer to r_G. Every biconnected component of a cactus is either a single edge or a simple cycle. Thus, it has a single **topmost vertex** and at most two **topmost edges**, which are exactly those adjacent to the topmost vertex.

Definition 1. *For each cycle in G, its two edges closest to r_G, i.e. topmost edges, form a* ***pair of associated edges****. Note that every edge belongs to at most one such pair.*

Although each vertex can be a topmost vertex in arbitrarily many biconnected components, it can belong to at most one without being its topmost vertex. Furthermore, all vertices except for the root belong to exactly one such biconnected component. Let us call it the main component of this vertex. For the root it is a special component, consisting only of itself (a single vertex). Our algorithm uses the tree of biconnected components rooted in this special component. Every other component is a child of the main component of its topmost vertex. Note that such a tree is unique. The tree of biconnected components of an example cactus graph is illustrated by Fig. 1.

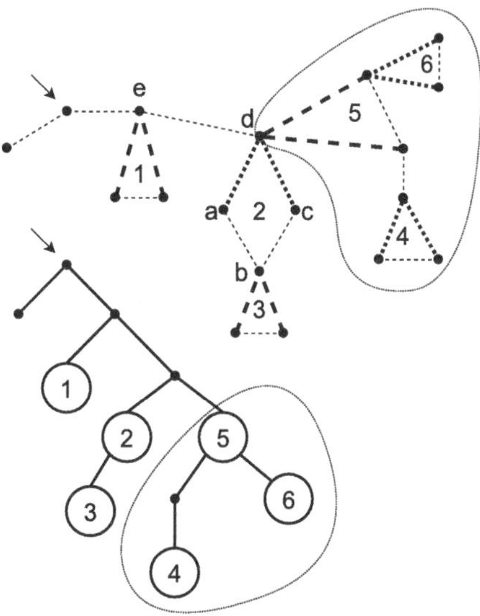

Fig. 1. An example cactus graph with marked pairs of associated edges and its tree of biconnected components. The arrows indicate respectively the root vertex and the root component. On both drawings the subtree graph of the cycle number 5 is marked out. Vertices a, b, c and d belong to biconnected component 2. It's also a main component of a, b and c. However, the main component of d is the edge e-d. Note that e belongs to three distinct biconnected components and it's main component is the edge between e and the root vertex.

Definition 2. *A* ***subtree graph of a component*** *C is the graph consisting of all the edges and vertices belonging to any descendant of C (inclusive) in the tree of biconnected components. The subtree graph of the root component is the whole graph G.*

2.2 Balanced Decomposition

Decompositions $D_1, D_2, \ldots D_L$ of G are defined recursively. Each of them is a family of edge-disjoint subgraphs, called fragments, which cover the graph G. In D_1 the whole graph G forms a single fragment, and in D_L each fragment consists of at most two edges. For all $j < L$ each fragment in partition D_j is split into a number of subgraphs, which become fragments in D_{j+1}.

Definition 3. *A vertex which belongs to multiple fragments in partition D_{j+1} is called a* ***border vertex*** *of j-th level. Furthermore, every vertex of G is considered to be a border vertex of L-th level.*

Note that a border vertex of j-th level is also a border vertex of $(j+1)$-th level.

Lemma 1. *Consider a family of decompositions $D_1, D_2, \ldots D_L$ of a cactus graph G satisfying the following invariants for each valid j:*

1. Each fragment in D_j is split into $\mathcal{O}(k)$ fragments in D_{j+1}.
2. The maximal number of edges in a fragment forming D_{j+1} is $\Omega(k)$ times smaller than in D_j.
3. Each fragment forming D_j contains at most $\mathcal{O}(k)$ border vertices of j-th level.
4. Each pair of associated edges belongs to the same fragment of D_j.
5. All fragments forming D_j are connected subgraphs of G.

For k being an unbounded and nondecreasing function of m (the number of edges in G), such a family can be found in polynomial time.

The second invariant ensures that the number of levels is bounded by $\mathcal{O}(\log_k m)$. By fixing $k = \left\lceil \log^{\frac{1}{2}} m \right\rceil$ we achieve an $\mathcal{O}\left(\frac{\log m}{\log \log m}\right)$ bound on L. We choose this value because some parts of the algorithm are exponential in k.

D_{j+1} is a refinement of D_j obtained by a two-phase procedure. In the first step each fragment is split into subparts of balanced size. The second phase refines this division in order to balance the number of border vertices in each resulting fragment. Precise description of this process and the proof of Lemma 1 can be found in the full version of the paper [13].

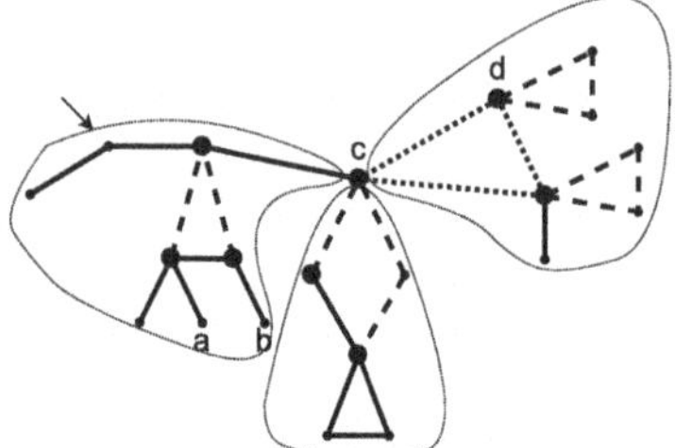

Fig. 2. Two levels of a recursive decomposition satisfying Lemma 1. Fragments from D_j are marked out and D_{j+1} are differentiated with line styles. The arrow indicates r_G, the root of G. Border vertices of j-th level are highlighted. A buyer wishing to purchase an a-b path is not assigned to j-th level, but a buyer interested in a-c paths is. Vertices c and d are connected at all levels of decomposition, so corresponding customers would be processed at its last level.

2.3 Classification of Buyers

A pair of vertices u and v is said to be connected in a decomposition D_j if there exists a path from u to v fully contained in a single fragment from D_j. Buyer $i \in B$ will be processed at the last level where u_i and v_i are connected. This way every buyer is assigned to a single level of decomposition. The decomposition process and the classification of buyers is illustrated by Fig. 2.

Remark 1. If j is the last level at which vertices u and v are connected, every u-v path in the whole graph contains a border vertex on j-th level.

3 Algorithm for a Single Decomposition

In the previous section buyers have been divided into subsets by a recursive graph decomposition. Now we focus on a single (j-th) level of decomposition. By exploiting its properties our algorithm constructs prices which achieve a constant factor approximation of revenue with respect to buyers assigned to this level (denoted B_j).

The main idea behind the algorithm for a single decomposition is to split the paths desired by buyers into smaller sections and handle them separately. In the following we define a partitioning of those paths and discuss that it suffices to be able to solve the natural two subcases.

3.1 The Skeleton

Definition 4. ***Skeleton on*** j***-th level****, denoted* SK_j*, is a minimal subgraph of* G *containing all simple paths between border vertices of* j*-th level. Equivalently, an edge belongs to the skeleton, i.e. is a* ***skeleton edge****, if and only if a simple path connecting two border vertices passes through it. A vertex adjacent to a skeleton edge is a* ***skeleton vertex****.*

Definition 5. *A* ***non-skeleton component on*** j***-th level*** *is a maximal connected subgraph of a fragment from* D_{j+1} *containing no edges from* SK_j*.*

Note that, by the definition of a border vertex, SK_{j+1} is always a superset of SK_j and $\mathrm{SK}_L = G$. The following lemma allows for a clear distinction between the paths inside the skeleton and outside it. The proof can be found in the full version of the paper [13].

Lemma 2. *Every simple path connecting two skeleton vertices passes only though skeleton edges.*

Corollary 1. *Each non-skeleton component contains exactly one skeleton vertex.*

Definition 6. *Let us define a* ***skeleton representative*** *of a vertex* v *on* j*-th level denoted by* $\mathrm{repr}_j(v)$*. If* v *is a skeleton vertex in* D_j*, then* $\mathrm{repr}_j(v) = v$*. Otherwise, the representative of* v *is the unique skeleton vertex in the non-skeleton component on* j*-th level containing* v*.*

Consider a buyer i from B_j wishing to buy the cheapest u_i-v_i path. Recall from Sect. 2 that each u_i-v_i path contains at least one border vertex of j-th level. By Corollary 1, each path from u_i to v_i contains vertices u_i, $\mathrm{repr}_j(u_i)$, $\mathrm{repr}_j(v_i)$, v_i. Although some of those four vertices may be equal, they are guaranteed to appear in this order. This allows us to split every such path into three parts (some of which may be empty):

- First **non-skeleton section** – a simple path from u_i to $\mathrm{repr}_j(u_i)$, which contains no skeleton edges.
- A **skeleton section** – a simple path from $\mathrm{repr}_j(u_i)$ to $\mathrm{repr}_j(v_i)$. By Lemma 2, it consists only of skeleton edges.
- Second **non-skeleton section** – a simple path from $\mathrm{repr}_j(v_i)$ to v_i. Similarly to the first one, it does not contain skeleton edges.

Note that the endpoints of individual sections do not depend on the choice of the particular u_i-v_i path. Our algorithm uses this property to handle both kinds of sections individually. An example partition of a path into skeleton and non-skeleton sections is illustrated by Fig. 3.

3.2 Splitting the Graph Into Two Independent Subproblems

The algorithm handles two subproblems: pricing the non-skeleton and skeleton edges to maximize revenue generated by respectively non-skeleton and skeleton sections.

The Skeleton Subproblem: Consider a buyer $i \in B_j$ wishing to purchase the cheapest u_i-v_i path. In this subproblem she buys a cheapest $\text{repr}_j(u_i)$-$\text{repr}_j(v_i)$ path as long as its cost is at most b_i (her original budget). This situation achieved by setting the price of all non-skeleton edges to zero.

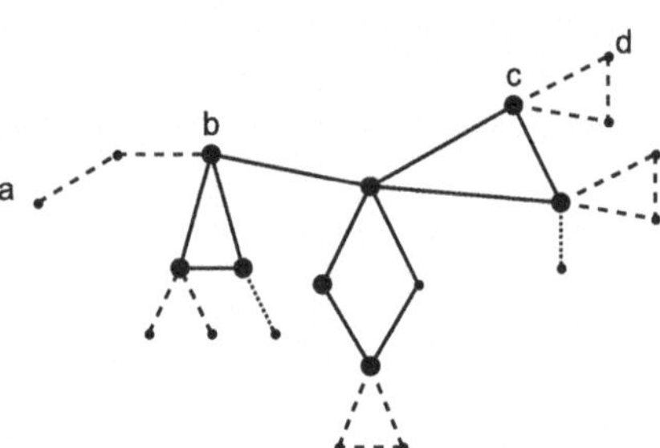

Fig. 3. The cactus from Fig. 2 with highlighted border vertices and the skeleton on j-th level. Each connected group of edges dotted in a same style forms a single non-skeleton component. Each a-d path is split into a skeleton section (a b-c path) and two non-skeleton sections: a-b and c-d paths.

The Non-skeleton Subproblem: In this case, we set the prices of all the skeleton edges to zero. Each buyer $i \in B_j$ will purchase the cheapest paths from u_i to $\text{repr}_j(u_i)$ and from v_i to $\text{repr}_j(v_i)$ if their total cost does not exceed b_i.

Let us introduce additional notation:

- Let OPT_j be the maximal revenue obtained by any price vector and envy-free assignment of paths to the buyers from B_j.
- Let SKOPT_j and NSKOPT_j be the maximal revenues for the skeleton and non-skeleton subproblem respectively.

Note that any envy-free solution for the whole graph immediately yields envy-free solutions for both subproblems. Thus, $\text{SKOPT}_j + \text{NSKOPT}_j \geq \text{OPT}_j$. The algorithm solves both subproblems independently. Then, the computed solutions are compared and the one with greater revenue is chosen. Sections 5 and 4 describe polynomial time approximation algorithms for the skeleton and non-skeleton subproblem respectively.

4 Non-skeleton Edges

This section describes an algorithm for solving the non-skeleton subproblem on j-th level, that is pricing the non-skeleton edges and maximizing revenue generated by non-skeleton sections of paths allocated to buyers from B_j. Prices of all edges in the skeleton on j-th level are set to zero. On the last level of decomposition the skeleton contains the whole graph G. Thus, we assume that $j < L$. The algorithm presented here finds prices generating at least $\frac{\text{NSKOPT}_j}{4}$ revenue.

4.1 The Rooted Case

Before describing the method for pricing non-skeleton edges, let us discuss an easier problem, solution to which is a subprocedure used by the final algorithm.

Definition 7. *Consider an instance of the tollbooth problem on cactus graphs defined by a cactus H and a set of buyers B_H. We will say that it is a* ***rooted instance*** *if there exists a vertex in H, called root, which is an endpoint of every path desired by the buyers.*

Definition 8. *Consider a buyer $i \in B_H$ in a rooted instance, who wishes to purchase a cheapest u_i-v_i path. Her* ***destination vertex*** *is the one of vertices u_i and v_i which is not the root.*

Lemma 3. *Any rooted instance of the tollbooth problem on cactus graphs can be solved in polynomial time. It is also true if the problem admits only those price assignments, under which the distances from the root to some vertices are equal to arbitrarily fixed constants.*

For every possible price assignment and envy-free allocation each buyer is assigned a shortest path from the root to her destination vertex as long as its cost does not exceed her budget. Thus, presenting a polynomial algorithm for finding optimal prices is sufficient to prove the above lemma.

The algorithm is based on dynamic programming whose subproblems mimic the structure of the tree of biconnected components of H rooted in r – the root from Definition 7. For each biconnected component C it calculates values $dp_{C,d}$, which are defined as the maximum revenue generated by buyers whose destination vertices are in the subtree graph of C (excluding its topmost vertex) under the assumption that the distance i.e., cost of a cheapest path, from r to C (its depth) equals d. Note that the distance from C to the root is in fact the distance between r and the topmost vertex of C. The following lemma allows us to consider only polynomially many values d.

Lemma 4. *For any rooted instance of the tollbooth problem on cactus graphs there exists an optimal solution, such that the distance from each vertex to the root belongs to the set $\mathcal{D}$ containing zero and buyers' budgets: $\mathcal{D} = \{0\} \cup \{b_i \mid i \in B_H\}$.*

In Sect. 5 the algorithm needs to find optimal prices given constraints on the distance from certain vertices to the root. The following corollary allows for handling such cases. Both the lemma and corollary are proven in the full version of the paper [13].

Corollary 2. *Consider a rooted instance of the tollbooth problem on cactus graphs and a subset S of vertices of H such that for each $v \in S$ required depth e_v of this vertex is given. Prices of edges in H are said to be feasible if the cost of a cheapest r-v path equals e_v for each $v \in S$. Let us assume that there is at least one such price assignment. Then, there exists a feasible price assignment maximizing revenue for which the distance from r to each vertex $v \notin S$ belongs to the set $\mathcal{D}'$:*

$$\mathcal{D}' = \{0\} \cup \{b_i \mid i \in B_H\} \cup \{e_i \mid i \in S\} \tag{2}$$

Solution for a Rooted Instance. The input to the procedure consists of a graph H, buyers B_H and a (possibly empty) set of constraints S from Corollary 2. For each vertex v we define a set of its possible depths $\mathcal{D}_v$ in the following way:

$$\mathcal{D}_v = \begin{cases} \{e_v\}, & v \in S \\ \{0\} \cup \{b_i \mid i \in B_H\} \cup \{e_u \mid u \in S\}, & v \notin S \end{cases} \tag{3}$$

For each biconnected component C the algorithm calculates the values of $dp_{C,d}$ for every $d \in \mathcal{D}_v$ where v is the topmost vertex of C. It is possible that some values of d inevitably lead to violation of the constraints on depths of vertices from S. In such a case we set $dp_{C,d} = -\infty$. For simplicity, we also assume that $dp_{C,d} = -\infty$ for each $d \notin \mathcal{D}_v$.

The biconnected components of H are processed bottom up based on the structure of the tree of biconnected components. The algorithm handles biconnected components differently depending on whether they consist of a single edge or a cycle. The root component R, which is the root of the tree of biconnected components, contains the whole H in its subtree graph and is treated in yet another way. Let us introduce useful notation:

- $cnt_{v,x}$ – the number of buyers whose budgets are at least x and whose destination vertex is v.
- $\mathcal{C}_v$ – the set of all biconnected components whose topmost vertex is v.

The Case of a Single Edge: Let us denote the lower vertex of the considered biconnected component C as v and the upper as u. Note that the subtree graph of C consists of the (u, v) edge and subtree graphs of biconnected components from $\mathcal{C}_v$, which are edge disjoint and share only the topmost vertex. All simple paths to the root from vertices contained in those subtree graphs pass though v. Furthermore, each simple path from v to the root must contain u. Basing on those observations, the algorithm calculates $dp_{C,d}$ for each $d \in \mathcal{D}_u$ according to the following formula:

$$dp_{C,d} = \max_{d' \in \mathcal{D}_v \,;\, d' \geq d} \left(cnt_{d',v} \cdot d' + \sum_{C' \in \mathcal{C}_v} dp_{C',d'} \right) \tag{4}$$

The optimal value of d' is stored along with $dp_{C,d}$ in order to find the prices after calculating optimal revenue ($(d' - d)$ is the price of the (u, v) edge).

The Case of the Root Component: The root must have depth 0, hence for this component only a single value ($dp_{R,0}$) is calculated: $dp_{R,0} = \sum_{C' \in \mathcal{C}_r} dp_{C',0}$.

The Case of a Cycle: Let us denote the considered cycle by C, its topmost vertex as v and its subtree graph by G_C. G_C consists of C and the subtree graphs of components from $\mathcal{C}_u$ for all $u \in C \setminus \{v\}$.

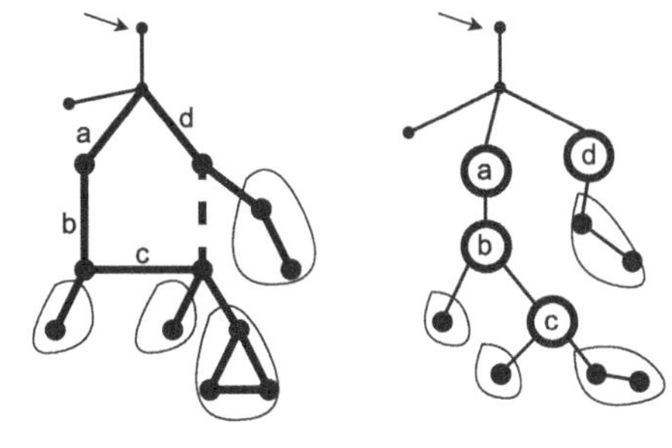

Let us examine the structure of the subproblem. All paths from vertices from G_C to the root pass through v. By construction of the tree of biconnected components each vertex s in $G_C \setminus \{v\}$ belongs either to C or to a subtree component of $C' \in \mathcal{C}_u$ for a unique $u \in C \setminus \{v\}$. In the latter case every path from s to the root also passes through u. Edges which form possible s-u paths belong to smaller subproblems, so given the depth of u, the optimal prices can be calculated. Thus, now we are only interested in the depths of vertices in C (more precisely, their distance to v, because it's depth is fixed). Note that these distances depend only on the edges in C.

Fig. 4. Processing a cycle for a fixed unused edge. Its subtree graph (without the topmost vertex) is highlighted by thicker edges and points both in the cactus graph (left) and the temporary tree of biconnected components (right). Solid lines enclose subproblems which already have been solved. Arrows indicate respectively the root of G and the root component. The algorithm calculates the values of dp^e for biconnected components c, b, a and d (in this order).

Consider any prices assigned to them. Let T be a shortest-path tree of C rooted in v. Exactly one edge from C does not belong to T, we will say that it is unused. After removing this edge, the cost of a cheapest path from any vertex in G_C to v does not change (Fig. 4).

The algorithm iterates over all edges in C fixing the current one, denoted e, to be the unused edge. In this step the algorithm finds an optimal solution among those price assignments which result in e being unused. First, the price of e is set to $b_{max} + 1$ ($b_{max} = \max\{b_i \mid i \in B_H\}$). This effectively removes e from the graph, as no buyer will ever purchase a path containing it.

Without e, C becomes a path and edges in $C \setminus \{e\}$ constitute individual biconnected components. The subtree graphs of components in $\mathcal{C}_u$ for $u \in C \setminus \{v\}$ remain intact, but now in the subtree graph they are descendants of the single-edge biconnected components from $C \setminus \{e\}$ instead of the cycle C. However, the solutions calculated for them for every valid depth still remain valid. Thus, the only biconnected components in $G_C \setminus \{e\}$ for which we need to calculate $dp_{C',d}$ are formed by single-edge biconnected components. The algorithm calculates these values in a bottom up manner as described previously. We assumed that e is an unused edge, so let us denote the results as dp^e.

Let C_1 and C_2 be the biconnected components formed by edges of C which are adjacent to v. If e, the unused edge, happens to be adjacent to v, there is only one such component. In this case C_2 is just a placeholder with an empty subtree graph and $dp^e_{C_2,d}$ equals zero for all $d \in \mathbb{R}$. Note that the union of subtree graphs of C_1 and C_2 contains the same vertices and edges (except for e) as G_C. Furthermore, the two subtree graphs can only share one vertex: v. Thus, if we admit only such solutions, where e is unused, then $dp_{C,d} = dp^e_{C_1,d} + dp^e_{C_2,d}$. Since for every possible price assignment there exists an optimal allocation where one edge of C is unused, it is enough to iterate over all possible edges $e \in C$:

$$dp_{C,d} = \max_{e \in C} \left(dp^e_{C_1,d} + dp^e_{C_2,d} \right) \tag{5}$$

Using the above formula the algorithm computes $dp_{C,d}$ for every $d \in \mathcal{D}_v$. Like previously, respective price assignments to edges of C are stored alongside the results.

Since all the above procedures run in polynomial time and each biconnected component is processed only once, the solution is found in polynomial time. Prices obtaining the computed maximal revenue can be easily calculated using additional information stored alongside the values of $dp_{C,d}$. This proves Lemma 3.

4.2 The Non-skeleton Subproblem

In order to solve the non-skeleton subproblem the algorithm utilizes a special structure of non-skeleton components on j-th level. For each of them, a rooted instance of the tollbooth problem on cactus graphs is created. Those subproblems are solved by the procedure described in Sect. 4.1. Resulting price assignments for individual non-skeleton components are merged by a probabilistic procedure which can, however, be derandomized.

Constructing Rooted Instances. Recall that each buyer $i \in B_j$ is defined by a triple (u_i, v_i, b_i), which means that she has a valuation of b_i for all u_i-v_i paths. Since the skeleton edges are given away for free, the algorithm only processes respective non-skeleton sections, which are modeled by two independent copies of i-th buyer: $(u_i, \mathrm{repr}_j(u_i), b_i)$ and $(v_i, \mathrm{repr}_j(v_i), b_i)$. Each of them is added to the instance associated with the non-skeleton component containing u_i and v_i respectively. If u_i or v_i is a skeleton vertex, the corresponding non-skeleton section is empty and can be ignored. Since $\mathrm{repr}_j(s)$ is the same for all vertices s within a single non-skeleton component (Corollary 1), all subproblems defined this way will indeed be rooted instances.

Remark 2. It follows from the classification of buyers, that if u_i and v_i are not in SK_j, the non-skeleton components containing u_i and v_i belong to the same fragment of D_j, but to different fragments from D_{j+1}.

Algorithm for a Single Fragment. The above observation allows us to treat all fragments in D_j independently. Let us consider a single fragment $H \in D_j$ and by $rev_{j,H}(p)$ let us denote the revenue generated by its selling non-skeleton edges for prices p to buyers from B_j. If for each buyer at most one non-skeleton section was non-empty, the rooted instance would be independent. We could apply their solutions verbatim – each buyer i present in an instance would always be able to spend b_i as assumed. However, it's not the case – for example a buyer i may be present in two non-skeleton components and both solutions to the corresponding rooted instances may require her to pay b_i for each non-skeleton section, in which case she would not buy anything.

We solve this issue by using a randomized procedure: each fragment $F \in D_{j+1}$ contained in H is independently and equiprobably colored black or white. Every non-skeleton component in a black fragment is priced according to the solution to the corresponding rooted instance. All edges in non-skeleton components in white fragments are given away for free.

Lemma 5. *Let p be the price vector found by the above randomized algorithm and q be any price vector feasible for the non-skeleton subproblem. Then, the following inequality holds:*

$$\mathbb{E}\left[\mathrm{rev}_{j,H}(p)\right] \geq \frac{1}{4}\mathrm{rev}_{j,H}(q) \tag{6}$$

This follows from the fact that for any non-skeleton section with probability at least $\frac{1}{4}$ it will be in the black fragment and the other non-skeleton section of the same buyer will be in a white one. The deterministic algorithm iterates over all possible colorings and chooses the best one, which will yield results at least as good as the expected value. Because there are at most $\mathcal{O}(k)$ $(\mathcal{O}(\sqrt{\log m}))$ fragments of the next level contained in H, this takes polynomial time. Detailed proofs of the lemma and the corollary can be found in the full version of the paper [13].

Corollary 3. *There exists a deterministic polynomial algorithm which for a non-skeleton subproblem on j-th level finds prices achieving at least $\frac{\mathrm{NSKOPT}_j}{4}$ revenue.*

5 Skeleton Edges

This section describes an algorithm solving the skeleton subproblem on a single, j-th level of decomposition. Non-skeleton edges do not influence envy-freeness of a solution because the are given away for free. Thus, a buyer $i \in B_j$ wishing to buy an u_i-v_i path can be thought of as a buyer with the same budget wishing to buy a shortest $\mathrm{repr}_j(u_i)$-$\mathrm{repr}_j(v_i)$ path. Our algorithm finds prices for edges in SK_j generating at least $\frac{\mathrm{SKOPT}_j}{2048}$ revenue. Here we present only the main ideas behind the polynomial time procedure achieving that. For the full description we refer the reader to the full version of the paper [13].

The algorithm first considers skeletons of individual fragments from D_j, i.e. minimal subgraphs of SK_j containing all paths between border vertices of j-th level lying in a given single fragment $F \in D_j$. One can think of the skeleton of a fragment as an extension of $F \cap \mathrm{SK}_j$ which includes all simple paths starting and ending in it. Prices of edges in the skeleton of each fragment are optimized independently. This is possible, because for each buyer $i \in B_j$ both $\mathrm{repr}_j(u_i)$

and $\text{repr}_j(v_i)$ lie in the same fragment $F \in D_j$. Hence, all paths between them are contained in the skeleton of F. Secondly, those solutions are combined to form a global solution for the whole SK_j.

5.1 Solution for a Skeleton of a Single Fragment

When finding solutions for the skeleton of an individual fragment $F \in D_j$, we take advantage of its simple structure. Since the skeleton of F is a cactus and a union of simple paths between $\mathcal{O}(k)$ vertices (there can be only that many border vertices in F), we can partition it into $\mathcal{O}(k)$ subgraphs, called **segments**, which can be thought of as edges. It is because each segment connects with the rest of the skeleton of F through at most two vertices. Thus for all buyers $i \in B_j$ for which $\text{repr}_j(u_i)$ and $\text{repr}_j(v_i)$ do not lie in a given segment, the segment will be either traversed from one endpoint to another or not at all. Hence, with regards to them, the algorithm can focus on pricing whole segments instead of individual edges. Here the price of a segment is the cost of traversing it from one endpoint to the other. This is helpful, because having only $\mathcal{O}(k)$ segments together with the observation that it is enough to consider only a polynomial number of possible prices for each segment, it allows us to exhaustively search through all such price assignments to the segments.

This yields a two-level algorithm, which first sets the prices of the entire segments and then distributes those costs (prices) among individual edges forming the segments. In each segment S the latter stage is done by optimizing for the revenue generated by buyers $i \in B_j$ for which $\text{repr}_j(u_i)$ and $\text{repr}_j(v_i)$ lie in S. As noted before, the other buyers can only traverse the whole segment from one endpoint to the other, so from their perspective the distribution of the cost into individual edges is irrelevant. By exploiting the structure of the cactus and properties of the decomposition, our polynomial time procedure (described in the full version of the paper [13]) computes the price distribution within a single segment which generates revenue within a constant factor of the optimal one.

The aforementioned observation that it is sufficient to consider polynomially many possible prices for each segment is based on the technique of price rounding. The main idea is that if we decrease the price of all segments by at most half, the revenue generated by the buyers will not decrease by more than a factor of two. If additionally, we round all the prices which are under a certain threshold t to zero, we can limit the prices to the set $\{0, t, 2t, 4t, \dots\ b_{max}\}$. For a proper choice of t we can show that the size of that set is polynomial in the number of edges and buyers while at the same time guaranteeing that the total revenue shrinks at most four times due to rounding.

5.2 Combining the Local Solutions

One of the major difficulties in combining the solutions for skeletons of individual fragments stems from the fact that G is a cactus graph and not necessarily a tree. If it was a tree, each fragment, which is, by construction, a connected subgraph, would contain all paths between border vertices belonging to it. Then, the skeleton of a fragment F would be fully contained in F and the skeletons of individual fragments would partition SK_j into edge-disjoint subgraphs. Hence, local solutions from the previous section would be completely independent and always compatible. Unfortunately, in the case of cactus graphs, although the fragments in D_j are edge-disjoint, their skeletons can overlap. Thus, solutions to the skeletons of two different fragments can also overlap, i.e. there may exist edges to which both solutions assign a price. Hence, the algorithm cannot just pick the most profitable solution for the skeleton of each fragment and combine them. Instead, it needs to find the most profitable combination of solutions which is compatible (i.e., if an edge is priced by multiple solutions, all of them assign the same price to it). We achieve that using an involved dynamic programming algorithm, which is presented in the full version of the paper [13].

6 Concluding Remarks

In Sects. 4 and 5 we have presented polynomial time constant factor approximation algorithms for the non-skeleton and skeleton subproblems. Thus, we have shown that prices achieving at least a constant fraction of optimal revenue can be found in polynomial time for each of the L levels of decomposition. Recall that by setting $k = \left\lceil \log^{\frac{1}{2}} m \right\rceil$ we ensure L to be $\mathcal{O}\left(\frac{\log m}{\log \log m}\right)$. Hence, our polynomial algorithm for the tollbooth problem on cactus graphs yields an $\mathcal{O}\left(\frac{\log m}{\log \log m}\right)$ approximation guarantee on revenue.

It remains an open question whether there exist polynomial time algorithms giving sublogarithmic guarantees on revenue for further generalizations of the tollbooth problem, for example for the cases where the underlying graphs are only assumed to have bounded treewidth.

References

1. Anshelevich, E., Kar, K., Sekar, S.: Envy-free pricing in large markets: approximating revenue and welfare. ACM Trans. Econ. Comput. **5**(3) (2017). https://doi.org/10.1145/3105786
2. Anshelevich, E., Sekar, S.: Price doubling and item halving: robust revenue guarantees for item pricing (2017)
3. Balcan, M.F., Blum, A., Mansour, Y.: Item pricing for revenue maximization. In: Proceedings of the 9th ACM Conference on Electronic Commerce, EC 2008, pp. 50–59. Association for Computing Machinery, New York (2008). https://doi.org/10.1145/1386790.1386802
4. Briest, P., Krysta, P.: Single-minded unlimited supply pricing on sparse instances. In: Proceedings of the 17th ACM-SIAM Symposium on Discrete Algorithms, pp. 1093–1102 (2006)
5. Cheung, M., Swamy, C.: Approximation algorithms for single-minded envy-free profit-maximization problems with limited supply. In: 2008 49th Annual IEEE Symposium on Foundations of Computer Science, pp. 35–44 (2008). https://doi.org/10.1109/FOCS.2008.15

6. Demaine, E.D., Feige, U., Hajiaghayi, M., Salavatipour, M.R.: Combination can be hard: approximability of the unique coverage problem. SIAM J. Comput. **38**(4), 1464–1483 (2008)
7. Elbassioni, K., Fouz, M., Swamy, C.: Approximation algorithms for non-single-minded profit-maximization problems with limited supply (2013)
8. Gamzu, I., Segev, D.: A sublogarithmic approximation for highway and tollbooth pricing. In: Abramsky, S., Gavoille, C., Kirchner, C., Meyer auf der Heide, F., Spirakis, P.G. (eds.) ICALP 2010. LNCS, vol. 6198, pp. 582–593. Springer, Heidelberg (2010). https://doi.org/10.1007/978-3-642-14165-2_49
9. Grandoni, F., Rothvoß, T.: Pricing on paths: a PTAS for the highway problem. SIAM J. Comput. **45**(2), 216–231 (2016)
10. Grandoni, F., Wiese, A.: Packing cars into narrow roads: ptass for limited supply highway. In: 27th Annual European Symposium on Algorithms (ESA 2019). Leibniz International Proceedings in Informatics (LIPIcs), vol. 144, pp. 54:1–54:14 (2019). https://doi.org/10.4230/LIPIcs.ESA.2019.54
11. Guruswami, V., Hartline, J.D., Karlin, A.R., Kempe, D., Kenyon, C., McSherry, F.: On profit-maximizing envy-free pricing. In: Proceedings of the Sixteenth Annual ACM-SIAM Symposium on Discrete Algorithms, SODA 2005, pp. 1164–1173. Society for Industrial and Applied Mathematics (2005)
12. Li, Y., Lu, P., Ye, H.: Revenue maximization with imprecise distribution (2019)
13. Turko, A., Byrka, J.: Sublogarithmic approximation for tollbooth pricing on a cactus (2023). https://arxiv.org/abs/2305.05405

To Regulate or Not to Regulate: Using Revenue Maximization Tools to Maximize Consumer Utility

Meryem Essaidi[1], Kira Goldner[2(✉)], and S. Matthew Weinberg[3]

[1] UC Berkeley, Berkeley, CA 94720, USA
essaidi.meryem@gmail.com
[2] Boston University, Boston, MA 02215, USA
goldner@bu.edu
[3] Princeton University, Princeton, NJ 08544, USA
smweinberg@princeton.edu

Abstract. We study a theoretical model inspired by regulated health insurance markets. The market regulator can choose to do nothing, running a Free Market, or can exercise her regulatory power by limiting the entry of providers (decreasing consumer *welfare* by limiting options, but also decreasing *revenue* via enhanced competition). We investigate whether limiting entry increases or decreases the *utility* (welfare minus revenue) of the consumers who purchase from the providers, specifically in settings where the outside option of "purchasing nothing" is prohibitively undesirable.

We focus primarily on the case where providers are symmetric. We propose a sufficient condition on the distribution of consumer values for (a) a unique symmetric equilibrium to exist in both markets and (b) utility to be higher with limited entry. (We also establish that these conclusions do not necessarily hold for all distributions, and therefore some condition is necessary.) Our techniques are primarily based on tools from revenue maximization, and in particular Myerson's virtual value theory. We also consider extensions to settings where providers have identical costs for providing plans, and to two providers with an asymmetric distribution.

Keywords: Limited Entry · Healthcare · Revenue · Virtual Values

1 Introduction

Consider a central planner who wishes to procure service from several *providers*, on behalf of a population of *consumers*. One common approach, especially in the domain of US healthcare, is for the planner to regulate an *exchange*. Specifically, providers are still in control of the price to charge for their service, but the regulator decides which providers get to actually enter the market (as a function of the prices charged).

G. Schäfer and C. Ventre (Eds.): SAGT 2024, LNCS 15156, pp. 315–332, 2024.
https://doi.org/10.1007/978-3-031-71033-9_18

Taking healthcare exchanges in the US as a running example, the following two paradigms are both common. In some cases, employers manage an exchange for their employees, and tend to *limit entry*, offering only a handful of providers access to the market. In other cases, governments manage an exchange (such as those created by the Affordable Care Act (ACA)), and instead allow any provider meeting some minimum coverage guarantees to enter the market. The tradeoff between these two approaches is that employer-managed exchanges tend to have significantly fewer options, while government-managed exchanges tend to have less competitive prices.[1] Depending on the consumer population at hand, either approach could optimize consumer utility.

Here is a motivating scenario to have in mind: perhaps there are multiple providers, but each serves some segment of the market significantly better than all others. For example, perhaps one provider has a strong network for diabetic consumers, another has a strong network for competitive athletes, etc. With little regulation, one should expect consumer *welfare* to be high (because there is a provider for each segment of the market), but also provider *revenue* to be high (because there is little competition within each segment).[2] With limited entry, one should expect consumer welfare to be lower (because some segments lose access to their preferred provider), but also provider revenue to be lower (due to increased competition just to enter the market in the first place). It is not immediately obvious which approach leads to higher consumer *utility* (which is the difference of the two). The main purpose of this paper is to provide theoretical tools to reason about this tradeoff, and understand under what circumstances consumer utility is improved.

Mandatory Purchase. Another important aspect of health insurance markets is that purchase is essentially mandatory. Employers may assign a default option if no option is selected, governments may assign a financial penalty for no selection (as was recently the case with the ACA's individual mandate), or the choice to purchase nothing may be prohibitively undesirable. We capture this aspect in our model by requiring all consumers to purchase an option (but we do not explicitly model the reason for why purchase is mandatory).

Our focus is the following: *when purchase is mandatory, under what conditions does limited entry (reducing prices via competition, but reducing welfare via reduced options) improve consumer utility?*

Observe that reasoning about this question requires a direct understanding of what prices the providers will choose to set. As such, our main techniques involve characterizing and analyzing equilibria, and understanding when they exist and are unique. We compare two settings: the *Free Market* setting, where the market regulator does not restrict entry at all, and the *Limited Entry* setting. Note, of

---

[1] For example in 2013, even with subsidies, ACA premiums represented between 4 and 9.5% of the median income [13]. In 2015, a survey found that without subsidies, average marketplace unsubsidized premiums were over 2.5× what most consumers claim is the maximum they can afford [19].

[2] In health insurance markets, lack of within-segment competition arises due to barriers to entry of building a strong network.

course, that while we use healthcare exchanges as a motivating example, the focus of our paper is to provide a model and theoretical tools to study procurement auctions. As such, our model is stylized and intended to capture one aspect of decision-making (whether to limit entry or not)—it is not intended to capture verbatim the full range of challenges facing regulators of healthcare exchanges. Still, we do emphasize that this is indeed a key decision facing regulators, and that theoretical tools to study this tradeoff were previously lacking.[3]

1.1 Model and Results

Our goal is to provide a clean model to reason about the impact of limiting entry. To this end, we study a model first posed by Perloff and Salop [17]. There are n providers, and a population D of consumers, each with a value v_i for the plan offered by provider i. For the majority of the paper, we consider the symmetric setting, where a random consumer from D has value v_i for plan i drawn i.i.d. from some single-variate distribution F (that is, individual consumers have different values for different providers, but providers are comparably desirable at the population-level view). We denote this as $D := F^n$. This i.i.d. assumption has two components: first, we assume that provider values are independent, and second, we assume that they are symmetrically distributed. Assuming that values are independent across providers is the natural analogue of the ubiquitous "independent items" assumption in multi-dimensional mechanism design [3,4,11]. The motivation for this assumption in our work is the same as in the vast literature of prior works: to enable tractability via Myersonian virtual-value theory [15]. Further assuming that values are symmetrically distributed simply means that we are restricting attention to cases where no provider is *a priori* stronger than another (e.g. providers have similarly-sized networks, but those networks are tailored to different medical needs, geographic preferences, etc.).

In Sect. 3, we study the Free Market setting, where each provider i simply sets a price p_i, and consumers drawn from D purchase the plan i which maximizes their utility $v_i - p_i$. We first identify sufficient conditions on F (involving a new condition we term "MHR$^+$") for the existence of a unique symmetric equilibrium, and we characterize the equilibrium prices in the Free Market in Theorem 1 below. Here, $h_2^n(F)$ denotes the expected hazard rate of the second-highest of n i.i.d. draws from F, and formal definitions of both MHR$^+$ and decreasing density appear in Sect. 2.

Theorem 1. *Let $D := F^n$, where F is MHR$^+$ and has decreasing density. Then the unique symmetric equilibrium in the Free Market setting is for each provider to set price $1/h_2^n(F)$.*

Observe that a characterization of a canonical equilibrium in the Free Market setting is necessary if we are to possibly analyze consumer utility. We include one vignette regarding our technical approach, which leverages machinery from

[3] Indeed, this direction was first posed to the authors by a researcher in applied markets for health insurance.

the revenue maximization literature [15]. Suppose all other providers are setting price p; what is a provider's best-response? To reason about this, consider instead a new distribution F_p^* defined such that $1 - F_p^*(q)$ is the probability that a consumer drawn from marginal F will purchase from this provider when he sets price q and all other providers set price p. The provider's best response is then to set q^* which maximizes his expected profit, $q \cdot (1 - F_p^*(q))$, and p is therefore a symmetric equilibrium if and only if $p = q^*$. Using this rephrasing, we then argue that if this distribution F_p^* happens to have a *monotone hazard rate* (MHR), then we are guaranteed the existence of a unique symmetric equilibrium. Of course, this distribution F_p^* is quite different from F itself (for example, F may be MHR and F_p^* may not even be regular!). We define a new distributional assumption, MHR$^+$, such that if F is MHR$^+$ then this implies that F_p^* is MHR. We note that MHR$^+$ is a strictly stronger condition than MHR, and that most common MHR distributions are also MHR$^+$ (e.g. exponential, uniform, Gaussian).

Next, in Sect. 4, we study the Limited Entry setting. Formally, each provider still sets a price p_i, but now only the $n-1$ providers[4] with lowest price enter the market (tie-breaking arbitrarily). That is, the providers participate in Bertrand competition to enter the market. Consumers again pick the plan maximizing $v_i - p_i$, but only among these $n-1$ providers. For symmetric instances, quickly observe that there is a unique (symmetric) equilibrium, and in it all providers set $p_i = 0$ (so in some sense, our model can be seen as "optimistic" towards the benefits of limiting entry). The main result of this section is a characterization of the precise condition on F that implies that consumer utility in the Limited Entry setting will be greater than in the Free Market setting; we call this the Limit-Entry condition.

Definition 1 (Limit-Entry Condition). *A distribution $D = F^n$ satisfies the* Limit-Entry Condition *if $H_1^n(F) \leq n/h_2^n(F)$. Here, $H_1^n(F)$ is the expected inverse hazard rate of the highest of n i.i.d. draws from F. (Recall that $h_2^n(F)$ is the expected hazard rate of the second-highest of n i.i.d. draws from F.)*

Theorem 2. *Let $D := F^n$ and admit a symmetric equilibrium in the Free Market setting. Then the expected consumer utility at the unique equilibrium in the Limited Entry setting exceeds that at the unique symmetric equilibrium in the Free Market setting* if and only if *F satisfies the Limit-Entry Condition.*

Finally, while the Limit-Entry condition is relatively clean, it is not obvious how it relates to more common distributional assumptions. Our final main result shows that the Limit-Entry condition is satisfied under standard distributional assumptions.

Theorem 3. *Let $D := F^n$, where F is MHR and has decreasing density. Then D satisfies the Limit-Entry Condition.*

[4] We consider an extension to any k and minimum price p in the Appendix of the full version.

Corollary 1. *Let $D := F^n$, where F is MHR$^+$ and has decreasing density. Then D has a unique symmetric equilibrium in the Free Market setting, and the expected consumer utility at this unique symmetric equilibrium is exceeded by the expected consumer utility in the Limited Entry setting.*

In the interest of completeness, we examine whether any of our assumptions can be relaxed to more standard assumptions (e.g. MHR$^+$ to MHR). In short, Proposition 1 establishes that the answer is no, suggesting that there is indeed a relevant aspect of our stronger assumptions as it relates to our conclusions. Its proof can be found in the Appendix of the full version.

Proposition 1. *(Different) distributions $D := F^n$ with the following properties all exist:*

- *F is MHR, but there exists a p for which F_p^* is not MHR (in fact, it is anti-MHR).*
- *F is MHR, but D has no symmetric equilibrium in the Free Market setting.*
- *D has a symmetric equilibrium in the Free Market, but does not satisfy the Limit-Entry Condition.*
- *D satisfies the Limit-Entry Condition, but does not have a symmetric equilibrium in the Free Market.*

Finally, we include some extensions. In Sect. 5 of the full version [9], we consider the case of asymmetric distributions $D = \times_i F_i$. We prove that if F_i is MHR$^+$, then the asymmetric analog, $F^*_{i,\mathbf{p}_{-i}}$, is MHR. We also show that for two providers, when $F^*_{i,\mathbf{p}_{-i}}$ is MHR, an equilibrium exists. In the Appendix of the full version, we extend our results to the setting where providers have identical costs for providing service, and also extend the Limit-Entry Condition to the Generalized Limited Entry setting where the regulator may choose parameters k and p: the k providers with lowest price enter the market, and a minimum price of p is imposed.

1.2 Related Work

As previously noted, the mathematical model we study (including that consumers are i.i.d. with mandatory purchase, and provider costs are identical as in the Appendix of the full version) is first posed in [17]. As a result, there is some technical overlap between our works, but minimal conceptual overlap. At a conceptual level, the entire goal of our paper is to understand the impact of limiting entry, which is not considered in [17] (they consider, for example, the impact of additional providers in the Free Market model). At a technical level, [17] also provides sufficient conditions for equilibria to exist, but these conditions are stated in terms of conditions on our F_p^* (rather than directly on F, as in our Theorem 1), and they do not provide an example of non-existence (our Proposition 1).[5] The

[5] Their conditions suffice to establish that when F is uniform or exponential, an equilibrium exists, but they do not give a classification in terms of direct conditions on F.

main technical overlap is that [17] also characterizes a Free Market symmetric equilibrium and proves that it is unique (under the assumption that it exists); however, their proof is via direct calculations, whereas ours derives intuition via revenue maximization techniques, and uniqueness follows for free. As far as motivation, [17] offers the same justification for the i.i.d. assumption: independence is a ubiquitous technical assumption to simplify analysis, and symmetry implies that the products are a priori of the same quality. Interestingly, they justify the mandatory purchase assumption as a tool for technical simplicity, whereas we propose mandatory purchase to better capture our running example.

The remaining literature pertains to procurement auctions *without* mandatory purchase for consumers. In order to compare with the literature on procurement auctions, we will call the providers "suppliers" below. The most relevant work is that of [18]. They also study procurement auctions with n heterogenous goods, each owned by a different supplier, a consumer population, and a mechanism designer whose objective is to maximize consumer utility. The simplest comparison between these works is that [18] studies a wide variety of different procurement models, whereas we focus in depth of one particular model. For example, our work fits into their "First-Price Auction" model (which is only one of many models they consider). But within this model, they consider only a two-supplier setting in a simple Hotelling game [12],[6] whereas we study this model in significantly more depth and generality.

Other procurement work [1,14] also studies optimal centralized allocations for consumer utility, where the designer chooses which suppliers allocate to which consumers. We do not study this form of allocation.

Most prior work studies the two-supplier case without mandatory purchase. [8] also uses a third-party mechanism designer apart from consumers and suppliers, but studies different objectives than us: (1) welfare (consumer utility plus supplier revenue) and (2) supplier revenue minimization. They do not, however, study consumer utility maximization.

Other works allow the consumers to act as the auctioneer [6,7] and investigate whether it is better for the consumers to have one or two suppliers; the answer differs depending on whether the consumers' information is private or not. These papers do not have a mechanism designer acting separately from the consumers; they also only study the stylized Hotelling model.

1.3 Brief Summary

We study consumer utility in a market with n providers under mandatory purchase. We find clean sufficient conditions for equilibria to exist (Theorem 1) in the Free Market, and establish that conditions like these are also necessary (Proposition 1). We also establish clean necessary and sufficient conditions for

[6] Specifically: consumers are uniformly distributed along the unit line. Each supplier offers an item with fixed value at the endpoint of the line, and consumers value the items at their value minus the distance.

consumer utility to improve with Limited Entry over the Free Market (Theorems 2 and 3, and Corollary 1).

We also wish to briefly note our technical highlights. Typically, establishing existence/uniqueness of market equilibria requires solving a system of non-linear equations (and establishing uniqueness). Of course, our proofs must also accomplish this, but we get a surprising amount of leverage via Myersonian virtual value theory. That is, we interpret equilibrium conditions as one price being revenue-maximizing for a related consumer distribution. Due to mandatory purchase, this interpretation (while mathematically involved) is conceptually fairly clean. This enables us to break down a complex mathematical proof into conceptually digestible chunks, and also provides insight into the right conditions to search for. We are optimistic that these tools will continue to be useful in extensions beyond those considered explicitly in this paper.

2 Notation and Preliminaries

We consider the following problem from the perspective of a market regulator. We use the language of healthcare providers throughout the paper (but we remind the reader that healthcare exchanges are just a motivating example for our stylized model). There are n providers, each of whom produces a single (distinct) plan. Each individual consumer in the market has a valuation vector $\boldsymbol{v} \in \mathbb{R}_+^n$ for the plans, with v_i denoting their value for plan i. The market consists of a continuum over valuations $\boldsymbol{v}$, which can alternatively be interpreted as a distribution D (over a random consumer drawn from the market).

We assume throughout the paper that D is a product distribution (that is, $D := \times_i D_i$ for single-dimensional D_i). We will use F_i to denote the CDF of D_i, and assume that each D_i also has a density function, or PDF, denoted by f_i. For our main results, we will also assume that D is symmetric (that is, $D_i = D_j$ for all i, j, or the valuations are identically drawn across providers). In Sect. 5 of the full version [9], we consider extensions to asymmetric distributions.

In our context, let's briefly elaborate on these assumptions. Assuming that each D_i admits a density function is extremely common in past literature (e.g. [15]), and is comparable to a "large market" assumption that no particular individual has an oversized role. The motivation for this assumption is purely technical, since it allows for clean closed-form definitions of conditions such as regularity or Monotone Hazard Rate. Assuming that D is a product distribution is also extremely common (e.g. [3,4,10,15]), and corresponds to the property that a consumer's value for one plan does not influence the probability of their value for another. While this assumption may initially appear restrictive, numerous works establish that results proved in this setting generally extend to richer settings as well. Indeed, our results immediately extend, for free, to the "common base-value" model of [5],[7] but we focus on the independent setting for ease of

[7] In the common base-value model, each consumer has a "base value" for all plans, and an idiosynchratic value for each plan separately. Their value for a plan sums these two together.

exposition (see Sect. 5 of the full version for details on this particular extension). Assuming that D is symmetric corresponds to the following: *individuals* may certainly have distinct values for distinct plans. The fact that D is symmetric simply means that *a priori* there is nothing special about one plan versus another.

Free Market Setting: In the Free Market setting, each provider i sets a price p_i on their plan. A consumer drawn from D purchases the plan $i^* = \arg\max_i\{v_i - p_i\}$. Importantly, notice that the consumer *must* purchase a plan, even if $v_i < p_i$ for all i. So provider i's payoff is equal to $p_i \cdot \Pr_{v \leftarrow D}[i = \arg\max_j\{v_j - p_j\}]$.[8] A best response of provider i to $\boldsymbol{p}_{-i}$, where $-i$ denotes all agents other than i, is the payoff-maximizing price in response to $\boldsymbol{p}_{-i}$. A price vector $\boldsymbol{p}$ is a pure equilibrium if each provider is simultaneously best responding. An equilibrium $\boldsymbol{p}$ is symmetric if $p_i = p_j$ for all i, j. Observe that when both D and $\boldsymbol{p}$ are symmetric, the payoff to each provider is just p_i/n.

Limited Entry Setting: The focus of this paper is contrasting the Free Market setting with a "Limited Entry" setting. In the Limited Entry setting, the regulator does not exert total control over the market (e.g. by directly setting prices), but simply restricts entry to the market. Specifically, each provider i first sets a price p_i on their plan, and then the market regulator allows the $n-1$ providers who set the lowest price (tie-breaking randomly, if necessary) to enter the market (refer to these providers as S). A consumer drawn from D purchases the plan $i^* = \arg\max_{i \in S}\{v_i - p_i\}$. Again, the consumer *must* purchase a plan in S, even if $v_i < p_i$ for all $i \in S$. Provider i's payoff is 0 if they are not selected to be in S, or equal to $p_i \cdot \Pr_{v \leftarrow D}[i = \arg\max_{j \in S}\{v_j - p_j\}]$ otherwise. A price vector $\boldsymbol{p}$ is again an equilibrium if each provider is simultaneously best responding, and symmetric if $p_i = p_j$ for all i, j. It is not hard to see that there is a unique equilibrium for the provider prices. Observation 1 follows simply as the providers participate in Bertrand competition (and e.g. would hold even if purchase were not mandatory).

Observation 1. *For all symmetric D, the unique equilibrium in the Limited Entry setting is $\boldsymbol{p} = \boldsymbol{0}$.*

Note that, in principle, a regulator could limit entry using other rules (e.g. they could allow only $k < n-1$ providers to enter). Taking our model literally, allowing $n-1$ providers to enter is optimal among all such rules (because of Observation 1, consumer utility is monotone in the providers who enter as long as it is $< n$). Still, we consider this generalization in the Appendix of the full version.

Consumer Utility: The focus of this paper is understanding the expected consumer *utility* in equilibrium for both settings. Specifically, the expected consumer *welfare* is equal to $\mathbb{E}_{v \leftarrow D}[v_{i^*},\ i^* = \arg\max_i\{v_i - p_i\}]$, where the argmax

[8] Observe also that because each D_i has a PDF, ties occur with probability 0, and there is always a unique arg max. As a result, we will not be careful between $\leq$ and $<$ when discussing preferences.

is taken over $i \in [n]$ in the Free Market setting, or $i \in S$ in the Limited Entry setting. That is, the expected welfare is simply the expected value a consumer receives for the plan they purchase. The expected *revenue* is simply the sum of the providers' payoffs. Consumer utility is then just welfare minus revenue. Recall that consumers *cannot* opt out, and some consumers may indeed get negative utility.

2.1 Distributional Properties

Symmetric equilibria do not always exist for symmetric distributions (Proposition 1), and limiting entry does not universally increase or decrease consumer utility compared to the Free Market setting (also Proposition 1). As such, the focus of this paper is in providing sufficient conditions for (e.g.) (1) equilibria to exist, and (2) limiting entry to improve consumer utility. Below are properties of single-variate distributions which we'll use. The first two are standard in the literature. MHR$^+$ is a new condition we introduce which is a proper subset of MHR (see Observation 2), and happens to be "the right" restriction of MHR for our setting. For all definitions below, "non-decreasing" means "non-decreasing over the support of F," and "for all x" means "for all x in the support of F." Modulo our new MHR$^+$, each of these conditions are common assumptions in past work (e.g. [2,16]). In typical applications, regularity (or at least MHR) suffice for desired positive results to hold. Proposition 1 establishes, perhaps surprisingly, that MHR doesn't suffice in our setting, motivating the MHR$^+$ definition.

Definition 2 (Regular). *A one-dimensional distribution with CDF F and PDF f is* regular *if for all x, $x - \frac{1-F(x)}{f(x)}$ is monotone non-decreasing.*

Definition 3 (Monotone Hazard Rate (MHR)). *A one-dimensional distribution with CDF F and PDF f is* MHR *if for all x, the hazard rate $h_F(x) := \frac{f(x)}{1-F(x)}$ is monotone non-decreasing.*

The following condition is new to this work. Note that $f'(x)$ denotes $\frac{d}{dx}f(x)$; similarly for $'$ throughout.

Definition 4 (MHR$^+$). *A one-dimensional distribution with CDF F and PDF f is* MHR$^+$ *if there exists a constant $c \geq 0$ such that $cf(x) \geq -f'(x)$ and $h_F(x) \geq c$ for all x.*[9]

Definition 5 (Decreasing Density). *A one-dimensional distribution with CDF F and PDF f has* decreasing density *if $f(\cdot)$ is non-increasing.*

The following observation provides several equivalent conditions for the above definitions. In particular, the second condition concerning MHR (4) and the second condition concerning MHR$^+$(6) establish how MHR$^+$distributions are MHR distributions "plus a little extra."

[9] Essentially every MHR distribution is also MHR$^+$except for those explicitly constructed so as not to be. See the Appendix of the full version.

Observation 2. *The definitions above are equivalent to the following conditions:*[10]

1. *A distribution is regular iff* $2f(x)^2 \geq -f'(x)(1-F(x))$ *for all* x.
2. *A distribution is regular iff* $2f(x)h_F(x) \geq -f'(x)$ *for all* x.
3. *A distribution is MHR iff* $f(x)^2 \geq -f'(x)(1-F(x))$ *for all* x.
4. *A distribution is MHR iff* $f(x)h_F(x) \geq -f'(x)$ *for all* x.
5. *A distribution is MHR*$^+$ *iff it is MHR and* $f(x)f(0) \geq -f'(x)$ *for all* x.
6. *A distribution is MHR*$^+$ *iff it is MHR and* $f(x)h_F(0) \geq -f'(x)$ *for all* x.
7. *A distribution is MHR*$^+$ *iff* $f(x)f(0) \geq -f'(x)$ *and* $h_F(x) \geq f(0)$ *for all* x.

Its proof can be found in the Appendix of the full version. We will use condition (3) for MHR and (7) for MHR$^+$several times throughout the proofs in Sects. 3 and 5 (of the full version). See Figure in the Appendix of the full version for examples in each class.

Finally, we'll use the following notation for many of our theorem statements.

Definition 6 (Expected Order Statistics). *For a single-variate distribution with CDF* F*, define:*

- $X_i^n(F)$ *to be a random variable which is the* i^{th} *highest of* n *i.i.d. draws from* F.
- $V_i^n(F)$ *to be the expected value of the* i^{th} *highest of* n *i.i.d. draws from* F. *That is,* $V_i^n(F) := \mathbb{E}[X_i^n(F)]$.
- $h_i^n(F)$ *to be the expected hazard rate of the* i^{th} *highest of* n *i.i.d. draws from* F. *That is,* $h_i^n(F) := \mathbb{E}\left[\frac{f(X_i^n(F))}{1-F(X_i^n(F))}\right]$. *Note that the definition inside the expectation is intentional: we first find the* i^{th} *highest sample, and then compute its hazard rate with respect to the original* F, f.
- $H_i^n(F)$ *to be the expected inverse hazard rate of the* i^{th} *highest of* n *i.i.d. draws from* F. *That is,* $H_i^n(F) := \mathbb{E}\left[\frac{1-F(X_i^n(F))}{f(X_i^n(F))}\right]$.

2.2 Virtual Value Preliminaries

A tool that we'll repeatedly use throughout our results is the Myersonian theory of virtual values [15].

Definition 7 (Virtual value). *For a single-dimensional distribution with CDF* F *and PDF* f*, define the* virtual value *with respect to* F *as* $\varphi_F(\cdot)$*, with* $\varphi_F(v) := v - \frac{1-F(v)}{f(v)}$.

Note also that $\varphi_F(v) = v - \frac{1}{h_F(v)}$. The inverse of $\varphi_F(v)$ is well-defined when $\varphi_F(v)$ is non-decreasing, and unique when strictly increasing.

Theorem 4 ([15]). *The following conditions hold*

[10] Let $\underline{x} = \inf \operatorname{supp}(F)$. If $\underline{x} \neq 0$, everything still holds replacing 0 with $\underline{x}$ in Observation 2 and the proof of Proposition 4.

- *Let F be regular. Then $\varphi_F(\cdot)$ is monotone non-decreasing, and* $\arg\max_p\{p \cdot (1 - F(p))\} = \varphi_F^{-1}(0)$. *Observe that $\varphi_F^{-1}(0)$ is the set of all v such that $v = 1/h_F(v)$.*
- *If F is not regular, it is still the case that $q :=$* $\arg\max_p\{p \cdot (1 - F(p))\}$ *satisfies $\varphi_F(q) = 0$.*[11]
- *Finally, for all n and F, $\mathbb{E}[X_2^n(F)] = \mathbb{E}[\varphi_F(X_1^n(F))]$.*[12]

Observation 3. *Let F be MHR. Then* $\arg\max_p\{p \cdot (1 - F(p))\}$ *is unique.*

Proof. Observe that v is strictly increasing in v, and $1/h_F(v)$ is weakly decreasing in v. Therefore, $v = 1/h_F(v)$ cannot have multiple solutions. □

3 Best Responding in the Free Market Setting

In this section, we expand on the mathematics behind what it means to best-respond in the Free Market setting. Importantly, recall that every consumer *must* select a provider, even if their utility for each is negative. Our focus is on the symmetric case (D is symmetric, searching for a symmetric equilibrium). In Sect. 5 of the full version [9] we consider extensions to the asymmetric case.

In the search for a symmetric equilibrium, the question we ask is "given that the $n-1$ other providers are setting price p, is p a best response for the remaining provider?" To resolve this, we first need to understand the payoff received by the remaining provider for setting price q while the other $n-1$ set price p.

Definition 8 (Star Operation). *Let $F_p^*(q)$ be such that when all providers $j \neq i$ are setting price p, and provider i sets price q, the probability that the consumer purchases from provider i is $1 - F_p^*(q)$. That is, the payoff to provider i in this circumstance is $q(1 - F_p^*(q))$.*

Our notation suggests that we will reason about best-responding as a single-item revenue problem, with the consumer's value distribution defined by F_p^*. Our goal will be to find sufficient conditions for a p to exist such that p itself is the revenue-maximizing price for the distribution F_p^*. Our plan is roughly as follows:

- Write an expression for $\varphi_{F_p^*}(\cdot)$.
- Observe that $\varphi_{F_p^*}(p) = 0$ is *necessary* for p to possibly be a symmetric equilibrium, by first-order conditions (Theorem 4). Show that this equation has a unique solution.

[11] For readers not familiar with this particular claim, it follows by observing that $\varphi_F(q) = 0$ is exactly the first-order condition for maximizing the revenue curve of F.

[12] This follows from the equivalence of expected virtual welfare and expected revenue. The expected revenue of the second-price auction with n bidders is the LHS, and the expected virtual value of the winner is the RHS.

- If F_p^* is regular or MHR, then first-order conditions suffice for $p = \varphi_{F_p^*}^{-1}(0)$ to be a best response, and therefore p is a symmetric equilibrium (but this is not *necessary* for p to be a symmetric equilibrium).
- Prove that if F is MHR$^+$ with decreasing density, then F_p^* is MHR.

Let's quickly highlight some aspects of this plan. Typically, finding a closed form for potential equilibria and establishing that sufficient conditions hold is a matter of solving systems of non-linear equations. Often, this process is mathematically engaging, but may not offer insight connecting the sufficient conditions to the conclusions. The final step of our outline (that MHR^+F implies MHR F_p^*) is still mostly "just math," but the rest of the outline leverages Myersonian virtual values to make the rest of the math more intuitive.

Let's now begin by writing an analytical expression for $F_p^*(q)$, $f_p^*(q)$, and $(f^*)'_p(q)$. This will let us (a) compute the virtual value $\varphi_{F_p^*}(q)$ and (b) check if F_p^* is regular or MHR. Recall that $1-F_p^*(q)$ is the probability that the consumer purchases from a provider priced at q when the other $n-1$ are priced at p. Below, $g_p(q,x)$ is the density of the maximum of $n-1$ draws from F after adding p and subtracting q.

Proposition 2. *Let* $g_p(q,x) := (n-1)f(x-q+p)(F(x-q+p))^{n-2}$. *Let also* $M := \max\{0, q-p\}$.

- $F_p^*(q) = \int_M^\infty F(x)g_p(q,x)dx$.
- $1 - F_p^*(q) = \int_M^\infty (1-F(x))g_p(q,x)dx + F(M+p-q)^{n-1}$.
- $f_p^*(q) = \int_M^\infty f(x)g_p(q,x)dx$.
- $(f_p^*)'(q) = \int_M^\infty (f'(x))g_p(q,x)dx + f(0)g_p(q,M)$.

Note that the definitions in Proposition 2 are referred to many times throughout the proofs of Propositions 3 and 4; the reader may want to keep them handy. Its proof appears in the Appendix of the full version.

We now describe a condition that, by first-order conditions, must be met in order to have a symmetric equilibrium in the Free Market. Note that this holds for *any* F, even those which are not MHR or regular.

Proposition 3. *Let* $D := F^n$. *The only possible symmetric equilibrium in the Free Market setting is* $p_F := \frac{1}{h_2^n(F)}$. *If* F_p^* *is regular, then* $\frac{1}{h_2^n(F)}$ *is a symmetric equilibrium.*

Proof. By Theorem 4 and the definition of F_p^*, in order for p to be a best response to all other providers setting price p, we must have $\varphi_{F_p^*}(p) = 0$. Note that this does not guarantee that p is a best response; this is just a necessary first-order condition.

Observe that, using Proposition 2, many of the terms in $F_p^*(q)$ (and $f_p^*(q)$) simplify when $p = q$, so:

$$\varphi_{F_p^*}(p) \quad = \quad p - \frac{1-F_p^*(p)}{f_p^*(p)} \quad = \quad p - \frac{\int_0^\infty (1-F(x))(n-1)f(x)F(x)^{n-2}dx}{\int_0^\infty f(x)(n-1)f(x)F(x)^{n-2}dx}.$$

Let's first examine the numerator. The numerator integrates over all x, the density $f(x)$, times the probability that exactly one of $n-1$ other draws from F exceeds x (this is $(n-1)(1-F(x))F(x)^{n-2}$). This is exactly the probability that one of n draws is the second-highest, which is just $1/n$. Another way to see that $1 - F_p^*(p) = 1/n$ is just that $1 - F_p^*(p)$, by definition, is the probability that a particular one of n providers is the consumer's favorite. But as D is symmetric, and the price p is the same for all providers, this is just $1/n$. So now we wish to examine the denominator, with an extra factor of n from the numerator:

$$\int_0^\infty n(n-1)f(x)^2 F(x)^{n-2}dx = \int_0^\infty n(n-1)h_F(x)f(x)F(x)^{n-2}(1-F(x))dx.$$

All we have done above is multiply and divide by $1 - F(x)$. But now the integral is interpretable: we are integrating over all x, the number of ways to choose an ordered pair (a, b) of n draws $(n(n-1))$, times the density of v_a at x $(f(x))$, times the probability that v_b exceeds x $(1-F(x))$, times the probability that the remaining $n-2$ items do not exceed x $(F(x)^{n-2})$, times the hazard rate at x. This is *exactly* computing the expected hazard rate of the second-highest of n draws from F! Therefore, we immediately conclude that:

$$\varphi_{F_p^*}(p) = p - \frac{1}{h_2^n(F)},$$

and therefore $\varphi_{F_p^*}(p) = 0$ if and only if $p = 1/h_2^n(F)$. Importantly, note that we have proven that *for all* p, even those which are not equilibria, or otherwise related to F, that $\varphi_{F_p^*}(p) = p - \frac{1}{h_2^n(F)}$. □

So at this point, we know the unique candidate for a symmetric equilibrium (because it is the only candidate which satisfies first-order conditions of Theorem 4). If we can find sufficient conditions for F_p^* to be regular (or MHR), then these first-order conditions suffice for $1/h_2^n(F)$ to indeed be a symmetric equilibrium. We identify such sufficient conditions below (remember that Proposition 1 establishes that MHR alone does not suffice, so some stronger conditions are necessary):

Proposition 4. *Let F be MHR$^+$ and have decreasing density. Then for all p, F_p^* is MHR.*

Due to space constraints, we defer the proof of Proposition 4 to the Appendix of the full version, but we provide some intuition here for why MHR$^+$ is a convenient condition for reasoning about the starred distribution. Observe that each of the starred CDF/PDF/etc. functions are convolutions of the original CDF/PDF/etc. with $g_p(q, x)$. Unfortunately, just knowing that, for example, $f'(x)(1-F(x)) \leq f(x)^2$ for all x (which is guaranteed by MHR from Observation 2 (3)) is not enough for us to reason about these convolutions. But MHR$^+$ buys us something stronger, which is exactly what's needed for the first half of the proof: not only is $-f'(x)(1-F(x)) \leq f(x)^2$ for all x, but in fact $-f'(x) \leq cf(x)$

everywhere, which does allow us to make direct substitutions into the convolution. The second step of the proof is dealing with the additional terms outside of the integral. Surprisingly, MHR$^+$ also turns out to be the right condition to reason transparently about these additional terms, although more creativity is required here than in step one.

And now, we can wrap up the proof of Theorem 1, which claims that whenever F is MHR$^+$, the unique symmetric equilibrium in the Free Market setting for $D := F^n$ is for each provider to set price $1/h_2^n(F)$.

Proof (Proof of Theorem 1). Proposition 3 establishes that $1/h_2^n(F)$ is a symmetric equilibrium as long as $F^*_{1/h_2^n(F)}$ is regular. Proposition 4 proves something even stronger: that F_p^* is MHR for all p, as long as F is MHR$^+$ with decreasing density. The two propositions together complete the proof. □

Note that Theorem 1 accomplishes several tasks:[13]

- It establishes that a symmetric equilibrium exists subject to MHR$^+$ (which is not generally true without some assumptions, Proposition 1).
- It provides a clean closed form for this symmetric equilibrium.
- It establishes uniqueness of this symmetric equilibrium (even stronger: this is the only possible symmetric equilibrium for all F). This is important because it lets us reason about "*the* utility in the Free Market setting" without needing to worry about exactly which equilibrium we should be analyzing.

4 Comparing Consumer Utilities

In this section, we derive a Limit-Entry condition, which dictates when consumer utility is higher in the Limited Entry setting versus the Free Market. Note that our condition is well-defined even when no symmetric equilibrium exists in the Free Market setting. Let's first recall the Limit-Entry condition from Sect. 1, which a distribution satisfies when

$$H_1^n(F) \leq n/h_2^n(F),$$

where again, $H_1^n(F)$ is the expected inverse hazard rate of the highest of n i.i.d. draws from F and $h_2^n(F)$ is the expected hazard rate of the second-highest of n i.i.d. draws from F. Recall that Theorem 2 states that consumer utility is higher in the Limited Entry setting versus the Free Market setting *if and only if* the Limit-Entry condition holds. The main result of this section is a proof of Theorems 2 and 3.

Let's first compute the expected consumer utility in the Limited Entry setting.

---

[13] The first two bullets below are *not* discussed in [17]: they do not provide direct conditions on F for equilibrium to exist, nor analyze a closed form for the unique potential equilibrium. The third bullet *was* also accomplished by [17], as they do show that a unique potential equilibrium exists.

Lemma 1. *The expected consumer utility at the unique Limited Entry equilibrium is* $V_1^{n-1}(F)$.

Proof. There are a total of $n-1$ providers, and recall from Observation 1 that the unique equilibrium has all prices set to 0. Therefore, the consumer's expected payment is zero. The consumer picks their favorite plan, with value simply the maximum of $n-1$ i.i.d. draws from F. Together, we see that the consumer's expected utility at the unique symmetric equilibrium of the Limited Entry setting is $V_1^{n-1}(F)$. □

Now, let's compute the expected consumer utility in the Free Market setting.

Lemma 2. *The expected consumer utility at the unique symmetric equilibrium (when it exists) in the Free Market setting is* $V_1^n(F) - 1/h_2^n(F)$.

Proof. There are a total of n providers, and the unique symmetric equilibrium (when it exists) has all prices set to $1/h_2^n(F)$. Therefore, the consumers expected payment is $1/h_2^n(F)$ (because the consumer must purchase a plan, even with negative utility for everything). The consumer's value for their favorite plan is the maximum of n i.i.d. draws from F. Therefore, the consumer's expected utility at the unique symmetric equilibrium of the Free Market setting is $V_1^n(F) - 1/h_2^n(F)$. □

With these two calculations in hand, we can prove Theorem 2.

Proof (Proof of Theorem 2). We observe first that the expected utility is higher in the Limited Entry setting versus Free Market if and only if $V_1^{n-1}(F) \geq V_1^n(F) - 1/h_2^n(F)$. The remainder of the proof is rewriting this condition, using Myersonian virtual value theory in yet another way. We produce the steps below, and justify each step afterwards. Two of the three steps follow from basic algebra or a coupling argument. The middle step (line three) makes use of virtual value theory.

$$
\begin{aligned}
V_1^{n-1}(F) &\geq V_1^n(F) - \frac{1}{h_2^n(F)} \\
\Leftrightarrow \frac{n-1}{n} V_1^n(F) + \frac{1}{n} V_2^n(F) &\geq V_1^n(F) - \frac{1}{h_2^n(F)} \\
\Leftrightarrow \frac{n-1}{n} V_1^n(F) + \frac{1}{n}\left(V_1^n(F) - H_1^n(F)\right) &\geq V_1^n(F) - \frac{1}{h_2^n(F)} \\
\Leftrightarrow H_1^n(F) &\leq \frac{n}{h_2^n(F)}.
\end{aligned}
$$

The first equivalence follows by a coupling argument. One way to draw the highest of $n-1$ draws from F, or $X_2^{n-1}(F)$, is to take n draws from F, remove one uniformly at random, and then examine the highest remaining. With probability $1/n$, the highest of the n draws is excluded, so the highest remaining is $X_2^n(F)$. The rest of the time, a different draw is excluded and the highest of n remains, giving $X_1^n(F)$. Hence in expectation, $V_1^{n-1}(F) = \frac{n-1}{n} V_1^n(F) + \frac{1}{n} V_2^n(F)$.

The second equivalence follows from Theorem 4, as $V_2^n(F) = \mathbb{E}[\varphi_F(X_1^n(F))]$. More familiarly, this is the fact that a second-price auction is revenue-maximizing, and that the revenue is equal to the virtual welfare of the highest-valued bidder in the i.i.d. setting. Recall that $\varphi_F(v) = v - 1/h_F(v)$; then $\mathbb{E}[\varphi_F(X_1^n(F))] = V_1^n(F) - H_1^n(F)$.

The final equivalence follows by subtracting $V_1^n(F)$ from both sides and multiplying by -1. □

Finally, we prove Theorem 3. Recall that Theorem 3 asserts that whenever F is MHR with decreasing density, it satisfies the Limit-Entry condition. Recall that MHR alone is not enough to guarantee that there is a symmetric equilibrium in the Free Market setting for the Limited Entry setting to improve over, but that the condition is well-defined anyway. When F is further MHR$^+$, there is a symmetric equilibrium in both settings, and the consumer utility is always higher with Limited Entry (Corollary 1).

Proof (Proof of Theorem 3). The proof will follow from the steps below (justification for each step is provided afterwards). If F is MHR with decreasing density, then:

$$\begin{aligned} f(0) &\geq \mathbb{E}[f(X_1^{n-1})] \\ \Leftrightarrow \frac{1}{f(0)} &\leq \frac{1}{\mathbb{E}[f(X_1^{n-1})]} \\ \Leftrightarrow \frac{1}{h_F(0)} &\leq \frac{n}{n\mathbb{E}[f(X_1^{n-1})]} \\ \Rightarrow H_1^n(F) &\leq \frac{n}{h_2^n(F)}. \end{aligned}$$

The first line follows because F has decreasing density. Therefore $f(0) \geq f(x)$ for all x, and certainly $f(0) \geq \mathbb{E}[f(X)]$ for any non-negative random variable X (including X_1^{n-1}). The second line follows by simple algebra. The third line makes two steps. On the LHS, we observe that $f(0) = h_F(0)$, so the left-hand sides are actually identical between the second and third lines. On the right-hand side, we just multiply the numerator and denominator by n. The final implication again makes two steps. On the left-hand side, we observe that as F is MHR, $1/h_F(0) \geq 1/h_F(x)$ for all $x \geq 0$. Therefore, $1/h_F(0) \geq \mathbb{E}[1/h_F(X)]$ for any non-negative random variable X (including X_1^n). On the right-hand side, we use the equality $n\mathbb{E}[f(X_1^{n-1})] = h_2^n(F)$, which follows from a technical lemma (Lemma 3, proved immediately after this proof).

The last line above completes the proof: we have shown that if F is MHR with decreasing density, then the Limit-Entry Condition is satisfied. □

The remaining task is to prove Lemma 3, below.

Lemma 3. $h_n^2(F) = \mathbb{E}[n \cdot f(X_1^{n-1}(F))]$.

Proof.

$$\begin{aligned} h_2^n(F) &= \int_0^\infty n(n-1)f(x)h_F(x)F(x)^{n-2}(1-F(x))dx \\ &= \int_0^\infty n(n-1)f(x)\cdot f(x)F(x)^{n-2}dx \\ &= \mathbb{E}[n\cdot f(X_1^{n-1}(F))]. \end{aligned}$$

The first line is simply the definition of $h_2^n(F)$. The second line just rewrites $h_F(x)(1-F(x)) = f(x)$. The third line observes that $(n-1)f(x)F(x)^{n-2}dx$ is the density of $X_1^{n-1}(F)$. Indeed, there are $n-1$ ways to choose a provider a from $n-1$, $f(x)$ is the density of v_a at x, and $F(x)^{n-2}$ is the probability that all $n-2$ other providers have $v_i \leq x$. So we are integrating the density of $X_1^{n-1}(F)$ at x, times $f(x)$ from 0 to ∞. This exactly computes the expected value of $f(X_1^{n-1})$. The extra factor of n is carried through. □

5 Conclusion

We study the impact of limiting entry on consumer surplus with mandatory purchase. We provide clean necessary and sufficient conditions for limiting entry to improve consumer surplus, as well as sufficient conditions for an equilibrium to exist in the Free Market setting.

Our model is of course stylized, so it is always important for future work to relax assumptions. While it is always interesting to relax technical assumptions (even if they do a reasonable job capturing practice), the most interesting direction would be to consider other models of limiting entry. For example, our model takes a pure view on competition, and implies that limiting entry to all but a single provider will drive prices down to the marginal cost of production. This seems to be the biggest deviation from our model and practice, as healthcare exchanges typically have much fewer options. It would be interesting to further refine our model to better match this aspect.

Acknowledgments. Kira Goldner was supported in part by NSF award DMS-1903037 and a Columbia Data Science Institute postdoctoral fellowship, and in part by a Shibulal Family Career Development Professorship. S. Matthew Weinberg was funded by NSF CAREER CCF-1942497.

We thank Mark Shepard for his expert guidance on health insurance markets, for initially posing this research direction, and for helping us to refine the model. We also thank Mark Braverman and Anna Karlin for helpful discussions in early stages of this work.

References

1. Anton, J.J., Gertler, P.J.: Regulation, local monopolies and spatial competition. J. Regul. Econ. **25**(2), 115–141 (2004)

2. Bulow, J., Klemperer, P.: Auctions vs. negotiations. Technical report, National Bureau of Economic Research (1994)
3. Chawla, S., Hartline, J.D., Kleinberg, R.: Algorithmic pricing via virtual valuations. In: Proceedings of the 8th ACM Conference on Electronic Commerce, pp. 243–251. ACM (2007)
4. Chawla, S., Hartline, J.D., Malec, D.L., Sivan, B.: Multi-parameter mechanism design and sequential posted pricing. In: Proceedings of the Forty-Second ACM Symposium on Theory of Computing, pp. 311–320. ACM (2010)
5. Chawla, S., Malec, D., Sivan, B.: The power of randomness in Bayesian optimal mechanism design. Games Econ. Behav. **91**, 297–317 (2015)
6. Chen, Y., Li, X.: Group buying commitment and sellers? Competitive advantages. J. Econ. Manag. Strategy **22**(1), 164–183 (2013)
7. Dana, J.D., Jr.: Buyer groups as strategic commitments. Games Econ. Behav. **74**(2), 470–485 (2012)
8. Engel, E., Fischer, R., Galetovic, A.: Competition in or for the field: which is better? Technical report, National Bureau of Economic Research (2002)
9. Essaidi, M., Goldner, K., Weinberg, S.M.: When to limit market entry under mandatory purchase. CoRR abs/2002.06326 (2020). https://arxiv.org/abs/2002.06326
10. Hart, S., Nisan, N.: Approximate revenue maximization with multiple items. In: The 13th ACM Conference on Electronic Commerce (EC) (2012)
11. Hart, S., Nisan, N.: Approximate revenue maximization with multiple items. J. Econ. Theory **172**, 313–347 (2017)
12. Hotelling, H.: Stability in competition. Econ. J. **39**(4), 57 (1929)
13. Johnson, E.J., Hassin, R., Baker, T., Bajger, A.T., Treuer, G.: Can consumers make affordable care affordable? The value of choice architecture. PloS One **8**(12) (2013)
14. McGuire, T.G., Riordan, M.H.: Incomplete information and optimal market structure public purchases from private providers. J. Public Econ. **56**(1), 125–141 (1995)
15. Myerson, R.B.: Optimal auction design. Math. Oper. Res. **6**(1), 58–73 (1981)
16. Pai, M.M., Vohra, R.: Optimal auctions with financially constrained buyers. J. Econ. Theory **150**, 383–425 (2014)
17. Perloff, J.M., Salop, S.C.: Equilibrium with product differentiation. Rev. Econ. Stud. **52**(1), 107–120 (1985)
18. Saban, D., Weintraub, G.Y.: Procurement mechanisms for assortments of differentiated products. Available at SSRN 3453144 (2019)
19. Williams, J.: The patient protection and affordable care act meets the 'persistently uninsured'. Soc. Policy Adm. **50**(4), 452–466 (2016)

Balancing Participation and Decentralization in Proof-of-Stake Cryptocurrencies

Aggelos Kiayias[1,5], Elias Koutsoupias[2,5], Francisco Marmolejo-Cossío[3,5](✉), and Aikaterini-Panagiota Stouka[4]

[1] University of Edinburgh, Edinburgh, UK
akiayias@inf.ed.ac.uk
[2] University of Oxford, Oxford, UK
elias.koutsoupias@cs.ox.ac.uk
[3] Harvard University, Cambridge, USA
fjmarmol@seas.harvard.edu
[4] Nethermind, London, UK
aikaterini-panagiota@nethermind.io
[5] Input Output(IOG), Singapore, Singapore

Abstract. Proof-of-stake blockchain protocols have emerged as a compelling paradigm for organizing distributed ledger systems. In proof-of-stake (PoS), a subset of stakeholders participate in validating a growing ledger of transactions. For the safety and liveness of the underlying system, it is desirable for the set of validators to include multiple independent entities as well as represent a non-negligible percentage of the total stake issued. In this paper, we study a secondary form of participation in the transaction validation process, which takes the form of stake delegation, whereby an agent delegates their stake to an active validator who acts as a stake pool operator. We study payment schemes that reward agents as a function of their collective actions regarding stake pool operation and delegation. Such payment schemes serve as a mechanism to incentivize participation in the validation process while maintaining decentralization. We observe natural trade-offs between these objectives and the total expenditure required to run the relevant payment schemes. Ultimately, we provide a family of payment schemes which can strike different balances between these competing objectives at equilibrium in a Bayesian game theoretic framework.

Keywords: delegation games · proof of stake · cryptocurrencies · decentralization

---

Full version of the paper: https://arxiv.org/abs/2407.08686.
A.-P. Stouka—Part of this work was conducted while Stouka was a research associate at the Edinburgh Blockchain Technology Lab.

G. Schäfer and C. Ventre (Eds.): SAGT 2024, LNCS 15156, pp. 333–350, 2024.
https://doi.org/10.1007/978-3-031-71033-9_19

1 Introduction

Proof-of-stake (PoS) blockchain protocols have emerged as a compelling paradigm for organizing distributed ledger systems. Unlike Proof-of-work (PoW), where computational resources are expended for the opportunity to append transactions to a growing ledger, PoS protocols designate the potential to update the ledger proportionally to the stake one has within the system. Common to both protocols is the fact that larger and more varied participation in the transaction validation process provides the system with increased security and liveness.

Although participating as a validator in a PoS protocol is computationally less intensive than doing so in a PoW protocol, it still demands some effort, e.g. that the validator be consistently online and maintain dedicated hardware and software, thus it is still not the case that every agent in the system decides to, or is even able to, do so. Given this, a compelling intermediate form of participation in the transaction validation process is stake delegation. In PoS systems with stake delegation, validators can be considered stake pool operators (SPOs), who activate pools controlling their own as well as delegated stake of others. Agents who prefer not to participate as validators have the opportunity to delegate their stake to active pools and gain rewards. In this paradigm, pools are chosen to update the ledger proportional to the combination of their "pledged stake" (i.e., stake they contribute) and externally delegated stake (stake contributed to them by others); in this way, delegation can be seen as a vetting of how frequently operators should be selected to participate. Furthermore, delegation is not borne out of good will alone, since the system provides additional payments to all agents as a function of the profile of pool operators and delegators in the system. The space of payment mechanisms provides for an interesting problem in balancing three objectives: increasing participation in the validation protocol of the system (via delegation), maintaining a decentralized validation creation process (in spite of added delegation), and balancing the budget of rewards to be given to operators and delegators.

1.1 Related Work and Motivation

Our work is most related to that of Brunjes et al. [2] which introduces a reward sharing scheme for stake pools as a mechanism to incentivize decentralization–a key objective shared with our work. The reward sharing scheme of their paper has been operational on the Cardano mainnet since July 2020.[1] In this existing Cardano reward sharing scheme, decentralization in the system is modulated by a system parameter, k, an integer representing the number of pools of equal size which are formed at equilibrium under the given payment scheme. This parametric formulation has the added benefit of preventing a single entity with low stake from controlling the majority of pools. Continuing with this line of work, the authors of [10] analyze the Nash dynamics of the Cardano reward sharing scheme and the decentralization that it offers through metrics similar to those

[1] https://roadmap.cardano.org/en/shelley/.

we employ to measure decentralization. In more detail, they use a variation of the Nakamoto coefficient [14] that takes into account not only the number of pools in the system, but also the overall composition of stake of the operators who run the pools. This metric can be loosely interpreted as a measure of the composition of "skin in the game" that SPOs have, by looking at the overall pledged stake from SPOs that may have enough cumulative stake (pledged and delegated) to perform an attack on the system. Multiple subsequent papers have proposed other metrics for decentralization of blockchain protocols (with applicability beyond PoS consensus), including [1,4,5,7–10,14,15].

Both [2] and [10] use in their analysis a framework for incentives called *non-myopic utility* that tries to predict how delegators will choose a pool when the system stabilizes at equilibrium. This analysis is essential because a key component of their reward mechanism is the *margin* of rewards an SPO keeps for themselves before further sharing rewards with delegators. We present a variation of the reward schemes of [2] in which the margin of the operators is implicitly set by the system. Most importantly, we study trade-offs between three competing objectives for the system: decentralization, overall participation, and the expenditure of the reward sharing scheme used. Furthermore, we study this performance in the presence of users who are only willing to delegate their stake if the reward they earn is lower bounded by an amount ϵ (i.e., users who may be "lazy", or who may have external sources of earnings for their stake).

Liquid Staking Protocols on Ethereum. We note that the framework of reward sharing schemes that we present is general enough to encompass key features of liquid staking protocols (LSPs) which are increasingly used in the Ethereum blockchain after its transition to PoS consensus [6]. At a high level, LSPs allow users to "stake" their cryptocurrency (such as ETH) to be used for validation even when their cryptocurrency held is below the 32 ETH threshold required to be a validator. Upon staking their assets, users receive a liquid token in return, representing the staked assets. These liquid tokens can be used in various decentralized finance (DeFi) applications, providing liquidity and earning additional rewards while the original stake continues to generate staking rewards when used to facilitate validation. Validators for LSPs are equivalent to SPOs in our model, and individuals who mint liquid stake tokens are similar to delegators in our model. Rewards are inherently generated by the Ethereum validation process and shared according to the specification of the corresponding LSP.

Currently on Ethereum, more than 25% of all ETH in circulation is staked to be used for PoS validation, of which more than 50% is attributed to 5 validators participating in LSPs (amounting overall to approximately 50 billion USD as of May 2024)[2]. Of all stakes in LSPs, the majority is related to the Lido LSP (approximately 29% of the total ETH in circulation), which facilitates staking ETH to a permissioned set of validators designated by the Lido DAO [13]. However, most relevant to our work is the permissionless validator setting, exemplified by Rocket Pool [12] and ether.fi [17] in which any user can become

[2] https://defillama.com/protocols/liquid%20staking/Ethereum.

a validator as long as they have enough ETH deposited as collateral, according to the specifications of the underlying LSP. Providing collateral is essential for aligning incentives of validators, for without locking collateral, they can mount attacks on the LSP by shorting liquid stake tokens with nothing to lose. In our work, SPOs choose how much stake to pledge to the pool they operate, and this quantity plays the same essential role as locked collateral in LSPs, forcing SPOs to have "skin in the game" in terms of the consensus validation process.

1.2 Overview

We consider a setting where a finite number of agents owns a publicly known amount of stake in a decentralized system. Agents at a high level are given three options: (i) They can create a stake pool, whereby they can be delegated stake from other players. Such agents are called stake pool operators (SPOs). To be an SPO, the agent must pledge whatever stake they own and, in addition, incur a private cost, $c > 0$, for the operation of the pool. (ii) They can delegate their stake to pools that are in operation. Such agents are called delegators. (iii) They can decide to abstain from participating in the protocol and remain idle, earning baseline utility $\epsilon > 0$.

Participation. We are interested in systems that encourage increased participation in the overall validation process. To prevent agents from abstaining from the protocol, they must at least be able to delegate in such a way as to earn more than ϵ, their baseline utility for remaining idle.

Rewards and Incentives. We consider reward schemes whereby pool operators and delegators are compensated as a function of which pools are active and whom delegators choose to delegate to. This creates a well-defined family of one-shot games that are played between all agents in the system, and we study the equilibria that result as a function of the reward scheme implemented.

Informal Design Objectives. Our goal is to create reward schemes that optimize for three distinct objectives: (i) Increasing participation in the system; (ii) Increasing Decentralization, i.e. preventing stake from overly accumulating (via delegation) in the hands of few SPOs; (iii) Minimizing the budget necessary to achieve the above.

1.3 Roadmap of Our Results

We consider the setting in which stakeholders of a PoS blockchain can either operate pools (receive delegation), delegate their stake, or abstain from the protocol, where each of these actions provides a certain reward from the system. Section 2 begins by introducing the notion of a delegation game, which is a general framework for encapsulating strategic considerations between stakeholders in this setting. At the end of Sect. 2, we introduce the notion of a uniform reward delegation game, which is a refinement of general delegation games by which all delegators in the system (roughly) earn a uniform reward per unit of stake that they delegate. Within the class of uniform delegation games, we further hone our focus on proper delegation games, which we define in such a way to exemplify relevant characteristics of existing reward sharing schemes deployed in practice. In Sect. 3 we provide sufficient conditions for pure Nash equilibria in proper delegation games. Section 4 introduces a Bayesian framework to proper delegation games and explores novel solution concepts intricately tied to ex post pure Nash equilibria. In Sect. 5 we introduce the main metrics by which we compare the equilibria of the Bayesian proper delegation game: participation, decentralization, and system expenditure. Section 6 provides details on the computational methods used to evaluate the performance of payment schemes in proper delegation games at equilibrium, together with experimental results. Finally, Sect. 7 provides a conceptual overview of the results obtained and provides future directions of work.

2 The Delegation Game

We formalize a general family of games modeling agent decisions regarding whether to create a pool, delegate their stake or remain idle. We consider $n > 0$ players, each with a publicly known stake, $s_i > 0$. We also assume that any agent who chooses to operate a pool incurs a cost of $c_i > 0$ and any agents who remain idle obtains a fixed utility ϵ_i. This can encompass the fact that an agent may find participating in stake delegation prohibitively complicated or that they prefer using their stake in other ways.

Player Strategies. For each player, $i \in [n]$, let $\mathcal{D}_i$ denote the set of functions $d_i : [n] \setminus \{i\} \to \mathbb{R}^+$ such that $\sum_{j \in [n] \setminus \{i\}} d_i(j) = s_i$. The action space of the i-th player corresponds to the set $\mathcal{A}_i = \{a_I\} \cup \{a_{SPO}\} \cup \mathcal{D}_i$. We further denote the space of all joint strategy profiles by $\mathcal{A} = \prod_i \mathcal{A}_i$. A joint strategy profile of the game is a vector $\mathbf{p} = (p_i)_{i=1}^n \in \mathcal{A}$, where $p_i \in \mathcal{A}_i$ denotes the action taken by the i-th agent. Furthermore, for a fixed agent $i \in [n]$, we let $\mathcal{A}_{-i}$ denote the action space of all players other than i, such that $\mathbf{p}_{-i} \in \mathcal{A}_{-i}$ denotes a specific collection of strategies for all players in $[n] \setminus \{i\}$, and $\mathbf{p} = (p_i, \mathbf{p}_{-i}) \in \mathcal{A}$ denotes a strategy profile that makes specific reference to the action $p_i \in \mathcal{A}_i$ played by the i-th player. There are 3 relevant cases for the values p_i can take and hence the actions that the i-th player can take:

- $p_i = a_I$ represents non-participation in delegation for the i-th agent. We say that the agent is *idle.*
- $p_i = a_{SPO}$ occurs when the i-th player chooses to operate their pool and be an SPO. They pledge their stake, s_i, to the pool and incur a pool operation cost of c_i. In this case, we say the i-th pool is active (otherwise it is inactive)
- $p_i = d_i \in \mathcal{D}_i$ occurs when the i-th player chooses to delegate their stake, s_i, to different pools operated by other agents. We call d_i the player's *delegation profile.* For each $j \in [n] \setminus \{i\}$, the player i delegates $d_i(j)$ stake to a pool operated by the j-th agent. We say that the agent is a *delegator.*

Rewards. For each agent, $i \in [n]$, we let $R_i : \mathcal{A} \to \mathbb{R}$ be their delegation game reward function. For $\mathbf{p} \in \mathcal{A}$, we denote $R_i(\mathbf{p})$ as the reward obtained by the i-th agent. We impose two constraints on R_i: (i) If $p_i = a_I$, then $R_i(\mathbf{p}) = \epsilon_i$; and (ii) If $p_i = d_i \in \mathcal{D}_i$, then $R_i(d_i, \mathbf{p}_{-i})$ can be further decomposed as the sum of $n-1$ delegation reward functions, $R_i(d_i, \mathbf{p}_{-i}) = \sum_{j \in [n] \setminus \{i\}} R_{i,j}(d_i(j), \mathbf{p}_{-i})$, with the constraints that $R_{i,j}(0, \mathbf{p}_{-i}) = 0$ for all $\mathbf{p}_{-i} \in \mathcal{A}_{-i}$ and $R_{i,j}(d_i(j), \mathbf{p}_{-i}) = 0$ if pool j is not active.

Utilities. For $i \in [n]$, we let $u_i : \mathcal{A} \to \mathbb{R}$, denote the i-th player's utility function, For $\mathbf{p} \in \mathcal{A}$, if $p_i = a_I$, then $u_i(\mathbf{p}) = \epsilon_i$. If $p_i = a_{SPO}$, then $u_i(\mathbf{p}) = R_i(\mathbf{p})) - c_i$. Finally, if $p_i \in \mathcal{D}_i$, then $u_i(\mathbf{p}) = R_i(\mathbf{p})$.

Definition 1 (The Delegation Game). *Suppose that we have n agents with publicly known stakes denoted by $\mathbf{s} = (s_i)_{i=1}^n$, privately known pool operation costs $\mathbf{c} = (c_i)_{i=1}^n$ and privately known idle utilities $\boldsymbol{\epsilon} = (\epsilon_i)_{i=1}^n$. In addition, suppose that $\mathbf{R} = (R_i)_{i=1}^n$ is a family of reward functions $R_i : \mathcal{A} \to \mathbb{R}^+$. We let $\mathcal{G}(\mathbf{R}, (\mathbf{s}, \mathbf{c}, \boldsymbol{\epsilon}))$ be the corresponding game with induced utilities $\mathbf{u} = (u_i)_{i=1}^n$ from above. This game is called the "Delegation Game" for $\mathbf{s}, \mathbf{c}, \boldsymbol{\epsilon}$, and $\mathbf{R}$.*

2.1 Games with Uniform Delegation Rewards

Given the large class of delegation games, we focus on a natural class of games similar to what is used on the Cardano blockchain [2]. Cardano implicitly rewards pools as a function of their pledge and accrued external delegation. SPOs in turn choose how much of this reward to proportionally share with delegators (by choosing what is called a *margin*). Going forward, we let λ_j be the operator pledge of pool j , given by $\lambda_j = s_j$ when $p_j = a_{PO}$, and $\lambda_j = 0$ otherwise. In addition, we let β_j be the external stake delegated to pool j under $\mathbf{p}$ given by $\beta_j = \sum_{i:p_i \in \mathcal{D}_i} d_i(j)$. Finally, we let $\sigma_j = \lambda_j + \beta_j$ be the size of the pool corresponding to player j.

Definition 2 (Pool Reward Function). *A pool reward function is given by $\rho : (\mathbb{R}^+)^2 \to \mathbb{R}^+$ that takes as input pool pledge and external delegation and returns pool rewards given by $\rho(\lambda, \beta)$.*

The Cardano pool reward function has the further property that rewards are capped, and rewards themselves can be decomposed into a specific algebraic form which we call separable:

Definition 3 (Capped Separable Pool Reward Function). *Let $\tau > 0$ and $a, b : \mathbb{R}^+ \to \mathbb{R}^+$ and define $\rho : (\mathbb{R}^+)^2 \to \mathbb{R}^+$ by $\rho(\lambda, \beta) = a(\lambda') + b(\lambda')\beta'$, where $\lambda' = \min\{\tau, \lambda\}$ and $\beta' = \min\{\tau - \lambda', \beta\}$. We say that ρ is a capped pool reward function with cap τ. We also say that ρ is separable into a and b, where a is the pledge reward component and b is the external delegation reward component.*

Delegation games, as per Definition 1, already exemplify an important point of departure from Cardano reward sharing schemes. Namely, they have a simpler action space for agents, amounting to the high-level choice of: being an SPO, being a delegator, and being idle. In Cardano, rewards have a more complicated action space whereby beyond the choice to become an SPO, agents can also pick the margin of rewards they wish to keep as SPOs. In [2], the authors study the parametric family of pool reward functions used in Cardano to show that when players are non-myopic, one can modulate the number of pools, k, which are formed at equilibrium. An important characteristic of these equilibria though is the fact that pool operators choose a margin such that delegators are indifferent amongst the k active pools in terms of the delegation reward they obtain from them. Rather than letting agents reach such an outcome at equilibrium, we study delegation games with the very property that delegators earn the same per-unit reward mostly irrespective of the pool to which they delegate. In order to do so, we introduce the notion of delegator rewards:

Definition 4. *A delegation reward function is given by* $r : \mathcal{A} \times (\mathbb{R}^+)^n \to \mathbb{R}^+$ *and takes as input* $\mathbf{p} = (p_i)_{i=1}^n \in \mathcal{A}$ *and* $\mathbf{s} = (s_i)_{i=1}^n$ *to output a fixed reward per unit of delegated stake given by* $r(\mathbf{p}, \mathbf{s})$.

We will shortly precisely define delegation games with uniform delegation rewards, but at a high level these games have reward functions that automatically enforce the fact that for a given strategy profile, delegators will receive $r(\mathbf{p}, \mathbf{s})$ rewards per unit of delegation. Continuing with the comparison with Cardano, at equilibrium, it is not the case that all pools have equal per-unit delegation rewards, but rather the k pools which offer the best per-unit delegation rewards to delegators. It can very well be the case that a suboptimal pool remain in operation, albeit offering lower per-unit rewards to potential delegators. In this spirit, we define the notion of pool feasibility to determine which pools are suboptimal, and use this to define delegation games with uniform rewards.

Definition 5 (Pool feasibility). *For a given joint strategy profile* $\mathbf{p}$*, we call the* i*-th pool* feasible *if* $p_i = a_{SPO}$ *and* $\rho(\lambda_i, \beta_i) \geq \sigma_i r(\mathbf{p}, \mathbf{s})$.

Definition 6 (Uniform Delegation Agent Rewards). *Suppose that we have* n *agents with stake distribution* $\mathbf{s}$*, participation costs* $\mathbf{c}$*, and idle utilities* $\boldsymbol{\epsilon}$*. Furthermore, suppose that* $\mathbf{p} \in \mathcal{A}$ *is such that* $p_i = d_i \in \mathcal{D}_i$*. If we let* $r = r(\mathbf{p}, \mathbf{s})$*, then the components of the reward function for the* i*-th agent are given by* $R_{i,j}(d_i(j), \mathbf{p}_{-i})$ *for each* $j \in [n]$*. This quantity is given by* $r \cdot d_i(j)$ *if the* j*-th pool is active and feasible,* $\frac{d_i(j)}{\sigma_j} \cdot \rho(\lambda_j, \beta_j)$ *if the* j*-th pool is active and infeasible, and 0 if the* j*-th pool is inactive. With this in hand, we can fully define the reward function for the* i*-th agent under arbitrary actions as follows:*

$$R_i(\mathbf{p}) = \begin{cases} \epsilon_i & \text{if } p_i = a_I \\ \rho(\lambda_i, \beta_i) - r \cdot \beta_i & \text{if } p_i = a_{SPO}, \text{ feasible} \\ \frac{\lambda_i}{\sigma_i} \cdot \rho(\lambda_i, \beta_i) & \text{if } p_i = a_{SPO}, \text{ infeasible} \\ \sum_{j \in [n] \setminus \{i\}} R_{i,j}(d_i(j), \mathbf{p}_{-i}) & \text{if } p_i = d_i \in \mathcal{D}_i \end{cases}$$

If a delegation game $\mathcal{G}$ *has uniform delegation rewards, we say it is a uniform delegation reward game.*

Narrowing Down Delegation Rewards. As seen before, Cardano rewards at equilibrium are specified by the most competitive agent who misses out on being an SPO. Similarly, we focus on a delegation reward function that is specified according to the "most competitive" delegator, with the property that once such a delegator is identified, all less competitive delegators do not deviate.

Definition 7. *For a given ρ, we let $\alpha : (\mathbb{R}^+)^2 \to \mathbb{R}^+$ given by $\alpha(s, c) = \frac{\rho(s,0)-c}{s}$ be the rewards per unit of stake that an SPO with stake s and pool cost c obtains for opening a pool without external delegation (a solo pool). We call $\alpha(s, c)$ the threat of deviation of a delegator with stake s and pool operation cost c.*

For a given $\mathbf{p} \in \mathcal{A}$, we would ideally set delegation rewards to be the maximum threat of deviation among delegators, to ensure no delegator has an incentive to deviate by becoming a solo pool. However, α depends on private c values, hence we suppose that $0 \leq c_{min} \leq c_i \leq c_{max}$ for any $i \in [n]$, where c_{min} and c_{max} are publicly known. With this, we define the max-delegate rewards:

Definition 8 (Max-delegate r). *Given ρ, we let $r_M : \mathcal{A} \times (\mathbb{R}^+)^n \to \mathbb{R}^+$ be $r_M(\mathbf{p}, \mathbf{s}) = \max_{i:p_i \in \mathcal{D}_i} \alpha(s_i, c_{min})$. If $\{i \in [n] \mid p_i \in \mathcal{D}_i\} = \emptyset$, then $r_M(\mathbf{p}, \mathbf{s}) = 0$.*

In what follows, we will consider pool reward functions ρ with the natural property that α is monotonically increasing in s as well. In this case, we can express the max-delegate reward function in a more simple and useful fashion by making use of the following:

Definition 9. *Suppose that $\mathcal{G}$ is a delegation game and $\mathbf{p} \in \mathcal{A}$. We let $s^* = \max_{i:p_i \in \mathcal{D}_i} s_i$ and call this quantity the pivotal delegation stake of $\mathbf{p}$. If $p_i \in \mathcal{D}_i$ and $s_i = s^*$, then we also say that the player is a pivotal delegate in $\mathbf{p}$.*

If the reward function, ρ, is such that α increases monotonically in s, then it follows that $r_M(\mathbf{p}, \mathbf{s}) = \alpha(s^*, c_{min})$. We focus on uniform delegation games with max-delegate rewards such that α is monotonically increasing in pledge. We give this class of games a specific name as the main focus of this paper:

Definition 10 (Proper delegation game). *Suppose that $\mathcal{G}$ is a uniform delegation game such that the following hold: (i) ρ is such that $\alpha(s, c)$ is monotonically increasing for $s \in [0, s_{max}]$, where $s_{max} = \max\{s_i\}$, (ii) ρ is capped and separable with $s_{max} < \tau$, and (iii) Delegation rewards are given by r_M. We say that ρ is a proper reward function and $\mathcal{G}$ is a proper delegation game. For a specific proper delegation game, we use notation $\mathcal{G}(\rho, \tau, (\mathbf{s}, \mathbf{c}, \epsilon))$.*

3 Equilibria in Proper Delegation Games

This section provides sufficient conditions for a joint strategy profile to be a pure Nash equilibrium. We use the shorthand $r = r_M(\mathbf{p}, \mathbf{s}) \in \mathbb{R}^+$ to refer to the per-unit reward for delegating to a feasible pool, and we begin by providing multiple structural results related to the best responses agents may have in a proper delegation game. Proofs can be found in the full version of the paper.

Lemma 1 (Feasible pool structural lemma). *Let $\mathcal{G}(\rho, \tau, (\mathbf{s}, \mathbf{c}, \boldsymbol{\epsilon}))$ be a proper delegation game with all agents playing $\mathbf{p} \in \mathcal{A}$. Let $p_i = a_{SPO}$ for an infeasible pool with pledge $\lambda_i = s_i$ and external delegation $\beta_i \geq 0$. Both delegators to the pool and the SPO can obtain strictly more utility by delegating to a feasible pool.*

Lemma 2 (Idle and Delegator Best Response). *Consider a proper delegation game $\mathcal{G}(\rho, \tau, (\mathbf{s}, \mathbf{c}, \boldsymbol{\epsilon}))$ and either $\mathbf{p} = (a_I, \mathbf{p}_{-i}) \in \mathcal{A}$ or $\mathbf{p} = (d_i, \mathbf{p}_{-i}) \in \mathcal{A}$, with the i-th player delegating to feasible pools. In both cases, deviating to become an SPO cannot be a strict best response for the i-th player.*

In what follows, we consider an SPO with pledge, pool operation cost, and idle utility given by (λ, c, ϵ). Moreover, we continue to let r be per-unit rewards for delegating to feasible pools. We call the following quantity the "Gap" of the given SPO: $G(\lambda, c, \epsilon, r) = \max\{\epsilon + c - a(\lambda), [r - \alpha(\lambda, c_{min})]^+ \cdot \lambda + (c - c_{min})\} > 0$, where we use the notational shorthand $[x]^+ = \max\{x, 0\}$. Furthermore, when the context is clear, we simply use G to refer to the gap of an SPO.

Lemma 3 *Suppose that an SPO has s stake, pool operation cost c, and idle utility ϵ. Furthermore, suppose that they operate a pool with pledge $\lambda = s$ and external delegation β. The SPO cannot benefit from unilaterally deviating from pool operation (by either becoming idle, becoming a delegator, or opening a new pool) if and only if: $b(\lambda)\beta' - r\beta \geq G(\lambda, c, r, \epsilon) > 0$.*

Definition 11 (Pool Deficit/Capacity). *Consider a proper pool delegation game given by $\mathcal{G}(\rho, \tau, (\mathbf{s}, \mathbf{c}, \boldsymbol{\epsilon}))$ where the pool reward function is given by $\rho(\lambda, \beta) = a(\lambda) + b(\lambda)\beta'$. Let $\mathbf{p} \in \mathcal{A}$ be such that per unit delegation reward is given by r, and the i-th player is an SPO with pledge $\lambda_i < \tau$ and pool operation cost c_i. We let $\beta_i^- = \beta^-(\lambda_i, c_i, \epsilon_i, r)$ and $\beta_i^+ = \beta^+(\lambda_i, c_i, \epsilon_i, r)$ denote the deficit and capacity, respectively, of the pool run by the i-th player as an SPO:*

$$\beta^-(\lambda_i, c_i, \epsilon_i, r) = \begin{cases} \frac{G(\lambda_i, c_i, \epsilon_i, r)}{b(\lambda_i) - r} & \text{if } (b(\lambda_i) - r)(\tau - \lambda_i) \\ & \geq G(\lambda_i, c_i, \epsilon_i, r) \\ \infty & \text{otherwise} \end{cases}$$

$$\beta^+(\lambda_i, c_i, \epsilon_i, r) = \begin{cases} \frac{b(\lambda_i)(\tau - \lambda_i) - G(\lambda_i, c_i, \epsilon_i, r)}{r} & \text{if } (b(\lambda_i) - r)(\tau - \lambda_i) \\ & \geq G(\lambda_i, c_i, \epsilon_i, r) \\ -\infty & \text{otherwise} \end{cases}$$

β_i^- and β_i^+ can take infinite values when no external delegation can prevent an SPO from deviating from stake pool operation. The following lemma formalizes how pool deficit and capacity serve as lower and upper bounds to the external delegation an SPO can bear while being content as an SPO.

Lemma 4. *Suppose that the i-th player is an SPO with pledge, λ_i, and pool operation cost, c_i, and that they are running a feasible pool under the joint strategy* *profile* $\mathbf{p}$ *with external delegation* β_i. *Furthermore, suppose that per-unit delegation rewards in* $\mathbf{p}$ *are given by* r. *The* i*-th player prefers operating their pool to becoming idle or becoming a delegator if and only if:* $0 < \beta_i^- \leq \beta_i \leq \beta_i^+$

Notice that $\beta_i^- \leq \beta_i \leq \beta_i^+$ also implies that the pool opened by the i-th player as an SPO is feasible. If this were not the case, then by Lemma 1 the SPO would prefer delegation, which is not possible due to Lemma 4. We summarize the collection of results from this section as a theorem:

Theorem 1. *Suppose that* $\mathcal{G}(\rho, \tau, (\mathbf{s}, \mathbf{c}, \boldsymbol{\epsilon}))$ *is a proper delegation game. Consider a joint strategy profile* $\mathbf{p}$ *that results in per-unit delegation rewards,* r. *The following are sufficient conditions for* $\mathbf{p}$ *to be a pure Nash equilibrium: i) Delegators only delegate to feasible pools; ii) If the* i*-th agent is not idle, they earn at least* ϵ_i *utility; iii) If the* i*-th agent is idle, their delegation utility is at most* ϵ_i*; iv) If the* i*-th agent is an SPO with pledge* $\lambda_i = s_i < \tau$ *and external delegation* β_i*, then* $\beta_i^- \leq \beta_i \leq \beta_i^+$.

4 The Bayesian Setting

In a proper delegation game, we let the *type* of the i-th player consist of their stake, pool operation cost and idle utility: (s_i, c_i, ϵ_i).

Definition 12 (Bayesian Proper Delegation Game (BPDG)). *A Bayesian proper delegation game requires four inputs: (i) A proper reward function,* ρ*, (ii) A pool cap,* τ*, (iii) A type distribution,* $\mathcal{X}$*, and (iv) The number of agents,* $n > 0$. *For such a game, player types are first drawn independently via* $(\mathbf{s}, \mathbf{c}, \boldsymbol{\epsilon}) \sim \mathcal{X}^n$*, and they subsequently play the proper delegation game* $\mathcal{G}(\rho, \tau, (\mathbf{s}, \mathbf{c}, \boldsymbol{\epsilon}))$. *We use the notation* $\mathcal{G}(\rho, \tau, \mathcal{X}, n)$ *to denote a specific Bayesian proper delegation game.*

In Bayesian games, one typically studies *ex ante* player strategies that consist of mappings from player types to actions taken. Agents in a proper delegation games, however, have a rich family of actions at their disposal and we are ultimately interested in the high-level decision taken by an agent whether to be an SPO, a delegator or idle. For this reason, we introduce the notion of a partial ex ante strategy which will be an important object of study of our paper.

Definition 13 (Partial Ex Ante Strategy). *A partial ex ante strategy for a BPDG is a function* $f : \mathbb{R}^3 \to \{0, 1\}$ *that dictates which players become SPOs. Under* f*, a player of type* (s, c, ϵ) *is an SPO if and only if* $f(s, c, \epsilon) = 1$.

We call such ex-ante strategies *partial* due to the fact that after drawing player types, there are multiple pure strategy profiles of the ex post proper delegation game which are consistent with f. For a given draw of player types, $(\mathbf{s}, \mathbf{c}, \boldsymbol{\epsilon})$, we let $\mathcal{A}_f(\mathbf{s}, \mathbf{c}, \boldsymbol{\epsilon})$ denote the set of pure strategy profiles of the ex post proper delegation game, $\mathcal{G}(\rho, \tau, (\mathbf{s}, \mathbf{c}, \boldsymbol{\epsilon}))$, which are consistent with f. In other words, $\mathbf{p} \in \mathcal{A}_f(\mathbf{s}, \mathbf{c}, \boldsymbol{\epsilon})$ when $p_i = a_{SPO} \iff f(s_i, c_i, \epsilon_i) = 1$. We are interested in strategies that can give rise to PNE ex post:

Definition 14 (Ex post SPO stable). *Suppose that f is a partial ex ante strategy for a BPDG $\mathcal{G}(\rho, \tau, \mathcal{X}, n)$. f is ex post SPO stable for the draw $(\mathbf{s}, \mathbf{c}, \boldsymbol{\epsilon}) \sim \mathcal{X}^n$ if there exists a joint strategy profile $\mathbf{p} \in \mathcal{A}_f(\mathbf{s}, \mathbf{c}, \boldsymbol{\epsilon})$ which is a PNE.*

The main result of this section provides useful sufficient conditions for a partial ex ante strategy, f, to be ex post SPO stable for a given draw of player types. The proof can be found in the full version of the paper. Before delving into the main theorem, though, we define some relevant quantities.

Definition 15 (Total Ex Post Stable Delegation). *Suppose that f is a partial ex ante strategy for a BPDG, $\mathcal{G}(\rho, \tau, \mathcal{X}, n)$ with player types given by $(\mathbf{s}, \mathbf{c}, \boldsymbol{\epsilon}) \sim \mathcal{X}^n$. Assume $s^* = \max\{i \in [n] \mid f(s_i, c_i, \epsilon_i) = 0$ and $\alpha(s_i, c_{min}) \geq \epsilon_i/s_i\}$ and $r = \alpha(s^*, c_{min})$,[3] we denote the total ex post stable delegation by $Del(f) = \sum_{i=1}^{n} s_i(1 - f(s_i, c_i, \epsilon))\mathbb{I}(rs_i \geq \epsilon_i)$, where $\mathbb{I}(\cdot)$ is an indicator function.*

Definition 16 (Total Ex Post Pool Deficit/Capacity). *Suppose that f is a partial ex ante strategy for a BPDG, $\mathcal{G}(\rho, \tau, \mathcal{X}, n)$ with player types given by $(\mathbf{s}, \mathbf{c}, \boldsymbol{\epsilon}) \sim \mathcal{X}^n$. Assume $s^* = \max\{i \in [n] \mid f(s_i, c_i, \epsilon_i) = 0$ and $\alpha(s_i, c_{min}) \geq \epsilon_i/s_i\}$ and $r = \alpha(s^*, c_{min})$, we denote the total ex post pool deficit/capacity by $Def(f)$ and $Cap(f)$ respectively where $Def(f) = \sum_{i=1}^{n} \beta_i^-(s_i, c_i, \epsilon_i, r) f(s_i, c_i, \epsilon_i)$ and $Cap(f) = \sum_{i=1}^{n} \beta_i^-(s_i, c_i, \epsilon_i, r) f(s_i, c_i, \epsilon_i)$.*

Theorem 2. *Suppose that f is a partial ex ante strategy for a BPDG $\mathcal{G}(\rho, \tau, \mathcal{X}, n)$ with player types given by $(\mathbf{s}, \mathbf{c}, \boldsymbol{\epsilon}) \sim \mathcal{X}^n$. The following is a sufficient condition for f to be ex post SPO stable: $0 < Def(f) \leq Del(f) \leq Cap(f)$.*

If f is ex post SPO stable for the draw $(\mathbf{s}, \mathbf{c}, \boldsymbol{\epsilon}) \sim \mathcal{X}^n$ there are generally multiple joint strategy profiles $\mathbf{p} \in \mathcal{A}_f(\mathbf{s}, \mathbf{c}, \boldsymbol{\epsilon})$ which give rise to PNE. In the following section we provide a means of distinguishing the performance different PNE which arise. We quantify the performance of a given joint strategy profile $\mathbf{p}$ using 3 key metrics: Participation, Expenditure, and Decentralization.

5 Design Objectives

If $\mathbf{p} \in \mathcal{A}$ gives rise to $k \geq 1$ active pools, we let $\boldsymbol{\lambda}$, $\boldsymbol{\beta}$ and $\boldsymbol{\sigma}$ be k-dimensional vectors encoding the pledge, external delegation and total size of each of the k active pools. We call $(\boldsymbol{\lambda}, \boldsymbol{\beta})$ the public pool profile of $\mathbf{p}$ with $\boldsymbol{\sigma} = \boldsymbol{\lambda} + \boldsymbol{\beta}$.

Definition 17 (Design Objectives). *Consider $\mathbf{p} \in \mathcal{A}$ which gives rise to public pool profile $(\boldsymbol{\lambda}, \boldsymbol{\beta})$. We define the participation objective as $O^P(\mathbf{p}) = \sum_{j=1}^{k}(\lambda_j + \beta_j) = \sum_{j=1}^{k} \sigma_j$. We define the expenditure objective as $O^E(\mathbf{p}) = \sum_{i=1}^{n} R_i(\mathbf{p})$. Finally, For $\ell \geq 0$, we let $P_\ell(\mathbf{p}) = \{S \subseteq [k] : \sum_{i \in S} \sigma_i \geq \ell O^P(\mathbf{p})\}$. We define the decentralization objective as $O_\ell^D(\mathbf{p}) = \min_{S \in P_\ell(\boldsymbol{\lambda}, \boldsymbol{\beta})} \sum_{i \in S} \lambda_i$.*[4]

[3] If $\{i \in [n] \mid \alpha(s_i, c_{min}) \geq \epsilon_i/s_i\} = \emptyset$, we let $r = 0$.

[4] O_ℓ^D can be seen as a measure of "skin in the game" for dominating coalitions. In a decentralized system this is high, and dominating coalitions stand to lose more rewards via penalties in case of undesirable behavior.

A system designer will seek to maximize participation, minimize expenditure and maximize decentralization. Simultaneously optimizing for each of these objectives is generally not possible, hence we use a framework inspired by multi-objective optimization to understand tradeoffs between all three.

6 Computational Methods and Results

Our main computational approach focuses on conceptualizing the performance of a partial ex ante strategy, f, for a given Bayesian proper delegation game $\mathcal{G}(\rho, \tau, \mathcal{X}, n)$. To do so, we measure the performance of f for a given $(\mathbf{s}, \mathbf{c}, \boldsymbol{\epsilon}) \sim \mathcal{X}^n$, in terms of the three objectives of Sect. 5. Our approach proceeds in two stages. First, we establish whether f satisfies the sufficient conditions set forth in Theorem 2 to be ex post SPO stable. Subsequently, If f is ex post SPO stable, then all $\mathbf{p} \in \mathcal{A}_f(\mathbf{s}, \mathbf{c}, \boldsymbol{\epsilon})$ which are PNE exhibit the same participation breakdown, and hence have equal values for O^P. This is not the case for O^E and O^D_ℓ, therefore, to study decentralization and expenditure, we construct a comprehensive set of ex post PNE, $\mathbf{p}^1, \ldots, \mathbf{p}^m \in \mathbf{P} \in \mathcal{A}_f(\mathbf{s}, \mathbf{c}, \boldsymbol{\epsilon})$ with different decentralization and expenditure performance to represent the potential spread of performance that can be achieved ex post for f.

6.1 Representative Ex Post PNE

We outline our methodology for building a representative set of PNE from $\mathcal{A}(\mathbf{s}, \mathbf{c}, \boldsymbol{\epsilon})$ to understand the potential decentralization and expenditure achieved by a given partial ex ante strategy, f, which is ex post SPO stable for a given draw of agent types. We consider a Bayesian proper delegation game $\mathcal{G}(\rho, \tau, \mathcal{X}, n)$ and a partial ex ante strategy f. Suppose that f is ex post SPO stable for $(\mathbf{s}, \mathbf{c}, \boldsymbol{\epsilon})$ where at least one agent is an SPO. We outline our methodology for building a representative set of PNE from $\mathcal{A}(\mathbf{s}, \mathbf{c}, \boldsymbol{\epsilon})$ for understanding the potential decentralization and expenditure achieved under f ex post.

Let $\lambda_{min} \leq \lambda_{max}$ represent the smallest and largest pledges made by SPOs under f and let $m \in \mathbb{N}$ be a resolution parameter that dictates the number of representative PNE from $\mathcal{A}_f(\mathbf{s}, \mathbf{c}, \boldsymbol{\epsilon})$ constructed. We construct an m-dimensional vector of *reference pledges*, $\bar{\boldsymbol{\lambda}} = (\bar{\lambda}_j)_{j=1}^m$, where $\bar{\lambda}_j = \lambda_{min} + (j-1)\frac{(\lambda_{max} - \lambda_{min})}{m-1}$. From $\bar{\lambda}_j$, we construct the j-th representative PNE from $\mathcal{A}_f(\mathbf{s}, \mathbf{c}, \boldsymbol{\epsilon})$ denoted by $\mathbf{p}^j$. As in Theorem 2, we fix the high-level actions of agents to ensure ex post SPO stability. All that remains to specify $\mathbf{p}^j$ is deciding where delegation goes to, for which we make use of the reference pledge, $\bar{\lambda}_j$. We do so by computing a delegation vector $\boldsymbol{\beta} = (\beta_i)_{i=1}^n$ first satisfying the deficit of all pools (using $Def(f) \leq Del(f)$ of the available delegation). Afterwards, we greedily fill pools with pledge closest to $\bar{\lambda}_j$ up to capacity using the remaining $Del(f) - Def(f)$ delegation at our disposal. The details of the greedy delegation allocation are provided in Algorithm 1. Given the target greedy delegation allocation, $\boldsymbol{\beta}$, we simply let $\mathbf{p}^j$ be any PNE which is consistent with the target delegation (since they all achieve the same expenditure and decentralization objectives).

Algorithm 1. Greedy Delegation Allocation

procedure GreedyDelegation($\bar{\lambda}_j, \boldsymbol{\beta}^-, \boldsymbol{\beta}^+, Del(f)$)
  $\boldsymbol{\beta} \leftarrow \boldsymbol{\beta}^-$ ▷ Satisfying pool deficit
  $X \leftarrow Del(f) - \sum_{i=1}^{n} \beta_i$ ▷ Remaining delegation
  $A \leftarrow \{i \in [n] \mid \beta_i < \beta_i^+\}$
  $j^* \leftarrow \text{argmin}_{i \in A} |\lambda_i - \bar{\lambda}_j|$ ▷ Ties broken lexicographically in argmin
  while $X \neq 0$ **do**
    $\beta_{j^*} \leftarrow \beta_{j^*} + \min\{X, (\beta_{j^*}^+ - \beta_{j^*})\}$
    $X \leftarrow Del(f) - \sum_{i=1}^{n} \beta_i$
    $A \leftarrow \{i \in [n] \mid \beta_i < \beta_i^+\}$
    $j^* \leftarrow \text{argmin}_{i \in A} |\lambda_i - \bar{\lambda}_j|$
  end while
  return $\boldsymbol{\beta}$
end procedure

Computing Design Objectives. Computing O^P and O^E for a given $\mathbf{p} \in \mathcal{A}$ in a proper delegation game, $\mathcal{G}(\rho, \tau, (\mathbf{s}, \mathbf{c}, \boldsymbol{\epsilon}))$, is straightforward.

We focus on the problem of computing the decentralization objective, O_ℓ^D. We can express this computational problem in terms of the public pool profile $(\boldsymbol{\lambda}, \boldsymbol{\beta})$ which arises from a given $\mathbf{p} \in \mathcal{A}$. The value of $O_\ell^D(\mathbf{p})$ is given by minimizing $\sum_{j=1}^{k} \lambda_j x_j$ subject to the constraints $\sum_{j=1}^{k} \sigma_j x_j \geq \ell O^P(\mathbf{p})$ and $x_j \in \{0,1\}$ for $j = 1, \ldots k$. This optimization problem is precisely an instance of the min-knapsack problem, hence it is NP-hard [3]. To approximate O_ℓ^D, we use typical knapsack approximation schemes as per [16].

6.2 Relevant Modeling Choices and Parameters

For a given Bayesian proper delegation game, $\mathcal{G}(\rho, \tau, \mathcal{X}, n)$, a partial ex ante strategy can be an arbitrary function from player types to whether they act as an SPO or not. In practice we expect larger players (with more stake) to be SPOs for multiple reasons (increased interest in the proper functioning of the underlying blockchain, potentially less frictions to operate as SPO, etc.). For this reason, we consider a simple class of partial ex ante strategies with agents operating as SPOs only if they exceed a stake threshold.

Definition 18 (Threshold Partial Ex Ante Strategy). *We let* $f_\alpha^t : \mathbb{R}^2 \rightarrow \{0,1\}$ *denote a threshold partial ex ante strategy with threshold* $\theta \geq 0$. *The strategy is specified by:* $f_\theta^t(s, c, \epsilon) = 1 \iff s \geq \theta$.

As is common in economic literature, We can assume that stake distributions obey a power law [11] (Pareto Distribution).

Definition 19 (Truncated Pareto Distribution). *For* $0 < L < H$, *we say* Z *is a truncated Pareto distribution over* $[L, H]$ *with inequality parameter* $\gamma > 0$ *if it has pdf* $\eta(x) = \left(\frac{\gamma L^\gamma}{1-(L/H)^\gamma}\right) x^{-\gamma-1}$ *for* $x \in [L, H]$ *and 0 otherwise. We write* $s \sim Pareto(L, H, \gamma)$ *when stake is distributed as a bounded Pareto distribution.*

In order to achieve marginal Pareto distributions on player stake, we consider type distributions $\mathcal{X}$ which result as product distributions over player stake, cost and idle utility respectively. Furthermore, without loss of generality, we normalize the value of stake with respect to the lower bound L, so we can let $L = 1$. In more detail, we consider type distributions parametrized by: H, γ, the upper bound and exponent in Pareto PDF for stake distribution, c_{min}, c_{max}, the minimal and maximal values of pool operation cost, and $\epsilon_{min}, \epsilon_{max}$, the minimal and maximal values of idle utility. The type distribution with these parameters is denoted $\mathcal{X}(H, \gamma, c_{min}, c_{max}, \epsilon_{min}, \epsilon_{max})$. In order to sample from the distribution, we independently sample each component: $s \sim Pareto(1, H, \gamma)$, $c \sim U[c_{min}, c_{max}]$ and $\epsilon \sim U[\epsilon_{min}, \epsilon_{max}]$.

6.3 Experimental Results

We provide some specific results for a proper Bayesian delegation game which demonstrate the flexibility of our approach in studying tradeoffs struck by payment schemes in proper delegation games. For extended experimental results, we refer the reader to the full version of the paper.

Baseline Parameter Settings. We begin by providing details regarding the family of ρ functions we explore in our experiments. Given we are modeling proper delegation games as per Definition 10, we are considering separable pool reward functions such that $\rho(\lambda, \beta) = a(\lambda) + b(\lambda)\beta'$, where $\beta' = \min\{\tau - \lambda, \beta\}$ for the cap τ, which we will specify shortly. In our experiments, we model $a(\lambda)$ and $b(\lambda)$ as polynomials of varying degree and positive coefficients (which is in fact similar to the formula for Cardano reward sharing schemes [2]). Our baseline formulas are given by $a(\lambda) = b(\lambda) = \lambda$.

For the marginal distribution of player stakes, we use a truncated Pareto distribution with lower bound $L = 1$, $H = 100$, and $\gamma = 1.5$. For SPO costs, we let lower and upper bounds for cost be $c_{min} = 0.4$ and $c_{max} = 0.6$ and for idle utilities, we simply assume that all players have the same $\epsilon = 0.01$. Finally, given the marginal stake distribution, we let $\tau = 200$ be the pool cap used for ρ. We begin by considering the threshold partial ex ante strategy f_θ^t with $\theta = 30$. Moreover, we consider a Bayesian proper delegation game with $n = 1000$ agents drawn from the type distribution described above. In addition, we create $m = 100$ representative ex post PNE as per Algorithm 1 whenever f_θ^t is ex post SPO stable, and use $\ell = 0.5$ for the decentralization objective O_ℓ^d. Finally, we repeat this process for $N = 500$ independent draws from $\mathcal{X}^n$.

Results from this parameteric setting can be seen within Fig. 1 for points corresponding to $\theta = 30$. The empirical frequency of ex post stability for the baseline setting was 496 of the $N = 500$ draws of player types. The figure provides a breakdown of participation achieved. We note that no players are idle in this setting. With regards to expenditure and decentralization, we can see that as delegation is sent to pools with higher pledge, the system achieves better decentralization, albeit at a higher expenditure.

Impact of Idle Utility. We modulate $\epsilon \in \{0.005, 0.1, 1.0, 5.0, 10.0\}$ of all players in the game, maintaining all other parameters as in the baseline setting. In Table 1 we see the empirical frequency of ex post stable PNE as we modulate ϵ values, and we see that there is no significant difference even as ϵ increases multiple orders of magnitude. We do however see significant differences in terms of the participation, decentralization and expenditure of ex post PNE as we change idle utilities. With regards to participation, Fig. 1 shows the changes in relative and absolute participation of agents as ϵ varies. As expected, with higher idle utilities, more agents prefer remaining idle over delegating. Moreover, this is in line with the fact that empirical frequencies for ex post stability do not change much, for if there is less delegation to go around, it can be easier to satisfy pool deficits and capacities. Of course, if too much delegation is idle, then there may not be enough delegation to satisfy pool deficits, and we may see a decrease in the empirical frequency of ex post SPO stability. Figure 1 also provides insight in terms of how decentralization and expenditure vary with ϵ. As expected, large values of ϵ result in lower expenditure, as the system needs to pay out less delegators. On the other hand, we also see that larger baseline utilities can increase decentralization, which also makes sense from the decreased delegation that occurs, as any dominating coalition of pools will necessarily have more skin in the game as they may have less external delegation.

Table 1. Frequency of ex post SPO stability for different parameter settings ($N = 500$ draws). Top rows modulate ϵ, middle rows θ, and bottom rows a, b in ρ.

Parameters	0.005	0.1	1.0	5.0	10.0	
Modulate ϵ	498	497	499	495	499	
Parameters	10	20	30	40	50	60
Modulate θ	500	500	496	478	428	344
Parameters	g_1	g_2	g_3	g_4	g_5	g_6
Modulate a	497	499	496	497	493	489
Modulate b	496	497	497	499	497	497
Modulate (a, b)	496	496	498	498	498	499

Impact of Reward Function. We modulate $\rho(\lambda, \beta) = a(\lambda) + b(\lambda)\beta'$. In addition we fix idle utilities to be larger than baseline at $\epsilon = 5$, where we've seen that agents can prefer to be idle over delegating. This way we glean insight regarding how different payment structures can foster participation. We modulate our payment scheme by varying, a and b. Going forward we consider setting the constituent functions of ρ with combinations of the following functions: $g_1(\lambda) = 0.5\lambda$, $g_2(\lambda) = \lambda$, $g_3(\lambda) = 2\lambda$, $g_4(\lambda) = \lambda + 0.005\lambda^2$, $g_5(\lambda) = \lambda + 0.01\lambda^2$, $g_6(\lambda) = \lambda + 0.05\lambda^2$. We modulate ρ in three different ways: unilaterally $a \in \{g_1, \ldots, g_6\}$; unilaterally $b \in \{g_1, \ldots, g_6\}$; and jointly $(a, b) \in \{(g_1, g_1) \ldots, (g_6, g_6)\}$. Empirical frequencies of ex post SPO stability can be found in Table 1.

In Fig. 2 we provide a detailed breakdown of how modulating a and b within ρ can impact the participation reached by the system at ex post PNE. First of

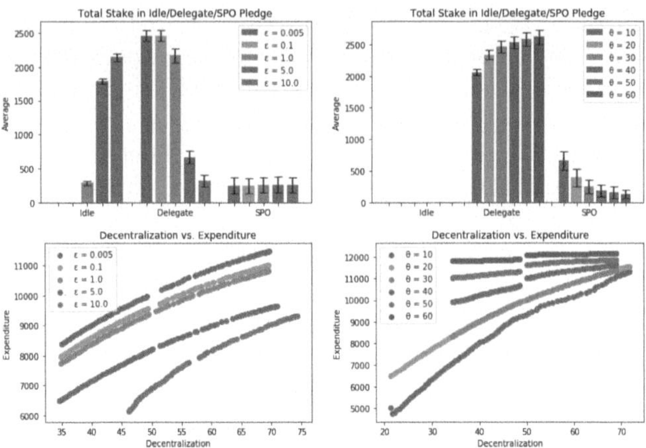

Fig. 1. The top images provide a breakdown of participation where bars give average values of absolute stake used by agents being idle, delegators or SPOs respectively for multiple draws in different parameter settings. The bottom image simultaneously plots the performance of representative ex post PNE in terms of average decentralization and expenditure over multiple draws of player types. Representative ex post PNE with larger reference pledge values exhibit both higher expenditure and decentralization. Left plots correspond to modulating ϵ and right plots to modulating θ.

all we see that unilaterally modulating $a \in \{g_1, \ldots, g_6\}$ accounts for much more change in participation over unilaterally modulating $b \in \{g_1, \ldots, g_6\}$. Moreover, when jointly modulating $(a, b) \in \{(g_1, g_1), \ldots, (g_6, g_6)\}$, changes in participation closely resemble those made by individually modulating a, which suggest that for the functional values chosen, changes in a account for the majority of differences in participation. This phenomenon largely results from the fact that the a functions we explore with larger quadratic coefficients in λ not only pay SPOs more, but they also increase values of $\alpha(s, c)$, which in turn increase delegation rewards. Increased delegation rewards in turn incentivize more players into being delegators over being idle. At the same time, this comes at an added expense, as can be seen in the same figure where higher degree expressions of λ result in much higher expenditure for the system. At the same time, these expensive ex post PNE also achieve large decentralization values, hence the system designer may find it beneficial to use such ρ functions if prioritizing participation and decentralization is more important than minimizing expenditure.

Impact of SPO Threshold in f_θ^t. We modulate the threshold for SPO operation in the ex ante strategy f_θ^t. We consider values $\theta \in \{10, 20, 30, 40, 50, 60\}$ and Table 1 shows the number of ex post SPO stable draws for each given threshold value. The first observation we can make is that the empirical probability that f_θ^t is ex post SPO stable is decreasing in θ. This makes sense for two reasons: (i) as θ increases, pivotal delegates become larger, which increases r, the per-unit delegator rewards, thus leaving less rewards for SPOs, and hence decreasing their

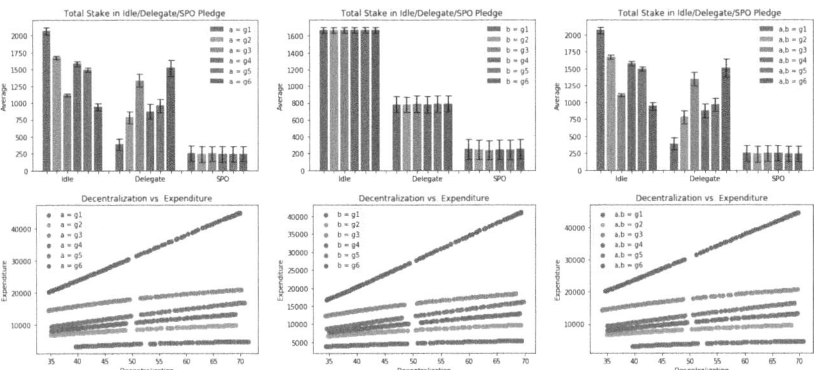

Fig. 2. The top images provide a breakdown of participation as ρ varies. Bars provide average values of absolute stake used by agents being idle, delegators or SPOs respectively for different ρ settings. The bottom images simultaneously plots the performance of representative ex post PNE in terms of decentralization and expenditure. Representative ex post PNE with larger reference pledge values exhibit both higher expenditure and decentralization. The left column corresponds to only modulating a, the middle column to only modulating b, and the right column to modulating both a and b.

pool capacity; (ii) an increased threshold also means that there is more delegation to go around, both from "large" delegates who lie just under the threshold, but also from agents who may have been idle, but with an increased r decide to delegate. These factors contribute to decreased empirical probability of being ex post SPO stable. Figure 1 also provides a more fine-grained perspective on how participation (and hence O^P) changes as a function of θ, where we see that increased thresholds decrease SPO operation and increase overall delegation.

For how decentralization and expenditure are affected by θ, we turn to Fig. 1. We see that as θ increases, decentralization and expenditure in general increase, and moreover they become more constant as a function of representative ex post PNE reference pledge. Furthermore, we see that the performance of the $\theta = 10$ threshold is better than others in terms of O_ℓ^D and O^E (but not O^P). All these points represent ex post PNE, hence depending on the threshold exhibited by players in an ex post PNE, the system can exhibit a multitude of decentralization and expenditure objective values (along all θ values).

7 Conclusion

In this work, we provide a multi-objective framework for studying tradeoffs inherent in delegation systems for PoS cryptocurrencies. We began by providing a broad game theoretic framework for incentives in delegation systems, and successively narrowed down the game at hand to both represent key characteristics of existing PoS delegation systems, and also be tractable to study in a Bayesian framework. We provide key sufficient conditions for equilibria in the one-shot and Bayesian setting and use this characterization to study the potential performance of various payment schemes with respect to three key objectives: participation, decentralization and expenditure. The computational tools we provide give us insight with respect to inherent tradeoffs system designers may face when attempting to maximize for these three natural objectives. We believe our work is a preliminary foray into these tradeoffs faced by system designers, and we hope that future work can further clarify the nuances in balancing these objectives via richer game theoretic models and reward sharing schemes.

References

1. Azouvi, S.: Levels of decentralization and trust in cryptocurrencies: consensus, governance and applications. Ph.D. thesis, UCL (2021)
2. Brünjes, L., Kiayias, A., Koutsoupias, E., Stouka, A.P.: Reward sharing schemes for stake pools. In: 2020 IEEE European Symposium on Security and Privacy (EuroS&p), pp. 256–275. IEEE (2020)
3. Csirik, J., Frenk, J.B.G.: Heuristics for the 0–1 min-knapsack problem. Acta Cybern. **10**(1–2), 15–20 (1991)
4. Gencer, A.E., Basu, S., Eyal, I., van Renesse, R., Sirer, E.G.: Decentralization in bitcoin and ethereum networks (2018)
5. Gersbach, H., Mamageishvili, A., Schneider, M.: Staking pools on blockchains. arXiv preprint arXiv:2203.05838 (2022)
6. Gogol, K., Velner, Y., Kraner, B., Tessone, C.: SoK: liquid staking tokens (LSTs). arXiv preprint arXiv:2404.00644 (2024)
7. Karakostas, D., Kiayias, A., Ovezik, C.: SoK: a stratified approach to blockchain decentralization (2022)
8. Lee, J., Lee, B., Jung, J., Shim, H., Kim, H.: DQ: two approaches to measure the degree of decentralization of blockchain. ICT Express **7**(3), 278–282 (2021)
9. Leonardos, N., Leonardos, S., Piliouras, G.: Oceanic games: centralization risks and incentives in blockchain mining. In: MaRBLe (2019)
10. Ovezik, C., Kiayias, A.: Decentralization analysis of pooling behavior in Cardano proof of stake. In: Proceedings of the Third ACM International Conference on AI in Finance, pp. 18–26 (2022)
11. Pareto, V.: Cours d'économie politique, vol. 1. Librairie Droz (1964)
12. Rugendyke, D.: Rocket pool: decentralized ethereum proof of stake network (2021). https://rocketpool.net/files/rocketpool-whitepaper.pdf. Accessed 20 May 2024
13. Shapovalov, V., Lomashuk, K., Fish, J., Harborne, W., Rasmussen, K.: Lido: a decentralized solution for liquid staking (2021). https://research.lido.fi/t/lido-whitepaper/2. Accessed 20 May 2024
14. Srinivasan, B., Lee, L.: Quantifying decentralization (2017). https://news.earn.com/quantifying-decentralization-e39db233c28eMedium
15. Stouka, A., Zacharias, T.: On the (de)centralization of fruitchains. In: 2023 IEEE 36th Computer Security Foundations Symposium (CSF) (CSF), Los Alamitos, CA, USA, pp. 299–314. IEEE Computer Society (2023). https://doi.org/10.1109/CSF57540.2023.00020
16. Tauhidul, I.M.: Approximation algorithms for minimum knapsack problem. Master's degree thesis, University of Lethbridge (2009)
17. Team, E.: Ether.fi: a decentralized, non-custodial liquid staking protocol (2023). https://etherfi.gitbook.io/etherfi/ether.fi-whitepaper. Accessed 20 May 2024

Matroid Theory in Game Theory

Price of Anarchy in Paving Matroid Congestion Games

Bainian Hao(✉) and Carla Michini

Department of Industrial and Systems Engineering, University of Wisconsin-Madison, Madison, WI, USA
{bhao8,michini}@wisc.edu

Abstract. Congestion games allow to model competitive resource sharing in various distributed systems. Pure Nash equilibria, that are stable outcomes of a game, could be far from being socially optimal. Our goal is to identify combinatorial structures that limit the inefficiency of equilibria. This question has been mainly investigated for congestion games defined over networks. Instead, we focus on symmetric matroid congestion games, where the strategies of every player are the bases of a given matroid. We derive new upper bounds on the Price of Anarchy (PoA) of congestion games defined over k-uniform matroids and paving matroids with delay functions in class $\mathcal{D}$. For both affine and polynomial delay functions, our bounds indicate that the inefficiency of pure Nash equilibria is limited by these combinatorial structures.

1 Introduction

Congestion games are a class of strategic games that provide an appealing paradigm to model resource sharing among selfish players. In a congestion game, a set of resources is given, and each player selects a feasible subset of the resources in order to minimize their cost function. The cost of a player's strategy is the sum of the delays of the resources selected by the player, and the delay of each resource is a function of the total number of players using it. The game is called *symmetric* if all players have the same strategy set. An example are *network congestion games* with single origin-destination pair, where the resources are the arcs of a given digraph and the strategies of each player are paths between the origin and the destination in the network. Congestion games are practically relevant for various problems related to resource sharing in distributed systems, e.g., routing, network design and scheduling.

A *pure Nash equilibrium* (PNE) is a configuration where no player can decrease their cost by unilaterally deviating to another strategy, and it represents a stable outcome of the game. However, since the players act selfishly and independently in a non-cooperative fashion, a PNE might be far from minimizing the social cost, which is commonly defined as the sum of all players' costs. Two classic metrics for quantifying the inefficiency of equilibria are the *Price of Anarchy* (PoA) [21] and the *Price of Stability* (PoS) [3].

G. Schäfer and C. Ventre (Eds.): SAGT 2024, LNCS 15156, pp. 353–370, 2024.
https://doi.org/10.1007/978-3-031-71033-9_20

Congestion games always admit a PNE [29]. However, the complexity of computing a PNE in a congestion game can be significantly affected by its combinatorial structure. While symmetric congestion games and asymmetric network congestion games are PLS-complete [14], Fabrikant et al. [14] gave a strongly polynomial-time algorithm to find a PNE in symmetric network congestion games, which was later extended to symmetric totally unimodular congestion games by Del Pia et al. [13].

Our main goal is to better understand how the combinatorial structure of a congestion game might affect the inefficiency of equilibria. For *nonatomic* congestion games, where each single player has a negligible impact on congestion, structure has no impact on the PoA. In fact, Roughgarden [31] proved that the worst-case PoA is equal to $\rho(\mathcal{D})$, a function that only depends on the class $\mathcal{D}$ of delay functions[1]. On the other hand, in *atomic* games, where each single player can affect the other players' decisions, there are structures that might reduce the inefficiency of equilibria. In the absence of structure, Awerbuch et al. [4,5] and Christodoulou and Koutsoupias [11] independently provided an upper bound of 5/2 on the PoA for general atomic congestion games with affine delays. This bound can be improved to $(5N-2)/(2N+1)$ if the game is symmetric [11], where N is the number of players. For atomic congestion games with polynomial delays of highest degree p, Aland et al. [2] obtained exact values for the worst-case PoA, see also [4,5,11]. These exact values admit a lower bound of $\lfloor \phi_p \rfloor^{p+1}$ and an upper bound of ϕ_p^{p+1}, where $\phi_p \in \Theta(p/\ln p)$ is the unique nonnegative real solution to $(x+1)^p = x^{p+1}$. In the general case, Bhawalkar et al. [6] proved that the worst-case PoA can be achieved in symmetric games.

However, in the symmetric case the PoA can significantly decrease if the players' strategy sets have a special structure. Most of the existing literature has focused on *graph* structures in network congestion games. Lücking et al. [23,24] studied symmetric congestion games on parallel links and proved that the PoA is 4/3 for linear delay functions. Fotakis [16] later extended this result to network congestion games defined over extension-parallel networks and proved that for these networks the worst-case PoA is equal to $\rho(\mathcal{D})$, if the delays belong to class $\mathcal{D}$. Recently, Hao and Michini explored a further extension to the larger family of series-parallel networks. For affine delays, they proved that the worst-case PoA is in $[27/19, 2]$ [18]; for polynomial delays of highest degree p they showed that the worst-case PoA is at most $2^{p+1}-1$ [17], which is significantly smaller than the worst-case PoA in general network congestion games.

In this paper we focus on another combinatorial structure, namely matroids. *Matroid congestion games* are congestion games where each player's strategy set is the set of bases of a given matroid. For this class of games, a PNE equilibrium can be efficiently computed, both in the symmetric and in the asymmetric case [1,13]. Concerning the inefficiency of equilibria, Kleer and Schäfer [20] showed that the PoS in general matroids is upper bounded by $\rho(\mathcal{D})$ when the delay functions belong to class $\mathcal{D}$. However, the PoA of matroid congestion games is not well understood. For affine delays, the worst-case PoA of general conges-

[1] The formal definition of $\rho(\mathcal{D})$ is recalled later in Eq. (2).

tion games, that is equal to 5/2, can be asymptotically achieved in asymmetric instances of singleton congestion games—that coincide with 1-uniform matroid congestion games—when the number of players goes to infinity [10]. In the symmetric case, the PoA of general matroid congestion games is still not completely understood. For graphic matroids and $N = 2, 3, 4$ or infinity the PoA can be as large as the worst-case PoA of symmetric congestion games, which is equal to $\frac{5N-2}{2N+1}$ [15]. However, for arbitrary N or different delay functions we don't know whether the worst-case PoA of symmetric congestion games can be achieved by symmetric matroid congestion games. Moreover, the worst-case PoA of k-uniform matroid congestion games with affine delays cannot exceed 1.4131 and it is equal 1.35188 when the number of players goes to infinity [12]. For k-uniform matroid congestion games with polynomial delays of highest degree p the worst-case PoA is in $O(2^{p(p+1)})$ and in $\Omega(2^p)$ [22]. Moreover, if $k = 1$ and the delays are all identical, then the worst-case PoA is in $\Theta((2 + o(1))^p)$ [7]. This indicates that the combinatorial structure of k-uniform matroids significantly limits the inefficiency of equilibria. However, k-uniform matroids are very special matroids, since every subset of the ground set of size at most k is independent. *Are there weaker matroid structures that affect the inefficiency of equilibria?* In this paper we focus on *paving matroids*, i.e., matroids whose circuits have cardinality greater than or equal to the matroid rank. Unlike k-uniform matroids, paving matroids exhibit a notable predominance within the enumeration of matroids. It has been conjectured that, in an asymptotic sense, the majority of matroids are paving matroids [25]. This conjecture holds if the ground set has size at most 9 [8,26]. Pendavingh and van der Pol [28] more recently proved that, as the size of the ground set goes to infinity, the ratio of logarithms between the total number of matroids and the number of sparse paving matroids, a subclass of paving matroids, converges to 1.

Our Contributions. First, we provide a lower bound of 13/9 on the worst-case PoA for symmetric paving matroid congestion games with affine delays. This ratio is worse than the previously known best upper bound ≈ 1.41 on the PoA of symmetric congestion games with affine delay functions over k-uniform matroids, which are a subclass of paving matroids. Thus, relaxing the structure of players' strategy sets from uniform matroids to paving matroids can increase the inefficiency of pure Nash equilibria.

Theorem 1. *The worst-case PoA of symmetric paving matroid congestion games with affine delay functions is at least* 13/9.

We next turn to the question of finding upper bounds on the PoA of symmetric paving matroid congestion games. Given the class of delay functions $\mathcal{D}$, we define the parameter $z(\mathcal{D})$ as

$$z(\mathcal{D}) = \sup_{d \in \mathcal{D},\ x \in \mathbb{N}^+} \frac{d(x+1)}{d(x)}.$$

Since the delay functions $d(x)$ are non-negative and non-decreasing, we have $z(\mathcal{D}) \geq 1$. Our first main result is an upper bound on the worst-case PoA in symmetric paving matroid congestion games with delay functions in class $\mathcal{D}$.

Theorem 2. *The PoA of symmetric paving matroid congestion games with delay functions in class $\mathcal{D}$ is at most $z(\mathcal{D})^2\rho(\mathcal{D})$.*

When $\mathcal{D}$ is the class of polynomial functions of maximum degree p, we have $z(\mathcal{D}) = 2^p$ and $\rho(\mathcal{D}) \in \Theta(p/\ln p)$. Thus, the worst-case PoA is in $O(4^p p/\ln p)$. For $p \geq 6$ our bound is smaller than the worst-case PoA that can be achieved in general symmetric congestion games, that is in $\Theta(p/\ln p)^{p+1}$ [2]. Thus, the worst-case PoA of symmetric congestion games *cannot* be achieved in paving matroids.

We also prove —with a substantially different approach— that this is the case for $p = 1$, i.e., when the delay functions are affine. In this case, the worst-case PoA for general symmetric congestion games is $5/2$.

Theorem 3. *The PoA of symmetric paving matroid congestion games with affine delay functions is at most* $17/7$.

Finally, the approach used to prove Theorem 2 also provides a new upper bound on the worst-case PoA in symmetric k-uniform matroid congestion games with delay functions in class $\mathcal{D}$.

Theorem 4. *The PoA of symmetric k-uniform matroid congestion games with delay functions in class $\mathcal{D}$ is at most $z(\mathcal{D})\rho(\mathcal{D})$.*

When $\mathcal{D}$ is the class of polynomial functions of maximum degree p, we obtain that the worst-case PoA is in $O(2^p p/\ln p)$. This significantly improves on the previously known upper bound of $O(2^{p(p+1)})$ [22] and partially closes the gap with the lower bound of $\Omega(2^p)$ [22].

Our Approach. Our approach is based on representing the "difference" between a PNE f and a social optimum o of a matroid congestion game as a flow on a complete directed graph, whose nodes correspond to the resources. Each unit of flow on arc (r, r') corresponds to a player replacing r with r' in their strategy. The *overloaded* resources (those with more players in f than in the o) act as supply nodes and the *underloaded* resources (those with more players in the o than in the f) act as demand nodes. If every path from supply u to demand v is such that the costs of u and v in the PNE are related through a constant α, then we can establish that the PoA is at most $\alpha\rho(\mathcal{D})$ (Theorem 6). When the delay functions are in class $\mathcal{D}$, we can determine values of α for the case where the matroid is k-uniform (Lemma 1) or paving (Lemma 3). These results allow us to establish Theorems 2 and 4. Note that our definition of flows generalizes the idea of the "augmenting paths" used by de Jong et al. [12], extending it from k-uniform matroids with affine delay functions to general matroids with delay functions in class $\mathcal{D}$.

For a paving matroid congestion game with affine delays we require a different approach in order to prove Theorem 3. Given f, o and the associated flow, we construct another congestion game with two states s and q such that $\frac{\text{cost}(f)}{\text{cost}(o)} \leq \frac{\text{cost}(s)}{\text{cost}(q)}$. We show that s and q and their associated flow satisfy some special properties, which are used to establish that $\text{cost}(s)/\text{cost}(q) \leq 17/7$ (Theorem 7).

2 Preliminaries

In this section, we first recall some basics of matroid theory and then we introduce some fundamental notions of congestion games.

Matroids. A *matroid* is a pair $(R, \mathcal{I})$ where the *ground set* R consists of a finite set of elements and $\mathcal{I}$ is a nonempty collection of subsets of R such that: (i) if $I \in \mathcal{I}$ and $J \subseteq I$, then $J \in \mathcal{I}$; and (ii) if $I, J \in \mathcal{I}$ and $|I| < |J|$, then $I \cup \{z\} \in \mathcal{I}$ for some $z \in J \setminus I$. Given a matroid $M = (R, \mathcal{I})$, a subset I of R is called *independent* if I belongs to $\mathcal{I}$, and dependent otherwise. A subset $B \subseteq R$ is called a *basis* if B is an inclusion-wise maximal independent subset. That is, $B \in \mathcal{I}$ and there is no $Z \in \mathcal{I}$ with $B \subset Z \subseteq R$. The common size of all bases is called the *rank* of the matroid, denoted by $r(M)$. A *circuit* of a matroid is an inclusion-wise minimal dependent set. For every basis B and every element x in $R \setminus B$, there is a unique circuit contained in $B \cup \{x\}$, that is called a *fundamental circuit*. Next, we introduce the *bijective basis-exchange property*:

Theorem 5 [9]. *Let $\mathcal{B}$ be the collection of bases of a matroid. For any $B, B' \in \mathcal{B}$, there is a bijection $\pi : B \to B'$ from B to B', such that for every $x \in B \setminus B'$, $B \setminus \{x\} \cup \{\pi(x)\}$ is a basis.*

A matroid is called *k-uniform* matroid if its independent sets are all the subsets of R of cardinality at most k, i.e. every $k+1$-element subset of R is a circuit. A matroid is called *paving* matroid if every circuit of M has cardinality $r(M)$ or $r(M) + 1$. The following proposition characterizes paving matroids in terms of their circuits.

Proposition 1 [27]. *Let $\mathcal{C}$ be a collection of non-empty subsets of a set R such that each member of $\mathcal{C}$ has size either t or $t + 1$. Let $\mathcal{C}' \subseteq \mathcal{C}$ consist only of the t-element members of $\mathcal{C}$. Then $\mathcal{C}$ is the set of circuits of a paving matroid on R of rank t if and only if*

1. *if two distinct members C_1 and C_2 of $\mathcal{C}'$ have $t - 1$ common elements, then every t-element subset of $C_1 \cup C_2$ is in $\mathcal{C}'$; and*
2. *$\mathcal{C} \setminus \mathcal{C}'$ consists of all the $(t+1)$-element subsets of R that contains no member of $\mathcal{C}'$.*

Congestion Games. We consider a congestion game with N players and resources set R. For $n \in \mathbb{N}$, we denote by $[n]$ the set $\{1, \ldots, n\}$. The set $X^i \subseteq R$ is the strategy set of player i. We call the game *symmetric* if all the players have the same strategy set, i.e. $X^i = X^j$ for all $i, j \in [N]$. A *state* of the game is a strategy profile $s = (s^1, \ldots, s^N)$ where $s^i \in X^i$ is the strategy chosen by player i, for $i \in [N]$. The set of states of the game is denoted by $X = X^1 \times \cdots \times X^N$.

For each $r \in R$ we have a nondecreasing delay function $d_r : [N] \to \mathbb{R}_{\geq 0}$. Given a state s we denote the number of players using resource r by s_r. Each player using r incurs a cost equal to $d_r(s_r)$, i.e., the cost of r depends on the total number of players that use r in s. Since d_r is a nondecreasing function, $d_r(j + 1) \geq d_r(j)$ for $j \in [N - 1]$, which models the effect of congestion. We denote the cost of a resource r with respect to state s by $\text{cost}_s(r) = d_r(s_r)$. We also define $\text{cost}_s^+(r) = d_r(s_r + 1)$. Finally, the *social cost* of state s is denoted by $\text{cost}(s) = \sum_{r \in R} s_r d_r(s_r) = \sum_{r \in R} s_r \text{cost}_s(r)$.

Matroid Congestion Games. A *matroid congestion game* is a congestion game where the strategy set of each player i is the set of bases $\mathcal{B}_i$ of a given matroid $M_i = (R_i \subseteq R, \mathcal{I}_i)$. For an arbitrary state s of the matroid congestion game, we denote by B_s^i the strategy of player i in s. A *paving* matroid congestion game is a matroid congestion game where M_i is a paving matroid for all $i \in [N]$. A *k-uniform* matroid congestion game is a congestion game where M_i is a k-uniform matroid for all $i \in [N]$ and $k \in [\min_i |R_i|]$.

Pure Nash Equilibria and Social Optima. A *pure Nash equilibrium* (PNE) is a state $s = (s^1, \dots, s^i, \dots, s^N)$ such that, for each $i \in [N]$ we have

$$\text{cost}_s(s^i) \leq \text{cost}_{\tilde{s}}(\tilde{s}^i) \qquad \forall \tilde{s} = (s^1, \dots, \tilde{s}^i, \dots, s^N) \in X.$$

A PNE represents a stable outcome of the game, since no player $i \in [N]$ can improve their cost if they select a different strategy $\tilde{s}^i$.

We are also interested in a *social optimum* (SO), which is a state that minimizes $\text{cost}(s)$ over all the states $s \in X$. The *Price of Anarchy* (PoA) is the maximum ratio $\frac{\text{cost}(f)}{\text{cost}(o)}$ such that o is a SO and f is a PNE. In other words, to compute the PoA we consider the "worst" PNE, i.e., a PNE whose social cost is as large as possible.

3 Upper Bounds on the PoA for Delays in Class $\mathcal{D}$

In this section, our goal is to prove Theorems 2 and 4. For a matroid congestion game over resource set R, we let $G = (R, E)$ be a complete directed graph, where the nodes correspond to the resources in R. Let s and q be two states of the congestion game. We define the following two sets:

$$R^-(s, q) = \{r \in R : s_r > q_r\} \qquad\qquad R^+(s, q) = \{r \in R : s_r < q_r\},$$

and we let $l = \sum_{r \in R^-(s,q)} (s_r - q_r) = \sum_{r \in R^+(s,q)} (q_r - s_r)$. In G, every node $r \in R^-$ has supply $s_r - q_r$, and every node $r \in R^+$ has demand $q_r - s_r$. A (single-commodity) flow $F \in \mathbb{Z}^{R \times R}$ in G is a non-negative vector such that for every node $r \in R$

$$F(\delta^-(r)) - F(\delta^+(r)) = q_r - s_r, \tag{1}$$

where $\delta^-(r)$ contains all the arcs whose head is r and $\delta^+(r)$ contains all the arcs whose tail is r. We call F an (s, q)-*difference flow*. Note that the above definitions can be applied to a generic congestion game. For a matroid congestion game, we can construct a special (s, q)-difference flow F, that we call (s, q)-*exchange flow*, as follows. According to Theorem 5, for each pair (B_s^i, B_q^i), there is a bijection $\pi^i(x) : B_s^i \to B_q^i$ such that for every $r \in B_s^i \setminus B_q^i$ there is a unique $\pi^i(r) \in B_q^i \setminus B_s^i$

and $B^i_s \setminus \{r\} \cup \{\pi^i(r)\} \in \mathcal{B}$. Starting from the zero vector, for every $i \in [N]$, $r \in B^i_s \setminus B^i_q$, we add one unit of flow to the arc $(r, \pi^i(r))$ to G in order to obtain F. We observe that F can be decomposed into l paths, each one starting from a node in $R^-(s,q)$ and ending at a node in $R^+(s,q)$, and carrying one unit of flow. Each path in the exchange flow can be interpreted as a sequence of resource exchanges such that each arc (r, r') in the path corresponds to some player replacing resource r with resource r' in their strategy.

In the next theorem, we consider an (f,o)-exchange flow. For any (u,v)-path from $R^-(f,o)$ to $R^+(f,o)$, if $\text{cost}_f(u)$ is equal to at least a fraction α of $\text{cost}^+_f(v)$, then we can upper bound the ratio between the social costs of f and o by $\alpha\rho(\mathcal{D})$. We recall that the function $\rho(\mathcal{D})$, initially introduced by Roughgarden [30], is defined as $\rho(\mathcal{D}) := \sup_{d\in\mathcal{D}} \rho(d)$, where

$$\rho(d) = \sup_{x \geq y \geq 0} \frac{xd(x)}{yd(y) + (x-y)d(x)}. \tag{2}$$

Theorem 6. *Let F be an (f,o)-exchange flow. Let $R^- = R^-(f,o)$ and $R^+ = R^+(f,o)$. For all paths p contained in F from $u \in R^-$ to $v \in R^+$, if $\alpha\, cost^+_f(v) \geq cost_f(u)$ for some $\alpha \geq 1$ then we have $cost(f) \leq \alpha\rho(\mathcal{D})cost(o)$.*

Proof. For every resource $r \in R^-$, inequality (2) and $\alpha \geq 1$ imply

$$\begin{aligned} f_r\text{cost}_f(r) = f_r d_r(f_r) &\leq \rho(\mathcal{D})(o_r d_r(o_r) + (f_r - o_r)d_r(f_r)) \\ &\leq \rho(\mathcal{D})(\alpha o_r d_r(o_r) + (f_r - o_r)d_r(f_r)). \end{aligned} \tag{3}$$

Let $\{p_1, \ldots, p_l\}$ be an arbitrary decomposition of the flow F, where each p_k is from r^-_k to r^+_k such that $r^-_k \in R^-$ and $r^+_k \in R^+$. We have

$$\begin{aligned} \sum_{r\in R^+} (o_r - f_r)\text{cost}_o(r) &= \sum_{k=1}^{l} \text{cost}_o(r^+_k) \geq \sum_{k=1}^{l} \text{cost}^+_f(r^+_k) \\ &\geq \sum_{k=1}^{l} \frac{1}{\alpha}\text{cost}_f(r^-_k) = \frac{1}{\alpha} \sum_{r\in R^-} (f_r - o_r)\text{cost}_f(r), \end{aligned} \tag{4}$$

where the equalities hold by the definition of F and equality (1), the first inequality holds because of the definition of R^+, and the second inequality holds by our assumption. Let $\bar{R} = \{r \in R : f_r = o_r\} = R \setminus (R^- \cup R^+)$.

$$\begin{aligned} \text{cost}(f) &= \sum_{r\in R^-} f_r\text{cost}_f(r) + \sum_{r\in R^+} f_r\text{cost}_f(r) + \sum_{r\in\bar{R}} f_r\text{cost}_f(r) \\ &\leq \sum_{r\in R^-} f_r\text{cost}_f(r) + \sum_{r\in R^+} f_r\text{cost}_o(r) + \sum_{r\in\bar{R}} o_r\text{cost}_o(r) \\ &\leq \rho(\mathcal{D}) \sum_{r\in R^-} \alpha o_r\text{cost}_o(r) + \rho(\mathcal{D}) \sum_{r\in R^-} (f_r - o_r)\text{cost}_f(r) \end{aligned}$$

$$+ \alpha\rho(\mathcal{D}) \sum_{r\in R^+} f_r \text{cost}_o(r) + \sum_{r\in\bar{R}} o_r \text{cost}_o(r)$$

$$\leq \rho(\mathcal{D}) \sum_{r\in R^-} \alpha o_r \text{cost}_o(r) + \rho(\mathcal{D}) \sum_{r\in R^+} \alpha(o_r - f_r)\text{cost}_o(r)$$

$$+ \alpha\rho(\mathcal{D}) \sum_{r\in R^+} f_r \text{cost}_o(r) + \sum_{r\in\bar{R}} o_r \text{cost}_o(r)$$

$$= \alpha\rho(\mathcal{D}) \sum_{r\in R^-\cup R^+} o_r \text{cost}_o(r) + \sum_{r\in\bar{R}} o_r \text{cost}_o(r) \leq \alpha\rho(\mathcal{D})\text{cost}(o).$$

The first inequality holds because of the definition of R^+ and $\bar{R}$; the second inequality holds because of inequality (3) and $\alpha \geq 1$, $\rho(\mathcal{D}) \geq 1$; the third inequality follows by applying (4); the last inequality follows because $\alpha \geq 1$, $\rho(\mathcal{D}) \geq 1$. □

We emphasize that the bound on the PoA provided by Theorem 6 is not restricted to the class of paving matroids. In fact, the assumption of the theorem involves an exchange flow, which is defined for any matroid, and a parameter α. Thus, for any matroid, if we are able to find such α, we are able to bound the PoA.

The next lemma implies that for k-uniform matroids $\alpha = z(\mathcal{D})$ satisfies the assumption of Theorem 6. This lemma is an extension of Lemma 5 in [12] from affine delay functions to general delay functions. Moreover, it can be verified that for polynomial delay functions the bound established in Lemma 1 is tight.

Lemma 1. *Suppose M is a k-uniform matroid. Let f be an arbitrary PNE state and q be an arbitrary state of the game. For every $u \in R^-(f,q)$ and $v \in R^+(f,q)$ we have $z(\mathcal{D})cost_f^+(v) \geq cost_f(u)$.*

Proof. Let u^* be the most expensive resource in $R^-(f,q)$, i.e., $\text{cost}_f(r) \leq \text{cost}_f(u^*)$ for every resource $r \in R^-(f,q)$. To prove the lemma, we will show that for every $v \in R^+(f,q)$ we have $z(\mathcal{D})\text{cost}_f^+(v) \geq \text{cost}_f(u^*)$. By contradiction, suppose there exists a resource $v \in R^+(f,q)$ such that

$$z(\mathcal{D})\text{cost}_f^+(v) < \text{cost}_f(u^*). \tag{5}$$

Since $q_v > f_v$, we have $f_v < N$, thus there exists at least one player i who does not use v in f, i.e., $v \notin B_f^i$. We claim that, for all $r \in B_f^i$, we have

$$\text{cost}_f(r) \leq \text{cost}_f^+(v). \tag{6}$$

This follows from the fact that, since M is a k-uniform matroid $B_f^i \setminus \{r\} \cup \{v\}$ is a basis of M for all $r \in B_f^i$. Thus, if (6) did not hold, player i could deviate from $r \in B_f^i$ to v to decrease their cost. As a consequence, $z(\mathcal{D})\text{cost}_f^+(v) < \text{cost}_f(u^*)$ implies that $v \notin B_f^i$. Moreover, recalling that $z(\mathcal{D}) \geq 1$, we have $\text{cost}_f^+(r) \leq z(\mathcal{D})\text{cost}_f(r)$ for all $r \in R$. Combining this with (5) and (6), we obtain that, for all $r \in B_f^i$

$$\text{cost}_f^+(r) < \text{cost}_f(u^*). \tag{7}$$

Note that (7) implies $u^* \notin B_f^i$. Since $u^* \in R^-(f,q)$, $f_{u^*} > o_{u^*} \geq 0$, thus there is at least one player j using u^* in f, i.e., $u^* \in B_f^j$. Since M is a k-uniform matroid, $B_f^j \setminus \{u^*\} \cup \{r\}$ is a basis of M for all $r \in B_f^i$. Moreover, since $u^* \notin B_f^i$ and $|B_f^i| = |B_f^j| = k$, we can conclude that $|B_f^i \setminus B_f^j| \geq 1$. I.e. there exists at least one resource $r^* \in B_f^i$ such that $r^* \notin B_f^j$. Thus, by (7), player j could deviate from u^* to r^* to decrease their cost. This contradicts the fact that f is a PNE. □

Applying Theorem 6 and Lemma 1, we can immediately derive Theorem 4.

Next, we show that for paving matroids $\alpha = z(\mathcal{D})^2$ satisfies the assumption of Theorem 6. To this purpose, we first introduce an auxiliary result.

Lemma 2. *Consider a symmetric matroid congestion game with delays in class $\mathcal{D}$. Let f be a PNE, and o a SO. Let v be a resource that is not used by player i in f and let C_v^i be the unique circuit in $B_f^i \cup \{v\}$. Then, for all $r \in C_v^i$ we have $cost_f^+(r) \leq z(\mathcal{D})cost_f(r) \leq z(\mathcal{D})cost_f^+(v)$.*

Proof. Assume that there exists a resource $r \in C_v^i$ such that $\text{cost}_f(r) > \text{cost}_f^+(v)$. Since C_v^i is the unique circuit that satisfies $C_v^i \setminus \{v\} \subseteq B_f^i$, we have that $B_f^i \setminus \{r\} \cup \{v\} \in \mathcal{B}$, i.e., exchanging r and v defines a feasible strategy for player i. By performing this exchange player i is able to lower their cost, thus contradicting the fact that f is a PNE. Thus, we can conclude that for each $r \in C_v^i$ we have $\text{cost}_f(r) \leq \text{cost}_f^+(v)$. This implies that $z(\mathcal{D})\text{cost}_f(r) \leq z(\mathcal{D})\text{cost}_f^+(v)$. Finally, by the definition of $z(\mathcal{D})$, thus we have $\text{cost}_f^+(r) \leq z(\mathcal{D})\text{cost}_f(r)$. □

For an arbitrary state q, consider an (f,q)-exchange flow F and any path contained in it starting from a node $u \in R^-(f,q)$ and ending at a node $v \in R^+(f,q)$. If the matroid is paving, the next lemma implies that $\text{cost}_f^+(v)$ cannot be smaller than a fraction of $\text{cost}_f(u)$.

Lemma 3. *Suppose M is a paving matroid with $r(M) = t \geq 1$. Let f be an arbitrary PNE state and q be an arbitrary state of the game and let F be an (f,q)-exchange flow. Let $R^- = R^-(f,q)$ and $R^+ = R^+(f,q)$. For all paths p contained in F from $u \in R^-$ to $v \in R^+$, and for every resource r in p we have $cost_f(r) \leq z(\mathcal{D})^2 cost_f^+(v)$.*

Proof. Let r^* be the most expensive resource of path p in f, i.e., $\text{cost}_f(r) \leq \text{cost}_f(r^*)$ for every resource r in p. Since $t \geq 1$ we know that r^* is used by at least one player in f. We will prove $\text{cost}_f(r^*) \leq z(\mathcal{D})^2\text{cost}_f^+(v)$. By contradiction, suppose

$$\text{cost}_f(r^*) > z(\mathcal{D})^2\text{cost}_f^+(v). \tag{8}$$

Define

$$S = \{r \in R : \text{cost}_f^+(r) < \text{cost}_f(r^*)\}, \quad \bar{S} = \{r \in R : z(\mathcal{D})\text{cost}_f^+(r) < \text{cost}_f(r^*)\}.$$

Since $z(\mathcal{D}) \geq 1$, we have $\bar{S} \subseteq S$. Moreover, we have the following property.

Claim 1. $|\bar{S}| \geq t$.

Proof of Claim. Since v is the last node in p, there exists a player j such that $v \notin B_f^j$. Let C_v^j be the fundamental circuit in $B_f^j \cup \{v\}$. By Lemma 2, for all $r \in C_v^j$ we have $\text{cost}_f^+(r) \leq z(\mathcal{D})\text{cost}_f^+(v)$. Thus:

$$z(\mathcal{D})\text{cost}_f^+(r) \leq z(\mathcal{D})^2\text{cost}_f^+(v) < \text{cost}_f(r^*),$$

where the last inequality comes from (8). This implies that $C_v^j \subseteq \bar{S}$. Since in a paving matroid of rank t every circuit has size at least t we obtain $|\bar{S}| \geq t$. ◇

Note that $v \in S$, since $z(\mathcal{D}) \geq 1$, and $r^* \notin S$. Since p traverses both r^* and v, there is an arc (a, b) in p such that $a \notin S$ and $b \in S$. Since (a, b) is contained in F there exists a player i such that $a \in B_f^i$, $b \notin B_f^i$ and $B_f^i \setminus \{a\} \cup \{b\} \in \mathcal{B}$.

First, $a \in B_f^i \setminus S = B_f^i \setminus (B_f^i \cap S)$. Thus $1 \leq t - |B_f^i \cap S|$. We have

$$|\bar{S} \setminus B_f^i| = |\bar{S}| - |\bar{S} \cap B_f^i| \geq t - |\bar{S} \cap B_f^i| \geq t - |S \cap B_f^i| \geq t + (1 - t) = 1,$$

where the first inequality follows from Claim 1. Thus $\bar{S} \setminus B_f^i \neq \emptyset$. Let $w \in \bar{S} \setminus B_f^i$. Let C_w^i be the fundamental circuit in $B_f^i \cup \{w\}$. By Lemma 2 for all $r \in C_w^i$ we have

$$\text{cost}_f^+(r) \leq z(\mathcal{D})\text{cost}_f^+(w) < \text{cost}_f(r^*),$$

where the last inequality holds because $w \in \bar{S}$.

This implies $C_w^i \subseteq S$. Recall that $C_w^i \setminus \{w\} \subseteq B_f^i$. Since the matroid is paving, $|C_w^i \setminus \{w\}| \geq t - 1$. Finally, as $a \in B_f^i \setminus S$ we can conclude that $B_f^i \setminus \{a\} = C_w^i \setminus \{w\} \subseteq S$. Since $b \in S$, we have $B_f^i \setminus \{a\} \cup \{b\} \subseteq S$. We now prove that every t-element subset of S is a circuit. This immediately contradicts the fact that $B_f^i \setminus \{a\} \cup \{b\}$ is a basis.

Claim 2. Every t-element subset of S is a circuit of the paving matroid M.

Proof of Claim. Let h be a player such that $r^* \in B_f^h$ and let r be an arbitrary resource in $S \setminus B_f^h$. We show that $B_f^h \setminus \{r^*\} \cup \{r\}$ is a circuit. Consider the fundamental circuit C_r^h in $B_f^h \cup \{r\}$. We argue that r^* is not in C_r^h. If that was the case, we would have $\text{cost}_f^+(r) \geq \text{cost}_f(r^*)$ by Lemma 2, which contradicts $r \in S$. Since we have a paving matroid $C_r^h \geq t$, thus $C_r^h = \{r\} \cup B_f^h \setminus \{r^*\}$. This proves that $B_f^h \setminus \{r^*\}$ forms a circuit with every resource $r \in S \setminus B_f^h$. By applying the first statement in Proposition 1 we can conclude that every t-element subset of $S \cup B_f^h \setminus \{r^*\}$ is a circuit. By the definition of S we have $r^* \notin S$, so $S \subseteq S \cup B_f^h \setminus \{r^*\}$ and every t-element subset of S is a circuit. ◇

□

Lemma 3 implies that for paving matroids $\alpha = z(\mathcal{D})^2$ satisfies the assumption of Theorem 6. Thus, Theorem 2 directly follows.

Remark 1. It can be verified that the bound of Lemma 3 is tight for polynomial delay functions, however we conjecture that the bound of Theorem 2 is *not* tight for the same class of delays. In fact, instances where the bound of Lemma 3 is tight can have PoA smaller than the upper bound of Theorem 2. An intuitive explanation is the following: when the bound in Lemma 3 is tight, in the PNE there is an "expensive" resource used by many players and a "cheap" resource used by few players. For this state to be a PNE, the circuits of the matroid must prevent single player deviations where the expensive resource is replaced by the cheap one. The existence of these circuits requires the existence of other resources with comparable costs both in the PNE and in the SO (this is implied by Lemma 2). As a result, the PoA in these instances will be lower than the upper bound of Theorem 2.

4 Lower Bound on the PoA of Paving Matroid Congestion Games with Affine Delays

In this section, we consider symmetric paving matroid congestion games with affine delays, i.e., we assume that the delay function of each resource $r \in R$ is of the form $d_r(x) = a_r x + b_r$ with $a_r \geq 0$ and $b_r \geq 0$. Our goal is to prove Theorem 1, stating that the worst-case PoA is at least 13/9. This lower bound is higher than the previously best known lower bound of ≈ 1.35, which is achieved in the symmetric k-uniform matroid congestion games [12]. Moreover, this lower bound indicates that the upper bound of ≈ 1.41 for symmetric k-uniform matroid congestion games does not hold for paving matroids.

Proof [Proof of Theorem 1]. We prove the theorem by constructing an instance of a symmetric paving matroid congestion game with affine delays that achieves the PoA of 13/9. Let $R = \{r_1\} \cup R_2 \cup R_3$, where $R_2 = \{r_2, r_3, r_4, r_5\}$ and $R_3 = \{r_6, \ldots, r_{13}\}$. Let

$$C_1 = \{\{r_1, r_{6+2i}, r_{6+2i+1}\} : \forall i \in \{0, 1, 2, 3\}\},$$
$$C_2 = \{S \subset R : |S| = 4 \text{ and } S' \not\subset S, \forall S' \in C_1\}.$$

Let $C = C_1 \cup C_2$. Using Proposition 1 with $\mathcal{C}' = C_1$ and $\mathcal{C} = C$ we can easily check that C is the set of circuits for a paving matroid of rank 3 defined over R.

Next we define a symmetric congestion game over $\mathcal{M}$. Let the delay function of r_1 be $d_{r_1}(x) = 1$, and for $i \in \{2, 3, \ldots, 13\}$ let $d_{r_i}(x) = x$. Let the number of players be $N = 6$. The strategy set of each player is the set of bases of the paving matroid. In a PNE, players 1 and 2 select resources $\{r_1, r_2, r_3\}$ and for $i \in \{3, 4, 5, 6\}$, player i selects resources $\{r_4, r_6, r_7\}$, $\{r_4, r_8, r_9\}$, $\{r_5, r_{10}, r_{11}\}$, $\{r_5, r_{12}, r_{13}\}$, respectively. Note that players will not deviate from r_4 or r_5 to r_1, since this would form a circuit in C_1. The social cost of this PNE state is 26. In the SO, each player $i \in [N]$ selects resources $\{r_1, r_{1+i}, r_{7+i}\}$. It can be easily checked that those strategies contain no circuit and the social cost is 18. Thus, the PoA of this instance is at least $26/18 = 13/9$. □

5 Upper Bound on the PoA of Paving Matroid Congestion Games with Affine Delays

In this section, we prove Theorem 3. Consider a symmetric matroid congestion game with N players over resource set R, and suppose that every delay function is affine. Let s and q be two arbitrary states of the game such that $\text{cost}(s) \geq \text{cost}(q)$, and let $R^- = R^-(s,q)$, $R^+ = R^+(s,q)$. We consider the graph G defined in Sect. 3, where each node $r \in R^-$ has supply $s_r - q_r$ and each node $r \in R^+$ has demand $q_r - s_r$, and we let Φ be an (s,q)-difference flow in G. The following theorem identifies some special properties of Φ that can be used to upper bound $\frac{\text{cost}(s)}{\text{cost}(q)}$. We sketch the proof of this theorem at the end of the section. The complete proof of the theorem appears in the full version of this paper [19].

Theorem 7. *Suppose that Φ is an acyclic (s,q)-exchange flow satisfying the following properties:*

1. *For every arc (u,v) with positive flow in Φ, $cost_s(u) \leq cost_s^+(v)$.*
2. *For every path p from $u \in R^-$ to $v \in R^+$, $cost_s^+(v) \geq \frac{1}{4} cost_s(u)$.*
3. *Let (v,w) be an arc with positive flow in Φ. If for every path to v starting at a node $u \in R^-$ we have $cost_s(v) \geq \frac{1}{2} cost_s(u)$, then $w \notin R^+$.*
4. *For all $r \in R^+$, $s_r = 0$ and $\Phi(\delta^+(r)) = 0$.*
5. *For all $r \notin R^+$, the delay function of r is linear.*

Then $cost(s)/cost(q) \leq 17/7$.

Now consider a symmetric paving matroid congestion game with N players over resource set R, and suppose that the delay functions $d = (d_r)_{r\in R}$ are affine. Let f and o be a PNE and a SO, respectively, that achieve the PoA. We consider an (f,o)-exchange flow F. We then apply five steps, to map $\mathcal{S} = (R,d,f,o,F)$ to a tuple $\mathcal{S}' = (R',d',s,q,\Phi)$ that defines a symmetric 1-uniform matroid congestion game over R' with affine delays $d' = (d'_r)_{r\in R}$, where s and q are two states of the game, and Φ is an (s,q)-exchange flow satisfying the assumptions in Theorem 7, and such that

$$\frac{\text{cost}(f)}{\text{cost}(o)} \leq \frac{\text{cost}(s)}{\text{cost}(q)}.$$

Then using Theorem 7 we can conclude that the worst-case PoA of symmetric paving matroid congestion games is at most 17/7.

Let $\mathcal{S}^0 = (R,d,f,o,F)$. F is an (f,o)-exchange flow of a matroid congestion game, thus for every arc (u,v) with positive flow in F there exists a player i who could replace resource u with resource v in their strategy. Since f is a PNE, player i is not able to decrease their cost by exchanging u and v, implying that F satisfies property 1. Moreover, since for affine delays $z(\mathcal{D}) = 2$, Lemma 2 implies that also property 7 is satisfied. We apply the following four steps, that preserve properties 1 and 2. Moreover, the construction guarantees $\sum_{r\in R} s_r = \sum_{r\in R} q_r$

in every step. This implies that in every step we can construct an instance of a symmetric 1-uniform matroid congestion game on resource set R where s and q are two states that are obtained by assigning players to resources so that for each $r \in R$ we have s_r players using r in s and q_r players using r in q. The corresponding (s,q)-exchange flow is redefined accordingly. Note that s and q are not necessarily a PNE and a SO of the game.

Step 1. First, we let $s = f$, $q = o$ and $\Phi = F$. We redefine (R, d, s, q, Φ) as follows. For every resource $v \in R^+(f, o)$ such that $f_v > 0$, we add a new resource v' with constant delay equal to $\text{cost}_o(v)$. We set $s_v = q_v = f_v$, $s_{v'} = 0$ and $q_{v'} = o_v - f_v > 0$. Note that $q_{v'} > s_{v'}$, i.e., $v' \in R^+(s,q)$, while $q_v = s_v$, i.e., $v \notin R^+(s,q)$. Moreover we define the flow Φ on arc (v, v') to be $o_v - f_v$. At the end, Φ is an (s,q)-exchange flow that satisfies property 4. Finally we show that $\frac{\text{cost}(f)}{\text{cost}(o)} \le \frac{\text{cost}(s)}{\text{cost}(q)}$ after Step 1. Denote the set of nodes we added in this step by V'. According to the construction in Step 1, we have

$$\text{cost}(s) = \sum_{r\in R} s_r \text{cost}_s(r) + \sum_{r\in V'} s_r \text{cost}_s(r) = \sum_{r\in R} f_r \text{cost}_f(r) + 0 = \text{cost}(f),$$

and

$$\begin{aligned}
\text{cost}(q) &= \sum_{r\in R\setminus V} q_r \text{cost}_q(r) + \sum_{r\in V} q_r \text{cost}_q(r) + \sum_{r\in V'} q_r \text{cost}_q(r) \\
&= \sum_{r\in R\setminus V} o_r \text{cost}_o(r) + \sum_{r\in V} f_r \text{cost}_q(r) + \sum_{r\in V} (o_r - f_r)\text{cost}_o(r) \\
&< \sum_{r\in R\setminus V} o_r \text{cost}_o(r) + \sum_{r\in V} f_r \text{cost}_o(r) + \sum_{r\in V} (o_r - f_r)\text{cost}_o(r) = \text{cost}(o),
\end{aligned}$$

where the inequality holds because $\text{cost}_q(r) = d_r(f_r) < d_r(o_r) = \text{cost}_o(r)$. By combining the above inequalities we obtain $\frac{\text{cost}(f)}{\text{cost}(o)} \le \frac{\text{cost}(s)}{\text{cost}(q)}$.

Step 2. For each resource $v \in R^+(s,q)$ receiving $t_1, \ldots, t_h$ units of flow from $h \ge 2$ resources $u_1, \ldots, u_h$ through arcs $(u_1, v), \ldots, (u_h, v)$ in Φ, we redefine (R, d) by replacing v with h new nodes $v_1, \ldots, v_h$, each having delay function d_v. We redefine (s,q) by setting $s_{v_i} = 0$ and $q_{v_i} = t_i$ for all $i \in [h]$. Next, we redefine Φ by replacing arc (u_i, v) with (u_i, v_i) having flow value t_i, for all $i \in [h]$. After this step, for each $v \in R^+(s,q)$ there is only one resource sending flow to v. Let $(s,q), (s',q')$ denote the input and output states of Step 2, respectively. We show that $\frac{\text{cost}(s)}{\text{cost}(q)} \le \frac{\text{cost}(s')}{\text{cost}(q')}$ holds after Step 2. For each $v \in R^+(s,q)$ that we selected in Step 2, we replaced it with $v_1, \ldots, v_h$. By the construction we have:

$$s_v \text{cost}_s(v) = \sum_{i=1}^{h} s'_{v_i} \text{cost}_{s'}(v_i) = 0,$$

and

$$q_v \text{cost}_q(v) = \sum_{i=1}^{h} q'_{v_i} \text{cost}_q(v) = \sum_{i=1}^{h} q'_{v_i} d_v(q_v) \ge \sum_{i=1}^{h} q'_{v_i} d_v(q'_v) = \sum_{i=1}^{h} q'_{v_i} \text{cost}_{q'}(v_i).$$

Thus, the social cost of s stays the same and the social cost of q decreases after Step 2, so we have $\frac{\text{cost}(s)}{\text{cost}(q)} \leq \frac{\text{cost}(s')}{\text{cost}(q')}$.

Step 3. For each resource $v \in R^+(s,q)$, let r^* be the most expensive resource in $R^-(s,q)$ that is connected to v along a path carrying at least one unit of flow in Φ. Let u be the only resource sending flow to v in Φ, and let h be the flow of Φ on arc (u,v). If $\text{cost}_s^+(v) > \frac{1}{2}\text{cost}_s(r^*)$, we redefine (R,d) by replacing v with h new nodes $v_1,\ldots,v_h$ having delay function $\frac{1}{2}\text{cost}_s^+(v)x$ for $i \in [h]$. Moreover, we add h new resources $w_1,\ldots,w_h$ with constant delay function $\frac{1}{2}\text{cost}_s^+(v)$ for $i \in [h]$. We redefine (s,q) by setting $s_{v_i} = 1$, $s_{w_i} = 0$ and $q_{v_i} = q_{w_i} = 1$ for $i \in [h]$. Thus, property 4 is preserved. Finally, we redefine Φ by setting to one the flow of arcs (u,v_i) and (v_i,w) for $i \in [h]$. We repeat this step until for all $v \in R^+(s,q)$ we have $\text{cost}_s^+(v) \leq \frac{1}{2}\text{cost}_s(r^*)$, thus achieving property 3. As in Step 2, let $(s,q),(s',q')$ denote the input and output states of each iteration in Step 3, respectively. We show that $\frac{\text{cost}(s)}{\text{cost}(q)} \leq \frac{\text{cost}(s')}{\text{cost}(q')}$ holds after each iteration of Step 3. Note that for each $v \in R^+(s,q)$ that we selected in an iteration of Step 3, v is replaced by $v_1,\ldots,v_h$ and $w_1,\ldots,w_h$. By our construction we have:

$$s_v\text{cost}_s(v) = 0 < \sum_{i=1}^{h}(v_i\text{cost}_{s'}(v_i) + w_i\text{cost}_{s'}(w_i)) = \sum_{i=1}^{h}\frac{1}{2}\text{cost}_s^+(v),$$

and

$$\begin{aligned} q_v\text{cost}_q(v) = h\text{cost}_q(v) \geq h\text{cost}_s^+(v) &= \sum_{i=1}^{h}\frac{1}{2}\text{cost}_s^+(v) + \sum_{i=1}^{h}\frac{1}{2}\text{cost}_s^+(v) \\ &= \sum_{i=1}^{h} q'_{v_i}\text{cost}_{q'}(v_i) + \sum_{i=1}^{h} q'_{w_i}\text{cost}_{q'}(w_i). \end{aligned}$$

The above inequalities imply that after each iteration of Step 3 the social cost of s increases and the social cost of q decreases, so we have $\frac{\text{cost}(s)}{\text{cost}(q)} \leq \frac{\text{cost}(s')}{\text{cost}(q')}$.

Step 4. For every resource $r \notin R^+(s,q)$, suppose $d_r(x) = ax + b$ where $a,b \geq 0$. We redefine the delay function of r as $\frac{\text{cost}_s(r)}{s_r}x = \frac{as_r+b}{s_r}x$. Next we show that $\frac{\text{cost}(s)}{\text{cost}(q)} \leq \frac{\text{cost}(s')}{\text{cost}(q')}$, where $(s,q),(s',q')$ are the input and output states of Step 4, respectively. According to the definition of the new delay functions, it is easy to conclude that $\text{cost}(s) = \text{cost}(s')$. For every resource $r \in R \setminus R^+(s',q')$, since we have $s_r = s'_r \geq q'_r = q_r$, then $\text{cost}_{q'}(r) \leq \text{cost}_q(r)$. For every resource $r \in R^+(s',q')$, since we did not change the associated delay function, we have $\text{cost}_{q'}(r) = \text{cost}_q(r)$. Thus, we can conclude that $\text{cost}(q') \leq \text{cost}(q)$, implying $\frac{\text{cost}(s)}{\text{cost}(q)} \leq \frac{\text{cost}(s')}{\text{cost}(q')}$.

Step 5. We delete all the cycles in Φ to make the flow acyclic. At the end, we set $\mathcal{S}' = \{R,d,s,q,\Phi\}$ and $\mathcal{S} = \mathcal{S}^0$. Thus, we achieve property 5.

Based on our discussion we obtain the following lemma.

Lemma 4. *S' satisfies the five assumptions in Theorem 7 and $\frac{cost(f)}{cost(o)} \leq \frac{cost(s)}{cost(q)}$.*

Remark 2. The construction that we use in the proof of Theorem 3 relies on Lemma 3 to satisfy property 2 in Theorem 7. As discussed in Remark 1, although there exist instances where the bound in Lemma 3 is tight, these instances might still have PoA smaller than the upper bound of Theorem 3. Thus, we conjecture that the upper bound of Theorem 3 is not tight.

We are now left with proving Theorem 7. Here we give a sketch of the proof. We refer the reader to the full version of this paper [19] for the complete proof.

Proof. [Proof Sketch of Theorem 7]
First, by property 4 and 5 we have

$$\frac{\text{cost}(s)}{\text{cost}(q)} = \frac{\sum_{r\in R\setminus R^+} s_r \text{cost}_s(r)}{\sum_{r\in R} q_r \text{cost}_q(r)} \leq \frac{\sum_{r\in R\setminus R^+} \frac{s_r^2}{s_r+1}\text{cost}_s^+(r)}{\sum_{r\in R\setminus R^+} \frac{q_r^2}{s_r+1}\text{cost}_s^+(r) + \sum_{r\in R^+} q_r \text{cost}_s^+(r)}. \tag{9}$$

Next, let $r \in R$. We define $\lambda(r) = \text{cost}_s(r) - \frac{1}{2}\text{cost}_s^+(r)$. Let $p = r_0, \ldots, r_k$ be a path in Φ carrying one unit of flow, where $r_0 \in R^-$ and $r_k \in R^+$. For each $i \in [k-1]$ we define:

$$\Omega(p,0) := \sum_{j=1}^{k-1} \left(\frac{1}{2}\right)^j \text{cost}_s(r_0), \quad \Omega(p,i) := \sum_{j=i+1}^{k-1} \left(\frac{1}{2}\right)^{j-i} \lambda(r_i). \tag{10}$$

Now let $P = \{p_1, \ldots, p_l\}$ be an arbitrary decomposition of the flow Φ where each path starts at a node in R^- and ends at a node in R^+ and carries one unit of flow. By property 4, $\Phi(\delta^+(r)) = 0$ for each $r \in R^+$, thus in every path $p \in P$ the only node in R^+ is the sink of the path, denoted by $t(p)$. Moreover, for each resource $r \in R$ we denote by $P(r)$ the paths in P that contain r and by $P^0(r)$ the paths in P starting at r. Finally, for each resource $r \in R$ and path $p \in P(r)$ we use the notation $p(r)$ to identify the position of r in p, precisely $p(r) = 0$ if r is the start node of p, and $p(r) = i$ if r is the i-th node appearing after the start node of p. Then we can rewrite (9) as:

$$\frac{\text{cost}(s)}{\text{cost}(q)} \leq \frac{\sum_{r\in R\setminus R^+} A_r}{\sum_{r\in R\setminus R^+} B_r} \leq \max_{r\in R\setminus R^+} \frac{A_r}{B_r}, \tag{11}$$

where

$$A_r = \frac{(s_r)^2}{s_r+1}\text{cost}_s^+(r) + \sum_{p\in P(r)} \Omega(p,p(r)) - \frac{\Phi(\delta^-(r))}{2}\text{cost}_s^+(r), \tag{12}$$

$$B_r = \frac{(q_r)^2}{s_r+1}\text{cost}_s^+(r) + \sum_{p\in P(r)} \Omega(p,p(r)) + \sum_{p\in P^0(r)} \text{cost}_s^+(t(p)) - \frac{\Phi(\delta^-(r))}{2}\text{cost}_s^+(r). \tag{13}$$

In the last part of the proof, we use properties 1, 2 and 3 to show that $\frac{A_r}{B_r} \leq \frac{17}{7}$ for all $r \in R \setminus R^+$. □

6 Conclusion

We have investigated the impact of matroid structures on the PoA of symmetric congestion games. In the symmetric case, the PoA of general matroid congestion games is still not completely understood. For graphic matroids and $N = 2, 3, 4$ or infinity with affine delay functions, the PoA can be as large as the worst-case PoA of symmetric congestion games [15], which is equal to $\frac{5N-2}{2N+1}$ [11]. However, for arbitrary N or different delay functions we don't know whether the worst-case PoA of symmetric congestion games can be achieved by symmetric matroid congestion games. Our results indicate that if we restrict to paving matroids, the worst-case PoA is significantly smaller than that of symmetric congestion games. A similar result had been previously established by de Jong et al. [12] for k-uniform matroids and affine delays. However, k-uniform matroids are only a mild generalization of singleton congestion games. Paving matroids, on the other hand, are a substantial generalization of k-uniform matroids, since they are conjectured to represent the vast majority of matroids. Since paving matroids are quite more complex than k-uniform matroids, it is not as easy to characterize the worst-case PoA. There is still a gap between our upper and lower bounds, and we conjecture that our upper bounds are not tight (see Remarks 1 and 2).

Our approach to bound the PoA relies on a constant α that we have quantified for both k-uniform matroids and paving matroids (Theorem 6). In particular, we can set $\alpha = z(\mathcal{D})$ for k-uniform matroids and $\alpha = z(\mathcal{D})^2$ for paving matroids. Since paving matroids of rank k contain circuits whose size is smaller than the circuit size of k-uniform matroids, this suggests that the difference between the sizes of bases and circuits might impact the PoA. Let δ be a parameter that is equal to the rank of the matroid minus the size of the smallest circuit in the matroid. We conjecture that for $\delta \geq 0$ we can satisfy the assumptions of Theorem 6 with $\alpha = z(\mathcal{D})^{2(\delta+1)}$. Thus, we would get an upper bound on the PoA which is equal to $\rho(\mathcal{D})z(\mathcal{D})^{2(\delta+1)}$. For polynomial delays of highest degree p, this bound is in $O((C^p)(p/\ln p))$, where $C = 4^{\delta+1}$. For fixed δ and large p this bound is still better than the PoA of general congestion games, that is in $O((p/\ln p)^{p+1})$. To summarize, it is possible that our approach could be extended to upper bound the PoA in arbitrary matroid congestion games where we have an upper bound on δ. On the other hand, our approach might fail to provide meaningful upper bounds for small values of p or when the circuits can be much smaller than the rank. Besides the *size* of the circuits, we suspect that the way in which the circuits overlap can affect the PoA. For example, circuits of k-uniform matroids are highly symmetric. When dealing with paving matroids, we observed that instances with highly symmetric circuits displayed a lower PoA. On the other hand, the paving matroid congestion game example in Sect. 4, whose PoA is larger than the worst-case PoA of uniform matroid congestion games, has circuits that more often overlap on a single resource. In conclusion, it is open to find lower and upper bounds of symmetric matroid congestion games that depend on the size of the matroid circuits and/or on their degree of symmetry.

Acknowledgments. Carla Michini gratefully acknowledges funding from DoD, Airforce, Award nb. FA9550-23-1-0487, which supported this work.

Disclosure of Interests. The authors have no competing interests to declare that are relevant to the content of this article.

References

1. Ackermann, H., Röglin, H., Vöcking, B.: On the impact of combinatorial structure on congestion games. J. ACM **55**(6), 25:1–25:22 (2008)
2. Aland, S., Dumrauf, D., Gairing, M., Monien, B., Schoppmann, F.: Exact price of anarchy for polynomial congestion games. In: Durand, B., Thomas, W. (eds.) STACS 2006. LNCS, vol. 3884, pp. 218–229. Springer, Heidelberg (2006). https://doi.org/10.1007/11672142_17
3. Anshelevich, E., Dasgupta, A., Kleinberg, J., Tardos, E., Wexler, T., Roughgarden, T.: The price of stability for network design with fair cost allocation. In: 45th Annual IEEE Symposium on Foundations of Computer Science, pp. 295–304 (2004)
4. Awerbuch, B., Azar, Y., Epstein, A.: The price of routing unsplittable flow. In: Proceedings of the Thirty-Seventh Annual ACM Symposium on Theory of Computing, STOC 2005, pp. 57–66. Association for Computing Machinery, New York (2005)
5. Awerbuch, B., Azar, Y., Epstein, A.: The price of routing unsplittable flow. SIAM J. Comput. **42**(1), 160–177 (2013)
6. Bhawalkar, K., Gairing, M., Roughgarden, T.: Weighted congestion games: price of anarchy, universal worst-case examples, and tightness. In: de Berg, M., Meyer, U. (eds.) ESA 2010. LNCS, vol. 6347, pp. 17–28. Springer, Heidelberg (2010). https://doi.org/10.1007/978-3-642-15781-3_2
7. Bilò, V., Vinci, C.: On the impact of singleton strategies in congestion games. In: Pruhs, K., Sohler, C. (eds.) 25th Annual European Symposium on Algorithms (ESA 2017). Leibniz International Proceedings in Informatics (LIPIcs), vol. 87, pp. 17:1–17:14. Schloss Dagstuhl–Leibniz-Zentrum fuer Informatik, Dagstuhl (2017)
8. Blackburn, J.E., Crapo, H.H., Higgs, D.A.: A catalogue of combinatorial geometries. Math. Comput. **27**(121), 155-s95 (1973)
9. Brualdi, R.A.: Comments on bases in dependence structures. Bull. Aust. Math. Soc. **1**(2), 161–167 (1969)
10. Caragiannis, I., Flammini, M., Kaklamanis, C., Kanellopoulos, P., Moscardelli, L.: Tight bounds for selfish and greedy load balancing. In: Bugliesi, M., Preneel, B., Sassone, V., Wegener, I. (eds.) ICALP 2006. LNCS, vol. 4051, pp. 311–322. Springer, Heidelberg (2006). https://doi.org/10.1007/11786986_28
11. Christodoulou, G., Koutsoupias, E.: The price of anarchy of finite congestion games. In: Proceedings of the Thirty-Seventh Annual ACM Symposium on Theory of Computing, STOC 2005, pp. 67–73. Association for Computing Machinery, New York (2005)
12. de Jong, J., Kern, W., Steenhuisen, B., Uetz, M.: The asymptotic price of anarchy for k-uniform congestion games. In: Solis-Oba, R., Fleischer, R. (eds.) WAOA 2017. LNCS, vol. 10787, pp. 317–328. Springer, Cham (2018). https://doi.org/10.1007/978-3-319-89441-6_23
13. Del Pia, A., Ferris, M., Michini, C.: Totally unimodular congestion games. In: Proceedings of the Twenty-Eighth Annual ACM-SIAM Symposium on Discrete Algorithms, pp. 577–588. SIAM (2017)

14. Fabrikant, A., Papadimitriou, C.H., Talwar, K.: The complexity of pure Nash equilibria. In: Proceedings of STOC 2004 (2004)
15. Fokkema, W.: The price of anarchy for matroid congestion games. In: ICALP 2023, Workshop on Congestion Games (2023)
16. Fotakis, D.: Congestion games with linearly independent paths: convergence time and price of anarchy. In: Monien, B., Schroeder, U.-P. (eds.) SAGT 2008. LNCS, vol. 4997, pp. 33–45. Springer, Heidelberg (2008). https://doi.org/10.1007/978-3-540-79309-0_5
17. Hao, B., Michini, C.: Inefficiency of pure nash equilibria in series-parallel network congestion games. In: Hansen, K.A., Liu, T.X., Malekian, A. (eds.) WINE 2022. LNCS, vol. 13778, pp. 3–20. Springer, Cham (2022). https://doi.org/10.1007/978-3-031-22832-2_1
18. Hao, B., Michini, C.: The price of anarchy in series-parallel network congestion games. Math. Program. 1–31 (2022)
19. Hao, B., Michini, C.: Price of anarchy in paving matroid congestion games. Optimization (2024)
20. Kleer, P., Schäfer, G.: Potential function minimizers of combinatorial congestion games: efficiency and computation. In: Proceedings of the 2017 ACM Conference on Economics and Computation, EC 2017, pp. 223–240. Association for Computing Machinery, New York (2017)
21. Koutsoupias, E., Papadimitriou, C.: Worst-case equilibria. In: Meinel, C., Tison, S. (eds.) STACS 1999. LNCS, vol. 1563, pp. 404–413. Springer, Heidelberg (1999). https://doi.org/10.1007/3-540-49116-3_38
22. Kraakman, Y.: The price of anarchy of symmetric and semi-symmetric uniform congestion games (2021)
23. Lücking, T., Mavronicolas, M., Monien, B., Rode, M.: A New Model for Selfish Routing. In: Diekert, V., Habib, M. (eds.) STACS 2004. LNCS, vol. 2996, pp. 547–558. Springer, Heidelberg (2004). https://doi.org/10.1007/978-3-540-24749-4_48
24. Lücking, T., Mavronicolas, M., Monien, B., Rode, M.: A new model for selfish routing. Theor. Comput. Sci. **406**(3), 187–206 (2008). Algorithmic Aspects of Global Computing
25. Mayhew, D., Newman, M., Welsh, D., Whittle, G.: On the asymptotic proportion of connected matroids. Eur. J. Combin. **32**(6), 882–890 (2011). Matroids, Polynomials and Enumeration
26. Mayhew, D., Royle, G.F.: Matroids with nine elements. J. Combin. Theory Ser. B **98**(2), 415–431 (2008)
27. Oxley, J.: Matroid Theory. Oxford University Press, Oxford (2011)
28. Pendavingh, R., van der Pol, J.: On the number of matroids compared to the number of sparse paving matroids. Electron. J. Combin. **22**(2) (2015)
29. Rosenthal, R.W.: A class of games possessing pure-strategy Nash equilibria. Internat. J. Game Theory **2**, 65–67 (1973)
30. Roughgarden, T.: The price of anarchy is independent of the network topology. J. Comput. Syst. Sci. **67**(2), 341–364 (2003). Special Issue on STOC 2002
31. Roughgarden, T., Tardos, E.: How bad is selfish routing? J. ACM **49**(2), 236–259 (2002)

Price of Anarchy for Graphic Matroid Congestion Games

Wouter Fokkema, Ruben Hoeksma, and Marc Uetz(✉)

Mathematics of Operations Research, University of Twente, Enschede, The Netherlands
{k.w.fokkema,r.p.hoeksma,m.uetz}@utwente.nl

Abstract. This paper analyzes the quality of pure-strategy Nash equilibria for symmetric Rosenthal congestion games with linear cost functions. For this class of games, the price of anarchy is known to be $(5N-2)/(2N+1)$, where N is the number of players. It has been open if restricting the strategy spaces of players to be bases of a matroid suffices to obtain stronger price of anarchy bounds. This paper answers this open question negatively. We consider graphic matroids, where each of the N players chooses a minimum cost spanning tree in a graph with linear cost functions on its edges. We provide constructions of graphs for $N = 2, 3, 4$ and for unbounded N, where the price of anarchy attains the known upper bounds $(5N-2)/(2N+1)$ and $5/2$, respectively. These constructions translate the tightness of algebraic constraints into combinatorial conditions which are necessary for tight lower bound instances. The main technical contribution lies in showing the existence of recursively defined graphs which fulfill these combinatorial conditions, and which are based on solutions of a bilinear Diophantine equation.

Keywords: Congestion Game · Minimum Spanning Tree · MST · Price of Anarchy · POA · Matroid

1 Introduction

Congestion games are strategic games where players select subsets of a finite set of resources, for example paths in a traffic network. Many players choosing the same resource may cause congestion, thereby increasing the players' costs. Studying resulting equilibria has a long history, specifically in the analysis of traffic networks. It is known since early works of Pigou [15] that equilibria of selfishly acting players may be inefficient with respect to total congestion.

The literature distinguishes non-atomic from atomic congestion games. The prime example of non-atomic congestion games are network routing games [18], where players need to route demand through a digraph in the form of a network flow. This is also known as the Wardrop traffic model [20], and a player, e.g., a car, may be thought of as an infinitesimally small part of the flow. For linear cost functions on the network edges, Roughgarden and Tardos [19] have shown

G. Schäfer and C. Ventre (Eds.): SAGT 2024, LNCS 15156, pp. 371–388, 2024.
https://doi.org/10.1007/978-3-031-71033-9_21

that for this model the total cost of an equilibrium, also known as Beckmann user equilibrium [3], can exceed the cost of an optimal solution minimizing total cost by at most a factor 4/3. This bound is tight. In other words, the price of anarchy [13] of linear, non-atomic network routing games equals 4/3. It is quite remarkable that this worst case bound is attained by routing one unit of flow on a very simple instance with only two vertices and two parallel edges, referred to as "Pigou example". And even for cost functions other than linear the worst case is attained on such simple Pigou-type instances [17]. That implies that for non-atomic congestion games, the combinatorial structure of the strategy spaces of the players has no meaningful impact on the price of anarchy.

The situation is different for the "combinatorial" version of congestion games, atomic congestion games. These were first studied by Rosenthal [16]. In the network routing connotation, it means that there is a finite number of players, N, each choosing a single path in the network. For any one player, a subset of edges is admissible for that player if it is a path from that player's origin to destination. More generally, there is a finite set of resources, and each player's strategy is to choose one subset of a collection of player-specific admissible subsets. The resulting game is a finite strategic form game, and the corresponding equilibrium concept is Nash equilibrium. In that setting, it is known that a pure-strategy Nash equilibrium always exists [16]. If the set of admissible subsets differ per player, the game is asymmetric, and otherwise symmetric.

The state of the art with respect to the price of anarchy for pure-strategy Nash equilibria is as follows. For asymmetric atomic congestion games with linear cost functions, the price of anarchy equals 2 for $N = 2$ players, and it equals 5/2 when $N \geq 3$ [2,5]. The price of anarchy remains 5/2 even if the players' strategy spaces are restricted to choosing just a single resource from two admissible resources [4]. For the symmetric case, the upper bound improves from 5/2 to $(5N - 2)/(2N + 1)$ [5]. This upper bound is attained also for the special case of symmetric network routing games [6].

The main motivation for this paper is to better understand what impact the combinatorial structure of players' strategy spaces has on the price of anarchy of symmetric and atomic congestion games. Arguably the most natural case to consider is that each player must select a subset that is a basis of a matroid. One specific example is the graphic matroid, where given a graph, each player needs to select a spanning tree in this graph. Generally speaking, matroid congestion games have favorable properties. E.g., when it comes to computation it is known that arbitrary best response sequences converge to pure-strategy Nash equilibria in polynomial time [1], while computing a pure-strategy Nash equilibrium is generally PLS hard already for network routing games [9]. With respect to the quality of equilibria in symmetric and atomic congestion games, and in contrast to non-atomic games, the combinatorial structure of strategy spaces does matter:

- Lücking et al. [14] and Fotakis [10] show that the price of anarchy drops to 4/3 when players select singleton resources. This corresponds to the trivial matroid of rank 1.

- Klimm, de Jong and Uetz [8] show that the price of anarchy is strictly larger than $4/3$ but at most $28/13 \approx 2.15$ for k-uniform matroid congestion games. This upper bound was subsequently improved to ≈ 1.4131 [7].
- Hao and Michini [12] show an upper bound of 2 for the special case of symmetric network routing games in series-parallel networks.
- Recently, Hao and Michini [11] proved an upper bound of $17/7 \approx 2.43$ for the case of a paving matroid.

It is has been open if the general assumption of matroid strategy spaces alone suffices to improve upon the known upper bound $(5N-2)/(2N+1)$. Even the case of a graphical matroid has remained open so far. In this paper we give a negative answer to this question, thereby settling the open question if general matroid congestion games allow improved price of anarchy bounds. Specifically, the paper gives constructions of graphs with corresponding linear cost functions on its edges when the number of players N equals $2, 3$, or 4, and for $N \to \infty$. For all cases, we show that the price of anarchy asymptotically reaches the upper bound $(5N-2)/(2N+1)$, respectively $5/2$.

Our main contribution lies in a systematic approach to design matching lower bound constructions. Indeed, while the construction for $N = 2$ players, achieving a price of anarchy 1.6 is still fairly simple and can be obtained more or less ad-hoc, our analysis provides the necessary insights into why this instance does the trick. Based on these insights, we tackle the more challenging cases with more than two players. These require more complex, recursive constructions of graphs, which differ per number of players. In a nutshell, the idea behind our construction is by translating the tightness of the algebraic constraints in the proof to obtain the upper bound $(5N-2)/(2N+1)$, into combinatorial conditions that the corresponding lower bound instances have to fulfill. The desired properties of such instances lead to a bilinear Diophantine equation. This equation has a solution for $N \in \{3, 4\}$, and in an asymptotic sense also for $N \to \infty$. Based on this, one still needs to show that the corresponding graphs can be constructed and admit solutions which correspond to the desired optimal and equilibrium solutions, respectively. We show that this is indeed possible.

2 Graphic Matroid Congestion Game

We consider symmetric and atomic congestion games with N players and linear cost functions on the set of resources. With slight abuse of notation, we also sometimes use N to denote the set of players. In matroid congestion games, the strategy space of each player is restricted to the bases of a matroid defined on the set of resources. The game is symmetric if that matroid is the same for all players. Specifically, in symmetric graphic matroid congestion games the resources are edges in a graph $G(V, E)$ and each player must choose a spanning tree in G. A solution is then denoted by a strategy profile $T = (T_1, T_2, \ldots, T_N)$, where T_i denotes the spanning tree that is chosen by player i. As we consider linear cost functions, each edge $e \in E$ has an associated cost or weight $w_e \in \mathbb{R}_{\geq 0}$. For given edge $e \in E$, denote by $n_e(T) = |\{i \in N \mid e \in T_i\}|$ the number of players that choose a spanning tree containing edge e. So if edge e is used by $n_e(T)$ players, each of these players incurs a cost of $w_e \cdot n_e(T)$ for this edge. The cost of player i is then given by

$$c_i(T) := \sum_{e \in T_i} w_e \cdot n_e(T),$$

and the total or social cost $c(T)$ is given by summing c_i over all players:

$$c(T) := \sum_{e \in E} w_e \cdot n_e^2(T).$$

We assume each player aims to minimize their cost, and are interested in the price of anarchy for pure-strategy Nash equilibria, measuring the relative loss of efficiency caused by selfish behaviour [13]. Using the notation $T = (T_i, T_{-i})$ for the strategy profile where player i chooses strategy T_i, and the remaining players T_{-i}, a solution T is a pure-strategy Nash equilibrium if for all players i, and all spanning trees T_i' that player i could choose, $c_i(T) \leq c_i(T_i', T_{-i})$.

For an instance I of the game, the (pure) price of anarchy $\text{PoA}(I)$ is given by largest cost of a pure-strategy Nash equilibrium $\text{NE}(I)$ relative to the cost of a solution $\text{OPT}(I)$ minimizing the total cost $c(T)$,

$$\text{PoA}(I) = \max \left\{ \frac{c(\text{NE}(I))}{c(\text{OPT}(I))} \,\middle|\, \text{NE}(I) \text{ is a Nash equilibrium for } I \right\},$$

and for a collection of instances $\mathcal{I}$, $\text{PoA}(\mathcal{I}) = \sup_{I \in \mathcal{I}} \text{PoA}(I)$. For the remainder of the paper, the instances will be graphic matroid congestion games with a given number of players N, hence we omit the dependence on I and $\mathcal{I}$. Moreover, Nash equilibrium and price of anarchy is always to be read as Nash equilibrium and price of anarchy in pure strategies. A Nash equilibrium solution is then denoted NE, and a solution that (approximately) minimizes total cost $c(T)$ is denoted OPT. Instance 1 provides a simple example.

Instance 1. *Consider the graph $G(V, E)$ with five vertices and eight edges as depicted in Fig. 1. All edges have unit cost.* △

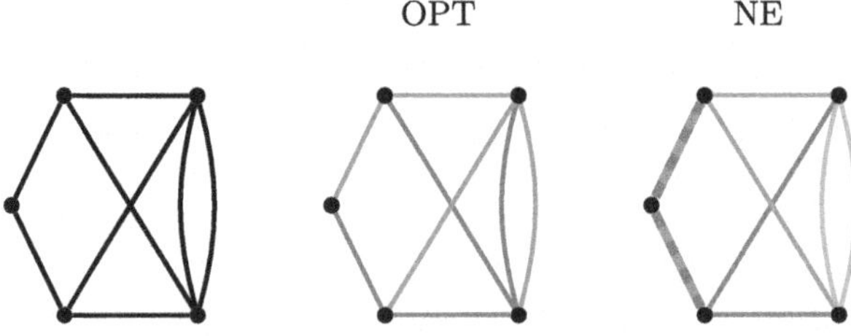

Fig. 1. Instance for $N = 2$ players, optimal solution OPT, and a Nash equilibrium NE. Blue edges are chosen by the blue player, vice versa for orange. Multicolored edges are chosen by both players, and light grey edges are not chosen by any player. (Color figure online)

In Instance 1, both players have a cost of 4 in the optimal solution OPT, which means that the total cost is 8. In the Nash equilibrium NE, both players have a cost of 6: the cost is 2 for the two edges which are used by both players. This gives a total cost of 12. Therefore, the PoA for $N = 2$ players is at least $12/8 = 1.5$.

3 Recap: Proof of the $(5N - 2)/(2N + 1)$ Bound

For linear and symmetric congestion games in general, Christodoulou and Koutsoupias proved that the price of anarchy for pure-strategy Nash equilibria is at most $(5N - 2)/(2N + 1)$ [5]. Because this proof lies at the core of our constructions to obtain matching lower bounds, we repeat it here.

Theorem 1 [5]. *The price of anarchy of linear and symmetric congestion games is at most* $(5N - 2)/(2N + 1)$.

Proof. Let $\mathrm{NE} = (\mathrm{NE}_1, \mathrm{NE}_2, \ldots, \mathrm{NE}_N)$ be a Nash equilibrium and let $\mathrm{OPT} = (\mathrm{OPT}_1, \mathrm{OPT}_2, \ldots, \mathrm{OPT}_N)$ be a solution that minimizes the total cost. In the Nash equilibrium, the players cannot decrease their cost by switching to another strategy. In particular, player i cannot decrease their cost by switching to OPT_j:

$$c_i(\mathrm{NE}) \leq c_i(\mathrm{OPT}_j, \mathrm{NE}_{-i}) = \sum_{e \in \mathrm{OPT}_j} w_e\,(n_e(\mathrm{NE}) + 1) - \sum_{e \in \mathrm{OPT}_j \cap \mathrm{NE}_i} w_e,$$

for all $i, j \in N$. Here, recall that $(\mathrm{OPT}_j, \mathrm{NE}_{-i})$ denotes the strategy profile NE where NE_i is replaced by OPT_j. We sum these inequalities over j to obtain

$$N \cdot c_i(\mathrm{NE}) \leq \sum_{e \in E} w_e(n_e(\mathrm{NE}) + 1) n_e(\mathrm{OPT}) - \sum_{e \in \mathrm{NE}_i} w_e n_e(\mathrm{OPT}), \qquad (1)$$

for all $i \in N$. Then we sum over i, obtaining

$$\begin{aligned} N \cdot c(\mathrm{NE}) &\leq N \sum_{e \in E} w_e(n_e(\mathrm{NE}) + 1) n_e(\mathrm{OPT}) - \sum_{e \in E} w_e n_e(\mathrm{NE}) n_e(\mathrm{OPT}) \\ &= (N - 1) \sum_{e \in E} w_e(n_e(\mathrm{NE}) + 1) n_e(\mathrm{OPT}) + \sum_{e \in E} w_e n_e(\mathrm{OPT}). \end{aligned}$$

Therefore,

$$c(\mathrm{NE}) \leq \frac{N-1}{N} \sum_{e \in E} w_e(n_e(\mathrm{NE}) + 1) n_e(\mathrm{OPT}) + \frac{1}{N} \sum_{e \in E} w_e n_e(\mathrm{OPT}). \qquad (2)$$

Now we use the fact that, for non-negative integers α and β, we have

$$(\alpha + 1)\beta \leq \frac{1}{3}\alpha^2 + \frac{5}{3}\beta^2, \qquad (3)$$

which we can use in (2) with $\alpha = n_e(\mathrm{NE})$ and $\beta = n_e(\mathrm{OPT})$. This gives the following upper bound on $c(\mathrm{NE})$

$$\frac{N-1}{3N}\sum_{e\in E} w_e n_e^2(\mathrm{NE}) + \frac{5(N-1)}{3N}\sum_{e\in E} w_e n_e^2(\mathrm{OPT}) + \frac{1}{N}\sum_{e\in E} w_e n_e(\mathrm{OPT}).$$

Finally, we bound $n_e(\mathrm{OPT})$ by $n_e^2(\mathrm{OPT})$, and substitute the summations by $c(\mathrm{NE})$ and $c(\mathrm{OPT})$, to obtain

$$c(\mathrm{NE}) \leq \frac{N-1}{3N}c(\mathrm{NE}) + \frac{5N-2}{3N}c(\mathrm{OPT}). \tag{4}$$

Now rearranging the terms yields the desired bound for the PoA. □

There are three steps where a strict inequality can occur. The first place is the Nash equilibrium constraints, which are aggregated in (1). The second is where we use inequality (3). Finally, where we bound $n_e(\mathrm{OPT})$ by $n_e^2(\mathrm{OPT})$ to obtain (4). This allows creating an expression for the price of anarchy where the difference with the desired upper bound is explained by the slacks in these inequalities. First, we define $s_1, s_2, s_3 \geq 0$, by the following expressions:

$$\begin{aligned} s_{ij} :&= c_i(\mathrm{OPT}_j, \mathrm{NE}_{-i}) - c_i(\mathrm{NE}) \\ &= \sum_{e\in \mathrm{OPT}_j} w_e\left(n_e(\mathrm{NE})+1\right) - \sum_{e\in \mathrm{OPT}_j\cap \mathrm{NE}_i} w_e - c_i(\mathrm{NE}), \end{aligned} \tag{5a}$$

$$s_1 := \frac{3}{2N+1}\sum_{i=1}^{N}\sum_{j=1}^{N} s_{ij}, \tag{5b}$$

$$s_2 := \frac{N-1}{2N+1}\sum_{e\in E} w_e\left(n_e^2(\mathrm{NE}) + 5n_e^2(\mathrm{OPT}) - 3n_e(\mathrm{OPT})(n_e(\mathrm{NE})+1)\right), \tag{5c}$$

$$s_3 := \frac{3}{2N+1}\sum_{e\in E} w_e\left(n_e^2(\mathrm{OPT}) - n_e(\mathrm{OPT})\right). \tag{5d}$$

Then it follows from the above upper bound proof that we can express the price of anarchy as follows.

Lemma 1. *The price of anarchy for linear and symmetric congestion games is given by*

$$\mathrm{PoA} = \frac{c(\mathrm{NE})}{c(\mathrm{OPT})} = \frac{5N-2}{2N+1} - \frac{s_1+s_2+s_3}{c(\mathrm{OPT})}.$$

To simplify the above expressions, we introduce the following notation.

Definition 1. *For $0 \leq i, j \leq N$, an (i,j)-*edge *is an edge that is used i times in* OPT *and j times in* NE. *Define the total cost of all (i,j)-edges by*

$$W_{ij} := \sum_{\{e\in E \mid n_e(\mathrm{OPT})=i,\, n_e(\mathrm{NE})=j\}} w_e. \tag{6}$$

Then we can rewrite s_2 and s_3:

$$s_2 = \frac{N-1}{2N+1} \sum_{0 \le i,j \le N} W_{ij} \left(j^2 + 5i^2 - 3i(j+1)\right) \ge 0, \tag{7a}$$

$$s_3 = \frac{3}{2N+1} \sum_{0 \le i,j \le N} W_{ij} \left(i(i-1)\right) \ge 0. \tag{7b}$$

Finally, define the penalties $P_{\text{switch}} := s_1$ and $P_{\text{edges}} = s_2 + s_3$, then we rewrite Lemma 1 to get:

Theorem 2. *For linear graphic matroid congestion games, the price of anarchy can be expressed as*

$$\text{PoA} = \frac{5N-2}{2N+1} - \frac{P_{\text{switch}} + P_{\text{edges}}}{c(\text{OPT})}. \tag{8}$$

The reason to define P_{switch} and P_{edges} is to give them a natural interpretation:

- P_{switch} denotes the penalty that is caused by the cost increase when players switch from their Nash equilibrium strategy NE to a spanning tree in OPT.
- P_{edges} denotes the penalty that is caused by having non-negligible cost in (i, j)-edges. Note that only $(0, 0)$-edges, $(1, 1)$-edges, $(1, 2)$-edges and edges with zero cost do *not* contribute to this penalty.

Penalty P_{edges} can be counter intuitive, because if many players use an edge in the Nash equilibrium, one might expect this edge to contribute to a high price of anarchy. However, the above tells us that in a tight lower bound instance, the cost of edges used by more than two players in the Nash equilibrium should be equal to zero, or should at least be negligible with respect to $c(\text{OPT})$.

4 Lower Bound Constructions

In this section we construct lower bounds for the price of anarchy of graphic matroid congestion games. When proving a lower bound, we need to provide

- a graph $G(V, E)$, with edge costs w_e,
- a solution OPT, which is an upper bound for the optimal solution,
- a solution NE, which is a pure-strategy Nash equilibrium.

When all conditions are satisfied, $c(\text{NE})/c(\text{OPT})$ is a lower bound for the price of anarchy. Alternatively, we can compute P_{switch} and P_{edges} and use Theorem 2.

To check whether a solution NE is indeed a Nash equilibrium, in principle we need to check for all players that there is no other spanning tree they can choose to give them a lower cost. However, we can simplify this by the following consequence from the strong exchange property for matroids; for the simple proof see the full version of the paper.

Lemma 2. *If no player can reduce their costs in* NE *by exchanging a single edge from their spanning tree to obtain another spanning tree, then* NE *is a Nash equilibrium.*

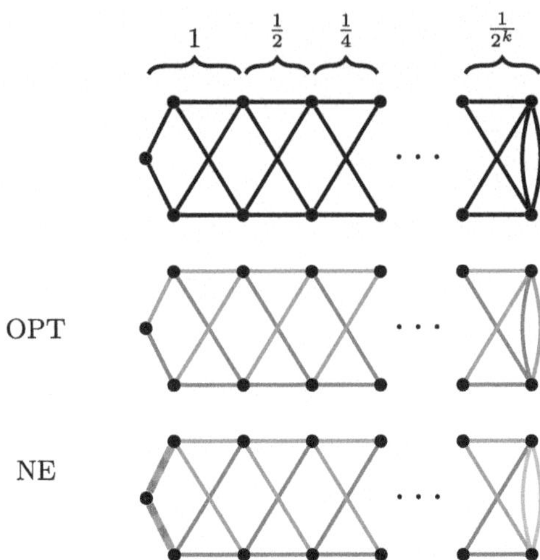

Fig. 2. Tight lower bound instance for $N = 2$ players. From top to bottom, $G(V, E)$ with the edge costs indicated above, OPT and NE. The two players are orange and blue, unused edges are light grey, and multicolored edges are used by both players. (Color figure online)

4.1 Tight Instance for 2 Players

We next give a simple construction showing that $\text{PoA} = (5N-2)/(2N+1) = 1.6$ for $N = 2$ players. The basic idea, which is the path the we take for the rest of this paper, is to construct instances where penalties P_{switch} and P_{edges} are either zero, or negligible asymptotically.

Instance 2 (Fence). *Consider the 2-player instance in Fig. 2, consisting of what we will also refer to as a* fence *construction, consisting of k times a $K_{2,2}$ "glued together", with (from the left) exponentially decreasing cost coefficients $1, 1/2, \ldots, 1/2^k$. The fence ends on the right with two parallel edges of cost $1/2^k$ each. The two left vertices of the leftmost $K_{2,2}$ are called "visible" for later reference. The fence is augmented with one additional vertex on the left, which is connected to the two visible vertices using two edges of cost 1.* △

Theorem 3. *The price of anarchy for linear graphic matroid congestion games with $N = 2$ players equals $(5N - 2)/(2N + 1) = 1.6$.*

Proof. Using Lemma 2, it is easy to verify that NE is indeed a Nash equilibrium, because no single edge exchange allows any one of the two players to improve. We compute the price of anarchy as 1.6 by taking $k \to \infty$ because:

$$c(\text{OPT}) = 2 + 4\sum_{i=0}^{k} \frac{1}{2^i} + \frac{2}{2^k} = 10 - \frac{2}{2^k},$$

$$c(\text{NE}) = 8 + 4\sum_{i=0}^{k} \frac{1}{2^i} = 16 - \frac{4}{2^k}.$$

□

Because the instance matches the upper bound for $k \to \infty$, we should have $P_{\text{switch}} = 0$ and $P_{\text{edges}} = 0$, at least asymptotically. To check that $P_{\text{edges}} = 0$, we note that we only have $(1,1)$-edges, $(1,2)$-edges and $(1,0)$-edges. Of these, only the two $(1,0)$-edges give a non-zero contribution to P_{edges}. However the combined cost of these two edges is $W_{10} = 1/2^{k-1} \to 0$ for $k \to \infty$, while $c(\text{OPT}) \to 10$. To see that indeed $P_{\text{switch}} = 0$, observe that the total cost of a player in NE equals $6 + 2(\sum_{i=1}^{k} 1/2^k) = 8 - 2/2^k$. One calculates the cost of switching to any of the two trees in OPT as $5 + 3(\sum_{i=1}^{k} 1/2^k) + 1/2^k = 8 - 2/2^k$.

4.2 Constructions for Three or More Players

The above instance for $N = 2$ players matches the upper bound 1.6 asymptotically. One may wonder if there exists a finite instance that does the job. This is not the case for any number of players N; the simple proof can be found in the full version of the paper.

Theorem 4. *There exists no finite graph for which a linear congestion game with N players has a price of anarchy that matches the value $(5N-2)/(2N+1)$.*

First Attempt for a Lower Bound. One can generalize the idea of Instance 2 to $N \geq 3$ players. However, this does not yet match the upper bound.

Instance 3 (N-fence). *Consider the N-player instances as illustrated in Fig. 3 for $N = 4$ players. It is the obvious generalization of the construction of Instance 2, only replacing $K_{2,2}$ by $K_{N,N}$, and keeping the same exponentially decreasing edge costs per $K_{N,N}$, namely $1, 1/2, \dots 1/2^k$. Define the costs of the edges that connect the additional vertex on the left to the N "visible" vertices of the N-fence as $2/N$.* △

Observe that in NE, each of the N edges with cost $2/N$ is used by all players, giving a total cost of 2 per player and per edge. One readily verifies that this is indeed a Nash equilibrium by considering possible edge exchanges. In OPT, depending on the value of N, it can be cheaper for multiple players to use these leftmost edges, rather than one player as illustrated in Fig. 3. Players then switch some edges of cost 1 for edges with cost $2/N$, as long as this is cheaper. If we assume that each edge with cost $2/N$ is used by q players in OPT, where $q \in \mathbb{N}$, there are $N(q-1)$ less of the N^2 edges with cost 1 being used, and we get the following costs:

$$c(\text{NE}) = 2N^2 + \sum_{i=0}^{k} \frac{N^2}{2^i} = \left(4 - \frac{1}{2^k}\right) N^2,$$

$$c(\text{OPT}) = \min_{q \in \mathbb{N}} \left[2q^2 - (q-1)N + \sum_{i=0}^{k} \frac{N^2}{2^i}\right] + \frac{N(N-1)}{2^k}$$

$$= \min_{q \in \mathbb{N}} \left[2\left(q - \frac{N}{4}\right)^2 + N\right] + \frac{15}{8}N^2 - \frac{N}{2^k}.$$

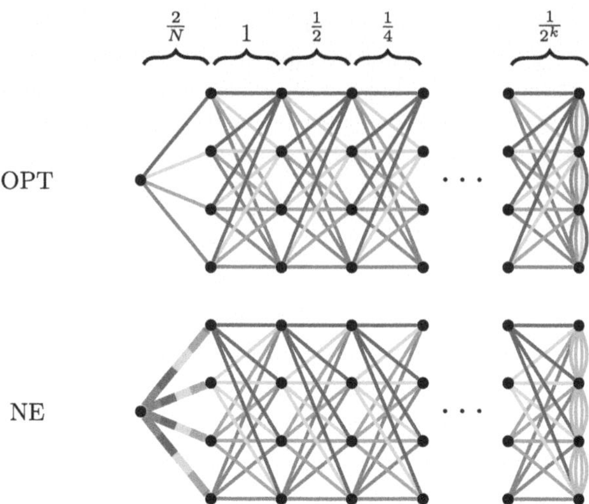

Fig. 3. Instance 3 for $N = 4$ players. At the top is (an approximate) OPT, and at the bottom is NE. The players are depicted by the colors red, orange, yellow and blue. Unused edges are colored light grey. Multicolored edges are used by all four players. (Color figure online)

To minimize for q, we note that the expression in brackets attains its minimum at $q = N/4$, so the optimal integer value for q is the integer nearest to $N/4$, which gives $q = \lfloor \frac{N+2}{4} \rfloor$.

Letting $k \geq N$ to make the negligible terms vanish, and then $N \to \infty$, one calculates the price of anarchy being $4/(15/8) = 32/15 \approx 2.13$. This is less than the desired upper bound 2.5. It turns out that the reason for the gap to the upper bound lies in the "heavily used" leftmost edges (well, together with roughly $N^2/4$ unused edges in the leftmost $K_{N,N}$ of the N-fence). Together these cause $P_{\text{edges}} \in \Theta(N^2)$, while $c(\text{OPT}) = \Theta(N^2)$, too. More specifically, the leftmost edges are used by N players in NE and q players in OPT, so that we have $W_{qN} = N \cdot 2/N = 2$. Moreover, because in OPT each player uses only $(3/4)N^2$ edges of the leftmost $K_{N,N}$, we also get that $W_{01} = N^2/4 + N(N-1)/2^k$. That said, we can compute the penalty term P_{edges}. One can verify that $P_{\text{edges}} = (9/16)N^2 + (1/8)N^2 + \text{o}(N^2) = (11/16)N^2 + \text{o}(N^2)$, while $c(\text{OPT}) = (15/8)N^2 + \text{o}(N^2)$. This exactly explains the gap between the price of anarchy 32/15 and the desired upper bound 5/2, which is equal to 11/30. In turn this implies that P_{switch} is negligible, which can also be verified manually.

Avoiding Heavily Used Edges. In an attempt to improve upon the above instance, one can avoid the penalty P_{edges} by replacing the leftmost $(1, N)$ edges of Instance 3: One simply replaces the singleton vertex on the left by a complete bipartite graph $K_{N-1,N}$, with edge costs all 1. This creates an instance with edges that are used exactly twice in NE and only once in OPT. This way, we get

$P_{\text{edges}} = 0$ asymptotically, yet at the expense of another issue, namely a non-negligible penalty P_{switch}. The resulting instance yields an even smaller lower bound in comparison to Instance 3, however.

4.3 Tight Instances for 3 or 4 Players

The idea to deal with the problem of non-vanishing penalty costs is to make their relative contribution negligible. To achieve that, we construct an instance recursively in "layers", where from layer to layer the edge costs decrease by a factor 3/2. After ℓ recursive layers, we continue with the standard N-fence construction as in the previous Instance 3 (and without the leftmost vertex). To simplify the corresponding algebraic expressions, these N-fences will from now on be assumed to be infinite, that is, in the limit $k \to \infty$, so that the contribution of such an N-fence to the total cost equal both in NE and in OPT. Indeed, the cost of the selected edges equals $\lim_{k\to\infty} \sum_{i=0}^{k} N^2/2^i = 2N^2$ in NE and $\lim_{k\to\infty}[N(N-1)/2^k + \sum_{i=0}^{k}(N^2/2^i)] = 2N^2$ in OPT. Note that in NE the visible vertices remain disconnected, while in OPT they are connected.

Instance 4. *First, define gadget B_0 to be the N-fence construction as in Instance 3, consisting of k times $K_{N,N}$ and with the same exponentially decreasing edge costs $1, 1/2^1, 1/2^2, \ldots,$ but without the leftmost vertex. Recursively define a gadget B_i by introducing N vertices that we again call "visible" vertices of gadget B_i, plus q copies of gadget B_{i-1}. That means that gadget B_i consists of $N + qN$ many vertices, the N visible vertices of gadget B_i, the qN visible vertices in the q copies of gadget B_{i-1}, and the edges among them. The edges will be discussed separately, keeping in mind that we intend to have only $(1, 2)$ edges so that penalty $P_{edges} = 0$. Define ℓ such recursive layers with gadgets B_ℓ, $B_{\ell-1}$, $\ldots B_0$. Note that this implies that we have $q^{\ell-i}$ many gadgets B_i. To finish the construction we connect a single vertex to all N visible vertices of gadget B_ℓ, using $(1, N)$-edges, just as in Instance 3. For convenience, this final leftmost vertex and its incident edges are called gadget $B_{\ell+1}$. Also define $\mathcal{B}_i$ as the union of B_i and recursively all gadgets B_{i-1}, B_{i-2}, $\ldots B_0$ contained in it. Figure 4 illustrates the idea behind the recursive construction for $N = 4$ and $q = 3$.* △

The exact definition of the edges inside each gadget B_i depends on the values q and N. There is some desiderata, however, to make the overall construction do the trick. For the case $N = 4$ and $q = 3$, the following items can best be inspected in Fig. 4.

(i) The intention is that every edge of every gadget B_i, $i \geq 1$, is used exactly once in OPT and exactly twice in NE.
(ii) In OPT, the N players select N disjoint spanning trees, such that, for all $i \geq 0$, each of these trees restricted to $\mathcal{B}_i$ is still a spanning tree for $\mathcal{B}_i$.
(iii) In NE, each player selects a spanning tree such that, for all $0 \leq i \leq \ell$, the subtree restricted to $\mathcal{B}_i$ consists of N connected components, each containing exactly one of the visible vertices of gadget B_i.

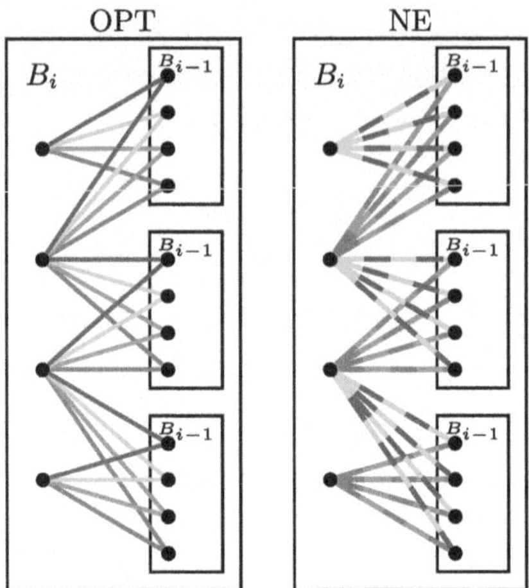

Fig. 4. Recursive building block for Instance 4, $N = 4$. On the left is OPT, and on the right NE. Gadget B_i contains $N = 4$ visible vertices on the left, $q = 3$ copies of B_{i-1} with qN visible vertices of gadgets B_{i-1} on the right, and $N(N + q - 1) = 24$ edges. Multicolored edges are used by two players.

(iv) In NE, for each player and each $1 \leq i \leq \ell$, within every gadget B_i, for each of its B_{i-1}-gadgets (q many), the N visible vertices are connected using edges from gadget B_i. Likewise, within gadget $B_{\ell+1}$ the N visible vertices of B_ℓ are connected.

Note that (ii) and (iii) are indeed also fulfilled for OPT and NE in B_0, as in Instance 3. The following lemma confirms that such a construction is indeed possible for the values $N = 3, 4$.

Lemma 3. *For $N = 3$ and $N = 4$ players, there exist graph constructions so that all above requirements (i)–(iv) can be fulfilled.*

Proof. First, the requirements imply the following conditions as to the number of edges of each gadget B_i.

In OPT: By (ii), inductively, the visible vertices of each gadget B_{i-1} can be assumed to be connected in $\mathcal{B}_{i-1}$. Hence to get (ii) for gadget B_i, it suffices to connect the N visible vertices of B_i with the q disconnected copies B_{i-1} using a spanning tree on $N + q$ vertices. This requires $N + q - 1$ edges per player. In Fig. 4, this is a path alternating between visible vertices of B_i and one vertex per copy of B_{i-1}. Since by (i) each edge is to be used by only one player in OPT, we therefore must have $N(N + q - 1)$ edges in layer B_i.

In NE: By (iii), the visible vertices in B_i remain disconnected in $\mathcal{B}_i$. At the same time, by (iv), for each player the N visible vertices of each B_{i-1} need to be connected within B_i. Together with (iii) that means that for each player the set of edges of B_i must connect $N + qN$ components to yield N components, which requires qN edges per player. In Fig. 4, this is the N-stars that connect the N

vertices of B_{i-1} with one vertex of B_i. Since by (i) the intention is that each edge shall be used twice in NE, this implies that we must have $qN^2/2$ edges in total in gadget B_i.

As a consequence, as long as we demand (i), that every edge of B_i is used exactly once in OPT and exactly twice in NE, this is possible only if $N(N+q-1)$ equals $qN^2/2$, so only if we find an integer solution to the bilinear Diophantine equation

$$qN = 2(N + q - 1). \tag{9}$$

This equation is fulfilled for $(q, N) = (3, 4)$. The existence of the construction fulfilling (i)–(iv) is testified by Fig. 4. The other integer solution is $(q, N) = (4, 3)$, and the existence of the construction is testified by Fig. 5. □

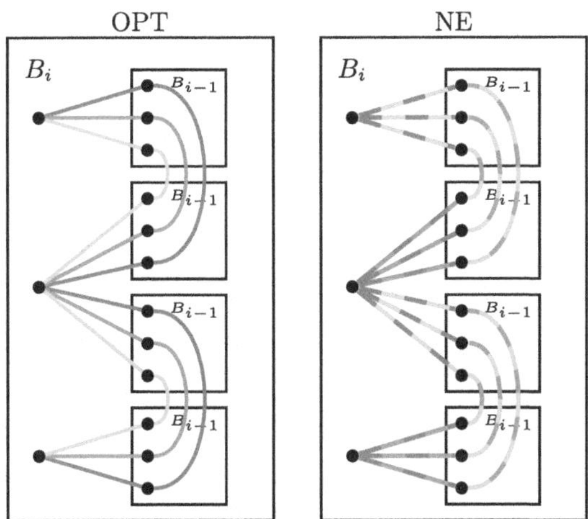

Fig. 5. Recursive building block for Instance 4 for $N = 3$ players. On the left is OPT, and on the right is NE. Gadget B_i contains $N = 3$ "visible" vertices on the left and $q = 4$ copies of B_{i-1} on the right. Multicolored edges are used by two players.

Next to the above solutions for $N = 3, 4$, there is a solution also for $N \to \infty$ while taking $q = 2$ which requires some additional adaptations because we cannot fulfill (i); we discuss it separately in Sect. 4.4.

In order to finish the description of the constructions for $N = 3, 4$, we still need to formally define edge costs. In gadgets B_i, $i \geq 1$, edge costs are defined to be equal to $(3/2)^{i-1}$. The cost of edges that connect the leftmost vertex to the visible vertices of B_ℓ are defined $3(3/2)^{\ell-1}/N$. Note that $(1, N)$ edges are in fact not optimal, meaning that the solution OPT where all leftmost edges are used by only one player in OPT is suboptimal, just as before in Instance 3. But taking the limit $\ell \to \infty$, the effect of a suboptimal OPT is negligible.

Lemma 4. *For the earlier described instances with $N = 3, 4$ players, the above described strategy profile* NE *constitutes a Nash equilibrium.*

Proof. We use Lemma 2. For the edges inside any of the N-fences B_0, we have already argued in Instance 3. So consider an edge in gadget B_i, $i \geq 1$. Players experience an edge cost equal to $2 \cdot (3/2)^{i-1}$. Clearly, edge exchanges within B_i, if feasible, cannot reduce the costs because all edges are equally loaded and have the same costs. Next, exchanging a B_i-edge with a B_{i-j}-edge with $j \geq 2$ is not eligible, because it creates a cycle by Property (iv). Now if a player were to exchange a B_1 edge (in NE used by 2 players) for an eligible B_0 edge that again yields a spanning tree (in NE used by 1 player), the switching cost is 0, as both edges have cost 1. If for $2 \leq i \leq \ell$ a player were to exchange a B_i-edge (in NE used by 2 players) for some B_{i-1}-edge (in NE used by 2 players), the switching cost is also 0, as the new edge costs $3 \cdot (3/2)^{i-2} = 2 \cdot (3/2)^{i-1}$. Finally, exchanging a $B_{\ell+1}$ edge (in NE used by N players) for some B_ℓ edge, the switching cost is again 0 as the new edge costs $3(3/2)^{\ell-1}$. □

Theorem 5. *The price of anarchy for linear graphic matroid congestion games with $N = 3$ players equals $(5N-2)/(2N+1) = 13/7$, and for $N = 4$ players it equals $(5N-2)/(2N+1) = 2$.*

Proof. The existence of instances for $N = 3$ and $N = 4$ players with the desired combinatorial properties of Nash equilibrium NE and solution OPT follows from Lemma 3, and by Lemma 4. As to the costs, recall that there are $q^{\ell-i}$ copies of gadget B_i, and that each gadget B_0 costs $2N^2$ both in OPT and NE. Moreover since each B_i has $N(N+q-1) = 6N$ edges for both $(q, N) = (3, 4)$ and for $(q, N) = (4, 3)$, we compute the total costs in NE and OPT as follows.

$$c(\mathrm{NE}) = 3\left(\frac{3}{2}\right)^{\ell-1} N^2 + \sum_{i=1}^{\ell}\left[q^{\ell-i}\left(\frac{3}{2}\right)^{i-1} \cdot 24\,N\right] + q^\ell \cdot 2N^2, \tag{10}$$

$$c(\mathrm{OPT}) = 3\left(\frac{3}{2}\right)^{\ell-1} + \sum_{i=1}^{\ell}\left[q^{\ell-i}\left(\frac{3}{2}\right)^{i-1} \cdot 6\,N\right] + q^\ell \cdot 2N^2. \tag{11}$$

Now for $(q, N) = (4, 3)$, we take out 4^ℓ from (10) and (11), and get:

$$c(\mathrm{NE}) = 4^\ell\left[18\left(\frac{3}{8}\right)^{\ell} + \frac{1}{4}\left(\frac{8}{5} - \sum_{i=\ell}^{\infty}\left(\frac{3}{8}\right)^{i}\right) 72 + 18\right],$$

$$c(\mathrm{OPT}) = 4^\ell\left[2\left(\frac{3}{8}\right)^{\ell} + \frac{1}{4}\left(\frac{8}{5} - \sum_{i=\ell}^{\infty}\left(\frac{3}{8}\right)^{i}\right) 18 + 18\right].$$

Then $\ell \to \infty$ yields $c(\mathrm{NE})/c(\mathrm{OPT}) \to ((2/5)72+18)/((2/5)18+18) = 13/7$. For $(q, N) = (3, 4)$, we take out 3^ℓ from (10) and (11), and get:

$$c(\mathrm{NE}) = 3^\ell\left[\frac{16}{2^{\ell-1}} + \left(2 - \frac{1}{2^{\ell-1}}\right) 32 + 32\right],$$

$$c(\mathrm{OPT}) = 3^\ell\left[\frac{1}{2^{\ell-1}} + \left(2 - \frac{1}{2^{\ell-1}}\right) 8 + 32\right].$$

Letting $\ell \to \infty$ yields that $c(\mathrm{NE})/c(\mathrm{OPT}) \to 96/48 = 2$. □

4.4 Tight Instance for Unconstrained N

We finally give a construction that, asymptotically, yields a lower bound 5/2 for the price of anarchy of linear matroid congestion games.

Instance 5. *The instance follows the same recursive design principle as Instance 4. By inspecting (9), observe that for large N, we must choose $q = 2$ in the corresponding recursive construction. To define the edges per gadget B_i, recall that by (i) we wish to have each edge used once in* OPT*, so by (ii) within B_i we need N disjoint spanning trees connecting N visible vertices and $q = 2$ copies of B_{i-1}, so the number of edges must be $N(N+2-1) = N^2+N$. In* NE*, on the other hand, by (iii) and (iv) for each player we need to use the edges of B_i to connect the N visible vertices of each gadget B_{i-1} with one visible vertex of B_i, which makes $2N$ edges per player, and since we wish that these edges are used by two players each, this is N^2 edges in total. However that means that we cannot demand that* all *of the $N^2 + N$ edges are used twice in* NE*, and we cannot fulfill condition (i). The solution is that from the $N^2 + N$ edges in each gadget B_i, only $N^2 - N$ shall be used by two players, while $2N$ edges are used by only one player in* NE*. This is the set of edges with unique red color in the* NE *solutions in Fig. 6. Because we will need to refer to these edges subsequently, call them "lonely" edges. For defining the edge costs per gadget B_i, we keep everything as before in Sect. 4.3, except for the costs of the lonely edges, which we define to be doubly as expensive, so they cost $2 \cdot (3/2)^{i-1}$ instead of $(3/2)^{i-1}$.* △

The feasibility of the resulting graph construction is exemplified for uneven number of players ($N = 5$) and even number of players ($N = 6$) in Fig. 6. For $N = 5$, gadget B_i has 30 edges, and for $N = 6$ it has 42 edges. Note that these construction have a simple symmetry, and it is not hard to see that the same constructions can be accomplished for any uneven and even number of players $N \geq 2$.

Lemma 5. *Given Instance 5 as defined above, for any $N \geq 2$ the solution* NE *is a Nash equilibrium.*

Proof. We again use Lemma 2. Denote for simplicity an edge from gadget B_i that in NE is loaded with 1 or 2 players as e_i^1 and e_i^2, respectively. We consider the following additional cases for potential edge exchanges involving lonely edges, which are not yet covered by the earlier proof of Lemma 4.

- exchange $e_i^1 \to e_i^2$: new cost $3 \cdot (3/2)^{i-1} > 2 \cdot (3/2)^{i-1}$,
- exchange $e_i^2 \to e_i^1$: new cost $2 \cdot 2(3/2)^{i-1} > 2 \cdot (3/2)^{i-1}$,
- exchange $e_i^1 \to e_{i-1}^1$: new cost $2 \cdot 2(3/2)^{i-2} > 3(3/2)^{i-2} = 2 \cdot (3/2)^{i-1}$,
- exchange $e_i^2 \to e_{i-1}^1$: new cost $2 \cdot 2(3/2)^{i-2} > 3(3/2)^{i-2} = 2 \cdot (3/2)^{i-1}$,
- exchange $e_1^1 \to e_0^1$: new cost $2 \cdot 1 \geq 2$.

Clearly, as all edges in layer B_ℓ have cost equal to or higher than $3(3/2)^{\ell-1}$, also none of the edges of layer $B_{\ell+1}$ can be beneficially exchanged in NE. This concludes the proof. □

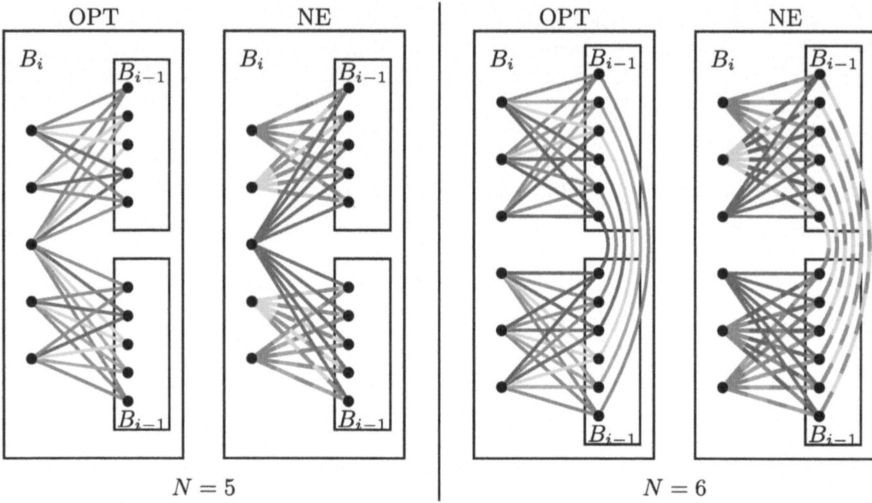

Fig. 6. Recursive building block for large N, here for $N = 5$ and $N = 6$ players. On the left is OPT, and on the right is NE. Gadget B_i contains the 5 (resp. 6) "visible" vertices on the left and $q = 2$ copies of gadget B_{i-1} on the right. The "lonely" edges are the edges with a unique color (red) in NE. Multicolored edges are used by two players. (Color figure online)

Theorem 6. *The price of anarchy for linear graphic matroid congestion games equals* 5/2 *when the number of players* N *is unconstrained.*

Proof. The previously given description yields a set of instances which are parametric in the number of players N, and Lemma 5 confirms the existence of a Nash equilibrium with the desired properties. It remains to compute the costs. In a nutshell, the so defined instance does the trick because the number of lonely edges is negligible for N large enough. Indeed, for any finite number of players and $\ell \to \infty$ the instance has penalty $P_{\text{edges}} = 0$ by construction, but a non-negligible penalty P_{switch}. But for $N \to \infty$, the latter penalty becomes negligible, too. Specifically, for NE we get:

$$\begin{aligned} c(\text{NE}) &= 3\left(\frac{3}{2}\right)^{\ell-1} N^2 + \sum_{i=1}^{\ell}\left[2^{\ell-i}\left(\left(\frac{3}{2}\right)^{i-1} \cdot 4(N^2-N) + 2\left(\frac{3}{2}\right)^{i-1} \cdot 2N\right)\right] \\ &\quad + 2^{\ell} \cdot 2N^2, \\ &= 3\left(\frac{3}{2}\right)^{\ell-1} N^2 + \sum_{i=1}^{\ell}\left[2^{\ell-i}\left(\frac{3}{2}\right)^{i-1} \cdot 4N^2\right] + 2^{\ell} \cdot 2N^2. \end{aligned}$$

Likewise, observing that the lonely edges have the same costs both in NE and in OPT, we get:

$$c(\text{OPT}) = \left(\frac{3}{2}\right)^{\ell-1} + \sum_{i=1}^{\ell}\left[2^{\ell-i}\left(\frac{3}{2}\right)^{i-1} \cdot (N^2+3N)\right] + 2^{\ell} \cdot 2N^2.$$

Taking out 2^ℓ we get

$$c(\mathrm{NE}) = 2^\ell \left[\frac{1}{2}N^2\left(\frac{3}{4}\right)^\ell + \frac{2}{3}\left(3 - \sum_{i=\ell}^{\infty}\left(\frac{3}{4}\right)^i\right)\cdot 4N^2 + 2N^2\right],$$

$$c(\mathrm{OPT}) = 2^\ell \left[\frac{1}{2}\left(\frac{3}{4}\right)^\ell + \frac{2}{3}\left(3 - \sum_{i=\ell}^{\infty}\left(\frac{3}{4}\right)^i\right)\cdot (N^2 + 3N) + 2N^2\right].$$

Now, $\ell \to \infty$ yields $c(\mathrm{NE})/c(\mathrm{OPT}) \to 10N^2/(4N^2 + 6N)$, and for $N \to \infty$, the limit is indeed $5/2$ as required. □

Note that for $N = 2, 3, 4, 5, 6, 7, 8$ players, Instance 5 also gives the lower bounds 1.428, 1.667, 1.818, 1.923, 2, 2.058, and 2.105.

5 Discussion

One idea behind the constructions in this paper is to use $(1, 2)$-edges in the recursive definition of graphs, which are edges that are used by one player in the optimal solution and by two players in the equilibrium solution. This yields combinatorial constraints for tight lower bound instances which are different per number of players, and so are the corresponding existence proofs. One may wonder how to generalize this idea to obtain tight lower bound constructions for each finite numbers of players $N \geq 5$. Obviously, that must be done so as to circumvent the bilinear Diophantine equation (9), which limits the constructions shown in this paper to the cases $N = 3, 4$, and $N \to \infty$. One promising idea is to vary the number of visible vertices and the number of copies of layers B_{i-1} within layer B_i. E.g. making them differ for even and odd index i. Additionally, the requirement to only use $(1, 2)$-edges might be relaxed, as we did for the instance for unbounded N. This gives more flexibility in the combinatorial design. However, two challenges remain. First, to provide an existence proof for the corresponding combinatorial designs, parametric in the number of players N. Second, to keep the resulting penalties negligible in those designs.

Acknowledgements. This research started around 2017 with discussions with our colleague and friend Walter Kern (1957–2021), who disproved an earlier conjecture, namely that the price of anarchy for matroid congestion games could be bounded from above by 2. For sure, Walter would have loved to see this problem finally solved.

References

1. Ackermann, H., Röglin, H., Vöcking, B.: On the impact of combinatorial structure on congestion games. J. ACM **55**, 1–22 (2008). https://doi.org/10.1145/1455248.1455249
2. Awerbuch, B., Azar, Y., Epstein, A.: The price of routing unsplittable flow. In: Proceedings of the 37th Annual ACM Symposium on Theory of Computing, pp. 57–66. ACM (2005). https://doi.org/10.1145/1060590.1060599

3. Beckmann, M., McGuire, C.B., Winsten, C.B.: Studies in the Economics and Transportation. Yale University Press, New Haven (1956)
4. Caragiannis, I., Flammini, M., Kaklamanis, C., Kanellopoulos, P., Moscardelli, L.: Tight bounds for selfish and greedy load balancing. Algorithmica **61**, 606–637 (2011). https://doi.org/10.1007/s00453-010-9427-8
5. Christodoulou, G., Koutsoupias, E.: The price of anarchy of finite congestion games. In: Proceedings of the 37th Annual ACM Symposium on Theory of Computing, pp. 67–73. ACM (2005). https://doi.org/10.1145/1060590.1060600
6. Correa, J., de Jong, J., de Keijzer, B., Uetz, M.: The inefficiency of Nash and subgame perfect equilibria for network routing. Math. Oper. Res. **44**(4), 1286–1303 (2019). https://doi.org/10.1287/moor.2018.0968
7. de Jong, J., Kern, W., Steenhuisen, B., Uetz, M.: The asymptotic price of anarchy for k-uniform congestion games. In: Solis-Oba, R., Fleischer, R. (eds.) WAOA 2017. LNCS, vol. 10787, pp. 317–328. Springer, Cham (2018). https://doi.org/10.1007/978-3-319-89441-6_23
8. de Jong, J., Klimm, M., Uetz, M.: Efficiency of equilibria in uniform matroid congestion games. In: Gairing, M., Savani, R. (eds.) SAGT 2016. LNCS, vol. 9928, pp. 105–116. Springer, Heidelberg (2016). https://doi.org/10.1007/978-3-662-53354-3_9
9. Fabrikant, A., Papadimitriou, C.H., Talwar, K.: The complexity of pure Nash equilibria. In: Babai, L. (ed.) Proceedings of the 36th Annual ACM Symposium on Theory of Computing, pp. 604–612. ACM (2004). https://doi.org/10.1145/1007352.1007445
10. Fotakis, D.: Stackelberg strategies for atomic congestion games. ACM Trans. Comput. Syst. **47**, 218–249 (2010). https://doi.org/10.1007/s00224-008-9152-8
11. Hao, B., Michini, C.: Price of anarchy in paving matroid congestion games. In: Schäfer, G., Ventre, C. (eds.) SAGT 2024, LNCS 15156, pp. 353–370. Springer, Cham (2024). https://doi.org/10.1007/978-3-031-71033-9_20
12. Hao, B., Michini, C.: The price of anarchy in series-parallel network congestion games. Math. Program. **203**, 499–529 (2024). https://doi.org/10.1007/s10107-022-01803-w
13. Koutsoupias, E., Papadimitriou, C.: Worst-case equilibria. Comput. Sci. Rev. **3**(2), 65–69 (2009). https://doi.org/10.1016/j.cosrev.2009.04.003
14. Lücking, T., Mavronicolas, M., Monien, B., Rode, M.: A new model for selfish routing. Theoret. Comput. Sci. **406**(3), 187–206 (2008). https://doi.org/10.1016/j.tcs.2008.06.045
15. Pigou, A.C.: The Economics of Welfare. Macmillan, London (1920)
16. Rosenthal, R.W.: A class of games possessing pure-strategy Nash equilibria. Internat. J. Game Theory **2**(1), 65–67 (1973). https://doi.org/10.1007/BF01737559
17. Roughgarden, T.: The price of anarchy is independent of the network topology. J. Comput. Syst. Sci. **67**, 341–364 (2002). https://doi.org/10.1016/S0022-0000(03)00044-8
18. Roughgarden, T.: Routing games. In: Nisan, N., Roughgarden, T., Tardos, E., Vazirani, V.V. (eds.) Algorithmic Game Theory, chap. 18, pp. 461–486. Cambridge University Press, Cambridge (2007). https://doi.org/10.1017/CBO9780511800481.020
19. Roughgarden, T., Tardos, É.: How bad is selfish routing? J. ACM **49**(2), 236–259 (2002). https://doi.org/10.1145/506147.506153
20. Wardrop, J.G.: Some theoretical aspects of road traffic research. Proc. Inst. Civ. Eng. **1**(3), 325–362 (1952). https://doi.org/10.1680/ipeds.1952.11259

Non-Adaptive Matroid Prophet Inequalities

Shuchi Chawla[1], Kira Goldner[2](✉), Anna R. Karlin[3], and J. Benjamin Miller[4]

[1] The University of Texas at Austin, Austin, TX 78701, USA
shuchi@cs.utexas.edu
[2] Boston University, Boston, MA 02215, USA
goldner@bu.edu
[3] University of Washington, Seattle, WA 98195, USA
karlin@cs.washington.edu
[4] Google, New York, NY 10011, USA
bmille@google.com

Abstract. We investigate non-adaptive algorithms for matroid prophet inequalities. Matroid prophet inequalities have been considered resolved since 2012 when [KW12] introduced thresholds that guarantee a tight 2-approximation to the prophet; however, this algorithm is adaptive. Other approaches of [CHMS10] and [FSZ16] have used non-adaptive thresholds with a feasibility restriction on the items that can be taken; however, this translates to adaptively changing an item's threshold to infinity when it cannot be taken with respect to the additional feasibility constraint, hence the algorithm is not truly non-adaptive. A major application of prophet inequalities is in auction design, where non-adaptive prices possess a significant advantage: they convert to order-oblivious posted pricings, and are essential for translating a prophet inequality into a truthful mechanism for multi-dimensional buyers. The existing matroid prophet inequalities do not suffice for this application. We present the first non-adaptive constant-factor prophet inequality for graphic matroids.

Keywords: Prophet Inequalities · Non-Adaptive Prices · Thresholds · Posted Pricings

1 Introduction

We study the classic prophet inequality problem introduced by Krengel and Sucheston [22]: n items arrive online in adversarial order. A gambler observes the value of each item as it arrives, and in that moment, must decide irrevocably whether to take the item or pass on it forever. He can accept at most one item. The gambler knows in advance the (independent) prior distribution of each item's value. What rule should he use to maximize the value of the item he accepts? In expectation, how does the maximum value that the gambler can guarantee

G. Schäfer and C. Ventre (Eds.): SAGT 2024, LNCS 15156, pp. 389–404, 2024.
https://doi.org/10.1007/978-3-031-71033-9_22

compare to the *prophet*, who knows all of the realized item values in advance and selects the highest valued one?

The prophet inequality is a standard model for online decision making in a stochastic/Bayesian setting and has many applications, particularly to mechanism design and pricing. Over the last few years many variants of the basic single-item setting have been studied. One natural generalization is to allow the gambler to accept more than one item, subject to a feasibility constraint. Formally, we can represent a feasibility constraint as a collection $\mathcal{S}$ of feasible sets. Then both the gambler and prophet can each select any feasible set of items $S \in \mathcal{S}$; in the single-item setting, the feasible sets are just all singletons. What is the gambler's best algorithm and guarantee?

A seminal result by Samuel-Cahn [25] showed that for the basic single-item setting, the online algorithm can obtain at least half of the prophet's value in expectation by determining a single threshold T and accepting the first item with value exceeding T. Further, this approximation factor is tight: there exist instances where the gambler can do no better than $\frac{1}{2}$ as well as the prophet. The threshold T is selected such that the probability that the value of *any* of the n items exceeds the threshold is exactly $\frac{1}{2}$.

In 2012[1], Kleinberg and Weinberg introduced an alternative approach for setting a single threshold: set $T = \frac{1}{2}\textsc{Opt}$. Here, $\textsc{Opt}$ is what the prophet can achieve in expectation, and this approach guarantees the same $\frac{1}{2}$-approximation for a single item. Kleinberg and Weinberg showed that this alternate approach generalizes also to *matroid* prophet inequalities: where both the gambler and the prophet are restricted to accepting independent sets in a given matroid. In this setting, the approach of Kleinberg and Weinberg still achieves a 2-approximation, matching the single-item lower bound.

There is a significant qualitative difference between Samuel-Cahn's approach for the single-item prophet inequality and Kleinberg and Weinberg's approach for the matroid setting. In particular, the former computes a single threshold that is used for the entire duration of the algorithm. The latter, however, recomputes thresholds after every decision. The threshold applied to the value of the second item, for example, depends on whether the first item was accepted by the algorithm or not, and thus also depends on the realized value of the first item. As a consequence, the KW algorithm is more complicated and involves more computation.

In this paper we address a natural problem exposed by this discussion: **Can an online algorithm compete against the prophet using static thresholds under a matroid feasibility constraint?**

There is an inherent connection between prophet inequalities and Bayesian mechanism design. The original problem by Krengel and Sucheston was formulated as an optimal stopping problem; it was Hajiaghayi et al. [18] that made the first connection to an economic welfare-maximization problem. Chawla et al. [7] studied this connection much more deeply, defining a truthful class of

[1] Their original result appeared in STOC 2012 [20], but we will cite their journal version from 2019 for the remainder of the paper.

simple mechanisms called "order-oblivious posted pricings". They show that one can translate a prophet inequality for n items with feasibility constraint $\mathcal{S}$ into an order-oblivious posted pricing for an n-unit setting with unit-demand buyers and a service feasibility constraint[2] corresponding to $\mathcal{S}$; this mechanism is truthful and it yields a revenue guarantee that matches its prophet inequality guarantee. When $\mathcal{S}$ is a matroid, by Kleinberg and Weinberg [21], the resulting mechanism yields $\frac{1}{2}$-approximation to the optimal expected revenue.

This reduction from truthful mechanisms to prophet inequalities crucially relies on the buyers being unit-demand. If we wish instead to translate the prophet inequality into a mechanism for a single constrained-additive buyer subject to feasibility constraint $\mathcal{S}$ over n heterogenous items—that is, the buyer is interested in buying *more* than one item—then adaptive thresholds will not translate to a truthful mechanism. Instead, they correspond to offering each item one-at-a-time to the buyer in any order, but prices change as a function of previous purchases. This update will not generally preserve truthfullness; that is, although the buyer may wish to purchase the first item offered when considered myopically, he may be better off declining, in order to avoid price increases on later items.

In order to fix this reduction for multi-parameter buyers beyond unit-demand, we must use only prophet inequalities with non-adaptive thresholds. This is our primary motivation: constructing non-adaptive prophet inequalities in order to expand the realm of settings where prophet inequalities can be used for truthful mechanism design (see related work for an understanding of how integral they are as a tool in this field). However, non-adaptive prophet inequalities possess numerous other attractive properties as well. For welfare-maximizing mechanisms, non-adaptive prophet inequalities correspond to prices that are not only order-oblivious, but also *anonymous*, using the same prices on each item regardless of the buyer. Additionally, since the thresholds (prices) are all computed before the items arrive and are never updated, there is much less computation required than for adaptive thresholds.

1.1 Our Contribution and Roadmap

We present the first non-adaptive thresholds that give a constant-factor prophet inequality for graphic matroids. We finish Sect. 1 with additional related work and in Sect. 2, we introduce mathematical preliminaries. In Sect. 3, we discuss why extending non-adaptive algorithms to graphic matroids is such a challenging objective, and why prior methods fail. Section 4 presents the ex-ante relation to the matroid polytope: a reduction from a given prophet inequality instance to an alternative setting with convenient properties for designing algorithms. Expert readers can safely skip this section. Then, in this context, Sect. 5 presents our construction for non-adaptive thresholds.

[2] A service feasibility constraint $\mathcal{S}$ says that the set of buyers that are served simultaneously must belong to some set $S \in \mathcal{S}$.

The ex-ante relaxation takes a given item i's value distribution and converts it into a Bernoulli distribution: with probability p_i, item i is "active", that is, non-zero, and takes on value t_i. Then, the threshold for item i is implicit (just the non-zero value t_i), and the only remaining questions are (1) with what probability should our algorithm consider this item, and (2) with what probability will the item be "unblocked", or feasible to accept, when the algorithm reaches it?

In order to obtain a constant-factor approximation, then for every item, the probabilities of both its selection and feasibility must be constant. In a graphic matroid, the elements are the edges and the independent sets are the forests—that is, any set of edges that does not contain a cycle. Depending on the given graph, an edge could be "blocked" by many different edges. Our main idea is to orient the graph to have good properties and then exploit them. For an edge (u, v), suppose it is oriented into vertex v. Notice that if no other edges incident to v are selected by the algorithm, then (u, v) will certainly be feasible to accept—it cannot possibly form a cycle by taking this edge. We orient the graph such that all edges directed into v have low enough probability mass such that with probability $\frac{1}{2}$, none are active. Then, our algorithm decides which edges to consider such that, with constant probability for every v, it will consider the edges into v and *not* the edges out of v. Hence, with no edges out of v and a good chance that no edges into v will be active, any edge into v can be accepted with constant probability. Our method for determining which edges to consider is simple: we take a random cut and consider only the edges in one direction across the cut. Since every edge is oriented into some vertex v, it will be both considered and unblocked with constant probability, as desired.

Unfortunately, this approach is quite specific to a graphic matroid. While some properties of the algorithm might extend to other matroids, we know that it cannot generalize to all matroids: [Kira: here] [16] prove a lower bound of $\Omega(\frac{\log n}{\log \log n})$ for prophet inequalities that use only non-adaptive thresholds for the class of general matroids. Their lower bound example is a gammoid.

We pose the following two remaining open (but likely very difficult) open questions for understanding how far non-adaptive constant-factor approximations reach between graphic matroids and the lower bound of a gammoid.

Open Problem 1. *What is the boundary within matroids for non-adaptive constant-factor approximations?*

Open Problem 2. *How do approximations decay for non-adaptive thresholds as matroids become more complex?*

1.2 Additional Related Work

Non-Adaptive Thresholds. As mentioned, the two predominant approaches for achieving $\frac{1}{2}$-approximation in the single-item setting are both non-adaptive [21, 25]. Chawla et al. [7] provide non-adaptive $\frac{1}{2}$-approximations to the prophet for both k-uniform and partition matroids; they also give a non-adaptive $O(\log r)$-approximation for general matroids, where r is the rank of the matroid. Recent work by Gravin and Wang [17] gives a non-adaptive algorithm that guarantees a 3-approximation to the prophet for online bipartite matching, which is the intersection of two matroids. Chawla et al. [6] optimize non-adaptive thresholds for the k-uniform settings depending on the range that k is in, improving existing guarantees in the $k < 20$ regime. Jiang et al. [19] provides a framework that optimizes for tight prophet inequalities in the k-uniform setting and recovers existing guarantees, such as the guarantee of [6], along with improving guarantees for the i.i.d. setting. Arnosti and Ma [1] study the question of non-adaptive thresholds in the k-uniform setting in the *prophet secretary* problem, where the order of the items is uniformly random rather than adversarial, as in the secretary problem (and the distributions of each item remain known), giving a $1 - e^k k^k / k!$-approximation, and their results are optimal for $k > 4$. Importantly, no non-adaptive algorithms are known beyond uniform and partition matroids and the special case of bipartite matching.

Constrained Non-Adaptive. Another class of algorithms uses non-adaptive thresholds and a restricted feasibility constraint. That is, given a feasibility constraint $\mathcal{S}$ and the prior distributions for n items, prior to the arrival of all items, the algorithm sets thresholds T_i for each item and a restricted feasibility constraint $\mathcal{S}'$ such that $\mathcal{S}' \subset \mathcal{S}$. Then, an item is accepted if it exceeds its threshold *and* is feasible with respect to the items already accepted and the subconstraint $\mathcal{S}'$. Notice that an item could exceed its threshold, be feasible with respect to previously accepted items and $\mathcal{S}$, and yet not be accepted because it is not feasible with respect to previously accepted items and $\mathcal{S}'$. In essence, imposing a subconstraint is equivalent to adaptively changing an item's threshold to $T_i = \infty$ if the item is not feasible with respect to the subconstraint.

Why is this different than when the gambler rejects an item that exceeds its threshold but is not feasible with respect to $\mathcal{S}$? We can interpret the gambler's value as constrained-additive with respect to $\mathcal{S}$, so the gambler does not have any marginal gain for items that are infeasible with respect to $\mathcal{S}$ and the items he has already accepted. Hence, he has no reason to take items with no positive marginal value to him. The original feasibility constrain is not a restriction on the algorithm, but rather a result of the gambler's valuation class.

Chawla et al. [7] first produced a $\frac{1}{3}$-approximation to the prophet for graphic matroids using non-adaptive thresholds with a partition matroid subconstraint. In a very elegant approach, Feldman et al. [15] produce an Online Contention Resolution Scheme (OCRS) that yields a $\frac{1}{4}$-approximation for all matroids using non-adaptive thresholds and a subconstraint built cleverly from the structure of the given matroid.

Prophet Inequalities Beyond Matroids. Prophet inequalities are well-studied and the literature is far too broad to cover; see [24] for an excellent survey. Note, however, that dynamic algorithms yield good approximations to the prophet in settings reaching beyond matroids. In addition to matroids, the approach of [15] also applies to matchings, knapsack constraints, and the intersections of each. Very recent work by Correa and Cristi [9] gives an algorithm guaranteeing a constant-factor approximation for the very general setting of multiple buyers with subadditive valuations.

Direct Applications to Pricing. The Chawla et al. [7] reduction from order-oblivious posted pricings to prophet inequalities was only the first of many pricing applications of prophet inequalities. Feldman et al. [14] considers the setting where buyers arrive online and face posted prices for items; non-adaptive anonymous prices are posted for each item equal to half its contribution to the optimal welfare. These prices guarantee $\frac{1}{2}$-approximation to the optimal welfare for fractionally subadditive valuations. Note that this is a prophet inequality when there is only one item. Dütting et al. [11,12] connect posted prices and prophet inequalities: they interpret the Kleinberg and Weinber [21] thresholds as "balanced prices" and derive an economic intuition for the proof. They extended these balanced prices to more complex settings, including a variety of feasibility constraints and valuation classes. The approach is to prove guarantees in the full information setting, where the realized values are known in advance. Then, via an extension theorem, they prove that the results hold for Bayesian settings too, where distributions are known but values are unknown. Note that their balanced prices result in non-adaptive anonymous prices for all settings they consider *except* for matroids feasibility constraints, where they remain adaptive and buyer-specific. The recent work of Dütting et al. [13] also implements posted prices for buyers with subadditive valuations, but rather than balanced prices, provides a weaker sufficient condition to get a tighter approximation, and shows the existence of such prices through a primal-dual approach. They also ask whether there is separation between prophet inequalities and posted pricings. Correa et al. [10] show that for single-dimensional agents and threshold-based algorithms, there is none. Banihashem et al. [2] answer this for more general settings and algorithms via a black-box reduction from prophet inequalities to posted prices, but this approach is only existential, and must be rectified with the known strong lower bounds for non-adaptive thresholds.

More Subtle Applications in Mechanism Design and Analysis. Beyond direct applications to pricing, prophet inequalities have also been used in to build more complex mechanisms and prove approximation guarantees. Chawla and Miller [8] design a two-part tariff mechanism to approximate optimal revenue for matroid-constrained buyers. Their benchmark is an ex-ante relaxation, and they use an OCRS [15] to achieve a constant fraction of that revenue. Cai and Zhao [4] prove that the better of a sequential posted price mechanism (where each buyer can only buy one item) and an anonymous sequential posted price mechanism with an entry fee yields a constant-approximation to the optimal revenue for multiple fractionally subadditive buyers (and $O(\log n)$-approximation for fully subadditive). In a specific case of their analysis that analyzes the core of the core (a double core-tail analysis follow the original of [23]), they use [14]. Work by Cai and Zhao [5] approximates the optimal profit—seller revenue minus cost—for constrained-additive buyers. Like [8], they also construct their benchmark using the ex-ante relaxation and use OCRS to bound a term here as well. Recent work by Cai et al. [3] studies gains from trade approximation in a two-sided market with a constrained-additive buyer and single-dimensional sellers—both the single-item prophet inequality of [21] and an OCRS are used to inspire prices for *both* the buyer and the sellers simultaneously and then show that enough gains from trade will be received to approximate one specific part of their benchmark.

2 Preliminaries

Definition 1. *A* matroid $M = (N, \mathcal{I})$ *is defined by a ground set of elements* N *(with* $|N| = n$*) and a set of independent sets* $\mathcal{I} \subseteq 2^N$*. It is a matroid if and only if it satisfies the following two properties:*

1. *Downward-closed: If* $I \subset J$ *and* $J \in \mathcal{I}$ *then* $I \in \mathcal{I}$*.*
2. *Matroid-exchange: For* $I, J \in \mathcal{I}$*, if* $|J| > |I|$ *then there exists some* $i \in J \setminus I$ *such that* $I \cup \{i\} \in \mathcal{I}$*.*

We review several standard notions for matroids:

- The *rank* of a set $\text{rank}(S)$ is the size of the largest independent set in S: $\max\{|I| \mid I \in \mathcal{I}, I \subseteq S\}$.
- The *span* of a set $\text{span}(S)$ is the largest set that contains S and has the same rank as S: $\{i \in N \mid \text{rank}(S \cup \{i\}) = \text{rank}(S)\}$.
- An element i is *spanned* by a set S when $i \in \text{span}(S)$.

We will informally use the language "blocked" (by a set S) to mean that an element is spanned (by the set S), and similarly "unblocked" to mean that an element is *not* spanned (by the set S).

For any matroid M, we have the *matroid polytope* $\mathcal{P}_M = \{\boldsymbol{p} \in \mathbb{R}^M_{\geq 0} \mid \forall S \in 2^N, \sum_{i \in S} p_i \leq \text{rank}(S)\}$. That is, $\mathcal{P}_M$ is the convex hull of the independent sets $\mathcal{I}$.

Definition 2. *A* Matroid Prophet Inequality *instance* $(\boldsymbol{X}, M)$ *is given by a matroid* $M = (N, \mathcal{I})$ *and distribution of values* $\boldsymbol{X}$ *for the* n *items that are the ground set* N*.* X_i *denotes the random variable representing the value for item* i*.*

For any given matroid prophet inequality instance, we let $\text{OPT}(\boldsymbol{X}, M)$ denote the value of the prophet's set in expectation of the value of the items. Formally, $\text{OPT}(\boldsymbol{X}, M) = \mathbb{E}\big[\max_{I \in \mathcal{I}} \sum_{i \in I} X_i\big]$. We omit the distributions $\boldsymbol{X}$ or matroid M when it is obvious from context.

Definition 3. *A* non-adaptive *threshold algorithm is given an instance* $(\boldsymbol{X}, M)$ *and determines thresholds* $\boldsymbol{T}$*. A threshold* T_i *for each item* i *is a function only of the random variables* $\boldsymbol{X}$ *(and, in particular, not as a function of any realizations of* $\boldsymbol{X}$ *or whether previous items have exceeded thresholds thus far).*

For any non-adaptive thresholds $\boldsymbol{T}$, we let $\text{ALG}(\boldsymbol{X}, M, \boldsymbol{T})$ denote the expected value obtained by the algorithm. Again, we omit the parameters when they are clear from context.

3 Where Straightforward Extensions Fail

Both of the non-adaptive single-item approaches—the probabilistic approach of Samuel-Cahn [25] and the $\frac{1}{2}$OPT approach of Kleinberg and Weinberg [21]—extend to the k-uniform matroid setting, in which any set of size at most k is feasible. We first see why these approaches work for k-uniform matroids yet break down for graphic matroids. Then, we attempt to use an idea for graphic matroids from Chawla et al. [7] to develop a non-adaptive algorithm, and again highlight where the approach breaks down.

We begin with the two generalizations to k-uniform methods. Note that we do not claim either as part of our contribution, although to the best of our knowledge, neither approaches' generalized thresholds and proof is written anywhere.

Formally, a k-uniform matroid is the matroid where, for any given ground set $N, \mathcal{I} = \{I \subset N : |I| \leq k\}$. Bear in mind that $k = 1$ returns to the single-item case.

The Probabilistic Approach. (Extension of Samuel-Cahn [25] single-item algorithm to non-adaptive thresholds for the k-uniform matroid.) Determine the thresholds T by setting $\Pr[< k \text{ item values exceed } T] = \Pr[\geq 1 \text{ slot empty}] = p = \frac{1}{2}$.

$$
\begin{aligned}
\textsc{Alg}(\boldsymbol{X}, T) &\geq \sum_i \Pr[i \text{ not blocked}]\mathbb{E}[(X_i - T)^+] + \Pr[\geq k \text{ above T}] \cdot kT \\
&\geq \Pr[< k \text{ above T}] \sum_i \mathbb{E}[(X_i - T)^+] + \Pr[\geq k \text{ above T}] \cdot kT \\
&\geq p\mathbb{E}\left[\max_{S:|S|\leq k} \sum_{i\in S}(X_i - T)^+\right] + (1-p)kT \\
&\geq p\mathbb{E}\left[\max_{S:|S|\leq k} \sum_{i\in S} X_i - kT\right] + (1-p)kT \\
&= \frac{1}{2}(\mathbb{E}\left[\max_{S:|S|\leq k} \sum_{i\in S} X_i\right]) - \frac{1}{2}kT + \frac{1}{2}kT \\
&= \frac{1}{2}\mathbb{E}\left[\max_{S:|S|\leq k} \sum_{i\in S} X_i\right] = \frac{1}{2}\textsc{Opt}(\boldsymbol{X}).
\end{aligned}
$$

For uniform matroids, a simple characterization based on size exists for sets that do not span *any* elements that have yet to arrive: they need only be of size strictly less than k. This property does not hold for more complex matroids.

The "Thresholds as Constant-Fraction of Prophet" Approach. (Extension of Kleinberg and Weinberg [21] single-item algorithm to non-adaptive thresholds for the k-uniform matroid; almost identical to those in Chawla et al. [7]). Set $T = \frac{1}{2k}\mathbb{E}\left[\max_{S:|S|\leq k} \sum_{i\in S} X_i\right] = \frac{1}{2k}\textsc{Opt}(\boldsymbol{X})$.

$$
\begin{aligned}
\text{ALG} &\geq \sum_i \Pr[i \text{ not blocked}]\mathbb{E}[(X_i - T)^+] + \Pr[\geq k \text{ above T}]kT \\
&\geq \Pr[< k \text{ above T}]\mathbb{E}[\sum_i (X_i - T)^+] + \Pr[\geq k \text{ above T}]kT \\
&\geq p\mathbb{E}\left[\max_{S:|S|\leq k} \sum_{i\in S}(X_i - T)^+\right] + (1-p)kT \\
&= p\left(\mathbb{E}\left[\max_{S:|S|\leq k} \sum_{i\in S} X_i\right] - kT\right) + (1-p)\frac{1}{2}\mathbb{E}\left[\max_{S:|S|\leq k} \sum_{i\in S} X_i\right] \\
&= \frac{1}{2}\mathbb{E}\left[\max_{S:|S|\leq k} \sum_{i\in S} X_i\right] = \frac{1}{2}\text{OPT}(\boldsymbol{X}).
\end{aligned}
$$

In uniform matroids, any element is exchangeable for any other element. Then so long as it contributes enough value, such as at least a constant fraction of the average contribution to the optimal basis, there is no reason not to accept an element. However, this does not hold for more complex matroids. A particular element, even if extremely high value, may cause so many other elements to be spanned that it is not worth taking.

One can imagine more nuanced extensions of either such approach: probabilistic thresholds for i according to how many elements it might block, or value-based thresholds for i based on the value of the sets it might block. However, any such extension would require a matroid-specific understanding of the relationship between elements, and element-specific thresholds.

Note that in addition to uniform matroids, both approaches easily extend to partition matroids by applying the approach to thresholds specific to the uniform matroid in each partition.

The Constrained Non-Adaptive Approach. Chawla et al. [7] construct non-adaptive thresholds for a graphic matroid that work *so long as* the algorithm can enforce an additional subconstraint. Specifically, they cleverly partition the graph such that, so long as at most one edge is accepted from each partition, then an independent set is guaranteed. Then as items arrive, they are accepted if and only if they exceed their threshold *and* are feasible with respect to the subconstraint—that is, no previous item from its partition has been accepted. This approach guarantees a $\frac{1}{3}$-approximation.

As discussed in the introduction, enforcing a subconstraint *is* in fact adaptive. But, we *could*, for example, randomly select one item from each partition in advance, defining our set for consideration C. Then, as items arrive, in each partition, we consider only the item in C, ignoring all other items from each partition. That is, we leave thresholds the same for all items in C and *a priori* set $T_i = \infty$ for all $i \notin C$. This ensures that we only consider a set that complies with our feasibility constraint *without* making any modifications online. Note that we can select items to be in the consideration set C with whatever probabilities we choose, even in a correlated fashion—as long as we make them prior to items arriving—thus setting all thresholds to T_i or ∞ in advance. Is there some clever way that we can implement our feasibility constraint, or any feasibility constraint, yet maintain a constant-factor approximation?

For the approach of CHMS, we might observe that a convenient property that bounds the probability mass of each partition could allow us to form a probability distribution over elements in each partition (i.e. place item i in C with probability $p_i/2$). However, this approach in fact reduces the probability too much, as it combines the probability that the element is active with the probability it is considered, and is no longer constant. If we use a constant probability, it would instead sell to too low of a quantile.

If such an approach *were* to work, we could convert *any* non-adaptive matroid prophet inequality to a prophet inequality, as a greedy OCRS exists for all matroids and constructs constrained non-adaptive thresholds all matroids [15]. However, Feldman et al. [16] also prove a super-constant lower bound of $\Omega(\frac{\log n}{\log \log n})$, so guarantees cannot possibly go through for every matroid. Thus, an interesting direction for future work is to characterize *when* an approach of converting constrained non-adaptive thresholds to a fully non-adaptive algorithm in this way would maintain good guarantees.

4 The Ex-Ante Relaxation to the Matroid Polytope

Reducing a given matroid prophet inequality instance to one with Bernoulli distributions that sits within the matroid polytope is "standard", and is used in [15]. It's "just" an ex-ante relaxation to the matroid polytope, and expert readers can safely skip this section. However, we present the reduction in detail for comprehensiveness and ease of reading, as we did not find it elsewhere.

First, given arbitrary independent random variables X_i, we reduce the problem to designing an algorithm for independent Bernoulli random variables X'_i:

$$X'_i = \begin{cases} t_i & \text{w.p. } p_i \\ 0 & \text{w.p. } 1 - p_i, \end{cases}$$

where $\boldsymbol{p} \in \mathcal{P}_M$ and there exist $t_i \in \mathbb{R}\, \forall i$.

Reducing to Bernoulli random variables gives two properties which greatly simplify the design of an algorithm:

1. Each element of the ground set is either *active* or *inactive*; and
2. There exists a worst-case total ordering of the elements.

The worst-case ordering is the typical greedy ordering. Assume $t_i \leq t_{i+1}$; then greedily selecting elements in order (maintaining independence and according to the rules of our algorithm) results in the lowest weight outcome over all orderings. For the rest of the paper, we assume $t_i \leq t_{i+1}$ for $1 \leq i < n$.

We now state our reduction formally.

Lemma 1. *Given a matroid $M = (N, \mathcal{I})$ and independent random weights X_i, $i \in N$, there exist independent Bernoulli weights X_i', where $X_i' = t_i$ w.p. p_i and $\boldsymbol{p} \in \mathcal{P}_M$, such that*

$$\textsc{Opt}(\boldsymbol{X}, M) \leq \sum_i p_i t_i.$$

Furthermore, for any algorithm Alg,

$$\textsc{Alg}(\boldsymbol{X}) \geq \textsc{Alg}(\boldsymbol{X'}).$$

Proof. First, rewrite the original optimal value as a sum over the ground set:

$$\begin{aligned}\textsc{Opt}(\boldsymbol{X}, M) &= \mathbb{E}\big[\max_{I \in \mathcal{I}} \sum_{i \in I} X_i\big] \\ &= \sum_{i \in N} \Pr[i \in I^*] \cdot \mathbb{E}[X_i \,|\, i \in I^*],\end{aligned}$$

where I^* is the maximum weight basis: $I^* = \text{argmax}_{I \in \mathcal{I}} \sum_{i \in \mathcal{I}} X_i$. Now let $p_i = \Pr[i \in I^*]$—the ex-ante probability that i is in the prophet's solution. Since $\boldsymbol{p}$ is a convex combination of basis vectors, then $\boldsymbol{p} \in \mathcal{P}_M$.

Now, observe that $\mathbb{E}[X_i \,|\, X_i \geq F_i^{-1}(1 - p_i)] \geq \mathbb{E}[X_i | H]$ for any event H with $\Pr[H] = p_i$. Let $t_i = \mathbb{E}[X_i \,|\, X_i \geq F_i^{-1}(1 - p_i)]$; then in particular $t_i \geq \mathbb{E}[X_i \,|\, i \in I^*]$. Hence

$$\textsc{Opt}(\boldsymbol{X}, M) \leq \sum_i p_i t_i.$$

Finally, to see that $\textsc{Alg}(\boldsymbol{X}) \geq \textsc{Alg}(\boldsymbol{X'})$, we simply couple $\boldsymbol{X}$ and $\boldsymbol{X'}$, so that $X_i \geq t_i$ if and only if $X_i' = t_i$. For any ordering of the elements, the algorithm applied to the original instance selects the same items as the algorithm applied to the Bernoulli instance. □

5 A Constant-Factor Approximation for Graphic Matroids

Given a Bernoulli instance from the matroid polytope, we show how to utilize it to obtain a constant-factor non-adaptive algorithm for graphic matroids.

A graphic matroid is defined by an undirected graph G with vertices V and edges E. The edges of the graph form the ground set, and the independent sets $\mathcal{I}$ are forests, i.e. cycle-free sets of edges: $\mathcal{I} = \{I \subseteq E : I \text{ contains no cycles}\}$; every spanning tree is a basis. In light of Lemma 1, each edge $i \in E$ has an associated weight t_i and is active (non-zero) with probability p_i, where $\boldsymbol{p} \in \mathcal{P}_G$, the matroid polytope for the graphic matroid G. The objective is then to select a maximum weight spanning tree. As discussed in the previous section, we assume without loss that the edges arrive in order with $t_i \leq t_{i+1}$ for all $1 \leq i \leq n-1$; this order obtains the worst-case performance.

Our approach works by considering only a subset of the edges which has the properties that (1) a significant fraction of the prophet's benchmark is accounted for and yet (2) with constant probability, elements selected earlier in the ordering do not block later elements.

Specifically, we do this in two steps. First, we show there exists a way to direct the edges such that every edge has at most a constant probability of being spanned by edges *except* for those leaving the vertex into which it is directed. Then, we take a random cut in the graph and allow our algorithm to select only edges crossing the cut in one direction, ensuring that for every vertex, the edges entering it are considered while the edges leaving it are not with constant probability.

Notation. We use $b_i(S)$ to denote the probability that element i is "blocked" or *spanned* by the active elements in a set S with respective to active probabilities $\boldsymbol{p} \in \mathcal{P}_M$. For $\boldsymbol{p} \in \mathcal{P}_M$, let $R_{\boldsymbol{p}}(S)$ be the random set containing $i \in S$ independently with probability p_i. We call this the "active" set. Formally, $b_i(S) = \Pr[i \in \text{span}(R_{\boldsymbol{p}}(S \setminus \{i\}))]$. Notice that even if $i \in S$, we do not worry that it would span itself.

One convenience of using the ex-ante relaxation is that, so long as each element is unblocked with constant probability, that is, $1 - b_i(S) \geq c$, we obtain a constant-factor approximation.

5.1 Directing the Graph

Lemma 2. *For $p \in \frac{1}{4}\mathcal{P}_G$, there exists a way to orient the edges of G such that for each vertex the total probability mass of incoming edges is at most* $1/2$.

Proof. Any vector from the graphic matroid polytope $\mathcal{P}_G$ is a convex combination bases, or spanning trees. The average vertex degree in any spanning tree is at most 2, so the average fractional degree in a convex combination of spanning trees is at most 2, and hence the average fractional degree under the scaled $\boldsymbol{p} \in \frac{1}{4}\mathcal{P}_G$ is at most $\frac{1}{2}$.

Let in-deg(v) denote the fractional in-degree of v in the constructed directed graph. That is, the sum of the "active" probabilities for the edges directed into v. We can find an orientation of the edges in the graph given probabilities $\boldsymbol{p}$ such that in-deg$(v) \leq \frac{1}{2}$ for all vertices v: because the average degree is at most $\frac{1}{2}$, there exists some vertex v with degree at most $\frac{1}{2}$. Orient all of the edges incident to v toward v, as in-deg$(v) \leq \frac{1}{2}$, and then recurse on the graph among the remaining vertices. □

Corollary 1. *Given a graph as guaranteed by Lemma 2, let* $\text{in}(v)$ *be the set of incoming edges to vertex v and let* $\text{out}(v)$ *be the outgoing edges. For any i, let v be the vertex such that $i \in \text{in}(v)$. Then for any $S \subseteq E$,*

$$b_i(S \setminus \text{out}(v)) \leq \frac{1}{2}.$$

Proof. Observe that for $i \in \text{in}(v)$, v cannot be spanned by a set that contains no other edges incident to v. Then in order for i to be spanned in $S \setminus \text{out}(v)$, at least one edge in $\text{in}(v)$ other than i must be active. By construction, $\sum_{i \in \text{in}(v)} p_i \leq \frac{1}{2}$. So the probability that no edges are active is at least $\frac{1}{2}$ by the union bound. □

5.2 Random Cut

Assume $\boldsymbol{p} \in \frac{1}{4}\mathcal{P}_G$, and direct the graph as described above. We consider a random cut of the graph $\widehat{S}$: let $A \subseteq V$ be a random set of vertices such that each vertex is included in A independently with probability $1/2$, and let $\bar{A} = B = V \setminus A$. Let $\widehat{S}$ be the set of directed edges across the cut from A to B, formally, $\widehat{S} = \{i : i \in \text{out}(u) \cap \text{in}(v), u \in A, v \in B\}$. Then the edges crossing the cut from A to B give a good approximation to our benchmark.

Claim.

$$\mathbb{E}_{\widehat{S}}\left[\sum_{i \in \widehat{S}} p_i t_i (1 - b_i(\widehat{S}))\right] \geq \frac{1}{8} \sum_{i \in E} p_i t_i.$$

Proof.

$$\begin{aligned}
\mathbb{E}_{\widehat{S}}\left[\sum_{i \in \widehat{S}} p_i t_i (1 - b_i(\widehat{S}))\right] &= \sum_{(u,v) \in E} p_{uv} t_{uv} \Pr[(u,v) \in \widehat{S}]\, \mathbb{E}\left[1 - b_{uv}(\widehat{S}) \middle| (u,v) \in \widehat{S}\right] \\
&= \sum_{(u,v) \in E} p_{uv} t_{uv} \Pr[u \in A] \Pr[v \in B]\, \mathbb{E}\left[1 - b_{uv}(\widehat{S}) \middle| u \in A, v \in B\right] \\
&= \frac{1}{4} \sum_{(u,v) \in E} p_{uv} t_{uv}\, \mathbb{E}\left[1 - b_{uv}(\widehat{S}) \middle| u \in A, v \in B\right] \\
&\geq \frac{1}{8} \sum_{(u,v) \in E} p_{uv} t_{uv}
\end{aligned}$$

where the last inequality follows from Corollary 1. □

5.3 Final Algorithm

For discrete random variables $\boldsymbol{X}$, our algorithm is constructive, albeit not efficient, because we can compute $\boldsymbol{p}$ and $\boldsymbol{t}$ as guaranteed by Lemma 1. (Of course, we can discretize continuous random variables to arbitrary approximation).

1: Compute $\boldsymbol{p}$ and $\boldsymbol{t}$ as guaranteed by Lemma 1.
2: Direct the graph as outlined in Lemma 2.
3: Choose a cut (A, B) uniformly at random; let $\widehat{S} = \{i : i \in \text{out}(u) \cap \text{in}(v), u \in A, v \in B\}$.
4: For all edges $i \in \widehat{S}$, set $T_i = t_i$.
5: For all edges $i \notin \widehat{S}$, set $T_i = \infty$.

Step 3 can be derandomized using the standard Max-Cut derandomization. Our main result is that this algorithm gives a $\frac{1}{32}$-approximation.

Theorem 1. *Let G be a graphic matroid with independent edge weights $\boldsymbol{X}$. Then*

$$32\,\mathbb{E}[\textsc{Alg}(G, \boldsymbol{X})] \geq \textsc{Opt}(G, \boldsymbol{X}).$$

Proof. Let $\boldsymbol{p} \in \mathcal{P}_G$ and $\boldsymbol{t}$ be the probabilities and values guaranteed by Lemma 1. Let $p'_i = \frac{1}{4}p_i$. Then our algorithm obtains $\textsc{Alg} = \mathbb{E}_{\widehat{S}}\left[\sum_{i \in \widehat{S}} p'_i t_i (1 - b_i(\widehat{S}))\right]$, which by our construction of $\widehat{S}$ and Claim 5.2, gives

$$\textsc{Alg} \geq \frac{1}{8} \sum_{(u,v) \in E} p'_{uv} t_{uv} = \frac{1}{32} \sum_{i \in E} p_i t_i.$$

□

Our approximation factor is, of course, a factor of 16 worse than the dynamic thresholds of [21] and a 10.67-factor worse than the constrained non-adaptive thresholds of [7]. However, our guarantee holds for fully non-adaptive thresholds, and thus will guarantee truthful mechanisms in multi-parameter mechanism design applications.

Acknowledgments. Shuchi Chawla was supported in part by NSF awards CCF-1617505, CCF-2008006, and CCF-2225259. Kira Goldner was supported in part by NSF award DMS-1903037 and a Columbia Data Science Institute postdoctoral fellowship, and in part by a Shibulal Family Career Development Professorship. Work by J. Benjamin Miller completed while at UW-Madison, and was supported in part by a Cisco graduate fellowship. Anna R. Karlin was supported by Air Force Office of Scientific Research grant FA9550-20-1-0212 and NSF grant CCF-1813135.

References

1. Arnosti, N., Ma, W.: Tight guarantees for static threshold policies in the prophet secretary problem. Oper. Res. **71**(5), 1777–1788 (2023)
2. Banihashem, K., Hajiaghayi, M., Kowalski, D.R., Krysta, P., Olkowski, J.: Power of posted-price mechanisms for prophet inequalities. In: Proceedings of the 2024 Annual ACM-SIAM Symposium on Discrete Algorithms (SODA), pp. 4580–4604. SIAM (2024)
3. Cai, Y., Goldner, K., Ma, S., Zhao, M.: On multi-dimensional gains from trade maximization. In: ACM-SIAM Symposium on Discrete Algorithms (SODA21) (2021)
4. Cai, Y., Zhao, M.: Simple mechanisms for subadditive buyers via duality. In: Proceedings of the 49th Annual ACM SIGACT Symposium on Theory of Computing, STOC 2017, pp. 170–183. ACM, New York (2017). https://doi.org/10.1145/3055399.3055465
5. Cai, Y., Zhao, M.: Simple mechanisms for profit maximization in multi-item auctions. In: Proceedings of the 2019 ACM Conference on Economics and Computation, EC 2019, p. 217–236. Association for Computing Machinery, New York (2019). https://doi.org/10.1145/3328526.3329616
6. Chawla, S., Devanur, N., Lykouris, T.: Static pricing for multi-unit prophet inequalities (2020)

7. Chawla, S., Hartline, J.D., Malec, D.L., Sivan, B.: Multi-parameter mechanism design and sequential posted pricing. In: Proceedings of the forty-second ACM symposium on Theory of Computing, pp. 311–320. ACM (2010). http://dl.acm.org/citation.cfm?id=1806733
8. Chawla, S., Miller, J.B.: Mechanism design for subadditive agents via an ex ante relaxation. In: Proceedings of the 2016 ACM Conference on Economics and Computation, EC 2016, pp. 579–596. ACM, New York (2016).https://doi.org/10.1145/2940716.2940756
9. Correa, J., Cristi, A.: A constant factor prophet inequality for online combinatorial auctions, STOC 2023, pp. 686–697. Association for Computing Machinery, New York (2023). https://doi.org/10.1145/3564246.3585151
10. Correa, J., Foncea, P., Pizarro, D., Verdugo, V.: From pricing to prophets, and back! Oper. Res. Lett. **47**(1), 25–29 (2019)
11. Dütting, P., Feldman, M., Kesselheim, T., Lucier, B.: Prophet inequalities made easy: Stochastic optimization by pricing non-stochastic inputs. In: Umans, C. (ed.) 58th IEEE Annual Symposium on Foundations of Computer Science, FOCS 2017, Berkeley, CA, USA, 15–17 October 2017, pp. 540–551. IEEE Computer Society (2017). https://doi.org/10.1109/FOCS.2017.56
12. Dütting, P., Feldman, M., Kesselheim, T., Lucier, B.: Prophet inequalities made easy: Stochastic optimization by pricing nonstochastic inputs. SIAM J. Comput. **49**(3), 540–582 (2020). https://doi.org/10.1137/20M1323850
13. Dütting, P., Kesselheim, T., Lucier, B.: An o(log log m) prophet inequality for subadditive combinatorial auctions. In: 61st IEEE Annual Symposium on Foundations of Computer Science, FOCS 2020 (2020). https://arxiv.org/abs/2004.09784
14. Feldman, M., Gravin, N., Lucier, B.: Combinatorial auctions via posted prices. In: Proceedings of the Twenty-Sixth Annual ACM-SIAM Symposium on Discrete Algorithms, pp. 123–135. SIAM (2015)
15. Feldman, M., Svensson, O., Zenklusen, R.: Online contention resolution schemes. In: Krauthgamer, R. (ed.) Proceedings of the Twenty-Seventh Annual ACM-SIAM Symposium on Discrete Algorithms, SODA 2016, Arlington, VA, USA, 10–12 January 2016, pp. 1014–1033. SIAM (2016). https://doi.org/10.1137/1.9781611974331.ch72
16. Feldman, M., Svensson, O., Zenklusen, R.: Online contention resolution schemes (2019)
17. Gravin, N., Wang, H.: Prophet inequality for bipartite matching: merits of being simple and non adaptive. In: Proceedings of the 2019 ACM Conference on Economics and Computation, EC 2019, pp. 93–109. Association for Computing Machinery, New York (2019). https://doi.org/10.1145/3328526.3329604
18. Hajiaghayi, M.T., Kleinberg, R., Sandholm, T.: Automated online mechanism design and prophet inequalities. In: Proceedings of the 22nd National Conference on Artificial Intelligence, AAAI 2007, vol. 1, pp. 58–65. AAAI Press (2007)
19. Jiang, J., Ma, W., Zhang, J.: Tightness without counterexamples: a new approach and new results for prophet inequalities. arXiv preprint arXiv:2205.00588 (2022)
20. Kleinberg, R., Weinberg, S.M.: Matroid prophet inequalities. In: Karloff, H.J., Pitassi, T. (eds.) Proceedings of the 44th Symposium on Theory of Computing Conference, STOC 2012, New York, NY, USA, 19–22 May 2012, pp. 123–136. ACM (2012). https://doi.org/10.1145/2213977.2213991
21. Kleinberg, R., Weinberg, S.M.: Matroid prophet inequalities and applications to multi-dimensional mechanism design. Games Econ. Behav. **113**, 97–115 (2019). https://doi.org/10.1016/j.geb.2014.11.002

22. Krengel, U., Sucheston, L.: Semiamarts and finite values. Bull. Am. Math. Soc. **83**(4), 745–747 (1977)
23. Li, X., Yao, A.C.C.: On revenue maximization for selling multiple independently distributed items. Proc. Natl. Acad. Sci. **110**(28), 11232–11237 (2013). https://doi.org/10.1073/pnas.1309533110
24. Lucier, B.: An economic view of prophet inequalities. SIGecom Exch. **16**(1), 24–47 (2017). https://doi.org/10.1145/3144722.3144725
25. Samuel-Cahn, E.: Comparison of threshold stop rules and maximum for independent nonnegative random variables. Ann. Probab. 1213–1216 (1984)

Matroid Bayesian Online Selection

Ian DeHaan and Kanstantsin Pashkovich(✉)

Department of Combinatorics and Optimization, University of Waterloo, Waterloo, Canada
{ijdehaan,kpashkovich}@uwaterloo.ca

Abstract. We study a class of Bayesian online selection problems with matroid constraints. Consider a vendor who has several items to sell, with the set of sold items being subject to some structural constraints, e.g., the set of sold items should be independent with respect to some matroid. Each item has an offer value drawn independently from a known distribution. Given distribution information for each item, the vendor wishes to maximize their expected revenue by carefully choosing which offers to accept as they arrive.

Such problems have been studied extensively when the vendor's revenue is compared with the offline optimum, referred to as the "prophet". In this setting, a tight 2-competitive algorithm is known when the vendor is limited to selling independent sets from a matroid [29]. We turn our attention to the online optimum, or "philosopher", and ask how well the vendor can do with polynomial-time computation, compared to a vendor with unlimited computation but with the same limited distribution information about offers.

We show that when the underlying constraints are laminar and the arrival of buyers follows a natural "left-to-right" order, there is a Polynomial-Time Approximation Scheme for maximizing the vendor's revenue. We also show that such a result is impossible for the related case when the underlying constraints correspond to a graphic matroid. In particular, it is PSPACE-hard to approximate the philosopher's expected revenue to some fixed constant $\alpha < 1$; moreover, this cannot be alleviated by requirements on the arrival order in the case of graphic matroids.

1 Introduction

In this paper, we study the problem of Bayesian online selection subject to structural constraints given by matroids. Let us consider a scenario where a vendor posts and updates prices in order to maximize their profit subject to structural constraints. This type of problems is omnipresent in our everyday life. Consider vendors, e.g., big e-commerce platforms or independent crafters, who sell items by posting prices. In order to maximize their profits, vendors usually calculate prices taking into account partial information about potential buyers and the constraints on inventory, transportation networks, legal regulations, etc.

I. DeHaan—Supported by an NSERC Canada Graduate Scholarship.
K. Pashkovich—Supported by NSERC Discovery Grants Program RGPIN-2020-04346.

G. Schäfer and C. Ventre (Eds.): SAGT 2024, LNCS 15156, pp. 405–422, 2024.
https://doi.org/10.1007/978-3-031-71033-9_23

One of the most prominent examples of this setting is the *single-item prophet inequality problem* [30]. In this problem, the vendor is selling one item, for which they observe a sequence of offers. The offers correspond to random variables $v_1, v_2, \ldots, v_n$ drawn independently from distributions known to the vendor. At each timestamp t, the vendor may choose to stop by selling their item and gaining the value v_t, or the vendor may choose to discard this offer and continue. The *prophet* in this problem represents a person who knows the realizations of all offers $v_1, v_2, \ldots, v_n$ ahead of time. Moreover, the expected gain of the vendor is evaluated using the maximum gain of the prophet as a benchmark. In this scenario, the vendor can do at least half as well as the prophet and no better [30,39]. This result generalizes to the setting of *matroid prophet inequalities*, in which the vendor is selling items from a matroid, and is limited to selling an independent set in this matroid [29].

The classical prophet inequality problems make the vendor compete with the prophet, where only the prophet knows the realizations of $v_1, v_2, \ldots, v_n$ ahead of time. Clearly, such a competition between the vendor and the prophet is very unfair. Indeed, in many cases the advantage of knowing all realizations cannot be alleviated through any efforts of the vendor. So let us change the benchmark and introduce the *philosopher*. The philosopher does not know the realizations of $v_1, v_2, \ldots, v_n$ but has unlimited computational power. The central question for our work is as follows. *How well can a vendor limited to polynomial time computation compete with the philosopher, where both know only the distributions of* $v_1, v_2, \ldots, v_n$ *but the latter has unlimited computational power?* Answering this question, we provide both positive and negative results for the ability of the vendor.

1.1 Our Results

In the single-item case, the vendor can achieve the same profit as the philosopher. Indeed, in this case the vendor can set a straightforward dynamic program that computes the optimal strategy.

Gupta posed the question of whether the vendor can achieve the same gain as the philosopher in the matroid setting, and more specifically in the graphic and laminar matroid settings [25]. We answer this question in the negative for graphic matroids. We show that for graphic matroids, it is PSPACE-hard for the vendor to approximate the expected gain of the philosopher up to some fixed constant. Moreover, for the graphic matroids there is no arrival order which "substantially increases competitiveness" of the vendor.

On the positive side, we provide a Polynomial-Time Approximation Scheme (PTAS) for all laminar matroids with "left-to-right" arrival orders, in which elements from each constraint arrive consecutively. Furthermore, the provided PTAS also holds for arrival orders that are "close" to "left-to-right" orders, i.e., to orders where each element is contained only in constantly many bins on which the arrival order is not "left-to-right". We note that our policy relies on the "left-to-right" order only in the analysis. We leave it as an open question to determine whether there are substantially different requirements that guarantee our policy to lead to a PTAS.

The defined order is called *left-to-right* because when the laminar matroid is drawn with all elements on one horizontal line, the elements can be arranged in order from left to right if and only if they are in a "left-to-right" ordering. These orderings are exactly those orderings that are obtained when the tree corresponding to the laminar family is explored with depth-first-search.

Let us provide an example of a Bayesian selection problem with a "left-to-right" arrival order. Consider a situation where items correspond to clients, and the vendor knows clients' arrival order and distributions for their offers. Let the parameter p represent some crucial resource and so determine restrictions on the number of clients the vendor can serve. In particular, let there be several critical thresholds p_1, p_2,... for p and limits γ_1, γ_2,.... For each i, if the value of p drops below p_i at some timestamp then we can serve at most γ_i clients between this timestamp and the next timestamp when the value of p is again at least p_i. In Fig. 1, one can see the example of how the parameter p changes over time and the corresponding laminar family. We note that the arrival order in the "production constrained Bayesian selection" from [5] corresponds to the case when p is a resource that is being delivered to us over time, see Fig. 2, plus one additional global constraint.

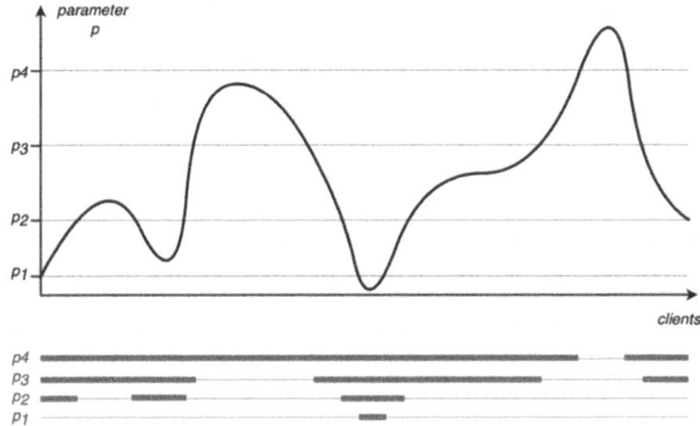

Fig. 1. Here, the horizontal axis is associated with clients and so with the timestamps, since their arrivals provide a measure for time. The vertical axis is associated with the value of the parameter p. The critical thresholds p_1, p_2, p_3 and p_4 for the value of p are depicted on the vertical axis. Below the picture of the graph, one can find the illustration for the corresponding laminar matroid. In particular, the picture below contains an interval for each critical threshold and the timestamps, when the value of p drops below the threshold and the next timestamp when the value of p reaches the value of the threshold.

Our PTAS holds for more general matroid Bayesian selection problems than the ones described above. We can handle the cases where the limits γ_1, γ_2,... are functions of timestamps when the value of p drops below p_i. Moreover, we allow

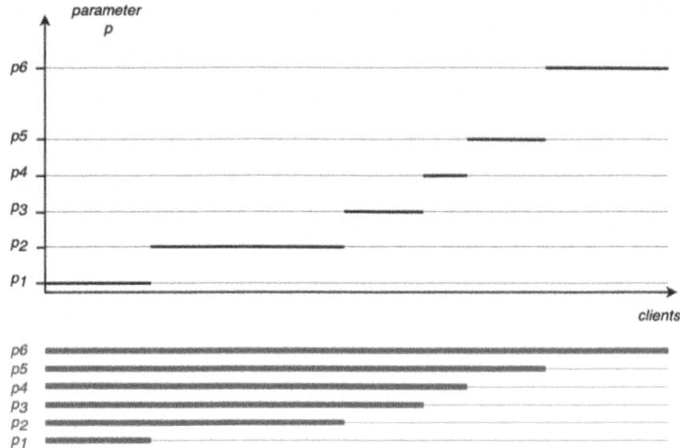

Fig. 2. Here the legend of the figure is the same as in Fig. 1. The structure of the function for the parameter p corresponds to the situation when p represents a resource that is being delivered over time, and is spent only on servicing clients. In this scenario, the vendor can serve only a certain number of clients until the next delivery of the resource.

the arrival order to be more complicated than just "left-to-right". One can imagine that the vendor runs several services in parallel, which mostly depend on different resources, however each type of service depends on at most a constant number of resources that are needed also for other types of service. In this situation, as long as the dependence on the "overarching" resources leads to a laminar matroid, our policy provides a PTAS for the arising Bayesian selection problem.

Thus our results provide a generalization and extension of the previous results for laminar matroids. The study of laminar Bayesian selection problems was initiated by Anari et al. [5]. They gave a PTAS for the special cases of bounded-depth laminar matroids and production constrained selection. Note that both of these special cases have required structure on the laminar family, with production constrained selection also having a special arrival order. In contrast, our results do not impose any requirements on the structure of the laminar matroids, but only on the arrival order.

1.2 Our Techniques

PTAS for Left-to-Right Laminar Bayesian Selection. We capture the optimal online policy for left-to-right laminar Bayesian selection with an exponential-sized linear program introduced in [5]. While this linear program is too large to solve efficiently, we create a polynomial-sized relaxation by partitioning bins into "big" and "small" based on their capacities. For each maximal "small" bin, we require the constraints from the exponential-sized linear program to hold exactly. But for each "big" bin, we only require its capacity constraints to "hold in expectation".

We show that by solving the linear program, we can efficiently obtain an online policy that is feasible for all "small" bins and has total expected gain equal to the optimal value of the linear program. In fact, the obtained online policies are optimal for the maximal "small" bins subject to changes in the value distributions. The argumentation about the above online policies goes generally along the lines of the analysis in [5]. One of our main technical contributions is showing that, with some pre-processing on capacities, "big" bins are unlikely to be violated by the obtained policies, see Lemma 7. To show this, we need concentration bounds on the number of items selected from each maximal "small" bin. The fact that the obtained policies act optimally on the maximal "small" bins does not guarantee us that selections of items are negatively correlated, see Fig. 3, and not even that the selection of an item is negatively correlated with the number of items selected before it, see Fig. 4. Thus, we cannot rely on the analysis from [5] and have to develop new tools. Nevertheless, we are able to show a more "global" version of negative dependence; we show that the number of items selected from each small bin is concentrated closely around the mean. Roughly speaking, we obtain this by showing that the selection of an item has a limited impact on the expected number of the items selected after it, see Lemma 5. This gives us the building blocks needed for Chernoff type results, see Lemma 6, which we use to bound the probability of "big" bins being violated in Lemma 7.

These concentration results are only possible for restricted arrival orders. We show in Theorem 2 that there exist laminar Bayesian online selection instances where the number of elements selected by the optimal policy is anti-concentrated. Due to the constructions done in the proof of Theorem 2, straightforward counting arguments show that given natural numbers r and $n \in \Omega(r^2)$, for a randomly chosen laminar matroid over n elements with rank r and a randomly chosen order, asymptotically almost surely one can choose value distributions and capacities such that the resulting instance exhibits anti-concentration for the number of elements selected by the optimal policy. This is a significant roadblock on the way to giving a good approximation for laminar Bayesian online selection problems with no restriction on arrival order. Most known results for problems of this type first break the problem into smaller pieces, solve the philosopher's problem optimally on each part, and then use some concentration result to show that combining the solution for these parts is unlikely to produce infeasible solutions for the global problem. Without strong concentration on the number of elements selected by the philosopher, this framework cannot work - further new ideas are needed to tackle the general problem.

Hardness of Graphic Matroid Bayesian Selection. To obtain `PSPACE`-hardness result for approximating Bayesian online selection for graphic matroids, we reduce from Stochastic MAX2SAT (`MAX-S2SAT`). The arrival order of edges is organized in three phases. In the first phase, the optimal policy for the constructed instance would need to make decisions that are equivalent to assigning the variables in `MAX-S2SAT` `True` and `False` values. In the second phase and

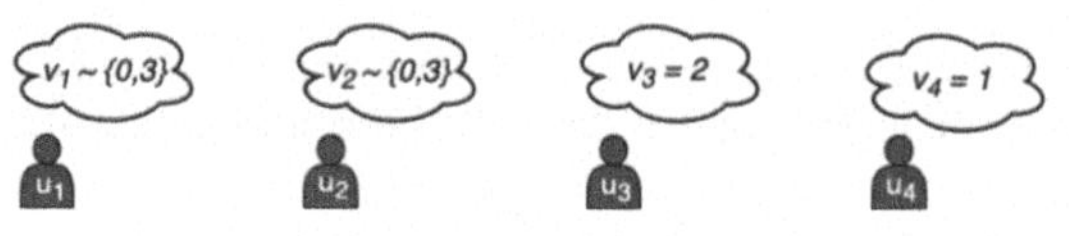

Fig. 3. Here, the vendor receives four clients u_1, u_2, u_3 and u_4 in the corresponding order, but can serve at most 2 clients. The distributions for the offers are as in the figure, e.g., u_2 offers 3 with probability 0.5 and otherwise offers 0, u_4 always offers 1. Let X_i, $i = 1, \dots, 4$ be the event (and the corresponding indicator variable) that u_i was served by the optimal online policy. We have $\mathbf{Pr}[X_3] = 3/4$, $\mathbf{Pr}[X_4] = 1/4$ while $\mathbf{Pr}[X_3 \wedge X_4] = 1/4$. Thus, we have $\mathbf{Cov}(X_3, X_4) = 1/16$.

Fig. 4. Here, the vendor receives six clients $u_1, \dots, u_6$ in the corresponding order, but can serve at most 3 clients. Moreover, among u_3 and u_4 the vendor can serve at most one client. The distributions for the offers are as on the figure, e.g. u_2 makes offer 3 with probability 0.5 and otherwise makes offer 0, u_5 always offers 1. Let X_i, $i = 1, \dots, 6$ be the event (and the corresponding indicator variable) that u_i was served by the optimal online policy. We have $\mathbf{Pr}[X_1] = \mathbf{Pr}[X_2] = 1/2$, $\mathbf{Pr}[X_3] = 1/4$, $\mathbf{Pr}[X_4] = 3/8$, $\mathbf{Pr}[X_1 \wedge X_4] = \mathbf{Pr}[X_2 \wedge X_4] = 1/4$ and $\mathbf{Pr}[X_3 \wedge X_4] = 0$. Thus, we have $\mathbf{Cov}(X_1 + X_2 + X_3, X_4) = 1/32$.

the third phase, the optimal policy would make greedy decisions, with expected payoff depending on the choices made in the first phase. Having shown that the behaviour of the optimal policy indeed follows the above rules, the expected gain of any such policy would be equal to the expected number of satisfied clauses plus a fixed term and a negligible error term. Our analysis to estimate the expected gain follows the lines of the work [34] on stochastic online matching. Later, we show that for graphic matroids, there are no orders like left-to-right orderings for laminar matroids. In particular, without knowing value distributions we cannot associate an arrival order to each graphic matroid such that the resulting class of instances admits a PTAS. To obtain this result, we show that we can embed our hardness instance into a sufficiently large complete graph for any arrival order on that complete graph.

1.3 Further Related Work

Study of prophet inequalities was initiated by [30] in which they considered adaptive algorithms for the single-item setting. Several years later, a simple single-threshold $\frac{1}{2}$-competitive algorithm was given in [39]. In the decades following, many variants of the classic prophet inequality problem have been considered.

We give a brief snapshot of work on variants related to this paper, and recommend reading surveys on the area for a more complete picture [11,28,31].

In recent years, there has been much work dedicated to the study of prophet inequalities under combinatorial constraints. Beginning with uniform matroid constraints [3,26], there has been exploration of general matroid constraints [29], knapsack constraints [20], matching constraints [4,17,24], downwards-closed constraints [36,37], and many more. Many other variations of the classic prophet inequality problem have been studied extensively. Some examples include settings where the arrival order of elements is random [7,12,15,16], where every element has its value drawn independently from the same distribution, [1,10,27], single sample prophet inequalities [6,38], and non-adaptive prophet inequalities [9,35].

Our work is far from the first to consider approximation of the online optimum for Bayesian online selection problems. Online stochastic weighted bipartite matching has been extensively studied in the last several years, with a series of improved approximation factors given in [8,32,34]. In a variant of the single-item prophet inequality problem where the order of elements is unknown and uniformly random, there is a Polynomial-Time Approximation Scheme [14]. [2] shows that even in the single-item prophet inequality problem, it is NP-hard to select the best order to observe elements in. Additional approximations on the online optimum in stochastic and Bayesian online selection problems have been given in a variety of settings [21–23,40]. More recently, there has been work comparing order-unaware algorithms to the philosopher benchmark [18,19].

During the preparation of our paper, we became aware also of an unpublished PSPACE-hardness result for graphic Bayesian online selection obtained independently by another group [41].

2 Problem Definition and Preliminaries

In this paper, we consider structural constraints defined by matroids. Each matroid $\mathcal{M}$ is defined by a ground set U and a collection of subsets of U, which are called *independent sets.* We work with two types of matroids: laminar and graphic. A matroid $\mathcal{M}$ over a ground set U is *laminar* if there is some laminar family of sets $\mathcal{L}$ over U and a capacity function $c : \mathcal{L} \to \mathbb{N}$ such that a set I is independent if and only if $|I \cap A| \leq c(A)$ for all $A \in \mathcal{L}$. In this case, we write $\mathcal{M} = (U, \mathcal{L}, c)$. Given a laminar matroid $\mathcal{M} = (U, \mathcal{L}, c)$, we call the sets in $\mathcal{L}$ *bins.* The *depth* of a bin $A \in \mathcal{L}$ is defined as $|\{B : A \subseteq B, B \in \mathcal{L}\}|$. A matroid $\mathcal{M}$ over a ground set $U := E$ is *graphic* if there is some graph $G = (V, E)$ such that a set $I \subseteq E$ is independent if and only if (V, I) has no cycles. For further reading about matroids and their properties, we refer to [33].

From now on, we refer to the vendor as a *gambler* and to the items as *elements.* In the matroid Bayesian online selection problem, the gambler is given the matroid $\mathcal{M}$ over a ground set $U = \{u_1, \ldots, u_n\}$, an arrival order of the elements in U, and for each element $u_i \in U$, a distribution F_i for its value v_i. Over the course of the game, the gambler maintains an independent set I which starts as the empty set. The set I consists of all elements selected by the gambler at the given timestamp.

Elements arrive one by one in the given order. When u_i arrives, the value $v_i \sim F_i$ is drawn from its distribution independently from values of other elements and presented to the gambler. The gambler may then choose to either select the element as long as independence is maintained, updating $I \leftarrow I \cup \{u_i\}$ and gaining the associated value v_i, or reject the element, gaining no value. This decision is final and may not be changed later in the game.

The goal of the gambler is to gain as much value as possible over the course of the game. Let $OPT_{\mathcal{M}}$ indicate the maximum expected gain that the gambler can achieve for the instance $\mathcal{M}$. Here, we abuse notation and associate the matroid Bayesian selection instance to the underlying matroid $\mathcal{M}$ when the value distributions and arrival order are not relevant or are clear from the context.

The Laminar Matroid Bayesian Selection (`LMBS`) problem is the matroid Bayesian online selection problem restricted to laminar matroids. The Graphic Matroid Bayesian Selection (`GMBS`) problem is the matroid Bayesian online selection problem restricted to graphic matroids.

To conserve space, most proofs are omitted and can be found in the preliminary version [13].

3 Left-to-Right Laminar Bayesian Selection

In this section, we show that the laminar Bayesian online selection problem with certain types of arrival orders admits a PTAS.

Definition 1. *Given a laminar matroid $\mathcal{M} = (U, \mathcal{L}, c)$ and an ordering u_1, u_2, $\dots$, u_n of the elements U, we say that this is a* left-to-right ordering *if for every bin $A \in \mathcal{L}$ of the laminar family, $A = \{u_i, u_{i+1}, \dots, u_j\}$ for some i, j.*

In other words, the left-to-right ordering captures the rule that once elements from some bin start to arrive, they must not stop until they have all arrived. We say that an `LMBS` *instance has a left-to-right arrival* order (or equivalently it is a *left-to-right `LMBS` instance*) if the elements arrive according to a left-to-right ordering.

In this section, we assume that each distribution in the input is atomic. Moreover, each distribution is given to us explicitly as the list of values and the corresponding probabilities. This allows us to efficiently represent all distributions and to perform computations on them in polynomial time.

3.1 Warm-Up: QPTAS

To design a QPTAS and PTAS, we rely on the concept of states. A *state* has an entry for each considered bin A in $\mathcal{L}$, and this entry equals $c(A)$ minus the number of currently selected elements from A. Informally, a state represents the remaining capacities of the bins under consideration.

In an LMBS instance, given a timestamp let us call a bin A *active* if at least one element of A already arrived and there are still some elements of A left to arrive. The next lemma is based on the fact that in an LMBS instance, we can use a dynamic programming algorithm that only keeps track of states for active bins.

Lemma 1. *Let $\mathcal{M} = (U, \mathcal{L}, c)$ be a left-to-right LMBS instance with depth at most L. Then the optimal gain and an optimal policy of the gambler in $\mathcal{M}$ can be computed in $n^{\mathcal{O}(L)}$ time.*

Corollary 1. *The optimal gain and an optimal policy of the gambler in a constant-depth left-to-right LMBS instance can be computed in polynomial time.*

Lemma 2. *Let $\mathcal{M} = (U, \mathcal{L}, c)$ be an LMBS instance and α be in $(0, 1)$. We can, in polynomial time, construct an LMBS instance $\mathcal{M}' = (U, \mathcal{L}', c')$ such that:*

- *$\mathcal{L}' \subseteq \mathcal{L}$,*
- *for all A, $B \in \mathcal{L}'$ with $A \subsetneq B$, we have $c'(A) \leq \lceil \alpha \cdot c'(B) \rceil$,*
- *all independent sets in $\mathcal{M}'$ are also independent in $\mathcal{M}$,*
- *$\alpha \cdot OPT_{\mathcal{M}} \leq OPT_{\mathcal{M}'} \leq OPT_{\mathcal{M}}$, and*
- *if $\mathcal{M}$ is a left-to-right instance then $\mathcal{M}'$ is also a left-to-right instance.*

Corollary 2. *There is a QPTAS for left-to-right LMBS through Lemma 1 and Lemma 2 with $\alpha = 1 - \epsilon$.*

3.2 Linear Programming Formulation and Rounding Algorithm

To design a PTAS, we build on a linear programming formulation from [5]. The linear program from [5] encodes dynamic programming ideas, allowing us to construct a relaxation by decomposing a given instance of LMBS into tractable parts and concentrating on optimal policies for each of these tractable parts. For the sake of completeness, we present the linear program and relevant results from [5] in this section.

Small and Big Bins. To decompose an instance of LMBS into tractable parts, we partition $\mathcal{L}$ into "big" and "small" bins based on their capacities. If a bin is small, we use a linear program based on dynamic programming to guarantee its "feasibility". If a bin is big, we enforce that its "feasibility" is guaranteed in expectation.

Consider a threshold $K \in \mathbb{N}$. Call a bin $A \in \mathcal{L}$ *big* if $c(A) \geq K$, and *small* otherwise. Let $\mathscr{B} \subseteq \mathcal{L}$ indicate the set of big bins and $\mathscr{S} \subseteq \mathcal{L}$ indicate the set of inclusion-wise maximal small bins, i.e., the set of small bins that are not contained in any other small bin.

Without loss of generality, we assume that each element of U is contained in some small bin. If we have an element not contained in any small bin, then we can add a bin with capacity 1 containing only this element. So, we can assume that $\mathscr{S}$ partitions U.

Feasibility on Small Bins. Let us consider a maximal small bin $B \in \mathscr{S}$. Let $\mathcal{L}^B = \{B' \in \mathcal{L} : B' \subseteq B\}$ be the set of bins contained in B. To capture states with respect to B at each timestamp, we use $\mathcal{S}^B \subseteq \mathbb{Z}^{\mathcal{L}^B}$ to indicate the set of feasible states. Again, each vector $\boldsymbol{s} \in \mathcal{S}^B$ has an entry for every $A \in \mathcal{L}^B$ which represents $c(A)$ minus the number of selected elements from A. Note that $|\mathcal{S}^B|$ is at most $n^{\mathcal{O}(c(B))}$ because no more than $c(B)$ elements can be selected from B by an online policy.

Let $\boldsymbol{d}_t \in \{0,1\}^{\mathcal{L}^B}$ be the indicator vector for the bins in $\mathcal{L}^B$ containing the element u_t. Let us define the set of *forbidden neighboring states*

$$\partial\mathcal{S}^B := \{\boldsymbol{f} \in \mathbb{Z}^{\mathcal{L}^B} \setminus \mathcal{S}^B \ : \text{there are } u_t \in B \text{ and } \boldsymbol{s} \in \mathcal{S}^B \text{ such that } \boldsymbol{s} = \boldsymbol{f} + \boldsymbol{d}_t\}\,.$$

In other words, $\partial\mathcal{S}^B$ contains the set of all states that can be reached by an online policy immediately after violating feasibility in addition to some irrelevant unreachable states.

We use *allocation variables* $\mathcal{X}_t(\boldsymbol{s}, v)$, which represent the probability that the gambler accepts the tth item u_t and the state upon its arrival is $\boldsymbol{s}$, conditioned on v being the realized value of v_t. We then use $\mathcal{X}_t(v)$ to represent the conditional probability that we accept u_t, conditioned on v being the realized value of v_t. The *state variables* $\mathcal{Y}_t(\boldsymbol{s})$ represent the probability that the gambler is at state $\boldsymbol{s}$ upon the arrival of u_t.

Now we introduce a polytope $\mathcal{P}^B$ to capture transitions between states through an application of an online policy on the bin B.

$$\begin{aligned}
&\mathcal{X}_t(v) = \sum_{\boldsymbol{s}\in\mathcal{S}^B} \mathcal{X}_t(\boldsymbol{s}, v) && \forall u_t \in B, v\\
&0 \le \mathcal{X}_t(\boldsymbol{s}, v) \le \mathcal{Y}_t(\boldsymbol{s}) && \forall \boldsymbol{s} \in \mathcal{S}^B, u_t \in B, v\\
&\mathcal{Y}_{t+1}(\boldsymbol{s}) = \mathcal{Y}_t(\boldsymbol{s}) - \mathbb{E}_{v_t}[\mathcal{X}_t(\boldsymbol{s}, v_t)] + \mathbb{E}_{v_t}[\mathcal{X}_t(\boldsymbol{s} + \boldsymbol{d}_t, v_t)] && \forall \boldsymbol{s} \in \mathcal{S}^B, u_t, u_{t+1} \in B\\
&\mathcal{Y}_{t_0}([c(A)]_{A\in\mathcal{L}^B}) = 1 && t_0 = \min\{t : u_t \in B\}\\
&\mathcal{Y}_t(\boldsymbol{s}) = 0 && \forall \boldsymbol{s} \in \partial\mathcal{S}^B, u_t \in B.
\end{aligned}$$

One can see that any online policy for the restriction of $\mathcal{M}$ on B induces a feasible point in $\mathcal{P}^B$. Let us show that the opposite also holds.

Proposition 1 (Proposition 2.1 from [5]). *Given a bin $B \in \mathscr{S}$ and a point $\{\mathcal{X}_t(\boldsymbol{s}, v), \mathcal{X}_t(v), \mathcal{Y}_t(\boldsymbol{s})\} \in \mathcal{P}^B$, there is an online policy for the restriction of $\mathcal{M}$ on B which guarantees the expected gambler's gain to be $\sum_{u_t\in B} \mathbb{E}_{v_t}[v_t \cdot \mathcal{X}_t(v_t)]$ and the expected number of selected elements to be $\sum_{u_t\in B} \mathbb{E}_{v_t}[\mathcal{X}_t(v_t)]$. Moreover, this policy can be found in polynomial time.*

Feasibility on Big Bins. For each maximal small bin $B \in \mathscr{S}$, we introduce a variable $\mathcal{N}_B$ to represent an upper bound on the expected number of elements selected from B. Using these variables we construct our final linear program.

$$
\begin{aligned}
\text{maximize} \quad & \sum_{t=1}^{n} \mathbb{E}_{v_t}[v_t \cdot \mathcal{X}_t(v_t)] && \text{(LP)} \\
\text{subject to} \quad & \sum_{A \in \mathscr{S} : A \subseteq B} \mathcal{N}_A \le c(B) && \forall B \in \mathscr{B} \\
& \sum_{u_t \in B} \mathbb{E}_{v_t}[\mathcal{X}_t(v_t)] \le \mathcal{N}_B && \forall B \in \mathscr{S} \\
& \{\mathcal{X}_t(\boldsymbol{s}, v), \mathcal{X}_t(v), \mathcal{Y}_t(\boldsymbol{s})\} \in \mathcal{P}^B && \forall B \in \mathscr{S}
\end{aligned}
$$

Note that in (LP) the first type of constraints are global ex-ante constraints. They enforce the "feasibility" of big bins in expectation. The second type of constraints enforces that for each $B \in \mathscr{S}$ the variable $\mathcal{N}_B$ is a correct upper bound on the expected number of selected elements. Finally, the third type of constraints guarantees "feasibility" on all small bins. One can see that any online policy induces a feasible solution for (LP).

Algorithm. Now we are ready to state the algorithm. We pre-processes an input LMBS instance as in Lemma 2 with $\alpha = (1 - \epsilon)$. Next, we decrease the capacities for big bins in order to introduce "slack" for our online policy to satisfy their constraints, and solve the resulting (LP).

ALGORITHM 1: ALG($\mathcal{M} = (U, \mathcal{L}, c)$, K, ϵ)

1 Obtain $\mathcal{M}' = (U, \mathcal{L}', c')$ from $\mathcal{M} = (U, \mathcal{L}, c)$ as in Lemma 2 based on $\alpha = (1 - \epsilon)$;
2 Compute maximal small bins $\mathscr{S}$ and big bins $\mathscr{B}$ for $\mathcal{M}'$ based on K ;
3 Obtain $\mathcal{M}'' = (U, \mathcal{L}', c'')$ from $\mathcal{M}' = (U, \mathcal{L}', c')$ by setting $c''(B) := (1 - \epsilon) \cdot c'(B)$ for $B \in \mathscr{B}$; and $c''(B) := c'(B)$ for $B \in \mathcal{L}' \setminus \mathscr{B}$;
4 Compute the optimal solution $\{\mathcal{X}_t^\star(\boldsymbol{s}, v_t), \mathcal{X}_t^\star(v_t), \mathcal{Y}_t^*(\boldsymbol{s}), \mathcal{N}_A^*\}$ for the linear program (LP) with respect to $\mathcal{M}''$;
5 Extract an online policy from $\{\mathcal{X}_t^\star(\boldsymbol{s}, v_t), \mathcal{X}_t^\star(v_t), \mathcal{Y}_t^*(\boldsymbol{s})\}$ for each $B \in \mathscr{S}$;
6 Independently run the obtained online policies on $B \in \mathscr{S}$. Select an element as long as the obtained policy suggests to select it and the selection is feasible with respect to $\mathcal{M}'$;

3.3 Analysis

In this subsection, we estimate the approximation guarantees achieved by Algorithm 1 with respect to an optimal online policy.

Losses from Transformations in Steps 1–3 **of Algorithm** 1. There are two steps in Algorithm 1 where we transform the original LMBS problem: Step 1 and Step 3. The next lemma captures how the optimal gain changes between these transformations. Define $LP_{\mathcal{M}''}$ to indicate the optimal objective value of (LP) with respect to $\mathcal{M}''$.

Lemma 3. *In Algorithm 1, we have that* $LP_{\mathcal{M}''} \ge (1 - \epsilon)^2 \cdot OPT_{\mathcal{M}}$.

Losses from Discarding Elements in Step 6 **of Algorithm** 1. The key challenge of analyzing Algorithm 1 is to estimate the expected value of elements that were not selected in Step 6 due to the feasibility restrictions of $\mathcal{M}'$. Lemma 7 shows that each element has a rather small probability of being not selected due to the feasibility restrictions of $\mathcal{M}'$. To show this, we first need some technical results about the concentration on the number of elements chosen in each small bin.

To obtain this concentration bound, we show that the online policy given by Algorithm 1 on any small bin B is actually an optimal algorithm for B with shifted values. Given this, we can analyze the behavior of the optimal algorithm on B.

Lemma 4. *Let $B \in \mathscr{S}$ be a maximal small bin. Then there exists λ^* such that the online policy for B computed in Step 5 is an optimal online policy for B with values of u_t, $t \in B$ being $v_t - \lambda^*$.*

Lemma 5. *Let $\widetilde{\mathcal{M}} = (\widetilde{U}, \widetilde{\mathcal{L}}, \widetilde{c})$ be a left-to-right* `LMBS` *instance with ordering of elements $\widetilde{u}_1, \widetilde{u}_2, \ldots, \widetilde{u}_n$. Consider the selections done by an optimal online policy when the policy encounters a fixed realization of values $\widetilde{v}_2, \widetilde{v}_3, \ldots, \widetilde{v}_n$. For this fixed realization, let μ_0 and μ_1 be the number of elements selected by an optimal online policy from $\{\widetilde{u}_2, \ldots, \widetilde{u}_n\}$ starting at $t = 2$ with S being $\varnothing$ and $\{\widetilde{u}_1\}$, respectively. Then we have that $\mu_1 \le \mu_0 \le \mu_1 + 1$.*

Lemma 6. *Given a maximal small bin $B \in \mathscr{S}$ and an element $u_t \in B$, let X_t be the event that the online policies computed in Step 5 of Algorithm 1 suggest to select u_t. Then, for every $\alpha > 0$ we have that*

$$\mathbb{E}[e^{\alpha \cdot X_B}] \le e^{(e^{\alpha}-1)\cdot \mathbb{E}[X_B]},$$

where $X_B := \sum_{u_t \in B} X_t$.

The Chernoff-like result of Lemma 6 then allows us to bound the failure probability of Step 5 and thus show that Algorithm 1 is a PTAS.

Lemma 7. *In left-to-right* `LMBS`*, for each $u_t \in U$, the probability that upon the arrival of u_t the online policies computed in Step 5 suggest to select u_t but it is not feasible with respect to $\mathcal{M}'$ is at most $3/(K \cdot \epsilon^3)$.*

Theorem 1. *Algorithm 1 is a $(1 - \epsilon)^3$ approximation for left-to-right* `LMBS` *instances when $K = 3\epsilon^{-4}$.*

We note that Lemmas 5 and 6 crucially rely on the left-to-right arrival order assumption. If this assumption is dropped, instances with essentially worst-possible concentration can be constructed.

Theorem 2. *For any $r \ge 1$ and $\epsilon > 0$, there exists an instance of laminar Bayesian online selection $\mathcal{M} = (U, \mathcal{L}, c)$ with rank$(\mathcal{M}) = r$ and $|U| \le (r+1)^2$ such that*

$$\boldsymbol{Pr}[\,|OPT(\mathcal{M})| = 0\,], \quad \boldsymbol{Pr}[\,|OPT(\mathcal{M})| = r\,] \in [\frac{1}{2} - \epsilon, \frac{1}{2} + \epsilon],$$

where $OPT(\mathcal{M}) \subseteq U$ is a random variable denoting the items selected by an optimal algorithm.

3.4 Extending to Arrivals "Close" to Left-to-Right

In this section, we extend the results for left-to-right LMBS instances to LMBS instances that are "close" to being left-to-right. Let $\mathcal{M} = (U, \mathcal{L}, c)$ be an LMBS instance and $B \in \mathcal{L}$ a bin. We say that B is a *left-to-right bin* if the instance $\mathcal{M}$ restricted to B is a left-to-right LMBS instance.

We are able to extend the results because we do not need every bin in laminar matroid to be left-to-right. For our analysis to go through, it is enough for every maximal small bin to be left-to-right. Given the above observation, we obtain the following theorem by using a modified definition of small and big bins, which is inspired by [5].

Theorem 3. *Let $L \in \mathbb{N}$ be a constant. There is a PTAS for LMBS instances where every element lies in at most L bins that are not left-to-right.*

Corollary 3 (Theorem 3.5 from [5]). *There is a PTAS for constant-depth LMBS instances.*

The production constrained Bayesian selection problem from [5] consists of LMBS instances in which all bins except the largest are left-to-right, plus some additional structure. This gives us the following corollary with $L = 1$.

Corollary 4 (Theorem 2.3 from [5]). *There is a PTAS for production constrained Bayesian selection.*

4 Graphic Bayesian Selection

LMBS can be viewed as a special case of the more general Matroid Bayesian Online Selection problem, where instead of a laminar matroid, we are given any matroid $\mathcal{M}$. Gupta questioned in 2017 whether the best strategy for Matroid Bayesian Online Selection is computationally hard to find [25]. We answer this question in the affirmative, even for the special case of graphic matroids.

Theorem 4. *There is an absolute constant $\alpha \in (0, 1)$ such that it is PSPACE-hard to approximate Graphic Matroid Bayesian Selection (GMBS) to a factor of α.*

Note that we can compute the optimal strategy for GMBS in polynomial space with a simple brute-force recursive algorithm, so GMBS is in fact PSPACE-complete. Also note that Theorem 4 says it is PSPACE-hard to approximate the *expected value* of an optimal strategy. This does not immediately imply that it is PSPACE-hard to act according to an approximate optimal strategy, but the proof of Theorem 4 shows this as well.

Definition 2. *We say that a class of matroids $\mathscr{M}$* admits a PTAS-compatible distribution-agnostic order *if there exists an order $\sigma(\mathcal{M})$ for every matroid $\mathcal{M} \in \mathscr{M}$ such that there is a Polynomial-Time Approximation Scheme for instances of Matroid Bayesian Online Selection with matroids from $\mathscr{M}$ and arrival orders given by σ.*

We show the following result by demonstrating that the hardness construction of Theorem 4 can be embedded in any arrival order of sufficiently large complete graphs.

Theorem 5. *The set of graphic matroids does not admit a PTAS-compatible distribution-agnostic order unless* `PSPACE` $=$ `P`*.*

As a corollary of Theorem 1, we have that the set of laminar matroids admits a PTAS-compatible distribution-agnostic order by taking $\sigma(\mathcal{M})$ to be any left-to-right ordering on $\mathcal{M}$. Theorem 5 provides a separation between graphic and laminar matroids.

4.1 Hardness for Graphic Bayesian Selection

`PSPACE`**-Hard Problem:** `MAX-S2SAT`. To prove Theorem 4, we first need some results about Stochastic Max 2SAT.

Definition 3. *In the* Stochastic Max 2SAT problem, *henceforth referred to as* `MAX-S2SAT`*, the input is a 2CNF formula ϕ over an ordered list of variables $(x_1, x_2, \ldots, x_n)$, where n is even and for every $i = 1, \ldots, n-1$, either x_i or x_{i+1} is contained in some clause of ϕ. We choose a value of* `True` *or* `False` *for x_1. Then, nature sets x_2 to either* `True` *or* `False` *with a probability of* 0.5*. We then get to choose the value of x_3, nature sets the value of x_4 to either* `True` *or* `False`*, and so on. Our goal is to maximize the expected number of satisfied clauses in ϕ after all the variables have been assigned a value. A variable is called a* random *variable if it is set by nature, and is called* deterministic *otherwise.*

Lemma 8. *There exist absolute constants $k \in \mathbb{N}$ and $\alpha \in (0, 1)$ so that it is* `PSPACE`*-hard to compute an α-approximation for a* `MAX-S2SAT` *instance ϕ satisfying the requirement that each variable appears in at most k clauses of ϕ.*

Reduction from `MAX-S2SAT` **to** `GMBS`**.** Now, we can state our reduction from `MAX-S2SAT` to `GMBS`. Let ϕ be an instance of `MAX-S2SAT` as in Lemma 8 with variables $(x_1, x_2, \ldots, x_n)$ and constant k. Let m indicate the number of clauses in ϕ. We construct an instance $\mathcal{I}_\phi$ of `GMBS` as follows.

The vertex set in the graph for $\mathcal{I}_\phi$ consists of a single central vertex w and two vertices for each variable x_i, $i = 1, \ldots, n$ in ϕ. We label the vertices for the variable x_i, $i = 1, \ldots, n$ as x_i and $\neg x_i$

$$V := \{w\} \cup \{x_i, \neg x_i : 1 \leq i \leq n\}.$$

The edges of $\mathcal{I}_\phi$ arrive in three distinct phases and have distinct value distributions. Before specifying $\mathcal{I}_\phi$ in full detail, let us explain the intuition behind the construction. The first phase simulates the selection of truth values for variables in the same order as they appear in ϕ. The second phase accounts for the number of clauses satisfied by the value selection of phase one. For this, the edges in the second phase correspond to clauses and attain very high values with very small probability, so that multiple clause edges attaining value is very unlikely. The third phase guarantees that for each variable at most one truth value is selected in the first phase by the optimal online policy.

First Phase. Respecting the order of variables as in ϕ, two edges wx_i and $w\neg x_i$ arrive for each variable x_i, $i = 1, \ldots, n$. The edge wx_i arrives immediately before $w\neg x_i$. If x_i is a deterministic variable, both edges wx_i and $w\neg x_i$ have deterministic value 1. If x_i is a random variable, the edge wx_i has value 2 with probability 0.5 and value 0 otherwise; the edge $w\neg x_i$ has deterministic value 1.

Second Phase. An edge $\neg\ell_1\neg\ell_2$ arrives for each clause $(\ell_1 \vee \ell_2)$ in ϕ. Each such edge has value $m^4/2k$ with probability m^{-4}, and value 0 otherwise. For each clause (ℓ) in ϕ, i.e., for each clause with a single literal, an edge $w\neg\ell$ arrives with the same value distribution.

Third Phase. An edge $x_i\neg x_i$ arrives for each variable x_i, $i = 1, \ldots, n$. The edge $x_i\neg x_i$ has deterministic value 2.

Note that within the first phase, the order of arrivals is important. In contrast, the order of edge arrivals within the second or third phase can be arbitrary.

Optimal Online Policy for $\mathcal{I}_\phi$. Let $OPT_{on}(\phi)$ refer to the expected value of the optimal online algorithm for MAX-S2SAT on ϕ. Let ALG_{opt} denote an optimal algorithm for GMBS. We can assume that ALG_{opt} selects only arrived edges. We say that an edge *arrived* if the value of the edge is nonzero.

Then the following properties hold.

- ALG_{opt} selects exactly one of wx_i, $w\neg x_i$ for each variable x_i.
- For a random variable x_i, ALG_{opt} selects wx_i if and only if wx_i arrives.
- After the first phase, ALG_{opt} selects every edge that arrives and can be selected without introducing a cycle.

Note that, given these properties, the first clause edge to arrive will be selected exactly when the truth assignment selected in phase 1 satisfies the clause. This along with the fact that the event in which more than 1 clause edge arrives is extremely unlikely can be used to show that the optimal algorithm obtains value

$$1.25n + 2n \cdot \gamma + 2n \cdot m^{-3} \cdot \gamma + \delta_A + OPT_{on}(\phi) \cdot \left(\frac{m^4}{2k} - 2\right) \cdot m^{-4} \cdot \gamma,$$

where $\gamma := (1 - m^{-4})^{m-1}$ and $\delta_A \in [0, 2m^{-1}]$. Finally, we can see that $OPT_{on}(\phi) \geq \frac{m}{2} \geq \frac{n}{8}$ due to the fact that a random assignment of variables will satisfy $\frac{m}{2}$ clauses in expectation. From here, standard techniques and the fact that MAX-S2SAT is PSPACE-hard to approximate finish the proof of Theorem 4.

4.2 No PTAS-Compatible Distribution-Agnostic Order

Let us show that for a graphic matroid on a complete graph there is no PTAS-compatible distribution-agnostic order. Let us be given a complete graph $G = (V, E)$ with $|V| = 3n + 2m + 1$, where n is even, and let us be given some arrival order for E. Let us show how one can construct a reduction for `MAX-S2SAT` analogously to Sect. 4.1.

Edges for Third Phase. In the reverse order of arrival, greedily construct a matching $A := \{a_1, \ldots, a_n\}$ such that $|A| = n$. Each edge in A has deterministic value 2.

Edges for Second Phase. In the reverse order of arrival, after constructing A, greedily keep adding edges to $B := \{b_1, \ldots, b_m\}$ such that $A \cup B$ is a matching and $|B| = m$. Each edge in B has value $m^4/2k$ with probability m^{-4}, and value 0 otherwise.

Edges for First Phase. Consider the vertex set U, which are the vertices not matched by the matching $A \cup B$. Consider a vertex $w \in U$ and let v_1, $\overline{v}_1$, v_2, $\overline{v}_2$, ..., v_n, $\overline{v}_n$ be such that the edges wv_1, $w\overline{v}_1$, wv_2, $w\overline{v}_2$, ..., wv_n, $w\overline{v}_n$ are the first n edges adjacent to w and their arrivals are in the above order. If i is odd, both edges wv_i and $w\overline{v}_i$ have deterministic value 1. If i is even, the edge wv_i has value 2 with probability 0.5 and value 0 otherwise; and the edge $w\overline{v}_i$ has deterministic value 1.

Auxiliary Edges. For each $i = 1, \ldots, n$ both edges $v_i\alpha_i$ and $\overline{v}_i\gamma_i$ have deterministic value 3, where α_i and γ_i are the endpoints of a_i. For each $j = 1, \ldots, m$ both edges $u_j\beta_j$ and $l_j\tau_j$ have deterministic value 3, where β_j, τ_j are the endpoints of b_j and l_j, u_j correspond to "negations" of the literals in the jth clause of ϕ.

Every remaining edge has deterministic value 0. Note, that all auxiliary edges arrive before the edges in the second and third phase due to the greedy construction of A and B. Also all auxiliary edges are selected by an optimal online policy. A similar analysis as the one presented in Sect. 4.1 then yields Theorem 5.

References

1. Abolhassani, M., Ehsani, S., Esfandiari, H., Hajiaghayi, M., Kleinberg, R., Lucier, B.: Beating 1-1/e for ordered prophets. In: Proceedings of the 49th Annual ACM SIGACT Symposium on Theory of Computing, pp. 61–71 (2017)
2. Agrawal, S., Sethuraman, J., Zhang, X.: On optimal ordering in the optimal stopping problem. In: Proceedings of the 21st ACM Conference on Economics and Computation, EC 2020, pp. 187–188. Association for Computing Machinery, New York (2020). https://doi.org/10.1145/3391403.3399484
3. Alaei, S.: Bayesian combinatorial auctions: Expanding single buyer mechanisms to many buyers. In: 2011 IEEE 52nd Annual Symposium on Foundations of Computer Science, pp. 512–521 (2011)https://doi.org/10.1109/FOCS.2011.90
4. Alaei, S., Hajiaghayi, M., Liaghat, V.: Online prophet-inequality matching with applications to ad allocation. In: Proceedings of the 13th ACM Conference on Electronic Commerce, pp. 18–35 (2012)
5. Anari, N., Niazadeh, R., Saberi, A., Shameli, A.: Nearly optimal pricing algorithms for production constrained and laminar Bayesian selection. In: Proceedings of the 2019 ACM Conference on Economics and Computation, pp. 91–92 (2019)

6. Azar, P.D., Kleinberg, R., Weinberg, S.M.: Prophet inequalities with limited information. In: Proceedings of the Twenty-Fifth Annual ACM-SIAM symposium on Discrete algorithms, pp. 1358–1377. SIAM (2014)
7. Azar, Y., Chiplunkar, A., Kaplan, H.: Prophet secretary: surpassing the 1-1/e barrier. In: Proceedings of the 2018 ACM Conference on Economics and Computation, pp. 303–318 (2018)
8. Braverman, M., Derakhshan, M., Molina Lovett, A.: Max-weight online stochastic matching: improved approximations against the online benchmark. In: Proceedings of the 23rd ACM Conference on Economics and Computation, EC 2022, pp. 967–985. Association for Computing Machinery, New York (2022). https://doi.org/10.1145/3490486.3538315
9. Chawla, S., Goldner, K., Karlin, A.R., Miller, J.B.: Non-adaptive matroid prophet inequalities. arXiv preprint arXiv:2011.09406 (2020)
10. Correa, J., Foncea, P., Hoeksma, R., Oosterwijk, T., Vredeveld, T.: Posted price mechanisms for a random stream of customers. In: Proceedings of the 2017 ACM Conference on Economics and Computation, pp. 169–186 (2017)
11. Correa, J., Foncea, P., Hoeksma, R., Oosterwijk, T., Vredeveld, T.: Recent developments in prophet inequalities. SIGecom Exch. **17**(1), 61–70 (2019). https://doi.org/10.1145/3331033.3331039
12. Correa, J., Saona, R., Ziliotto, B.: Prophet secretary through blind strategies. Math. Program. **190**(1–2), 483–521 (2021)
13. DeHaan, I., Pashkovich, K.: Matroid Bayesian online selection. arXiv preprint arXiv:2406.00224 (2024)
14. Dütting, P., Gergatsouli, E., Rezvan, R., Teng, Y., Tsigonias-Dimitriadis, A.: Prophet secretary against the online optimal. In: Proceedings of the 24th ACM Conference on Economics and Computation, EC 2023, pp. 561–581. Association for Computing Machinery, New York (2023). https://doi.org/10.1145/3580507.3597736
15. Ehsani, S., Hajiaghayi, M., Kesselheim, T., Singla, S.: Prophet secretary for combinatorial auctions and matroids. In: Proceedings of the Twenty-Ninth Annual ACM-SIAM Symposium on Discrete Algorithms, pp. 700–714. SIAM (2018)
16. Esfandiari, H., Hajiaghayi, M., Liaghat, V., Monemizadeh, M.: Prophet secretary. SIAM J. Discret. Math. **31**(3), 1685–1701 (2017)
17. Ezra, T., Feldman, M., Gravin, N., Tang, Z.G.: Online stochastic max-weight matching: prophet inequality for vertex and edge arrival models. In: Proceedings of the 21st ACM Conference on Economics and Computation, pp. 769–787 (2020)
18. Ezra, T., Feldman, M., Gravin, N., Tang, Z.G.: who is next in line? On the significance of knowing the arrival order in Bayesian online settings. In: Proceedings of the 2023 Annual ACM-SIAM Symposium on Discrete Algorithms (SODA), pp. 3759–3776. SIAM (2023)
19. Ezra, T., Garbuz, T.: The importance of knowing the arrival order in combinatorial Bayesian settings. In: Garg, J., Klimm, M., Kong, Y. (eds.) WINE 2023. LNCS, vol. 14413, pp. 256–271. Springer, Cham (2023). https://doi.org/10.1007/978-3-031-48974-7_15
20. Feldman, M., Svensson, O., Zenklusen, R.: Online contention resolution schemes. In: Proceedings of the twenty-seventh annual ACM-SIAM Symposium on Discrete Algorithms, pp. 1014–1033. SIAM (2016)
21. Feng, Y., Niazadeh, R., Saberi, A.: Two-stage stochastic matching with application to ride hailing. In: Proceedings of the 2021 ACM-SIAM Symposium on Discrete Algorithms (SODA), pp. 2862–2877. SIAM (2021)

22. Feng, Y., Niazadeh, R., Saberi, A.: Near-optimal Bayesian online assortment of reusable resources. In: Proceedings of the 23rd ACM Conference on Economics and Computation, pp. 964–965 (2022)
23. Fu, H., Li, J., Xu, P.: A PTAS for a class of stochastic dynamic programs. In: Chatzigiannakis, I., Kaklamanis, C., Marx, D., Sannella, D. (eds.) Proceedings of the 45th International Colloquium on Automata, Languages, and Programming (ICALP 2018), Prague, Czech Republic, pp. 1–56 (2018)
24. Gravin, N., Wang, H.: Prophet inequality for bipartite matching: merits of being simple and non adaptive. In: Proceedings of the 2019 ACM Conference on Economics and Computation, pp. 93–109 (2019)
25. Gupta, A.: Lecture Notes. IPCO Summer School (2017)
26. Hajiaghayi, M.T., Kleinberg, R., Sandholm, T.: Automated online mechanism design and prophet inequalities. In: Proceedings of the 22nd National Conference on Artificial Intelligence, AAAI 2007, vol. 1, p. 58–65. AAAI Press (2007)
27. Hill, T.P., Kertz, R.P.: Comparisons of stop rule and supremum expectations of I.I.D. random variables. Ann. Probab. **10**(2), 336–345 (1982). http://www.jstor.org/stable/2243434
28. Hill, T.P., Kertz, R.P.: A survey of prophet inequalities in optimal stopping theory. Contemp. Math. **125**(1), 191–207 (1992)
29. Kleinberg, R., Weinberg, S.M.: Matroid prophet inequalities. In: Proceedings of the Forty-Fourth Annual ACM Symposium on Theory of Computing, pp. 123–136 (2012)
30. Krengel, U., Sucheston, L.: Semiamarts and finite values. Bull. Am. Math. Soc. **83**(4), 745–747 (1977)
31. Lucier, B.: An economic view of prophet inequalities. SIGecom Exch. **16**(1), 24–47 (2017). https://doi.org/10.1145/3144722.3144725
32. Naor, J., Srinivasan, A., Wajc, D.: Online dependent rounding schemes (2023)
33. Oxley, J.G.: Matroid Theory. Oxford Graduate Texts in Mathematics. Oxford University Press (2006). https://books.google.ca/books?id=puKta1Hdz-8C
34. Papadimitriou, C., Pollner, T., Saberi, A., Wajc, D.: Online stochastic max-weight bipartite matching: beyond prophet inequalities. In: Proceedings of the 22nd ACM Conference on Economics and Computation, EC 2021, pp. 763–764. Association for Computing Machinery, New York (2021). https://doi.org/10.1145/3465456.3467613
35. Pashkovich, K., Sayutina, A.: Non-adaptive matroid prophet inequalities. arXiv preprint arXiv:2301.01700 (2023)
36. Rubinstein, A.: Beyond matroids: secretary problem and prophet inequality with general constraints. In: Proceedings of the Forty-Eighth Annual ACM Symposium on Theory of Computing, pp. 324–332 (2016)
37. Rubinstein, A., Singla, S.: Combinatorial prophet inequalities. In: Proceedings of the Twenty-Eighth Annual ACM-SIAM Symposium on Discrete Algorithms, pp. 1671–1687. SIAM (2017)
38. Rubinstein, A., Wang, J.Z., Weinberg, S.M.: Optimal single-choice prophet inequalities from samples. arXiv preprint arXiv:1911.07945 (2019)
39. Samuel-Cahn, E.: Comparison of threshold stop rules and maximum for independent nonnegative random variables. Ann. Probab. 1213–1216 (1984)
40. Segev, D., Singla, S.: Efficient approximation schemes for stochastic probing and prophet problems. In: Proceedings of the 22nd ACM Conference on Economics and Computation, pp. 793–794 (2021)
41. Wajc, D.: Personal Communication (2023)

Information Sharing and Decision Making

Prediction-Sharing During Training and Inference

Yotam Gafni[1(✉)], Ronen Gradwohl[2], and Moshe Tennenholtz[3]

[1] Weizmann Institute of Science, Rehovot, Israel
yotam.gafni@gmail.com
[2] Ariel University, Ariel, Israel
roneng@ariel.ac.il
[3] Technion - Israel Institute of Technology, Haifa, Israel
moshet@ie.technion.ac.il

Abstract. Two firms are engaged in a competitive prediction task. Each firm has two sources of data—labeled historical data and unlabeled inference-time data—and uses the former to derive a prediction model and the latter to make predictions on new instances. We study data-sharing contracts between the firms. The novelty of our study is to introduce and highlight the differences between contracts to share prediction models only, contracts to share inference-time predictions only, and contracts to share both.

Our analysis proceeds on three levels. First, we develop a general Bayesian framework that facilitates our study. Second, we narrow our focus to two natural settings within this framework: (i) a setting in which the accuracy of each firm's prediction model is common knowledge, but the correlation between the respective models is unknown; and (ii) a setting in which two hypotheses exist regarding the optimal predictor, and one of the firms has a structural advantage in deducing it.

Within these two settings we study optimal contract choice. More specifically, we find the individually rational and Pareto-optimal contracts for some notable cases, and describe specific settings where each of the different sharing contracts is optimal. Finally, on the third level of our analysis we demonstrate the applicability of our concepts in a synthetic simulation using real loan data.

Keywords: Data Sharing · Strategic Machine Learning · Strategic Classification · Information Sharing

1 Introduction

Machine learning (ML) is becoming a highly distributed endeavor. Data is spread among different firms, each of whom may have their own ML capabilities and economic utilities. In many cases, one firm's data and prediction capabilities are complemented by those available to a competing firm, and each firm would benefit from access to the other's predictions. For example, two investment banks

G. Schäfer and C. Ventre (Eds.): SAGT 2024, LNCS 15156, pp. 425–442, 2024.
https://doi.org/10.1007/978-3-031-71033-9_24

that attempt to predict loan defaults could each improve their respective predictions by accessing the other's predictions. Indeed, this is in the spirit of one of the most fundamental ideas in ML—aggregating weak learners into strong ones [9]. However, the distributed nature introduces a major obstacle: Why, and under what conditions, would firms willingly share their predictions with competitors? And what would equilibrium behavior look like, given such sharing?

Our main innovation in this paper is the observation that this obstacle actually consists of two separate questions: Would firms share the labels they have in the training phase? And would firms share their predictions for unlabeled instances?

In order to tackle this question of training/inference-stage prediction-sharing, we proceed on three levels. First, we develop a general Bayesian model that captures the two kinds of sharing. The Bayesian model specifies the informational environment, while a utility model specifies the economic implications. In the Bayesian model, each firm obtains a *training signal* that represents the *prediction model* (a.k.a. classifier) learned by that firm via its labeled historical data. The firm also obtains an *inference-time signal* that represents the classifier's prediction on unlabeled inference-time data. In the utility model we associate a real number with each outcome quadrant: True-positive, true-negative, false-positive, and false-negative predictions. We moreover assume that if both firms arrive at the same outcome, then the associated utility is split between them.

In the second level of our analysis, we apply our model to a game-theoretic study of two natural settings. In the first setting, the accuracy of each firm's prediction model is common knowledge, but the correlation between the respective models is unknown. As for utilities, firms have a safe prediction with utility zero (whether right or wrong), and a risky prediction. An example is a firm predicting a customer's trustworthiness to decide if to issue a loan. If a loan is provided, the firm's utility depends on the accuracy of the trustworthiness prediction, and whether or not the customer has other offers. If no loan is provided, the firm's utility is fixed at 0. In the second setting we study, there are two hypotheses regarding the optimal predictor, and one of the firms has a structural advantage in deriving it. Furthermore, firms' utilities are symmetric across prediction types (unlike the first setting), and depend only on the predictions' correctness. An example is a firm recommending a movie to a viewer, where the firm's utility depends on whether or not it accurately predicts the viewer's tastes.

Finally, in the third level of our analysis, we demonstrate the applicability of our ideas in a synthetic simulation using real loan data. This is intended to provide an accessible, practical recasting of our abstract model's results. In broad terms, if we take a single firm's perspective, the *no-sharing* contract allows it to build a classifier based on its own historical data. Then, based on its assessment (prior) of the competitor, it decides whether or not to act in accordance with the classifier's prediction (signal). An example of choosing to ignore the classifier's signal would be if the firm knows that its competitor can perfectly predict whether a loan would be repaid. Then, all the benefit of issuing a good loan is split (e.g., by the random decision of the consumer as to which of the offered loans to accept). However, since the firm knows that its own classifier is imperfect, it knows it will also end up issuing some bad loans. If the cost of bad loans outweighs the benefit of splitting the profit from good loans, the firm would decide to ignore its classifier and not issue any loans. Expanding on this example, the *train-sharing* contract can allow the firm to make a more refined decision: Based on seeing how the other firm predicts on the historical data, it can assess whether or not to follow its own classifier. The *full-sharing* contract allows even more intricate decision rules: They can depend both on what the firm learns about the competitor's predictions on historical data, and also on the competitor's prediction on each specific consumer. Lastly, the *infer-sharing* contract does not reveal the competitor's predictions on historical data, so it must maintain its prior over the other firm's classifier, but can use the competitor's prediction on the real-time consumer to decide whether to follow its own classifier's prediction. In our practical implementation of Sect. 5 we examine the performance of the optimal decision rules under different contracts, and show that each of no-sharing, full-sharing, and train-sharing is uniquely optimal for some set of parameters.

The emphasis of our work in game theoretic terms is to require that a contract is both *individually rational* and *Pareto-optimal* (IRPO). This follows the assumption that the natural state of affairs is that no contract is signed (no-sharing). Thus, for the firms to agree for any kind of prediction-sharing, it must be that for each of them, the expected utility under the prediction-sharing contract is at least as good as under no-sharing. We refer to this property as the contract being *individually rational*. Moreover, the contract must be *Pareto-optimal* with respect to the four possible contracts. That is, if the utilities under full-sharing dominate these under train-sharing, even if train-sharing is by itself individually rational, it would make sense that the firms choose to sign the Pareto-optimal contract rather than a Pareto dominated one. As we will see, there are different settings so that each of the contract types may be the unique IRPO contract.

Lastly, we note that in order for the firms to share their predictions, they need a way to match records. Facing this issue is common in the industry and there are companies that specialize in this task.[1] This type of prediction-sharing is valuable, even if done for identifiers both firms hold, as different firms may be exposed to different properties of the same identifier. As an example, think of firms that know different social and financial features associated with the same social security number. In this case, there is a difference between sharing each firm's binary prediction regarding the user, or the entire data it holds for that identifier. Importantly, our model assumes that firms share their training and inference-time *signals*, and not their entire data. In practice, in the training stage the signals come in the form of true labels in the historical data, and in the inference stage in the form of the classifier's predictions. The fact that this still proves to be useful is by itself interesting, as it suggests a path to data sharing that protects both the firm's intellectual property (in terms of both data and models used in training), and possibly the users' privacy.

[1] E.g., in advertising, identifying the same user on different devices is called cross-device targeting, enabled by "attribution providers" such as AppsFlyer and Singular.

1.1 Our Contribution

In Sect. 2, we provide the first model to reason about contracts that may involve sharing prediction both in the training and inference stage. In Sects. 3 and 4 we then focus on two natural sub-models of the general model we present:

1. *A Correlation Model:* Both firms know their own and their competitor's prediction accuracy, but not the correlation between the two prediction models. We characterize the uniquely individually rational and Pareto-optimal contracts for some notable cases. We also show that all contracts except inference-sharing can be optimal in this setting.
2. *A Two Hypotheses Model:* One firm is able to determine the correct hypothesis during training, while the other has information about customers that is valuable during inference. Here, we show that inference-sharing can be the unique individually rational and Pareto-optimal contract.

Overall, we conclude that each of the four train/inference contracts can be optimal:

- *No-sharing* is IRPO when the cost of making a wrong prediction is equal to the reward of making a correct prediction (Lemma 1, Theorem 1). Then, (1) Under full-sharing, when the two firms share their inference-time signals, the firms will simply follow the primary firm signal. This is because a negative primary firm signal overshadows a positive secondary firm signal. (2) Given the first insight, the primary firm is only set to lose by sharing its signal.
- *Full-sharing* is IRPO when the firms can use both signals to "amplify" or mitigate their individual signal (Theorem 2), and when train-sharing/infer-sharing have symmetric equilibria, whereas the symmetric full-sharing equilibrium is more informed (Lemma 4, Lemma 2).
- *Train-sharing* is IRPO when the two firms benefit from reaching different equilibria given a different correlation between their signals. In particular, the firms may prefer to each follow its signal when the correlation is low but have the secondary firm 'yield' to the primary firm when the correlation is high and exit the market.
- *Infer-sharing* is never uniquely IRPO under the correlation model (Lemma 2) but can appear under other natural models (Theorem 4). It is also used to amplify the individual signal, even without knowing what the other firm's signal is based off.

Beyond the existence results detailed above, which help provide intuition into the different types of prediction-sharing contracts, the theorems of Sect. 3 also provide a partial characterization of our correlation model in several important cases such as symmetric utilities or symmetric prediction-accuracy. In Sect. 5, we demonstrate how our abstract Bayesian model may be put into practice and implemented, using a real loan dataset.

1.2 Related Work

Strategic Collaborative ML. *Federated Learning* [17] is a popular framework where agents run training locally on data they have access to, and share it to create an optimized global model. A known issue is the possible problem of *free-riding* [8], where agents may benefit from other agents' inputs but keep the improvements from their own data to themselves. More generally, agents participating in collaborative learning tasks against competitors may seek to actively mislead others, while still benefiting from their data. [10] shows bounds on when collaborative learning is safe, depending on the number of fake identities an attacker may control, and the parameters of the learning task. [5] suggest distributing (different) noisy versions of the global model, and using budget-balanced payment schemes, as methods to ensure honest collaborative behavior. [2] study competition in regression, where instead of optimizing for the best model fit, agents try to be the best fit for as large a market share as possible. They find a sample and time efficient algorithm to calculate a best response in the game. [7] study how firms may choose the distribution of errors resulting from their learning algorithm in a competitive environment. They find that firms would prioritize minimizing bias rather than variance, even at the cost of having a higher total error rate.

Information Sharing and Selling. The economic literature on information sharing [3] studies how signals can be structured out of existing information, and be sold at an appropriate price. [1] finds that unlike when selling traditional items, achieving good welfare may require multiple rounds of communication, and large back and forth monetary transfers. [19] find that when outsourcing a learning task, it is best to use a "threshold contract" where the firm is paid only if it achieves a certain level of accuracy. [15] study a model where a firm pre-trains a model, to be revenue-shared with a firm that fine-tunes it. When firms' preferences are uni-modal, revenue sharing is Pareto-optimal within the interval between each of the firms' preferred revenue share, but no longer necessarily so with multiple fine-tuners, and characterize free-riding in their model.

The above works focus on different aspects of the machine learning pipeline, but mostly have a downstream perspective, where the information flows from the agent holding it towards the firm benefiting and paying for it. Our work belongs to the economic literature on data *sharing*, among different firms that hold parts of the information. [13] studies data aggregation between competitors. In this model, segmentation information about consumers is split between firms, and firms decide whether or not to share their part of the data with others during the *inference* phase. The work assumes that the segmentation scheme itself is known (which we would consider as something that is learned during the *training* phase). [12] studies competition in serving consumers within a Hotelling model. It finds that fully sharing the data on existing consumers increases competition and harms the firms, but that partial sharing schemes exist that not only benefit both firms but the consumers as well. Both works focus on contracts that involve only information sharing, without monetary transfers, as we do as well.

[20] study a model where competing firms share data samples. They find that sharing is more prevalent when the competition model is milder (Cournot vs. Bertrand competition), or when the learning task is harder (requires more samples to achieve good accuracy). There are many differences from our work: We employ a Bayesian model for our reasoning, we focus on simple binary predictions, and we assume the firms share their predictions and not the underlying data.

Lastly, we note that some works have observed the important role of information as a *coordination device* [4]. In this line of reasoning, the importance of the information that different firms see may not be in and of itself, but the fact that other firms see the same information and may use it to coordinate. We find a similar phenomenon in the train-sharing case, where the firms benefit from knowing that one of them may only lose in expectation when participating in the market. Then, the firm yields, which benefits both the firm itself, and its competitor which effectively becomes a monopoly.

2 Model

Informational Environment. There are two firms engaged in a competitive prediction task. Each firm obtains data in two phases: training and inference. In the training phase, examples with binary labels are drawn at random, and each firm learns a respective prediction model (i.e., classifier). The training phase may consist of one example, multiple examples, or "infinitely many" examples. In the inference phase firms use their learned model in order to predict the label of a new example. Firm 1's prediction is either A or B and firm 2's prediction is either a or b, where the former indicates that the firm's prediction model believes the label is 1 and the latter indicates the label is 0. We model this interaction in an abstract Bayesian framework using the *rich signal spaces* of [14] and [11]. We next describe the formal model, and highlight the main elements and interpretations.

A *world model* w consists of a prior distribution π_w over $\{0, 1\}$, as well as two *signal spaces*, one for each firm. For every true label $t \in \{0, 1\}$, each signal space partitions $[0, 1]$ into two sets, representing the probabilities associated with firms' prediction models, given true label t.[2] For the first firm, the first set is denoted $A^t_w \subset [0, 1]$, and the second is denoted $B^t_w = [0, 1] \setminus A^t_w$. For the second firm, the two sets are denoted a^t_w and b^t_w. Given w, a random example is modeled as a label t drawn from $\{0, 1\}$ according to π_w, as well as ζ drawn from $[0, 1]$ uniformly at random.[3] Firm 1's signal (i.e., its model's suggested prediction under w) on this example is then 1 if $\zeta \in A^t_w$ under t, and 0 otherwise; firm 2's signal is 1 if $\zeta \in a^t_w$ under t, and 0 otherwise. In words, ζ chooses a "location" on the interval [0,1]. This location decides some signal for Firm 1 (according to the way it partitions the interval $[0, 1]$), and similarly for Firm 2 (possibly with a different partition). Sampling ζ uniformly at random from $[0, 1]$ is in a sense similar to sampling a

[2] Formally, each signal space is a Lebesgue measurable bi-partition of $[0, 1] \times \{0, 1\}$.
[3] The uniformity assumption here is without loss of generality.

random feature vector that is used to train the firms' prediction models/requires a prediction at inference time.

In general, firms may not know the true w. Instead, let W be a possibly infinite set of possible world models, and suppose there is a commonly known prior π over them. An example of this framework is illustrated in Fig. 1.

Given this informational environment, the interaction proceeds as follows. In stage 0, Nature chooses an element w of W according to π. Then:

1. In the *training* stage, each firm i obtains a *training signal* w_i about the realized world model w. Each w_i is a function of firm i's respective signal space under w. Given signal w_1 (respectively, w_2) and the prior over W, each firm i uses Bayesian updating to posterior beliefs π_i over world models W.
2. In the *inference* stage, ζ is drawn from $[0,1]$ uniformly at random, and a label t is drawn from $\{0,1\}$ according to π_w. Firm 1 obtains the *inference-time signal* $X \in \{A, B\}$ that satisfies $\zeta \in X^t_{w'}$, where $w' \sim \pi_1$; firm 2 obtains the *inference-time signal* $x \in \{a, b\}$ that satisfies $\zeta \in x^t_{w'}$, where $w' \sim \pi_2$.[4]
3. In the *action* stage, each firm i takes an action $a_i \in \{0,1\}$. Utilities depend on both firms' actions, and true label t.

Next, we consider different contracts for prediction sharing. Under *no-sharing*, the interaction proceeds as above. Under *train-sharing*, there is an additional stage between 1 and 2:

1b. Firms share their respective training signals w_1 and w_2.

Under *infer-sharing*, an additional stage between 2 and 3:

2b. Firms share respective inference-time signals X and x.

Finally, under *full-sharing* both 1b and 2b take place.

Summary and Interpretation. We now summarize the model elements and interpretation:

- The *world model* w is an information-theoretically optimal pair of classifiers.
- The *training signal* w_i implies a posterior π_i over world models, which we interpret as the actual classifier firm i is able to train. We interpret the signal w_i as firm i's predictions on its labeled historical data. In practical terms, the training signal can be interpreted as the model that best fits the training data, out of all possible models. The firms can then share these signals (i.e., the functions or code representing their best models given their data), without sharing the data itself.
- The *inference-time signal* is the prediction ($X \in \{A, B\}$ for firm 1, $x \in \{a, b\}$ for firm 2) made by the classifier on an unlabeled inference-time example.
- Under *train-sharing*, firms share w_1 and w_2, their predictions on labeled data.

[4] Notice that the inference-time signal is drawn according to the firm's *posterior*, rather than according to some specific true possible world. This is since we are interested in calculating the firms' equilibrium behaviors, which follow their Bayesian perspective.

- Under *infer-sharing*, firms share X and x, their respective predictions on the unlabeled inference-time example.

This formulation can capture a wide range of scenarios. The prior over W implies a prior over the relative share π_w of each label, a prior over the accuracy of each firm's model, and a prior over the correlation between the predictions of firms' models. The framework is illustrated in Fig. 1. See also Fig. 1 and Fig. 2 in [11].

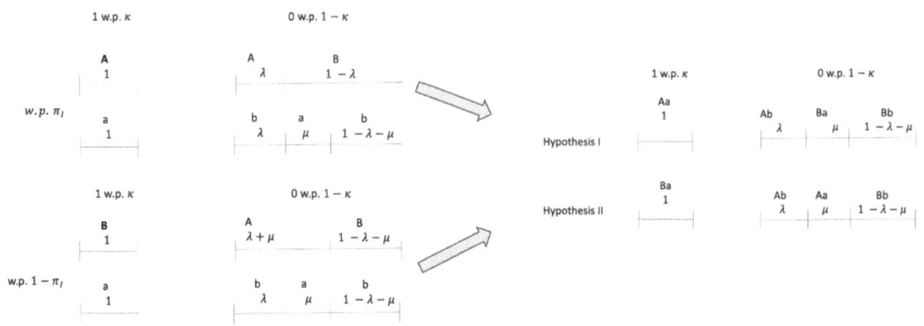

Fig. 1. There are two world models, represented by the top two and bottom two pairs of intervals, respectively. For both world models, $\pi_w = Pr[t = 1] = \kappa$. In the first world, $A^1_w = [0, 1]$ and $A^0_w = [0, \lambda]$. Thus, if $t = 1$ firm 1 always obtains signal A, and if $t = 0$ firm 1 obtains signal A with probability λ—i.e., whenever $\zeta \in [0, \lambda]$—and signal B with probability $1 - \lambda$. Furthermore, $a^1_w = [0, 1]$ and $a^0_w = [\lambda, \lambda + \mu]$. Thus, if $t = 1$ firm 2 always obtains signal a, and if $t = 0$ obtains signal a with probability μ—i.e., whenever $\zeta \in [\lambda, \lambda + \mu]$—and signal b with probability $1 - \mu$. Finally, the bottom two pairs of line segments represent the firms' signal spaces in the second world model, which differs from the first only in firm 1's signal under $t = 1$, namely, $A^1_w = \emptyset$ and $B^1_w = [0, 1]$. The interval structure of each of the firms results in a joint interval structure (and an induced joint probability over firm 1 signal A/B, firm 2 signal a/b, and the true realization 0/1), shown on the rhs of the figure. In the infinite data model, where each of the firms learns its own interval structure with certainty, firm 1 is able to deduce the correct world model just by knowing its own interval structure. On the other hand, firm 2 does not learn (in a Bayesian sense) anything from its own interval structure. This example captures our "Two Hypotheses" model of Sect. 4.

Strategies. A strategy s_i of firm i in the action stage is a mapping from the firm's signals to a distribution over actions $a_i \in \{0, 1\}$. The firm's signals depend on the contract: under no-sharing, the respective signals are $\sigma_1^{ns} = (w_1, X)$ for firm 1 and $\sigma_2^{ns} = (w_2, x)$ for firm 2. Under train-sharing, they are $\sigma_1^{ts} = (w_1, w_2, X)$ and $\sigma_2^{ts} = (w_1, w_2, x)$. Under infer-sharing, they are $\sigma_1^{is} = (w_1, X, x)$ and $\sigma_2^{is} = (w_2, X, x)$. And under full sharing, both firms obtain signals $\sigma_i^{fs} = (w_1, w_2, X, x)$.

Utility Model. As noted above, utility $u_i(p, t, p')$ of firm i depends on 3 variables: The firm's action p, the true label t, and the other firm's action p'. For a given example, action p is *correct* if it matches the example's label t. Given a training signal w_i, a contract $ct \in \{ns, ts, is, fs\}$, and a pair of strategies (s_1, s_2), the expected utility of firm i is

$$u_i^{ct}(w_i, s_1, s_2) = E\Big[u_i\Big(s_1\left(\sigma_1^{ct}\right), t, s_2\left(\sigma_2^{ct}\right)\Big)\Big], \tag{1}$$

where the expectation is over the draw of w from W according to $\pi|w_i$, the draw of t according to π_w, the draw of w_j under w, the draws of inference-time signals X and x under w, and the distributions of firms' randomization over actions.

We make some simplifying assumptions about utilities. First, we assume that $u_1 = u_2$. Second, we assume $u_i(p, t, p) = \frac{1}{2}u_i(p, t, \neg p)$, i.e., that if the two firms take the same action, the utility (whether positive or negative) is divided between them, in the sense that:

$$\sum_{i=1}^{2} u_i(p, t, p) = \sum_{i=1}^{2} \frac{1}{2}u_i(p, t, \neg p) = u_1(p, t, \neg p).$$

To emphasize the notation, $u_i(p, t, p)$ is the utility when the other firm's prediction p' is equal to p, and $u_i(p, t, \neg p)$ is the utility when the other firm's prediction p' is *different* than p. Thus, the ex-post utility is determined by four numbers: $R_0 = u_1(0, 0, 1), R_1 = u_1(1, 1, 0), C_0 = u_1(0, 1, 1), C_1 = u_1(1, 0, 0)$, where for example R_0 is the reward from correctly taking action 0 while the other firm takes action 1. We assume that $R_0, R_1 \geq 0$ and $C_0, C_1 \leq 0$.

In the paper, we largely focus on two specific utility models that capture important settings. In Sect. 3, we focus on a utility model we call *significant-action* utilities. In this model, there is a significant action—w.l.o.g., the action 1. For example, this action may be choosing to issue a loan. When taking the other, safe action, both reward and cost satisfy $R_0 = C_0 = 0$. If a firm takes a correct significant action exclusively, meaning that the other firm takes the safe action, it gets the full reward R_1. On the other hand, if a firm takes an incorrect significant action exclusively, it pays a cost C_1. When $C_1 = 1$, we call this the *symmetric* significant-action utility model.

In Sect. 4, we focus on a utility model we call *matching recommendations*, as in, e.g., [13]. In this model, there are no costs to a mistake—formally, $C_0 = C_1 = 0$—and there is a symmetric reward for any correct action—formally, $R_0 = R_1 = 1$. E.g., consider a firm that chooses between two possible recommendations to a user, and, if it correctly recommends what the user is looking for, the user will make a purchase.

Equilibrium, Individual Rationality, and Pareto Optimality. Given training signals w_1 and w_2 and a contract $ct \in \{ns, ts, is, fs\}$, a pair of strategies $s = (s_i, s_{\neg i})$ form a *Nash equilibrium* at (w_1, w_2) if for each i and each strategy s_i',

$$u_i^{ct}(w_i, s_i, s_{\neg i}) \geq u_i^{ct}(w_i, s_i', s_{\neg i}). \tag{2}$$

An equivalent and perhaps more useful formulation of the equilibrium condition takes the perspective of the agent together with her beliefs, see the discussion in Chap. 9 of [16]. We define the utility from taking action $p_i \in \{0,1\}$ given the collection of signals σ_i^{ct} and the other firm's strategy $s_{\neg i}$ as

$$\tilde{u}_i^{ct}(\sigma_i^{ct}, p_i, s_{\neg i}) = E\Big[u_i\Big(p_i, t, s_{\neg i}\left(\sigma_{\neg i}^{ct}\right)\Big) \mid \sigma_i^{ct}\Big], \tag{3}$$

where the expectation is over the conditional draw of $\sigma_{\neg i}^{ct}$ and t given signals σ_i^{ct}. We then say that s is an equilibrium for firm i if for every belief σ_i^{ct} and every possible action $p_i' \in \{0,1\}$,

$$\tilde{u}_i^{ct}(\sigma_i^{ct}, s_i(\sigma_i^{ct}), s_{\neg i}) \geq \tilde{u}_i^{ct}(\sigma_i^{ct}, p_i', s_{\neg i}). \tag{4}$$

Next, a contract *ct Pareto dominates* contract ct' at (w_1, w_2) if there exists an equilibrium s under ct such that, for every equilibrium s' under ct',

$$u_1^{ct}(w_1, s) \geq u_1^{ct'}(w_1, s') \quad \text{and} \quad u_2^{ct}(w_2, s) \geq u_2^{ct'}(w_2, s'). \tag{5}$$

If at least one of the inequalities is strict then the Pareto dominance is *strict.* Contract *ct Pareto dominates* contract ct' if it Pareto dominates ct' at every (w_1, w_2), and in this case we write $ct \succeq ct'$. If $ct \succeq ct'$ and $ct' \succeq ct$, we write $ct = ct'$, and say that the two contracts are equivalent. Contract *ct strictly Pareto dominates* ct' if $ct \succeq ct'$ but $ct' \not\succeq ct$, and in this case we write $ct \succ ct'$.

Contract *ct* is *individually rational (IR)* at (w_1, w_2) either if it is the no-sharing contract (which we consider the default contract), or if *ct* Pareto dominates the no-sharing contract at (w_1, w_2). Contract *ct* is *always IR* if it is IR at every (w_1, w_2), namely, $ct \succeq ns$. Contract *ct* is *Pareto optimal* if it is not Pareto dominated by any other contract, *Pareto-optimal IR* (IRPO) if it is both Pareto optimal and always IR, and *uniquely IRPO* if it is the only contract that is both Pareto optimal and always IR. Although our model is general, and can handle both mixed and pure Bayesian equilibria, our results in Sect. 3 onwards are for *pure* Bayesian equilibria.

3 Contracts for Prediction-Sharing with Unknown Correlation

In this section we focus on the first of two specific settings within our framework. We assume firms have significant-action utilities and "infinite data". The latter assumption means that each firm's prediction model is in some sense an optimal classifier given the data features it is able to see. We believe that this is the most natural assumption to closely approximate massive data sets.[5] The main caveat is that neither firm knows the correlation between the firms' classifiers, even after learning its own classifier. We further assume that the prediction accuracy of each firm's classifier—formally, $Pr_{\pi_w}[1|X = A]$ and $Pr_{\pi_w}[1|x = a]$—are common

[5] See also our analysis of a finite data case in Sect. 4.1.

knowledge.[6] Finally, for simplicity we assume that $Pr[0] = Pr[1] = \frac{1}{2}$, and that the false-positive and false-negative rates are the same for each of the firms:

$$\alpha \stackrel{\text{def}}{=} Pr[A|1] = Pr[B|0] \quad \text{and} \quad \beta \stackrel{\text{def}}{=} Pr[a|1] = Pr[b|0]. \tag{6}$$

Still, the full **joint** distribution of the firms' pair of signals together with the true realizations under w is unknown. As we see later in Sect. 3.2, this is equivalent to both firms not knowing how the signals of the two firms are correlated under the no-sharing contract, regardless of (w_1, w_2). Firms also do not know the true label of the outcome they are trying to predict in the inference phase. We assume w.l.o.g. that $\alpha \geq \beta \geq \frac{1}{2}$.

At one extreme, it is possible that the firms' signals are independent. At the other extreme, it is possible that they are fully correlated. In the full version of this paper, we show how this model can be formulated using our general model from Sect. 2.

3.1 Warm-Up: Known Correlation

We start our investigation with a simple model in which the correlation between the firms is known.

When the precision accuracy of both firms is common knowledge, as we assume throughout this section, then the *correlation* between the firms' predictions fully determines the joint distribution of the pair of signals under label t. We show that formally in Claim 3.1. By correlation we mean the Pearson correlation of the signals, namely

$$\theta_t = \frac{Pr[X = A \wedge x = a|t] - \alpha\beta}{\sqrt{\alpha(1-\alpha)\beta(1-\beta)}},$$

where t is the true label realization. Notice that the two Bernoulli variables are the two firms' signals given the true realization. For simplicity, we assume that $\theta_1 = \theta_0$, and denote the correlation simply by θ.

Claim. In the correlation model, knowing correlation θ determines the joint distribution of Firm1's signal A/B, Firm2's signal a/b, and true realization $0/1$.

Proof. To see that the correlation determines the joint distribution in our setting, recall that for the Bernoulli variables in our settings, the Pearson correlation under label 1 satisfies $\theta = \frac{Pr[X=A\wedge x=a|1]-\alpha\beta}{\sqrt{\alpha(1-\alpha)\beta(1-\beta)}}$. Thus, given θ, α, and β we have

$$Pr[X = A \wedge x = a|1] = \sqrt{\alpha\beta}\left(\sqrt{\alpha\beta} + \theta \cdot \sqrt{(1-\alpha)(1-\beta)}\right). \tag{7}$$

This then determines $Pr[X = A \wedge x = b|1] = \alpha - Pr[X = A \wedge x = a|1]$, $Pr[X = B \wedge x = a|1] = \beta - Pr[X = A \wedge x = a|1]$, and $Pr[X = B \wedge x = b|1] = 1 - Pr[X =$

[6] We use $Pr[1]$ as shorthand for $Pr[t = 1]$, and omit π_w, X, x when clear from context.

$A \wedge x = a|1] - Pr[X = A \wedge x = b|1] - Pr[X = B \wedge x = a|1]$. That is, it fully determines the joint distribution. For example, when $\alpha = \beta$ and $\theta = 0$ (i.e., the signals are conditionally independent), we have $Pr[X = A \wedge x = a|1] = \alpha^2$, and when $\alpha = \beta$ and $\theta = 1$, we have $Pr[X = A \wedge x = a|1] = \alpha$. Finally, a symmetric argument holds under label 0.

When the correlation is known, there is no added value in sharing w_i, since the world model w is already known to both firms. Therefore, no-sharing is equivalent to train-sharing, and infer-sharing is equivalent to full-sharing. The only question is, which of these contracts, if any, is IRPO?

Lemma 1. *With known correlation and symmetric significant-action utilities, only no-sharing and the equivalent train-sharing are IRPO. The unique equilibrium under these contracts has two regimes: A high β regime where both firms play by their inference-time signals, and a low β regime where Firm 2 "gives in" and always takes action* 0*, while Firm 1 matches its action to its inference-time signal.*

This matches what we learned to expect in practice: Firms develop their own classification models, and, assuming they are accurate enough, predict according to them. In Sect. 3.2 we show that once the correlation is not known with certainty, this conclusion may change, and full-sharing or train-sharing contracts may be uniquely IRPO. We also note that the threshold that separates the high and low β regime is itself dependent on α. The higher α is, the higher the threshold for the high beta regime, where Firm 2 follows its prediction signal. I.e., fixing Firm 2's prediction accuracy β, the firm is more likely to give in the higher Firm 1's prediction accuracy α is.

Lemma 1 deals with symmetric significant-utilities. In the full version of this paper, we study the *asymmetric* case. We show that with a higher cost for a mistake in the significant action C_1, but not so high as to prohibit ever taking a significant action altogether, the firms would prefer full-sharing, which enables them to take the significant action only when both receive positive signals.

3.2 Unknown Correlation

So far we have considered *known* correlations. However, a more natural model is that the correlation is *unknown*, and only some distribution over it is known. As we will see, this model can give rise to train-sharing as uniquely IRPO.

We begin with some preliminary lemmas. First, we show that within the specification of this subsection, full-sharing always Pareto dominates infer-sharing.

Lemma 2. *For any distribution π_θ over correlations and any R_1 and C_1, $fs \succeq is$.*

Proof. In the correlation model, the private signal w_i a firm gets during the training phase does not impact its posterior regarding the correlation θ, which follows the distribution Θ. Thus under infer-sharing, where each firm i only sees

w_i, we can ignore it, and we have $\sigma_1^{is} = \sigma_2^{is} = Xx$ for some pair of inference-time signal X, x.

Thus, we can conclude that the infer-sharing equilibrium is symmetric between the firms. That is since as we argue above, the posterior for both firms after the training phase stays the same as the common prior. In the inference phase, both firms share their signals, and so both firms end up with the exact same information. Both firms' equilibrium strategy is to predict 1 if and only if

$$E_{\theta\sim\Theta}[Pr[1|X = x_1, x = x_2, \theta] - C_1 \cdot Pr[0|X = x_1, x = x_2, \theta]] \geq 0.$$

Under full-sharing, a similar argument shows that for every pair of signals Xx and correlation θ (which both firms learn during the training phase), the symmetric equilibrium strategy is to predict 1 if and only if $Pr[1|X = x_1, x = x_2, \theta] - C_1 \cdot Pr[0|X = x_1, x = x_2, \theta]$.

We can thus write, for the symmetric equilibrium strategies $s \stackrel{def}{=} s_1 = s_2$ of the infer-sharing contract,

$$\begin{aligned} u_i^{is} &= E[u_i(s(X, x), t, s(X, x)] \\ &= E_{\theta\sim\Theta}[E[u_i(s(X, x), t, s(X, x)|\theta]] \\ &\leq E_{\theta\sim\Theta}[E[\max_s u_i(s(X, x), t, s(X, x)|\theta]] = u_i^{fs}. \end{aligned}$$

Next, we see that train-sharing and no-sharing contracts are equivalent under sufficient symmetry.

Lemma 3. *If $R_1 = C_1$ then $ts = ns$.*

Proof. Suppose first that, under train-sharing, the firms follow the same equilibrium strategies s_1, s_2 for any realization $\theta \sim \Theta$. Then, it must be that, under no-sharing, s_1, s_2 is also an equilibrium: This is immediate since the IC conditions of Eq. 4 under no-sharing follow immediately if the more granular IC conditions of the same equation under train-sharing are satisfied.

Now, we know by Lemma 1 that for any fixed θ the equilibrium strategies under train-sharing (the same as the strategies for no-sharing given we know that the correlation is θ) depend only on the values of α, β, and so are independent of θ. Thus, the same equilibrium strategies are played for any θ.

Lemma 4. *If $\alpha = \beta$, then for any distribution π_θ over correlations and any R_1 and C_1, $fs \succeq ts$.*

Finally, we can use the lemmas above to identify IRPO contracts. The two theorems below show that, with some symmetry, only full-sharing or no-sharing are such contracts.

Theorem 1. *If $R_1 = C_1$ then no-sharing is uniquely IRPO.*

Theorem 2. *If $\alpha = \beta$ then full-sharing is either uniquely IRPO, or $fs = ns$ are the only IRPO.*

However, outside the symmetries in Theorems 1 and 2, train-sharing can emerge as uniquely optimal.

Theorem 3. *Train-sharing is uniquely IRPO for an open subset of parameters* π_θ, α, β, R_1, C_1.

The intuition underlying the construction in the proof of Theorem 3 is the following. Under no-sharing, the firms play the same equilibrium regardless of their train-phase signals w_1 and w_2. Under train-sharing, the equilibrium may depend on w_1 and w_2, and so in some cases may improve both firms' utilities relative to the no-sharing equilibrium. This happens particularly when the firms learn that their signals are highly correlated, which results in Firm 2 not taking a significant action (e.g., not issue a loan). This saves Firm 2 from attaining negative utility, and allows Firm 1 to extract the full utility.

4 A "Two Hypotheses" Model

In Sect. 3 we showed that all contracts except for infer-sharing can be uniquely optimal. In this section we complete the picture by describing a setting where infer-sharing is uniquely optimal. We focus on the second setting described in the introduction, which is summarized in Fig. 1 of Sect. 2. We assume that firms have "infinite data" and matching-recommendations utilities: $C_0 = C_1 = 0$ and $R_0 = R_1 = 1$. This setting captures a variety of natural circumstances, such as multi-factorial genetic disease and chemical testing. Consider a genetic disease that only manifests itself with some environmental cause. Firm 1 performs genetic testing and knows (i) what genes cause the disease (ii) for a specific person, whether these genes are present. Firm 2 has users' behavioral data (e.g., credit card histories) and can identify the environmental cause. However, it does not understand the underlying genes that enable the disease.

Formally, let $t = 1$ denote the presence of the disease, inference-time signals A and B denote the presence of two different gene mutations in the population, and inference-time signals a and b denote the presence and absence of the environment cause, respectively. There are two hypotheses: (I) the disease is caused by mutation A and the environmental cause, and (II) the disease is caused by mutation B and the environmental cause. Thus, Hypothesis I (resp., Hypothesis II) is that firms see inference-time signals Aa (resp., Ba) if and only if $t = 1$. The following are common knowledge:

- Hypothesis I is correct w.p. π_I, Hypothesis II w.p. $1 - \pi_I$;
- without the environmental cause, the disease remains dormant ($Pr[0|b] = 1$);
- the incidence rate of the disease in the general population is $Pr[1] = \kappa$; and
- the incidence rates of the two different gene mutations in the general population are $Pr[A] = \kappa + (1 - \kappa) \cdot \lambda$ and $Pr[B] = 1 - Pr[A]$.

Our main result is that, within this setting, there are instances where infer-sharing is uniquely optimal.

Theorem 4. *Infer-sharing is uniquely IRPO for an open subset of parameters* π_I, κ, λ, *and* μ.

The intuition underlying the construction in the proof of Theorem 4 is the following. Generally, in the two hypotheses model, Firm 1 has the ability to deduce the correct world model during training, even with only its own signal. In the cases we identify, train/full-sharing makes it lose this advantage, and thus cannot be beneficial for it. We are left with no/infer-sharing as possible individually rational contracts. Since generally in the two hypotheses model, the signal of Firm 1 by itself is not enough to decide the user classification with certainty, infer-sharing helps in that Firm 1 can both determine the correct hypothesis and has the pair of signals that determines the true realization, and thus it always predicts correctly. In the cases we identify, the behavior of Firm 2 remains the same under both contracts, because of the fact that it cannot deduce the correct world model during training. Hence, infer-sharing allows Firm 1 a "free information meal", similar to the example, given for a model that only captures inference stage sharing, without consideration of the training stage, in [13].

4.1 Beyond the Infinite-Data Model

So far, we focused on the infinite-data model, where the training signal allows the firm to deduce the marginal distribution over its signal and the true realization. We conjecture that with enough data, the results are similar to the idealized infinite case that we analyze. However, with few samples, the results may change significantly. To demonstrate how the analysis may lead to different results when there is only little historical data, we consider the setting of Sect. 4, but when only one labeled example of past data is available to the firms. Thus, after the hypothesis (world) is drawn (Hypothesis I w.p. π_I, and otherwise Hypothesis II), a sample is drawn from the joint distribution over the pair of signals and true realizations, and each firm sees its own signal and the true realization. I.e., if the true hypothesis is Hypothesis I, then Firm 1 sees $(A, 0)$ w.p. α, $(A, 1)$ w.p. β, and $(B, 1)$ w.p. $1-\alpha-\beta$. The firms then update a Bayesian posterior over the world models. Under train-sharing and full-sharing, when historical predictions are shared, both firms see the entire sample, i.e., the pair of signals and the true realization.

In the full version of this paper, we prove that, in the two hypotheses model with the parameters used for Theorem 4 but with a single labeled example, the statement of Theorem 4 breaks down, as do some of the properties of equilibria derived in the theorem's proof. In particular:

Theorem 5. *Under the parameters of Theorem 4 but with one sample, no-sharing and train-sharing are not necessarily equivalent, no-sharing is IRPO (rather than infer-sharing), and Firm 1 has lower equilibrium expected utility than Firm 2.*

5 Implementation for a Real Data-Set

To see how our ideas may be put to practice, we use the peer-to-peer loan data of LendingClub, available publicly at Kaggle [18], to conduct a synthetic simulation. We take a random subset of 25% of the features and assign it to Firm 1[7]. We take another subset of 10% of the features (possibly overlapping) and assign it to Firm 2. Vertically, we split the data into train, test and validation sets. We let each of the firms train a neural net over the training data (that includes only its features). Each neural net was trained for 20 epochs on a 8-GB RAM M1 MacBook Pro, which takes about half an hour. The training signal consists of the neural net's predictions on whether loans are good or bad. The firms use the test data to learn the signal performance, which we assume then becomes common knowledge. Depending on the contract, the firms choose their equilibrium strategies based on the performance in the test data: under train-sharing and full-sharing they also see the other firm's predictions on the test data (rather than only knowing the aggregate performance measurements). The firms then use their models to get a signal for every example in the validation data. Under infer-sharing and full-sharing they see the other firm's signals on the validation set, and may use it to alter their final actions. We evaluate their actions on the validation data, under significant action utilities with $R_1 = 1$ and cost C_1.

We find that the results generally follow the lines of our discussion in Sect. 3: Varying by cost (going from $C_1 = 0$ to $C_1 = 2.5$ in 0.05 steps), as summarized in Fig. 2, we find regimes where either full-sharing, no-sharing, or train-sharing are uniquely IRPO. While full-sharing is almost always a Pareto optimal contract, there are significant regimes where it is not IR for firm 1, which results in the no-sharing and train-sharing regimes. In almost all cost values of the simulation, infer-sharing is Pareto dominated by full-sharing, as predicted by Lemma 2.

The behavior of no-sharing and train-sharing is of particular interest. With low values of C_1, both contracts have the two firms issue a loan regardless of the signal. Then, with higher values of C_1, the firms move to an equilibrium where each acts according to its signal, and later to an equilibrium where firm 1 predicts its signal while firm 2 does not issue any loans. At each such equilibrium shift, there is a discontinuity for firm 1's utility. For example, moving from each firm predicting its own signal to Firm 2 not issuing loans, allows it to get the full utility of its action instead of half.

6 Discussion

The analysis of incentives is a crucial aspect of the general effort to encourage data sharing, as recognized by the European Commission: "In spite of the economic potential, data sharing between companies has not taken off at sufficient scale. This is due to a lack of economic incentives (including the fear of losing a competitive edge)" [6]. This paper introduces a novel element of data

[7] In the full version, we include robustness tests where we vary the choice of features, and explain how the practical implementation corresponds to our formal model.

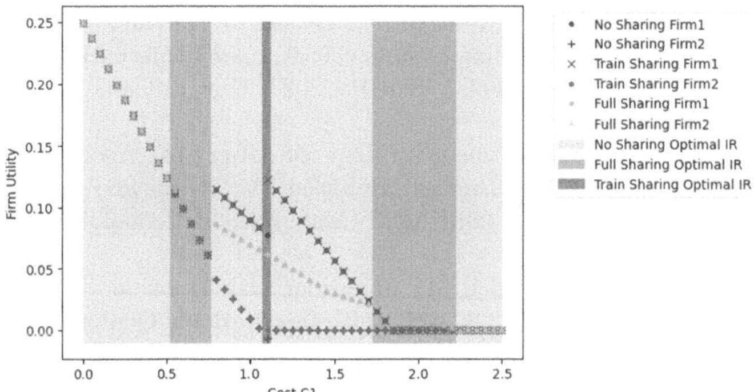

Fig. 2. No Sharing, Train Sharing and Full Sharing contracts performance for both firms and different costs. We do not include the infer sharing contract utility as they are very similar to (and dominated by) full sharing. We mark regimes where each contract is the optimal-welfare IR contract.

sharing—the distinction between sharing during training and inference—and demonstrates its importance to understanding firms' data-sharing incentives. Some natural questions arise as a result of our work:

- We have assumed a common prior over priors for the firms. What if the firms have different beliefs? How robust is the emergence of uniquely optimal contracts to small differences in the epistemic models of the firms?
- Our work is set within the framework of mechanism design without money, i.e., we suppose that firms share data based on mutual gain, rather than based on monetary compensation. In some cases it is natural to consider that one of the firms may compensate the other as part of the data sharing process. This could be interesting as future work and may build on the framework and insights we develop.

References

1. Babaioff, M., Kleinberg, R., Paes Leme, R.: Optimal mechanisms for selling information. In: Proceedings of the 13th ACM Conference on Electronic Commerce, EC 2012, pp. 92–109. Association for Computing Machinery, New York (2012)
2. Ben-Porat, O., Tennenholtz, M.: Best response regression. In: Advances in Neural Information Processing Systems 30: Annual Conference on Neural Information Processing Systems 2017, 4–9 December 2017, Long Beach, CA, USA, pp. 1499–1508 (2017)
3. Bergemann, D., Bonatti, A.: Markets for information: an introduction. Annu. Rev. Econ. **11**, 85–107 (2019)
4. Bimpikis, K., Crapis, D., Tahbaz-Salehi, A.: Information sale and competition. Manage. Sci. **65**(6), 2646–2664 (2019)
5. Dorner, F.E., Konstantinov, N., Pashaliev, G., Vechev, M.T.: Incentivizing honesty among competitors in collaborative learning and optimization. In: Oh, A.,

Naumann, T., Globerson, A., Saenko, K., Hardt, M., Levine, S. (eds.) Advances in Neural Information Processing Systems 36: Annual Conference on Neural Information Processing Systems 2023, NeurIPS 2023, New Orleans, LA, USA, 10–16 December 2023 (2023)
6. European Commission: A European strategy for data (2020. Accessed 13 May 2021
7. Feng, Y., Gradwohl, R., Hartline, J., Johnsen, A., Nekipelov, D.: Bias-variance games. In: Proceedings of the 23rd ACM Conference on Economics and Computation, pp. 328–329 (2022)
8. Fraboni, Y., Vidal, R., Lorenzi, M.: Free-rider attacks on model aggregation in federated learning. In: AISTATS 2021 - 24th International Conference on Artificial Intelligence and Statistics (2021)
9. Freund, Y., Schapire, R.E.: A decision-theoretic generalization of on-line learning and an application to boosting. J. Comput. Syst. Sci. **55**(1), 119–139 (1997)
10. Gafni, Y., Tennenholtz, M.: Long-term data sharing under exclusivity attacks. In: EC 2022: The 23rd ACM Conference on Economics and Computation, Boulder, CO, USA, 11–15 July 2022, pp. 739–759. ACM (2022)
11. Gentzkow, M., Kamenica, E.: Bayesian persuasion with multiple senders and rich signal spaces. Games Econom. Behav. **104**, 411–429 (2017)
12. Gradwohl, R., Tennenholtz, M.: Pareto-improving data-sharing. In: Proceedings of the 2022 ACM Conference on Fairness, Accountability, and Transparency, FAccT 2022, pp. 197-198. Association for Computing Machinery, New York (2022)
13. Gradwohl, R., Tennenholtz, M.: Coopetition against an amazon. J. Artif. Intell. Res. **76**, 1077–1116 (2023)
14. Green, J.R., Stokey, N.L.: Two representations of information structures and their comparisons. Decis. Econ. Financ. **45**(2), 541–547 (2022). Originally circulated as IMSSS Technical Report No. 271, Stanford University, 1978
15. Laufer, B., Kleinberg, J., Heidari, H.: Fine-tuning games: bargaining and adaptation for general-purpose models (2023)
16. Maschler, M., Solan, E., Zamir, S.: Game Theory. Cambridge University Press, Cambridge (2013)
17. McMahan, B., Moore, E., Ramage, D., Hampson, S., y Arcas, B.A.: Communication-efficient learning of deep networks from decentralized data. In: Proceedings of the 20th International Conference on Artificial Intelligence and Statistics, AISTATS 2017, vol. 54, pp. 1273–1282. PMLR (2017)
18. George, N.: All lending club loan data (2007). Accessed 13 May 2023
19. Saig, E., Talgam-Cohen, I., Rosenfeld, N.: Delegated classification. In: Oh, A., Naumann, T., Globerson, A., Saenko, K., Hardt, M., Levine, S. (eds.) Advances in Neural Information Processing Systems 36: Annual Conference on Neural Information Processing Systems 2023, NeurIPS 2023, New Orleans, LA, USA, 10–16 December 2023 (2023)
20. Tsoy, N., Konstantinov, N.: Strategic data sharing between competitors. In: Oh, A., Naumann, T., Globerson, A., Saenko, K., Hardt, M., Levine, S. (eds.) Advances in Neural Information Processing Systems, vol. 36, pp. 16483–16514. Curran Associates, Inc. (2023)

Calibrated Recommendations for Users with Decaying Attention

Jon Kleinberg, Emily Ryu(✉), and Éva Tardos

Cornell University, Ithaca, NY 14853, USA
{kleinberg,eva.tardos}@cornell.edu, eryu@cs.cornell.edu

Abstract. There are many settings, including ranking and recommendation of content, where it is important to provide diverse sets of results, with motivations ranging from fairness to novelty and other aspects of optimizing user experience. One form of diversity of recent interest is *calibration*, the notion that personalized recommendations should reflect the full distribution of a user's interests, rather than a single predominant category—for instance, a user who mainly reads entertainment news but also wants to keep up with news on the environment and the economy would prefer to see a mixture of these genres, not solely entertainment news. Existing work has formulated calibration as a subset selection problem; this line of work observes that the formulation requires the unrealistic assumption that all recommended items receive equal consideration from the user, but leaves as an open question the more realistic setting in which user attention decays as they move down the list of results.

In this paper, we consider calibration with decaying user attention under two different models. In both models, there is a set of underlying genres that items can belong to. In the first setting, where items are coarsely binned into a single genre each, we surpass the $(1 - 1/e)$ barrier imposed by submodular maximization and provide a novel bin-packing analysis of a 2/3-approximate greedy algorithm. In the second setting, where items are represented by fine-grained mixtures of genre percentages, we provide a $(1 - 1/e)$-approximation algorithm by extending techniques for constrained submodular optimization. Our work thus addresses the problem of capturing ordering effects due to decaying attention, allowing for the extension of near-optimal calibration from recommendation *sets* to recommendation *lists*.

Keywords: Calibration · Recommendations · Submodularity · Ranking

1 Introduction

Recommendation systems, now a ubiquitous feature of online platforms, have also been a long-standing source of fundamental theoretical problems in computing. Based on a model derived from a user's past behavior, such systems suggest relevant pieces of content that they predict the user is likely to be interested

G. Schäfer and C. Ventre (Eds.): SAGT 2024, LNCS 15156, pp. 443–460, 2024.
https://doi.org/10.1007/978-3-031-71033-9_25

in. This is typically achieved by optimizing for an objective function based on a model of the user's interests (such as relevance or utility), and such questions lead to a number of interesting optimization questions. Often, these basic formulations try to capture relevance in aggregate without considering the diversity of the results produced; they also generally treat lists of recommended results as an unordered sets, while in reality they are ordered sequences (reflecting the key influence of position and rank on the amount of attention a piece of content receives). Considering these two directions in conjunction leads to new and interesting theoretical questions, which form the focus of this paper.

In particular, a recurring concern with algorithmic recommendations is that optimizing for relevance risks producing results that are too homogeneous; it can easily happen that all the most relevant pieces of content are similar to one another, and that they collectively correspond to only one facet of a user's interests at the expense of other facets that go unrepresented [12]. To address such concerns, a long-standing research paradigm seeks recommendation systems whose results are not only relevant but also *diverse*, reflecting the range of a user's interests. Explicitly pursuing diversity in recommendations has been seen as a way to help mitigate the homogenizing effects that might otherwise occur [2,5].

Calibrated Recommendations. Within this area, an active line of research has pursued *calibration* as a means of optimizing for diversity [18]. In this formalism, we want to present a list of k recommended items to a single user (*e.g.*, movies on an entertainment site, or articles on a news site), and there is a set of underlying *genres* that the items belong to. The user has a *target distribution* over genres that reflect the extent to which they want to consume each genre in the long run. A natural goal is that the average distribution induced by the list of recommendations should be "close," or *calibrated*, to the user's target distribution. (For example, a user who likes both documentaries and movies about sports might well be dissatisfied with recommendations that were always purely about sports and contained no documentaries; this set of recommendations would be badly calibrated to the user's target distribution of genres.)

In a user study that systematically varied the quality of results from a recommender system, the researchers reported significant differences in users' evaluations of the system based on a single session lasting only 15 min [10]. Considering calibration an important aspect of quality (asserted by multiple papers including [9,11]), it is thus critical to achieve calibration within each single session rather than to simply hope for recommendations to eventually "average out" in the long run, lest the user become so dissatisfied after a sufficiently miscalibrated session that they decide to abandon the system altogether.

Prior work by [18] showed that for natural measures of distributional similarity, the selection of a set of k items to match the user's target distribution can be formulated as the maximization of a submodular set function. Because of this, the natural greedy algorithm produces a set of k items whose distributional similarity to the user's target distribution is within a $(1 - 1/e)$ factor of optimal. In this way, the work provided an approximately optimal calibrated *set* of recommendations.

Decaying Attention. This same work observed a key limitation at the heart of approximation algorithms for this and similar objectives: it necessarily treats the k recommendations as a set, for which the order does not matter. In contrast, one of the most well-studied empirical regularities in the social sciences is decaying attention as a user reads through a list of results. Results at the top of a list get much more attention than results further down—this phenomenon has been documented not only through traditional content engagement metrics, but also directly through eye-tracking and other behavioral studies [8,15,21]. Given this, the average genre distribution induced by a list of k recommendation results is really a *weighted average* over the genres of these items, with the earlier items in the list weighted more highly than the later ones.

Once we introduce the crucial property of decaying user attention, the formalism of set functions—and hence of submodularity, which applies to set functions—is no longer available to us. Moreover, it is no longer clear how to obtain algorithms for provably near-optimal calibration. There exist formalisms that extend the framework of submodular functions, in restricted settings, to handle inputs that are ordered sequences [3,4,6,13,19,23], but none of these formalisms can handle the setting of calibrated recommendations with decaying attention that we have here. It has thus remained an open question whether non-trivial approximation guarantees can be obtained for this fundamental problem.

The Present Work: Calibrated Recommendations with Decaying Attention. In this paper, we address this question by developing algorithms that produce lists of recommendations with provably near-optimal calibration for users with decaying attention. We provide algorithms for two models of genres: the *discrete* model in which each item comes from a single genre, and the *distributional* model in which each item is described by a distribution over genres. (For example, in this latter version, a documentary about soccer in Italy is a multi-genre mixture of a movie about sports, a movie about Italy, and a documentary.)

As noted above, a crucial ingredient in these models is to measure the similarity between the user's desired target distribution over genres and the distribution of genres present in the results we show them. In Sect. 3 we make concrete what it means for these distributions to be similar through the notion of an *overlap measure*, which we define to unify in a simple way standard measures of distributional similarity. Our results apply to a large collection of overlap measures including a large family of f-divergence measures, including overlap measures derived from the well-known *Hellinger distance*. These were also at the heart of earlier approaches that worked without decaying attention, where these measures gave rise to non-negative submodular set functions [1,14,17].[1]

[1] It is useful to note that the KL-divergence—arguably the other most widely-used divergence along with the Hellinger distance—is not naturally suited to our problem, since it can take both positive and negative values, and hence does not lead to well-posed questions about multiplicative approximation guarantees. This issue is not specific to models with decaying user attention; the KL-divergence is similarly not well-suited to approximation questions in the original unordered formalism, where the objective function could be modeled as a set function.

Overview of Results. In our two models of genres, we offer technical results of two distinct flavors in Sects. 4 and 5. First, the discrete genre model takes a completely new approach to the analysis of the greedy algorithm: to the best of our knowledge, our bin-packing argument is entirely novel; we also highlight that it allows us to surpass the barrier imposed by traditional submodularity arguments and achieve a stronger approximation guarantee.

For both versions of the problem, direct attempts at generalizing the methods of submodular maximization from unordered items to ordered items face a natural approximation barrier at $(1 - 1/e)$, simply because this is the strongest approximation guarantee we can obtain if we know only that the underlying function is submodular, and a special case of decaying attention is the case in which all weights are the same, which recovers the traditional submodular case. For the discrete version of the model, however, we are able to break through this $(1 - 1/e)$ barrier via a different technique based on a novel type of bin-packing analysis; through this approach, we are able to obtain a 2/3-approximation to the optimal calibration for overlap measures based on the Hellinger distance. We find this intriguing, since the problem is NP-hard and amenable to submodular maximization techniques; but unlike other applications of submodular optimization (including hitting sets and influence maximization) where $(1-1/e)$ represents the tight bound subject to hardness of approximation, here it is possible to go further by using a greedy algorithm combined with a careful analysis in place of submodular optimization.

To do this, we begin by observing that the objective function over the ordered sequence of items selected satisfies a natural inequality that can be viewed as an analogue of submodularity, but for functions defined on sequences rather than on sets. We refer to this inequality as defining a property that we call *ordered submodularity*, and we show that ordered submodularity by itself guarantees that the natural greedy algorithm for sequence selection provides a 1/2-approximation to the optimal sequence. This bound is not as strong as $1 - 1/e$; but unlike the techniques leading to the $1 - 1/e$ bound, ordered submodularity provides a direction along which we are able to obtain an improvement. In particular, for the discrete problem we can think of each genre as a kind of "bin" that contains items belonging to this genre, and the problem of approximating a desired target distribution with respect to the Hellinger measure then becomes a novel kind of load-balancing problem across these bins. Using a delicate local-search analysis, we are able to maintain a set of inductive invariants over the execution of a greedy bin-packing algorithm for this problem and show that it satisfies a strict strengthening of the general ordered submodular inequality; and from this, we are able to show that it maintains a 2/3-approximation bound.

Subsequently, in Sect. 5 for the distributional genre model we build on an existing line of work on constrained submodular maximization by introducing a new transformation technique to allow for position-based weights, which were not previously handled. A separate line of work has posed, but left open, the question of the effect of such position-based weights on achieving near-optimal diversity in recommender systems. Our work unites these two bodies of research by developing new methods from the former line of work to answer questions from the latter, and thereby provide a deeper fundamental understanding of the effects of weights and ordering on approximate submodular maximization.

For the case of distributional genres, we begin by noting that if we were to make the unrealistic assumption of repeated items (i.e. availability of many items with the exact same genre distribution q), then we could apply a form of submodular optimization with matroid constraints of [7] to obtain a $(1 - 1/e)$-approximation to the optimal calibration with decaying attention. This approach is not available to us, however, when we make the more reasonable assumption that items each have their own specific genre distribution. Instead, we construct a more complex laminar matroid structure, and we are able to show that with these more complex constraints, a continuous greedy algorithm and pipage rounding produces a sequence of items within $(1 - 1/e)$ of optimal.

2 Related Work

The problem of calibrated recommendations was defined by [18], in which *calibration* is proposed as a new form of diversity with the goal of creating recommendations that represent a user's interests. In this model, items represent distributions over genres, and weighting each item's distribution according to its rank induces a genre distribution for the entire recommendation list. Calibration is then measured using a *maximum marginal relevance* objective function, a modification of the KL divergence from this induced distribution to the user's desired distribution of interests. In the case where all items are weighted equally, the maximum marginal relevance function is shown to be monotone and submodular, and thus $(1 - 1/e)$-approximable by the standard greedy algorithm. However, when items have unequal weights (such as with decaying user attention), the function becomes a sequence function rather than a set function, and the tools of submodular optimization can no longer be applied. Further, the use of KL divergence with varying weights results in a mixed-sign objective function (refer to online Appendix C for an example), so formal approximation guarantees are not even technically well-defined in this setting. Hence, [18]'s approximation results are limited to only the equally-weighted (essentially unordered) case.

Since then, there has been recent interest in improving calibration in recommendation systems via methods such as greedy selection using statistical divergences directly or other proposed metrics [14,17] and LP-based heuristics [16], but this line of work largely focuses on empirical evaluation of calibration heuristics rather than approximation algorithms for provably well-calibrated lists. To the best of our knowledge, our work provides the first nontrivial approximation guarantees for calibration with unequal weights due to decaying attention.

Within the recommendation system literature, there is a long history of modeling calibration and other diversity metrics as submodular set functions, and leaving open the versions where ordering matters because user engagement decays over the course of a list (*e.g.*, [2,5,18]). Although numerous approaches to extending the notion of submodularity to have sequences have been proposed (*e.g.*, [3,4,6,13,19,20,22,23]), none is designed to handle these types of ordering effects. For a detailed survey of general theories of submodularity in sequences and a discussion of how they do not model our problem of calibration with decaying user attention, we refer the reader to Appendix A.

3 Problem Statement and Overlap Measures

[18] considers the problem of creating calibrated recommendations using the language of *movies* as the items with which users interact, and *genres* as the classes of items. Each user has a preference distribution over genres that can be inferred from their previous activity, and the goal is to recommend a list of movies whose genres reflect these preferences (possibly also incorporating a "quality" score for each movie, representing its general utility or relevance). In our work, we adopt [18]'s formulation of distributions over genres and refer to items as movies (although the problem of calibrated recommendations is indeed more general, including also news articles and other items, as discussed in the introduction). We describe the formal definition of our problem next.

3.1 Item Genres and Genre of Recommendation Lists

Consider a list of recommendations π for a user u. Let $p(g)$ be the distribution over genres g preferred by the user (possibly inferred from previous history). Given our focus on a single user u, we keep the identity of the user implicit in the notation. For simplicity of notation, we will label the items as the elements of $[K]$, and say that item i has genre distribution q_i.

Following the formulation of [18], we define the distribution over genres $q(\pi)$ of a recommendation list $\pi = \pi_1\pi_2\dots\pi_k$ as $q(\pi)(g) := \sum_{j=1}^{k} w_j \cdot q_{\pi_j}(g)$, where w_j is the weight of the movie in position j, and we assume that the weights sum to 1: $\sum_{j=1}^{k} w_j = 1$.[2] Note that the *position-based weights* make the position of each recommendation important, so this is no longer a subset selection problem.

To model attention decay, we assume that the weights are weakly decreasing in rank (i.e., $w_a \geq w_b$ if $a < b$). We also assume that the desired length of the recommendation list is a fixed constant k. This assumption is without loss of generality, even with the more typical cardinality constraint that the list may have length *at most* k—we simply consider each possible length $\ell \in [1, k]$, renormalize so that the first ℓ weights sum to 1, and perform the optimization. We then take the maximally calibrated list over all k length-optimal lists.

The goal of the *calibrated recommendations* problem is to choose π such that $q(\pi)$ is "close" to p. To quantify closeness between distributions, we introduce the formalism of *overlap measures*.

[2] Various weights are possible; [18] suggests "Possible choices include the weighting schemes used in ranking metrics, like in Mean Reciprocal Rank (MRR) or normalized Discounted Cumulative Gain (nDCG)." Alternatively, given empirical measurements of attention decay such as in [15], one might use numerically estimated weights.

3.2 Overlap Measures

For the discussion that follows, we restrict to finite discrete probability spaces Ω for simplicity, but the concepts generalize to continuous probability measures.

A common tool for quantitatively comparing distributions is statistical divergences, which measure the "distance" from one distribution to another. A divergence D has the property that $D(p, q) \geq 0$ for any two distributions p, q, with equality attained if and only if $p = q$. This means that divergences cannot directly be used to measure calibration, which we think of as a non-negative metric that is uniquely *maximized* when $p = q$. Instead, we define a new but closely related tool that we call *overlap*, which exactly satisfies the desired properties.

Our definition is also more general in two important ways. First, we do not limit ourselves to the KL divergence, so that other divergences and distances with useful properties may be used (such as the Hellinger distance, $H(p,q) = \frac{1}{\sqrt{2}}||\sqrt{p} - \sqrt{q}||_2$, which forms a bounded metric and has a convenient geometric interpretation using Euclidean distance). Second, in our definition q may be any *subdistribution*, a vector of probabilities summing to *at most* 1. This is crucial because it enables the use of algorithmic tools such as the greedy algorithm – which incrementally constructs q from the 0 vector by adding a new movie (weighted by its rank), and thus in each iteration must compute the overlap between the true distribution p and the partially constructed subdistribution q.

Definition 1 (Overlap measure). *An* ***overlap measure*** *G is a function on pairs of distributions and subdistributions (p, q) with the properties that (i) $G(p, q) \geq 0$ for all distributions p and subdistributions q, (ii) for any fixed p, $G(p, q)$ is uniquely maximized at $q = p$.*

Definition 2 (Distance-based overlap measure). *Let $d(p, q)$ be a bounded distance function on the space of distributions p and subdistributions q with the property that $d(p, q) \geq 0$, with $d(p, q) = 0$ if and only if $p = q$. Denote by d^* the maximum value attained by d over all pairs (p, q). Then, the* ***d-overlap measure*** *G_d is defined as $G_d(p, q) := d^* - d(p, q)$.*

Now, it is clear that G_d indeed satisfies both properties of an overlap measure (Definition 1): property (i) follows from the definition of d^*, and property (ii) follows from the unique minimization of d at $q = p$.

For an overlap measure G and a recommendation list $\pi = \pi_1 \pi_2 \ldots \pi_k$, we define $G(\pi) := G(p, q(\pi)) = G(p, \sum_{i=1}^{k} w_i q_{\pi_i})$.

3.3 Constructing Families of Overlap Measures

An important class of distances between distributions are f-divergences. Given a convex function f with $f(1) = 0$, the f-divergence from distribution q to distribution p is $D_f(p, q) := \sum_{x \in \Omega} f\left(\frac{p(x)}{q(x)}\right) q(x)$. One such f-divergence is the KL divergence, which [18] uses to define a *maximum marginal relevance* objective function similar to an overlap measure. However, this proposed function has issues with mixed sign (see Appendix C for an example), so it does not admit well-specified formal approximation guarantees. Instead, we consider a broad class of overlap measures based on f-divergences for all convex functions f. As a concrete example, consider the squared Hellinger distance (obtained by choosing $f(t) = (\sqrt{t} - 1)^2$ or $f(t) = 2(1 - \sqrt{t})$), which is of the form $H^2(p, q) = \frac{1}{2}\sum_{x\in\Omega}(\sqrt{p(x)} - \sqrt{q(x)})^2 = 1 - \sum_{x\in\Omega}\sqrt{p(x)\cdot q(x)}$. This divergence is bounded above by $d^* = 1$; the resulting H^2-overlap measure is $G_{H^2}(p, q) = \sum_{x\in\Omega}\sqrt{p(x)\cdot q(x)}$.

Inspired by the squared Hellinger-based overlap measure, we also construct another general family of overlap measures based on non-decreasing concave functions. Given any nonnegative non-decreasing concave function h, we define the overlap measure $G^h(p, q) = \sum_{x\in\Omega}\frac{h(q(x))}{h'(p(x))}$. For instance, taking $h(x) = x^\beta$ for $\beta \in (0, 1)$ gives $\frac{1}{h'(x)} = \frac{1}{\beta}x^{1-\beta}$, which produces the (scaled) overlap measure $G^{x^\beta}(p, q) = \sum_{x\in\Omega} p(x)^{1-\beta}q(x)^\beta$. Observe that the natural special case of $\beta = \frac{1}{2}$ gives $h(x) = \frac{1}{h'(x)} = \sqrt{x}$, providing an alternate construction that recovers the squared Hellinger-based overlap measure.

3.4 Monotone Diminishing Return (MDR) Overlap Measures

Many classical distances, including those discussed above, are originally defined on pairs of distributions (p, q) but admit explicit functional forms that can be evaluated using the values of $p(x)$ and $q(x)$ for all $x \in \Omega$. This allows us to compute $d(p, q)$, and consequently $G_d(p, q)$, when q is not a distribution (*i.e.*, the values do not sum to 1), which will be useful in defining algorithms for finding well-calibrated lists. Using this extension, we can take advantage of powerful techniques from the classical submodular optimization literature when a certain extension of the overlap measure G_d is monotone and submodular, properties satisfied by most distance-based overlap measures.

Consider an extension of an overlap measure $G(p, q)$ to a function on the ground set $V = \{(i, j)\}$, where $i \in [K]$ is an item and $j \in [k]$ is a position. For a set $R \subseteq V$, define $R^{\leq j}$ be the set of items assigned to position j or earlier; that is, $R^{\leq j} := \{i \in [K] \mid \exists \ell \leq j \text{ s.t. } (i, \ell) \in R\}$.

Assuming the overlap measure G is well-defined as long as q is non-negative (but not necessarily a probability distribution), we define the set function

$$F_G(R) := G\left(p, \sum_{j=1}^{k} w_j \left(\sum_{i\in R^{\leq j}\setminus R^{\leq j-1}} q_i\right)\right).$$

With this definition, we can define monotone diminishing return (MDR) and strongly monotone diminishing return (SMDR) overlap measures:

Definition 3 ((S)MDR overlap measure). *An overlap measure G is* ***monotone diminishing return (MDR)*** *if its corresponding set function F_G is monotone and submodular. If, in addition, G is non-decreasing with respect to all $q(x)$, we say G is* ***strongly monotone diminishing return (SMDR)****.*

For any bounded monotone f-divergence, the corresponding overlap measure satisfies the MDR property, where by monotone we mean that if subdistribution q_2 coordinate-wise dominates subdistribution q_1, then $D_f(p, q_1) \geq D_f(p, q_2)$ for all p. Further, all overlap measures G^h defined above by concave functions h satisfy the SMDR property. A detailed technical discussion is deferred to Appendix B.1, but at a high level, since D_f is negated in the construction of G_{D_f}, the convexity of f (since D_f is negated) and the concavity of h result in concave overlap measures (corresponding to diminishing returns).

Theorem 1. *Given any bounded monotone f-divergence D_f with maximum value $d^* = \max_{(p,q)} D_f(p, q)$, the corresponding D_f-overlap measure $G_{D_f}(p, q) = d^* - D_f(p, q)$ is MDR.*

Theorem 2. *Given any nonnegative non-decreasing concave function h, the overlap measure $G^h(p, q) = \sum_{x \in \Omega} \frac{h(q(x))}{h'(p(x))}$ is SMDR.*

Observe that D_f-overlap measures are not necessarily *SMDR*, but many D_f-overlap measures based on common f-divergences are not only bounded and monotone, but also increasing in $q(x)$. This includes the squared Hellinger distance, so the resulting $G_{H^2}(p, q)$ is indeed both MDR and SMDR.

4 Calibration in the Discrete Genre Model

In this section we consider the version of the calibration model with *discrete genres*, in which each item is classified into a single genre. In this model, we allow the list of items to contain repeated genres, since it is natural to assume that the universe contains many items of each genre, and that a recommendation list may display multiple items of the same genre.

We start by thinking about a solution to the problem in this model as a sequence of choices of genres, and we study how the value of the objective function changes as we append items to the end of the sequence being constructed. In particular, we show that as we append items, the value of the objective function changes in a way that is governed by a basic inequality, that intuitively can be viewed as an analogue of monotonicity and submodularity but for sequences rather than sets. We pursue this idea by defining any function on sequences to be *ordered-submodular* if it satisfies this basic inequality; in particular, the Hellinger measures of calibration for our problem (as well as more general families based on the overlap measures defined earlier) are ordered-submodular in this sense.

As a warm-up to the main result of this section, we start by showing that for *any* ordered-submodular function, the natural greedy algorithm that iteratively adds items to maximally increase the objective function achieves a factor 1/2-approximation to the optimal sequence. Note that this approximation guarantee is weaker than the $(1 - 1/e)$ guarantee obtained by classical (constrained) submodular optimization, but we present it because it creates a foundation for analyzing the greedy algorithm which we can then strengthen to break through the $(1-1/e)$ barrier and achieve a 2/3-approximation for the problem of calibration with discrete genres. (In contrast, the techniques achieving $1-1/e$ appear to be harder to use as a starting point for improvements, since they run up against tight hardness bounds for submodular maximization.)

To start, we make precise exactly how the greedy algorithm works for approximate maximization of a function f over sequences. The greedy algorithm initializes $A_0 = \emptyset$ (the empty sequence), and for $\ell = 1, 2, \ldots, k$, it selects A_ℓ to be the sequence that maximizes our function $f(A)$ over all sequences obtained by appending an element to the end of $A_{\ell-1}$. In other words, it iteratively appends elements to the sequence A one by one, each time choosing the element that leads to the greatest marginal increase in the value of f.

For two sequences A and B we use $A||B$ to denote their concatenation. For a single element s, we use $A||s$ to denote s added at the end of the list A.

4.1 Ordered-Submodular Functions and the Greedy Algorithm

Let f be a function defined on a sequences of elements from some ground set; we say that f is *ordered-submodular* if for all sequences of elements $s_1 s_2 \ldots s_k$, the following property holds for all $i \in [k]$ and all other elements $\bar{s}_i$:

$$f(s_1 \ldots s_i) - f(s_1 \ldots s_{i-1}) \geq f(s_1 \ldots s_i \ldots s_k) - f(s_1 \ldots s_{i-1} \bar{s}_i s_{i+1} \ldots s_k). \quad (1)$$

Notice that if f is an ordered-submodular function that takes sequences as input but does not depend on their order (that is, it produces the same value for all permutations of a given sequence), then it follows immediately from the definition that f is a monotone submodular set function. In this way, monotone submodular set functions are a special case of our class of functions.

A standard algorithmic inductive argument shows that the greedy algorithm described earlier attains a 1/2-approximation to the optimal sequence. Next, observe that the MDR property defined in Sect. 3.3 directly implies ordered submodularity (via submodularity and monotonicity of $\hat{F}_G$), and hence the greedy algorithm is a 1/2-approximation algorithm for these calibration problems. Full proofs of both claims above as well as the theorem are given in Appendix B.3.

Theorem 3. *The greedy algorithm for nonnegative ordered-submodular function maximization over sets of cardinality k outputs a solution whose value is at least $\frac{1}{2}$ times that of the optimum solution.*

Theorem 4. *Any MDR overlap measure G is ordered-submodular. Thus, the greedy algorithm provides a 1/2-approximation for calibration heuristics using MDR overlap measures.*

4.2 Improved Approximation for Calibration with Discrete Genres

Next, we focus on calibration using the squared Hellinger-based overlap measure, which has several useful properties: (1) it is SMDR, and thus the approximation guarantee is directly comparable to the $(1-1/e)$ guarantee in the distributional model that we discuss next in Sect. 5; (2) its mathematical formula is amenable to genre-specific manipulations; (3) perhaps most importantly, it is well-motivated by frequent use in the calibration literature (*e.g.*, [1,14,17]). (We note that our techniques apply generally to many overlap measures, such as the second family based on concave functions described in Sect. 3.3, but the quantitative 2/3 bound is specific to the squared Hellinger-based overlap measure.[3]) We prove this improved approximation result using the concrete form of the Hellinger distance to establish a stronger version of ordered submodularity.

Given that each item belongs to a single genre, and that we have many copies of items for each genre, the question we ask in each step of the greedy algorithm is now: at step i, which genre should we choose to assign weight w_i to? We can think of this problem as a form of "bin-packing" problem, packing the weight w_i into a bin corresponding to a genre g. Since every item represents a single discrete genre, we can interpret a recommendation list as an assignment of *slots* to *genres*. Then, using $s_i = g$ to denote that a sequence S assigns slot i to genre g, we can write the squared Hellinger-based overlap measure as

$$f(S) = \sum_{\text{genres } g} \sqrt{p(g)} \sqrt{\sum_{i \in [k]: s_i = g} w_i}. \tag{2}$$

The main technical way we rely on the Hellinger distance is the following Lemma, which strengthens inequality (3) from Appendix B.3 but does not assume that the sequence T^i or $\bar{T}^{i+1}$ is coming from the optimal sequence, or that they are identical except for their first element.

Lemma 1. *With calibration defined via the Hellinger distance, for all sequences A_{i-1} and T^i, and the greedy choice of extending A_{i-1} with the next element a_i, there exists a sequence $\bar{T}^{i+1}$ such that $f(A_i||\bar{T}^{i+1}) \geq f(A_{i-1}||T^i) - \frac{1}{2}\left(f(A_i) - f(A_{i-1})\right)$.*

Before we prove the lemma, we show that it yields a 2/3-approximation:

Theorem 5. *For calibration with discrete genres, the greedy algorithm provides a 2/3-approximation for the squared Hellinger-based overlap measure.*

Proof. Define $S^{(1)}$ to be the optimal sequence S, and using Lemma 1 with $T^i = S^{(i)}$, define inductively $S^{(i+1)} = \bar{T}^{i+1}$. We show via induction that for all i, $f(A_i||S^{(i+1)}) \geq OPT(k) - \frac{1}{2} f(A_i)$.

[3] In particular, our bin-packing analysis of the greedy algorithm relies on concavity along the direction of improvement, so it applies to other overlap measures such as those in the G^{x^β} family, but the numerical constant of $\frac{1}{2}$ in Lemma 1 (and thus the final approximation guarantee of $\frac{2}{3}$ in Theorem 5) would change. Here, we focus on the particular case of $\beta = \frac{1}{2}$, as the induced Hellinger-based overlap measure is one that is commonplace in practice.

For the base case of $i = 0$, $f(A_0||S^{(1)}) = f(S) = OPT(k) \geq OPT(k) - \frac{1}{2}f(A_0)$. So suppose the claim holds for some i, and observe that by Lemma 1 and the fact that $f(A_{i+1}) \geq f(A_i||s_{i+1})$ by definition of the greedy algorithm,

$$\begin{aligned} f(A_{i+1}||S^{(i+2)}) &\geq f(A_i||S^{(i+1)}) - \frac{1}{2}(f(A_{i+1}) - f(A_i)) \\ &\geq OPT(k) - \frac{1}{2}f(A_i) - \frac{1}{2}(f(A_{i+1}) - f(A_i)) \\ &= OPT(k) - \frac{1}{2}f(A_{i+1}), \end{aligned}$$

completing the induction. Finally, setting $i = k$ establishes $ALG(k) \geq \frac{2}{3}OPT(k)$.

Remark. This approximation guarantee is fairly robust to settings in which we do not have perfectly accurate information about the preferences and genres, but a small degree of error or noise to within a multiplicative factor of $(1 + \varepsilon)$. We still maintain a $\frac{2/3}{(1+\varepsilon)^2}$-approximation; for more details, see Appendix E.

Next, we outline the proof of Lemma 1; the full analysis is in Appendix B.4.

Proof (Proof outline of Lemma 1). Consider the sequence $A_{i-1}||T^i$ and the greedy choice a_i, and let t_i be the first element of T^i. Recall that each of these items is a genre, and the term multiplying $\sqrt{p(g)}$ in the Hellinger distance (2) is the sum of the weights of all positions where a given genre g is used. We define notation for the total weight of positions that have a genre g in A_{i-1} and in T^{i+1} respectively, skipping the genre in the ith position. Since this lemma focuses on a single position i, we keep i implicit in some of the notation.

Let $\alpha(g) := \sum_{\{j \in [i-1], a_j = g\}} w_j$ denote the total weight of the slots assigned to genre g by A_{i-1}. Let $\tau(g) := \sum_{\{j \in [i+1,k], t_j = g\}} w_j$ denote the total weight assigned to genre g by T^{i+1}. Say that the greedy algorithm assigns slot i to genre $a_i = g'$, but in T^i the first genre (corresponding to slot i) is $t_i = g^*$.

Next, notice that for the squared Hellinger-based overlap measure (2), there are only two genres in which $f(A_i||T^{i+1})$ and $f(A_{i-1}||T^i)$ differ: the genre $a_i = g'$ chosen by the greedy algorithm, and the genre $t_i = g^*$ of the first item of the sequence T^i. For all other genres, the sum of assigned weights in the definition of the Hellinger distance is unchanged.

First, writing T^{i+1} to denote simply dropping the first item assignment from T^i, and denoting the blank in position i by $\_$, we get

$$\begin{aligned} f(A_{i-1}||a_i||T^{i+1}) - f(A_{i-1}||\_||T^{i+1}) &= \sqrt{p(g')}\left(\sqrt{\alpha(g') + w_i + \tau(g')} - \sqrt{\alpha(g') + \tau(g')}\right), \\ f(A_{i-1}||t_i) - f(A_{i-1}) &= \sqrt{p(g^*)}\left(\sqrt{\alpha(g^*) + w_i} - \sqrt{\alpha(g^*)}\right), \\ f(A_{i-1}||t_i||T^{i+1}) - f(A_{i-1}||\_||T^{i+1}) &= \sqrt{p(g^*)}\left(\sqrt{\alpha(g^*) + w_i + \tau(g^*)} - \sqrt{\alpha(g^*) + \tau(g^*)}\right). \end{aligned}$$

Using these expressions and the monotonicity/convexity of the square root:

$$\begin{aligned}f(A_{i-1}||T^i) - f(A_i||T^{i+1}) &= \sqrt{p(g^*)}\left(\sqrt{\alpha(g^*) + w_i + \tau(g^*)} - \sqrt{\alpha(g^*) + \tau(g^*)}\right)\\ &\quad - \sqrt{p(g')}\left(\sqrt{\alpha(g') + w_i + \tau(g')} - \sqrt{\alpha(g') + \tau(g')}\right)\\ &\leq \sqrt{p(g^*)}\left(\sqrt{\alpha(g^*) + w_i + \tau(g^*)} - \sqrt{\alpha(g^*) + \tau(g^*)}\right)\\ &\leq \sqrt{p(g^*)}\left(\sqrt{\alpha(g^*) + w_i} - \sqrt{\alpha(g^*)}\right)\\ &= f(A_{i-1}||t_i) - f(A_{i-1})\\ &\leq f(A_i) - f(A_{i-1})\end{aligned}$$

If $f(A_i||T^{i+1}) \geq f(A_{i-1}||T^i)$ (*e.g.*, if $g' = g^*$), then it suffices to take $\bar{T}^{i+1} = T^{i+1}$ and the inequality holds trivially. Hence, we assume that $g' \neq g^*$ and $f(A_i||T^{i+1}) \leq f(A_{i-1}||T^i)$. Now, we may need to modify T^i to get $\bar{T}^{i+1}$, depending on the size of $\tau(g')$ relative to w_i.

Case 1: $\tau(g') \geq \frac{1}{2}w_i$. Intuitively, since the greedy algorithm added w_i to g' instead of g^*, we should not assign so much additional weight to g'. To create $\bar{T}^{i+1}$, we start from T^{i+1} (the part of T^i without the first item), but make an improvement by reassigning some subsequent items from g' to g^*.

Because $\tau(g')$ is the sum of weights each of which is at most w_i (as the weights of positions are in decreasing order), we can move some weight z satisfying $\frac{1}{2}w_i \leq z \leq w_i$ from g' to g^*. Now, consider the function

$$c(x) = \sqrt{p(g')}\sqrt{\alpha(g') + \tau(g') + w_i - x} + \sqrt{p(g^*)}\sqrt{\alpha(g^*) + \tau(g^*) + x},$$

representing the contribution from genres g' and g^* towards f, after moving an amount x from g' to g^* (the change in f will only be due to these two genres, since all others are unchanged). Observe that $x = 0$ corresponds to $f(A_i||T^{i+1})$ and $x = w_i$ corresponds to $f(A_{i-1}||T^i)$, so $c(0) \leq c(w_i)$. Further, c is concave in x. As depicted in the figure below, a correction that is at least $\frac{1}{2}w_i$ increases f by at least half the amount that a full correction of w_i would have achieved (Fig. 1).

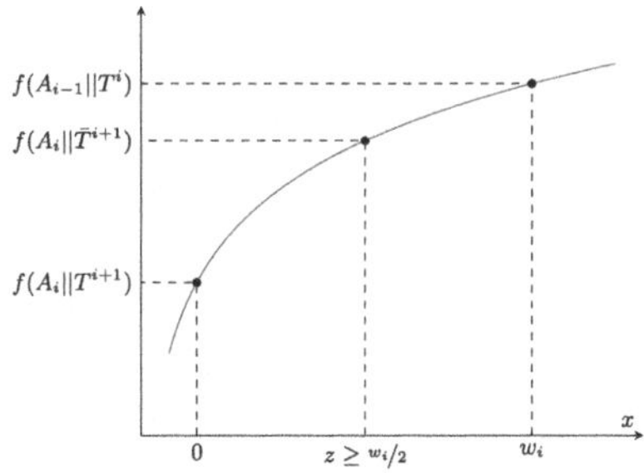

Fig. 1. Change in $f(A_i||\bar{T}^{i+1})$ as we move x weight from g' to g^*.

Then, the remaining amount is at most half the uncorrected difference; i.e. $f(A_{i-1}||T^i) - f(A_i||\bar{T}^{i+1}) \leq \frac{1}{2}(f(A_{i-1}||T^i) - f(A_i||T^{i+1}))$. Combining this with the form of inequality (3) re-established at the beginning of the proof yields $f(A_i||T^i) - f(A_i||\bar{T}^{i+1}) \leq \frac{1}{2}(f(A_i) - f(A_{i-1}))$, which we rearrange to give the desired inequality: $f(A_i|\bar{T}^{i+1}) \geq f(A_{i-1}|T^i) - \frac{1}{2}(f(A_i) - f(A_{i-1}))$.

Case 2: $\tau(g') < \frac{1}{2}w_i$. Observe that any greedy misstep is due to the fact that the greedy algorithm must choose based only on $\alpha(g')$, with no knowledge of $\tau(g')$. If there is a large $\tau(g')$ that the greedy algorithm does not know about, then the choice to fill g' may have been overly eager, and ultimately ends up being less helpful than expected after the remaining items are assigned.

But here, the fact that $\tau(g')$ is small means that this is *not* the case – the greedy algorithm was not missing a large piece of information, so the choice based only on $\alpha(g')$ was actually quite good. In particular, it cannot turn out to be much worse than g^*, meaning that the difference between $f(A_{i-1}||T^i)$ and $f(A_i||T^{i+1})$ is fairly small. In fact, the greedy algorithm's lack of knowledge is most harmful when $\tau(g')$ is large and $\tau(g^*)$ is small. So the *worst* possible outcome for this case occurs when $\tau(g') = \frac{1}{2}w_i$ and $\tau(g^*) = 0$, for which we have

$$\begin{aligned}
&f(A_{i-1}||T^i) - f(A_i||T^{i+1})\\
&\quad= \sqrt{p(g')}\left(\sqrt{\alpha(g') + w_i/2} - \sqrt{\alpha(g') + 3w_i/2}\right) + \sqrt{p(g^*)}\left(\sqrt{\alpha(g^*) + w_i} - \sqrt{\alpha(g^*)}\right)\\
&\quad= \sqrt{p(g')}\left(\sqrt{\alpha(g') + w_i/2} - \sqrt{\alpha(g') + 3w_i/2}\right) + f(A_{i-1}||t_i) - f(A_{i-1})\\
&\quad\leq \sqrt{p(g')}\left(\sqrt{\alpha(g') + w_i/2} - \sqrt{\alpha(g') + 3w_i/2}\right) + f(A_i) - f(A_{i-1}).
\end{aligned}$$

Then, this gives

$$\begin{aligned}
\frac{f(A_{i-1}||T^i) - f(A_i||T^{i+1})}{f(A_i) - f(A_{i-1})} &\leq 1 - \frac{\sqrt{p(g')}\left(\sqrt{\alpha(g') + 3w_i/2} - \sqrt{\alpha(g') + w_i/2}\right)}{f(A_i) - f(A_{i-1})}\\
&= 1 - \frac{\sqrt{\alpha(g') + 3w_i/2} - \sqrt{\alpha(g') + w_i/2}}{\sqrt{\alpha(g') + w_i} - \sqrt{\alpha(g')}}.
\end{aligned}$$

This final expression is minimized when $\alpha(g') = 0$, for which

$$\frac{f(A_{i-1}||T^i) - f(A_i||T^{i+1})}{f(A_i) - f(A_{i-1})} \leq 1 - \sqrt{2 - \sqrt{3}} \leq \frac{1}{2},$$

which rearranges to $f(A_i||T^{i+1}) \geq f(A_{i-1}||T^i) - \frac{1}{2}(f(A_i) - f(A_{i-1}))$. Thus, simply taking $\bar{T}^{i+1} = T^{i+1}$ suffices to give the desired result.

5 Calibration in the Distributional Genre Model

In this section we consider the general calibrated recommendations problem with a model of *distributional genres*, in which each item has a specific distribution over genres (as in Sect. 3.1), or a fine-grained breakdown of all the genres represented by that item. Note that if we permitted our list to include repeats, then a $(1 - 1/e)$-approximation would be possible using a reduction to submodular maximization over a partition matroid (Appendix B.2). But realistically, genre mixtures are too specific to have multiple items with identical distributions, and recommendation lists should not show the same item repeatedly. Our main result addresses this, providing a $(1-1/e)$-approximation for calibrated recommendation lists *without* repeated elements using SMDR overlap measures.[4]

To begin, we view a list as an assignment of (at most) one item to each position, and consider the ground set of all *item-position pairs* $\{(i,j) \mid i \in [K], j = \ell\}$ ("item i in position j"). Define the laminar family of sets $D_\ell := \{(i,j) \mid i \in [K], j \leq \ell\}$, and the laminar matroid $\mathcal{M} = (V, \mathcal{I})$, where $R \subset V$ is an independent set in $\mathcal{I}$ if and only if $|R \cap D_\ell| \leq \ell$ for all $\ell \in [k]$ (*i.e.*, R assigns at most ℓ items to the first ℓ positions, corresponding to a "valid" list). Now, there is a correspondence between recommendation lists and laminar matroid bases: any list assigns exactly ℓ items to the first ℓ slots for all $\ell \in [k]$ (and is thus a basis); any basis can also be converted into a list solely by promoting items upwards (and by strong monotonicity, this preserves the value of the calibration objective). Then, it suffices to optimize over matroid bases using the continuous greedy algorithm and pipage rounding technique of [7], then convert the approximately-optimal basis to an approximately-optimal list.

Proposition 1. *Given a basis $R \in \mathcal{I}$, we can construct a length-k list π such that $G(\pi) \geq F_G(R)$.*

Proof. For every item i, define $\ell_R(i)$ to be the first position that i occurs in R; *i.e.*, $\ell_R(i) := \min\{j \in [k] | (i,j) \in R\}$, or $\ell_R(i) = k+1$ if no such j exists. Also denote $w_{k+1} = 0$. Sort the items in increasing order of $\ell_R(\cdot)$ (breaking ties arbitrarily), and call this sequence π. We claim that $G(\pi) \geq F_G(R)$.

Consider an arbitrary item j. By definition of the laminar matroid, $|R \cap D_{\ell_R(i)}| = \sum_{y=1}^{k} \sum_{x=1}^{\ell_R(i)} \mathbb{1}_{[(x,y) \in R]} \leq \ell_R(i)$. The summation is an upper bound on the number of items x with $(x,y) \in R$ for some $y \leq \ell_R(i)$. But these are exactly the items with $\ell_R(x) \leq \ell_R(i)$ (including i itself), and therefore the items that can appear before i in π. So the position at which i appears in π, denoted $\pi^{-1}(i)$, is less than or equal to $\ell_R(i)$. This implies $w_{\pi^{-1}(i)} \geq w_{\ell_R(i)}$ for all i. Now, observe that $R^{\leq j} \backslash R^{\leq j-1}$ is exactly the set of items which appear for the *first* time in position j; thus $\ell_R(i) = j$ for all $i \in R^{\leq j} \backslash R^{\leq j-1}$. Additionally, $R^{\leq 1} \subseteq R^{\leq 2} \subseteq \cdots \subseteq R^{\leq k} \subseteq [K]$. Then, for any genre g, we have

[4] One might hope that it would suffice to take a solution with repeats and convert it to a solution without repeats simply by showing items in the order of the first time they appear. Unfortunately, this approach may destroy the submodular structure of the original function, so that the continuous greedy algorithm no longer provides a near-optimal approximation guarantee. Further details are in Appendix B.2.

$$\sum_{j=1}^{k} w_j \left(\sum_{i \in R^{\le j} \setminus R^{\le j-1}} q_i(g) \right) = \sum_{j=1}^{k} \sum_{i \in R^{\le j} \setminus R^{\le j-1}} w_{\ell_R(i)} q_i(g) = \sum_{i \in R^{\le k}} w_{\ell_R(i)} q_i(g)$$

$$\le \sum_{i \in [K]} w_{\pi^{-1}(i)} q_i(g) = \sum_{j=1}^{k} w_j q_{\pi(j)}(g).$$

Then, since G is non-decreasing w.r.t. all $q(g)$, we have $G(R) \le F_G(\pi)$.

Proposition 2. $\max_{R \in \mathcal{I}} F_G(R) \ge \max G(\pi)$.

Proof. Let $\pi^* := \arg\max_\pi G(\pi)$, and define $R^* := \{((\pi^*)^{-1}(j), j) \mid j \in [k]\}$ as the set of item-position pairs in π^*. By construction, $F_G(R^*) = G(\pi^*)$, and $R^* \in \mathcal{I}$. By monotonicity the the maximum over independent sets is attained by a basis, and $\max_{R \in \mathcal{I}} F_G(R) \ge F_G(R^*) = G(\pi^*) = \max G(\pi)$.

Theorem 6. *There exists a $(1 - 1/e)$-approximation algorithm for the calibration problem with distributional genres using any SMDR overlap measure G.*

Proof. Since G is an SMDR overlap measure, F_G is a monotone submodular function. Then, the continuous greedy algorithm and pipage rounding technique of [7] finds an independent set $\bar{R} \in \mathcal{I}$ such that $F_G(\bar{R}) \ge (1 - 1/e) \max_{R \in \mathcal{I}} F_G(R)$. We can assume $\bar{R}$ is a basis. By Proposition 2, $F_G(\bar{R}) \ge (1 - 1/e) \max G(\pi)$. Using Proposition 1, convert $\bar{R}$ into a sequence $\bar{\pi}$ such that $G(\bar{\pi}) \ge F_G(\bar{R})$. Now, $G(\bar{\pi}) \ge (1 - 1/e) \max G(\pi)$, so we take $\bar{\pi}$ to be our output.

6 Conclusion

In this paper, we have studied the problem of calibrating a recommendation list to match a user's interests, where user attention decays over the course of the list. We have introduced the notion of overlap measures to quantify calibration under two different models of genre distributions. In the first model, where every item belongs to a single *discrete genre*, by defining an ordered submodularity property and utilizing a careful bin-packing argument, we have shown that the greedy algorithm is a 2/3-approximation. In the second model of *distributional genres*, where each item has a fine-grained mixture of genre percentages, we have extended tools from constrained submodular optimization to supply a $(1 - 1/e)$-approximation algorithm. Prior work had highlighted the importance of the order of items due to attention decay but had left open the question of provable guarantees for calibration on these types of sequences; this prior work obtained guarantees only under the assumption that the ordering of items does not matter. Now, our work has provided the first performance guarantees for near-optimal calibration of recommendation *lists*, working within the models of user attention that form the underpinnings of applications in search and recommendation.

Finally, we highlight a number of directions for further work. First, it is interesting to consider the greedy algorithm for calibration with discrete genres and ask whether the approximation bound of 2/3 is tight, or if it can be sharpened using an alternative analysis technique. Additionally, we ask whether $(1 - 1/e)$ and 2/3 are the best possible approximation guarantees possible for the distributional and discrete genre models, respectively, or if there exists a polynomial time approximation algorithm that achieves a stronger constant factor under either model. As noted earlier, both models with decaying attention are amenable to the general framework of submodular optimization, but these tools are limited to an approximation guarantee of $(1 - 1/e)$. In the discrete genre model, by using different techniques we surpass this barrier and obtain a stronger guarantee; might the same be possible in the distributional genre model?

To further investigate the performance of our algorithms, it may be useful to parametrize worst-case problem instances, since we found through computational simulations that the greedy solution tends to be very close to optimal across many randomly generated instances (Appendix F). Another potential direction is constructing additional families of overlap measures, or deriving a broader characterization of functional forms that satisfy the (S)MDR properties so that they may be used with our algorithms. As personalized recommendations become increasingly commonplace and explicitly optimized, the answers to these questions will be essential in developing tools to better understand the interplay between relevance, calibration, and other notions of diversity in these systems.

References

1. Abdollahpouri, H., Mansoury, M., Burke, R., Mobasher, B.: The connection between popularity bias, calibration, and fairness in recommendation. In: Proceedings of the 14th ACM Conference on Recommender Systems, pp. 726–731. RecSys 2020, Association for Computing Machinery, New York, NY, USA (2020). https://doi.org/10.1145/3383313.3418487
2. Agrawal, R., Gollapudi, S., Halverson, A., Ieong, S.: Diversifying search results. In: Proceedings of the Second ACM International Conference on Web Search and Data Mining, pp. 5–14 (2009). https://doi.org/10.1145/1498759.1498766
3. Alaei, S., Makhdoumi, A., Malekian, A.: Maximizing sequence-submodular functions and its application to online advertising (2019)
4. Asadpour, A., Niazadeh, R., Saberi, A., Shameli, A.: Sequential submodular maximization and applications to ranking an assortment of products. In: Proceedings of the 23rd ACM Conference on Economics and Computation, p. 817. EC 2022, Association for Computing Machinery, New York, NY, USA (2022). https://doi.org/10.1145/3490486.3538361
5. Ashkan, A., Kveton, B., Berkovsky, S., Wen, Z.: Optimal greedy diversity for recommendation. In: Twenty-Fourth International Joint Conference on Artificial Intelligence (2015)
6. Bernardini, S., Fagnani, F., Piacentini, C.: A unifying look at sequence submodularity. Artif. Intell. **297**, 103486 (2021). https://doi.org/10.1016/j.artint.2021.103486
7. Calinescu, G., Chekuri, C., Pal, M., Vondrák, J.: Maximizing a monotone submodular function subject to a matroid constraint. SIAM J. Comput. **40**(6), 1740–1766 (2011)
8. Fessenden, T.: Scrolling and attention (2018). https://www.nngroup.com/articles/scrolling-and-attention/. Accessed 04 Feb 2022

9. Kowald, D., Mayr, G., Schedl, M., Lex, E.: A study on accuracy, miscalibration, and popularity bias in recommendations (2023)
10. Liao, M., Sundar, S.S., Walther, J.B.: User trust in recommendation systems: a comparison of content-based, collaborative and demographic filtering. In: Proceedings of the 2022 CHI Conference on Human Factors in Computing Systems. CHI 2022, Association for Computing Machinery, New York, NY, USA (2022). https://doi.org/10.1145/3491102.3501936
11. Lin, K., Sonboli, N., Mobasher, B., Burke, R.: Calibration in collaborative filtering recommender systems: a user-centered analysis. In: Proceedings of the 31st ACM Conference on Hypertext and Social Media, pp. 197–206. HT 2020, Association for Computing Machinery, New York, NY, USA (2020). https://doi.org/10.1145/3372923.3404793
12. McNee, S.M., Riedl, J., Konstan, J.A.: Being accurate is not enough: how accuracy metrics have hurt recommender systems. In: CHI 2006 Extended Abstracts on Human Factors in Computing Systems, pp. 1097–1101. CHI EA 2006, Association for Computing Machinery, New York, NY, USA (2006). https://doi.org/10.1145/1125451.1125659
13. Mitrovic, M., Feldman, M., Krause, A., Karbasi, A.: Submodularity on hypergraphs: from sets to sequences. In: Storkey, A., Perez-Cruz, F. (eds.) Proceedings of the Twenty-First International Conference on Artificial Intelligence and Statistics. Proceedings of Machine Learning Research, vol. 84, pp. 1177–1184. PMLR (2018). https://proceedings.mlr.press/v84/mitrovic18a.html
14. Naghiaei, M., Rahmani, H.A., Aliannejadi, M., Sonboli, N.: Towards confidence-aware calibrated recommendation (2022). arXiv:2208.10192
15. Pan, B., Hembrooke, H., Joachims, T., Lorigo, L., Gay, G., Granka, L.: In google we trust: users' decisions on rank, position, and relevance. J. Comput.-Mediat. Commun. **12**(3), 801–823 (2007)
16. Seymen, S., Abdollahpouri, H., Malthouse, E.C.: A constrained optimization approach for calibrated recommendations. In: Proceedings of the 15th ACM Conference on Recommender Systems, pp. 607–612. RecSys 2021, Association for Computing Machinery, New York, NY, USA (2021). https://doi.org/10.1145/3460231.3478857
17. da Silva, D.C., Durão, F.A.: Introducing a framework and a decision protocol to calibrate recommender systems (2022). arXiv:2204.03706
18. Steck, H.: Calibrated recommendations. In: Proceedings of the 12th ACM Conference on Recommender Systems, pp. 154–162. RecSys 2018, Association for Computing Machinery, New York, NY, USA (2018). https://doi.org/10.1145/3240323.3240372
19. Tschiatschek, S., Singla, A., Krause, A.: Selecting sequences of items via submodular maximization. In: Thirty-First AAAI Conference on Artificial Intelligence (2017)
20. Udwani, R.: Submodular order functions and assortment optimization (2021). arXiv:2107.02743
21. Williams, H.E.: Clicks in search (2012). https://hughewilliams.com/2012/04/12/clicks-in-search/. Accessed 04 Feb 2022
22. Zhang, G., Tatti, N., Gionis, A.: Ranking with submodular functions on a budget. Data Min. Knowl. Disc. **36**(3), 1197–1218 (2022)
23. Zhang, Z., Chong, E.K.P., Pezeshki, A., Moran, W.: String submodular functions with curvature constraints. IEEE Trans. Autom. Control **61**(3), 601–616 (2016). https://doi.org/10.1109/TAC.2015.2440566

Matrix Rationalization via Partial Orders

Agnes Totschnig[(✉)], Rohit Vasishta, and Adrian Vetta

McGill University, Montreal, Canada
{agnes.totschnig,rohit.vasishta}@mail.mcgill.ca,
adrian.vetta@mcgill.ca

Abstract. A *preference matrix* M has an entry for each pair of candidates in an election whose value p_{ij} represents the proportion of voters that prefer candidate i over candidate j. The matrix is *rationalizable* if it is consistent with a set of voters whose preferences are total orders. A celebrated open problem asks for a concise characterization of rationalizable preference matrices. In this paper, we generalize this matrix rationalizability question and study when a preference matrix is consistent with a set of voters whose preferences are partial orders of width α. The width (the maximum cardinality of an antichain) of the partial order is a natural measure of the rationality of a voter; indeed, a partial order of width 1 is a total order. Our primary focus concerns the *rationality number*, the minimum width required to rationalize a preference matrix. We present two main results. The first concerns the class of half-integral preference matrices, where we show the key parameter required in evaluating the rationality number is the chromatic number of the undirected *unanimity graph* associated with the preference matrix M. The second concerns the class of integral preference matrices, where we show the key parameter now is the dichromatic number of the directed *voting graph* associated with M.

1 Introduction

At the heart of macroeconomics is the concept of *choice* [2]. For example, what should a consumer (respectively, producer) demand (respectively, supply) given market prices? More generally, given a set of options to choose from an agent selects an option(s). The agent is considered *rational* if it always selects the best choice among the given options. Specifically, a rational agent has a *total order* over the entire collection of options and, presented with a subset of the options, always chooses the option highest in the ordering.

1.1 Rationalizable Choice Data

But how can we evaluate whether or not agents are rational? The simple answer is to test the data. Is a collection of observational choice data consistent with

A. Totschnig—supported by FRQNT Grant 332481.
R. Vasishta—supported by a McCall MacBain Scholarship.
A. Vetta—supported by NSERC Discovery Grant 2022-04191.

G. Schäfer and C. Ventre (Eds.): SAGT 2024, LNCS 15156, pp. 461–479, 2024.
https://doi.org/10.1007/978-3-031-71033-9_26

decision-making by a group of rational agents? A classical way to model this is via an election. Assume there are n candidates over which we have pairwise choice data; that is p_{ij} is the proportion of voters that prefer candidate i over candidate j. This induces a non-negative *preference matrix* $M = (p_{ij})_{i,j\in[n]}$, where $p_{ii} = 0$ for each candidate i and $p_{ij} + p_{ji} = 1$ for every pair of candidates i and j. Is this preference matrix compatible with an electorate of rational voters, where each voter ranks the candidates via a total order?

THE MATRIX RATIONALIZABILITY PROBLEM: Given a preference matrix M, does there exist a set of rational voters such that, for any pair of candidates, a random voter prefers i over j with probability exactly p_{ij}?[1]

The fundamental value of this problem is highlighted by its importance in a wide range of disciplines. The problem and its variants have been studied in depth in mathematical psychology (primarily with respect to human decision making) [9,22], economics (w.r.t. econometrics and behavioural economics) [6,18,19], operations research (w.r.t. consumer choice and advertising) [4,19], combinatorial optimization (w.r.t. geometry and integrality in mathematical programming) [11,15], and theoretical computer science [1].

A good characterization of rationalizable preference matrices has eluded the research community for over 60 years. Consequently, the objectives of this paper are more modest, but more general. Rather than study preference matrices that are compatible with a collection of total orders, we study preference matrices that are compatible with a collection of *partial orders*. There are two major advantages to this approach. First, from a practical perspective, the problem of non-existence is avoided. Every preference matrix is compatible with a set of voters with partial order preferences (see Observation 1). Second, and more substantially, the use of partial orders induces a natural measure of approximate rationality. Specifically, a total order corresponds to a poset of width 1, where the width is the maximum cardinality of an antichain in the poset. More generally, the smaller the width of a partial order the closer it is to a total order. Intuitively, the smaller the width the more "decisive" and rational the voter. In contrast, the poset of a voter with higher width has a higher number of linear extensions; the voter is thus more ambiguous and less decisive. We say that a voter whose preferences are given by a partial order of width at most α is α-rational. Thus, a 1-rational voter is rational. Further, we say that a preference matrix M is α-rationalizable if it can be explained by a set of voters who are α-rational.

1.2 The Model

A (strict) *partial order* $\succ$ over a set $[n] = \{1, 2, \ldots, n\}$ of candidates satisfies, for any $i, j, k \in [n]$, the following three properties:

`Irreflexivity`: NOT $i \succ i$.
`Asymmetry`: If $i \succ j$, then NOT $j \succ i$.
`Transitivity`: If $i \succ j$ and $j \succ k$, then $i \succ k$.

[1] For irrational matrices one may ask if the matrix is compatible with a probability distribution over total orders.

We say that a pair of candidates i and j are *comparable* if either $i \succ j$ or $j \succ i$; else they are *incomparable.* A partial order is a *total order* if every pair of candidates is comparable.

We assume each voter v has a personal set of preferences given by a partial order $\succ_v$ over the candidates, where the voter *strongly prefers* candidate i over j if $i \succ_v j$ (we omit the subscript and write $i \succ j$ if the context is clear). We say the voter *weakly prefers* i over j if either $i \succ j$ (strongly prefers) or if i and j are incomparable (voter is indifferent). Recall that a partial order $\succ$ induces a *poset* $\mathcal{P}$ over the candidates. A *chain* is a subset of candidates in $\mathcal{P}$ that induces a total order. An *antichain* in $\mathcal{P}$ is a subset of pairwise incomparable candidates. We say that the voter is *α-rational* if the maximum cardinality of an antichain in its partial order is at most α.

A preference matrix is a non-negative matrix $M = (p_{ij})$, where $p_{ii} = 0$ and $p_{ij} + p_{ji} = 1$ for all $i, j \in [n]$. A preference matrices M is α-rationalizable if there exists a set $\mathcal{V}$ of α-rational voters such that, for any pair of candidates i and j,

(i) at least a p_{ij} fraction of the voters weakly prefer i over j,
(ii) at least a p_{ji} fraction of the voters weakly prefer j over i.

One way to see that this definition accords with M being compatible with the set of voters is via sampling. Suppose we take a large sample of the voters and ask them if they prefer i or j. If the voter prefers i over j (namely, $i \succ j$) or vice versa then we insist the voter must declare truthfully. If the voter is indifferent between i and j then we allow the voter to choose either of them. Then, in the limit, it is feasible that exactly a p_{ij} fraction of the voters (from the sample) state a preference for i over j if and only if (i) and (ii) hold.

Observe that since $p_{ij} + p_{ji} = 1$, a set of α-rational voters is **not** compatible with M if more than a $p_{ij} = 1 - p_{ji}$ fraction of the voters strongly prefer i over j, or if more than a $p_{ji} = 1 - p_{ij}$ fraction of the voters strongly prefer j over i. Consequently, we can encode (i) and (ii) as the following *rationality constraints* for all $i \neq j \in [n]$:

$$\frac{\#\{v \in \mathcal{V} : v \text{ strongly prefers } i \text{ over } j\}}{|\mathcal{V}|} \leq p_{ij} \leq \frac{\#\{v \in \mathcal{V} : v \text{ weakly prefers } i \text{ over } j\}}{|\mathcal{V}|} \quad (*)$$

If $\mathcal{V}$ satisfies $(*)$ for a preference matrix M then we say $\mathcal{V}$ is *consistent* or *compatible* with M. Note that we impose no restriction on the cardinality of $\mathcal{V}$, we simply desire a voting set of any cardinality that α-rationalizes M.

We remark that if every voter only has strong preferences then (i) and (ii) are equivalent to exactly a p_{ij} fraction of the voters (strongly) preferring i over j (in accordance with The Matrix Rationalizability Problem). Such voters have total order preferences and thus are 1-rational and, in this case, M would be 1-rationalizable. Naturally, in our general setting, we then desire the minimum α such that M is α-rationalizable; we call this minimum the *rationality number* of M and denote it by $\alpha(M)$. This induces the following decision problem:

The Rationality Number Problem: Given a preference matrix M and a positive integer k, is the rationality number $\alpha(M)$ at most k?

1.3 Examples

`Example I:` Consider the integral preference matrix M shown in Fig. 1. Observe that there is a simple way to represent an integral preference matrix by a *voting graph*, $D_M = (V, A)$, whose vertices are the candidates and there is an arc from i to j if and only if $p_{ij} = 1$. Thus, as illustrated, the voting graph for M is simply the directed 3-cycle. More generally an integral preference matrix corresponds to a *tournament*, an orientation of the complete graph on n vertices.

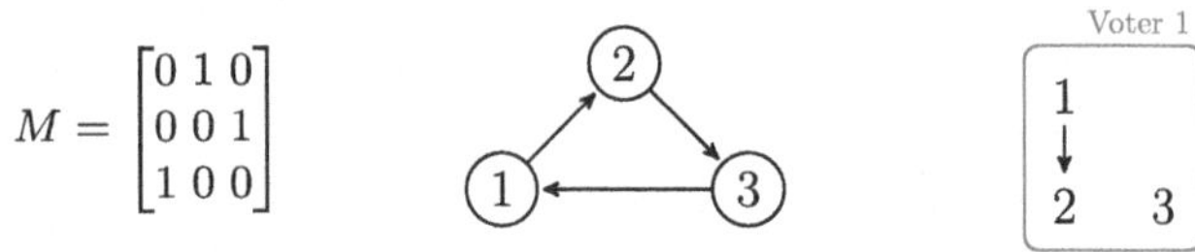

Fig. 1. An integral preference matrix (and its voting graph D_M) that is 2-rationalizable using a single voter.

The matrix M is not rationalizable (1-rationalizable), as any voter with a total order preference list would strongly prefer j over i, for some arc $(i, j) \in G$, which is incompatible with the fact that $p_{ij} = 1$. However, M is 2-rationalizable. Indeed it is compatible with a single voter with partial order of width 2 as shown by the red voter in Fig. 1. Specifically, the voter prefers candidate 1 over 2 but is indifferent between candidates 1 and 3 and between candidates 2 and 3.

Let us verify that this voter does 2-rationalize M. We need to prove that the *rationality constraints* are satisfied for every pair of voters. Observe $1 \leq p_{12} = 1 \leq 1$. Here the lower bound holds because the fraction of voters that strongly prefer 1 over 2 is one (as there is a single voter!). Furthermore $0 \leq p_{13} = 0 \leq 1$ because the voter is indifferent between 1 and 3. Thus the fraction of voters that strongly prefer 1 over 3 is zero and the fraction that weakly prefer 1 over 3 is one. Similarly $0 \leq p_{23} = 1 \leq 1$. We remark that if conditions $(*)$ hold for p_{ij} then they hold for p_{ji}. Thus the rationality constraints $(*)$ are satisfied for every pair of candidates and M is 2-rationalizable.

`Example II:` Consider the half-integral preference matrix M shown in Fig. 2. Again, we represent a half-integral preference matrix by a *voting graph*, $D_M = (V, A)$, where there is an arc from i to j if and only if $p_{ij} = 1$. Thus if $p_{ij} = \frac{1}{2}$ there is no arc (in either direction) between i and j. In Fig. 2, the voting graph for M is illustrated with a dashed line for the absence of arcs. We similarly define an undirected *unanimity graph* $G_M = (V, E)$, which has an edge between i and j whenever $p_{ij} = 1$ or $p_{ji} = 1$. Thus it contains an edge for each pair of candidates for which the voters cannot strongly disagree. Note that it corresponds to the undirected version of the voting graph D_M.

The matrix M is not rationalizable (1-rationalizable). Again, this is because the voting graph contains a directed cycle C on the six candidates. Thus, any voter with a total order preference list would strongly prefer j over i, for at least one arc $(i, j) \in C$, which is incompatible with the fact that $p_{ij} = 1$.

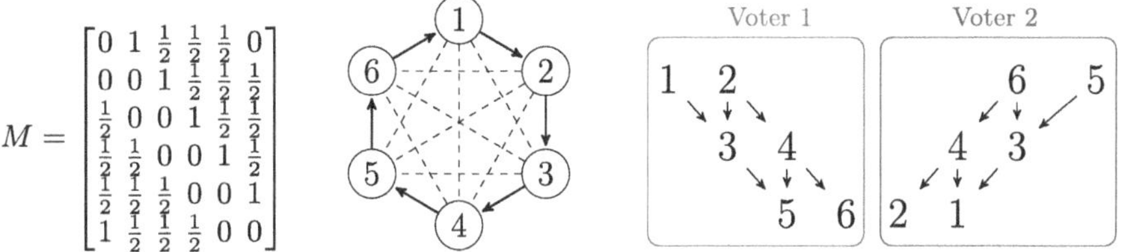

Fig. 2. A half-integral preference matrix (and its voting graph D_M) that is 2-rationalizable using two voters.

However, M is 2-rationalizable. But, unlike in Example I, this requires at least two voters. To see that one voter is insufficient, consider the three candidates $\{1, 3, 5\}$. For any 2-rational voter v, at least two of these must be comparable in its partial order; otherwise they form an antichain of cardinality 3. Without loss of generality, let $1 \succ_v 3$. So if v is the only voter then the proportion of voters that prefer 1 over 3 is one. This contradicts condition $(*)$ because $p_{13} = \frac{1}{2}$.

On the other hand, M is compatible with two voters that are each 2-rational as illustrated by the red and blue voters in Fig. 2. For example, both the red and blue voters are indifferent between candidates 1 and 2. Thus the fraction of voters that strongly prefer 1 over 2 is zero and the fraction that weakly prefer 1 over 2 is one. Thus condition $(*)$ holds for this pair as $0 \leq p_{12} = 1 \leq 1$. The reader may verify that the conditions $(*)$ also hold for every other pair of candidates. Hence M is 2-rationalizable, as claimed.

`Example III:` Finally consider the generic preference matrix M with three candidates shown in Fig. 3. M is 3-rationalizable using a single voter whose partial order is an antichain on all the candidates. Observe that since the voter has no strict preference, the fraction of voters that strongly prefer i over j is zero, for any pair of candidates. Similarly, the fraction of voters that weakly prefer i over j is one. Thus the rationality constraints $(*)$ are simply $0 \leq p_{ij} \leq 1$ which are trivially satisfied. Thus M is 3-rationalizable.

$$M = \begin{bmatrix} 0 & p_{12} & p_{13} \\ 1 - p_{12} & 0 & p_{23} \\ 1 - p_{13} & 1 - p_{23} & 0 \end{bmatrix}$$

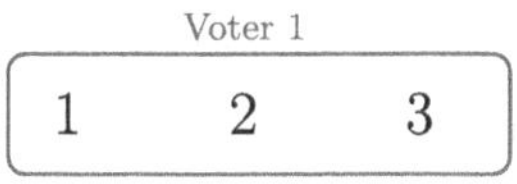

Voter 1

1 2 3

Fig. 3. A generic preference matrix that is 3-rationalizable using a single voter.

Of course, this example, trivially generalizes to any number n of candidates. A single voter whose partial order is an antichain of size n will n-rationalize any preference matrix with n candidates.

Observation 1. *A preference matrix with n candidates is n-rationalizable.*

Observation 1 confirms the existence of a set of voters that α-rationalize a preference matrix M. Of course, our interest is whether or not the matrix is α-rationalizable for some α much smaller than n.

1.4 Our Results

In Sect. 2 we present structural results concerning α-rationalizable preference matrices. Then in Sect. 3, we focus on the important class $\mathcal{M}^{\frac{1}{2}}$ of half-integral preference matrices. Our first main result is that, for this class, the rationality number $\alpha(M)$ is bounded by the chromatic number of the (undirected) unanimity graph G_M. Specifically, we prove:

Theorem 1.1. *Let $\mathcal{M}^{\frac{1}{2}}$ be the class of half-integral preference matrices. Then*

$$\frac{1}{5}\chi(G_M) \;\leq_{\exists}\; \alpha(M) \;\leq_{\forall}\; \chi(G_M)$$

In order to concisely formulate our results, we use the notation $\leq_{\exists}$ and $\leq_{\forall}$. Here $\leq_{\exists}$ means that for every $k \in \mathbb{N}$ *there exists* a preference matrix M in that class (here, $\mathcal{M}^{\frac{1}{2}}$) with chromatic number k such that the inequality holds, and $\leq_{\forall}$ means that the inequality holds *for every* preference matrix M in that class.

Next, in Sect. 4, we strengthen our results for the class $\mathcal{M}^{0/1}$ of integral preference matrices. For this class, we prove that the rationality number $\alpha(M)$ is equal to the dichromatic number of the (directed) voting graph D_M. We use this to give even more precise bounds on the rationality number for the class of integral matrices. Specifically our second main result is:

Theorem 1.2. *Let $\mathcal{M}^{0/1}$ be the class of integral preference matrices. Then*

$$\frac{n}{2\log n + 1} \;\leq_{\exists}\; \alpha(M) \;\leq_{\forall}\; \frac{3n}{\log n}$$

Note that for an integral preference matrix M its unanimity graph G_M is the complete graph and thus has chromatic number exactly $\chi(G_M) = n$.

We conclude, in Sect. 5, by showing that the rationality number problem is NP-complete, even for the class of integral matrices.

1.5 Literature Review

The matrix rationalizability problem asks when a binary probability system (that is, a preference matrix) corresponds to a distribution over total orders. This problem dates back to the 1950s and the works of Guilbaud [16] and Marschak [18]. Early works proved that the triangle inequality[2] is a necessary and sufficient condition for a preference matrix to be rationalizable in the case of five or fewer candidates [8,11]. But this fails for six or more candidates [8,11], leading to a search for other necessary conditions [15,19]. Obtaining a concise characterization for matrix rationalizability remains an outstanding open problem.

Systems of choice probabilities have been widely studied in mathematical psychology. Specifically, in stochastic choice behaviour, decision-makers must select an item i when presented with a subset S of the items. Binary probability systems are the special case where the subsets considered have size two.

[2] The triangle inequality states that $p_{ij} \leq p_{ik} + p_{kj}$, for any three candidates i, j and k.

Falmagne [9] showed a complete system of choice probabilities to be induced by rankings whenever it satisfies the Block-Marschak conditions [3] and normalization equalities, the proof of which was simplified by Fiorini [10].

Closely related to the matrix rationalizability problem is the *linear ordering problem*, which asks for the total order that best approximates a given binary probability system. For standard measures this problem is NP-hard [14], but understanding the geometry of the polytope of rationalizable choice matrices aids in the development of heuristic and approximation algorithms [12], [?]. Also closely related to the matrix rationalizability problem is the classic combinatorial *majority digraph problem*. Here a directed graph models the pairwise majority relation of a voter profile; that is, the arc ij is included in the digraph whenever the majority of voters prefer candidate i over candidate j. McGarvey [20] showed that every asymmetric digraph corresponds to the majority relation of some voter profile. Various relaxations of rationalizability have been studied. One popular relaxation is *regularity* [6,22] which asks if a system of binary probabilities can be extended to a system of choice probabilities under which the probability of selecting an item from a set does not increase when that set expands.

2 Preliminaries

We begin with a monotonicity property.

Lemma 2.1. *Let M be a preference matrix consistent with m voters with preferences $\{\succ_1, \dots, \succ_u, \dots, \succ_m\}$. Then M is consistent with $\{\succ_1, \dots, \succ'_u, \cdots \succ_m\}$, where $\succ'_u$ is identical to $\succ_u$ except that voter u prefers x over y in $\succ_u$ but is indifferent between x and y in $\succ'_u$.*

Proof. So, for any pair of candidates i and j, $\{\succ_1, \dots, \succ_u, \dots, \succ_m\}$ satisfies the conditions $(*)$. Namely

$$\frac{\#\{v : v \text{ strongly prefers } i \text{ over } j\}}{m} \leq p_{ij} \leq \frac{\#\{v : v \text{ weakly prefers } i \text{ over } j\}}{m}$$

These conditions trivially still hold with respect to $\{\succ_1, \dots, \succ'_u, \cdots \succ_m\}$ for any pair except $\{x, y\}$ and $\{y, x\}$. Let us verify that $(*)$ still holds for these two cases as well. As u prefers x over y in $\succ_u$ but is indifferent between x and y in $\succ'_u$, the number of voters that strongly prefer x over y has fallen by one (namely, u) while the number of voters that weakly prefer x over y is the same. Thus the lower bound has fallen whilst the upper bound is identical. Hence $(*)$ holds for $\{x, y\}$. Similarly, $(*)$ holds for $\{y, x\}$, as the lower bound is the same whilst the upper bound has increased. Hence M is consistent with $\{\succ_1, \dots, \succ'_u, \cdots \succ_m\}$. □

Take a finite poset $\mathcal{P} = (S, \succ)$ on a set S of elements with partial order $\succ$. A *chain decomposition* of $\mathcal{P}$ is a partition of the elements of the poset into disjoint chains. The cardinalility of a chain decomposition is the number of chains in the decomposition. A famous result of Dilworth [7] states that the width of $\mathcal{P}$ is the minimum cardinality of a chain decomposition.

Theorem 2.2. *[7] Let $\mathcal{P} = (S, \succ)$ be a finite poset. The maximum cardinality of an antichain of $\mathcal{P}$ equals the minimum cardinality of a chain decomposition.* □

Theorem 2.2 allows us to restrict our attention to voter preferences composed of disjoint chains.

Theorem 2.3. *Let M be α-rationalizable by $\{\succ_1, \ldots, \succ_u, \ldots, \succ_m\}$. Then M is α-rationalizable by $\{\succ_1, \ldots, \succ'_u, \cdots \succ_m\}$, where $\succ'_u$ consists of at most α disjoint chains.*

Proof. Let $\succ'_u$ correspond to a minimum cardinality chain decomposition of $\mathcal{P} = ([n], \succ_u)$. Thus, by Theorem 2.2, $\succ'_u$ consists of at most α disjoint chains. Thus the width of $\succ_u$ and $\succ'_u$ are the same, and voter u is still α-rational.

Next we know M is consistent with $\{\succ_1, \ldots, \succ_u, \cdots \succ_m\}$. We can now use the monotonicity property. Repeatedly applying Lemma 2.1, we conclude that M is consistent with $\{\succ_1, \ldots, \succ'_u, \cdots \succ_m\}$, as desired. □

Of course, repeated application of Theorem 2.3 implies we can assume that every voter has a partial order that is a collection of disjoint chains.

Corollary 2.4. *If M is α-rationalizable then it is consistent with a set of α-rational voters whose partial orders are a collection of (at most) α chains.* □

For an application, re-consider `Example II`. Recall the preference matrix M is 2-rationalizable using two voters. By Corollary 2.4, it must be consistent with two 2-rational voters whose partial orders each consist of two disjoint chains. To do this we find a minimum chain decomposition of partial orders for the red and blue voters in `Example II`. This gives us the two voters illustrated in Fig. 4.

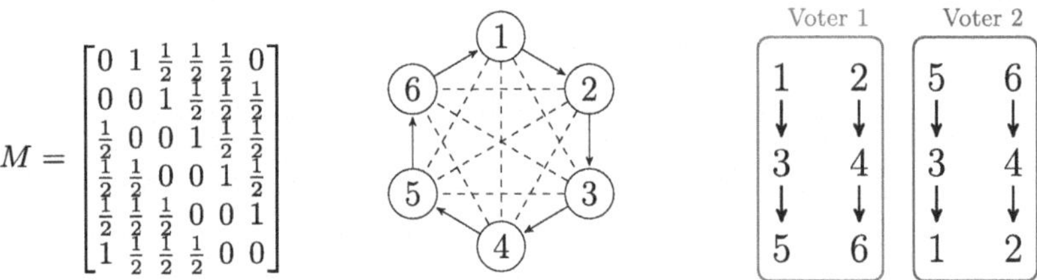

Fig. 4. A 2-rationalizable matrix consistent with two voters whose partial orders are disjoint chains.

3 Half-Integral Preference Matrices

We now restrict attention to the class $\mathcal{M}^{\frac{1}{2}}$ of half-integral matrices. This class is important in the field as it is the simplest class of preference matrices for which no characterization of rationality is known.[3] Recall, associated with a half-integral preference matrix is a directed *voting graph* D_M and an undirected *unanimity graph* G_M. In this section, we will see that understanding the chromatic number of the unanimity graph is critical in understanding the α-rationalizability of the class of half-integral preference matrices.

3.1 Rationality and the Unanimity Graph

We begin with a simple reduction that allows us to restrict the computation of the rationality number of M to its computation in a collection of submatrices based upon the structure of the unanimity graph G_M.

Theorem 3.1. *Given a half-integral preference matrix M with connected components $G_1, G_2, \ldots, G_t$ in the unanimity graph G_M. For each, $1 \leq \ell \leq t$, let M_ℓ be the sub-matrix of M induced by G_ℓ. Then $\alpha(M) = \max_\ell \alpha(M_\ell)$.*

Proof. Given a set of α-rational voters that satisfy the conditions $(*)$ for every pair of candidates in M. Then, for each $1 \leq \ell \leq t$, the same set of voters trivially satisfy $(*)$ for each pair of candidates in V_ℓ, the set of candidates restricted to the sub-matrix M_ℓ. Thus $\alpha(M) \geq \alpha(M_\ell)$ and, hence, $\alpha(M) \geq \max_\ell \alpha(M_\ell)$.

So it remains to prove the harder direction, that $\alpha(M) \leq \max_\ell \alpha(M_\ell)$. Assume, for each $1 \leq \ell \leq t$, that M_ℓ is consistent with a collection of m_ℓ voters who are each α-rational. We claim that M is consistent with a collection of $2 \cdot \prod_{\ell=1}^{t} m_\ell$ voters who are each α-rational.

To prove this, we create two new voters, L^S and R^S, for every set $S = \{v_1, v_2, \ldots, v_t\}$ of voters, where v_ℓ is one of the n_ℓ voters used to α-rationalize M_ℓ. Note that voter v_ℓ has a partial order on the candidates in V_ℓ. Both voters L^S and R^S will copy the partial order v_ℓ has on the set V_ℓ, for all $1 \leq \ell \leq t$. That is if candidates i and j are both in V_ℓ then L^S and R^S comparatively rank i and j exactly how v_ℓ does.

But what if candidate $i \in V_\ell$ and candidate $j \in V_\gamma$, where $\ell \neq \gamma$? Imagine an ordering of the sets of candidates $\{V_1, V_2, \ldots, V_t\}$ from left to right. Then voter L^S will prefer sets from left to right, and voter R^S will prefer sets from right to left. That is, if $\gamma < \ell$, then L^S prefers i over j and R^S prefers j over i.

So in total we have created $2 \cdot \prod_{\ell=1}^{t} m_\ell$ new voters. Moreover each of these new voters is α-rational. This is because both L^S and R^S have a strict preference for any pair of candidates for $i \in V_\ell$ and candidate $j \in V_\gamma$, where $\ell \neq \gamma$. Hence, any antichain of cardinality greater than one can only contain candidates within the same set V_ℓ. Thus, the maximum size of an antichain in the partial order of L^S (or R^S) is equal to the maximum size of an antichain in any of the partial orders for the set of voters $S = \{v_1, \ldots, v_t\}$, which by definition is at most α. So L^S and R^S are both α-rational, for any set S.

[3] There is a simple characterization for the class $\mathcal{M}^{0/1}$ of integral matrices. An integral preference matrix M is rationalizable (1-rationalizable) if and only if its voting graph D_M is acyclic.

Finally it remains to prove that the constraints (∗) hold for every pair of candidates using the $2 \cdot \prod_{\ell=1}^{t} m_\ell$ new voters. First, take a pair of candidates $i \in V_\ell$ and candidate $j \in V_\gamma$, where $\ell \neq \gamma$. Then since G_ℓ and G_γ separate components in the unanimity graph G_M it follows that $p_{ij} = \frac{1}{2}$. Moreover for any set $S = \{v_1, v_2, \ldots, v_t\}$ L^S and R^S rank i and j in the opposite way. Thus exactly half the voters strongly prefer i over j and half strongly prefer j over i. It follows that (∗) holds for this pair of candidates. Second, take a pair of candidates $i, j \in V_\ell$. Recall there are m_ℓ α-rational voters consistent with the submatrix M_ℓ. Therefore $f_1 \leq p_{ij} \leq f_2$, where f_1 is the fraction of these m_ℓ voters that strongly prefer i over j, and f_2 is the fraction of these voters that weakly prefer i over j. But each voter in M_ℓ is selected to be v_ℓ in $S = \{v_1, v_2, \ldots, v_t\}$ with exactly the same probability, namely $\frac{1}{m_\ell}$. It immediately follows that among the $2 \cdot \prod_{\ell=1}^{t} m_\ell$ new voters exactly an f_1 fraction of them strongly prefer i over j, and exactly an f_2 fraction of them weakly prefer i over j. Thus (∗) holds, and M is indeed α-rationalizable. So $\alpha(M) \geq \max_\ell \alpha(M_\ell)$. □

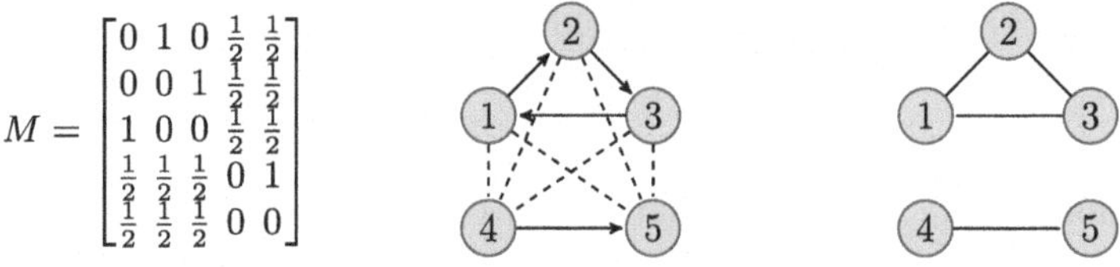

$M = \begin{bmatrix} 0 & 1 & 0 & \frac{1}{2} & \frac{1}{2} \\ 0 & 0 & 1 & \frac{1}{2} & \frac{1}{2} \\ 1 & 0 & 0 & \frac{1}{2} & \frac{1}{2} \\ \frac{1}{2} & \frac{1}{2} & \frac{1}{2} & 0 & 1 \\ \frac{1}{2} & \frac{1}{2} & \frac{1}{2} & 0 & 0 \end{bmatrix}$
1
2
3
4
5
1
2
3
4
5

Fig. 5. A half-integral preference matrix with its voting graph and unamimity graph.

`Example IV`. Consider the half-integral preference matrix M shown in Fig. 5, along with its voting graph D_M and unanimity graph G_M. Observe that the unanimity graph has exactly two components on the candidate sets $V_1 = \{1, 2, 3\}$ and $V_2 = \{4, 5\}$. We can prove that M is 2-rationalizable by applying the method of Theorem 3.1. The submatrix M_1 induced by V_1 simply corresponds to the 3-cycle we saw in `Example I`. Thus M_1 is 2-rationalizable with $m_1 = 1$ voter with preference 1 over 2 and an indfference between candidate 3 and the other two candidates. The submatrix M_2 induced by V_2 corresponds to an arc $(4, 5)$. This

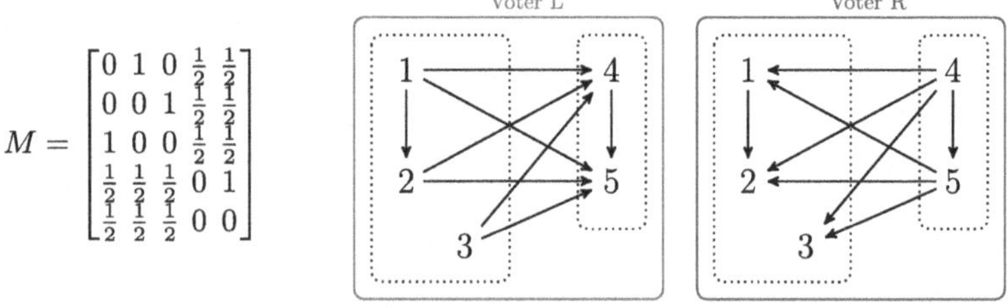

Fig. 6. The matrix M is 2-rationalizable using two voters.

is trivially 1-rationalizable (and, hence, 2-rationalizable) with $m_2 = 1$ voter with preference 4 over 5. So we only need $2 \cdot m_1 \cdot m_2 = 2$ new voters to 2-rationalize M. Let these voters be L and R. Following the proof of Theorem 3.1, let L and R copy the preferences within each of $V_1 = \{1, 2, 3\}$ and $V_2 = \{4, 5\}$. Now L prefers any vertex in V_1 over any vertex in V_2, but R has the opposite preference. Thus we obtain the two voters shown in Fig. 6.

3.2 An Upper Bound on the Rationality Number of Half-Integral Matrices

We can also use the unanimity graph to bound the rationality number of half-integral matrices. First we give an upper bound. The rationality number $\alpha(M)$ is upper bound by the chromatic number $\chi(G_M)$ of its unanimity graph.

Lemma 3.2. *If M is a half-integral preference matrix then $\alpha(M) \leq \chi(G_M)$.*

Proof. Given M let $\chi(G_M) = k$ be the chromatic number of its unanimity graph G_M. Take any k-coloring of the candidates, namely the vertices in G_M. Let C_ℓ be the set of candidates receiving color ℓ, for $1 \leq \ell \leq k$. We will show that M is k-rationalizable using just two voters. The construction is simple. Both voters will have k chains in their partial order and are thus k-rational. There is a chain for each color class C_ℓ. Voter 1 places the candidates in C_ℓ in an arbitrary total order to generate its ℓth chain, for $1 \leq \ell \leq k$. Voter 2 does the same thing, except it chooses exactly the opposite total order for the candidates in C_ℓ to generate its ℓth chain.

Let us verify that this construction satisfies the rationality constraints $(*)$. Take two candidates i and j. There are two cases to consider. First, assume that both candidates belong to the same color class, $i, j \in C_\ell$. As each color class is an independent set we have that $(i, j) \notin G_M$ and so $p_{ij} = \frac{1}{2}$. As the chains are reversals of each other, half the voters strongly prefer i over j and half the voters weakly prefer i over j. So $\frac{1}{2} \leq p_{ij} = \frac{1}{2} \leq \frac{1}{2}$ and $(*)$ holds for these candidates.

Second, assume the candidates belong to different color classes, $i \in C_\ell$ and $j \in C_\gamma$, where $\ell \neq \gamma$. This means i and j are in different chains in the partial orders of both voter 1 and voter 2. Thus, the fraction of voters that strongly prefer i over j is zero and the fraction of voters that weakly prefer i over j is one. Thus, regardless of the value of p_{ij}, we have $0 \leq p_{ij} = \frac{1}{2} \leq 1$ and $(*)$ holds for this pair of candidates. Thus M is k-rationalizable and so $\alpha(M) \leq \chi(G_M)$. □

`Example V.` Consider the half-integral preference matrix M shown in Fig. 7. This has a two-colorable unanimity graph, $\chi(G_M) = 2$. It has color classes green and pink with $V_{green} = \{2, 3, 5\}$ and $V_{pink} = \{1, 4\}$. Let voter 1 order the two corresponding chains by preferring lower numbered candidates, and let voter prefer higher numbered candidates. This produces the partial orders illustrated in Fig. 7. These two voters prove that M is 2-rationalizable.

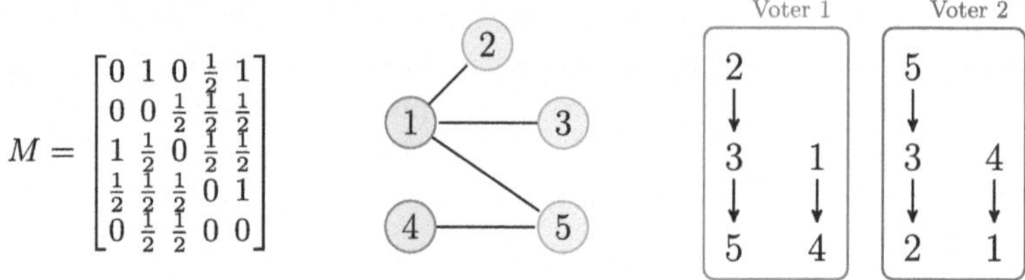

Fig. 7. A 2-chromatic unanimity graph inducing two 2-rational voters consistent with its half-integral preference matrix M.

3.3 A Lower Bound on the Rationality Number of Half-Integral Matrices

Let us now show that the upper bound in Lemma 3.2 is tight (to within a constant factor) for the class of half-integral preference matrices.

Take any integral preference matrix M. Its voting graph is a *tournament*. If this tournament is acyclic then the matrix is rationalizable, that is, it is $\alpha(M) = 1$. But its unanimity graph is a clique so has chromatic number $\chi(G_M) = n$.

Because an integral matrix is trivially half-integral, this example shows that it is **not** the case that $\alpha(M) = \Omega(\chi(G_M))$ for every half-integral matrix. However, the upper bound is indeed tight in the following sense:

Lemma 3.3. *Given $k \in \mathbb{N}$, there exists a half-integral matrix M with $\chi(G_M) = k$ and $\alpha(M) \geq \frac{1}{5} \cdot k$.*

Proof. Take any $k \in \mathbb{N}$. We build a unanimity graph G_M on n vertices composed of k disjoint independent sets $\{C_1, C_2, \ldots, C_k\}$, where $|C_\ell| = \frac{n}{k}$, for all $1 \leq \ell \leq k$. (Note we will determine below how large n needs to be.) In G_M every pair of candidates belonging to different independent sets is connected by an edge. Thus we have that $\chi(G_M) = k$. We now need to construct a voting graph D_M and a half-integral preference matrix M corresponding to this unanimity graph G_M. To do this we apply the probabilistic method. We uniformly at random orient each edge in G_M independently to produce D_M (and thus M).

We prove the following key property. With non-zero probability, every subset $S \subseteq V(D_M)$ of size $5\frac{n}{k}$ contains a directed triangle. Let S be an arbitrary subset of the vertices of size $5\frac{n}{k}$ and let $S_\ell = S \cap C_\ell$, for all $1 \leq \ell \leq k$. Set $m_\ell = |S_\ell|$, the number of vertices of S contained in C_ℓ. We desire a lower bound on the number of edge-disjoint triangles contained within S. To do this we apply the following "merge" operation. Given S_ℓ and S_γ, where $\ell \neq \gamma$, we set create an independent set $S_\ell \cup S_\gamma$ by removing the arcs between vertices in S_ℓ and S_γ. Clearly, by removing arcs we cannot increase the number of triangles. So any lower bound we obtain after applying this operation applies to the original instance.

Our goal is to obtain (at least) two independent sets of size at least $\frac{n}{k}$. So if $m_1 < \frac{n}{k}$, we merge other sets into S_1 until $m_1 \geq \frac{n}{k}$. Observe that at this point we must have $m_1 < 2 \cdot \frac{n}{k}$. We then repeat this process on another set, say S_2, until

$\frac{n}{k} \leq m_2 < 2 \cdot \frac{n}{k}$. There are now at least $5\frac{n}{k} - m_1 - m_2 \geq \frac{n}{k}$ vertices remaining. So $m_3 + \cdots + m_k \geq \frac{n}{k}$.

We can now show that there are at least $\left(\frac{n}{k}\right)^2$ edge-disjoint triangles in S. First observe that every vertex in S_1 is adjacent to every vertex in S_2. By deleting vertices we may assume $m_1 = m_2 = \frac{n}{k}$. Then, by repeatedly applying Hall's theorem, we can find $\frac{n}{k}$ edge-disjoint perfect matchings between S_1 and S_2. Note that each of these perfect matchings has cardinality $\frac{n}{k}$.

Next, by deleting vertices, we may assume $\bar{S} = S_3 \cup \cdots S_k$ contains exactly $\frac{n}{k}$ vertices, that is, $\sum_{i \geq 3} m_i = \frac{n}{k}$. Now pair each perfect matching with a different vertex v from $\bar{S}$. Since v is adjacent in G_M to each vertex in $S_1 \cup S_2$ this creates $\frac{n}{k}$ disjoint triangles. Because there are $\frac{n}{k}$ perfect matchings paired to $\frac{n}{k}$ distinct vertices in $\bar{S}$, this gives a total of $(\frac{n}{k})^2$ triangles in total, as claimed.

But each of these triangles is a directed 3-cycle with probability $\frac{1}{4}$. Furthermore, as each of these triangles are edge-disjoint, these are independent events. So the probability that S contains no directed cycle, which is less than the probability that none of our triangles are directed, is upper bounded as follows.

$$\mathbb{P}\{S \text{ has no directed cycle}\} \leq \mathbb{P}\{\text{none of the triangles are directed}\} = \left(\frac{3}{4}\right)^{\left(\frac{n}{k}\right)^2}$$

Thus, by the union bound, we can show that the probability of the existence of such a subset S without directed cycle is bounded away from 1. Specifically

$$\mathbb{P}\{\exists S \text{ with no directed cycle}\} \leq \sum_{S} \mathbb{P}\{S \text{ has no directed cycle}\} \leq N \times \left(\frac{3}{4}\right)^{\left(\frac{n}{k}\right)^2}$$

where N is the number of subsets of $V(G_M)$ of size $5 \cdot \frac{n}{k}$. Thus

$$N = \binom{n}{\frac{5n}{k}} = \frac{n(n-1)\cdots(n-\frac{5n}{k}+1)}{\left(\frac{5n}{k}\right)!} \leq \frac{n^{\frac{5n}{k}}}{\left(\frac{5n}{ke}\right)^{\frac{5n}{k}}} = \left(\frac{ke}{5}\right)^{\frac{5n}{k}}$$

where in the denominator we used the fact that $t! \geq \left(\frac{t}{e}\right)^t$ for any integer t, by Stirling's formula. Hence:

$$N \times \left(\frac{3}{4}\right)^{\left(\frac{n}{k}\right)^2} \leq \left(\frac{ke}{5}\right)^{\frac{5n}{k}} \cdot \left(\frac{3}{4}\right)^{\left(\frac{n}{k}\right)^2}$$

which is strictly less than 1 for n large enough. To see this, raising this expression to the power $\frac{k}{n}$ we have $\left(\frac{ke}{5}\right)^5 \cdot \left(\frac{3}{4}\right)^{\frac{n}{k}}$ which goes to zero as $n \to \infty$. Thus, we conclude that

$$\mathbb{P}\{\text{every subset } S \text{ has a cycle}\} = 1 - \mathbb{P}\{\text{there exists subset } S \text{ with no cycle}\} > 0.$$

Therefore, by the probabilistic method, there exists an orientation of G_M for which every subset $S \subseteq V(G_M)$ of size $5 \cdot \frac{n}{k}$ contains a directed triangle. This implies that the longest chain any voter can have in its partial order is of length at most $5\frac{n}{k}$. Consequently each voter must have at least $\frac{k}{5}$ chains in their partial order. This proves that this orientation D_M of G_M requires that each voter be at best $\frac{k}{5}$-rational. Hence, for this voting graph D_M, we have $\alpha(M) \geq \frac{k}{5}$. □

Together Lemma 3.2 and Lemma 3.3 prove our first main result, Theorem 1.1. Observe the lower and upper bounds are tight up to a constant factor of 5. It is an intriguing combinatorial problem to completely close the gap between the lower and upper bounds.

4 Integral Preference Matrices

Stronger results can be obtained when the preference matrix M is integral, that is, $p_{ij} \in \{0, 1\}$ for all i, j.

4.1 One Voter Suffices

In general, if M is α-rational more than one voter might be required to α-rationalize the matrix. However, if M is integral, then one voter suffices.

Theorem 4.1. *Let M be an integral preference matrix. If M is α-rational then it is consistent with a single α-rational voter.*

Proof. Take an integral preference matrix M. Consider any pair of candidates, i and j. Without loss of generality, $p_{ij} = 1$, i.e. the fraction of voters that weakly prefer i over j must be one, by the rationality constraints $(*)$. That is, *every* voter must either prefer i over j or must be indifferent between i and j. In particular, if $i \succ_v j$ in the partial order of voter v then it must be the case that $p_{ij} = 1$, as $p_{ij} = 0$ would violate $(*)$. Thus any strict preference in the partial order $\succ_v$ must agree with p_{ij}, whereas any indifference imposes no constraint.

Now assume M is consistent with a collection of m α-rational voters. By the above argument, each of these m voters must satisfy $(*)$ on its own! So each such voter suffices to α-rationalize M. Therefore, if M is α-rational then it is consistent with a single α-rational voter. □

4.2 The Dichromatic Number

Using Theorem 4.1, we can characterize the rationality number of an integral preference matrix M in terms of a coloring of its voting graph D_M. Now D_M is a directed graph, so what do we mean by a coloring of a directed graph? A *chromatic coloring* of an undirected graph G is a partition of the vertices into independent sets. A *dichromatic coloring* of a directed graph D is a partition of the vertices into acyclic sets. A dichromatic coloring can be viewed as a generalisation of a chromatic coloring of an undirected graph. To see this, if we bidirect every edge in an undirected graph G then an acyclic set in the resultant directed graph D is an independent set in the original undirected graph. Analogous to the chromatic number of an undirected graph, Neumann-Lara [21] defined the *dichromatic number* $\overrightarrow{\chi}(D)$ of a digraph D to be the minimum number of colors required in any dichromatic coloring of D. It is particularly important in the study of the voting graph D_M of an integral preference matrix, as shown below:

Theorem 4.2. *Let M be an integral preference matrix. Then its rationality number is equal to the dichromatic number of its voting graph: $\alpha(M) = \overrightarrow{\chi}(D_M)$.*

Proof. Let M be an integral preference matrix. Then its voting graph D_M is a tournament. First assume that $\overrightarrow{\chi}(D_M) = k$. Then we can partition the vertices of D_M into k acyclic subgraphs $\{C_1, C_2, \dots, C_k\}$. Take any C_ℓ, for $1 \leq \ell \leq k$. Then C_ℓ is itself an tournament. Because it it is acyclic it has an acyclic ordering. Furthermore, this ordering is unique as C_ℓ is a tournament. We use acyclic ordering as total order to induce a chain on C_ℓ. In this way we have a partial order that consists of k disjoint chains on $\{C_1, C_2, \dots, C_k\}$. This partial order corresponds to a single α-rational voter. Moreover if $i \succ j$ in this partial order then $p_{ij} = 1$. Thus by the argument of Theorem 4.1 this single voter is consistent with M in satisfying the constraints $(*)$. Thus $\alpha(M) \leq k$.

Second assume that $\alpha(M) = k$. Then, by Theorem 4.1, there is a single α-rational voter v that is consistent with M. Let $\succ$ be the partial order of voter v. Then, by Corollary 2.4, we can assume the partial order $\succ$ consists of exactly k disjoint chains, $\{C_1, C_2, \dots, C_k\}$. We claim that each chain induces an acyclic subgraph in the voting graph D_M. Suppose not, then there exist candidates i and j such that $i \succ j$ and $p_{ij} = 0$. But this contradicts $(*)$ since the fraction of voters that strongly prefer i over j is one. So voter v is not consistent with M, a contradiction. So $\{C_1, C_2, \dots, C_k\}$ are a partition of the vertices in D_M into acyclic subgraphs. Thus $\overrightarrow{\chi}(D_M) \leq k$.

Putting this together we have $\alpha(M) = \overrightarrow{\chi}(D_M)$ as desired. □

`Example VI.` Consider the integral preference matrix M in Fig. 8. Its voting graph has dichromatic number 3 and is consistent with a single 3-rational voter.

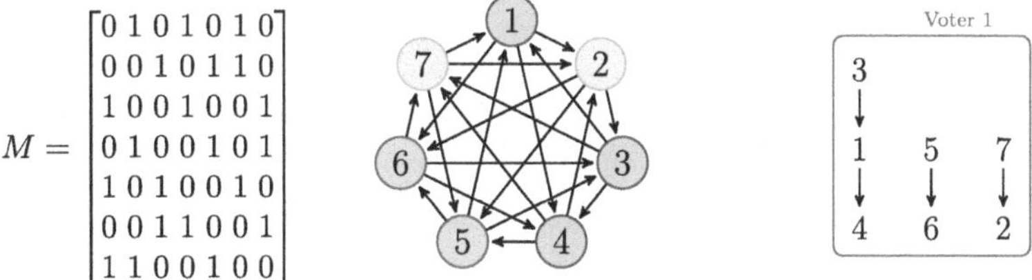

Fig. 8. An integral preference matrix M, its 3-dichromatic voting graph D_M, and a corresponding 3-rational voter.

4.3 An Upper Bound on the Rationality Number of Integral Matrices

For the class $\mathcal{M}^{0/1}$ of integral preference matrices we can strengthen the bounds in Theorem 1.1.

Lemma 4.3. *If M is an integral preference matrix then $\alpha(M) \leq \frac{3n}{\log n}$.*

Proof. By Theorem 4.2, to analyze the rationality number of an integral preference matrix M we must consider dichromatic colorings of the voting graph D_M. In particular, any algorithm to dichromatic color the voting graph will give an upper bound on the rationality number of M. We now present a greedy algorithm which gives tight bounds on the rationality number over the class of integral preference matrices.

The algorithm selects a color class C_1 as follows. It picks the vertex v_1 with the highest out-degree in D_M and adds it to C_1. It then selects the vertex v_2 with the highest out-degree in the subgraph $V(D_M) \cap \Gamma^+(v_1)$, induced by the set of out-neighbours of v_1, and adds v_2 to C_1. It then selects the vertex v_3 with the highest out-degree in the subgraph $V(D_M) \cap \Gamma^+(v_1) \cap \Gamma^+(v_2)$, induced by the set of out-neighbours of both v_1 and v_2, and adds v_3 to C_1. This process terminates with $C_1 = \{v_1, v_2, \ldots, v_t\}$ when $\bigcap_{i=1}^{t} \Gamma^+(v_i) \cap V(D_M) = \emptyset$. The vertices in C_1 are given the color 1. The algorithm is then repeated on $D_M \backslash C_1$ to select the second color class C_2, etc. The algorithm terminates when every vertex has been colored.

Let this greedy algorithm output the color classes $\{C_1, C_2, \ldots, C_k\}$. We claim this is a valid dichromatic coloring of D_m. This is true because each C_ℓ, for $1 \leq \ell \leq k$, is acyclic. In particular, by construction, if the $C_\ell = \{v_1, v_2, \ldots, v_r\}$, then v_j is an out-neighbour of v_i, for any $i < j$. Thus C_ℓ contains no directed cycle. Furthermore, we claim that $k \leq \frac{3n}{\log n}$. That is, the greedy algorithm gives a k-dichromatic coloring of D_M using at most $\frac{3n}{\log n}$, where n is the number of vertices (candidates). Since, $\chi(G_M) = n$, by Theorem 4.2, this will prove that $\alpha(M) \leq \frac{3n}{\log n} = \frac{3\chi(G_M)}{\log n}$. So let us verify this claim.

To do this, observe that if there are at least $\frac{n}{2^i}$ vertices remaining when the algorithm begins to construct a new color class C_ℓ, then the cardinality of C_ℓ will be at least $\log \frac{n}{2^i} = \log n - i$. Indeed, as the graph is a tournament, at any time there is a vertex whose out-degree is at least (the floor of) half the number of vertices under consideration. In particular, consider the first time we have at most $\frac{n}{2^{i-1}}$ vertices remaining. Then the number of color classes we find until the number of remaining vertices is at most $\frac{n}{2^i}$ is upper bounded by $\frac{\frac{n}{2^i}}{\log n - i}$.

If there are less than $\frac{n}{\log n}$ vertices remaining then the number of color classes the greedy algorithm finds from that point on is trivially upper bounded by $\frac{n}{\log n}$. Thus the total number of color classes the greedy algorithm finds in coloring every vertex is at most

$$\frac{n}{\log n} + \sum_{i=1}^{\log \log n} \frac{\frac{n}{2^i}}{\log n - i} = \frac{n}{\log n} + n \cdot \sum_{i=1}^{\log \log n} \frac{1}{2^i \cdot (\log n - i)} \leq \frac{3n}{\log n}$$

This gives our upper bound on the rationality number $\alpha(M)$. □

4.4 A Lower Bound on the Rationality Number of Integral Matrices

Let us now show that the upper bound in Lemma 4.3 is tight (to within a constant factor) for the class of integral preference matrices.

Lemma 4.4. *There exists an integral matrix M with $\alpha(M) \geq \frac{n}{2\log n+1}$.*

Proof. We claim there is a tournament with dichromatic number at least $\frac{n}{2\log n+1}$. To prove this we again apply the probabilistic method. Take a random tournament on n vertices. Next select any subset S of k vertices. There are $k!$ acyclic orderings of a tournament on k vertices and $2^{\binom{k}{2}}$ ways to orient the arcs in the tournament. Thus, the probability that S is a acyclic is exactly $\frac{k!}{2^{\binom{k}{2}}}$. Furthermore there are $\binom{n}{k}$ ways to choose S. So, by the union bound, the probability that at least one of them induces an acyclic tournament is at most

$$\binom{n}{k} \cdot \frac{k!}{2^{\binom{k}{2}}} < \frac{n^k}{2^{\binom{k}{2}}} = \frac{n^k}{2^{\frac{1}{2}(k-1)k}} = \left(\frac{n}{2^{\frac{1}{2}(k-1)}}\right)^k$$

But this is less than 1 if $2^{\frac{1}{2}(k-1)} \geq n$. That is, if $k \geq 2\log n + 1$. This implies there exists a tournament D_M on n vertices that contains no acyclic subgraphs of cardinality greater than $2\log n + 1$. For this tournament, every dichromatic color class has cardinality at most $2\log n + 1$. Thus, its dichromatic number is at least $\frac{n}{2\log n+1}$. Now D_M is the voting graph of an integral preference matrix M. So, by Theorem 4.2, the rationality number of M is $\alpha(M) = \vec{\chi}(D_M) \geq \frac{n}{2\log n+1}$. □

Together Lemma 3.2 and Lemma 3.3 prove our second main result, Theorem 1.2. Again, closing the factor 6 gap between the lower and upper bounds is an interesting open problem.

5 Computational Complexity

Theorem 5.1. *The rationality number problem is NP-complete for $k \geq 2$, even for the case of integral preferences matrices.*

Proof. First note that the problem is in NP, as a set of voter preferences gives a certificate to the rationality number, which can be verified in polynomial time.

Given an integral preference matrix M, by Theorem 4.2, determining whether $\alpha(M) \leq k$ is equivalent to deciding whether the tournament D_M has dichromatic number k. Consider first the case of $k = 2$. Now a tournament T has dichromatic number 2 if and only if the vertices of T can be partitioned into two feedback vertex sets, since the complement of a feedback vertex set induces, by definition, an acyclic graph. Using a reduction from Not-All-Equal-3SAT, Chen, Hu and Zang [5] proved that determining if a tournament can be partitioned into two feedback vertex sets is NP-complete. Next consider $k \geq 3$. Fox et al. [13] gave a reduction from $(k-1)$-dicolorability to k-dicolorability, hence proving

NP-completeness for all $k \geq 2$. The reduction is simple. Given a tournament T, construct a new tournament $\hat{T}$ consisting of two identical copies of T and an extra vertex z, connected in the order $T_1 \to T_2 \to z \to T_1$. It can easily be verified that $\vec{\chi}(T) = k - 1$ if and only if $\vec{\chi}(\hat{T}) = k$. The theorem follows. □

This hardness result indicates that it may be fruitful to search for approximation algorithms for the rationality number of a preference matrix. We remark that for the special case of integral preference matrices with $\alpha(M) = 2$ a 5-approximation can be derived from the work of Klingelhoefer and Newman [17].

Acknowledgements. We thank Sophie Spirkl for showing us a reduction from MONOTONE-NOT-ALL-EQUAL-3-SAT to the problem of deciding if a tournament has dichromatic number 2 and Gerardo Berbeglia for discussions.

References

1. Bachmeier, G., et al.: k-majority digraphs and the hardness of voting with a constant number of voters. J. Comput. Syst. Sci. **105**, 130–157 (2019)
2. Batchelder, W., Colonius, H., Dzhafarov, E., Myung, J.: New Handbook of Mathematical Psychology, vol. 1. Cambridge University Press, Foundations and Methodology (2016)
3. Block, H., Marschak, J.: Random orderings and stochastic theories of response. In: Olkin, I. (ed.) Contributions to Probability and Statistics. Stanford University Press (1960)
4. Cameron, T., Charmot, S., Pulaj, J.: On the linear ordering problem and the rankability of data. Found. Data Sci. **3**(2), 133–149 (2021)
5. Chen, X., Hu, X., Zang, W.: A min-max theorem on tournaments. SIAM J. Comput. **37**, 923–937 (2007)
6. Dasgupta, I., Pattanaik, P.: 'regular' choice and the weak aximom of stochastic revealed preference. Econ. Theor. **31**, 35–50 (2007)
7. Dilworth, R.: A decomposition theorem for partially ordered sets. Ann. Math. **51**(1), 161–166 (1950)
8. Dridi, T.: Sur les distribution binaires associées à des distribution ordinales. Math. Sci. Hum. **69**, 15–31 (1980)
9. Falmagne, J.C.: A representation theorem for finite random scale systems. J. Math. Psychol. **18**, 52–72 (1978)
10. Fiorini, S.: A short proof of a theorem of Falmagne. J. Math. Psychol. **48**, 80–82 (2004)
11. Fishburn, P.: Decomposing weighted digraphs into sums of chains. Discret. Appl. Math. **16**(3), 223–238 (1987)
12. Fishburn, P.: Induced binary probabilities and the linear ordering polytope: a status report. Math. Soc. Sci. **23**(1), 67–80 (1992)
13. Fox, J., Gishboliner, L., Shapira, A., Yuster, R.: The removal lemma for tournaments. J. Comb. Theory, Ser. B **136**, 110–134 (2017)
14. Garey, M.R., Johnson, D.S.: Computers and Intractability; A Guide to the Theory of NP-Completeness. W. H. Freeman & Co., USA (1979)
15. Grötschel, M., Jünger, M., Reinelt, G.: Optimal triangulation of large real world input-output matrices. Stat Hefte **25**, 261–295 (1983)

16. Guilbaud, G.: Sur une difficulté de la théorie du risque. Colloques Internationaux Du CNRS, Econométrie **40**, 19–25 (1953)
17. Klingelhoefer, F., Newman, A.: Coloring tournaments with few colors: algorithms and complexity. In: Proceedings of the 31st Annual European Symposium on Algorithms (ESA), pp. 71:1–71:14 (2023)
18. Marschak, J.: Binary-choice constraints and random utility indicators. In: Arrow, K., Karlin, S., Suppes, P. (eds.) Economic Information, Decision, and Prediction, pp. 312–329. Stanford University Press (1960)
19. Martí, R., Reinelt, G.: The Linear Ordering Problem: Exact and Heuristic Methods in Combinatorial Optimization. Applied Mathematical Sciences, Springer, Berlin Heidelberg (2011). https://doi.org/10.1007/978-3-642-16729-4
20. McGarvey, D.: A theorem on the construction of voting paradoxes. Econometrica **21**(4), 608–610 (1953)
21. Neumann-Lara, V.: The dichromatic number of a digraph. J. Comb. Theory, Ser. B **33**(3), 265–270 (1982)
22. Suck, R.: Regular choice systems: a general technique to represent them by random variables. J. Math. Psychol. **75**, 110–117 (2016)

Computational Complexity and Resource Allocation

k-Times Bin Packing and its Application to Fair Electricity Distribution

Dinesh Kumar Baghel[1(✉)], Alex Ravsky[2], and Erel Segal-Halevi[1]

[1] Ariel University, 40700 Ariel, Israel
dinkubag21@gmail.com, erelsgl@gmail.com
[2] Pidstryhach Institute for Applied Problems of Mechanics and Mathematics of National Academy of Sciences of Ukraine, Lviv, Ukraine
alexander.ravsky@uni-wuerzburg.de

Abstract. Given items of different sizes and a fixed bin capacity, the bin-packing problem is to pack these items into a minimum number of bins such that the sum of item sizes in a bin does not exceed the capacity. We define a new variant called *k-times bin-packing (kBP)*, where the goal is to pack the items such that each item appears exactly k times, in k different bins. We generalize some existing approximation algorithms for bin-packing to solve kBP, and analyze their performance ratio.

The study of kBP is motivated by the problem of *fair electricity distribution*. In many developing countries, the total electricity demand is higher than the supply capacity. We prove that every electricity division problem can be solved by k-times bin-packing for some finite k. We also show that k-times bin-packing can be used to distribute the electricity in a fair and efficient way. Particularly, we implement generalizations of the First-Fit and First-Fit Decreasing bin-packing algorithms to solve kBP, and apply the generalizations to real electricity demand data. We show that our generalizations outperform existing heuristic solutions to the same problem.

Due to space constraints, several parts of the paper were moved to appendices. All appendices are available in the full version [1].

Keywords: Approximation algorithms · bin-packing · First-Fit · First-Fit Decreasing · Next-Fit · fair division · Karmarkar-Karp algorithms · Fernandez de la Vega-Lueker algorithm · electricity distribution · utilitarian metric · egalitarian metric · utility difference

1 Introduction

This work is motivated by the problem of *fair electricity distribution*. In developing countries, the demand for electricity often surpasses the available supply [17]. Such countries have to come up with a fair and efficient method of allocating of the available electricity among the households.

Formally, we consider a power-station that produces a fixed supply S of electricity. The station should provide electricity to n households. The demands

G. Schäfer and C. Ventre (Eds.): SAGT 2024, LNCS 15156, pp. 483–500, 2024.
https://doi.org/10.1007/978-3-031-71033-9_27

of the households in a given period are given by a (multi)set D. Typically, $\sum_i D[i] > S$ (where $D[i]$ is the electricity demand of a household i), so it is not possible to connect all households simultaneously. Our goal is to ensure that each household is connected the same amount of time, and that this amount is as large as possible. We assume that an agent gains utility only if the requested demand is fulfilled; otherwise it is zero. Practically it can be understood as follows: Suppose at some time, a household i is running some activity that requires $D[i]$ kilowatt of electricity to operate; in the absence of that amount, the activity will not function. Therefore, an allocation where demands are fractionally fulfilled is not relevant.

A simple approach to this problem is to partition the households into some q subsets, such that the sum of demands in each subset is at most S, and then connect the agents in each subset for a fraction $1/q$ of the time. To maximize the amount of time each agent is connected, we have to minimize q. This problem is equivalent to the classic problem of *bin-packing*. In this problem, we are given some n items, of sizes given by a multiset of positive numbers numbers D, and a positive number S representing the capacity of a bin. The goal is to pack items in D into the smallest possible number of bins, such that the sum of item sizes in each bin is at most S. The problem is NP-complete [9], but has many efficient approximation algorithms.

However, even an optimal solution to the bin-packing problem may provide a sub-optimal solution to the electricity division problem. As an example, suppose we have three households x, y, z with demands $2, 1, 1$ respectively, and the electricity supply $S = 3$. Then, the optimal bin-packing results in 2 bins, for instance, $\{x, y\}$ and $\{z\}$. This means that each agent would be connected $1/2$ of the time. However, it is possible to connect each agents $2/3$ of the time, by connecting each of the pairs $\{x, y\}, \{x, z\}, \{y, z\}$ for $1/3$ of the time, as each agent appears in 2 different subsets. More generally, suppose we construct q subsets of agents, such that each agent appears in exactly k different subsets. Then we can connect each subset for $1/q$ of the time, and each agent will be connected k/q of the time.

1.1 The k-Times Bin-Packing Problem

To study this problem more abstractly, we define the *k-times bin-packing problem (or kBP)*. The input to kBP is a set of n items of sizes given by a multiset D, a positive number S representing the capacity of a bin, and an integer $k \geq 1$. The goal is to pack items in D into the smallest possible number of bins, such that the sum of item sizes in each bin is at most S, and each item appears in k different bins, where each item occurs at most once in a bin. In the above example, $k = 2$. It is easy to see that, in the above example, 2-times bin-packing yields the optimal solution to the electricity division problem.

Our first main contribution (**Section** 4) is to prove that, for every electricity division problem, there exists some finite k for which the optimal solution to the kBP problem yields the optimal solution to the electricity division problem.

We note that kBP may have other applications beyond electricity division. For example, it could be to create a backup of the files on different file servers [14].

We would like to store k different copies of each file, but obviously, we want at most one copy of the same file on the same server. This can be solved by solving kBP on the files as items, and the server disk space as the bin capacity.

Motivated by these applications, we would like to find ways to efficiently solve kBP. However, it is well-known that kBP is NP-hard even for $k = 1$. We therefore look for efficient approximation algorithms of kBP.

1.2 Using Existing Bin-Packing Algorithms for kBP

Several existing algorithms for bin-packing can be naturally extended to kBP. However, it is not clear whether the extension will have a good approximation ratio.

As an example, consider the simple algorithm called *First-Fit (FF)*: process the items in an arbitrary order; pack each item into the first bin it fits into; if it does not fit into any existing bin, open a new bin for it. In the example $D = [10, 20, 11], S = 31$, the FF would pack two bins: $\{10, 20\}$ and $\{11\}$. This is clearly optimal. The extension of FF to kBP would process the items as follows: for each item x_r in the list (in order), suppose that b bins have been used thus far. Let j be the lowest index ($1 \leq j \leq b$) such that (a) bin j can accommodate x_r and (b) bin j does not contain any copy of x_r, should such j exist; otherwise open a new bin with index $j = b + 1$. Place x_r in bin j.

There are two ways to process the input. One way is by processing each item k times in sequence. In the above example, with $k = 2$, FF will process the items in order $[10^1, 10^2, 20^1, 20^2, 11^1, 11^2]$, where the superscript specifies the instance to which an item belongs. This results in four bins: $\{10^1, 20^1\}, \{10^2, 20^2\}, \{11^1\}, \{11^2\}$, which simply repeats k times the solution obtained from FF on D. However, the optimal solution here is 3 bins: $\{10^1, 20^1\}, \{11^1, 10^2\}, \{20^2, 11^2\}$.

Another way is to process the whole sequence D, k times. In the above example, FF will process the sequence $D_2 = DD = [10^1, 20^1, 11^1, 10^2, 20^2, 11^2]$. Applying the FFk algorithm to this input instance will result in three bins $\{10^1, 20^1\}, \{11^1, 10^2\}, \{20^2, 11^2\}$, which is optimal. Thus, while the extension of FF to kBP is simple, it is not trivial, and it is vital to study the approximation ratio of such algorithms in this case.

As another example, consider the approximation schemes by de la Vega and Lueker [24] and Karmarkar and Karp [16]. These algorithms use a linear program that counts the number of bins of each different *configuration* in the packing (see subsection 6.1 for the definitions) One way in which these algorithms can be extended, without modifying the linear program, is to give D_k as the input. But then, a configuration might have more than one copy of an item in D, which violates the kBP constraint. Another approach is to modify the constraint in the configuration linear program, to check that there are k copies of each item in the solution, while keeping the same configurations as for the input D. Doing so will respect the kBP constraint. Again, while the extension of the algorithm is straightforward, it is not clear what the approximation ratio would be; this is the main task of the present paper.

The most trivial way to extend existing algorithms is to run an existing bin-packing algorithm, and duplicate the output k times. However, this will not let us enjoy the benefits of kBP for electricity division (in the above example, this method will yield 4 bins, so each agent will be connected for $2/4 = 1/2$ of the time). Therefore, we present more elaborate extensions, that attain better performance. The algorithms we extend can be classified into two classes:

1. Fast constant-factor approximation algorithms **(Section** 5). Examples are First-Fit (FF) and First-Fit-Decreasing (FFD). For bin-packing, these algorithms find a packing with at most $1.7 \cdot OPT(D)$ and $\frac{11}{9} \cdot OPT(D) + \frac{6}{9}$ bins respectively [5–7] We adapt these algorithms by running them on an instance made of k copies, $DD \ldots D$ (k copies of D), which we denote by D_k. We show that, for $k > 1$, the extension of FF to kBP (which we call FFk) finds a packing with at most $\left(1.5 + \frac{1}{5k}\right) \cdot OPT(D_k) + 3 \cdot k$ bins. For any fixed $k > 1$, the asymptotic approximation ratio of FFk for large instances (when $OPT(D_k) \to \infty$) is $(1.5 + \frac{1}{5k})$, which is better than that of FF, and improves towards 1.5 when k increases.

We also prove that the lower bound for $FFDk$ (the extension of FFD to kBP) is $\frac{7}{6} \cdot OPT(D_k) + 1$, and conjecture by showing on simulated data that $FFDk$ solves kBP with at most $\frac{11}{9} \cdot OPT(D_k) + \frac{6}{9}$ bins which gives us an asymptotic approximation ratio of at most 11/9.

We also show that the extension of NF (next-fit algorithm) to kBP (we call this extension as NFk) has the asymptotic ratio of 2.

2. Polynomial-time approximation schemes **(Section** 6). Examples are the algorithms by Fernandez de la Vega and Lueker [24] and Karmarkar and Karp [16]. We show that the algorithm by Fernandez de la Vega and Lueker can be extended to solve kBP using at most $(1+2\cdot\epsilon)OPT(D_k)+k$ bins for any fixed $\epsilon \in (0, 1/2)$. For every $\epsilon > 0$, Algorithm 1 of Karmarkar-Karp algorithms [16] solves kBP using bins at most $(1 + 2 \cdot k \cdot \epsilon)OPT(D_k) + \frac{1}{2\cdot\epsilon^2} + (2 \cdot k + 1)$ bins, and runs in time $O(n(D_k) \cdot \log n(D_k) + T(\frac{1}{\epsilon^2}, n(D_k))$, where $n(D_k)$ is the number of items in D_k, and T is a polynomially-bounded function. Algorithm 2 of Karmarkar-Karp algorithms [16] generalized to solve kBP using at most $OPT(D_k) + O(k \cdot \log^2 OPT(D))$ bins, and runs in time $O(T(\frac{n(D)}{2}, n(D_k)) + n(D_k) \cdot \log n(D_k))$.

Electricity Distribution **(Section** 7). The fair electricity division problem was introduced by Oluwasuji and Malik and Zhang and Ramchurn [19,20] under the name of "fair load-shedding". They presented several heuristic as well as ILP-based algorithms, and tested them on a dataset of 367 households from Nigeria. We implement the FFk and $FFDk$ algorithms for finding approximate solutions to kBP, and use the solutions to determine a fair electricity allocation. We test the performance of our allocations on the same dataset of Oluwasuji and Malik and Zhang and Ramchurn [19]. We compare our results on the same metrics used by Oluwasuji and Malik and Zhang and Ramchurn [19]. These metrics are utilitarian and egalitarian social welfare and the maximum utility difference between agents. We compare our results in terms of hours of connection to supply on average, utility delivered to an agent on average, and electricity supplied on average, along with their standard deviation. We find that our results surmount their results in all the above parameters. FFk and $FFDk$ run in time that is nearly linear in the number of agents. We conclude that using kBP can provide a practical, fair and efficient solution to the electricity division problem.

In Sect. 8, we conclude with a summary and directions for future work.

2 Related Literature

Due to space constraints, some related work was removed; it can be found in the full version [1].

First-Fit. We have already defined the working of FF in Subsect. 1.2. Denote by FF the number of bins used by the First-Fit algorithm, and by OPT the number of bins in an optimal solution for a multiset D. An upper bound of $FF \leq 1.7OPT + 3$ was first proved by Ullman in 1971 [23]. The additive term was first improved to 2 by Garey and Graham and Ullman [8] in 1972. In 1976, Garey and Graham and Johnson and Yao [10] improved the bound further to $FF \leq \lceil 1.7OPT \rceil$, equivalent to $FF \leq 1.7OPT + 0.9$ due to the integrality of FF and OPT. This additive term was further lowered to $FF \leq 1.7OPT + 0.7$ by Xia and Tan [27]. Finally, in 2013 Dosa and Sgall [5] settled this open problem and proved that $FF \leq \lfloor 1.7OPT \rfloor$, which is tight.

First-Fit Decreasing. Algorithm *First-Fit Decreasing* (FFD) first sorts the items in non-increasing order, and then implements FF on them. In 1973, in his doctoral thesis [15], D. S. Johnson proved that $FFD \leq \frac{11}{9}OPT + 4$. Unfortunately, his proof spanned more than 100 pages. In 1985, Baker [2] simplified their proof and improved the additive term to 3. In 1991 Minyi [28] further simplified the proof and showed that the additive term is 1. Then, in 1997, Li and Yue [18] narrowed the additive constant to 7/9 without formal proof. Finally, in 2007 Dosa [6] proved that the additive constant is 6/9. They also gave an example which achieves this bound.

Next-fit. The algorithm next-fit works as follows: It keeps the current bin (initially empty) to pack the current item. If the current item does not pack into the currently open bin then it closes the current bin and opens a new bin to pack the current item. Johnson in his doctoral thesis [15] proved that the asymptotic performance ratio of next-fit is 2.

Efficient Approximation Schemes. In 1981, Fernandez de la Vega and Lueker [24] presented a polynomial time approximation scheme to solve bin-packing. Their algorithm accepts as input an $\epsilon > 0$ and produces a packing of the items in D of size at most $(1 + \epsilon)OPT + 1$. Their running time is polynomial in the size of D and depends on $1/\epsilon$. They invented the *adaptive rounding* method to reduce the problem size. In adaptive rounding, they initially organize the items into groups and then round them up to the maximum value in the group. This results in a problem with a small number of different item sizes, which can be solved optimally using the linear configuration program. Later, Karmarkar and Karp [16] devised several PTAS for the bin-packing problem. One of the Karmarkar-Karp algorithms solves bin-packing using at most $OPT + O(\log^2 OPT)$ bins. Other Karmarkar–Karp algorithms have different additive approximation guarantees, and they all run in polynomial time. This additive approximation was further improved to $O(\log OPT \cdot \log \log OPT)$ by Rothvoss [21]. They used a "glueing" technique wherein they glued small items to get a single big item. In 2017, Hoberg and Rothvoss [13] further improved the additive approximation to a logarithmic term $O(\log OPT)$.

Jansen [14] has proposed a FPTAS for the generalization of the bin-packing problem called as *bin-packing with conflicts*. The input instance for their algorithm is the conflict graph. Its vertices are the items and any two items are adjacent provided they cannot be packed into the same bin. In particular, kBP can be considered as the bin-packing with conflicts, where the conflict graph D_k is a disjoint union of copies of a complete graph K_k. Their bin-packing problem with conflicts is restricted to $q-$inductive graphs. In a $q-$inductive graph the vertices are ordered from $1, \ldots, n$. Each vertex in the graph has at most q adjacent lower numbered vertices. Since the degree of each vertex of D_k equals $k-1$, D_k is a $k-1$-inductive graph. In their method first they obtain an instance of large items from the given input instance. Let this instance be J_k. They apply the linear grouping method of Fernandez de la Vega and Lueker [24] to.obtain a constant number of different item sizes. Next they apply the Karmarkar and Karp algorithm [16] to obtain an approximate packing of the large items. The bins in this approximate packing may have conflicts, so they use the procedure called COLOR which places each conflicted item into a new bin. In the worst case it may happen that all the items in each bin have conflict and hence each one of them is packed into a separate bin. Finally, after removing the conflicts, they packed the small items into the existing bins, respecting conflicts among items. In doing so, new bins are opened if necessary. In this paper we focus on a special kind of conflicts, and for this special case, we present a better approximation ratio. Their algorithm solves the kBP using at most $(1 + 2 \cdot \epsilon)OPT(D_k) + \frac{2 \cdot k - 1}{4 \cdot \epsilon^2} + 3 \cdot k + 1$ bins, whereas our extension to Algorithm 1 and 2 of Karmarkar-Karp algorithms solves the kBP using at most $(1 + 2 \cdot \epsilon)OPT(D_k) + \frac{1}{2 \cdot \epsilon^2} + 2 \cdot k + 1$ and $OPT(D_k) + O(k \cdot \log^2 OPT(D))$ bins respectively.

Gendreau and Laporte and Semet [11] propose six heuristics named H1 to H6 for bin-packing with item-conflicts, represented by a general conflict graph. The heuristic H1 is a variant of FFD which incorporates the conflicts, whereas H6 is a combination of a maximum-clique procedure and FFD. They show that H6 is better than H1 for conflict graphs with high density, whereas H1 performs marginally better for low density conflict graphs (where density is defined as the ratio of the number of edges to the number of possible edges). kBP can be represented by duplicating each item k times, and constructing a conflict graph in which there are edges between each two copies of the same item. The density of this graph is $(k-1)/(kn-1)$, which becomes smaller for large n. This suggests that H1 is a better fit for kBP. But in their adoption of kBP the vertices of the conflict graphs are ordered as blocks corresponding to the items, in such a way that the their sizes are non-decreasing. We have already seen that when we change such item order we can obtain a better packing.

Recently, Doron-Arad and Kulik and Shachnai [4] has solved in polynomial time a more general variant of bin-packing, with partition matroid constraints. Their algorithm packs the items in $OPT + O\left(\frac{OPT}{(\ln \ln OPT)^{1/17}}\right)$ bins. Their algorithm can be used to solve the kBP: for each item in D, define a category that contains k items with the same size. Then, solve the bin-packing with the constraint that each bin can contain at most one item from each category (it is a special case of a partition-matroid constraint). However, in the present paper we focus on the special case of kBP. This allows us to attain a better running-time (with FFk and $FFDk$), and a better approximation ratio (with the de la Vega–Lueker and Karmarkar–Karp algorithms).

Cake Cutting and Electricity Division. The electricity division problem can also be modeled as a classic resource allocation problem known as 'cake cutting'. The problem was first proposed by Steinhaus [22]. A number of cake-cutting protocols have been discussed in [3,25]. In cake cutting, a cake is a metaphor for the resource. Like previous approaches, a time interval can be treated as a resource. A cake-cutting protocol then allocates this divisible resource among agents who have different valuation functions (or preferences) according to some fairness criteria. The solution to this problem differs from the classic cake-cutting problem in the sense that at any point in time, t, the sum of the demands of all the agents, should respect the supply constraint, and several agents may share the same piece.

3 Definitions and Notation

3.1 Electricity Division Problem

The input to Electricity Division consists of:

- A number $S > 0$ denoting the total amount of available supply (e.g. in kW);
- A number n of households, and a list $D = D[1], \ldots, D[n]$ of positive numbers, where $D[i]$ represents the demand of households i (in kW);
- An interval $[0, T]$ representing the time in which electricity should be supplied to the households.

The desired output consists of:

- A partition $\mathcal{I}$ of the interval $[0, T]$ into sub-intervals, $I_1, \ldots, I_p$;
- For each interval $l \in [p]$, a set $A_l \subseteq [n]$ denoting the set of agents that are connected to electricity during interval l, such that $\sum_{i \in A_l} D[i] \leq S$ (the total demand is at most the total supply).

Throughout most of the paper, we assume that the utility of agent i equals the total time agent i is connected: $u_i(\mathcal{I}) = \sum_{l:i\in A_l} |I_l|$ (we will consider other utility functions at Sect. 7).

The optimization objective is $\max_{\mathcal{I}} \min_{i\in[n]} u_i(\mathcal{I})$, where the maximum is over all partitions that satisfy the demand constraints.

3.2 *k*-Times Bin Packing

We denote the bin capacity by $S > 0$ and the multiset of n items by D.

Let $n(D)$ and $m(D)$ denote the number of items and the number of different item sizes in D, respectively. We denote these sizes by $c[1], \ldots, c[m(D)]$. Moreover, for each natural $i \leq m(D)$ let $n[i]$ be the number of items of size $c[i]$. The *size* of a bin is defined as the sum of all the item sizes in that bin. Given a multiset B of items, we assume that its *size* $V(B)$ equals the sum of the sizes of all items of B.

We denote k copies of D by $D_k := DD\ldots D$. We denote the number of bins used to pack the items in D_k by the optimal and the considered algorithm by $OPT(D_k)$ and $bins(D_k)$, respectively.

Note that each item in D_k is present at most once in each bin, so it is present in exactly k distinct bins. Consider the example in Sect. 1. There are three items x, y, z with demand $2, 1, 1$ respectively. Let $k = 2$ and $S = 3$. Then, $\{x, y\}, \{y, z\}, \{z, x\}$ is a valid bin-packing. Note that each item is present twice overall, but at most once in each bin. In contrast, the bin-packing $\{x, y\}, \{y, z, z\}, \{x\}$ is not valid, because there are two copies of z in the same bin.

4 On Optimal *k* for *k*-times Bin-Packing

In this section we prove that, for every electricity division instance, there exists an integer k such that kBP yields the optimal electricity division. Moreover, we give an upper bound on k as a function of the number of agents.

Let X be a nonempty set. We denote by $\mathbb{R}^X$ the linear space of all functions from X to the real numbers $\mathbb{R}$. So elements of $\mathbb{R}^X$ have the form $(w_\alpha)_{\alpha\in X}$, where for each element $\alpha \in X$, w_α is the corresponding real number.

For each nonempty subset Y of X, let $\pi_Y : \mathbb{R}^X \to \mathbb{R}^Y$ be the natural projection, which maps each element $(w_\alpha)_{\alpha\in X} \in \mathbb{R}^X$ to the element $(w_\alpha)_{\alpha\in Y} \in \mathbb{R}^Y$.

We shall need the following lemmas (all proofs are in the full version [1]).

Lemma 1. *Let $W' \subseteq \mathbb{R}^X$ be a nonempty finite linearly independent set. Then there exists a subset Y of X with $|Y| = |W'|$ such that the set $\pi_Y(W')$ is linearly independent.* □

Given a finite set Y let $\|\cdot\|_Y$ be the Euclidean norm on the linear space $\mathbb{R}^Y$, that is for each $w = (w_\alpha)_{\alpha\in Y} \in \mathbb{R}^Y$ we have $\|w\|_Y = \sqrt{\sum_{\alpha\in Y} w_\alpha^2}$.

Lemma 2. *Let $W \subset \mathbb{Z}^X$ be a nonempty finite linearly dependent set of nonzero vectors. Let $p = |W| - 1$ and $K = \sup\{\|\pi_Y(w)\|_Y : w \in W, Y \subset X, |Y| = p\}$. Then there exist integers $(\Delta_w)_{w \in W}$ which are not all zeros such that $|\Delta_w| \leq K^p$ for each $w \in W$ and $\sum_{w \in W} \Delta_w \cdot w = 0$. That is if* some *nontrivial linear combination of W equals 0, then there exists such a linear combination in which the coefficients are all integers bounded by K^p.* □

Given an input set of items D, let $OPT(D_k)$ denote the optimal number of bins in k-times bin-packing of the items in D.

Theorem 1. *Given the bin size S and D the input set of items, there exists a $k \leq n^{n/2}$ such that $\frac{k}{OPT(D_k)}$ is the maximum possible connection-time per agent. This time can be attained by solving kBP on D and allocating a fraction $\frac{1}{OPT(D_k)}$ of the time to each bin in the optimal solution.* □

We do not know if the upper bound $n^{n/2}$ for k is tight. Proving tightness, or finding a better bound, remains an intriguing open question.

5 Fast Approximation Algorithms

The results of the previous section are not immediately applicable to fair electricity division, as kBP is known to be an NP-hard problem. However, they do hint that good approximation algorithms for kBP can provide good approximation for electricity division. Therefore, in this section, we study several fast approximation algorithms for kBP.

5.1 *FFk*—First-Fit for *k*BP

The k-times version of the First-Fit bin-packing algorithm packs each item of D_k in order into the first bin where it fits and does not violate the constraint that each item should appear in a bin at most once. If the item to pack does not fit into any currently open bin, FFk opens a new bin and packs the item into it. For example: consider $D = \{10, 20, 11\}, k = 2, S = 31$. FFk will result the bin-packing $\{10, 20\}, \{11, 10\}, \{20, 11\}$. It is known that the asymptotic approximation ratio of FF is 1.7 [5]. For any fixed $k > 1$, the asymptotic approximation ratio of FFk for large instances (when $OPT(D_k) \to \infty$) is better, and it improves when k increases.

Theorem 2. *For every input D and $k \geq 1$, $FFk(D_k) \leq \left(1.5 + \frac{1}{5k}\right) \cdot OPT(D_k) + 3 \cdot k$.* □

Approximation Ratio Lower Bound. In the full version [1] we have shown that the approximation ratio lower bound for FFk for $k = 2$ is 1.375. We conjecture that for $k > 1$ the *absolute* approximation ratio for FFk is 1.375.

5.2 *FFDk*

The k-time version of the First-Fit Decreasing bin-packing algorithm first sorts D in non-increasing order. Then it constructs D_k using k consecutive copies of the sorted D, and then implements FFk on D_k. In contrast to FFk, we could not prove an upper bound for $FFDk$ that is better than the upper bound for FFD; we only have a lower bound.

Lemma 3. $FFDk(D_k) \geq \frac{7}{6} \cdot OPT(D_k) + 1.$ □

Based on experimental results presented in the full version [1], we conjecture that the upper bound for $FFDk$ is $\frac{11}{9} \cdot OPT(D_k) + \frac{6}{9}$, as for the case $k = 1$.

5.3 *NFk*

Given the input D_k, the algorithm NFk works as follows: like NF, NFk always keeps a single bin open to pack items. If the current item does not pack into the currently open bin then NFk closes the current bin and opens a new bin to pack the item.

We can assume that $V(D) > S$, otherwise there is a trivial solution with k bins. While processing input D_k, NFk holds only one open bin, and it cannot contain a copy of each item of D. In fact, the open bin always contains a part of some instance of D, and possibly a part of the next instance of D, with no overlap. Therefore, if the current item x is not packed into the current open bin, the only reason is that x does not fit, as there is no previous copy of x in the current bin (all previous copies, if any, are in already-closed bins).

Theorem 3. *For every input D_k and $k \geq 1$, the asymptotic ratio of $NFk(D_k)$ is 2.* □

6 Polynomial-Time Approximation Schemes

6.1 Some General Concepts and Techniques

The basic idea behind generalizing Fernandez de la Vega-Lueker and all the Karmarkar-Karp algorithms to solve kBP is similar. It consists of three steps: 1. Keeping aside the small items, 2. Packing the remaining large items, and 3. Packing the small items in the bins that we get from step 2 (opening new bins if necessary) to get a solution to the original problem.

In step 3, the main difference from previous work is that, in kBP, we cannot pack two copies of the same small item into the same bin, so we may have to open a new bin even though there is still remaining room in some bins. The following lemma analyzes the approximation ratio of this step.

Lemma 4. *Let D_k be an instance of the kBP problem, and $0 < \epsilon \leq 1/2$. We say that the item is* large, *if its size is bigger than $\epsilon \cdot S$ and* small *otherwise. Assume that the large items are packed into L bins. Consider an algorithm which* *starts adding the small items into the L bins respecting the constraint of kBP, but whenever required, the algorithm opens a new bin. Then the number of bins required for the algorithm to pack the items in D_k is at most* $\max\{L, (1 + 2 \cdot \epsilon) \cdot OPT(D_k) + k\}$. □

The analysis of all methods discussed in this section have been moved to the full version [1]. Step 2 is done using a linear program based on *configurations.*

Definition 1. A *configuration* (or a *bin type*) is a collection of item sizes which sums to, at most, the bin capacity S.

For example [26]: suppose there are 7 items of size 3, 6 items of size 4, and $S = 12$. Then, the possible configurations are $[3, 3, 3, 3], [3, 3, 3], [3, 3], [3]$, $[4, 4, 4], [4, 4], [4]$, $[3, 3, 4], [3, 4, 4], [3, 4]$.

Enumerate all possible configurations by the natural numbers from 1 to t. Let $A = \|a_{ij}\|$ be a $m(D) \times t$ matrix, such that for each natural $i \leq m(D)$ and $j \leq t$ the entry a_{ij} is the number of items of size $c[i]$ in the configuration j. Let $\mathbf{n}$ be a $m(D)$-dimensional vector such that for each natural $i \leq m(D)$ its ith entry is $n[i]$ (the number of items of size $c[i]$). Let $\mathbf{x}$ be a t-dimensional vector such that for each natural $j \leq t$ we have that $x[j]$ is the number of bins filled with configuration j, and $\mathbf{1}$ be a t-dimensional vector whose each entry is 1. Consider the following linear program

$$(C_1) \qquad \begin{aligned} \min \quad & \mathbf{1} \cdot \mathbf{x} \\ \text{such that} \quad & A\mathbf{x} = \mathbf{n} \\ & \mathbf{x} \geq 0 \end{aligned}$$

When $\mathbf{x}$ is restricted to integer entries ($\mathbf{x} \in \mathbb{Z}^t$), the solution of this linear program defines a feasible bin-packing. We denote by F_1 the fractional relaxation of the above program, where $\mathbf{x} \in \mathbb{R}^t$.

Recall that in kBP, each item of D has to appear in k distinct bins. *One can observe that kBP uses the same configurations as in the bin-packing, to ensure that each bin contains at most one copy of each item.* Therefore, the configuration linear program C_k for kBP is as follows, where A, $\mathbf{n}$, and $\mathbf{x}$ are the same as in C_1 above (for $k = 1$ it is the same as in [24]):

$$(C_k) \qquad \begin{aligned} \min \quad & \mathbf{1} \cdot \mathbf{x} && (1) \\ \text{such that} \quad & A\mathbf{x} = k\mathbf{n} && (2) \\ & \mathbf{x} \geq 0 && (3) \end{aligned}$$

Lemma 5. *Every integral solution of C_k can be realised as a feasible solution of kBP.* □

Let F_k be the fractional bin-packing problem corresponding to C_k. Step 2 involves grouping. Grouping reduces the number of different item sizes, and thus reduces the number of constraints and configurations in the fractional linear program F_k.

To solve the configuration linear program efficiently, both Fernandez de la Vega-Lueker algorithm and Algorithm 1 of Karmarkar Karp use a *linear grouping* technique. In linear grouping, items are divided into groups (of fixed cardinality, except possibly the last group), and each item size (in each group) increases to the maximum item size in that group.

Our extension of the Fernandez de la Vega-Lueker and the Karmarkar-Karp algorithms to kBP differs from their original counterparts in mainly two directions. First, in the configuration linear program (see the constraint 2 in C_k), and hence the obtained solution to this configuration linear program is not necessarily the k times copy of the original solution of BP. Second, in greedily adding the small items, see lemma 4. In extension of Karmarkar-Karp algorithm 1 to kBP we have also shown that getting an integer solution from $\mathbf{x}$ by rounding method may require at most $(k-1)/2$ additional bins. We discuss extensions to the Fernandez de la Vega-Lueker and Karmarkar-Karp algorithms and their analyses in Subsects. 6.2 and 6.3, respectively.

The inputs to the extension of the algorithms by Fernandez de la Vega-Lueker and Algorithm 1 and Algorithm 2 of Karmarkar-Karp are an input set of items D, a natural number k, and an approximation parameter $\epsilon \in (0, 1/2]$. Algorithm 2 of Karmarkar-Karp, in addition, accepts an integer parameter $g > 0$.

6.2 Fernandez de la Vega-Lueker Algorithm to kBP

Fernandez de la Vega and Lueker [24] published a PTAS which, given an input instance D and $\epsilon \in (0, 1/2]$, solves a bin-packing problem with, at most, $(1+\epsilon) \cdot OPT(D) + 1$ bins. They devised a method called "adaptive rounding" for this algorithm. In this method, the given items are put into groups and rounded to the largest item size in that group. This resulting instance will have fewer different item-sizes. This resulting instance can be solved efficiently using a configuration linear program C_k.

Theorem 4. *Generalizing the Fernandez de la Vega-Lueker algorithm to kBP will require $bins(D_k) \leq (1 + 2 \cdot \epsilon) \cdot OPT(D_k) + k$ bins.* □

We give a detailed proof of Theorem 4, along with the runtime analysis of the Fernandez de la Vega-Lueker algorithm to kBP in the full version [1].

6.3 Karmarkar-Karp Algorithms to kBP

Karmarkar and Karp [16] improved the work done by Fernandez de la Vega and Lueker [24] mainly in two directions: (1) Solving the linear programming relaxation of C_1 using a variant of the GLS method [12] and (2) using a different grouping technique. These improvements led to the development of three algorithms. Their algorithm 3 is a particular case of the algorithm 2; we will discuss the generalization of algorithms 1 and 2 of Karmarkar-Karp algorithms to solve kBP. Let $LIN(F_k)$ denote the optimal solution to the fractional linear program F_k for kBP. Helpful results which bounds from above the number of bins needed to pack the items in an optimal packing of some instance D_k, and result which concerns obtaining an integer solution from a basic feasible solution of the fractional linear program have been discussed in full version [1].

Before moving further, we would like to mention that if we use some instance (or group) without subscript k, we are talking about the instance when $k = 1$.

All Karmarkar-Karp algorithms use a variant of the ellipsoid method to solve the fractional linear program. So, we will talk about adapting this method to kBP.

Solving the Fractional Linear program: Solving the fractional linear program F_k involves a variable for each configuration. This results in a large number of variables. The fractional linear program F_k has the following dual D_F.

$$\max \quad k \cdot \mathbf{n} \cdot \mathbf{y} \tag{4}$$

$$\text{such that} \quad A^T \mathbf{y} \leq \mathbf{1} \tag{5}$$

$$\mathbf{y} \geq 0 \tag{6}$$

The above dual linear program can be solved to any given tolerance h by using a variant of the ellipsoid method that uses an approximate separation oracle [16]. The running time of the algorithm is $T(m(D_k), n(D_k)) = O\left(m(D)^8 \cdot \ln m(D) \cdot \ln^2\left(\frac{m(D)\cdot n(D)}{\epsilon \cdot S \cdot h}\right) + \frac{m(D)^4 \cdot k \cdot n(D) \cdot \ln m(D)}{h} \ln \frac{m(D)\cdot n(D)}{\epsilon \cdot S \cdot h}\right)$.

Karmarkar-Karp Algorithm 1 Extension To *k*BP: Algorithm 1 of the Karmarkar-Karp algorithms uses the linear grouping technique as illustrated in Subsect. 6.1.

Theorem 5. *Let $bins(D_k)$ denote the number of bins produced by Karmarkar-Karp Algorithm 1 extension to kBP. Then, $bins(D_k) \leq (1 + 2 \cdot k \cdot \epsilon) OPT(D_k) + \frac{1}{2\cdot\epsilon^2} + (2 \cdot k + 1)$.* □

Karmarkar-Karp Algorithm 2 Extension To *k*BP. Algorithm 2 of the Karmarkar-Karp algorithms uses the *alternative geometric grouping technique.* Let J be some instance and $g > 1$ be some integer parameter, then, alternative geometric grouping partitions the items in J into groups such that each group contains the necessary number of items so that the size of each group but the last (i.e. the sum of the item sizes in that group) is at least $g \cdot S$.

Theorem 6. *Let $bins(D_k)$ denote the number of bins produced by Karmarkar-Karp Algorithm 2 extension to kBP. Then, $bins(D_k) \leq OPT(D_k) + O(k \cdot \log^2 OPT(D))$.* □

7 Experiment: *FFk* and *FFDk* for Fair Electricity Distribution

In this section, we describe an experiment checking the performance of the kBP adaptations of FFk and $FFDk$ to our motivating application of fair electricity distribution.

7.1 Dataset

We use the same dataset of 367 Nigerian households described in [20].[1] This dataset contains the hourly electricity demand for each household for 13 weeks (2184 hours). In addition, they estimate for each agent and hour, the *comfort* of that agent, which is an estimation of the utility the agent gets from being connected to electricity at that hour. For more details about the dataset, readers are encouraged to refer to the papers [19,20]. The electricity demand of agents can vary from hour to hour. We execute our algorithms for each hour separately, which gives us essentially 2184 different instances.

As in [20], we use the demand figures in the dataset as mean values; we determine the actual demand of each agent at random from a normal distribution with a standard deviation of 0.05 (results with a higher standard deviation are presented in the full version [1]).

As in [20], we compute the supply capacity S for each day by averaging the hourly estimates of agents' demand for that day. We run nine independent simulations (with different randomization of agents' demands). Thus, the supply changes in accordance with the average daily demand, but cannot satisfy the maximum hourly demand.

7.2 Experiment

For each hour, we execute the FFk and $FFDk$ algorithms on the households' demands for that hour. We then use the resulting packing to allocate electricity: if the packing returns q bins, then each bin is connected for $1/q$ of an hour, which means that each agent is connected for k/q of an hour.

The authors of [20] measure the efficiency and fairness of the resulting allocation, not only by the total time each agent is connected, but also by more complex measures. In particular, they assume that each agent i has a utility function, denoted u_i, that determines the utility that the agent receives from being connected to electricity at a given hour. They consider three different utility models:

1. The simplest model is that u_i equals the amount of time the agent i is connected to electricity (this is the model we mentioned in the introduction).
2. The value u_i can also be equal to the total amount of electricity that the agent i receives. For each hour, the amount of electricity given to i is the amount of time i is connected, times i's demand at that hour.
3. They also measure the "comfort" of the agent i in time t by averaging their demand over the same hour in the past four weeks, and normalizing it by dividing by the maximum value.

For each utility model, they consider three measures of efficiency and fairness:

- Utilitarian: the sum $\sum_i u_i(x)$ (or the average) of all agents' utilities u_i.
- Egalitarian: the minimum utility $\min_i u_i(x)$ of a single agent,
- The maximum difference $\max_{i,j}\{|u_i(x) - u_j(x)|\}$ of utilities between each pair of agents.

[1] We are grateful to Olabambo Oluwasuji for sharing the dataset with us.

7.3 Results

For the comparison with the results from [20], we show our results for FFk and $FFDk$ for $k = 9$ along with the results in [20] in tables[2][3] 1,2, and 3 . We highlight the best results in bold. As can be seen in Tables 1, 2 and 3, FFk and $FFDk$ outperform the previous results for all 9 combinations of utility models and social welfare metrics, even though they were only designed for the egalitarian welfare of the time-based utility function. More results have been discussed in the full version [1].

Table 1. Comparing results of FFk and $FFDk$ for $k = 9$ with the results in [20] in terms of hours of connection to supply on the average, along with their standard deviation (SD) within parenthesis. In the third column, we have shown the average number of hours an agent is connected to the supply.

Algorithm	Utilitarian: sum(SD)	Utilitarian: average	Egalitarian(SD)	MUD
FFk	**723982.0121(401.4496)**	**1972.7030**	**1972.7030 (1.0939)**	**0.0(0.0)**
$FFDk$	**723804.0727(654.8798)**	**1972.2182**	**1972.2182 (1.7844)**	**0.0(0.0)**
CM	717031(3950)	1953.7629	1920(3.24)	123(2.09)
SM	709676(3878)	1933.7221	1922(3.41)	71(2.04)
GA	629534(4178)	1715.3515	1609(4.69)	695(3.28)
CSA1	647439(3063)	1764.1389	1764(2.27)	1(0.00)
RSA	643504(4094)	1753.4169	1753(4.33)	1(0.00)
CSA2	641002(3154)	1746.5995	1746(2.38)	1(0.00)

8 Conclusion and Future Directions

We have shown that the existing approximation algorithms, like the First-Fit and the First-Fit Decreasing, can be extended to solve kBP. We have proved that, for any $k \geq 1$, the asymptotic approximation ratio for the FFk algorithm is $\left(1.5 + \frac{1}{5k}\right) \cdot OPT(D_k) + 3 \cdot k$. We have also proved that the asymptotic approximation ratio for the NFk algorithm is 2. We have also demonstrated that the generalization of efficient approximation algorithms like Fernandez de la Vega-Lueker and Karmarkar Karp algorithms solves kBP in $(1 + 2 \cdot \epsilon) \cdot OPT(D_k) + k$ and $OPT(D_k) + O(k \cdot \log^2 OPT(D))$ bins respectively in polynomial time. We have also shown the practical efficacy of FFk and $FFDk$ in solving the fair electricity distribution problem.

Given the usefulness of k-times bin-packing to electricity division, an interesting open question is how to determine the optimal value of k—the k that

[2] In Tables 1-3 CM, SM, GA, CSA1, RSA, CSA2 stands for: The Comfort Model, The Supply Model, Grouper Algorithm, Consumption-Sorter Algorithm, Random-Selector Algorithm, Cost-Sorter Algorithm respectively [19,20].

[3] MUD stands for Maximum Utility Difference in all the Tables

Table 2. Comparing results of FFk and $FFDk$ for $k = 9$ with the results in [20] in terms of electricity supplied on the average, along with their standard deviation (SD) within parenthesis.

Algorithm	Utilitarian(SD)	Egalitarian(SD)	MUD
FFk	**1444617.5746(793.4190)**	**0.8957 (0.0005)**	**0.0179 (0.0008)**
$FFDk$	**1444242.4530(1333.086)**	**0.8954 (0.0009)**	**0.01771 (0.0010)**
CM	1340015(8299)	0.78(0.01)	0.17(0.02)
SM	1347801(8304)	0.83(0.01)	0.11(0.02)
GA	1297020(11264)	0.35(0.04)	0.58(0.03)
CSA1	1296939(7564)	0.66(0.02)	0.28(0.02)
RSA	1344945(11284)	0.68(0.03)	0.25(0.03)
CSA2	1345537(7388)	0.63(0.03)	0.30(0.02)

Table 3. Comparing results of FFk and $FFDk$ for $k = 9$ with the results in [20] in terms of comfort delivered on the average, along with their standard deviation (SD) within parenthesis.

Algorithm	Utilitarian(SD)	Egalitarian(SD)	MUD
FFk	**314035.2667 (171.7323)**	**0.8975 (0.0005)**	**0.0109 (0.0004)**
$FFDk$	**313967.0364 (268.3526)**	**0.8974 (0.0007)**	**0.0108 (0.0004)**
CM	303217(3447)	0.81(0.01)	0.13(0.02)
SM	292135(3802)	0.83(0.01)	0.09(0.02)
GA	291021(5198)	0.38(0.04)	0.56(0.03)
CSA1	291909(3201)	0.67(0.02)	0.25(0.02)
RSA	268564(5106)	0.65(0.04)	0.28(0.03)
CSA2	270262(3112)	0.64(0.02)	0.28(0.02)

maximizes the fraction of time each agent is connected—the fraction $\frac{k}{OPT(D_k)}$. Note that this ratio is not necessarily increasing with k. For example, consider the demand vector $D = \{11, 12, 13\}$:

- For $k = 1$, $OPT(D_k) = 2$, so any agent is connected $\frac{1}{2}$ of the time.
- For $k = 2$, $OPT(D_k) = 3$, so any agent is connected $\frac{2}{3}$ of the time.
- For $k = 3$, $OPT(D_k) = 5$, so any agent is connected only $\frac{3}{5} < \frac{2}{3}$ of the time.

Some other questions left open are

1. To bridge the gap in the approximation ratio of FFk, between the conjectured lower bound 1.375 and the upper bound $\left(1.5 + \frac{1}{5k}\right) \cdot OPT(D_k) + 3 \cdot k$, which converges to 1.5.
2. To prove or disprove that the conjectured bound $\frac{11}{9}OPT + \frac{6}{9}$ is tight for $FFDk$.

Acknowledgments. We thank the reviewers for their constructive comments. This research is partly funded by the Israel Science Foundation grant ISF 712/20.

Disclosure of Interests. The authors have no competing interests.

References

1. Baghel, D.K., Ravsky, A., Segal-Halevi, E.: k-times bin packing and its application to fair electricity distribution (2024). arXiv:2311.16742
2. Baker, B.S.: A new proof for the first-fit decreasing bin-packing algorithm. J. Algorithms **6**(1), 49–70 (1985). https://linkinghub.elsevier.com/retrieve/pii/0196677485900185
3. Brams, S.J., Taylor, A.D.: Fair Division: From Cake-Cutting to Dispute Resolution. Cambridge University Press, Cambridge (1996). https://doi.org/10.1017/CBO9780511598975
4. Doron-Arad, I., Kulik, A., Shachnai, H.: Bin packing with partition matroid can be approximated within $o(OPT)$ bins (arXiv:2212.01025) (2022). http://arxiv.org/abs/2212.01025, arXiv:2212.01025 [cs]
5. Dosa, G., Sgall, J.: First Fit bin packing: a tight analysis. In: Leibniz International Proceedings in Informatics, LIPIcs, vol. 20, pp. 538–549 (2013). https://doi.org/10.4230/LIPIcs.STACS.2013.538, iSBN: 9783939897507
6. Dósa, G.: The Tight Bound of First Fit Decreasing Bin-Packing Algorithm Is F F D (I) $\leq$ 11 / 9 OP T (I) + 6 / 9 **2007**(2), 1–11 (2007)
7. Dósa, G., Li, R., Han, X., Tuza, Z.: Tight absolute bound for first fit decreasing bin-packing: FFD(L) $\leq$ 11/9OPT(L) + 6/9. Theoret. Comput. Sci. **510**(11101065), 13–61 (2013). https://doi.org/10.1016/j.tcs.2013.09.007
8. Garey, M.R., Graham, R.L., Ullman, J.D.: Worst-case analysis of memory allocation algorithms. In: Proceedings of the Fourth Annual ACM Symposium on Theory of Computing, pp. 143–150. STOC '72, Association for Computing Machinery, New York, NY, USA (1972). https://doi.org/10.1145/800152.804907
9. Garey, M.R., Johnson, D.S.: Computers and Intractability; A Guide to the Theory of NP-Completeness. W. H. Freeman & Co., USA (1990)
10. Garey, M., Graham, R., Johnson, D., Yao, A.C.C.: Resource constrained scheduling as generalized bin packing. J. Comb. Theory, Ser. A **21**(3), 257–298 (1976). https://www.sciencedirect.com/science/article/pii/0097316576900017
11. Gendreau, M., Laporte, G., Semet, F.: Heuristics and lower bounds for the bin packing problem with conflicts. Comput. Oper. Res. **31**(3), 347–358 (2004). https://doi.org/10.1016/S0305-0548(02)00195-8
12. Grötschel, M., Lovász, L., Schrijver, A.: The ellipsoid method and its consequences in combinatorial optimization. Combinatorica **1**(2), 169–197 (1981). http://link.springer.com/10.1007/BF02579273
13. Hoberg, R., Rothvoss, T.: A logarithmic additive integrality gap for bin packing. In: Proceedings of the Twenty-Eighth Annual ACM-SIAM Symposium on Discrete Algorithms, pp. 2616–2625. Society for Industrial and Applied Mathematics (2017). http://epubs.siam.org/doi/10.1137/1.9781611974782.172
14. Jansen, K.: An approximation scheme for bin packing with conflicts. In: Arnborg, S., Ivansson, L. (eds.) SWAT 1998. LNCS, vol. 1432, pp. 35–46. Springer, Heidelberg (1998). https://doi.org/10.1007/BFb0054353
15. Johnson, D.S.: Near-Optimal Bin Packing Algorithms. Thesis, p. 400 (1973)

16. Karmarkar, N., Karp, R.M.: Efficient approximation scheme for the one-dimensional bin-packing problem. In: Annual Symposium on Foundations of Computer Science - Proceedings, pp. 312–320 (1982). https://doi.org/10.1109/sfcs.1982.61
17. Kaygusuz, K.: Energy for sustainable development: a case of developing countries. Renew. Sustain. Energy Rev. **16**(2), 1116–1126 (2012). https://doi.org/10.1016/j.rser.2011.11.013
18. Li, R., Yue, M.: The proof of FFD(L) < -OPT(L) + 7/9. Chin. Sci. Bull. **42**(15), 1262–1265 (1997). https://doi.org/10.1007/BF02882754
19. Oluwasuji, O.I., Malik, O., Zhang, J., Ramchurn, S.D.: Algorithms for fair load shedding in developing countries. In: Proceedings of the Twenty-Seventh International Joint Conference on Artificial Intelligence, pp. 1590–1596. International Joint Conferences on Artificial Intelligence Organization, Stockholm, Sweden (2018). https://www.ijcai.org/proceedings/2018/220
20. Oluwasuji, O.I., Malik, O., Zhang, J., Ramchurn, S.D.: Solving the fair electric load shedding problem in developing countries. Auton. Agents Multi-Agent Syst. **34**(1), 12 (2020). http://link.springer.com/10.1007/s10458-019-09428-8
21. Rothvoss, T.: Approximating bin packing within O(log opt · log log opt) bins. In: 2013 IEEE 54th Annual Symposium on Foundations of Computer Science, pp. 20–29. IEEE, Berkeley, CA, USA (Oct 2013). https://ieeexplore.ieee.org/document/6686137/
22. Steinhaus, H.: Sur la division pragmatique. Econometrica **17**, 315–319 (1949). http://www.jstor.org/stable/1907319
23. Ullman, J.D.: The Performance of a Memory Allocation Algorithm. Technical report (Princeton University. Department of Electrical Engineering. Computer Sciences Laboratory), Princeton University (1971). https://books.google.co.il/books?id=gnwNPwAACAAJ
24. Fernandez de la Vega, W., Lueker, G.S.: Bin packing can be solved within $1 + \epsilon$ in linear time. Combinatorica **1**(4), 349–355 (1981). https://doi.org/10.1007/BF02579456
25. Webb, Jack Robertson, W.: Cake-Cutting Algorithms: Be Fair if You Can. A K Peters/CRC Press, New York (1998). https://doi.org/10.1201/9781439863855
26. Wikipedia contributors: Configuration linear program — Wikipedia, the free encyclopedia (2023). https://en.wikipedia.org/w/index.php?title=Configuration_linear_program&oldid=1139054649. Accessed 16 March 2023
27. Xia, B., Tan, Z.: Tighter bounds of the first fit algorithm for the bin-packing problem. Discr. Appl. Math. **158**(15), 1668–1675 (2010). https://doi.org/10.1016/j.dam.2010.05.026
28. Yue, M.: A simple proof of the inequality FFD(l) $\leq$ 11/9OPT(L)+ 1,$\forall$ L for the FFD bin-packing algorithm. Acta Math. Appl. Sin. **7**(4), 321–331 (1991)

Condorcet Markets

Stéphane Airiau[1], Nicholas Kees Dupuis[4], and Davide Grossi[2,3](✉)

[1] University Paris-Dauphine, Paris, France
stephane.airiau@dauphine.fr
[2] University of Groningen, Groningen, The Netherlands
d.grossi@rug.nl
[3] University of Amsterdam, Amsterdam, The Netherlands
[4] Berkeley, USA

Abstract. The paper studies information markets concerning single events from an epistemic social choice perspective. Within the classical Condorcet error model for collective binary decisions, we establish equivalence results between elections and markets, showing that the alternative that would be selected by weighted majority voting (under specific weighting schemes) corresponds to the alternative with highest price in the equilibrium of the market (under specific assumptions on the market type). This points to the possibility, in principle, of implementing specific weighted majority elections, which are known to have superior truth-tracking performance, by means of information markets without needing to elicit voters' competences.

Keywords: Information markets · Jury theorems · Crowd-wisdom

1 Introduction

Information, or prediction, markets are markets of all-or-nothing contracts (so-called Arrow securities) that pay one unit of currency if a designated event occurs and nothing otherwise (see [1,5,14] for models of such markets). Under the view, inspired by [12], that markets are good aggregators of the information dispersed among traders, proponents of information markets have argued that equilibrium prices are accurate estimates of the probability of the designated event. Recent research—theoretical and empirical—has probed this interpretation of prices in information markets, finding that equilibrium prices successfully track the traders' average belief, under several utility models [16,19].

In this paper we address a closely related, but different question: *if a decision maker takes a decision based on the information they extract from the equilibrium price of the market, how accurate would the decision be?* Therefore, rather than relating equilibrium prices to the aggregation of traders' beliefs, we relate them directly to the quality of the decision they would support. We frame the above question within the standard binary choice framework of epistemic social choice, stemming from the Condorcet jury theorem tradition [6,11,20] and the maximum-likelihood estimation approach to voting [7,10,11,17].

G. Schäfer and C. Ventre (Eds.): SAGT 2024, LNCS 15156, pp. 501–519, 2024.
https://doi.org/10.1007/978-3-031-71033-9_28

Contribution. To answer the above question, we study information markets when traders' beliefs are obtained by Bayesian update from a private independent signal with accuracy known to the trader, just like in the classic jury theorems setting. In other words, we study 'jurors' as if they were 'traders' who, instead of relaying their vote to a central mechanism, trade in an information market. In taking this perspective, we compare the decisions that would be taken based on the equilibrium price of an information market, with the decisions that would be taken by specific weighted majority elections, whose truth-tracking behavior is already well-understood [11]. Specifically, we aim at identifying correspondences between classes of markets and of weighted majority elections which are equivalent from a decision-making point of view. That is, if agents vote according to the event they believe more likely and aggregate these votes by weighted majority, then they identify the alternative whose Arrow security would have highest price in the equilibrium of the market in which the same agents trade based on their beliefs. Figure 1 depicts such relationship via a commutative diagram. This type of results point to the possibility (in principle) of implementing weighted majority voting with proven truth-tracking performance without needing to know (or estimate [3]) jurors' competences.

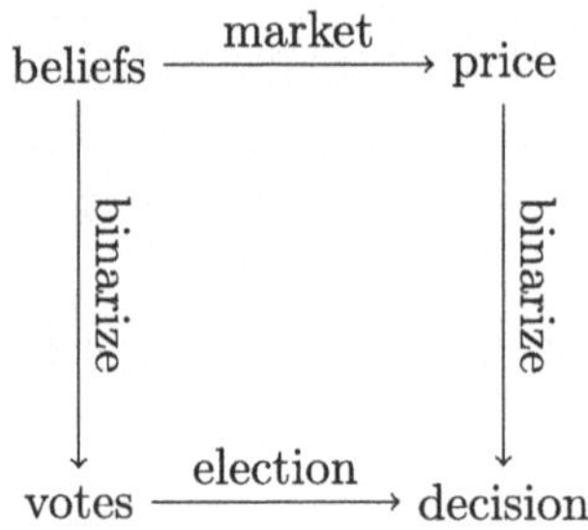

Fig. 1. Elections and information markets commute.

Paper Outline. Section 2 introduces the standard binary truth-tracking framework and presents our model of information markets. Section 3 presents results on equilibrium prices in two of the three types of markets we consider (Naive and Kelly markets) and Sect. 4 proves 'Fig. 1-type' results for those markets. Section 5 then shows how such results could be lifted even to the case of majority voting where jurors are weighted perfectly according to their competence. Section 6 discusses two examples illustrating our framework and analysis. Section 7 outlines future research directions. Auxiliary results and omitted proofs are available in a longer version of the paper at https://arxiv.org/abs/2306.05028.

2 Preliminaries

2.1 Collective Truth-Tracking

Collective Decisions. We are concerned with a finite set of agents $N = \{1, \ldots, n\}$ who have to decide collectively on the correct state of the world $x \in \{A, B\}$. A prior probability $P(x = A) = \pi = 0.5$ is given, that the correct state is A. Each agent i observes a private independent signal $y_i \in \{A, B\}$ that has quality $q_i \in (0.5, 1]$. Each q_i represents the competence or accuracy of i, which is assumed to satisfy $q_i = P(y = A \mid x = A) = P(y = B \mid x = B)$. We call each vector $\mathbf{q} = (q_1, \ldots, q_n)$ of individual accuracies an *accuracy* or *competence profile* of the group. Having observed her private signal, each agent then forms a posterior belief $b_i = P(x = A \mid y = A)$ about state $x = A$ by Bayes rule. By Bayes' rule and the condition on the prior, we have that either $b_i = q_i > 0.5$ (the belief in A equals q_i) or $b_i = 1 - q_i < 0.5$ (the belief in A equals $1 - q_i$). This gives us, for all $i \in N$:

$$b_i = \mathbb{1}(b_i > 0.5) \cdot (2q_i - 1) + (1 - q_i), \tag{1}$$

where $\mathbb{1}$ denotes the indicator function. Individual beliefs are then collected in a *belief profile* $\mathbf{b} = (b_1, \ldots, b_n) \in [0,1]^n$. Given an accuracy profile $\mathbf{q}$, the set of possible belief profiles is denoted $\mathcal{B}_\mathbf{q} = \{\mathbf{b} \in [0,1]^n \mid P(\mathbf{b} \mid \mathbf{q}) > 0\}$. Observe that the size of this set equals 2^n: the number of all signal realizations.

Based on a profile $\mathbf{b}$ of individual beliefs, the group then takes a decision by mapping the profile to A or to B. In this process, agents may have different weights, which are collected in a *weight profile* $\mathbf{w} = (w_1, \ldots, w_n) \in [0,1]^n$. We refer to $\mathbf{1} = (1, \ldots, 1)$ as the *egalitarian weight profile* in which all agents have equal weight. Assuming a weight profile $\mathbf{w}$, we call an *aggregator* any function

$$\mathcal{A}^\mathbf{w} : [0,1]^n \to 2^{\{1,0\}} \setminus \{\emptyset\}, \tag{2}$$

mapping belief profiles to alternatives, where $\{1\}$ denotes A; $\{0\}$ denotes B; and $\{1,0\}$ denotes a tie.

Types of Aggregators. We will study two classes of mechanisms to implement aggregators. In the first class, agents cast binary ballots based on their beliefs and these ballots are submitted to a voting mechanism. The winning alternative is the outcome of the aggregation process. In the second class, agents' trade in special types of securities, based on their beliefs. The equilibrium price of this securities market is then used as a proxy for the group's belief in the probability of state A. In this case, it is the alternative favored by this collective belief to be the outcome of the aggregation process.

Let us make the above notions more precise. First of all, a belief $b \in [0,1]$ is translated into binary opinions, or *votes*, for A or B via the binarization function $\widehat{\ }: [0,1] \to 2^{\{1,0\}} \setminus \emptyset$ defined as follows:

$$\widehat{b} = \begin{cases} \{1\} & \text{if } b > 0.5, \\ \{0\} & \text{if } b < 0.5, \\ \{0,1\} & \text{otherwise.} \end{cases} \tag{3}$$

That is, agents are assumed to vote in accordance to their posterior belief (this is sometimes referred to as sincere voting [2]). A binarized belief profile $\widehat{\mathbf{b}} = (\widehat{b}_1, \ldots, \widehat{b}_n)$ is therefore a binary vector and we will referred to such vectors also as *voting profiles* and denote them by $\mathbf{v} = (v_1, \ldots, v_n)$.[1]

Given a weight profile $\mathbf{w}$, a (belief) merger is a function $F^\mathbf{w} : [0,1]^n \to [0,1]$ taking as input a belief profile and outputting a group belief. A choice function is a function $f^\mathbf{w} : \{1,0\}^n \to 2^{\{0,1\}} \setminus \emptyset$ taking as input a voting profile

[1] As individual beliefs cannot equal 0.5, the reduction function always outputs a singleton $\{0\}$ or $\{1\}$ on individual beliefs. We will see, however, that this is not the case for collective beliefs.

and outputting a possibly tied choice between 1, i.e., A, and 0, i.e., B. We will study aggregators of the type $f^{\mathbf{w}} \circ \widehat{\ }$ (voting) and $\widehat{\ } \circ F^{\mathbf{w}}$ (trading), where $\circ$ denotes function composition. A voting mechanism is a choice function f^{w} which, applied to a binarized belief profile $\widehat{\mathbf{b}}$, yields a collective choice $f^{\mathbf{w}}(\widehat{\mathbf{b}})$ (under the weight profile $\mathbf{w}$). A market mechanism is a belief aggregation function F that, once applied to a belief profile $\mathbf{b}$, yields a collective belief $F^{\mathbf{w}}(\mathbf{b})$ whose binarization $\widehat{F^{\mathbf{w}}(\mathbf{b})}$ yields a collective choice (under the weight profile $\mathbf{w}$).

We are concerned with the truth-tracking performance of aggregators. The accuracy of an aggregator $\mathcal{A}^{\mathbf{w}}$ under the accuracy profile $\mathbf{q}$, is the conditional probability that the outcome of the aggregator is x if the state of the world is x. What we outlined describes an epistemic social choice setting involving a maximum-likelihood estimation task in a dichotomous choice situation (see [10]).

2.2 Voting and Market Mechanisms

We turn now to the description of the mechanisms we are concerned with.

Voting Mechanisms. After observing their private signal, agents decide whether to vote for A or B according to Eq. (3). A weighted majority rule is then applied to these votes to determine the group's choice:

$$M^{\mathbf{w}}(\mathbf{v}) = \begin{cases} \{1\} & \text{if } \sum_{i\in N} w_i v_i > \frac{\sum_{i\in N} w_i}{2}, \\ \{0\} & \text{if } \sum_{i\in N} w_i v_i < \frac{\sum_{i\in N} w_i}{2}, \\ \{0,1\} & \text{otherwise.} \end{cases} \tag{4}$$

We will be working in particular with three variants of Eq. (4) defined by three different weight profiles: the egalitarian weight profile $\mathbf{1}$; the weight profile allocating to each agent i a weight proportional to $q_i - 0.5$; the weight profile allocating to each agent i a weight proportional to $\log \frac{q_i}{1-q_i}$. The first weight profile defines the *simple majority* rule. The second weight profile simulates decision-making according to the mean belief of the group. The latter weight profile can be inferred from Bayes theorem and induces the weighted majority rule which we refer to as *perfect majority*, and which has been proven to optimize the truth-tracking ability of the group.

Theorem 1 ([11]). *For any competence profile $\mathbf{q}$, the accuracy of $M^{\mathbf{w}}$ given $\mathbf{q}$ is maximal if $\mathbf{w}$ is such that $w_i \propto \ln\left(\frac{q_i}{1-q_i}\right)$ for all $i \in N$.*

Markets. The market model we use is borrowed from [5,14]. Two symmetric Arrow securities are traded: securities of type A, which cost $p_A \in [0,1]$ and pay 1 unit of currency if $x = A$, and 0 otherwise; securities of type B, which cost $p_B \in [0,1]$ and pay 1 unit if $x = B$ and 0 otherwise. After observing their private signal, agents decide what fraction of their endowment to invest in which securities. The endowment is fixed to 1 for all agents. When the true state of the world is revealed, the market resolves and payouts based on the agents' investments are distributed. We refer to tuples $\mathbf{s}^A = (s_1^A, \ldots, s_n^A)$ (respectively, $\mathbf{s}^B = (s_1^B, \ldots, s_n^B)$) as investment profiles in A-securities (respectively, B-securities). We refer to a pair $\mathbf{s} = (\mathbf{s}^A, \mathbf{s}^B)$ as an *investment profile*. We assume that agents *invest in at most one* of these securities, so if $s^A > 0$ then $s^B = 0$ and vice versa. In our setting, this assumption can be shown to be without loss of generality.[2] We call agents investing in A, A-*traders* and agents investing in B, B-*traders*.

Market Mechanism. When the market opens, all purchasing orders for each security are executed by the market operator. The market operator sells all requested securities to agents when the market opens and pays the winning securities out immediately when the market resolves, that is, when either A or B turns out to be the case. We further assume that the operator makes no profits and incurs no losses. So, for every A-security sold at price p^A, a B-security is sold at price $p^B = 1 - p^A$ and vice versa. It follows that there are as many A-securities as B-securities and the price of the risk-less asset consisting of one of each security is $p_A + p_B = 1$. In this way the operator finances the payout of any bet by the pay-in of the opposite bet.

Under the above assumptions, the market clears[3] when the total amount of individual wealth invested in A-securities, divided by the price of A-securities (demand of A-securities) matches the amount of individual wealth invested in B-securities, divided by the price of B-securities (demand of B-securities):[4]

$$\frac{1}{p^A} \sum_{i \in N} s_i^A = \frac{1}{1 - p^A} \sum_{i \in N} s_i^B. \tag{5}$$

It follows that, given an investment profile $\mathbf{s}$, solving Eq. (5) for p^A, yields the clearing price $\frac{\sum_{i \in N} s_i^A}{\sum_{i \in N} s_i^A + \sum_{i \in N} s_i^B}$, denoted $p^A(\mathbf{s})$. Note that the price is undefined if $p^A = 0$ or $p^A = 1$. We come back to this issue in Remark 2.

When the market resolves, each agent receives a different payout depending on how much of each security she owns, how the market resolves, and how much of her endowment is not invested. The payout, that is, the amount of wealth

[2] This auxiliary result can be found in the long version of the paper at https://arxiv.org/abs/2306.05028. We are indebted to Marcus Pivato for bringing this issue to our attention.

[3] A market is said to clear when supply and demand match. In our model, supply and demand are implicit in the following way: demand for an A-security at price p^A implies supply for a B-security at price $p^B = 1 - p^A$ and vice versa. The same applies to supply and demand for B securities.

[4] It may be worth observing that by the above design we are effectively treating the operator as an extra trader in the market, who holds a risk-less asset consisting of $\frac{1}{p^A} \sum_{i \in N} s_i^A$ A-securities and $\frac{1}{1-p^A} \sum_{i \in N} s_i^B$ B-securities. We are indebted to Marcus Pivato for this observation.

obtained by i with a given strategy s_i^A investing in A under a price p^A, is:

$$z(p^A, s_i^A) = \begin{cases} \frac{s_i^A}{p^A} & A \text{ is correct,} \\ 1 - s_i^A & \text{otherwise,} \end{cases} \tag{6}$$

where $\frac{s_i^A}{p^A}$ equals the amount of A-securities that i has purchased. The payout for an investment in B-securities is defined in the same manner.

Remark 1. For simplicity, in what follows we will refer to the price of A-securities as p instead of p^A and to the price of B-securities as $1 - p$ instead of p^B.

Utility. We study price p by making assumptions on how much utility agents extract from their payout at that price. We consider two types of utility functions:

Naive. Given a price $p \in [0, 1]$, the naive utility function of an A-trader i is $u(p, s_i^A) = z(p, s_i^A)$ Similarly, for a B-trader, it is $u(1 - p, s_i^B) = z(1 - p, s_i^B)$. The expected utility for investment in A-securities is then:

$$U_i^A(p, s_i^A) = \mathbb{E}[u(p, s_i^A)] = b_i \left(\frac{s_i^A}{p} - s_i^A + 1 \right) + (1 - b_i)(1 - s_i^A). \tag{7}$$

The expected utility for investment in B-securities is, correspondingly, $b_i(1 - s_i^B) + (1 - b_i) \left(\frac{s_i^B}{1-p} - s_i^B + 1 \right)$. We will refer to markets under a naive utility assumption as *Naive markets.*

Kelly. Given a price $p \in [0, 1]$, the Kelly [13] utility function of an A-trader i is $u(p, s_i^A) = \ln(z(p, s_i^A))$, and mutatis mutandis for B-traders. The expected Kelly utility for an A-trader is therefore:

$$U_i^A(p, s_i^A) = \mathbb{E}[u(p, s_i^A)] = b_i \ln \left(\frac{s_i^A}{p} - s_i^A + 1 \right) + (1 - b_i) \ln(1 - s_i^A). \tag{8}$$

Correspondingly, the expected utility of investment s_i^B for a B-traders is $b_i \ln(1 - s_i^B) + (1 - b_i) \ln \left(\frac{s_i^B}{1-p} - s_i^B + 1 \right)$. We will refer to markets under such logarithmic utility assumption as *Kelly markets.* Investing with a logarithmic utility function is known as Kelly betting and is known to maximize bettor's wealth over time [13]. Information market traders with Kelly utilities have been studied, for instance, in [5].

Equilibria. For each of the above models of utility we will work with the notion of equilibrium known as competitive equilibrium [16]. This equilibrium assumes that agents optimize the choice of their investment strategy s_i under the balancing assumption of Eq. (5), while not considering the effect of their choice on the price (they behave as 'price takers').

Definition 1. (Competitive equilibrium). *Given a belief profile* $\mathbf{b}$, *an investment profile* $\mathbf{s}$ *is in competitive equlibrium for price* $p^\star$ *if and only if:*

1. *Equation* (5) *holds, that is,* $p^\star = p(\mathbf{s})$,
2. *for all* $i \in N$, *if* i *is a* t*-trader in* $\mathbf{s}$, *then* $s_i^t \in arg\ max_{x\in[0,1]} U_i^t(p^t, x)$, *for* $t \in \{A, B\}$.

So, when the investment profile $\mathbf{s}$ is in equilibrium with respect to the A-securities $p^\star$, no agent would like to purchase more securities of any type given their beliefs. If $\mathbf{s}$ is in equilibrium for price $p(\mathbf{s})$, then we say that $\mathbf{s}$ is an *equilibrium*. If equilibria always exist, and are such for one same price, then the equilibrium price can be interpreted as the market's belief that the state of the world is A, given the agents' underlying beliefs $\mathbf{b}$. We can therefore view a market as a belief merger $F^{\mathbf{w}} : [0,1]^n \to [0,1]$, mapping beliefs to the equilibrium price.

Remark 2. (Null price). Under Eq. (5) a price $p = 0$ (respectively, $p = 1$) implies that there are no A-traders (respectively, no B-traders). In such cases Eqs. (6), (7) (Naive utility) and (8) (Kelly utility) would be formally undefined. Such situations, however, cannot occur in equilibrium because as p approaches 0 (respectively, 1), the utility for $s_i^A > 0$ (respectively, $s_i^B > 0$) approaches ∞ under both utility models. No investment profile can therefore be in equilibrium with respect to prices $p = 0$ or $p = 1$.

3 Equilibrium Price in Naive and Kelly Markets

In order to see markets as belief aggregators we need to show that the above market types always admit equilibria and, ideally, that equilibrium prices are unique, thereby making the aggregator resolute. We do so in this section.

3.1 Equilibrium p in Naive Markets is the $(1-p)$-Quantile Belief

Let us start by observing that, under naive utility, agents maximize their utility by investing all their wealth, unless their belief equals the price, in which case any level of investment would yield the same utility to them in expectation.

Lemma 1. *In Naive markets, for any competence profile* $\mathbf{q}$, *belief profile* $\mathbf{b} \in \mathcal{B}_{\mathbf{q}}$, *and price* $p \in [0,1]$ *we have that, for any* $i \in N$:

$$\underset{x\in[0,1]}{arg\ max} U_i^A(p,x) = \begin{cases} \{1\} & \text{if } p < b_i, \\ \{0\} & \text{if } p > b_i, \\ [0,1] & \text{otherwise,} \end{cases}$$

$$\underset{x\in[0,1]}{arg\ max} U_i^B(p,x) = \begin{cases} \{1\} & \text{if } (1-p) < (1-b_i), \\ \{0\} & \text{if } (1-p) > (1-b_i), \\ [0,1] & \text{otherwise.} \end{cases}$$

Proof. We reason for A. The argument for B is symmetric. Observe first of all that Eq. (7) can be rewritten as $U_i^A(p, s_i^A) = \frac{b_i}{p}(s_i^A(1-p)+p) + (1-b_i)(1-s_i^A)$. So, the expected utility for strategy $s_i^A = 1$ is $\frac{b_i}{p}$ and for $s_i^A = 0$ is 1. If $\frac{b_i}{p} > 1$,

$U_i^A(p, s_i^A) \in [1, \frac{b_i}{p}]$ and so $s_1^A = 1$ maximizes Eq. (7). By our assumptions, it follows that $s_i^B = 0$. If $\frac{b_i}{p} < 1$ instead $U_i^A(p, s_i^A) \in [\frac{b_i}{p}, 1]$ and $s_1^A = 0$ maximizes Eq. (7). The agent then takes the opposite side of the bet and maximizes $U_i^B(p, s_i^B)$ by setting $s_i^B = 1$. Finally, if $\frac{b_i}{p} = 1$, all investment strategies yield expected utility 1. □

The above result tells us that if $\mathbf{s}$ is in competitive equilbrium with respect to price $p(\mathbf{s})$ in a Naive market, then for each agent i: $s_i^A = 1$ if $b_i > p(\mathbf{s})$, $s_i^A = 0$ if $b_i < p(\mathbf{s})$, and $s_i \in [0, 1]$ if $b_i = p(\mathbf{s})$. The same holds, symmetrically, for s_i^B.

Let $NC(\mathbf{b})$ be the set of investment profiles in competitive equilibrium (under naive utilities) given $\mathbf{b}$. We show that such equilibria always exist and are unique.

Lemma 2. *In Naive markets, for any competence profile* $\mathbf{q}$ *and belief profile* $\mathbf{b} \in \mathcal{B}_{\mathbf{q}}$, $|NC(\mathbf{b})| \geq 1$.

Proof. We prove the claim by construction via Algorithm 1, by showing that the algorithm outputs an investment profile which is a competitive equilibrium.

The algorithm consists of two routines: lines 1–7, and lines 8–21. We first show that, via these two routines, the algorithm always yields an output: if the first routine does not return an output, the second one does. The two routines compare entries in two vectors: the n-long vector of beliefs $(b_1, \ldots, b_n)$, assumed to be ordered by decreasing values (thus, stronger beliefs first); the $n+1$-long vector $(0, \frac{1}{n}, \frac{2}{n}, \ldots, \frac{n}{n})$, ordered therefore by increasing values. The two vectors define two functions from $\{0, \ldots, n\}$ to $[0, 1]$ (we postulate $b_0 = 1$). Because the first function is non-increasing, and the second one is increasing and its image contains both 0 and 1, there exists $i \in \{0, \ldots n\}$ such that the two segments $[b_{i+1}, b_i]$ and $[\frac{i}{n}, \frac{i+1}{n}]$ intersect. There are two cases: $\frac{i}{n}$ lies in $[b_{i+1}, b_i]$, in which case the condition of the first routine applies; or b_{i+1} lies in $[\frac{i}{n}, \frac{i+1}{n}]$, in which case the condition of the second loop applies.

It remains to be shown that the outputs of the two routines are equilibria. The output of the first routine is an investment profile $\mathbf{s} = (\mathbf{s}^A, \mathbf{s}^B)$ where i agents fully invest in A and the remaining agents fully invest in B, yielding a price $p(\mathbf{s}) = \frac{i}{n} \in [b_i, b_{i+1}]$. By Lemma 1 such a profile is an equilibrium. The output of the second routine is an investment profile $\mathbf{s}$ where $i-1$ agents fully invest in A, $n-i$ agents fully invest in B and agent i, whose belief equals the price, invests partially in either A or B in order to meet the clearing Eq. (5). By Lemma 1 we conclude that the profile is in equilibrium for b_i. □

Observe that the price constructed by Algorithm 1 lies in the $[b_i, b_{i+1}]$ interval.

Lemma 3. *In Naive markets, for any competence profile* $\mathbf{q}$ *and belief profile* $\mathbf{b} \in \mathcal{B}_{\mathbf{q}}$, $|NC(\mathbf{b})| \leq 1$.

Proof. Assume towards a contradiction there exist $\mathbf{s} \neq \mathbf{t} \in NC(\mathbf{b})$. It follows that $p(\mathbf{s}) \neq p(\mathbf{t})$. Assume w.l.o.g. that $p(\mathbf{s}) < p(\mathbf{t})$. By Eq. 5 and the definition of competitive equilibrium, it follows that $\sum_{i \in N} s_i^A \leq \sum_{i \in N} t_i^A$ (larger A-investment in $\mathbf{t}$). By Lemma 1, there are more agents i such that $b_i > p(\mathbf{t})$ rather than $b_i > p(\mathbf{s})$, and therefore $p(\mathbf{t}) < p(\mathbf{s})$. A contradiction follows. □

Algorithm 1: Competitive equilibria in Naive markets

---

input : A belief profile $\mathbf{b} = (b_1, \ldots, b_n)$ ordered from highest to lowest beliefs
output: An investment profile $\mathbf{s} = (\mathbf{s}^A, \mathbf{s}^B)$

$\mathbf{s}^A \leftarrow (0, \ldots, 0)$; /* We start by assuming no agent invests in A */
for $1 \leq i < n$ **do**
  if $b_i \geq \frac{i}{n} \geq b_{i+1}$ **then**
    $\mathbf{s}^A \leftarrow (\underbrace{1, \ldots, 1}_{i \text{ times}}, 0, \ldots, 0)$ and $\mathbf{s}^B \leftarrow (\underbrace{0, \ldots, 0}_{i \text{ times}}, 1, \ldots, 1)$;
    return $(\mathbf{s}^A, \mathbf{s}^B)$ and exit ; /* profile with price $\frac{i}{n}$ */
  end
end
for $1 \leq i < n$ **do**
  if $\frac{i-1}{n} < b_i < \frac{i}{n}$ **then**
    $x \leftarrow$ solve $\frac{1}{b_i}((i-1) + x) = \frac{1}{1-b_i}(n-i)$; /* partial A investment */
    if $x \geq 0$ **then**
      $s_i^A \leftarrow x$;
      $\mathbf{s}^A \leftarrow (\underbrace{1, \ldots, 1}_{i-1 \text{ times}}, s_i^A, 0, \ldots, 0)$ and $\mathbf{s}^B \leftarrow (\underbrace{0, \ldots, 0}_{i-1 \text{ times}}, 0, 1, \ldots, 1)$;
      return $(\mathbf{s}^A, \mathbf{s}^B)$ and exit
    else
      $x \leftarrow$ solve $\frac{1}{b_i}(i-1) = \frac{1}{1-b_i}((n-i) + x)$; /* partial B investment */
      $s_i^B \leftarrow x$;
      $\mathbf{s}^B \leftarrow (\underbrace{0, \ldots, 0}_{i-1 \text{ times}}, s_i^B, 1, \ldots, 1)$ and $\mathbf{s}^A \leftarrow (\underbrace{1, \ldots, 1}_{i-1 \text{ times}}, 0, 0, \ldots, 0)$;
      return $(\mathbf{s}^A, \mathbf{s}^B)$ and exit ; /* profile with price b_i */
    end
  end
end

We can thus conclude that in Naive markets there exists exactly one competitive equilibrium and, therefore, only one equilibrium price.

Theorem 2. *In Naive markets, for any competence profile* $\mathbf{q}$ *and belief profile* $\mathbf{b} \in \mathcal{B}_{\mathbf{q}}$, $NC(\mathbf{b})$ *is a singleton.*

Proof. The result follows directly from Lemmas 2 and 3. □

We will refer to the equilibrium profile as $\mathbf{s}_{NC}(\mathbf{b})$ and to its price as $p_{NC}(\mathbf{b})$. An interesting consequence of the above results is that the equilibrium price splits $\mathbf{b}$ into segments roughly proportional to the price.

Corollary 1. *In Naive markets, for any competence profile* $\mathbf{q}$ *and belief profile* $\mathbf{b} \in \mathcal{B}_{\mathbf{q}}$, *there are* $n \cdot p(\mathbf{s})$ *agents* i *such that* $b_i \geq p_{NC}(\mathbf{b})$ *and there are* $n \cdot (1 - p(\mathbf{s}))$ *agents* i *such that* $b_i \leq p_{NC}(\mathbf{b})$.

The equilibrium price $p_{NC}(\mathbf{b})$ corresponds to the $(1 - p_{NC}(\mathbf{b}))$-quantile of $\mathbf{b}$.[5]

3.2 The Average Belief is the Equilibrium Price in Kelly Markets

The two following lemmas are known results from the betting [13] and the information markets literature [5], which we restate here for completeness.

Lemma 4 ([13]). *In Kelly markets, for any $b_i \in [0, 1]$ and $p \in [0, 1]$:*

$$\underset{x\in[0,1]}{arg\ max} U_i^A(p, x) = \begin{cases} \frac{b_i - p}{1-p} & \text{if } p < b_i, \\ 0 & \text{otherwise,} \end{cases}$$

$$\underset{x\in[0,1]}{arg\ max} U_i^B(p, x) = \begin{cases} \frac{p - b_i}{p} & \text{if } (1-p) < (1-b_i), \\ 0 & \text{otherwise.} \end{cases}$$

So, a strategy profile $\mathbf{s}$ is in Kelly competitive equilibrium with respect to price $p(\mathbf{s})$ whenever Eq. (5) is satisfied together with the 'Kelly conditions' of Lemma 4. Unlike in the case of Naive markets it is easy to see that such equilibrium is unique. So, for a given belief profile $\mathbf{b}$, let us denote by $\mathbf{s}_{KC}(\mathbf{b})$ such competitive equilibrium and by $p_{KC}(\mathbf{b})$ the price at such equilibrium.

Lemma 5 ([5]). *For any $\mathbf{q}$ and $\mathbf{b} \in \mathcal{B}_{\mathbf{q}}$, $p_{KC}(\mathbf{b}) = \frac{1}{|N|} \sum_{i \in N} b_i$.*

4 Truth-Tracking via Equilibrium Prices

In this section we show how competitive equilibria in Naive and Kelly markets correspond to election by simple majority and, respectively, by a majority in which agents carry weight proportional to their competence minus 0.5.

4.1 Simple Majority and Naive Markets

The following result shows that simple majority is implemented in competitive equilibrium by a Naive market: for any belief profile $\mathbf{b}$ induced by independent individual competences in $(0.5, 1]$, the diagram on the right commutes. That is, the outcome of simple majority always consists of the security that the $(1-p)$-quantile belief (where p is the equilibrium price) would invest in equilibrium when the market is naive. So we can treat NC as an aggregator $[0, 1]^n \to [0, 1]$ mapping beliefs to equilibrium prices (Fig. 2).

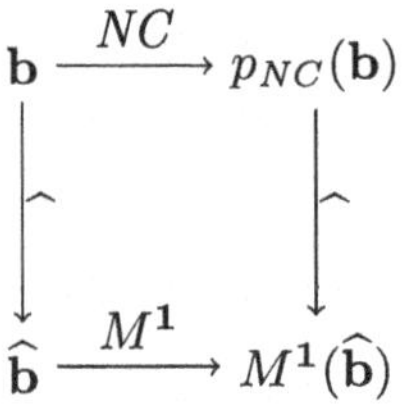

$$\begin{array}{ccc} \mathbf{b} & \xrightarrow{NC} & p_{NC}(\mathbf{b}) \\ \downarrow \widehat{} & & \downarrow \widehat{} \\ \widehat{\mathbf{b}} & \xrightarrow{M^1} & M^1(\widehat{\mathbf{b}}) \end{array}$$

Fig. 2. Simple majority and Naive markets commute.

[5] A similar observation, but for a continuum of players ($N = [0, 1]$) and for subjective beliefs, is made in [15].

Theorem 3. *In Naive markets, for any competence profile* $\mathbf{q}$ *and* $\mathbf{b} \in \mathcal{B}_{\mathbf{q}}$*:*

$$M^1(\widehat{\mathbf{b}}) = \widehat{p_{NC}(\mathbf{b})}.$$

Proof. The claim follows from the observation that, by Corollary 1, $p_{NC}(\mathbf{b}) > 0.5$ if and only if there exists a majority of traders whose beliefs are higher than the price. From which we conclude that $\hat{\mathbf{b}}$ contains a majority of votes for A. □

Remark 3. Note that, by Theorem 3, known extensions of the Condorcet Jury Theorem with heterogeneous competences [11] directly apply to Naive markets in competitive equilibrium. In particular with $N \to \infty$ the probability that $p_{NC}(\mathbf{b})$ is correct approaches 1 for any $\mathbf{b}$ induced by a competence profile.

4.2 Weighted Majority and Kelly Markets

A similar result to Theorem 3 can be obtained for the weighted majority rule with individual weights proportional to $q_i - 0.5$, for each individual i. Such a rule is implemented in competitive equilibrium by Kelly markets. Intuitively, such markets then implement a majority election where individuals' weights are proportional to how better the individual is compared to an unbiased coin.

Theorem 4. *In Kelly markets, for any competence profile* $\mathbf{q}$*, and* $\mathbf{b} \in \mathcal{B}_{\mathbf{q}}$*:*

$$M^{\mathbf{w}}(\widehat{\mathbf{b}}) = \widehat{p_{KC}(\mathbf{b})},$$

where $\mathbf{w}$ *is such that for all* $i \in N$*,* $w_i \propto 2q_i - 1$*.*

Proof. By the assumed weight profile, the normalized total weight of votes for A is $\frac{\sum_{i\in N} \mathbb{1}(b_i>0.5)(2q_i-1)}{\sum_{i\in N}(2q_i-1)}$. For A (respectively, B) to be chosen, this value should exceed (resp., fall short of) $\frac{1}{2}$. This is the case if and only if $\frac{\sum_{i\in N}(2q_i-1)}{n} \cdot \left(\frac{\sum_{i\in N} \mathbb{1}(b_i>0.5)(2q_i-1)}{\sum_{i\in N}(2q_i-1)} - \frac{1}{2}\right) + \frac{1}{2}$ exceeds $\frac{1}{2}$. Let us denote this value $\rho(\mathbf{b})$. The following series of equivalences shows that $\rho(\mathbf{b})$ equals the average belief in $\mathbf{b}$.

$$\begin{aligned}
\rho(\mathbf{b}) &= \frac{\sum_{i\in N}(2q_i-1)}{n} \cdot \left(\frac{\sum_{i\in N} \mathbb{1}(b_i > 0.5)(2q_i-1)}{\sum_{i\in N}(2q_i-1)} - \frac{1}{2}\right) + \frac{1}{2} \\
&= \frac{\sum_{i\in N}(2q_i-1)}{n} \cdot \left(\frac{2\sum_{i\in N} \mathbb{1}(b_i > 0.5)(2q_i-1) - \sum_{i\in N}(2q_i-1)}{2\sum_{i\in N}(2q_i-1)}\right) + \frac{1}{2} \\
&= \frac{\sum_{i\in N} \mathbb{1}(b_i > 0.5)(2q_i-1) - \sum_{i\in N}(q_i - \frac{1}{2}) + \frac{n}{2}}{n} \\
&= \frac{\sum_{i\in N} \mathbb{1}(b_i > 0.5)(2q_i-1) - \sum_{i\in N} q_i + \frac{n}{2} + \frac{n}{2}}{n} \\
&= \frac{\sum_{i\in N} \mathbb{1}(b_i > 0.5)(2q_i-1) + \sum_{i\in N}(1-q_i)}{n} \\
&= \frac{\sum_{i\in N} \mathbb{1}(b_i > 0.5)(2q_i-1) + (1-q_i)}{n} \qquad \text{(recall Equation (1))} \\
&= \frac{\sum_{i\in N} b_i}{n}.
\end{aligned}$$

From this, the definition of $M^{\mathbf{v}}(\widehat{\mathbf{b}})$ (Eq. (4)), and Lemma 5 we obtain $M^{\mathbf{v}}(\widehat{\mathbf{b}}) = \widehat{\frac{\sum_{i \in N} b_i}{n}} = \widehat{p_{KC}(\mathbf{b})}$, as desired. □

Intuitively, the theorem tells us that by implementing a weighted average of the beliefs of the traders, the competitive equilibrium price in markets with Kelly utilities behaves like a weighted majority where agents' weights are a linear function of their individual competence (specifically, $2q_i - 1$). So, for any belief profile $\mathbf{b}$ induced by a competence profile $\mathbf{q}$ and by weights $w_i = 2q_i - 1$, we again a realization of Fig. 1 depicted in the commutative diagram on the right (Fig. 3).

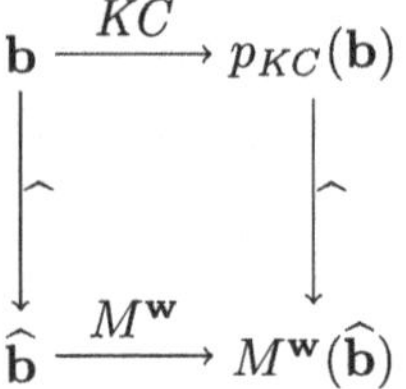

Fig. 3. Weighted majority with weights $q_i - 0.5$ and Kelly markets commute.

5 Markets for Perfect Elections

In this section we show how, by introducing a specific tax scheme, we can modify Kelly markets to make their equilibrium price implement a perfect weighted majority. Recall that we refer to perfect majority voting as weighted majority voting in which the weight of each individual is proportional to the natural logarithm of their competence ratio (recall Theorem 1). The intuition of our approach is the following: Theorem 4, has shown that Kelly markets correspond to elections where individuals are weighted proportionally to their competence in excess of 0.5. In order to bring such weights closer to the ideal values of Theorem 1 we need therefore to allow more competent agents to exert substantially more influence on the equilibrium price; we do so by designing a tax scheme which achieves such effect asymptotically in one parameter of the scheme.

5.1 Taxing Payouts

We modify Eq. (8) by building in the effects of a tax scheme T as follows:

$$U_i^A(p, s_i) = b_i \ln T\left(s_i \frac{1-p}{p} + 1\right) + (1 - b_i)\ln(1 - s_i), \tag{9}$$

where

$$T(x) = \frac{1 - e^{-kx\frac{p}{1-p}}}{k\frac{p}{1-p}}, \tag{10}$$

with $k \in \mathbb{R}_{>0}$. Observe that, as parameter k approaches 0, $T(x)$ approaches x and null taxation is therefore approached.

To gain an intuition of the working of function T, it is useful to observe its effects on the agent's optimal investment strategy supposing the price $p = 0.5$. For $p = 0.5$ the optimal strategy of a Kelly trader is $2b_i - 1$ (Lemma 4). Function

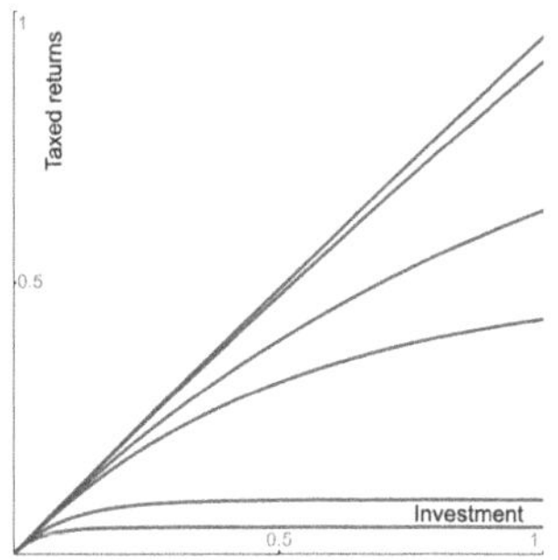

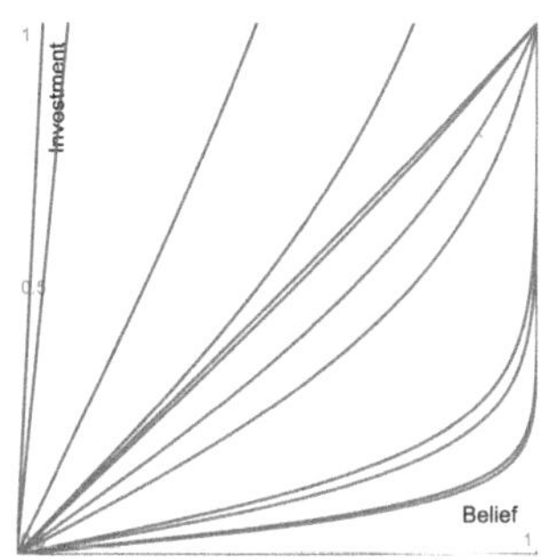

Fig. 4. Left: returns after taxation by T as a function of investment (Equation (9)). Right: investment strategy (red) approximating $\ln\left(\frac{b_i}{1-b_i}\right)\frac{1}{k}$ (blue) as k grows when price equals 0.5. Functions plotted for $k \in \{0.1, 0.2, 1, 2, 10, 20\}$.

T makes that strategy asymptotically proportional to $\ln\left(\frac{b_i}{1-b_i}\right)$ (Fig. 4) as k grows.

We call markets under the utility in Eq. (9) *taxed markets* and denote their equilibrium prize by $p_{TC}(\mathbf{b})$ for any belief profile $\mathbf{b}$.

5.2 Equilibria in Taxed Kelly Markets

Like for Naive and Kelly markets, we first determine the optimal strategy of the traders. We do that for A-traders, as the lemma for B-traders is symmetric.

Lemma 6. *In Taxed markets, for any $i \in N$, if $b_i > p$, then as $k \to \infty$,*

$$\underset{x\in[0,1]}{arg\ max} U_i^A(p,x) \propto \ln\left(\frac{1-p}{p}\cdot\frac{b_i}{1-b_i}\right).$$

Proof. We start from i's utility, given by Eq. (9). By setting $\frac{dU_i^A}{ds_i} = 0$ (first order condition) we obtain:

$$\frac{bT'(s\frac{1-p}{p})\frac{1-p}{p}}{1+T(s\frac{1-p}{p})} = \frac{1-b_i}{1-s_i} \qquad (11)$$

If we replace Eq. (10) into Eq. (11), we obtain:

$$\frac{be^{-ks_i}\frac{1-p}{p}}{1+\frac{1-e^{-ks_i}}{k\frac{p}{1-p}}} = \frac{1-b_i}{1-s_i}. \qquad (12)$$

and therefore

$$\frac{kbe^{-ks_i}}{k\frac{p}{1-p}+1-e^{-ks_i}} = \frac{1-b}{1-s_i}. \qquad (13)$$

As k approaches infinity, s_i approaches zero. For this reason we rescale strategies by k and consider a value $y = sk$. This allows us to understand the form to which strategies tend as they approach zero. We thus obtain

$$\frac{kbe^{-y}}{k\frac{p}{1-p} + 1 - e^{-y}} = \frac{1-b}{1-\frac{y}{k}}. \tag{14}$$

As k approaches infinity this approaches

$$\frac{be^{-y}}{\frac{p}{1-p}} = (1-b), \tag{15}$$

which can be rewritten in turn as

$$y = \ln\left(\frac{1-p}{p}\frac{b}{1-b}\right), \tag{16}$$

from which we conclude $s_i = \frac{1}{k}\log(\frac{1-p}{p}\frac{b}{1-b})$, as desired. □

As k tends to infinity, the optimal investment strategy will tend to 0 for all agents. However, it will do so in such a way that as k grows, the optimal investment strategy tends to be proportional to $\ln(\frac{1-p}{p} \cdot \frac{b_i}{1-b_i})$ as desired.

So, as k grows large, a strategy profile $\mathbf{s}$ is in competitive equilibrium in a taxed market with respect to price $p(\mathbf{s})$ whenever Eq. (5) is satisfied together with the condition identified by Lemma 6. We denote by $\mathbf{s}_{TC}(\mathbf{b})$ such competitive equilibrium and by $p_{TC}(\mathbf{b})$ the price at such equilibrium. We then obtain the following lemma.

Lemma 7. *In Taxed markets, for any profile* $\mathbf{q}$ *and* $\mathbf{b} \in \mathcal{B}_{\mathbf{q}}$*, as* $k \to \infty$*,*

$$\ln\left(\frac{p_{TC}(\mathbf{b})}{1-p_{TC}(\mathbf{b})}\right) \propto \sum_i^n \ln\left(\frac{b_i}{1-b_i}\right).$$

Proof. To lighten notation we write p for $p_{TC}(\mathbf{b})$. From the equilibrium condition (Eq. (5)) and Lemma 6 we have that

$$\frac{1}{p}\sum_{i\in N^A} \ln\frac{b_i}{1-b_i} = \frac{1}{1-p}\sum_{i\in N^B}\frac{1-b_i}{b_i}, \tag{17}$$

where $N^A = \{i \in N \mid b_i > p\}$ and $N^B = \{i \in N \mid b_i < p\}$. From the above we obtain

$$0 = \sum_i^N \ln\left(\frac{1-p}{p}\frac{b_i}{1-b_i}\right), \tag{18}$$

which rewrites to

$$\ln\left(\frac{p}{1-p}\right) = \frac{1}{N}\sum_i^N \ln\left(\frac{b_i}{1-b_i}\right), \tag{19}$$

as desired. □

That is, the equilibrium price ratio between A and B securities in a taxed market tends to be proportional, in logarithmic scale, to the average belief ratio.

Theorem 5. *In Taxed markets, for any profile* $\mathbf{q}$*,* $\mathbf{b} \in \mathcal{B}_{\mathbf{q}}$ *and as* $k \to \infty$*,*

$$M^{\mathbf{w}}(\widehat{\mathbf{b}}) = \widehat{p_{TC}(\mathbf{b})},$$

where $\mathbf{w}$ *is such that for all* $i \in N$ $w_i \propto \ln \frac{q_i}{1-q_1}$*.*

Proof. First of all, observe that: $\widehat{p_{TC}(\mathbf{b})} = \{1\}$ iff $\ln\left(\frac{p_{TC}(\mathbf{b})}{1-p_{TC}(\mathbf{b})}\right) > 0$; $\widehat{p_{TC}(\mathbf{b})} = \{0,1\}$ iff $\ln\left(\frac{p_{TC}(\mathbf{b})}{1-p_{TC}(\mathbf{b})}\right) = 0$; and $\widehat{p_{TC}(\mathbf{b})} = \{0\}$ iff $\ln\left(\frac{p_{TC}(\mathbf{b})}{1-p_{TC}(\mathbf{b})}\right) < 0$. Then, by Lemma 7, Eq. (1) and some algebra we obtain the following relations:

$$\begin{aligned}
\ln\left(\frac{p_{TC}(\mathbf{b})}{1-p_{TC}(\mathbf{b})}\right) &\propto \sum_i^n \ln\left(\frac{b_i}{1-b_i}\right) \\
&= \sum_i^n \mathbb{1}(b_i > 0.5) \cdot \ln\left(\frac{q_i}{1-q_i}\right) \\
&= \sum_{i:b_i>0.5} \ln\left(\frac{q_i}{1-q_i}\right) + \sum_{i:b_i<0.5} \ln\left(\frac{1-q_i}{q_i}\right) \\
&= \sum_{i:b_i>0.5} \ln\left(\frac{q_i}{1-q_i}\right) - \sum_{i:b_i<0.5} \ln\left(\frac{q_i}{1-q_i}\right).
\end{aligned}$$

The last expression is: positive whenever weighted voting with optimal weights returns $\{1\}$; negative whenever it returns $\{0\}$; and 0 whenever it returns $\{0,1\}$ (Eq. (4)). □

This last result shows that elections that are perfect from a truth-tracking perspective (Theorem 1) can be implemented increasingly faithfully by markets with Kelly utilities, once the taxation scheme T is applied and the taxation parameter k in Eq. (10) grows larger and, therefore, that taxation grows. So, for any belief profile $\mathbf{b}$ induced by a competence profile $\mathbf{q}$ and weights $w_i = \frac{q_i}{1-q_i}$, we obtain a realization of Fig. 1 consisting of the commutative diagram on the right, under the assumption that k tends to infinity (Fig. 5).

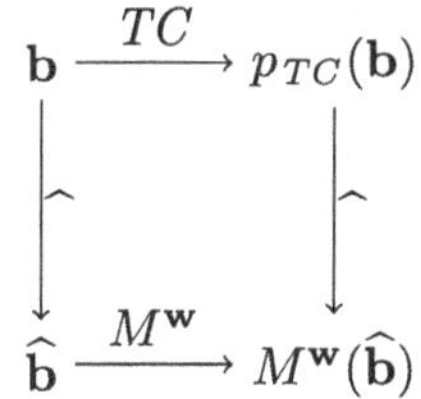

Fig. 5. As tax parameter $k \to \infty$, perfect majority and taxed Kelly markets commute.

6 Numerical Examples

Assume $N = \{1, \ldots, 5\}$ with competence profile $\mathbf{q} = (0.9, 0.7, 0.6, 0.6, 0.6)$. Assume further that only the first and last agent receive signal A while the rest receives signal B. This gives us the following belief profile by Bayesian update: $\mathbf{b} = (0.9, 0.3, 0.4, 0.4, 0.6)$. These beliefs result in the voting profile $\mathbf{v} = (1, 0, 0, 0, 1)$, from which we obtain:

- $M^1(\mathbf{v}) = \{0\}$, that is, standard majority selects B;
- $M^{\mathbf{w}}(\mathbf{v}) = \{1\}$, where $\mathbf{w} = (0.8, 0.4, 0.2, 0.2, 0.2)$ (weight profile given by $2q_i - 1$), as $0.8 + 0.2 - (0.4 + 0.2 + 0.2) > 0$, that is, the sum of weights of the first and last agents are larger then the sum of weights of the others;
- $M^{\mathbf{w}}(\mathbf{v}) = \{1\}$, where $\mathbf{w} = (\ln \frac{0.9}{0.1}, \ln \frac{0.7}{0.3}, \ln \frac{0.6}{0.4}, \ln \frac{0.6}{0.4}, \ln \frac{0.6}{0.4})$ (optimal weights), as the following expression is positive:

$$\ln \frac{0.9}{0.1} + \ln \frac{0.6}{0.4} - \left(\ln \frac{0.7}{0.3} + 2 \cdot \ln \frac{0.6}{0.4} \right). \tag{20}$$

We move now to the choices made by the markets based on equilibrium prices. We have that: by Algorithm 1, $p_{NC}(\mathbf{b}) = \frac{2}{5}$ (Naive market equilibrium) where the two agents who received the A signal invest all their endowment in A-securities, and the remaining agents invest all their endowment in B-securities; $p_{KC}(\mathbf{b}) = \frac{2.6}{5}$ (Kelly market equilibrium) corresponding to the mean belief in $\mathbf{b}$. So, a Naive market given the above beliefs selects B while the Kelly market selects A by a very small margin. As to the taxed markets, our results do not give us a closed expression for $p_{TC}(\mathbf{b})$ but rather determine whether the price favors A- or B-securities based on the logarithm of the ratio between the two prices, which is proportional to the logarithm of the weighed support for A and for B when the taxation parameter k tends to infinity (Theorem 5). In this example, we thus have that $\ln \frac{p_{TC}(\mathbf{b})}{1-p_{TC}(\mathbf{b})}$ is proportional to Eq. (20) and therefore points to security A.

Assume $N = \{1, \ldots, 4\}$ with competence profile $\mathbf{q} = (0.8, 0.6, 0.6, 0.6)$ and that only the first agent receives signal A while the rest receives signal B. This gives us the following belief profile: $\mathbf{b} = (0.8, 0.4, 0.4, 0.4)$. These beliefs result in the voting profile $\mathbf{v} = (1, 0, 0, 0)$, from which we obtain:

- $M^1(\mathbf{v}) = \{0\}$, that is, standard majority selects B
- $M^{\mathbf{w}}(\mathbf{v}) = \{0, 1\}$ where $\mathbf{w} = (0.6, 0.2, 0.2, 0.2)$ (weight profile given by $2q_i - 1$) as $0.6 - (0.2 + 0.2 + 0.2) = 0$. That is, we have a split weighted majority.
- $M^{\mathbf{w}}(\mathbf{v}) = \{1\}$ where $\mathbf{w} = (\ln \frac{0.8}{0.2}, \ln \frac{0.6}{0.4}, \ln \frac{0.6}{0.4}, \ln \frac{0.6}{0.4})$ (optimal weights) as

$$\ln \frac{0.8}{0.2} - 3 \cdot \ln \frac{0.6}{0.4} \tag{21}$$

is positive.

As to equilibrium prices, by applying Algorithm 1, we have that also in this case $p_{NC}(\mathbf{b}) = \frac{2}{5}$ (Naive market equilibrium). This price equals the posterior beliefs of the three agents that receive signal B. By the algorithm, the agent receiving signal A invests all its wealth in A, one of the agents receiving signal B invests $\frac{1}{3}$ of their wealth in A (to guarantee market clearing at that price, line 10 of Algorithm 1), and the remaining agents invest all their endowment in B-securities. The equilibrium price in Kelly markets is in this case 0.5 (mean belief). So, a Naive market given the above beliefs selects B while the Kelly market remains undecided. In taxed markets, we have that $\ln \frac{p_{TC}(\mathbf{b})}{1-p_{TC}(\mathbf{b})}$ is proportional to Eq. (21) and therefore points to security A.

7 Discussion and Outlook

Our paper is the first one to establish a formal link between voting and information markets from an epistemic social choice perspective. The link consists specifically of correspondence results between weighted majority voting on the one hand, and information markets under three types of utility on the other. Such results open up the possibility, in principle, to implement weighted majority voting with strong epistemic guarantees even without having access to individual competences, because such information becomes indirectly available in the market via the equilibrium price. Notice that, in particular, while it may be difficult to elicit truthful weights from agents, investment strategies are subject to the natural incentive of maximizing investment returns. Whether this can prove advantageous also in practice, for instance in the setting of classification markets [4] or voting-based ensembles [8], should be object of future research.

The study we presented is subject to at least four main limitations. First, our analysis inherits all assumptions built into standard jury theorems, in particular: jurors' independence; homogeneous priors; equivalence of type-1 and type-2 errors in jurors' competences; binary events. Future research should try to lift our correspondence to more general settings relaxing the above assumptions (see [9] for a recent overview, and [17,18] for more general frameworks for epistemic social choice). Second, our study limited itself to one-shot interactions. However, markets and specifically Kelly betting make most sense in a context of iterated decisions. Extending our results to the iterated setting, along the lines followed for instance in [5], is also a natural avenue for future research. Third, our market model makes use of the notion of competitive equilibrium. Although such notion of equilibrium is standard in information markets, it responds to the intuition that individuals operate in a large group and, therefore, behave as price takers. We consider it interesting to study how different notions of equilibrium that do not make such assumption (e.g., Nash equilibrium), would behave within our framework. Fourth, our analysis assumes very stylized utility functions which are identical for all agents. This is of course highly restrictive and our results would be substantially strengthened if lifted to more general, and possibly heterogeneous, classes of utilities.

Acknowledgments. This research was (partially) funded by the Hybrid Intelligence Center, a 10-year programme funded by the Dutch Ministry of Education, Culture and Science through the Netherlands Organisation for Scientific Research, https://hybrid-intelligence-centre.nl, grant number 024.004.022. Davide Grossi wishes to also thank Université Paris Dauphine and the Netherlands Institute for Advanced Studies (NIAS), where parts of this research were completed. We are grateful to Marcus Pivato, Feline Lindeboom and the anonymous reviewers of COMSOC'23 and SAGT'24 for many useful comments on earlier versions of the paper.

References

1. Alós-Ferrer, C., Ania, A.: The asset market game. J. Math. Econ. **41**, 67–90 (2005)
2. Austen-Smith, D., Banks, J.: Information aggregation, rationality, and the Condorcet jury theorem. Am. Polit. Sci. Rev. **90**, 34–45 (1996)
3. Baharad, E., Goldberger, J., Koppel, M., Nitzan, S.: Beyond condorcet: optimal aggregation rules using voting records. Bar-Ilan University Department of Economics Research Paper (2010-20) (2010)
4. Barbu, A., Lay, N.: An introduction to artificial prediction markets for classification. J. Mach. Learn. Res. **13**, 2177–2204 (2012)
5. Beygelzimer, A., Langford, J., Pennock, D.M.: Learning performance of prediction markets with Kelly bettors. In: van der Hoek, W., Padgham, L., Conitzer, V., Winikoff, M. (eds.) International Conference on Autonomous Agents and Multiagent Systems, AAMAS 2012, Valencia, Spain, 4–8 June 2012 (3 Volumes), pp. 1317–1318. IFAAMAS (2012)
6. de Condorcet, M., de C., M.J.A.N.: Essai sur l'Application de l'Analyse à la Probabilité des Décisions Rendues à la Pluralité des Voix. Imprimerie Royale, Paris (1785)
7. Conitzer, V., Sandholm, T.: Common voting rules as maximum likelihood estimators. In: UAI 2005, Proceedings of the 21st Conference in Uncertainty in Artificial Intelligence, Edinburgh, Scotland, 26–29 July 2005, pp. 145–152. AUAI Press (2005)
8. Cornelio, C., Donini, M., Loreggia, A., Pini, M.S., Rossi, F.: Voting with random classifiers (VORACE): theoretical and experimental analysis. In: Faliszewski, P., Mascardi, V., Pelachaud, C., Taylor, M.E. (eds.) 21st International Conference on Autonomous Agents and Multiagent Systems, AAMAS 2022, Auckland, New Zealand, 9–13 May 2022, pp. 1929–1931. International Foundation for Autonomous Agents and Multiagent Systems (IFAAMAS) (2022)
9. Dietrich, F., Spiekermann, K.: Jury theorems. Stanford Encycl. Philos. (2021)
10. Elkind, E., Slinko, A.: Rationalizations of voting rules. In: Handbook of Computational Social Choice, pp. 169–196 (2016)
11. Grofman, B., Owen, G., Feld, S.: Thirteen theorems in search of the truth. Theor. Decis. **15**(3), 261–278 (1983)
12. Hayek, F.: The use of knowledge in society. Am. Econ. Rev. **35**, 519–530 (1945)
13. Kelly, J.L.: A new interpretation of information rate. Bell Syst. Tech. J. **35**(4), 917–926 (1956)
14. Kets, W., Pennock, D.M., Sethi, R., Shah, N.: Betting strategies, market selection, and the wisdom of crowds. In: Proceedings of the Twenty-Eighth AAAI Conference on Artificial Intelligence, 27–31 July 2014, Québec City, Québec, Canada, pp. 735–741 (2014)
15. Manski, C.F.: Interpreting the predictions of prediction markets. Econ. Lett. **91**(3), 425–429 (2006)
16. Pennock, D.M.: Aggregating Probabilistic Beliefs: Market Mechanisms and Graphical Representations (1999)
17. Pivato, M.: Voting rules as statistical estimators. Soc. Choice Welfare **40**(2), 581–630 (2013)

18. Pivato, M.: Epistemic democracy with correlated voters. J. Math. Econ. **72**, 51–69 (2017)
19. Wolfers, J., Zitzewitz, E.: Prediction markets. J. Econ. Perspect. **18**(2), 107–126 (2004)
20. Young, H.P.: Condorcet's theory of voting. Am. Polit. Sci. Rev. **82**(4), 1231–1244 (1988)

Complexity of Round-Robin Allocation with Potentially Noisy Queries

Zihan Li[1], Pasin Manurangsi[2], Jonathan Scarlett[1], and Warut Suksompong[1](✉)

[1] National University of Singapore, Singapore, Singapore
lizihan@u.nus.edu, {scarlett,warut}@comp.nus.edu.sg
[2] Google Research, Bangkok, Thailand
pasin@google.com

Abstract. We study the complexity of a fundamental algorithm for fairly allocating indivisible items, the round-robin algorithm. For n agents and m items, we show that the algorithm can be implemented in time $O(nm\log(m/n))$ in the worst case. If the agents' preferences are uniformly random, we establish an improved (expected) running time of $O(nm + m\log m)$. On the other hand, assuming comparison queries between items, we prove that $\Omega(nm + m\log m)$ queries are necessary to implement the algorithm, even when randomization is allowed. We also derive bounds in noise models where the answers to queries are incorrect with some probability. Our proofs involve novel applications of tools from multi-armed bandit, information theory, as well as posets and linear extensions.

1 Introduction

A famous computer science professor is retiring soon, and she wants to distribute the dozens of books that she has accumulated in her office over the years as a parting gift to her students. As one may expect, the students have varying preferences over the books, depending on their favorite authors or the branches of computer science that they specialize in. How can the professor take the students' preferences into account and distribute the books in a fair manner?

The problem of fairly allocating scarce resources has long been studied under the name of *fair division* [12,35] and received significant interest in recent years [1,3,31,36,37]. A simple and well-known procedure for allocating discrete items—such as books, clothes, or household items—is the *round-robin algorithm.* In this algorithm, the agents take turns picking their most preferred item from the remaining items according to a cyclic agent ordering, until all items have been allocated. The algorithm is sometimes known as the *draft* mechanism for its use in allocating sports players to teams [14]. Despite its simplicity, the allocation chosen by the round-robin algorithm satisfies a surprisingly strong fairness guarantee called *envy-freeness up to one item (EF1)* provided that the agents have additive utilities over the items. EF1 means that if an agent envies another agent, this envy can be eliminated by removing some item from the latter agent's bundle. Furthermore, round-robin can be implemented using only the agents' *ordinal*

G. Schäfer and C. Ventre (Eds.): SAGT 2024, LNCS 15156, pp. 520–537, 2024.
https://doi.org/10.1007/978-3-031-71033-9_29

rankings over individual items. This stands in stark contrast to other important fair division algorithms such as *envy-cycle elimination*, which requires eliciting the agents' rankings over *sets* of items, or *maximum Nash welfare*, for which the agents' *cardinal* utilities for items must be known.[1]

It is clear that the round-robin algorithm can be implemented in time polynomial in the number of agents and items. However, despite being one of the very few basic algorithms in discrete fair division and used, adapted, and extended numerous times to provide various fairness guarantees, its complexity has not been analyzed in detail to our knowledge. A moment of thought reveals two sensible approaches for implementing round-robin.[2] Firstly, we can find each agent's ranking over all individual items—this allows us to determine the item picked in every turn, no matter who the picker is. Since sorting m numbers can be done in $O(m \log m)$ time, this approach takes $O(nm \log m)$ time, where n and m denote the number of agents and items, respectively. Secondly, we can instead, for the picking agent at each turn, find the agent's most valuable item from the remaining items. As there are m turns and finding the maximum among m numbers takes $O(m)$ time, the time complexity of this approach is $O(m^2)$. Hence, the first approach is better when n grows at a slower rate than $m/\log m$, while the second approach is more efficient if n has a higher growth rate. Are these two approaches already the best possible, or are there faster ways—that is, faster than $O(m \cdot \min\{n \log m, m\})$ time—to implement the fundamental round-robin algorithm?

1.1 Overview of Results

Fix the agent ordering $1, 2, \ldots, n$, and assume that each agent has a strict ranking over the m items, where $m \geq n$. Hence, the round-robin allocation is uniquely defined, and the task of an algorithm is to output this allocation. Note that we do not require the algorithm to output the item that each agent picks in each turn—this makes our lower bounds stronger, while for the upper bounds, our algorithms can also return this additional information. We consider two models for eliciting agent preferences. In the *comparison query* model, an algorithm can find out with each query which item an agent prefers between a pair of items; in the *value query* model, it can find out the utility of an agent for an item.

In Sect. 3, we consider the *noiseless setting*, where the answers to queries are always accurate. We present a deterministic algorithm that runs in time $O(nm \log(m/n))$ in both query models. Since $nm \log(m/n)$ is less than both $nm \log m$ and m^2, our algorithm is more efficient than both of the approaches mentioned earlier, for any asymptotic relation between n and m. We then show that when the preferences are uniformly random—meaning that each agent has a uniformly random and independent ranking of the items—we can obtain an improved (expected) running time of $O(nm + m \log m)$. We complement these

[1] For descriptions of these algorithms, we refer to the survey by Amanatidis et al. [1].

[2] We remain informal with the query model for this discussion, but will make this precise later.

Table 1. Summary of our results on the query complexity of round-robin allocation. In the noisy setting, δ is the allowed error probability. The uniformly random lower bounds hold because the proof of Theorem 3 (and Corollary 3) uses uniformly random preferences. Our algorithms also offer running time guarantees that match the comparison query upper bounds.

Noise	Preferences	Queries	Upper Bound	Lower Bound
Noiseless	Worst-case	Comparison	$O(nm\log(m/n))$	$\Omega(nm + m\log m)$
		Value	$O(nm)$	$\Omega(nm)$
	Uniformly random	Comparison	$O(nm + m\log m)$	$\Omega(nm)$
		Value	$O(nm)$	$\Omega(nm)$
Noisy	Worst-case	Comparison	$O(nm\log(m/\delta))$	$\Omega(nm\log(1/\delta) + m\log(m/\delta))$
		Value	$O(nm\log(m/\delta))$	$\Omega(nm\log(1/\delta) + m\log(m/\delta))$

results by establishing lower bounds: $\Omega(nm + m\log m)$ and $\Omega(nm)$ queries are necessary in the comparison and value model, respectively. Since the entire preferences can be elicited via $O(nm)$ queries in the value model, the latter bound is tight. Our lower bounds hold even against algorithms that may fail with some constant probability; proving the former bound entails leveraging results on posets and linear extensions.

In Sect. 4, we turn our attention to the *noisy setting*. For comparison queries, we assume that the answer to each query is incorrect with probability ρ independently of all other queries, where $\rho \in (0, 1/2)$ is a given constant. For value queries, the answer to each query is the true utility with probability $1-\rho$ and an arbitrary value with probability ρ, where this value can be chosen adversarially. We focus on algorithms that are correct with probability at least $1-\delta$ for a given parameter[3] δ. We show that for both types of queries, there exists a deterministic algorithm running in time $O(nm\log(m/\delta))$. On the other hand, we provide a lower bound of $\Omega(nm\log(1/\delta) + m\log(m/\delta))$ on the number of queries even for randomized algorithms;[4] the proof involves novel applications of tools from multi-armed bandit and information theory and may be of independent interest to researchers in different areas.

A summary of our results can be found in Table 1.

1.2 Related Work

The fair division literature typically considers fairness guarantees (such as EF1) that are feasible in different settings along with algorithms that achieve these guarantees [1,12,31]. We take a central algorithm in the literature—the round-robin algorithm—and analyze its complexity. Query complexity in discrete fair

[3] We assume that $\delta \in (0, 1/2 - c)$ for some constant $c > 0$.

[4] Note that if $\delta \in O(m^{-d})$ for some constant $d > 0$, the upper and lower bounds asymptotically match.

division has previously been studied by Plaut and Roughgarden [33] and Oh et al. [32]. For instance, Oh et al. showed that for two agents, it is possible to compute an EF1 allocation using $O(\log m)$ queries. However, all of these authors assumed a query model such that with each query, an algorithm can find out an agent's utility for any *set* of items. Since determining this value for a large set can be quite demanding, our query models, in which each query only involves one or two items, are arguably more realistic.

The vast majority of work in fair division assumes that accurate information on the agent preferences is available to the algorithms. Nevertheless, in reality there may be noise or uncertainty in these preferences, possibly due to the limited time or high cost for determining the true preferences. Aziz et al. [4] and Li et al. [28] investigated uncertainty models for item allocation. For example, in their "compact indifference model", each agent reports a ranking over items that may contain ties; the ties indicate that the agent is uncertain about her preferences among the tied items. Note that this approach to uncertainty is very different from ours; in our approach, the uncertainty does not appear in the initial input to the problem, but instead arises as noise in answers to queries. Our noisy comparison model has been used for several algorithmic problems including searching, sorting, and selection [8,13,20–22], and our noisy value model has also been studied for similar problems [16], though the use of these models in fair division is new to the best of our knowledge.

Not surprisingly given its wide applicability, the round-robin algorithm has been examined from various angles, including strategic considerations [5,10], equilibrium properties [2], and monotonicity guarantees [15].

2 Preliminaries

Let $N = [n]$ be the set of agents and $M = [m]$ be the set of items, where $[k] := \{1, 2, \ldots, k\}$ for any positive integer k. Denote by $u_i(j) \geq 0$ the *utility* (also referred to as the *value*) of agent i for item j, and assume for convenience that $u_i(j) \neq u_i(j')$ for all $i \in N$ and distinct $j, j' \in M$. For $X, Y \subseteq M$, we write $X \succ_i Y$ if $u_i(x) > u_i(y)$ for all $x \in X$ and $y \in Y$. The *round-robin allocation* is the allocation that results from letting agents take turns picking items in the order $1, 2, \ldots, n, 1, 2, \ldots,$ where in each turn, the picking agent picks the item for which she has the highest utility. Since each agent has distinct utilities for all items, this allocation is uniquely defined. We sometimes refer to each sequence $1, 2, \ldots, n$ as a *round*, where the last round might not include all agents. Assume without loss of generality that $m \geq n$ (otherwise, we may simply ignore the agents who do not receive any item) and $n \geq 2$.

We consider two models of how an algorithm can discover information about agents' utilities. In the *comparison query* model, an algorithm can specify an agent $i \in N$ and a pair of distinct items $j, j' \in M$, and find out whether agent i prefers item j to j'. In the *value query* model, an algorithm can specify $i \in N$ and $j \in M$, and find out the value of $u_i(j)$. We assume that each query takes constant time and the algorithm can be adaptive. In the *noiseless setting*

(Sect. 3), the answer to each query is always accurate. Observe that with no noise, each comparison query can be simulated using two value queries, so a lower bound for value queries implies a corresponding one for comparison queries with an extra factor of 1/2. In the *noisy setting* (Sect. 4), for comparison queries, there is a constant $\rho \in (0, 1/2)$, and the answer to each query is incorrect with probability ρ, independently of all other queries (including those with the same i, j, j').[5] For value queries, the answer to each query is the true utility with probability $1 - \rho$ and an arbitrary value with probability ρ, independently of all other queries (including those with the same i, j). The arbitrary value can be chosen by an adversary, who may choose the value based not only on the algorithm's past queries (and their answers), but also on the entire set of utilities $u_i(j)$ for $i \in N$ and $j \in M$.

By *uniformly random preferences*, we refer to the setting where we associate each agent i with a permutation $\sigma_i : [m] \to [m]$ chosen uniformly and independently at random, and let i prefer j to j' if and only if $\sigma_i(j) > \sigma_i(j')$.

All omitted proofs can be found in the full version of our paper [27].

2.1 Selection and Quantiles Algorithms

Selection. We will use the following classic linear-time algorithm for the so-called *selection* problem.

Lemma 1 (Selection algorithm [9]). *For every $i \in N$, $S \subseteq M$, and $\ell \in [|S|]$, there is an $O(|S|)$-time deterministic algorithm $\textsc{Sel}_{i,\ell}(S)$ that makes $O(|S|)$ comparison queries and outputs a partition $(S^\uparrow, S^\downarrow)$ of S such that $S^\uparrow \succ_i S^\downarrow$ and $|S^\uparrow| = \ell$.*

Quantiles. We next consider the (m, n)-*quantiles* problem, where we want to partition a set of m items into subsets of size at most n each, so that every item in the first set is preferred to every item in the second set, which is in turn preferred to every item in the third set, and so on. This problem can be solved in $O(m \log(m/n))$ time (in contrast to $O(m \log m)$ for sorting).

Lemma 2 (Quantiles algorithm). *For every $i \in N$, there is an $O(m \log (m/n))$-time deterministic algorithm $\textsc{Quant}_i$ that makes $O(m \log(m/n))$ comparison queries and outputs a partition $(S_1^i, \ldots, S_k^i)$ of M for some $k \in \mathbb{N}$, with the property that $S_1^i \succ_i \cdots \succ_i S_k^i$ and $|S_1^i|, \ldots, |S_k^i| \le n$.*

Similar results are known in the literature; see, e.g., Exercise 9.3-6 of Cormen et al. [17]. For completeness, we provide the proof of Lemma 2.

---

[5] While one could consider an alternative model in which comparison faults are *persistent*, there is no hope of getting a reasonable success rate for our problem under that model. For example, even with *one* persistent error, if that error is on agent 1's comparison between the top two items and some other agent has the same favorite item as agent 1, then the output will be wrong.

Proof. The algorithm QUANT_i on input $S \subseteq M$ works as follows:

- If $|S| \leq n$, terminate.
- Otherwise, let $(S^{\uparrow}, S^{\downarrow}) \leftarrow \text{SEL}_{i,\lfloor |S|/2 \rfloor}$ (where SEL is from Lemma 1). Then, recurse on $S^{\uparrow}$ and $S^{\downarrow}$.

We start with $S = M$. Observe that the recurrence relation for both the running time and the query complexity is

$$T(\ell) \leq \begin{cases} O(1) & \text{if } \ell \leq n; \\ O(\ell) + T(\lfloor \ell/2 \rfloor) + T(\lceil \ell/2 \rceil) & \text{otherwise,} \end{cases}$$

where ℓ denotes the size of the set S. One can verify using standard methods for solving recurrence relations (e.g., a recursion tree) that $T(\ell) \leq O(\ell \log(\ell/n))$. □

3 Noiseless Setting

We begin in this section by considering the noiseless setting, where every query always receives an accurate answer.

3.1 Upper Bounds

First, we show that it is possible to achieve a running time of $O(nm \log(m/n))$ for comparison queries.

Theorem 1. *Under the noiseless comparison query model, there exists a deterministic algorithm that outputs the round-robin allocation using $O(nm \log(m/n))$ queries and $O(nm \log(m/n))$ time.*

Proof. The algorithm is presented as Algorithm 1. Its correctness is due to the guarantee of Lemma 2 that $S^i_1 \succ_i \cdots \succ_i S^i_k$, which implies that in each turn, the picking agent picks her most preferred item among the remaining items.

As for the number of comparison queries, note that there are only two places that require comparisons: (i) when we call QUANT_i, and (ii) when we find the best item for i in $S^i_{b_i} \cap S$ (Line 10). For (i), Lemma 2 ensures that the number of queries is $O(m \log(m/n))$ for each i, resulting in a total of $O(nm \log(m/n))$ across all $i \in N$. For (ii), since $|S^i_{b_i}| \leq n$, we can find the most preferred item of agent i in $S^i_{b_i} \cap S$ using $O(n)$ queries. Since Line 10 is invoked m times, the number of queries for this part is $O(nm)$. It follows that the total number of queries used by the algorithm is $O(nm \log(m/n))$.

Apart from Line 8, it is clear that the running time of the rest of the algorithm is $O(nm \log(m/n))$. As for Line 8, let us fix $i \in N$. Note that we can check whether $S^i_{b_i} \cap S$ is non-empty in time $O(|S^i_{b_i}|) \leq O(n)$. Since we increment b_i each time the check fails and in each round the check passes only once, the total running time of this step (for this agent i) is at most $O\big(\sum_{b \in [k]} |S^i_b| + \lceil m/n \rceil \cdot n\big) = O(m)$. Therefore, in total, the running time of this step across all agents is $O(nm)$. We conclude that the total running time of the entire algorithm is $O(nm \log(m/n))$, as desired. □

Algorithm 1. For worst-case preferences

Input: Set of agents N, set of items M, utilities $u_i(j)$ for $i \in N, j \in M$
Output: Round-robin allocation $(A_1, \dots, A_n)$

```
for i ∈ N do
    A_i ← ∅
    (S^i_1, ..., S^i_k) ← QUANT_i(M)                 {See Lemma 2}
    b_i ← 1                                 {First b such that S^i_b ≠ ∅}
S ← M                                           {Set of remaining items}
for r = 1, ..., ⌈m/n⌉ do
    for i = 1, ..., min{n, m − n(r − 1)} do
        while S^i_{b_i} ∩ S = ∅ do
            b_i ← b_i + 1
        j ← best item for agent i in S^i_{b_i} ∩ S
        A_i ← A_i ∪ {j}
        S ← S \ {j}
return (A_1, ..., A_n)
```

For value queries, we can query all nm values and run the algorithm from Theorem 1.

Corollary 1. *Under the noiseless value query model, there is a deterministic algorithm that outputs the round-robin allocation using $O(nm)$ queries and $O(nm \log(m/n))$ time.*

Next, we consider uniformly random preferences, which constitute a standard stochastic model in fair division [19,25,29,30]. For these preferences, we present an improvement over the worst-case bound. We remark that the round-robin algorithm on random preferences has been studied by Manurangsi and Suksompong [30, Thm. 3.1] and Bai and Gölz [7, Prop. 2]. Since both of these papers used preference models that imply uniformly random ordinal preferences,[6] our result can be applied for the running time analysis of their algorithms.

Theorem 2. *For uniformly random preferences, under the noiseless comparison query model, there exists a deterministic algorithm that outputs the round-robin allocation using expected $O(nm + m \log m)$ queries and expected $O(nm + m \log m)$ time.*

Proof. The algorithm (Algorithm 2) proceeds by keeping a sorted list L_i of the best remaining items for each agent i, which is initially empty. At i's turn, if L_i is non-empty, i picks the best item j from L_i, and item j is removed from every other set $L_{i'}$ that contains it. On the other hand, if L_i is empty, the algorithm

[6] Specifically, Manurangsi and Suksompong [30] assumed that all agents' utilities for all items are drawn independently from the same (non-atomic) distribution, while Bai and Gölz [7] allowed each agent's utilities for items to be drawn independently from an agent-specific (non-atomic) distribution.

Algorithm 2. For uniformly random preferences

Input: Set of agents N, set of items M, utilities $u_i(j)$ for $i \in N, j \in M$
Output: Round-robin allocation $(A_1, \ldots, A_n)$

```
1: for i ∈ N do
2:     A_i ← ∅
3:     L_i ← ∅                              {Sorted list of i's best remaining items}
4: S ← M                                                 {Set of remaining items}
5: for r = 1, ..., ⌈m/n⌉ do
6:     for i = 1, ..., min{n, m − n(r − 1)} do
7:       if L_i = ∅ then
8:         ℓ ← ⌈|S|/n⌉
9:         (S↑, S↓) ← SEL_{i,ℓ}(S)                                {See Lemma 1}
10:        Find the sorted list L_i of S↑ for agent i, using a standard sorting algorithm
           (e.g., merge sort).
11:      j ← best item for agent i in L_i
12:      A_i ← A_i ∪ {j}
13:      S ← S \ {j}
14:      for i' ∈ N do
15:        L_i' ← L_i' \ {j}
16: return (A_1, ..., A_n)
```

first finds the best $1/n$ fraction of the remaining items (rounded up) for i using Lemma 1, sorts these items, and inserts them into L_i.

The correctness of the algorithm is again trivial. Furthermore, the running time does not add extra asymptotic terms on top of the query complexity, so we will only establish the latter. In particular, we will show that the expected number of queries made by the first agent is at most $O(m + \frac{m}{n} \log m)$; the proof is similar for the other agents, and the desired statement follows from summing this up across all agents.

Let $Q_n(m)$ denote the number of queries made by the first agent when there are m items and n agents. Fix $n \in \mathbb{N}$. We will show by induction on m that

$$Q_n(m) \leq C \cdot \left(m + \left\lceil \frac{m}{n} \right\rceil \log m\right) \tag{1}$$

where $C > 0$ is a sufficiently large constant. Specifically, let C_2 be a constant such that the re-initialization of L_i between Lines 7 and 10 of the algorithm takes at most $C_2 \cdot (|S| + \ell \log \ell)$ comparison queries. Then, we let $C = 100C_2$.

For the base case where $m \leq 2n$, note that there are at most two rounds and each round only takes at most $C_2 \cdot (m + \lceil m/n \rceil \log m)$ comparisons for the first agent.

Next, we address the induction step. Suppose that for some $m^* > 2n$, inequality (1) holds for all $m < m^*$; we will show that it also holds for $m = m^*$.

Consider running the algorithm for $m = m^*$. Let $R^* = \lceil m^*/n \rceil \geq 3$ be the total number of rounds to be run. Let $r \geq 1$ denote the first round such that L_1 becomes empty after the end of the round; for notational convenience, we let $r = m^*/n$ instead of $\lceil m^*/n \rceil$ in the case that this happens in the last round.

Note that r is a random variable. Observe that from round $r+1$ onward, the expected number of queries made by the agent is the same as if the algorithm is run on $m^* - rn$ items—this is because, conditioned on the items selected so far by all agents, the remaining items admit uniformly random preferences. In other words, we have

$$Q_n(m^*) \leq C_2 \cdot (m^* + R^* \log m^*) + \mathbb{E}_r[Q_n(m^* - rn)],$$

where we use the convention $Q_n(0) = 0$.

Plugging the inductive hypothesis into the inequality above, we get that $Q_n(m^*)$ is at most

$$\begin{aligned} &C_2 \cdot (m^* + R^* \log m^*) + \mathbb{E}_r\left[C \cdot ((m^* - rn) + (R^* - r)\log m^*)\right] \\ &= C \cdot (m^* + R^* \log m^*) + C_2 \cdot ((m^* - 100nX) + (R^* - 100X)\log m^*), \end{aligned}$$

where $X := \mathbb{E}[r]$. Hence, to show (1) for $m = m^*$, it suffices to prove that $X \geq R^*/100$. In turn, to prove this, it is sufficient to show that $\mathbb{P}[r \geq R^*/50] \geq 0.5$, or equivalently, $\mathbb{P}[r < R^*/50] \leq 0.5$.

To show that $\mathbb{P}[r < R^*/50] \leq 0.5$, let $r_0 := \lfloor R^*/50 \rfloor$. Let $X_{i',r'}$ be an indicator variable that equals 1 if and only if *both* of the following are true: (i) agent i' takes an item from L_1 in round r', and (ii) L_1 has not been re-initialized by round r'. Notice that

$$\mathbb{P}\left[r < \frac{R^*}{50}\right] = \mathbb{P}\left[R^* = \sum_{r' \in [r_0]} \sum_{i' \in N} X_{i',r'}\right] \leq \frac{1}{R^*} \cdot \mathbb{E}\left[\sum_{r' \in [r_0]} \sum_{i' \in N} X_{i',r'}\right],$$

where the equality follows from the fact that we start with L_1 of size R^*, and the inequality follows from Markov's inequality. Thus, to show that $\mathbb{P}[r < R^*/50] \leq 0.5$, it suffices to show that $\mathbb{E}\left[\sum_{r' \in [r_0]} \sum_{i' \in N} X_{i',r'}\right] \leq 0.5R^*$.

To calculate this expectation, let us make the following observations on $X_{i',r'}$. If $i' \neq 1$, then since each agent i' removes an item from S most preferred by her at that point, the probability that this item belongs to L_1 is exactly $\frac{|L_1|}{|S|} \leq \frac{R^*}{|S|}$, where the inequality follows from the fact that L_1 starts with size R^*. In other words, we have $\mathbb{E}[X_{i',r'}] \leq \frac{R^*}{|S|}$ when $i' \neq 1$. Note also that for $r' \leq r_0$, we always have $|S| \geq m^*/3$. Indeed, since there are $R^* - r_0 > R^*/3 + 1$ rounds remaining (not including the current round), the number of items left is at least $n \cdot R^*/3 \geq m^*/3$. From this, we can derive

$$\begin{aligned} \mathbb{E}\left[\sum_{r' \in [r_0]} \sum_{i' \in N} X_{i',r'}\right] &\leq r_0 + \mathbb{E}\left[\sum_{r' \in [r_0]} \sum_{i' \in N \setminus \{1\}} \frac{R^*}{m^*/3}\right] \leq r_0 \left(1 + \frac{3nR^*}{m^*}\right) \\ &\leq r_0 \cdot 10 \leq 0.5R^*, \end{aligned}$$

which concludes our proof. □

Similarly to Corollary 1, for value queries, we can query all nm values and run the algorithm from Theorem 2.

Corollary 2. *For uniformly random preferences, under the noiseless value query model, there exists a deterministic algorithm that outputs the round-robin allocation using $O(nm)$ queries and expected $O(nm + m \log m)$ time.*

3.2 Lower Bounds

We now turn to lower bounds. First, we present a lower bound of $\Omega(nm)$ for comparison queries.

Theorem 3. *Under the noiseless comparison query model, any (possibly randomized) algorithm that outputs the round-robin allocation with probability at least $2/3$ makes $\Omega(nm)$ queries in expectation.*

We remark that all of our lower bounds in this section hold even when $2/3$ is replaced by any constant strictly larger than $1/2$.

Before we proceed, we introduce some additional notation. Let Alg be an algorithm for the round-robin problem. A triplet $a = (i, j, j')$ represents a query in which Alg asks if agent i prefers item j to j'. Let $\mathcal{A}$ be the set of all such triplets, with $|\mathcal{A}| = n\binom{m}{2}$; since querying (i, j_1, j_2) is equivalent to querying (i, j_2, j_1) and flipping the answer, we omit such "duplicate" triplets from $\mathcal{A}$ without loss of generality.

Define an *instance* ν of our problem to be a setting of the agent preferences, and let L_ν be the correct round-robin allocation when the instance is ν. For any instance ν and agent i, let $\mathcal{J}_i(\nu)$ be the set of all items j such that in the correct round-robin procedure, item j is not allocated to any of the agents $1, 2, \dots, i-1$ in the first round and is not allocated to agent i in any round. We will use the following lemma.

Lemma 3. *Let ν be an arbitrary instance of our round-robin problem. Then, $\sum_{i \in N} |\mathcal{J}_i(\nu)| \geq nm/4$.*

Proof. We write $k_i := 1 + \lfloor (m - i)/n \rfloor$ to denote the number of items allocated to agent i across all rounds. Since $i - 1$ items are allocated to agents $1, \dots, i-1$ in the first round, and k_i items are allocated to agent i across all rounds, we have $|\mathcal{J}_i(\nu)| \geq m - (i-1) - k_i$. Summing this over all $i \in N$, we get

$$\sum_{i \in N} |\mathcal{J}_i(\nu)| \geq nm - \frac{n(n-1)}{2} - m = (n-1)\left(m - \frac{n}{2}\right) \geq \frac{n}{2} \cdot \frac{m}{2} = \frac{nm}{4},$$

as desired. □

For brevity, we say that an algorithm Alg is *α-correct* if $\mathbb{P}_{\mathsf{Alg}}[\mathsf{Alg}(\nu) = L_\nu] \geq \alpha$ for any instance ν, where the probability is taken over the randomness of Alg. Moreover, for a distribution $\mathcal{D}$ over instances, we say that Alg is *$(\alpha, \mathcal{D})$-correct* if $\mathbb{P}_{\nu \sim \mathcal{D}, \mathsf{Alg}}[\mathsf{Alg}(\nu) = L_\nu] \geq \alpha$, where the probability is taken over both the random instance ν drawn from $\mathcal{D}$ and the randomness of Alg. We will also use the following lemma, which is in the spirit of Yao's principle. The proof of this lemma and the next can be found in the full version of our paper [27].

Lemma 4. *If there exists a 2/3-correct algorithm using at most q queries in expectation for some $q \in \mathbb{R}^+$, then for any distribution $\mathcal{D}$ of instances, there exists a deterministic algorithm that makes $O(q)$ queries in the worst case and is $(0.99, \mathcal{D})$-correct.*

Given Lemma 4, to prove Theorem 3, the following lower bound against deterministic algorithms is sufficient.

Proposition 1. *Under the noiseless comparison query model, there exists a distribution $\mathcal{D}$ over instances such that any deterministic $(0.99, \mathcal{D})$-correct algorithm makes $\Omega(nm)$ queries in the worst case.*

Proof. Let $\mathcal{D}$ be the distribution based on uniformly random preferences. Suppose for contradiction that there exists a deterministic algorithm Alg that is $(0.99, \mathcal{D})$-correct and makes at most $q := 0.01nm$ queries in the worst case.

For any instance ν, agent $i \in N$, and item $j \in M$, let $q_{i,j}(\nu)$ be the indicator variable of whether the pair (i, j) is involved in any query made by Alg when run on ν. Furthermore, let $\nu^{i,j}$ denote the instance that is the same as ν except that item j is made the most preferred item of agent i. To help with the proof of Proposition 1, we will use the following lemma.

Lemma 5. *For any algorithm Alg, it holds that $q_{i,j}(\nu) = 1$ or $j \notin \mathcal{J}_i(\nu)$ or $\mathsf{Alg}(\nu) \neq L_\nu$ or $\mathsf{Alg}(\nu^{i,j}) \neq L_{\nu^{i,j}}$.*

Next, observe that if we pick $\nu \sim \mathcal{D}$, $i \in N$, and $j \in M$ uniformly and independently at random, then $\nu^{i,j}$ has the same distribution as $\mathcal{D}$, due to symmetry. Hence, picking ν, i, j in this way, we have

$$\begin{aligned}
2\mathbb{P}_\nu[\mathsf{Alg}(\nu) \neq L_\nu] &= \mathbb{P}_\nu[\mathsf{Alg}(\nu) \neq L_\nu] + \mathbb{P}_{\nu,i,j}[\mathsf{Alg}(\nu^{i,j}) \neq L_{\nu^{i,j}}] \\
&= \mathbb{E}_{\nu,i,j}\left[\mathbf{1}\left[\mathsf{Alg}(\nu) \neq L_\nu\right] + \mathbf{1}\left[\mathsf{Alg}(\nu^{i,j}) \neq L_{\nu^{i,j}}\right]\right] \\
&\geq \mathbb{E}_{\nu,i,j}[1 - q_{i,j}(\nu) - \mathbf{1}[j \notin \mathcal{J}_i(\nu)]] \\
&= \mathbb{E}_{\nu,i,j}\left[\mathbf{1}[j \in \mathcal{J}_i(\nu)] - q_{i,j}(\nu)\right] \\
&\geq \frac{1}{nm}\mathbb{E}_\nu\left[\sum_{i \in N} |\mathcal{J}_i(\nu)|\right] - \frac{2q}{nm} \geq \frac{1}{4} - 0.02 > 0.2,
\end{aligned}$$

where the first and third inequalities follow from Lemma 5 and Lemma 3, respectively, and the factor of 2 in the second inequality arises because each query (i, j, j') can contribute to both $q_{i,j}(\nu)$ and $q_{i,j'}(\nu)$. This contradicts our assumption that Alg is $(0.99, \mathcal{D})$-correct. □

The proof of an analogous bound for value queries is essentially the same. Note that since nm value queries are clearly sufficient, this bound cannot be improved.

Corollary 3. *Under the noiseless value query model, any (possibly randomized) algorithm that outputs the round-robin allocation with probability at least 2/3 makes $\Omega(nm)$ queries in expectation.*

Next, we prove a bound of $\Omega(m \log m)$ for comparison queries—by Theorem 1, this bound is tight for constant n.

Theorem 4. *Under the noiseless comparison query model, any (possibly randomized) algorithm that outputs the round-robin allocation with probability at least* $2/3$ *makes* $\Omega(m \log m)$ *queries in expectation.*

Given Lemma 4, to prove Theorem 4, it suffices to show the following bound against deterministic algorithms.

Proposition 2. *Under the noiseless comparison query model, there is a distribution* $\mathcal{D}$ *over identical-preference instances such that any deterministic* $(0.99, \mathcal{D})$*-correct algorithm makes* $\Omega(m \log m)$ *queries in the worst case.*

Since the proof of Proposition 2 only uses identical preferences, we do not distinguish between queries for different agents and view each comparison simply as a tuple (j, j') of items. Further, we represent an identical-preference instance ν by a permutation $\sigma : [m] \to [m]$, where item j is preferred to j' (by all agents) exactly when $\sigma(j) > \sigma(j')$. Let L_σ be the correct round-robin allocation when the instance is σ. Let $\mathcal{R}$ denote the complete set of comparison results, i.e., $\mathcal{R} = \{(j, j', r) \mid 1 \le j < j' \le m, r \in \{0, 1\}\}$, where $(j, j', 1)$ means that j is preferred to j' and $(j, j', 0)$ indicates the opposite preference. For any set $R \subseteq \mathcal{R}$ of comparison query results, let $\mathcal{X}(R)$ be the set of all permutations on $[m]$ that are compatible with R. We write $\sigma \sim \mathcal{X}(R)$ to signify a permutation drawn uniformly at random from $\mathcal{X}(R)$. Notice that $|\mathcal{X}(R)| = 1$ if and only if the comparison results in R completely determine the ordering of items. Our main lemma is that, unless this is the case, we cannot find an allocation that agrees with almost all permutations in $\mathcal{X}(R)$.

Lemma 6. *For any* $R \subseteq \mathcal{R}$*, if* $|\mathcal{X}(R)| > 1$*, then for any allocation* A*, we have* $\mathbb{P}_{\sigma \sim \mathcal{X}(R)}[L_\sigma \neq A] \ge 3/44$.

The proof of Lemma 6 involves showing that we can find an item $j \in M$ such that, for a random $\sigma \sim \mathcal{X}(R)$, the value $\sigma(j)$ is sufficiently random. We show this by leveraging results from the theory of posets and linear extensions [23,34]. The full proof is deferred to the full version of our paper [27].

Lemma 6 implies that we have to determine σ with sufficiently high probability using the comparison queries. Without the "sufficiently high probability" part, this is exactly the sorting problem, for which it is well-known that $\Omega(m \log m)$ queries are required. To establish Proposition 2, we show that a similar number of queries is still necessary even with the "high probability" relaxation.

Proof (of Proposition 2*).* Let $\mathcal{D}$ be the distribution based on identical preferences such that the preference order of items is uniformly random. Suppose for contradiction that there exists a deterministic algorithm Alg that is $(0.99, \mathcal{D})$-correct and makes at most $q := 0.1\, m \log_2 m$ queries in the worst case. We use the standard representation of Alg as a binary decision tree:[7] each internal node of the tree corresponds to a comparison query (j, j'), and the left and right children correspond to the query answer being 0 and 1, respectively. Let Λ denote the set of leaves of the tree; since Alg makes at most q queries, $|\Lambda| \leq 2^q$. Each leaf $\lambda \in \Lambda$ corresponds to the algorithm's termination, at which point it outputs some allocation A^λ. We use $R_\lambda \subseteq \mathcal{R}$ to denote the set of comparison results leading to the leaf λ. Finally, let $\lambda(\sigma)$ denote the leaf that Alg ends up in when run on σ.

We can now bound the probability that Alg is incorrect on a random $\sigma \sim \mathcal{D}$ as follows:

$$\begin{aligned}
\mathbb{P}_{\sigma\sim\mathcal{D}}[L_\sigma \neq \mathsf{Alg}(\sigma)] &= \mathbb{P}_{\sigma\sim\mathcal{D}}[L_\sigma \neq A^{\lambda(\sigma)}] \\
&= \sum_{\lambda\in\Lambda} \mathbb{P}_{\sigma\sim\mathcal{D}}[\lambda(\sigma) = \lambda] \cdot \mathbb{P}_{\sigma\sim\mathcal{D}}[L_\sigma \neq A^\lambda \mid \lambda(\sigma) = \lambda] \\
&= \sum_{\lambda\in\Lambda} \frac{|\mathcal{X}(R_\lambda)|}{m!} \cdot \mathbb{P}_{\sigma\sim\mathcal{X}(R_\lambda)}[L_\sigma \neq A^\lambda] \\
&\geq \sum_{\lambda\in\Lambda} \frac{|\mathcal{X}(R_\lambda)|}{m!} \cdot \frac{3}{44} \cdot (1 - \mathbf{1}[|\mathcal{X}(R_\lambda)| = 1]) \\
&= \frac{3}{44} - \frac{3}{44} \sum_{\lambda\in\Lambda} \frac{\mathbf{1}[|\mathcal{X}(R_\lambda)| = 1]}{m!} \\
&\geq \frac{3}{44}\left(1 - \frac{|\Lambda|}{m!}\right) \geq \frac{3}{44}\left(1 - \frac{2^q}{m!}\right) > 0.01,
\end{aligned}$$

where the first inequality follows from Lemma 6 and the last inequality from $q = 0.1\, m \log_2 m$. This contradicts our assumption that Alg is $(0.99, \mathcal{D})$-correct. □

4 Noisy Setting

In this section, we turn our attention to the noisy setting. Because of the noise, we cannot expect algorithms to always output the correct answer. Therefore, we will instead require them to be correct with probability at least $1 - \delta$, for a given parameter δ. Throughout the section, we adopt the mild assumption that $\delta \in (0, 1/2 - c)$ for some constant $c > 0$, and we treat the noise parameter ρ as a fixed constant in $(0, 1/2)$ (not scaling with n and m).

4.1 Upper Bounds

We start with a simple upper bound for comparison queries.

Theorem 5. *Under the noisy comparison query model, there exists a deterministic algorithm that outputs the round-robin allocation with probability at least* $1 - \delta$ *using* $O(nm \log(m/\delta))$ *queries and* $O(nm \log(m/\delta))$ *time.*

[7] See, e.g., Sect. 8.1 of Cormen et al. [17].

Proof. For each agent, sort the items according to her preferences using a noisy sorting algorithm; then, allocate the items using the n resulting sorted lists. Noisy sorting is a well-studied problem, and for our proof, it suffices to use the algorithm of Feige et al. [20, Thm. 3.2], which requires $O(m \log(m/\delta_0))$ queries and time[8] to correctly sort m items with probability at least $1 - \delta_0$, for any $\delta_0 \in (0, 1/2)$.

Since we have n lists to be sorted, we set $\delta_0 = \delta/n$, so that a union bound yields an overall success probability of $1 - \delta$. Hence, the overall complexity is $O(nm \log(nm/\delta))$, which is equivalent to $O(nm \log(m/\delta))$ due to the fact that $m/\delta \leq nm/\delta \leq m^2/\delta^2$ (recall our assumption $n \leq m$). □

For value queries, we query each agent's utility for each item a sufficient number of times and run our noiseless algorithm based on the majority values.

Theorem 6. *Under the noisy value query model, there exists a deterministic algorithm that outputs the round-robin allocation with probability at least* $1 - \delta$ *using* $O(nm \log(m/\delta))$ *queries and* $O(nm \log(m/\delta))$ *time.*

While the proofs of our upper bounds are simple, based on our current understanding, it is conceivable that these bounds are already optimal. In the full version of our paper [27], we discuss some challenges that we faced when trying to improve the bounds.

4.2 Lower Bounds

Next, we shift our focus to lower bounds. We derive a bound of $\Omega(nm \log(1/\delta))$ for comparison queries.

Theorem 7. *Under the noisy comparison query model, any (possibly randomized) algorithm that outputs the round-robin allocation with probability at least* $1 - \delta$ *makes* $\Omega(nm \log(1/\delta))$ *queries in expectation.*

Proof. We use the same notation as in Sect. 3.2. We interpret the query model as a multi-armed bandit problem [26] (with a highly unconventional objective) in which each $a \in \mathcal{A}$ is an "arm" or "action". If the t-th query made is $a_t = (i, j, j')$, then the resulting observation is denoted by y_t, and is drawn from the distribution Bernoulli$(1 - \rho)$ if i prefers j to j', and Bernoulli(ρ) otherwise. Let P_a denote the (Bernoulli) distribution associated with action a, i.e., it holds for any query index t and $y \in \{0, 1\}$ that $P_a(y) = \mathbb{P}[y_t = y \mid a_t = a]$.

Let $\mathbb{P}^\nu$ and $\mathbb{E}^\nu$ denote the probability and expectation (with respect to the randomness in the algorithm and/or the query answers), respectively, when the underlying instance is ν. By assumption, we have $\mathbb{P}^\nu[\mathsf{Alg}(\nu) = L_\nu] \geq 1 - \delta$ for all ν. Note that the number of queries taken when Alg terminates is a random variable, as this may depend on the observed y_t values (which are themselves random) and moreover Alg itself may be randomized.

[8] Feige et al. [20] did not make an explicit claim on time. However, one can observe that the time complexity of their algorithm is the same as its query complexity.

With the actions $\{a_t\}$ and query responses $\{y_t\}$ being interpreted under a bandit framework as above, we can make use of a highly general result from the bandit literature that translates into an information-theoretic lower bound on the number of times certain actions are played (in expectation). These bounds are expressed in terms of the *KL divergence* $D(P\|Q) = \sum_x P(x) \log \frac{P(x)}{Q(x)}$ for probability mass functions P, Q, and its binary version $d(a,b) = a \log \frac{a}{b} + (1-a) \log \frac{1-a}{1-b}$ for real numbers $a, b \in [0,1]$ (i.e., $d(a,b)$ is the KL divergence between Bernoulli(a) and Bernoulli(b) distributions). This result and similar variations have been used in numerous bandit works—for example, see Lemma 1 of Kaufmann et al. [24] and Exercise 15.7 of Lattimore and Szepesvári [26].

Lemma 7. *Let ν and ν' be any two bandit instances defined on the same (finite) set of arms $\mathcal{A}$, with corresponding observation distributions $\{P_a\}_{a\in\mathcal{A}}$ and $\{P'_a\}_{a\in\mathcal{A}}$. Let τ be the (random) total number of queries made when the algorithm terminates, and let $\mathcal{E}$ be any probabilistic event that can be deduced from the resulting history $(a_1, y_1, \dots, a_\tau, y_\tau)$, possibly with additional randomness independent of that history. Then, we have*

$$\sum_{a\in\mathcal{A}} \mathbb{E}^\nu[T_a] D(P_a\|P'_a) \geq d(\mathbb{P}^\nu[\mathcal{E}], \mathbb{P}^{\nu'}[\mathcal{E}]), \tag{2}$$

where T_a is the (random) number of times action a is queried up to the termination index τ.

Intuitively, the right-hand side of (2) identifies an event $\mathcal{E}$ that (ideally) occurs with significantly different probabilities under the two instances (e.g., the algorithm outputting L_ν when $L_{\nu'} \neq L_\nu$). The left-hand side indicates that in order to permit such a difference in probabilities, actions with sufficient distinguishing power (i.e., high $D(P_a\|P'_a)$) must be played sufficiently many times (i.e., high $\mathbb{E}^\nu[T_a]$).

Let ν be any instance of our round-robin problem. Recall Lemma 3 (and the notation $\mathcal{J}_i(\nu)$), which asserts that

$$\sum_{i\in N} |\mathcal{J}_i(\nu)| \geq \frac{nm}{4}. \tag{3}$$

Now, for fixed i and $j \in \mathcal{J}_i(\nu)$, consider a different instance ν' in which j is made the most preferred item for agent i, and all other preferences remain unchanged. This means that j is allocated to i in $L_{\nu'}$, in particular implying that $L_\nu \neq L_{\nu'}$. Moreover, in (2), we observe the following:

- Unless the action a corresponds to agent i and item j (along with some other arbitrary item), the quantity $D(P_a\|P'_a)$ is zero; this is due to our construction of ν' and the fact that $D(P\|P) = 0$ for any P.
- For *any* action a, the quantity $D(P_a\|P'_a)$ is either zero or $d(\rho, 1-\rho)$ (which is equal to $d(1-\rho, \rho)$), since our observation distributions are always Bernoulli(ρ) or Bernoulli($1-\rho$).

– Let $\mathcal{E}$ be the event that Alg outputs $L_{\nu'}$. The success condition on Alg implies that $\mathbb{P}^{\nu}[\mathcal{E}] \leq \delta$ and $\mathbb{P}^{\nu'}[\mathcal{E}] \geq 1-\delta$. Since $\delta < 1/2$, this in turn implies $d(\mathbb{P}^{\nu}[\mathcal{E}], \mathbb{P}^{\nu'}[\mathcal{E}]) \geq d(\delta, 1-\delta)$ by a standard monotonicity property of $d(a,b)$ [24].

Combining the above findings, and defining $\widetilde{\mathcal{A}}_{ij}$ to be the set of all actions involving agent i and item j, we obtain $\sum_{a\in\widetilde{\mathcal{A}}_{ij}} \mathbb{E}^{\nu}[T_a] d(\rho, 1-\rho) \geq d(\delta, 1-\delta)$, or equivalently, $\sum_{a\in\widetilde{\mathcal{A}}_{ij}} \mathbb{E}^{\nu}[T_a] \geq \frac{d(\delta,1-\delta)}{d(\rho,1-\rho)}$. Since this holds for all pairs (i,j) such that $j \in \mathcal{J}_i(\nu)$, we can sum over all such pairs and apply (3) to obtain

$$\sum_{i\in N} \sum_{j\in\mathcal{J}_i(\nu)} \sum_{a\in\widetilde{\mathcal{A}}_{ij}} \mathbb{E}^{\nu}[T_a] \geq \frac{nm}{4} \cdot \frac{d(\delta,1-\delta)}{d(\rho,1-\rho)}. \tag{4}$$

Next, we claim that the left-hand side of (4) is upper-bounded by $2\mathbb{E}^{\nu}[\tau]$, where τ is the (random) total number of queries. To see this, we upper-bound the summation $\sum_{j\in\mathcal{J}_i(\nu)}$ by $\sum_{j\in M}$, apply linearity of expectation, and observe that $\sum_{i\in N}\sum_{j\in M}\sum_{a\in\widetilde{\mathcal{A}}_{ij}} T_a$ is exactly 2τ; the factor of 2 arises because each query (i, j_1, j_2) is counted twice (once when $j = j_1$ and once when $j = j_2$). It follows that $\mathbb{E}^{\nu}[\tau] \geq \frac{nm}{8} \cdot \frac{d(\delta,1-\delta)}{d(\rho,1-\rho)}$. The proof is completed by recalling that ρ is a fixed constant in $(0, 1/2)$, and noting that $d(\delta, 1-\delta) \in \Omega(\log(1/\delta))$ since $\delta \leq 1/2 - c$. □

Next, we establish an analogous result for value queries.

Theorem 8. *Under the noisy value query model, any (possibly randomized) algorithm that outputs the round-robin allocation with probability at least* $1-\delta$ *makes* $\Omega(nm\log(1/\delta))$ *queries in expectation.*

Finally, we derive a lower bound of $\Omega(m\log(m/\delta))$ for both query models.

Theorem 9. *Under the noisy comparison query model, for two agents, any (possibly randomized) algorithm that outputs the round-robin allocation with probability at least* $1-\delta$ *makes* $\Omega(m\log(m/\delta))$ *queries in expectation. The same holds for the noisy value query model.*

5 Conclusion and Future Directions

In this paper, we have analyzed the round-robin algorithm, one of the most widespread algorithms in the fair division literature, and presented several bounds on its complexity in the potential presence of noise. Besides tightening the bounds themselves, our work opens up a number of appealing conceptual directions. First, it would be interesting to explore the complexity of other fair division algorithms that rely only on ordinal rankings of items [6,11]; such algorithms are cognitively less demanding for agents than algorithms that require either cardinal utilities or ordinal information on sets of items. In addition, one could consider questions on fulfilling certain fairness notions in the presence of noise. Another intriguing avenue is to consider other noise models, for example, a comparison query model in which the probability of error depends on how differently the relevant agent ranks the two queried items [18].

Acknowledgments. We thank the anonymous reviewers for their valuable feedback, and acknowledge support by the Singapore Ministry of Education under grant numbers T1 251RES2218 and MOE-T2EP20221-0001 and by an NUS Start-up Grant.

References

1. Amanatidis, G., et al.: Fair division of indivisible goods: recent progress and open questions. Artif. Intell. **322**, 103965 (2023)
2. Amanatidis, G., Birmpas, G., Fusco, F., Lazos, P., Leonardi, S., Reiffenhäuser, R.: Allocating indivisible goods to strategic agents: pure Nash equilibria and fairness. In: Proceedings of the 17th International Conference on Web and Internet Economics (WINE), pp. 149–166 (2021)
3. Aziz, H.: Developments in multi-agent fair allocation. In: Proceedings of the 34th AAAI Conference on Artificial Intelligence (AAAI), pp. 13563–13568 (2020)
4. Aziz, H., Biró, P., de Haan, R., Rastegari, B.: Pareto optimal allocation under uncertain preferences: uncertainty models, algorithms, and complexity. Artif. Intell. **276**, 57–78 (2019)
5. Aziz, H., Bouveret, S., Lang, J., Mackenzie, S.: Complexity of manipulating sequential allocation. In: Proceedings of the 31st AAAI Conference on Artificial Intelligence (AAAI), pp. 328–334 (2017)
6. Aziz, H., Gaspers, S., Mackenzie, S., Walsh, T.: Fair assignment of indivisible objects under ordinal preferences. Artif. Intell. **227**, 71–92 (2015)
7. Bai, Y., Gölz, P.: Envy-free and Pareto-optimal allocations for agents with asymmetric random valuations. In: Proceedings of the 31st International Joint Conference on Artificial Intelligence (IJCAI), pp. 53–59 (2022)
8. Ben-Or, M., Hassidim, A.: The Bayesian learner is optimal for noisy binary search (and pretty good for quantum as well). In: Proceedings of the 49th Annual IEEE Symposium on Foundations of Computer Science (FOCS), pp. 221–230 (2008)
9. Blum, M., Floyd, R.W., Pratt, V.R., Rivest, R.L., Tarjan, R.E.: Time bounds for selection. J. Comput. Syst. Sci. **7**(4), 448–461 (1973)
10. Bouveret, S., Lang, J.: Manipulating picking sequences. In: Proceedings of the 21st European Conference on Artificial Intelligence (ECAI), pp. 141–146 (2014)
11. Brams, S.J., Kilgour, D.M., Klamler, C.: Two-person fair division of indivisible items: an efficient, envy-free algorithm. Not. AMS **61**(2), 130–141 (2014)
12. Brams, S.J., Taylor, A.D.: Fair Division: From Cake-Cutting to Dispute Resolution. Cambridge University Press, Cambridge (1996)
13. Braverman, M., Mao, J., Weinberg, S.M.: Parallel algorithms for select and partition with noisy comparisons. In: Proceedings of the 48th Annual ACM Symposium on Theory of Computing (STOC), pp. 851–862 (2016)
14. Caragiannis, I., Kurokawa, D., Moulin, H., Procaccia, A.D., Shah, N., Wang, J.: The unreasonable fairness of maximum Nash welfare. ACM Trans. Econ. Comput. **7**(3), 12:1–12:32 (2019)
15. Chakraborty, M., Schmidt-Kraepelin, U., Suksompong, W.: Picking sequences and monotonicity in weighted fair division. Artif. Intell. **301**, 103578 (2021)

16. Cohen-Addad, V., Mallmann-Trenn, F., Mathieu, C.: Instance-optimality in the noisy value-and comparison-model. In: Proceedings of the 31st Annual ACM-SIAM Symposium on Discrete Algorithms (SODA), pp. 2124–2143 (2020)
17. Cormen, T.H., Leiserson, C.E., Rivest, R.L., Stein, C.: Introduction to Algorithms, 3rd edn. MIT Press, Cambridge (2009)
18. Davidson, S., Khanna, S., Milo, T., Roy, S.: Top-k and clustering with noisy comparisons. ACM Trans. Database Syst. **39**(4), 35:1–35:39 (2014)
19. Dickerson, J.P., Goldman, J., Karp, J., Procaccia, A.D., Sandholm, T.: The computational rise and fall of fairness. In: Proceedings of the 28th AAAI Conference on Artificial Intelligence (AAAI), pp. 1405–1411 (2014)
20. Feige, U., Raghavan, P., Peleg, D., Upfal, E.: Computing with noisy information. SIAM J. Comput. **23**(5), 1001–1018 (1994)
21. Gan, J., Wirth, A., Zhang, X.: An almost optimal algorithm for unbounded search with noisy information. In: Proceedings of the 18th Scandinavian Symposium and Workshops on Algorithm Theory (SWAT), pp. 25:1–25:15 (2022)
22. Gu, Y., Xu, Y.: Optimal bounds for noisy sorting. In: Proceedings of the 55th Annual ACM Symposium on Theory of Computing (STOC), pp. 1502–1515 (2023)
23. Kahn, J., Saks, M.: Balancing poset extensions. Order **1**(2), 113–126 (1984)
24. Kaufmann, E., Cappé, O., Garivier, A.: On the complexity of best-arm identification in multi-armed bandit models. J. Mach. Learn. Res. **17**(1), 1:1–1:42 (2016)
25. Kurokawa, D., Procaccia, A.D., Wang, J.: When can the maximin share guarantee be guaranteed? In: Proceedings of the 30th AAAI Conference on Artificial Intelligence (AAAI), pp. 523–529 (2016)
26. Lattimore, T., Szepesvári, C.: Bandit Algorithms. Cambridge University Press, Cambridge (2020)
27. Li, Z., Manurangsi, P., Scarlett, J., Suksompong, W.: Complexity of round-robin allocation with potentially noisy queries. arXiv preprint arXiv:2404.19402 (2024)
28. Li, Z., Bei, X., Yan, Z.: Proportional allocation of indivisible resources under ordinal and uncertain preferences. In: Proceedings of the 38th Conference on Uncertainty in Artificial Intelligence (UAI), pp. 1148–1157 (2022)
29. Manurangsi, P., Suksompong, W.: When do envy-free allocations exist? SIAM J. Discret. Math. **34**(3), 1505–1521 (2020)
30. Manurangsi, P., Suksompong, W.: Closing gaps in asymptotic fair division. SIAM J. Discret. Math. **35**(2), 668–706 (2021)
31. Moulin, H.: Fair division in the internet age. Annu. Rev. Econ. **11**, 407–441 (2019)
32. Oh, H., Procaccia, A.D., Suksompong, W.: Fairly allocating many goods with few queries. SIAM J. Discret. Math. **35**(2), 788–813 (2021)
33. Plaut, B., Roughgarden, T.: Almost envy-freeness with general valuations. SIAM J. Discret. Math. **34**(2), 1039–1068 (2020)
34. Stanley, R.P.: Two combinatorial applications of the Aleksandrov-Fenchel inequalities. J. Combin. Theory Ser. A **31**(1), 56–65 (1981)
35. Steinhaus, H.: The problem of fair division. Econometrica **16**(1), 101–104 (1948)
36. Suksompong, W.: Constraints in fair division. ACM SIGecom Exchanges **19**(2), 46–61 (2021)
37. Walsh, T.: Fair division: the computer scientist's perspective. In: Proceedings of the 29th International Joint Conference on Artificial Intelligence (IJCAI), pp. 4966–4972 (2020)

Abstracts

The Computational Complexity of the Housing Market

Edwin Lock[1,3(✉)], Zephyr Qiu[2,3], and Alexander Teytelboym[1,3]

[1] Department of Computer Science, University of Oxford, Oxford, UK
edwin.lock@cs.ox.ac.uk , alexander.teytelboym@economics.ox.ac.uk
[2] Department of Economics, University of Oxford, Oxford, UK
zephyr.qiu.cn@gmail.com
[3] ETH Zürich, Zürich, Switzerland

Keywords: housing market · indivisible goods · competitive equilibrium · PPAD · computational complexity · query complexity

We prove that the classic problem of finding a(n approximate) competitive equilibrium in an exchange economy with indivisible goods, money, and unitdemand agents is PPAD-complete. In this "housing market", agents have general preferences over the house and amount of money they end up with; they can experience income effects and their willingness to pay for a house might depend on their level of wealth. Surprisingly, a competitive equilibrium allocation always exists, a result shown (under various assumptions) by Quinzii (1984), Gale (1984) and Svensson (1984).

Our results contrast with the existence of polynomial-time algorithms for related problems: Top Trading Cycles for the "housing exchange" problem in which there are no transfers, and the Hungarian algorithm for the "housing assignment" problem in which agents' utilities are linear in money. We show that the housing market problem is computationally equivalent to the Rainbow-KKM problem, a total search problem based on a generalization by Gale (1984) of the Knaster-Kuratowski-Mazurkiewicz (KKM) lemma, and then prove that Rainbow-KKM is PPAD-complete. Our reductions also imply exponential lower bounds on the query complexity of finding equilibria with four or more agents.

We leave open several avenues for further work including the complexity of finding equilibria under stronger existence assumptions (e.g., Quinzii (1984)), as well as connections to envy-free cake-cutting.

A full version of this paper can be found at https://arxiv.org/abs/2402.08484.

This project has received funding from the European Research Council (ERC) under the European Union's Horizon 2020 research and innovation programme (grant agreement No. 949699). This material is based upon work supported by the National Science Foundation under Grant No. DMS-1928930 and by the Alfred P. Sloan Foundation under grant G-2021-16778, while Teytelboym was in residence at the Simons Laufer Mathematical Sciences Institute (formerly MSRI) in Berkeley, California, during the Fall 2023 semester. We are grateful to Alex Hollender and Aviad Rubinstein for their comments.

G. Schäfer and C. Ventre (Eds.): SAGT 2024, LNCS 15156, p. 541, 2024.
https://doi.org/10.1007/978-3-031-71033-9

Ex-Post Stability Under Two-Sided Matching: Complexity and Characterization

Haris Aziz[1(✉)], Peter Biro[2,3], Gergely Csáji[2,4], and Ali Pourmiri[1]

[1] UNSW Sydney, Sydney, Australia
haris.aziz@unsw.edu.au

[2] Institute of Economics, HUN-REN KRTK, Budapest, Hungary
biro.peter@krtk.hu

[3] Corvinus University of Budapest,Budapest, Hungary

[4] Eötvös Lóránd Science University, Budapest, Hungary

Abstract. A probabilistic approach to the stable matching problem has been identified as an important research area with several important open problems. When considering random matchings, ex-post stability is a fundamental stability concept. A prominent open problem is characterizing ex-post stability and establishing its computational complexity. We investigate the computational complexity of testing ex-post stability. Our central result is that when either side has ties in the preferences/priorities, testing ex-post stability is NP-complete. The result even holds if both sides have dichotomous preferences. On the positive side, we give an algorithm using an integer programming approach, that can determine a decomposition with a maximum probability of being weakly stable. We also consider stronger versions of ex-post stability (in particular robust ex-post stability and ex-post strong stability) and prove that they can be tested in polynomial time.

Keywords: Matching theory · Stability Concepts · Fairness · Random Assignment

Acknowledgment. The authors thank Onur Kesten, Isaiah Iliffe, Tom Demeulemeester, and M. Utku Ünver for comments. Aziz acknowledges the support by the NSF-CSIRO grant on 'Fair Sequential Collective Decision-Making' (Grant number RG230833). Biró and Csáji acknowledge the financial support by the Hungarian Academy of Sciences, Momentum Grant No. LP2021-2, and by the Hungarian Scientific Research Fund, OTKA, Grant No. K143858. Csáji acknowledges support by the Ministry of Culture and Innovation of Hungary from the National Research, Development and Innovation fund, financed under the KDP-2023 funding scheme (grant number C2258525).

G. Schäfer and C. Ventre (Eds.): SAGT 2024, LNCS 15156, p. 542, 2024.
https://doi.org/10.1007/978-3-031-71033-9

Approval-Based Committee Voting Under Uncertainty

Hariz Aziz[1(✉)], Venkateswara Rao Kagita[2], Baharak Rastegari[3], and Mashbat Suzuki[1]

[1] University of New South Wales, Sydney, Australia
{haris.aziz,mashbat.suzuki}@unsw.edu.au
[2] National Institute of Technology, Warangal, India
venkat.kagita@nitw.ac.in
[3] University of Southampton, Southampton, UK
b.rastegari@soton.ac.uk

Abstract. We study approval-based committee voting in which a target number of candidates are selected based on voters' approval preferences over candidates. In contrast to most of the work, we consider the setting where voters express uncertain approval preferences and explore four different types of uncertain approval preference models. For each model, we study the problems such as computing a committee with the highest probability of satisfying axioms such as justified representation.

Keywords: Approval preferences · Committee voting · ABC voting

We initiate work on problems where voters' uncertain approval preferences are taken into account to compute desirable committees that satisfy representation with high probability. We consider four different types of uncertain approval preferences: *Joint Probability model*, *Lottery model*, and *Candidate-Probability model*. We also consider a restricted version of the latter model. For each of the uncertain approval models, we consider problems such as computing a committee with the highest probability of being JR. We undertake a detailed computational complexity analysis of several problems with respect to the four preference uncertainty models.

Acknowledgments. This work was supported by the NSF-CSIRO grant on "Fair Sequential Collective Decision-Making" (Grant No. RG230833). Venkateswara Rao Kagita is grateful to Science and Engineering Research Board (SERB) for providing financial support for this research through the SERB-SIRE project (Project Number: SIR/2022/001217). Mashbat Suzuki is supported by the ARC Laureate Project FL200100204 on "Trustworthy AI". Baharak Rastegari is grateful to UNSW Sydney's Faculty of Engineering for supporting her research visit through Diversity in Engineering Academic Visitor Funding Scheme.

G. Schäfer and C. Ventre (Eds.): SAGT 2024, LNCS 15156, p. 543, 2024.
https://doi.org/10.1007/978-3-031-71033-9

Author Index

G. Schäfer and C. Ventre (Eds.): SAGT 2024, LNCS 15156, pp. 545–546, 2024.
https://doi.org/10.1007/978-3-031-71033-9

SPRINGER NATURE

GPSR Compliance

The European Union's (EU) General Product Safety Regulation (GPSR) is a set of rules that requires consumer products to be safe and our obligations to ensure this.

If you have any concerns about our products, you can contact us on ProductSafety@springernature.com

In case Publisher is established outside the EU, the EU authorized representative is:

Springer Nature Customer Service Center GmbH
Europaplatz 3
69115 Heidelberg, Germany

The manufacturer's authorised representative in the EU is Springer Nature Customer Service Centre GmbH, Europaplatz 3, 69115 Heidelberg, Germany. If you have any concerns regarding our products, please contact ProductSafety@springernature.com

Printed and bound by CPI Group (UK) Ltd, Croydon, CR0 4YY

29/04/2026

02099532-0019